COLLECTED PAPERS OF

SALOMON BOCHNER

Courtesy of Princeton University Press

Salomon Bochner
August 20, 1899–May 2, 1982

COLLECTED PAPERS OF SALOMON BOCHNER

Part 3

Robert C. Gunning
Editor

American Mathematical Society
Providence, Rhode Island

1991 *Mathematics Subject Classification.*
Primary 01A75, 00B60.

Library of Congress Cataloging-in-Publication Data

Bochner, S. (Salomon), 1899–1982.
[Works. 1991]
Collected papers of Salomon Bochner/Robert C. Gunning, editor.
p. cm.
Includes bibliographical references.
1. Mathematics. 2. Bochner, S. (Salomon), 1899–1982. I. Gunning, R. C. (Robert Clifford), 1931– . II. Title.
QA3.B5719 1991 91-31045
510—dc20 CIP
ISBN 0-8218-0161-9 (set: alk. paper)
ISBN 0-8218-0174-0 (pt. 1: alk. paper)
ISBN 0-8218-0175-9 (pt. 2: alk.paper)
ISBN 0-8218-0176-7 (pt. 3: alk. paper)
ISBN 0-8218-0177-5 (pt. 4: alk. paper)

A complete list of acknowledgments can be found at
the back of this publication.

The paper used in this book is acid-free and falls within the guidelines
established to ensure permanence and durability. ♾
10 9 8 7 6 5 4 3 2 1 97 96 95 94 93 92

Contents of Part 3

Preface

The present volumes contain the collected papers of Salomon Bochner. He worked in a variety of different areas of mathematics in a long and productive career. For the convenience of those using these volumes, the papers are grouped in eleven different categories: Almost Periodic Functions, Fourier Analysis, Real Analysis (topics in real analysis that do not easily fit into the categories of almost periodic functions and Fourier analysis), Complex Analysis (general and special analytic functions of one complex variable), Several Complex Variables, Differential Geometry, Lie Groups, Differential Equations (both ordinary and partial), Probability, History, and Miscellaneous (including a few papers that did not fit naturally anywhere else together with obituary articles). The choice of these categories and the assignment of papers to a particular category are of course both rather arbitrary. Bochner saw the interconnections between topics very clearly and was often led to results in one area by his work in other areas, so many papers properly fall between categories or alternatively within two or more categories.

This collection does not contain the many reviews of other papers and books that Bochner wrote. He was a long-time reviewer for *Mathematical Reviews* and wrote some review articles for the *Bulletin of the American Mathematical Society* as well. Although his reviews do convey his own opinions and evaluations of a good deal of the mathematics of his time, they contain little of his own characteristic contributions to the subject. However, not included in this collection are some entries he wrote for the *Dictionary of the History of Ideas* (Scribner, 1973), the entries on "Space," "Continuity and Discontinuity," "Infinity," "Symmetry and Asymmetry," "Mathematics in Cultural History," and the article "Analysis" in the *Encyclopedia Britannica* (1961). These are undoubtedly of interest to the general reader, but less so to those who are interested in Bochner's own work.

In addition to his mathematical papers, Bochner wrote seven books on mathematical and historical topics. The books are listed in the summaries at the beginning of the appropriate categories as well as in the chronological bibliography included in this collection, but are not reproduced here.

Robert C. Gunning

Chronological Bibliography

The notation to the right of a paper indicates the part and section where it is located in this collection. For example, 1:A means part 1, section A. The absence of notation or the word "book" means that the work has not been included.

Orthgonale Systeme analytischer und harmonischer Funktionen, Deutsche Math.-Ver. 30 (1921), 100–101. 1:A

Über orthogonale Systeme analytischer Funktionen, Math. Zeit. 14 (1922), 180–207. 1:A

Sur les fonctions presque périodiques de Bohr, C. R. Acad. Sci. Paris 180 (1925), 1156–1158. 2:C

Properties of Fourier series of almost periodic functions, Proc. London Math. Soc. 26 (1926), 433–452. 2:C

Remarks on the theorems of Picard-Landau an Picard-Schottky, J. London Math. Soc. 1 (1926), 100–103. 2:E

Properties of Fourier series of almost periodic functions, J. London Math. Soc. 1 (1926), 131–133. 2:C

Notes on two theorems of Norbert Wiener (with G. H. Hardy), J. London Math. Soc. 1 (1926), 240–244. 1:B

Über Fourier-Reihen von fastperiodischen Funktionen, Sitzungsberi chte Berliner Math. Ges. 26 (1927), 49–56. 2:C

Beiträge zur Theorie der fastperiodischen Funktionen I. *Teil. Funktionen einer Variablen,* Math. Ann. 96 (1927), 119–147. 2:C

Beiträge zur Theorie der fastperiodischen Funktionen I. *Teil. Funktionen mehrerer Variablen,* Math. Ann. 96 (1927), 383–409. 2:C

Darstellung reellvariabler und analytischer Funktionen durch verallgemeinerte Fourier- und Laplace-Integrale, Math. Ann. 97 (1927), 635–662. 1:A

Arithm. Satz der Röntgenkristallog (with H. Seyfarth), Probl. Mod. Physki (1928), 8 pp. 4:J

Hauptlösungen von Differenzengleichungen, Acta. Math. 51 (1928), 1–21. 1:B

Über die Struktur von Fourierreihen fastperiodischer Funktionen, Sitzungsberichte Math.-Nat. Abteil. Bayerischen Akad. Wiss. (1928), 181–190. 2:C

Konvergenzsäte für Fourierreihen grenzperiodischen Funktionen, Math. Zeit. 27 (1928), 187–211. 2:C

Fortsetzung Riemannscher Flächen, Math. Ann. 98 (1928), 406–421. 2:E

Über einen arithmetischen Satz der Rontgenkristallographie (with H. Seyfarth), Probleme der Moderne Physik, Leipzig, (1928), 88–94. 4:J

Über Factorenfolgen für Fouriersche Reihen, Acta Litt. Ac. Sci. Univ. Franc.-Josephinae, 4 (1929), 5 pp. 1:A

Über Sturm-Liouvillesche Polynomsysteme, Math. Zeit. 29 (1929), 730–736. 3:H

Ein Satz über lineare Operationen, Math. Zeit 29 (1929), 737–743. 1:B

Über gewisse Differential- und allgemeinere Gleichungen, deren lösungen fastperiodisch sind. I. *Existenzsatz.* Math. Ann. 102 (1929), 489–504. 2:C

Über eine Klasse singulärer Integralgleichungen, Sitzungsberichte Preuss. Akad. Wiss. (1930), 9 pp. 1:B

Beitrag zur absoluten Konvergenz fastperiodischer Funktionen, Jber. Deutschen Math. Ver. 39 (1930), 52–54. 2:C

Über gewisse Differential- und allgemeinere Gleichungen, deren lösungen fasterperiodisch sind. II. Beschranktheitsatz Math. Ann. 103 (1930), 588–597. 2:C

Der Dimensionsbegriff für Punktmengen in kartesischen Räumen, Math. Ann. 102 (1930), 768–784. 1:B

Kurven endlicher Länge, Jber. Dtsch. Math. Ver. 40 (1931), 245–249. 1:B

Über gewisse Differential- und allgemeinere Gleichungen, deren lösungen fastperiodisch sind. III. *Systeme von Gleichungen,* Math. Ann. 104 (1931), 579–587. 2:C

Allgemeine lineare Differenzengleichungen mit asymptotisch konstanten Koeffizienten, Math. Zeit. 33 (1931). 426–450. 1:B

Ein Konvergensatz für mehrvariablige Fouriersche Integrale, Math. Zeit. 34, 440–447. 1:A

Die Poissonsche Summationsformel in mehreren Veränderlichen, Math. Ann. 106 (1932), 56–63. 1:A

Limitierung mehrfacher Folgen nach dem Verfahren der aritmetischen Mittel, Math. Zeit. 35 (1932), 122–126. 1:B

Vorlesungen über Fouriersche Integrale, Leipzig, Akademische Verlagsgesellschaft, 1932. Book

Umkehrsätze für allgemeine Limitierungsverfahren, S.-B. preuss. Akad. Wiss. (1933), 126–144. 1:B

Spektralzerlegung linearer Scharen unitärer Operatoren, S.-B. preuss Akad. Wiss. (1933), 371–376. 1:B

Abstrakte Funktionen und die Besselsche Ungleichung, Nachr. Ges. Wiss. Göttingen (1933), 178–180. 1:A

Integration von Funktionen, deren Werte die Elemente eines Vektorraumes sind, Fundam. Math. 20 (1933), 262–276. 1:B

Eine Bemerkung zum Satz von Fubini, Fundam. Math. 20 (1933), 277–280. 1:B

Absolut-additiv abstrakte Mengenfunktionen, Fundam. Math. 21 (1933), 211–213. 1:B

Remark on the integration of almost periodic functions, J. London Math. Soc. 8 (1933), 250–254. 2:C

Homogeneous systems of differential equations with almost periodic coefficients, J. London Math. Soc. 8 (1933), 283–288. 2:C

Abstrakte fastperiodische Funktionen, Acta Math. 61 (1933), 149–184. 2:C

Monotone Funktionen, Stieltjessche Integrale und harmonische Analyse, Math. Ann. 108 (1933), 378–410. 1:A

Ein Satz von Landau und Ikehara, Math. Zeit 37 (1933), 1–90. 2:E

Simultankonvergenz von mehrfachen Fourierschen Reihen und Integralen, Math. Zeit 37 (1933), 566–571. 1:A

Average distribution of arbitrary masses under group translations, Proc. Nat. Acad. Sci. U.S.A. 20 (1934), 206–210. 1:B

A theorem on Fourier-Stieltjes integrals, Bull. Amer. Math. Soc. 40 (1934), 270–276. 1:A

Remark on the Euclidean algorithm, J. London Math. Soc. 9 (1934), 4. 4:J

An extension of a Tauberian theorem on series with positive terms, J. London Math. Soc. 9 (1934), 141–148. 1:A

Fastperiodische Lösungen der Wellengleichung, Acta. Math. 62 (1934), 227–237. 2:C

Inversion formulae and unitary transformations, Ann. of Math. 35 (1934), 111–115. 1:A

Analytic functions with almost periodic coefficients (with F. Bohnenblust), Ann. of Math. 35 (1934), 152–161. 2:C

Distribution functions and positive definite functions (with B. Jessen), Ann. of Math. 35 (1934), 252–257. 1:A

Summation of multiple Fourier series by spherical means, Proc. Nat. Acad. Sci. U.S.A. 21 (1935), 353–355. 1:A

Almost periodic functions in groups, II Trans. Amer. Math. Soc. 37 (1935), 21–50. 2:C

On compact solutions of operational-differential equations, I (with J. v. Neumann), Ann. of Math. 36 (1935), 255–291. 2:C

Summation of double Fourier series by spherical means (abstract), Bull. Amer. Math. Soc. 41 (1935), 332. 1:A

Summation of multiple Fourier series by spherical means, Trans. Amer. Math. Soc. 40 (1936), 175–207. 1:A

Summation of derived Fourier series. An application to Fourier expansions on compact Lie groups, Ann. of Math. 37 (1936), 345–356. 1:A

A converse of Poisson's theorem in the theory of probability, Ann. of Math. 37 (1936), 816–822. 2:D

Some general convergence theorems (with S. Izumi), Tôhoku Math. J. 42 (1936), 191–194. 1:A

On general Fourier series with gaps, Prace mat.-fiz 43 (1936), 63–79. 1:A

Bounded analytic functions in several variables and multiple Laplace integrals, Amer. J. Math. 59 (1937), 732–738. 3:F

Singularities of composite functions in several variables (with W. T. Martin), Ann. of Math. 38 (1937), 298–302. 3:F

Remark on the theorem of Green, Duke Math. J. 3 (1937), 334–338. 3:G

Analytic mappings of compact Riemann spaces into Euclidean space, Duke Math. J. 3 (1937), 339–354. 3:G

Completely monotone functions of the Laplace operator for torus and sphere, Duke Math. J. 3 (1937), 488–502. 2:D

Stable laws of probability and completely monotone functions, Duke Math. J. 3 (1937), 726–728. 2:D

A theorem on analytic continuation of functions in several variables, Ann. of Math. 39 (1938), 14–19. 3:F

Closure theorems for translations (with R. P. Boas), Ann. of Math 39 (1938), 287–300. 2:C

Linear functionals on certain spaces of abstractly-valued functions (with A. E. Taylor), Ann. of Math. 39 (1938), 913–944. 1:B

Differentiable and Riemann metric, Duke Math. J. 4 (1938), 51–54. 3:G

Functions of integrable square in several complex variables, Duke Math. J. 4 (1938), 635–639. 3:F

Additive set functions on groups, Proc. Nat. Acad. Sci. U.S.A. 25 (1939), 158–160. 1:B

On a theorem of M. Riesz for Fourier series (with R. P. Boas), J. London Math. Soc. 14 (1939), 62–73. 1:A

Some theorems on quasi-analyticity for functions of several variables (with A. E. Taylor), Amer. J. Math. 61 (1939), 303–329. 3:F

Additive set functions on groups, Ann. of Math. 40 (1939), 769–799. 1:B

A uniqueness theorem for analytic almost-periodic functions, Duke Math. J. 5 (1939), 937–940. 2:C

Integration and differentiation in partially ordered spaces, Proc. Nat. Acad. Sci. U.S.A. 26 (1940), 29–31. 1:B

Harmonic surfaces in Riemann metric, Trans. Amer. Math. Soc. 47 (1940), 146–154. 3:G

Finitely additive integral, Ann. of Math. 41 (1940), 495–504. 1:B

Hadamard's theorem for Dirichlet series, Ann. of Math. 41 (1940), 711–714. 2:E

A generalization of Poisson's summation formula, Duke Math. J. 6 (1940), 229–234. 1:A

Additive set functions and vector lattices (with R. S. Phillips) Ann. of Math. 42 (1941), 316–324. 1:B

Hilbert distances and positive definite functions, Ann. of Math. 42 (1941), 647–656. 1:A

On a theorem on Tannaka and Krein, Ann. of Math. 43 (1942), 56–58. 2:C

Absolutely convergent Fourier expansions for non-commutative normed rings (with R. S. Phillips), Ann. of Math. 43 (1942), 409–418. 1:A

Completely monotone functions in partially ordered spaces, Duke Math. J. 9 (1942), 519–526. 1:B

Analytic and meromorphic continuation by means of Green's formula, Ann. of Math. 44 (1943), 652–673. 3:F

Group invariance of Cauchy's formula in several variables, Ann. of Math. 45 (1944), 686–707. 3:F

Boundary values of analytic functions in several variables and of almost periodic functions, Ann. of Math 45 (1944), 708–722. 3:F

Compact groups of differentiable transformations, Ann. of Math. 46 (1945), 372–381. 4:I

Groups of differentiable and real or complex analytic transformations (with D. Montgomery), Ann. of Math. 46 (1945), 685–694. 4:I

Finitely additive set functions and stochastic processes, Proc. Nat. Acad. Sci. U.S.A. 32 (1946), 259–261. 2:D

Bloch's theorem for real variables, Bull. Amer. Math. Soc. 52 (1946), 715–719. 3:G

Vector fields and Ricci curvature, Bull. Amer. Math. Soc. 52 (1946), 776–797. 3:G

Formal Lie groups, Ann. of Math. 47 (1946), 192–201. 4:I

Linear partial differential equations with constant coefficients, Ann. of Math. 46 (1946), 202–212. 3:H

Locally compact groups of differentiable transformations (with D. Montgomery), Ann. of Math. 47 (1946), 639–653. 4:I

Curvature in Hermitian metric, Bull. Amer. Math. Soc. 53 (1947), 179–195. 3:G

On compact complex manifolds, J. Indian Math. Soc. 11 (1947), 1–21. 3:G

Distributive order-preserving operations in partially ordered vector sets (with Ky Fan), Ann. of Math. 48 (1947), 168–179. 1:B

Groups on analytic manifolds (with D. Montgomery), Ann. of Math 48 (1947), 659–669. 4:I

Stochastic processes, Ann. of Math. 48 (1947), 1014–1061. 2:D

A homogeneous differentiable system of infinite order with nonvanishing solution (with D.V. Widder), Bull. Amer. Math. Soc. 54 (1948), 409–415. 3:H

On spherical partial sums of multiple Fourier series, Rev. Ci., Lima, 50 (1948), 85–104. 1:A

Curvature and Betti numbers, Ann. of Math. 49 (1948), 379–390. 3:G

On the localization property for multiple Fourier series (with K. Chandrasekharan), Ann. of Math. 49 (1948), 966–978. 1:A

Summations over lattice points in K-space, (with K. Chandrasekharan), Quart. J. Math. 19 (1948), 238–248. 1:A

Several complex variables, (with W. T. Martin). Princeton University Press, 1948. Book

Diffusion equation and stochastic processes, Proc. Nat. Acad. Sci. U.S.A. 35 (1949), 368–370. 2:D

Gauss summability of trigonometric integrals (with K. Chandrasekharan), Amer. J. Math. 71 (1949), 50–59. 1:A

Curvature and Betti numbers, II, Ann. of Math. 50 (1949), 77–93. 3:G

Fourier series of L_2-functions (with K. Chandrasekharan), Duke Math. J. 16 (1949), 579–583. 1:A

Fourier transforms (with K. Chandrasekharan), Princeton University Press, 1949; Annals of Math. Studies 19. Book

Partial ordering in theory of stochastic processes, Proc. Nat. Acad. Sci. U.S.A. 36 (1950), 439–443. 2:D

Su un teorema di Frobenius per le funzioni di Jacobi, Atti Accad. Naz. Lincei, Rend., Cl. Sci. fis. mat. natur. 8 (1950), 327–331. 2:E

Lattice points and Fourier expansions (with K. Chandrasekharan), Acta Sci. Math. 12B (1950), 1–15. 1:A

Quasi-analytic functions, Laplace operator, positive kernels, Ann. of Math. 51 (1950), 68–91. 1:A

Euler–Poincaré characteristic for locally homogeneous and complex spaces, Ann. of Math. 51 (1950), 241–261. 3:G

Vector fields on complex and real manifolds, Ann. of Math. 52 (1950), 642–649. 3:G

Summations over lattice points in K-space (A supplementary note) (with K. Chandrasekharan), Quart. J. Math. 1 (1950), 80. 1:A

Localization of best approximation, Contrib. Fourier Analysis, Ann. of Math. Studies 25 (1950), 3–23. 1:A

Dirichlet problem for domains bounded by spheres, Contrib. Fourier Analysis, Ann. of Math. Studies 25 (1950), 24–25. 1:A

Complex spaces with transitive commutative groups of transformations, Proc. Nat. Acad. Sci. U.S.A. 37 (1951), 356–359. 3:G

Tensor fields and Ricci curvature in Hermitian metric, Proc. Nat. Acad. Sci. U.S.A. 37 (1951), 704–706. 3:G

Theta relations with spherical harmonics, Proc. Nat. Acad. Sci. U.S.A. 37 (1951), 804–808. 2:E

Some properties of modular relations, Ann. of Math. 53 (1951), 332 363. 2:E

Tensor fields with finite bases, Ann. of Math. 53 (1951), 400–411. 3:G

A new viewpoint in differential geometry, Canadian J. Math. 3 (1951), 460–470. 3:G

Laplace operator on manifolds, Proc. Int. Congr. Math., Cambridge, Mass., 1950, 2 (1952), 189–201. 3:G

Partial differential equations and analytic continuations, Proc. Nat. Acad. Sci. U.S.A. 38 (1952), 227–230. 3:H

Local transformations with fixed points on complex spaces with singularities (with W.T. Martin), Proc. Nat. Acad. Sci. U.S.A. 38 (1952), 726–732. 3:F

Harald Bohr, April 22, 1887–January 22, 1951. Bull. Amer. Math. Soc. 58 (1952), 72–75. 4:J

On the addition theorem for multiply periodic functions, Proc. Amer. Math. Soc. 3 (1952), 99–106. 3:F

Remarks on Gaussian sums and Tauberian theorems, J. Indian Math. Soc. 15 (1952), 97–104. 1:A

Algebraic and linear dependence of automorphic functions in several variables, J. Indian Math. Soc. 16 (1952), 1–6. 3:F

Connection between functional equations and modular relations and functions of exponential type, J. Indian Math. Soc. 16 (1952), 99–102. 2:E

Hartogs' theorem in complex spaces with singularities (with W.T. Martin) J. Indian Math. Soc. 16 (1952), 137–146. 3:F

Tensor-fields in non-symmetric connections (with K. Yano), Ann. of Math. 56 (1952), 504–519. 3:G

Bessel functions and modular relations of higher type and hyperbolic differential equations, Meddel. Lunds Univ. Mat. Sem. Suppl.-band M. Riesz (1952), 12–20. 2:E

Stochastic processes with finite and non-finite variance, Proc. Nat. Acad. Sci. U.S.A. 39 (1953), 190–197. 2:D

Fourier transforms of time series, Proc. Nat. Acad. Sci. U.S.A. 39 (1953), 302–307. 2:D

Closure classes originating in the theory of probability, Proc. Nat. Acad. Sci. U.S.A. 39 (1953), 1082–1088. 2:D

The theorem of Morera in several variables, Ann. Mat. Pura Appl. 34 (1953), 27–39. 3:F

Zeta functions and Green's functions for linear partial differential operators of elliptic type with constant coefficients, Ann. of Math. 57 (1953), 32–56. 3:H

Length of random paths on general homogeneous spaces, Ann. of Math. 57 (1953), 309–313. 2:D

Complex spaces with singularities (with W.T. Martin), Ann. of Math. 57 (1953), 490–516. 3:F

Curvature and Betti numbers (with K. Yano), Princeton University Press, 1953; Annals of Math. Studies 32. Book

Structure of complex spaces, Ann. of Math. Studies 30 (1953), 189–201. 3:F

Boundedness and stationary of time series, Proc. Nat. Acad. Sci. U.S.A. 40 (1954), 289–294. 2:D

Limit theorems for homogeneous stochastic processes, Proc. Nat. Acad. Sci. U.S.A. 40 (1954), 699–703. 2:D

Positive zonal functions on spheres, Proc. Nat. Acad. Sci. U.S.A. 40 (1954), 1141–1147. 1:A

Green's formula and analytic continuation, Ann. of Math. Studies 33 (1954), 1–14. 3:F

Stationary, boundedness, almost periodicity of random-valued functions, Proc. Third Berkeley Symp. Math. Stat. and Prob. 2 (1954), 7–27. 2:D

Contributions to the theory of partial differential equations, (Edited with L. S. Bers and F. John), Ann. of Math. Studies 33, Princeton University Press, 257.

Stationary space-time in general relativity, Proc. Nat. Acad. Sci. U.S.A. 41 (1955), 485–490. 3:G

Existence of functionally independent automorphic functions (with R.C. Gunning), Proc. Nat. Acad. Sci. U.S.A. 41 (1955) 746–752. 3:F

Curvature and Betti numbers in real and complex vector bundles, Univ. e Politec. Torino, Rend. Sem. Mat. 15 (1955), 225–253. 3:G

Partial ordering in the theory of martingales, Ann. of Math. 62 (1955), 162–169. 2:D

Green-Goursat theorem, Math. Zeit. 63 (1955), 230–242. 3:F

Functions in one complex variable as viewed from the theory of functions in several variables, Lect. on functions of a complex variable, (University of Michigan, Ann Arbor, 1955), 315–333. 3:F

Harmonic analysis and the theory of probability, Berkeley, University of California Press, 1955. Book

Gamma factors in functional equations, Proc. Nat. Acad. Sci. U.S.A. 42 (1956), 86–89. 2:E

On Rieman's functional equation (with K. Chandrasekharan), Ann. of Math. 63 (1956), 336–360. 2:E

Weak solutions of linear partial differential equations, J. Math. Pures Appl. 35 (1956), 193–202. 3:H

Systèmes des équations aux dérivées partielles qui se rattachent aux formules de Green, Th. Eqns. aux Der. Part. Coll. Int. C.N.R.S. 71, Paris (1956), 19–27. 3:H

Sturm–Liouville and heat equations whose eigenfunctions are ultraspherical polynomials or associated Bessel functions, Proc. Conf. on Diff. Eqns., University of Maryland, 1956, 23–48. 3:H

Theorems on analytic continuation which occur in the study of Riemann's functional equation, J. Indian Math. Soc. 21 (1957), 127–147. 2:E

Krivizna i Chisla Betti (with K. Yano), Moscow, 1957, (translated from the English). (Translation of Curvature and Beth Numbers) Book

John von Neumann, Nat. Acad. Sci. U.S.A. Biographical Memoirs 32 (1958), 438–457. 4:J

On Riemann's functional equation with multiple Gamma factors, Ann. of Math. 67 (1958), 29–41. 2:E

A special integral transformation in Euclidean space (with T. Kawata), Ann. of Math. 68 (1958), 150–158. 2:D

A limit theorem for the periodogram (with T. Kawata), Ann. Math. Stat. 29 (1958), 1198–1208. 2:D

On identical vanishing of holomorphic functions in several complex variables, Proc. Nat. Acad. Sci. U.S.A. 45 (1959), 46–47. 3:F

Linear and algebraic dependance of functions on compact complex spaces with singularities, Proc. Nat. Acad. Sci. U.S.A. 45 (1959), 47–49. 3:F

Generalized conjugate and analytic functions without expansions, Proc. Nat. Acad. Sci. U.S.A. 45 (1959), 855–857. 2:E

Strong law of large numbers for sequences of almost periodic functions (with S. Izumi), Ann. of Math. 69 (1959), 718–732. 2:D

Lectures on Fourier integrals, with an Author's Supplement on Monotonic Functions, Stieltjes Integrals, and Harmonic Analysis. Princeton University Press, 1959; Annals of Math. Studies 42. (Translated from the original by Morris Tenenbaum and Harry Pollard.) (Translation of Vorlesungen über Fouriersche Integrale) Book

Classes of holomorphic functions of several variables in circular domains, Proc. Nat. Acad. Sci. U.S.A. 46 (1960), 721–723. 3:F

Almost periodic solutions of the inhomogeneous wave equation, Proc. Nat. Acad. Sci. U.S.A. 46 (1960), 1233–1236. 2:C

Mathematics, McGraw-Hill Encyc. of Sci. and Tech. 8 (1960), 175–180. 4:K

Hartogs' theorem in Euclidean space and a related Cousin the theorem on the torus, Contributions to Function Theory, Tata Institute of Fundamental Research, Bombay, 12–19 January 1960, 79–114. 3:F

Uniform convergence of monotone sequences of functions, Proc. Nat. Acad. Sci. U.S.A. 47 (1961), 582–585. 1:B

Vector fields on Riemannian spaces with boundary, Ann. Mat. Pura Appl. 53 (1961), 57–62. 3:G

General analytical setting for the central limit theorem of probability, Calcutta Math. Soc. Golden Jubilee Comm. Vol. part 1, Calcutta Math. Soc. (1961), 111–128. 2:D

Analysis. Encyc. Britannica 1 (1961), 864–864F.

Subordination of non-Gaussian stochastic processes, Proc. Nat. Acad. Sci. U.S.A. 48 (1962), 19–22. 2:D

A new approach to almost periodicity, Proc. Nat. Acad. Sci. U.S.A. 48 (1962), 2039–2043. 2:C

Entire functions in several variables with constant absolute values on a circular uniqueness set, Proc. Amer. Math. Soc. 13 (1962), 117–120. 3:F

Infinite linear pseudogroups of transformations (with R.C. Gunning), Ann. of Math. 75 (1962), 93–104. 3:G

The role of mathematics in the rise of mechanics, American Scientist 50 (1962), 294–311. 4:K

Lektsu ob integralakh Fur'l. Moscow, 1962 (translated from the English). (Translation of Vorlesungen über Fouriersche Integrale) Book

A generalization of bounded Haar measure, Proc. Nat. Acad. Sci. U.S.A. 50 (1963), 293–297. 1:B

Potential-theoretic approach to analytic Bohr almost periodic functions in half-planes, Math. Ann. 150 (1963), 150–155. 2:C

Function rings and analytic measures, J. de Math. Pures et Appl. 42 (1963), 209–221. 1:B

The significance of some basic mathematical conceptions for physics, Isis 54 (1963), 179–205. 4:K

Revolutions in physics and crises in mathematics, Science 141 (1963), 408–411. 4:K

Interpolation of general bounded and of almost periodic sequences by functions of stratified exponential type, Proc. Nat. Acad. Sci. U.S.A. 51 (1964), 164–168. 2:C

Continuous mappings of almost automorphic and almost periodic functions, Proc. Nat. Acad. Sci. U.S.A. 52 (1964), 907–910. 2:C

Analytic measures on compact Bohr groups, Scripta Math. 27 (1964), 5–26. 1:A

Aristotle's physics and today's physics, Inter. Phil. Quar. 4 (1964), 217–244. 4:K

Aristotle's notion of place (topos) in physics, Acts of the Tenth International Congress of the History of Science VIII, (Ithaca,1962), (1965), 471–474. 4:K

Why mathematics grows, Jour. Hist. Ideas 26 (1965), 3–24. 4:K

Intrinsic analytic continuation and envelopes of holomorphy, Proc. Nat. Acad. Sci. U.S.A. 53 (1965), 904–907. 3:F

Plato and the One-ness of Knowledge, University Magazine 32 (Spring 1967), Princeton University Press, p. 18. 4:K

The role of mathematics in the rise of science, Princeton University Press, 1966; Japanese edition, 1970. Book

The size of the Universe in Greek thought, Scientia 103 (1968), 511–531. 4:K

The rise of functions, Complex Analysis II: Proceedings of the Conference on Complex Analysis, Rice University, 26–29 March 1969, Rice University Studies, 56, No. 2 (1970), 3–23. 4:K

Space, Dictionary of the History of Ideas, Charles Scribner's Sons, 1973, New York.

Continuity and discontinuity in nature and knowledge, Dictionary of the History of Ideas, Charles Scribner's Sons, 1973, New York.

Infinity, Dictionary of the History of Ideas, Charles Scribner's Sons, 1973, New York.

Symmetry and asymmetry, Dictionary of the History of Ideas, Charles Scribner's Sons, 1973, New York.

Mathematics in Cultural History, Dictionary of the History of Ideas, Charles Scribner's Sons, 1973, New York.

Eclosion and synthesis, perspectives on the history of knowledge, W. A. Benjamin, Inc., 1969. Book

Singularities and discontinuities, Proceedings of the Conference on Complex Analysis, Rice University, March 1972, Rice University Studies, 59, No. 2 (1973). 3:F

Mathematical reflections, The American Mathematical Monthly 81, No. 8, (1974), 827–852. The first part of the article appeared as "Mathematical Americana" in Graduate Studies, Texas Tech University No. 13, October 1976, 87–97. 4:K

"Henry Bateman," Dictionary of American Biography, Supplement Four 1949–1950, 1974, 57–58.

A formal approach to stochastic stability, Zeitschrift für Wahrscheinlichkeitstheorie 31 (1975), 187–198. 2:D

Mathematical background space in astronomy and cosmology, Vistas in Astronomy 19, part 2 (1975), 133–161. 4:K

Note on Kepler's contribution to mathematical analysis (Kepler-Four Hundred Years), Vistas in Astronomy 18, (1975). 4:K

General almost automorphy, Proc. Nat. Acad. Sci. U.S.A. 72 (1975). 2:C

"Commentary" on Curtis A. Wilson's "Rheticus, Ravetz, and the 'necessity' of Copernicus' Innovation." In The Copernican Achievement (Robert S. Westman, Editor). Univ. of California Press (1975), 40–48. 4:K

Review of "Genevieve Guitel, Histoire Comparee des numeration écrites." Paris, Flammarion, 1975. American Scientist 65 (1977), 105.

Kepler: A personal Footnote, Vistas in Astronomy, 22, (1978), 19–20. 4:K

The Emergence of Analysis in the Renaissance and After, Rice University Studies, 64 Nos. 2 and 3 (1978), 11–56. 4:K

Fourier Series came first, The American Mathematical Monthly, 86 (1979), 197–199. 4:K

Einstein between Centuries, Rice University Studies, Vol. 65, No.3, Summer 1979, 1–54. 4:K

Positivity of the Heat Kernel for Ultraspherical Polynomials and Similar Functions, Archive of Rational Mechanics and Analysis, 70 (1979), 211–217. 1:A

Review of Gustar Herglotz, Gesammelte Schritten, 1979 in Bulletin of the American Mathematical Society, Vol. 1, 1020–1022.

Section F

Functions of Several Complex Variables

Bochner's interest in functions of several complex variables began with their Fourier analysis, leading to his characterization of the envelopes of holomorphy of tube domains. He later wrote on generalizations of Cauchy's integral formula for functions of several variables (including what is known as the Bochner-Martinelli integral formula) and applications of these formulas to analytic continuation on singularities of analytic spaces and on conditions for the analytic and linear dependence of complex analytic functions in various cases. The book *Several Complex Variables* (Princeton, 1948), written jointly with W. T. Martin, summarized much of his earlier work and his own outlook on the subject.

BOUNDED ANALYTIC FUNCTIONS IN SEVERAL VARIABLES AND MULTIPLE LAPLACE INTEGRALS.*

By S. Bochner.

Introduction. We shall consider the k-dimensional Euclidean space of the real points $x = (x_1, \cdots, x_k)$ and its complex extension consisting of the complex points $z = (z_1, \cdots, z_k)$; $z_\kappa = x_\kappa + iy_\kappa$. Any point set S of the real space is the basis of a *tube* $T = T_S$ of the complex space. The latter is the point set of the complex space consisting of all k-dimensional planes

$$x_\kappa = x_\kappa^0; \qquad -\infty < y_\kappa < \infty, \qquad (\kappa = 1, \cdots, k),$$

where $(x_1^0, \cdots, x_k^0)$ is an arbitrary point of S. A tube is an (open) domain if and only if its basis is a domain. We shall say that a tube T' lies within a tube T if the closure of T' is part of the interior of T.

The transformation

$$(1) \qquad w_\kappa = e^{-z_\kappa} \qquad (\kappa = 1, \cdots, k)$$

transforms a tube of the z-space into a so-called Reinhardt set of the w-space and, conversely, the image of a Reinhardt set in the w-space is a tube of the z-space.[1] For shortness, we shall term a Reinhardt region *convex* if the corresponding tube is convex in the usual sense. Any function analytic in a Reinhardt region R has a Laurent expansion

$$(2) \qquad \sum \cdot \overset{\infty}{\underset{-\infty}{\cdot}} \cdot \sum a_{n_1 \ldots n_k} w_1^{n_1} \cdots w_k^{n_k}$$

which is absolutely convergent in R. On the other hand the largest domain of absolute convergence of a series of the form (2) is a convex Reinhardt region.[2] Therefore, if a function is analytic in a Reinhardt region R it also exists in the smallest convex Reinhardt region enclosing R. In z-coördinates we obtain the following result. *If a function $f(z) = f(z_1, \cdots, z_k)$ is analytic in a tube T and has the period $2\pi i$ in each variable then it also exists in the convex closure $\bar{T}$ of T and has the period $2\pi i$ in each variable.*

Similar theorems hold for many classes of analytic functions other than

* Received June 14, 1937.

[1] H. Behnke und P. Thullen, "Theorie der Funktionen mehrerer komplexer Veränderlichen," *Ergebnisse der Mathematik*, vol. 3 (1934), pp. 33-34.

[2] Behnke-Thullen, *ibid.*, pp. 37-38.

periodic. As an illustration we shall discuss the class of bounded functions by two different methods. In the first proof we shall represent functions of integrable square by multiple Laplace integrals of the form

$$\int_{-\infty}^{\infty}\cdots\int \exp(-z_1t_1-\cdots-z_kt_k)\cdot\phi(t)\,dv_t, \tag{3}$$

the symbol dv_t denoting the Euclidean volume element $dt_1\cdots dt_k$.

Functions of integrable square. A function $f(z)$ which is analytic in T_S is of integrable square in T_S if the function

$$f_x(y)\equiv f(x_1+iy_1,\cdots,x_k+iy_k) \tag{4}$$

belongs to the Lebesgue class L_2 over the y-space, for every $x\subset S$, and moreover,

$$\int_{-\infty}^{\infty}\cdots\int |f_x(y)|^2\,dv_y\leq K\equiv K_S. \tag{5}$$

Of course, if (5) holds only for every domain S' within S then $f(z)$ is of integrable square within T.[3]

Let $\phi(t)=\phi(t_1,\cdots,t_k)$ be a measurable function of the real t-space. If, for every $x\subset S$, the function

$$\phi_x(t)=\exp(-x_1t_1-\cdots-x_kt_k)\cdot\phi(t) \tag{6}$$

belongs to L_2 in the t-space then the integral (3) is absolutely and uniformly convergent within T_S and the sum function $f(z)$ is of integrable square within T_S. In proving this we may assume, applying the Heine-Borel theorem, that S is an interval

$$a_\kappa\leq x_\kappa\leq b_\kappa \qquad (\kappa=1,\cdots,k). \tag{7}$$

Furthermore we may split the range of integration in (3) into the 2^k octants $t_1\gtrless 0,\cdots,t_k\gtrless 0$. For instance, if we restrict ourselves to the octant $t_1>0,\cdots,t_k>0$, then (3) has the value

$$\int_{-\infty}^{\infty}\cdots\int \exp[-(y_1t_1+\cdots+y_kt_k)i]\cdot\phi_x(t)\,dv_t. \tag{8}$$

[3] A function has a property within a tube T if it has the property in every tube T' within T.

The absolute and uniform convergence of (8) follows immediately from the relation

$$\left[\int_0^\infty\cdots\int |\phi_{x+\xi}(t)|\, dv_t\right]^2$$
$$\leq \int_0^\infty\cdots\int \exp(-2\xi_1 t_1 - \cdots - 2\xi_k t_k)\, dv_t \cdot \int_0^\infty\cdots\int |\phi_x(t)|^2\, dv_t$$

which is a consequence of Schwarz' inequality. The sum function of (8) is therefore analytic in $z_1, \cdots, z_k$, and relation (5) follows immediately from the Parseval equation

$$\int_{-\infty}^\infty\cdots\int |f_x(y)|^2\, dv_y = (2\pi)^{k/2}\int_0^\infty\cdots\int |\phi_x(t)|^2\, dv_t.$$

Any other octant can be treated in a similar fashion or can be reduced to the first octant by changing the signs of the negative t-components.

Conversely, any function which is of integrable square in S has a unique representation of the form (3). In fact, by Plancherel's theorem, the function $f_x(y)$ has a representation

$$f_x(y) \sim \int_{-\infty}^\infty\cdots\int \exp[-(y_1 t_1 + \cdots + y_k t_k)i]\cdot \phi_x(t)\, dv_t, \tag{9}$$

and we have to prove that (6) holds. Integrating (5) with respect to x over the domain S we obtain

$$\int_T |f_x(y)|^2\, dv_y dv_x < \infty.$$

Hence there exists a constant A such that for any polycylinder

$$P_\zeta : |z_\kappa - \zeta_\kappa| \leq r_\kappa$$

which lies entirely in T_S the relation

$$\int_{P_\zeta} |f(z)|^2\, dv_x dv_y \leq A$$

holds. Expanding $f(z)$ in a power series around ζ we easily obtain

$$|f(\zeta)|^2 \leq \frac{1}{\pi^k r_1^2 \cdots r_k^2}\int_{P_\zeta} |f(z)|^2\, dv_x dv_y.$$

Thus $f(z)$ is bounded within S. With the Fejer kernel

$$F_n(\eta) = \prod_{\kappa=1}^{k} \frac{(\sin (n/2)\eta_\kappa)^2}{\pi n \eta_\kappa{}^2}$$

we form the sequence of functions

$$(10)\quad f^n(z) = \int_{-\infty}^{\infty} \cdots \int f(x_1 + iy_1 + i\eta_1 + ,\cdots, x_k + iy_k + i\eta_k) F_n(\eta_1,\cdots,\eta_k)\, dv_\eta.$$

They have the following properties: α) they are analytic and of integrable square within T_S, β) for each (x, y)

$$\lim_{n\to\infty} f_x{}^n(y) = f_x(y),$$

γ) the Fourier transform $\phi_x{}^n(t)$ of $f_x{}^n(t)$, see (9), has the value

$$\phi_x(t) \prod_{\kappa=1}^{k} \left(1 - \frac{|t_\kappa|}{n}\right),$$

if $|t_x| \leq 1$ and the value 0 if some t_κ is > 1.[4] As a consequence of these properties it is sufficient to prove relation (6) for functions $\phi_x(t)$ vanishing outside a finite interval of the t-space. We pick any special point $x^0 \subset S$ and introduce the functions

$$(11)\qquad \phi(t) = \exp(x_1{}^0 t_1 + \cdots + x_k{}^0 t_k) \cdot \phi_{x^0}(t)$$

$$g(z) = \int_{-\infty}^{\infty} \cdots \int \exp(-z_1 t_1 - \cdots - z_k t_k) \cdot \phi(t)\, dv_t.$$

Since $\phi(t)$ vanishes outside a finite interval the function $g(z)$ is analytic. By (11), it is equal to the given function $f(z)$ on the linear manifold $x_1 = x_1{}^0, \cdots, x_k = x_k{}^0$. Therefore $f(z) = g(z)$ identically, and this completes the proof of (6).

Finally, the point set S for which $\phi_x(t)$ belongs to L_2 in the t-space is convex. In fact, if x^1 and x^2 are two points of S, and $x = \alpha x^1 + (1-\alpha)x^2$, $0 < \alpha < 1$, is a point of the segment joining these points, then the quantities

$$E_1 = \exp(-x_1{}^1 t_1 - \cdots - x_k{}^1 t_k), \qquad E_2 = \exp(-x_1{}^2 t_1 - \cdots - x_k{}^2 t_k)$$
$$E = \exp(-x_1 t_1 - \cdots - x_k t_k)$$

---

[4] S. Bochner, *Vorlesungen über Fouriersche Integrale*, 1932, p. 191, Satz 57.

satisfy the relation

$$E = E_1{}^a E_2{}^{1-a} \leq \max(E_1, E_2) \leq E_1 + E_2.$$

Hence we conclude that *a function which is analytic and of integrable square within a tube exists and is of integrable square within the convex closure of the tube.*

Bounded functions. From the last statement we shall conclude that *a function which is analytic and bounded within a tube exists and is bounded within its convex closure.*

For the proof we may assume that S is a bounded domain whose closure is contained in the octant $x_1 > 0, \cdots, x_k > 0$. This being the case, the function

$$g(z) = \frac{f(z)}{z_1 \cdots z_k} \tag{12}$$

is of integrable square within T and therefore it exists and is bounded within $\tilde{T}$. Thus $f(z)$ is bounded within T and $O(z_1 \cdots z_k)$ within T. Given any two points x^1 and x^2 of S and any real numbers $\eta_1, \cdots, \eta_k$ we form the function of the one complex variable $w = u + iv$,

$$h(w) = f(x_1{}^1 + w(x_1{}^2 - x_1{}^1) + i\eta_1, \cdots, x_k + w(x_k{}^2 - x_k{}^1) + i\eta_k).$$

It is defined in the strip $0 \leq u \leq 1$, it is bounded on the boundary $u = 0$, $u = 1$, and it is $O(v^k)$ as $v \to \pm\infty$. By Phragmen-Lindelöf, $h(w)$ is bounded in the strip, and any upper bound for its values on the boundary is a bound for its values in the interior of the strip. For appropriate values of x^1, x^2, η_k, and w every point of $\tilde{T}$ can be reached, and hence we easily conclude that $f(z)$ is bounded within $\tilde{T}$.

Functions in octant-shaped tubes. If a Reinhardt region contains the origin and is star-shaped with respect to the origin then the corresponding tube is "octant-shaped": if $(x_1{}^0, \cdots, x_k{}^0)$ is a point of its base S then the whole octant

$$x_\kappa \leq x_\kappa{}^0 \qquad (\kappa = 1, \cdots, k) \tag{13}$$

is contained in S. The convex closure of an octant-shaped tube is again octant-shaped.

If $f(z)$ is analytic in an octant-shaped tube T it also exists in the convex closure $\tilde{T}$ of T; if it is bounded within T it is also bounded within $\tilde{T}$.

Let $x' = (x'_1, \cdots, x'_k)$, $x'' = (x'_1, \cdots, x''_k)$ be any two points of S. We want to show that $f(z)$ is analytic for the points

$$x_\kappa \leq \alpha x'_\kappa + (1 - \alpha) x''_\kappa, \qquad 0 \leq \alpha \leq 1 \tag{14}$$

of the basis S of T.

For $\epsilon > 0$ sufficiently small and $A > 0$ sufficiently large the two polycylinders

$$\begin{aligned} P'&: \quad | z_\kappa + A | \leq x'_\kappa + 2\epsilon + A \\ P''&: \quad | z_\kappa + A | \leq x''_\kappa + 2\epsilon + A \end{aligned} \tag{15}$$

are both contained in T. The smallest convex Reinhardt region (*not* tube) of the z-space which is concentric with, and contains P' and P'', contains the polycylinders

$$| z_\kappa + A | \leq (x'_\kappa + \epsilon + A)^\alpha (x''_\kappa + \epsilon + A)^{1-\alpha}. \tag{16}$$

The sum of these point-sets for variable values of A contains the points of S for which

$$x_\kappa < \alpha(x'_\kappa + \epsilon) + (1 - \alpha)(x''_\kappa + \epsilon)$$

and therefore $f(z)$ is analytic in (14). The complex variables $z_1, \cdots, z_k$ can be altered by arbitrary fixed quantities $iy_1^0, \cdots, iy_k^0$; therefore $f(z)$ is analytic for all points whose real parts satisfy relation (14). The points x', x'' being arbitrary points of S we thus conclude that $f(z)$ is analytic throughout $\tilde{T}$. This proves the first part of the theorem. The second part is a consequence of the following remark.[5] If $|f(z)|$ is $\leq M$ in (15) then it is $\leq M \cdot C$ in (16) where C depends only on x'_κ, x''_κ, ϵ, A.

Application to bounded functions. We shall give a new proof of our previous statement concerning bounded functions, and we shall again assume that our function is a function of integrable square, see (12).

If a function $f(z)$ of one variable is of integrable square in the closed interval

$$a \leq x \leq b$$

then Cauchy's formula

$$f(z) = \frac{1}{2\pi i} \int_{a+i\infty}^{a-i\infty} \frac{f(s)}{s - z}\, dx + \frac{1}{2\pi i} \int_{b-i\infty}^{b+i\infty} \frac{f(s)}{s - z}\, ds$$

[5] The remark can be proved easily by the reasoning in H. Tietze, "Über den Bereich absoluter Konvergenz von Potenzreihen mehrerer Veränderlichen," *Mathematische Annalen* 99 (1928), p. 181.

leads to a decomposition of $f(z)$ into a sum of two functions $f_1(z)$, $f_2(z)$ of which $f_1(z)$ is analytic and bounded within $x > a$ and $f_2(z)$ is analytic and bounded within $x < b$. Up to additive constants such a decomposition is unique.[6] More generally, if $f(z) = f(z_1, \cdots, z_k)$ is of integrable square in the tube (7), it can be written as a sum of 2^k functions $f_\nu(z)$, $(\nu = 1, \cdots, 2^k)$, each of which is analytic and bounded within an octant containing the interior of the tube (7). For example, if $k = 2$, the four octants are 1) $x_1 > a_1$, $x_2 > a_2$, 2) $x_1 > a_1, x_2 < b_2$, 3) $x_1 < b_1, x_2 > a_2$, 4) $x_1 < b_1, x_2 < b$. The functions $f_\nu(z)$ are again uniquely determined up to additive constants.

We cover the basis S by a denumerable number of octants

$$a_\kappa^n \leq x_\kappa \leq b_\kappa^n, \qquad (n = 1, 2, \cdots).$$

By the uniqueness property the functions $f_\nu^n(z)$ can be so normalized that for fixed ν they are analytic continuations of each other. Therefore $f(z)$ is a sum of functions $f_\nu(z)$ each of which is analytic and bounded within an octant-shaped tube T_ν containing T. By the result of the previous section each $f_\nu(z)$ is analytic, and bounded within the convex closure $\tilde{T}_\nu$ of T_ν. But the intersection of the domains $\tilde{T}_\nu$ contains the convex closure $\tilde{T}$ of T.

PRINCETON UNIVERSITY.

[6] Harald Bohr, "Fastperiodische Funktionen III," *Acta Mathem.*, Bd. 47 (1926), pp. 250-251.

Annals of Mathematics
Vol. 38, No. 2, April, 1937

SINGULARITIES OF COMPOSITE FUNCTIONS IN SEVERAL VARIABLES

By S. Bochner and W. T. Martin[1]

(Received May 1, 1936)

In the present note we shall give a generalization to functions in several variables, of the following classical theorem of Hadamard.[2]

A. *Let*

$$a(z) = \sum_0^\infty a_n z^n, \qquad b(z) = \sum_0^\infty b_n z^n.$$

be any two power series converging in the neighborhood of the origin, and let $\pi(z)$ *be the composite series*

$$\pi(z) = \sum_0^\infty a_n b_n z^n.$$

If $\xi_\varphi e^{i\varphi}$, $\eta_\varphi e^{i\varphi}$, $\zeta_\varphi e^{i\varphi}$, $0 \leqq \varphi < 2\pi$, *are the vertices of the stars of* $a(z)$, $b(z)$, $\pi(z)$ *respectively, then*

$$\zeta_\varphi \geqq \underset{0 \leqq \vartheta < 2\pi}{\text{g.l.b.}} \{\xi_\vartheta \eta_{\varphi-\vartheta}\}.$$

The reader will find a proof of this classical theorem in an annex to this note.

The proof of the generalization will be simple enough but its formulation was by no means evident. In fact, the classical theorem itself will emanate in the following alternative form.

I. *Let*

$$a(x) = \sum_0^\infty a_n x^n, \qquad b(y) = \sum_0^\infty b_n y^n$$

be any two power series converging in the neighborhood of the origin, and let $\pi(x, y)$ *be the composite series in two complex variables*

$$\pi(x, y) = \sum_0^\infty a_n x^n \cdot b_n y^n.$$

If x^0, y^0 *are any two numbers, then* $\pi(x, y)$ *is analytic along the segment*

$$(\rho x^0, \rho y^0), \qquad 0 \leqq \rho^2 < \underset{0 \leqq \vartheta < 2\pi}{\text{g.l.b.}} \{\xi_\vartheta(x^0)\eta_{-\vartheta}(y^0)\},$$

[1] National Research Fellow.

[2] J. Hadamard, "Théorème sur les séries entières," *Acta Mathematica*, 22 (1898), 55–64; P. Dienes, *The Taylor Series*, Oxford (1931), esp. 346–348; S. Mandelbrojt, *Les singularités des fonctions analytiques representées par une Séries de Taylor*, Fascicule 54, Paris, (1932), 18–19.

where $\xi_\varphi(x^0)e^{i\varphi}$, $\eta_\varphi(y^0)e^{i\varphi}$ *are vertices of the stars of the functions* $a(x^0z)$, $b(y^0z)$, *these functions being considered as functions of the complex variable* z.

The transition from one to several variables will be based on the following key theorem of Hartogs.[3]

B. *Let* $f(x_1, \cdots, x_\mu; z)$ *be analytic in* z *in a fixed domain* T *containing the origin for every set* $(x_1, \cdots, x_\mu)$ *from the region*

$$I_\delta(x^0): | x_1 - x_1^0 | + \cdots + | x_\mu - x_\mu^0 | < \delta,$$

and analytic in $(x_1, \cdots, x_\mu; z)$ *in the region* $(I_\delta(x^0); | z | < r_0)$ *for some* $r_0 > 0$.
Then $f(x_1, \cdots, x_\mu; z)$ *is analytic in* $(x_1, \cdots, x_\mu; z)$ *in the region* $(I_\delta(x^0); T)$.

The corresponding generalization of Hurwitz's theorem[4] will involve notions that were introduced recently.[5]

Hadamard composite series. *Let* $a(x_1, \cdots, x_\mu)$, $b(y_1, \cdots, y_\nu)$ *be two power series, each in several complex variables. Grouping their terms in homogeneous polynomials, we put*

$$(1)\quad a(x_1, \cdots, x) = \sum_{n=0}^{\infty} A_n(x_1, \cdots, x_\mu), \qquad b(y_1, \cdots, y_\nu) = \sum_{n=0}^{\infty} B_n(y_1, \cdots, y_\nu).$$

Our composite series will be a function in all occurring variables,

$$(2)\qquad \pi(x; y) = \sum_{n=0}^{\infty} A_n(x_1, \cdots, x_\mu)\, B_n(y_1, \cdots, y_\nu).$$

We observe that this composition is invariant under any affine transformation in each group of variables. A composition of the coefficients themselves would fail in this and other respects.

Circular regions. We shall first make a very simple statement concerning the regions of uniform convergence of the three series (1), (2). These regions will be denoted by G_a, G_b, G_π respectively. Each of them is a circular region.[6] A region in the $(x_1, \cdots, x_\mu)$-space is circular if with any point $(x_1, \cdots, x_\mu)$ it contains all the points $(kx_1, \cdots, kx_\mu)$, $| k | \leqq 1$.

[3] F. Hartogs, "Zur Theorie der analytischen Funktionen mehrerer unabhängiger Veränderlichen, etc.," *Math. Annalen*, 62 (1906), 1–88, esp. 12–20.

[4] A. Hurwitz, "Sur un theorème de M. Hadamard," *Comptes Rendus*, Paris, 128 (1899), 350–353; S. Pincherle, "A priposita di un recente teorema del Sig. Hadamard," *Rendiconti de l'Acc. du Sc. de Bologn,a* 3 (1898–99), 67–74; S. Mandelbrojt, *loc. cit.*[2], 21–22.

[5] W. T. Martin, "Special regions of regularity of functions of several complex variables," *Proc. Nat. Acad. Sc.*, 22 (1936), 33–35.

[6] For a discussion of this and other fundamental notions see the tract by H. Behnke and P. Thullen, "Theorie der Funktionen mehrerer komplexen Veränderlichen," *Ergebnisse der Mathematik und ihrer Grenzgebiete*, Berlin (1934), 34, 41–47.

G_a is described by[7]

$$H_a(x_1, \cdots, x_\mu) < 1,$$

where

$$(3) \qquad H_a(x_1, \cdots, x_\mu) = \limsup_{x'=x} \left\{\limsup_{n=\infty} | A_n(x_1', \cdots, x_\mu') |^{1/n}\right\}.$$

The functions H_a, H_b, H_π satisfy the relation

$$H_\pi(x; y) = \limsup_{x'=x, y'=y} \left\{\limsup_{n=\infty} | A_n(x')B_n(y') |^{1/(2n)}\right\}$$

$$\leqq \{H_a(x_1, \cdots, x_\mu)H_b(y_1, \cdots, y_\nu)\}^{\frac{1}{2}},$$

hence:

II. *The composite function* $\pi(x; y)$ *is certainly analytic within*

$$H_a(x_1, \cdots, x_\mu)\, H_b(y_1, \cdots, y_\nu) < 1.$$

This is a circular region contained in the circular region G_π; *indeed, the point* $(x_1, \cdots, x_\mu; y_1, \cdots, y_\nu)$ *belongs to* G_π *if numbers* α, β *with* $| \alpha\beta | = 1$ *exist such that* $(\alpha x_1, \cdots, \alpha x_\mu)$ *belongs to* G_a *and* $(\beta y_1, \cdots, \beta y_\nu)$ *belongs to* G_b.

Stars. The star S_f of $f(x_1, \cdots, x_\mu)$ is defined as follows:[8] a point $(x_1^0, \cdots, x_\mu^0)$ belongs to S_f if $f(x)$ is regular along the segment $\rho x_1^0, \cdots, \rho x_\mu^0$, $0 \leqq \rho \leqq 1$.

We shall need the following lemma which is based on Hartog's key theorem.

C. *Let* $f(x_1, \cdots, x_\mu)$ *be regular at the origin. Corresponding to any* $x_1^0, \cdots, x_\mu^0$ *let* $\xi(x^0)$,

$$0 < \xi(x^0) \leqq \infty,$$

be the vertex of S_f *in the direction* $x_1^0, \cdots, x_\mu^0$, *that is, let* $\xi(x^0)$ *be the greatest number such that* $f(x_1, \cdots, x_\mu)$ *is regular along the segment* $\rho x_1^0, \cdots, \rho x_\mu^0$, $0 \leq \rho < \xi(x^0)$. *Then*

$$(4) \qquad \xi(x^0) = \liminf_{x=x^0} \xi^*(x)$$

where $\xi^*(x)$ *is the vertex on the positive real axis in the z-plane of the star of* $f(x_1 z \cdots, x_\mu z)$, *this function being considered as a function of z.*

For the proof of this lemma, let ξ' denote any positive number less than $\liminf_{x=x^0} \xi^*(x)$ *for fixed* $x_1^0, \cdots, x_\mu^0$. There exists an $\epsilon = \epsilon(\xi'; x^0) > 0$ such that

$$\xi^*(x) \geqq \xi' \qquad \text{for } x_1, \cdots, x_\mu \text{ in } I_\epsilon(x^0).$$

---

[7] Compare W. T. Martin, "Special regions of regularity of functions of several complex variables." To appear in *Annals of Math.*

[8] See footnote 5 and B. Almer "Sur quelques problèmes de la théorie des fonctions analytiques de deux variables complexes," *Arkiv for Mathmatik, Astronomi och Fysik*, 17 (1922–23).

There exists a δ, depending only on ϵ and x^0 such that for any x from $I_\delta(x^0)$ and any real φ satisfying: $|\varphi| < \delta$ the new point $xe^{i\varphi}$ falls into $I_\epsilon(x^0)$.

The vertices in the z-plane of the star of $f(x_1 z, \cdots, x_\mu z)$ as a function of z are obviously $\xi^*(xe^{i\varphi})e^{i\varphi}$. Hence for x from $I_\delta(x_0)$ the function $f(x_1 z \cdots, x_\mu z)$ is regular in the sector

$$|z| < \xi', \qquad |\arg z| < \delta.$$

On the other hand, for x from $I_\delta(x_0)$ and $|z|$ less some $r_0 > 0$ the function $f(x_1 z, \cdots, x_\mu z)$ is regular in all variables. Defining T as the region consisting of the points of the sector

$$|z| < \xi', \qquad |\arg z| < \delta$$

and of the points of the circle $|z| < r_0$, we may now apply theorem B. We find that $f(x_1 z, \cdots, x_\mu z)$ is analytic in $(I_\delta(x^0); T)$. Hence $f(x_1, \cdots, x_\mu)$ is analytic along the segment $\rho x_1^0, \cdots, \rho x_\mu^0$ for $\rho < \xi'$. Hence $\xi' \leqq \xi(x^0)$. But ξ' was any number smaller than the quantity on the right side of (4). Therefore

$$\xi(x^0) \geqq \liminf_{x=x^0} \xi^*(x).$$

But the greater sign cannot hold. Otherwise there would exist numbers ξ'' and ξ''' and a $I_\delta(x^0)$ such that

$$\xi(x^0) > \xi'' > \xi''' > \xi^*(x) \qquad \text{for } x \text{ from } I_\delta(x^0).$$

Since $f(x_1, \cdots, x_\mu)$ is regular along the segment

$$x_1^0\rho, \cdots, x_\mu^0\rho, \qquad \rho < \xi(x^0),$$

there exists and $I_{\epsilon'}(x^0)$ such that $f(x_1, \cdots, x_\mu)$ is regular along the segments

$$x_1\rho, \cdots, x_\mu\rho, \qquad \rho < \xi'', \qquad x \text{ from } I_{\epsilon'}(x^0).$$

Hence there exist an ϵ and a φ_0 such that $f(x_1 z, \cdots, x_\mu z)$ is regular for x from $I_\epsilon(x^0)$ and z from the sector

$$|z| \leqq \xi''', \qquad |\arg z| \leqq \varphi_0 .$$

Let $(x_1', \cdots, x_\mu')$ be a point common to $I_\delta(x^0)$ and $I_\epsilon(x^0)$. By the definition of $\xi^*(x')$ we have $\xi^*(x') \geqq \xi'''$ and this contradicts the assumption $\xi''' > \xi^*(x')$

This completes the proof of lemma C and the theorem proper will follow easily.

III. *Let*

$$a(x) = \sum_0^\infty A_n(x_1, \cdots, x_\mu), \qquad b(y) = \sum_0^\infty B_n(y_1, \cdots, y_\nu)$$

be any two power series, each in several complex variables, converging in the neighborhood of the origin, and let $\pi(x; y)$ *be the composite series in all occurring variables,*

$$\pi(x; y) = \sum_{n=0}^\infty A_n(x_1, \cdots, x_\mu)\, B_n(y_1, \cdots, y_\nu).$$

If $(x^0; y^0)$ *is any composite point then* $\pi(x; y)$ *is analytic along the segment*

$$(\rho x^0; \rho y^0), \qquad 0 \leqq \rho^2 < \operatorname*{g.l.b.}_{0 \leqq \vartheta < 2\pi} \{\xi(x^0 e^{i\vartheta})\, \eta(y^0 e^{-i\vartheta})\}$$

where

$$x^0 e^{i\varphi}\, \xi(x^0 e^{i\varphi}); \qquad\qquad y^0 e^{i\varphi}\, \eta(y^0 e^{i\varphi})$$

are vertices of the stars of the given functions $a(x_1, \cdots, x_\mu)$, $b(y_1, \cdots, y_\nu)$.

The vertices of the stars of the functions

$$a(xz) = \sum_0^\infty A_n(x) z^n, \qquad b(yz) = \sum_0^\infty B_n(x) z^n,$$

as functions in z, may be written in the form

$$e^{i\varphi}\, \xi^*(xe^{i\varphi}), \qquad\qquad e^{i\varphi}\, \eta^*(ye^{i\varphi}).$$

By Hadamard's classical theorem (see A)

$$\sum_0^\infty A_n(x) B_n(y) w^n$$

is analytic along the segment

$$0 \leqq \Re(w) < \operatorname*{g.l.b.}_{0 \leqq \vartheta < 2\pi} \{\xi^*(xe^{i\vartheta}) \eta^*(ye^{-i\vartheta})\}, \qquad \Im(w) = 0.$$

Hence

$$\pi(xz; yz) \equiv \sum_0^\infty A_n(x) B_n(y) z^{2n}$$

is analytic along the segment

$$z = \rho, \qquad 0 \leqq \rho^2 < \operatorname*{g.l.b.}_{0 \leqq \vartheta < 2\pi} \{\xi^*(xe^{i\vartheta}) \eta^*(ye^{-i\vartheta})\}.$$

Introducing the quantities

$$\zeta(x^0; y^0), \qquad\qquad \zeta^*(x; y)$$

in analogy to $\xi(x^0)$, $\xi^*(x)$ and $\eta(y^0)$, $\eta^*(y)$ puts the last relation into the form

$$[\zeta^*(x; y)]^2 \geqq \operatorname*{g.l.b.}_{0 \leqq \vartheta \leqq 2\pi} \{\xi^*(xe^{i\vartheta}) \eta^*(ye^{-i\vartheta})\}.$$

Since $\xi^*(xe^{i\vartheta})$ is for fixed x the vertex distance of a function in one variable, it is continuous from below in the variable ϑ. The same holds for $\eta^*(ye^{-i\vartheta})$, and consequently for $\xi^*(xe^{i\vartheta})\eta^*(ye^{-i\vartheta})$. But such a function attains a minimum on a closed set. Therefore we have

$$[\zeta^*(x; y)]^2 \geqq \xi^*(xe^{i\vartheta(x,y)}) \eta^*(ye^{-i\vartheta(x,y)}).$$

There exists sequences $x^n \to x^0$; $y^n \to y^0$, $\vartheta_n = \vartheta(x^n, y^n) \to \vartheta_0$ such that

$$\liminf_{x; y = x^0; y^0} \xi^*(xe^{i\vartheta(x,y)}) \eta^*(ye^{-i\vartheta(x,y)}) = \lim_{n=\infty} \xi^*(x^n e^{i\vartheta_n}) \cdot \lim_{n=\infty} \eta^*(y^n e^{-i\vartheta_n}).$$

In consequence of lemma C the last expression is

$$\geqq \xi(x^0 e^{i\vartheta_0}) \cdot \eta(y^0 e^{-i\vartheta_0}).$$

But

$$\liminf_{x;\, y = x^0;\, y^0} \zeta^*(x; y) = \zeta(x^0; y^0),$$

hence

$$[\zeta(x^0, y^0)]^2 \geqq \xi(x^0 e^{i\vartheta_0}) \cdot \eta(y^0 e^{-i\vartheta_0}).$$

Hence

$$[\zeta(x^0; y^0)]^2 \geqq \underset{0 \leqq \vartheta < 2\pi}{\text{g.l.b.}} \{\xi(x^0 e^{i\vartheta}) \cdot \eta(y^0 e^{-i\vartheta})\}$$

and this is precisely the statement of our theorem.

The Hurwitz composition. With (1) we form

$$c(x; y) = \sum_{n=0}^{\infty} C_n(x; y) \tag{5}$$

where

$$C_n(x; y) = \sum_{k=0}^{n} \binom{n}{k} A_k(x) B_{n-k}(y).$$

If circular regions of convergence of our three series are characterized by $H_a(x) < 1$, $H_b(y) < 1$, $H_c(x, y) < 1$ then, compare (3),

$$H_c(x; y) \leqq H_a(x) + H_b(y). \tag{6}$$

A more refined relation, analogous to (6), is

$$h_c(x; y) \leqq h_a(x) + h_b(y) \tag{7}$$

where $h_a(x)$, $h_b(y)$, $h_c(x; y)$ describe the diagrams of the series $a(x)$, $b(y)$, $c(x; y)$. The relation follows from

$$h_a(x^0) = \limsup_{x = x^0} \left\{ \limsup_{\rho = \infty} \frac{\log | A(x\rho) |}{\rho} \right\}$$

where

$$A(x) = \sum_0^{\infty} \frac{A_n(x)}{n!}$$

and from the analogous formulas for $h_b(y)$, $h_c(x; y)$, see footnote 5.

If $\alpha(x^0)$, $\beta(y^0)$, $\gamma(x^0; y^0)$ are the reciprocal values of the vertex distances of our three functions then

$$\gamma(x^0; y^0) \leqq \text{l.u.b.} \{\alpha(x^0), \beta(y^0), \alpha(x^0 e^{-i\varphi}) e^{i\varphi} + \beta(y^0 e^{-i\psi}) e^{i\psi}\}$$

taken over such real values of φ and ψ for which the expression in which they occur is positive.

A generalization. Let $k_1, \cdots, k_\mu; l_1, \cdots, l_\nu$ be any positive integers and suppose that the given expressions $A_n(x)$, $B_n(y)$ occurring in the Hadamard and Hurwitz composites are polynomials of weight n in the variables $x_1, \cdots, x_\mu;$ $y_1, \cdots, y_\nu$, if these variables are given the weights $k_1, \cdots, k_\mu; l_1, \cdots, l_\nu$. Then all assertions remain valid, if the circular regions of uniform convergence are the corresponding Cartan circular regions pertaining to the indices $(k_1, \cdots, k_\mu)$, $(l_1 \cdots, l_\nu)$ and $(k_1, \cdots, k_\mu; l_1, \cdots, l_\nu)$ respectively and the stars and diagrams are similarly generalized, see footnote 5.

An illustration. Let $X_p(x)$, $p = 1, 2, \cdots$ denote a sequence of homogeneous polynomials in $x_1^{\kappa_1}, \cdots, x_\mu^{\kappa_\mu}$, let $\| X_p \|$ be the maximum of $| X_p(x) |$ for $| x_1 |^2 + \cdots + | x_\mu |^2 = 1$; and let α_p be a sequence of numbers. If the sequence $\| X_p \|$ is bounded and the series

$$\sum_{p=1}^{\infty} \frac{| \alpha_p |}{\| X_p \|}$$

is convergent, then the series

$$\sum_{p=1}^{\infty} \frac{\alpha_p}{1 - X_p(x)}$$

is uniformly absolutely convergent in the neighborhood of the origin, and defines a function which will be our $a(x)$. Its expansion $\sum_0^\infty A_n(x)$, where

$$A_n(x) = \sum_{p=1}^{\infty} \{X_p(x)\}^n$$

is its Cartan diagonal expansion for the indices

$$k_j = \frac{\kappa_1 \cdots \kappa_\mu}{\kappa_j} = \frac{\kappa}{\kappa_j}, \qquad j = 1, \cdots, \mu$$

and converges in the Cartan circular region

$$(8) \qquad | X_p(x) | < 1, \qquad p = 1, 2, \cdots,$$

see footnote 7. It does not converge in a larger Cartan region since it has a singularity on the curve in which the analytic manifold

$$x_1 = x_1^0 z^{k_1}, \cdots, x_\mu = x_\mu^0 z^{k_\mu}$$

intersects the boundary of (8); in fact, the function in z

$$a(x^0 z^k) = \sum_{p=1}^{\infty} \frac{\alpha_p}{1 - X_p(x^0) z^\kappa}$$

has a singularity on the corresponding circle in the z-plane.[9] Hence its Cartan region is

[9] See, for example, A. Pringsheim, "Über bemerkenswerte Singularitätenbildung bei gewissen Partialbruchreihen," *Sitzungsberichte der Bayerischen Akademie* (1927), pp. 145–164, where also various references are given.

(9) $$\underset{p}{\text{l.u.b.}} \mid X_p(x) \mid < 1.$$

Let

$$b(y) = \sum_{q=1}^{\infty} \frac{\beta_q}{1 - Y_q(y)} = \sum_{n=0}^{\infty} B_n(y_1, \cdots, y_\nu)$$

be a similar expression in the y-space with indices $\lambda_1, \cdots, \lambda_\nu$; $l_1, \cdots, l_\iota$; $\lambda = \lambda_1 \cdots \lambda_\nu$. Its Cartan region is

$$\mid Y_q(y) \mid < 1, \qquad q = 1, 2, \cdots,$$

or

(10) $$\underset{q}{\text{l.u.b.}} \mid Y_q(y) \mid < 1.$$

The Hurwitz composite $c(x; y)$ is the expression

$$\sum_{p,q=1}^{\infty} \frac{\alpha_p \beta_q}{1 - X_p(x) - Y_q(y)}.$$

If we add the assumption $\kappa = \lambda$ then it has properties analogous to those of $a(x)$, $b(y)$. In particular, its Cartan region is

$$\underset{p,q}{\text{l.u.b.}} \mid X_p(x) + Y_q(y) \mid < 1.$$

In agreement with our previous assertion we have

$$\underset{p,q}{\text{l.u.b.}} \mid X_p(x) + Y_q(y) \mid \leqq \underset{p}{\text{l.u.p.}} \mid X_p(x) \mid + \underset{q}{\text{l.u.b.}} \mid Y_q(y) \mid,$$

the equality *not* holding as a rule. The Hadamard composite is

$$\pi(x; y) = \sum_{p,q=1}^{\infty} \frac{\alpha_p \beta_q}{1 - X_p(x) Y_q(y)},$$

the resulting Cartan region being

(11) $$\underset{p,q}{\text{l.u.b.}} \mid X_p(x) Y_q(y) \mid < 1.$$

Thus in this case we even have the equality

$$\underset{p,q}{\text{l.u.b.}} \mid X_p(x) Y_q(y) \mid = \underset{p}{\text{l.u.b.}} \mid X_p(x) \mid \cdot \underset{q}{\text{l.u.b.}} \mid Y_q(y) \mid;$$

in particular, the inequalities (9), (10) imply the inequality (11), in accordance with our general theorem.

The stars of our functions do not lend themselves to a similarly simple analytic representation. But the result relating to stars can be partially illustrated. In general, the star contains the Cartan region; in fact the radial distance $\rho(x)$ of the Cartan region and the vertex distance $\xi(x)$ of the star satisfy the relation

$$\rho(x) = \underset{\varphi}{\text{g.l.b.}}\ \xi(xe^{i\varphi}).$$

From the self explanatory argument

$$\begin{aligned}\tau(x;y) &= \operatorname*{g.l.b.}_{\varphi} \zeta(xe^{i\varphi}; ye^{i\varphi}) \\ &= \operatorname*{g.l.b.}_{\varphi} \left\{\operatorname*{g.l.b.}_{\vartheta} \xi(xe^{i\varphi}e^{i\vartheta})\eta(ye^{i\varphi}e^{-i\vartheta})\right\} \\ &\geqq \operatorname*{g.l.b.}_{\vartheta} \left\{\operatorname*{g.l.b.}_{\varphi} \xi(xe^{i\vartheta}e^{i\varphi}) \operatorname*{g.l.b.}_{\varphi} \eta(ye^{-i\vartheta}e^{i\varphi})\right\} \\ &\geqq \operatorname*{g.l.b.}_{\vartheta} \{\rho(xe^{i\vartheta})\sigma(ye^{-i\vartheta})\} \\ &\geqq \rho(x)\sigma(y)\end{aligned}$$

it follows that a point on the boundary of the Cartan region of $\pi(x;y)$ cannot be singular (that is a star vertex) unless it is a composite of such points of the functions $a(x)$, $b(y)$. And this statement can be verified for the Cartan regions (9), (10), (11).

To return to the Hurwitz composite if the constants $\kappa_1, \cdots, \kappa_\mu; \lambda_1, \cdots, \lambda_\nu$ are all 1, then the indicator functions $h_a(x)$, $h_b(y)$, $h_c(x;y)$ have the values

$$\begin{aligned}h_a(x) &= \operatorname*{l.u.b.}_{p} \{\text{real part } X_p(x)\} \\ h_b(y) &= \operatorname*{l.u.b.}_{q} \{\text{real part } Y_q(y)\} \\ h_c(x;y) &= \operatorname*{l.u.b.}_{p,q} \{\text{real part } [X_p(x) + Y_q(y)]\},\end{aligned}$$

and hence precisely

$$h_c(x;y) = h_a(x) + h_b(y).$$

Annex. Proof of theorem A. Let $e^{i\varphi}\rho(\varphi)$ denote the star vertices of a function $f(z)$ which is regular around the origin. Since $f(z)$ is regular along the segment $z = e^{i\varphi}\rho$, $0 \leqq \rho < \rho(\varphi)$ it follows that $\rho(\varphi)$, as a function of φ, is continuous from below.

Since both given functions $\bar{\xi}_\vartheta$, $\bar{\eta}_{\varphi-\vartheta}$ are continuous from below, their product, as a function in ϑ, is also continuous from below. Such a function attains its minimum on every closed set. Let $\alpha_n(\varphi)$, $\beta_n(\varphi)$ be sequence of continuous (periodic) functions converging increasingly towards $\bar{\xi}_\varphi$, $\bar{\eta}_\varphi$ as $n \to \infty$. Then, for every φ the sequence $\alpha_n(\varphi)\beta_n(\varphi - \vartheta)$ converges increasingly towards $\bar{\xi}_\vartheta \bar{\eta}_{\varphi-\vartheta}$. Since the functions of the sequence are continuous and converge increasingly, and their limit function attains its minimum, we have

$$\lim_{n=\infty} \operatorname*{g.l.b.}_{0 \leqq \vartheta < 2\pi} \{\alpha_n(\vartheta)\beta_n(\varphi - \vartheta)\} \geqq \operatorname*{g.l.b.}_{0 \leqq \vartheta < 2\pi} \{\bar{\xi}_\vartheta \bar{\eta}_{\varphi-\vartheta}\}.$$

The appromating functions $\alpha_n(\varphi)$, $\beta_n(\varphi)$ can be so chosen that the stars $e^{i\varphi}\alpha_n(\varphi)$, $e^{i\varphi}\beta_n(\varphi)$ are bounded by rectifiable curves. Hence it is sufficient to prove the following statement.

If $a(z)$ is analytic in the closed star $e^{i\varphi}\alpha(\varphi)$ and $b(z)$ in the closed star $e^{i\varphi}\beta(\varphi)$ and the star boundaries are rectifiable then $\pi(z)$ is analytic in the interior of the composite star

$$(12) \qquad z = e^{i\varphi}\rho \qquad 0 \leqq \rho < \min_{0 \leqq \vartheta < 2\pi} \{\alpha(\vartheta)\beta(\varphi - \vartheta)\}.$$

If t is a point of the contour $e^{-i\vartheta}/\beta(\vartheta)$,—we shall denote this contour by C,—and z is a point in (12) then the contour zt is interior to the star $e^{i\psi}\alpha(\psi)$. In fact, zt has the form $e^{i(\varphi-\vartheta)}\rho/\beta(\vartheta)$ and $\rho/\beta(\vartheta)$ is less than $\alpha(\varphi - \vartheta)$. Hence we can form the integral

$$(13) \qquad \frac{1}{2\pi i}\int a(tz)b\left(\frac{1}{t}\right)\frac{dt}{t}$$

for all z in (12). Furthermore if z runs over a closed set S interior to (12) and t runs over C, then the totality of points tz is closed, and thus a closed subset of the interior of the star in which $a(z)$ is being considered. Therefore for all t on C and z from S, $a(tz)$ is uniformly continuous in (t, z) and analytic in z. Hence the function (13) is analytic in (12). Let α denote the minimum of $\alpha(\varphi)$ and β the minimum of $\beta(\varphi)$. Then $a(tz)$ is analytic for $| t | \leqq \alpha/ | z |$ and $b(1/t)$ for $| t | \geqq 1/\beta$. Let $| z |$ be so small that C lies in the annulus $(1/\beta, \alpha/ | z |)$. Then, in the integral (13), C can be replaced by any concentric circle in the annulus. On this circle we can substitute $\sum_0^\infty a_n t^n z^n$ for $a(tz)$ and $\sum_0^\infty b_n t^{-n}$ for $b(1/t)$, and thus we identify the integral (13) with the function $\pi(z)$.

PRINCETON, N. J.

Annals of Mathematics
Vol. 39, No. 1, January, 1938

A THEOREM ON ANALYTIC CONTINUATION OF FUNCTIONS IN SEVERAL VARIABLES

By S. Bochner

(Received July 23, 1937)

Introduction. As in a previous paper[1] we shall consider the k-dimensional Euclidean space of the real points $x = (x_1, \cdots, x_k)$ and its complex extension consisting of the complex points $z = (z_1, \cdots, z_k)$; $z_\kappa = x_\kappa + iy_\kappa$. Any point set S of the real space is the basis of a *tube* $T = T_S$ of the complex space. The latter is the point set of the complex space consisting of all k-dimensional planes

$$x_\kappa = x_\kappa^0; \qquad -\infty < y_\kappa < \infty, \quad \kappa = 1, \cdots, k,$$

where $(x_1^0, \cdots, x_\kappa^0)$ is an arbitrary point of S. A tube is an (open) domain if and only if its basis is a domain. We shall say that a tube T' lies *within* a tube T if the closure of T' is part of the interior of T. The convex closure (in the usual sense) of the tube T will be denoted by $\tilde{T}$. Obviously $\tilde{T}$ is again a tube and its base $\tilde{S}$ is the convex closure of S.

In the paper loc. cit. it was shown that if a function $f(z) = f(z_1, \cdots, z_k)$ is analytic and bounded within T it also exists and is bounded within $\tilde{T}$. In the present paper we shall establish the theorem omitting the property of boundedness.

Theorem. *Any function which is analytic in a tube T is analytic in its convex closure $\tilde{T}$.*

On the other hand, any convex tube $\tilde{T}$ is the natural domain of analyticity for some function. In fact, there exists functions of this kind which have the period $2\pi i$ in each variable z_κ. Combining this fact with our theorem we are led to the following statement: *the envelope of regularity (Regularitätshülle)*[2] *of T is $\tilde{T}$.*

From the viewpoint of the theory of analytic functions of real variables our theorem may be formulated as follows. A function $f(x_1, \cdots, x_k)$ which is analytic over a domain S has, by definition, an extension

$$f(x_1 + iy_1, \cdots, x_k + iy_k)$$

into a complex neighborhood of S. If, now, the extended neighborhood is the whole tube T_S, then $f(x_1, \cdots, x_k)$ has an analytic continuation into the convex closure $\tilde{S}$ of S (and its tube $\tilde{T}$).

---

[1] *Bounded analytic functions in several variables and multiple Laplace integrals.* American Jour. of Math., vol. 59 (1937), pp. 732–738.

[2] H. Behnke und P. Thullen, *Theorie der Funktionen mehrerer komplexer Veränderlichen,* Ergebnisse der Mathematik, vol. 3 (1934), p. 70.

Our method of proof seems to be new. We shall approximate to the infinite tube T by finite regions with elliptic cross-sections and expand $f(z)$ in multiple series of Legendre polynomials. In this way we shall prove the analyticity of $f(z)$ in a tube T^* containing T and, finally, we shall show that T^* is identical with $\tilde{T}$.

Expansions in multiple Legendre series. Denoting by $C(r_\kappa) \equiv C(r_1, \cdots, r_k)$ the circular polycylinder

$$(1) \qquad | z_\kappa | < r_\kappa \qquad \kappa = 1, \cdots, k,$$

if $f(z)$ is analytic in a neighborhood of the origin containing the polycylinders $C(r'_\kappa)$, $C(r''_\kappa)$, it also is analytic in each polycylinder $C(r_\kappa)$, where

$$(2) \qquad \log r_\kappa = \alpha \log r'_\kappa + (1 - \alpha) \log r''_\kappa, \qquad 0 \leqq \alpha \leqq 1.$$

This statement on the analyticity of $f(z)$ follows immediately from the corresponding statement on the absolute convergence of its power series,[3] ,

$$(3) \qquad \sum a_{n_1 \cdots n_k} z_1^{n_1} \cdots z_k^{n_k}.$$

Replacing (1) by

$$(4) \qquad | z_\kappa + (z_\kappa^2 - 1)^{\frac{1}{2}} | < r_\kappa \qquad \kappa = 1, \cdots, k$$

we obtain a polycylinder $E(r_\kappa)$ whose components are ellipses with foci at the points $z_\kappa = +1, -1$, the quantity r_κ being the sum of the two semi-axes of the κ-th ellipse.

Lemma. *If $f(z)$ is analytic in a neighborhood of the rectangle*

$$-1 \leqq x_\kappa \leqq +1, \qquad \kappa = 1, \cdots, k$$

containing the polycylinders $E(r'_\kappa)$, $E(r''_\kappa)$, it also is analytic in each polycylinder $E(r_\kappa)$ whose radii belong to the family (2).

Proof. We consider the Legendre polynomials $P_n(z)$ and the Legendre functions of the second kind $Q_n(\zeta)$, and we write

$$w = z + (z^2 - 1)^{\frac{1}{2}}, \qquad \omega = \zeta + (\zeta^2 - 1)^{\frac{1}{2}}.$$

It is known that[4] for fixed $\delta > 0$, $M > 0$,

$$\frac{1}{\zeta - z} = \sum_{n=0}^{\infty} (2n + 1) P_n(z) Q_n(\zeta),$$

uniformly in

$$| w | \leqq | \omega | - \delta \leqq M,$$

[3] Behnke-Thullen, l. c., p. 37, pp. 74–75.

[4] E. W. Hobson, *Spherical and ellipsoidal harmonics*, Cambridge (1931), pp. 60-62.

and that

$$|P_n(z)| \leqq |w|^n \tag{5}$$

$$|Q_n(\zeta)| \leqq \left(\frac{\pi}{n}\right)^{\frac{1}{2}}\left(1 - \frac{1}{|\omega|^2}\right)^{-\frac{1}{2}} \cdot \frac{1}{|\omega|^{n+1}}$$

(6)

$$Q_0(\zeta) = \tfrac{1}{2} \log \frac{\zeta + 1}{\zeta - 1}.$$

If $f(z)$ is analytic in the closure of $E(\rho_\kappa)$, we can apply Cauchy's formula

$$f(z_1, \cdots, z_k) = \frac{1}{(2\pi i)^k} \int \cdots \int \frac{f(\zeta_1, \cdots, \zeta_k)\, d\zeta_1 \cdots d\zeta_k}{(\zeta_1 - z_1) \cdots (\zeta_k - z_k)}$$

in which the integrations extend over the boundaries of the ellipses. The substitution

$$\frac{1}{\zeta_\kappa - z_\kappa} = \sum_{n=0}^{\infty} (2n + 1) P_n(z_\kappa) Q_n(\zeta_\kappa)$$

leads to an expansion

$$f(z) = \sum a_{n_1 \cdots n_k} P_{n_1}(z_1) \cdots P_{n_k}(z_k) \tag{7}$$

which is absolutely and uniformly convergent in every closed subset of $E(\rho_\kappa)$; and, on account of (6), the coefficients satisfy an inequality

$$|a_{n_1 \cdots n_k}| \leqq K \cdot \rho_1^{-n_1} \cdots \rho_k^{-n_k}$$

in which the constant K is independent of the indices $n_1, \cdots, n_k$.

As in the case of power series (3), an expansion of the form (7) which converges absolutely and uniformly in an elliptic polycylinder is unique. Hence, under the assumptions of our lemma, there exists an expansion (7) such that, for fixed $\epsilon > 0$,

$$|a_{n_1 \cdots n_k}| \leqq K \cdot (r_1' - \epsilon)^{-n_1} \cdots (r_k' - \epsilon)^{-n_k}$$

$$|a_{n_1 \cdots n_k}| \leqq K \cdot (r_1'' - \epsilon)^{-n_1} \cdots (r_k'' - \epsilon)^{-n_k}.$$

If a system of coefficients satisfies these two inequalities simultaneously, the series

$$\sum |a_{n_1 \cdots n_k}| r_1^{n_1} \cdots r_k^{n_k}$$

converges for any system of radii for which

$$\log r_\kappa < \alpha \log (r_\kappa' - \epsilon) + (1 - \alpha) \log (r_\kappa'' - \epsilon).$$ [5]

[5] H. Tietze, *Über den Bereich absoluter Konvergenz von Potenzreihen mehrerer Veränderlichen*, Math. Annalen, vol. 99 (1928), pp. 181–182.

Letting $\epsilon \to 0$, the assertion of our lemma becomes an immediate consequence of property (5).

In the following section we shall replace the polycylinder $E(r_\kappa)$ by the polycylinder $E_l(r_\kappa)$ whose components are ellipses with foci at the points $z_\kappa = +il$, $-il$, the quantity r_κ being again the sum of the two semi-axes of the k-th ellipse. The quantity l is a fixed positive number. It is easily seen that the lemma remains valid.

A special case. We shall first prove our theorem for a tube whose basis S is the sum of two (k-dimensional) rectangles

$$|x_\kappa| < a'_\kappa \qquad \kappa = 1, \cdots, k \tag{8}$$

$$|x_\kappa| < a''_\kappa \qquad \kappa = 1, \cdots, k. \tag{9}$$

Since a linear transformation of the x-coordinates can be interpreted as a linear transformation of the z-coordinates and carries tubes into tubes we are actually dealing with the case of a basis S being the sum of two rectangles which are concentric and coaxial.

The tubes (8), (9) contain the polycylinders $E_l(r'_\kappa)$, $E_l(r''_\kappa)$ for which

$$r'_\kappa = a'_\kappa + (a'^2_\kappa + l^2)^{\frac{1}{2}}, \qquad r''_\kappa = a''_\kappa + (a''^2_\kappa + l^2)^{\frac{1}{2}}.$$

By our lemma $f(z)$ is analytic also in the polycylinders $E_l(r_\kappa)$ whose small semi-axes a_κ satisfy the relations

$$\log (a_\kappa + (a^2_\kappa + l^2)^{\frac{1}{2}} = \alpha \log (a'_\kappa + (a'^2_\kappa + l^2)^{\frac{1}{2}} + (1 - \alpha) \log (a''_\kappa + (a''^2_\kappa + l^2)^{\frac{1}{2}}$$

or, what is the same,

$$\operatorname{arcsinh} \frac{a_\kappa}{l} = \alpha \operatorname{arcsinh} \frac{a'_\kappa}{l} + (1 - \alpha) \operatorname{arcsinh} \frac{a''_\kappa}{l}, \qquad 0 \leqq \alpha \leqq 1. \tag{10}$$

Letting $l \to \infty$, relation (10) goes over into

$$a_\kappa = \alpha a'_\kappa + (1 - \alpha) a''_\kappa,$$

and the polycylinder $E_l(r_\kappa)$ converges to the corresponding tube

$$|x_\kappa| < a_\kappa .$$

The totality of these tubes for all values α is the convex closure of the tubes (8), (9) and our theorem is proved in this case.

For general tubes we easily conclude that if $f(z)$ is analytic in the tube T it also exists and is analytic in the tube T^* whose basis S^* is the smallest (connected) neighborhood of S having the following property: if a subdomain σ of S^* is the sum of two rectangles which are concentric and coaxial, then the convex closure $\tilde{\sigma}$ of σ is likewise a subdomain of S.

A characterization of convex domains. Finally we have to show that the domain S^* is convex in the usual sense. Any two points A, A_n of S can be connected by a (rectilinear) polygon $AA_1A_2 \cdots A_n$ and we have to prove that the polygon can be replaced by the segment AA_n. Applying induction it will be sufficient to prove that the segment AA_2 lies in S. We shall assume the contrary and derive a contradiction.[6] In order to simplify the argument we shall first treat the two-dimensional case $k = 2$. Let X be a point of A_1A_2 moving from A_1 to A_2 and let X_0 be the first value of X for which the segment AX_0 is not contained in S^*. Let P be the first point of AX_0, counting from A, which is a boundary point of S^*, and let B be any point of AX_0 beyond P such that PB is shorter than AP. We can now construct in S^* (compare the figure) two rectangles $ABCD$, $abcd$, which are concentric and coaxial. Their convex

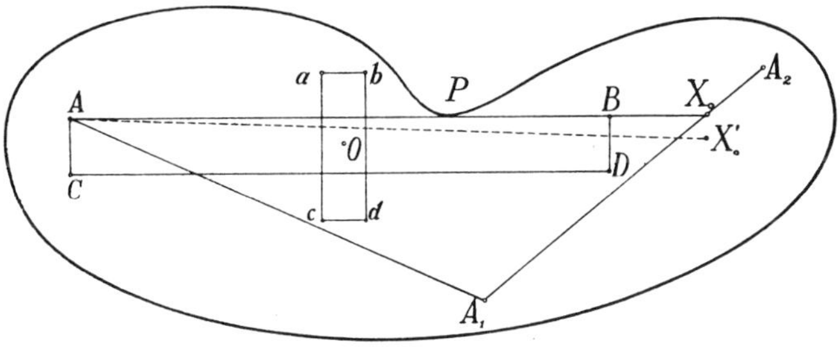

closure is the octagon $A\ a\ b\ B\ D\ d\ c\ C$, of which P is an interior point. Therefore P cannot be a boundary point of S^*.

If $k \geqq 3$ we draw the same two-dimensional figure as before and rotate it around A by a small angle which moves the line AX_0 into a new position AX_0' nearer the point A_1. The two rectangles go over into new rectangles which are lying within S^* together with their boundaries, and, provided, the angle of rotation is small enough, the enclosing octagon will again contain the point P, although P is no longer located on the boundary of the rectangle $ABCD$. The two rectangles can be made the cross-sections of two k-dimensional rectangles whose remaining $k - 2$ dimensions are sufficiently small. In this way we can construct in S^* two k-dimensional rectangles which are concentric and coaxial and whose convex closure again contains the boundary point P.

Remark. Our convex tube $\tilde{T}$ is a star region with respect to any of its points. In fact if it contains the points c_κ, z_κ, then, for any positive exponents σ_κ, the segment

$$\zeta_\kappa = c_\kappa + (z_\kappa - c_\kappa)t^{\sigma_\kappa}, \qquad 0 \leqq t \leqq 1,$$

is likewise contained in $\tilde{T}$. Hence, by a theorem of B. Almer,[7] if

$$f(z) = \sum a_{n_1 \cdots n_k}(z_1 - c_1)^{n_1} \cdots (z_k - c_k)^{n_k}$$

[6] Compare H. Tietze, *Über Konvexheit im kleinen und im grossen und über gewisse den Punkten einer Menge zugeordnete Dimensionszahlen*, Math. Zeitschrift, vol. 28 (1928), pp. 704–707.

[7] *Sur quelques problèmes de la théorie des fonctions analytiques de deux variables complexes.* Ark. Mat. Astron. Fys. vol. 17 (1922).

is the power series expansion of our function around the point c_κ, then for all points in $\tilde{T}$,

$$f(z) = \lim_{\epsilon \to 0} \sum \frac{a_{n_1 \cdots n_k}(z_1 - c_1)^{n_1} \cdots (z_k - c_k)^{n_k}}{\Gamma(1 + \epsilon\sigma_1 n_1 + \cdots + \epsilon\sigma_k n_k)},$$

In particular, since the point c_κ can be chosen as a point in S, we obtain a formula giving in one step the extension of our function $f(x)$ from the original domain S into its convex closure $\tilde{S}$.

PRINCETON UNIVERSITY.

Reprinted from DUKE MATHEMATICAL JOURNAL
Vol. 4, No. 4, December, 1938

FUNCTIONS OF INTEGRABLE SQUARE IN SEVERAL COMPLEX VARIABLES

BY S. BOCHNER

As in two previous notes[1] we consider in the space C_k of k complex variables

$$z = (z_1, \cdots, z_k), \qquad z_\kappa = x_\kappa + iy_\kappa,$$

point sets of a special nature which we call *tubes*. A point set T of C_k is a tube, if there exists a point set S in the space R_k of real variables $x = (x_1, \cdots, x_k)$ such that T consists of all k-dimensional planes

$$(1) \qquad x_\kappa = x_\kappa^0 \qquad (-\infty < y_\kappa < \infty;\ \kappa = 1, \cdots, k)$$

for which $(x_1^0, \cdots, x_k^0)$ is any point of S. The set S is called the basis of T, and we also denote T more explicitly by T_S. The tube T_S is open or closed in C_k if and only if S is open or closed in R_k; it is convex if and only if S is convex, and the convex hull[2] $\tilde{T}$ of a tube T is again a tube whose basis $\tilde{S}$ is the convex hull of S.

We say that a function $f(z) = f(z_1, \cdots, z_k)$ is of integrable square in T if the function

$$f_x(y) = f(x_1 + iy_1, \cdots, x_k + iy_k)$$

belongs to the Lebesgue class L_2 over the y-space, for every $x \subset S$, and if moreover there exists a constant K such that

$$(2) \qquad \int_{-\infty}^{\infty}\cdots\int |f_x(y)|^2 dv_y \leqq K,$$

for all $x \subset S$, the symbol dv_y denoting the Euclidean volume element $dy_1 \cdots dy_k$.

In our first note we proved the following theorem. If $f(z)$ is analytic and of integrable square in an open tube T, then it also exists and is of integrable square in $\tilde{T}$. In the present paper we shall extend this theorem to the case of tubes which are not necessarily open.

ASSUMPTIONS. (1) *The basis S is such that any two points P, Q of S have a finite Euclidean distance $D(P, Q)$ on S in the following sense. Corresponding to*

Received May 6, 1938.

[1] S. Bochner, *Bounded analytic functions in several variables and multiple Laplace integrals*, American Journal of Math., vol. 59(1937), pp. 731–738; *A theorem on analytic continuation of functions in several variables*, Annals of Math., vol. 39(1938), pp. 14–19. I am indebted to H. Behnke for pointing out to me that the theorem of the second note can be proved in a much simpler fashion. See K. Stein, *Zur Theorie der Funktionen mehrerer komplexer Veränderlichen*, Math. Annalen, vol. 114(1937), p. 557.

[2] This is the smallest convex set containing T; it is not necessarily closed.

any $\epsilon > 0$ *there exist points* $P_0, P_1, \cdots, P_n$ *in* S, *with the following properties:* (α) $P_0 = P$, $P_n = Q$; (β) *the Euclidean distance* $\overline{P_\nu P_{\nu+1}}$ *is* $< \epsilon$, *for* $\nu = 0, 1, \cdots, n-1$; *and*

$$(\gamma) \qquad \overline{P_0P_1} + \overline{P_1P_2} + \cdots + \overline{P_{n-1}P_n} \leqq D(P, Q).$$

(2) *The convex hull* $\tilde{S}$ *of* S *is* k-*dimensional.*

(3) *The function* $f(z)$ *is analytic and bounded in some* ($2k$-*dimensional, open*) *domain*[3] U *of* C_k *containing* T_S, *and is of integrable square in* T_S.

ASSERTION. *The function* $f(z)$ *exists and is of integrable square in the convex hull* $\tilde{T}$ *of* T. *Also, it satisfies relation* (2) *in* $\tilde{S}$ *for the same constant* K *as in* S.

This theorem includes our previous theorem bearing on open tubes. In fact, if T is open, assumptions (1) and (2) are trivially fulfilled, and assumption (3) is satisfied for $U = T$, subject to the inessential qualification that $f(z)$ is not necessarily bounded in the whole tube T but is so in every closed tube interior to it.

For the proof of the theorem we may assume that S is bounded. Otherwise we take a fixed point O of S, a sequence of spheres Σ_n in C_k with centers at O whose radii tend to infinity, and, for each n, the set S_n consisting of those points of S which, in the sense of assumption (1), can be connected with the point O within Σ_n. Obviously S_n tends to S, and T_{S_n} tends to T_S. Thus it is sufficient to prove our theorem for $S = S_n$.

Another specialization is less trivial, and will be justified afterwards; we assume that

$$(3) \quad f(z_1, \cdots, z_k) = O((y_1^2 + \cdots + y_k^2)^{-\frac{1}{2}(k+1)}) \qquad \text{as } y_1^2 + \cdots + y_k^2 \to \infty,$$

uniformly for $z \subset U$.

As a consequence of (3), there exists a constant A such that

$$(4) \qquad f(z_1, \cdots, x_\kappa \pm ia, \cdots, z_k) \leqq Aa^{-k-1}$$

for $\kappa = 1, \cdots, k$, and

$$(z_1, \cdots, x_\kappa \pm ia, \cdots, z_k) \subset U.$$

Corresponding to any $a > 0$ there exists in R_k an (open) neighborhood S_a of S such that the point set

$$(x_1, \cdots, x_k) \subset S_a \qquad (-a < y_\kappa < a; \kappa = 1, \cdots, k)$$

of C_k is contained in U. We may assume that S_a is contained in a sphere

$$(5) \qquad x_1^2 + \cdots + x_k^2 \leqq \alpha^2$$

whose radius α is independent of a. We now introduce the function

$$\varphi_a(x, t) = \int_{-a}^{a} \cdots \int_{-a}^{a} f_x(y) \cdot \exp\left[\sum_{\kappa=1}^{k} (x_\kappa + iy_\kappa)t_\kappa\right] dv_y$$

[3] Which need not contain an open tube.

for $x \subset S_a$ and arbitrary real points $t = (t_1, \cdots, t_k)$. We are interested in the partial derivative of this function with respect to x_1. The integrand on the right is an analytic function of $x_1 + iy_1$, hence its derivative with respect to x_1 is, but for the factor $-i$, its partial derivative with respect to y_1, thus

$$\frac{\partial \varphi_a(x, t)}{\partial x_1} = -i \int_{-a}^{a} \cdots \int_{-a}^{a} \frac{\partial}{\partial y_1} \left\{ f_x(y) \exp \left[\sum_{\kappa=1}^{k} (x_\kappa + iy_\kappa) t_\kappa \right] \right\} dv_y .$$

Carrying out the integration with respect to y_1, we obtain

$$-i \int_{-a}^{a} \cdots \int_{-a}^{a} f_x(a, y_2, \cdots, y_k) \exp \left[(x_1 + ia)t_1 + \sum_{\kappa=2}^{k} (x_\kappa + iy_\kappa)t_\kappa \right] dy_2 \cdots dy_k$$

and a similar term with $-a$ instead of a. Using (4), we obtain, for $x \subset S_a$,

$$\left| \frac{\partial \varphi_a(x, t)}{\partial x_1} \right| \leqq A(\alpha, t) a^{-1}$$

and in general,

$$\left| \frac{\partial \varphi_a(x, t)}{\partial x_\kappa} \right| \leqq A(\alpha, t) a^{-1} \qquad (\kappa = 1, \cdots, k),$$

where $A(\alpha, t)$ is independent of a. If $x' = (x_1', \cdots, x_k')$ and $x'' = (x_1'', \cdots, x_k'')$ are any two points of S, and if the connecting segment

$$x_\kappa = \rho x_\kappa' + (1 - \rho) x_\kappa'' \qquad (0 \leqq \rho \leqq 1)$$

lies in S_a, then

$$| \varphi_a(x', t) - \varphi_a(x'', t) | \leqq \sum_{\kappa=1}^{k} \max_{0 \leqq \rho \leqq 1} \left| \frac{\partial \varphi_a(x, t)}{\partial x_\kappa} \right| \cdot | x_\kappa' - x_\kappa'' |$$
$$\leqq kA(\alpha, t) \left[\sum_{\kappa=1}^{k} (x' - x'')^2 \right]^{\frac{1}{2}} \cdot a^{-1}.$$

Consequently, by assumption (1),

$$| \varphi_a(x', t) - \varphi_a(x'', t) | \leqq k \cdot A(\alpha, t) D(x', x'') \cdot a^{-1},$$

for any two points x', x'' of S. Letting $a \to \infty$, we conclude that the integral

$$\int_{-\infty}^{\infty} \cdots \int f_x(y) \exp \left[\sum_{\kappa=1}^{k} (x_\kappa + iy_\kappa) t_\kappa \right] dv_y$$

which we can set up for all points $x \subset S$ is actually independent of x. We denote it by $\varphi(t)$.

By Plancherel's formulas we have, for $x \subset S$,

$$f_x(y) \sim \frac{1}{(2\pi)^k} \int_{-\infty}^{\infty} \cdots \int \varphi(t) \exp \left[- \sum_{\kappa=1}^{k} (x_\kappa + iy_\kappa) t_\kappa \right] dv_t \tag{6}$$

and

$$\int_{-\infty}^{\infty} \cdots \int | f_x(y) |^2 dv_y = \frac{1}{(2\pi)^k} \int_{-\infty}^{\infty} \cdots \int | \varphi(t) |^2 \exp \left[-2 \sum_{\kappa=1}^{k} x_\kappa t_\kappa \right] dv_t . \tag{7}$$

Thus, by (2),

$$(8)\qquad \int_{-\infty}^{\infty}\cdots\int |\varphi(t)|^2 \exp\left(-2\sum_{\kappa=1}^{k} x_\kappa t_\kappa\right) dv_t \leqq (2\pi)^k K$$

for $x \subset S$. The function in (8) is the limit, as $a \to \infty$, of

$$\int_{-a}^{a}\cdots\int_{-a}^{a} |\varphi(t)|^2 \exp\left(-2\sum_{\kappa=1}^{k} x_\kappa t_\kappa\right) dv_t .$$

The latter function is convex along any straight line in R_k. Therefore it is $\leqq (2\pi)^k K$ for all $x \subset \tilde{S}$. We hence conclude that (8) holds for all $x \subset \tilde{S}$. Therefore we can construct the analytic function

$$(9)\qquad g_x(y) \equiv g(z) = \frac{1}{(2\pi)^k}\int_{-\infty}^{\infty}\cdots\int \varphi(t) \exp\left[-\sum_{\kappa=1}^{k}(x_\kappa + iy_\kappa)t_\kappa\right] dv_t ,$$

the integral converging absolutely and uniformly in every closed tube in the interior of $\tilde{T}$. Also

$$\int_{-\infty}^{\infty}\cdots\int |g_x(y)|^2\, dv_y \leqq K \qquad \text{for } x \subset \tilde{S}.$$

The next step is to identify the functions $f(z)$ and $g(z)$. Comparing (6) and (9), we easily see that they are identical if some point x^0 of S is an inner point of $\tilde{S}$. Otherwise, if S lies on the boundary of $\tilde{S}$, there exist $k + 1$ points of S on the boundary of $\tilde{S}$ such that the rectilinear simplex with these points as vertices is k-dimensional and part of $\tilde{S}$.

Any non-singular linear transformation

$$Az_\kappa = a_{\kappa 1} z_1 + \cdots + a_{\kappa k} z_k$$

with real coefficients $a_{\kappa\lambda}$ carries a tube into a tube, and the basis of the one tube into the basis of the other. In particular, we can choose the transformation in such a way that (after the transformation) our simplex contains the basis of a tube

$$(10)\qquad x_\kappa^0 < x_\kappa \leqq x_\kappa^0 + \eta, \qquad -\infty < y_\kappa < \infty,$$

the point x_κ^0 being a point of S. Our function $g(z)$ is analytic inside (10) and defined by (9), whereas $f(z)$ is analytic in a neighborhood of (1) and defined by (6) on (1), the function $\varphi(t)$ being the same in both cases. As in the case $k = 1$, we now readily conclude

$$(11)\qquad \lim_{x_\kappa \to x_\kappa^0} \int_{-\infty}^{\infty}\cdots\int |f_{x^0}(t) - g_x(t)|^2\, dv_t = 0.^{4}$$

But the function $f_x(y)$ is continuous in a neighborhood of (1), hence by (11), for every finite $a > 0$,

$$(12)\qquad \lim_{x_\kappa \to x_\kappa^0} \int_{-3a}^{3a}\cdots\int_{-3a}^{3a} |\delta_x(t)|\, dv_t = 0,$$

[4] E. C. Titchmarsh, *Introduction to the Theory of Fourier Integrals*, Oxford, 1937, p. 130, Theorem 97.

where

$$\delta(z) = f(z) - g(z)$$

is defined in the point set

$$x_\kappa^0 < x_\kappa \leqq x_\kappa^0 + \eta, \qquad 0 \leqq | y_\kappa | \leqq 3a,$$

for η sufficiently small (depending on a). For $0 < \epsilon < a$, the function

$$\delta^\epsilon(z) = \frac{1}{\epsilon^k} \int_0^\epsilon \cdots \int^\epsilon \delta(x_1 + iy_1 + iu_1, \cdots, x_k + iy_k + iu_k)\, dv_u$$

is analytic and bounded in

$$x_\kappa^0 < x_\kappa \leqq x_\kappa^0 + \eta, \qquad 0 \leqq | y_\kappa | \leqq 2a, \tag{13}$$

and on account of (12) it tends to 0 as $x_\kappa \to x_\kappa^0$. For any fixed numbers y_κ, η_κ such that

$$0 \leqq | y_\kappa | \leqq a, \qquad 0 \leqq \eta_\kappa \leqq \max(1, \eta),$$

the function

$$\lambda_\epsilon(w) = \delta^\epsilon(x_1^0 + iy_1 + \eta_1 w, \cdots, x_k^0 + iy_k + \eta_k w)$$

of the complex variable $w = u + iv$ is analytic and bounded in the rectangle

$$0 < u \leqq 1, \qquad -a \leqq v \leqq a$$

and tends to 0 as $u \to 0$. Hence $\lambda_\epsilon(w) = 0$. In particular $\lambda_\epsilon(1) = 0$, or

$$\delta^\epsilon(x_1^0 + \eta_1 + iy_1, \cdots, x_k^0 + \eta_k + iy_k) = 0$$

for all values (13). Thus $\delta^\epsilon(z)$ vanishes on a $2k$-dimensional set and is therefore identically zero. But $\delta(z) = \lim_{\epsilon \to 0} \delta^\epsilon(z)$, and hence $\delta(z) = 0$, or $f(z) = g(z)$.

This completes the proof of our theorem under the additional assumption (3). If $f(z)$ is bounded,

$$f^\sigma(z) = f(z) \exp [\sigma(z_1^2 + \cdots + z_k^2)]$$

will satisfy (3) for $0 < \sigma < 1$, and since, for (5),

$$\int_{-\infty}^{\infty} \cdots \int | f_x^\sigma(y) |^2 dv_y \leqq A(\alpha) \int_{-\infty}^{\infty} \cdots \int | f_x(y) |^2 dv_y$$

and

$$\lim_{\sigma \to 0} \int_{-\infty}^{\infty} \cdots \int | f_x^\sigma(y) |^2 dv_y = \int_{-\infty}^{\infty} \cdots \int | f_x(y) |^2 dv_y,$$

our theorem holds for $f^\sigma(z)$, and for $f(z)$ itself.

PRINCETON UNIVERSITY.

SOME THEOREMS ON QUASI-ANALYTICITY FOR FUNCTIONS OF SEVERAL VARIABLES.*

By S. BOCHNER and A. E. TAYLOR.[1]

Introduction. The notion of quasi-analyticity may be described roughly as follows. A class of functions defined in a fixed domain is said to be quasi-analytic if within the class an individual function is completely determined by its behavior in an arbitrarily small sub-region of its domain of definition, or, what is sometimes the same thing, if the function is determined by the knowledge of its value and that of all its derivatives at a single point.

In the present paper we shall exhibit various ways of realizing, for definite classses of functions of several variables, this notion of quasi-analyticity. All of our results depend upon the following theorem of Denjoy and Carleman, which has to do with functions of one real variable.[2]

THEOREM D. *Let f be a real function of the real variable x, defined and possessing derivatives of all orders on the interval $a \leqq x \leqq b$. Then if $f^{(n)}(a) = 0$ $(n = 0, 1, \cdots)$ and if for all values of x the inequalities*

$$| f^{(n)}(x) | \leqq m_n \qquad (n = 0, 1, \cdots)$$

are satisfied, where the constants m_n are such that the series

$$\Sigma\, m_n^{-1/n}$$

is divergent, then f is identically zero.

It is evident from this that for a fixed sequence of positive constants $\{m_n\}$ such that $\Sigma m_n^{-1/n}$ is divergent the class of functions f defined on (a, b) and such that

$$| f^{(n)}(x) | \leqq A^n m_n \qquad (n = 0, 1, \cdots)$$

where A is a constant depending only on f, is quasi-analytic in the sense that if f and g are in the class, and $f^{(n)}(a) = g^{(n)}(a)$, then $f \equiv g$.

When one seeks to establish some similar result for functions of several variables a number of different modes of procedure present themselves. If we

* Received July 28, 1938.

[1] National Research Fellow.

[2] T. Carleman, *Les Fonctions Quasi Analytiques*, Paris (1926), p. 20.

consider, to begin with, Euclidean spaces only, we may, with a fixed domain R, restrict the magnitude of the function and all its partial derivatives in R. Or we may perhaps require restrictions on certain combinations of these derivatives only, that is to say, on certain differential operators applicable to f in R—the successive iterates of the Laplacian, for example. Again, the conditions $f^{(n)}(a) = 0$ may have as their analogues in the generalization conditions on the function and its derivatives at a single point, or they may be replaced by similar conditions relating to a point set in R with the property that a function which is analytic in R and vanishes on the set must vanish identically. We have made use of each of these alternatives, and it is on the basis of them that the paper is divided into two parts, the second part dealing with certain differential operators, especially the Laplacian. The theorems of the second part, it will be seen, are essentially theorems on functions of several variables, being genuinely different in character from those of the first part.

We have also considered Riemannian spaces; in this case likewise we have resorted to generalizations of two different kinds. Theorems 3, 4, 5, 7 (Part I) and Theorem 10 (Part II) deal with Riemannian spaces. In particular, Theorem 10 is concerned with functions defined on a space of constant positive curvature, the unit sphere

$$x_1^2 + \cdots + x^2_{k+1} = 1$$

in Euclidean space of $k+1$ dimensions, $k \geqq 2$. In its formal statement Theorem 10 differs from Theorem D merely by the substitution of $\Delta^n f$ for $f^{(n)}(x)$, Δ being the Laplace-Beltrami operator for the sphere, and the substitution of a point set of specified character for the single point $x = a$.

In order to avoid repetitions in the statement of our theorems, all of which have the same general form, we shall agree at the outset on the following conventions: a) f shall denote a real, continuous functions defined in some domain (= open set) of a space; it shall possess continuous derivatives of all orders with respect to the coördinates used. b) $\{m_n\}$ $n = 0, 1, \cdots$ shall denote a sequence of positive constants, arbitrary except for the conditions imposed in the statement of the theorem in which it occurs. c) Statements involving the index n shall hold for $n = 0, 1, \cdots$ unless the contrary is explicitly asserted. d) If $(x_1, \cdots, x_k)$ are the coördinates of a point, we shall frequently refer to it as the point x; if f is a function of $x_1, \cdots, x_k$ we shall often express this by the notation $f(x)$. e) E_k shall denote the Eucildean arithmetic space of k dimensions; V_k shall denote a Riemannian space of k dimensions. We shall require merely that the components of the fundamental metric tensor g_{ij} for the space V_k be of class C^∞; that is, they admit continuous partial derivatives of all orders. We stress the fact that the g's need not be analytic.

Part I.

1. If f is defined over a domain of E_k we introduce the symbols

$$(1.1)\qquad \begin{aligned} D_0(f,x) &= |f(x)| \\ D_n(f,x) &= \left\{ \sum \left| \frac{\partial^n f}{\partial x_{\alpha_1} \cdots \partial x_{\alpha_n}} \right|^2 \right\}^{1/2} \qquad (n = 1, 2, \cdots) \end{aligned}$$

where $\alpha_1, \cdots, \alpha_n$ are summed independently from 1 to k.

THEOREM 1. *Let f be defined in a connected domain R in E_k, and let x^0 be an interior point of R. Then the conditions*

a) $D_n(f,x) \leqq m_n \quad x$ in R

b) $D_n(f,x^0) = 0$

c) $\sum m_n^{-1/n} = \infty \quad$ (the series is divergent)

imply that f is zero throughout R.

Proof. Selecting an arbitrary straight line emanating from x^0 and a point x' on this line such that the segment x^0x' lies wholly in R, we have as a parametric representation of this segment

$$x_i(s) = a_i s + x_i^0 = \frac{(x'_i - x_i^0)}{d} s + x_i^0$$

where d is the distance from x^0 to x' and s is the distance from x^0 to $x(s)$, so that

$$(1.2)\qquad \sum_1^k a_i^2 = 1.$$

Define

$$F(s) = f(x_1(s), \cdots, x_k(s)) \equiv f(x(s)).$$

Then clearly

$$F^{(n)}(s) = \sum a_{\alpha_1} \cdots a_{\alpha_n} f_{\alpha_1 \ldots \alpha_n}(x(s))$$

where

$$f_{\alpha_1 \ldots \alpha_n}(x) = \frac{\partial^n f}{\partial x_{\alpha_1} \cdots \partial x_{\alpha_n}}.$$

By (1.2) and the Cauchy-Schwarz inequality we find

$$|F^{(n)}(s)| \leqq D_n(f, x(s)) \leqq m_n \qquad 0 \leqq s \leqq d$$

and so also $F^{(n)}(0) = 0$. Therefore by Theorem D, $F(s) = 0$ and in particular $f(x') = 0$.

Since R is connected it is obvious that a finite number of applications of the above reasoning allows us to conclude, as in the process of analytic continuation, that f is zero at all points of R.

Theorem 1 is perhaps the most obvious generalization of Theorem D. The generalization may be pushed even further, to include functions defined on an arbitrary Banach space[3] E. For this purpose we introduce the following considerations. If E is a Banach space with elements $x, y, \cdots$, and f is a real function defined in a domain R in E, we shall say that f is differentiable arbitrarily often along all lines in R if the quantities

$$\delta^n f(x; y) = \left[\frac{\partial^n f(x + sy)}{\partial s^n}\right]_{s=0} \tag{1.3}$$

exist for each x in R and each y in E. It is readily seen that if E is the Euclidean space E_k we have

$$\delta^n f(x; y) = \sum \frac{\partial^n f}{\partial x_{a_1} \cdots \partial x_{a_n}} y_{a_1} \cdots y_{a_n}$$

and consequently

$$| \delta^n f(x; y) | \leqq D_n(f, x) \{ \sum y_i^2 \}^{n/2}.$$

This leads us at once to the statement of our next theorem.

THEOREM 2. *Let f be defined in a connected domain R of a Banach space E, and let f be differentiable arbitrarily often along all lines in R. Then if x^0 is a point in R the conditions*

$$\begin{array}{ll} \text{a)} & | \delta^n f(x; y) | \leqq m_n \| y \|^n \quad x \text{ in } R,\ y \text{ in } E \\ \text{b)} & \delta^n f(x^0; y) = 0 \\ \text{c)} & \sum m_n^{-1/n} = \infty \end{array}$$

imply that f is zero throughout R.

Proof. If x' is chosen as in the proof of Theorem 1, and we put

$$F(s) = f(x^0 + s(x' - x^0)) \qquad 0 \leqq s \leqq 1$$

then

$$F^{(n)}(s) = f(x^0 + s(x' - x^0); x' - x^0)$$

and by conditions a)–c) and Theorem D we conclude that $F(s) \equiv 0$. The rest of the proof is then obvious.

2. In order to formulate theorems of the desired type for Riemannian spaces it is first of all necessary to devise something corresponding to the

[3] For the definition of such spaces see Benach, *Opération Linéaires,* Warsaw (1932), Chapter V.

quantities $D_n(f, x)$ entering in Theorem 1. It is desirable to define $D_n(f, x)$ in invariant form; covariant differentiation naturally suggests itself. However, if we keep formulae (1.1) in mind and think in terms of normal coördinates, it is another process which occurs most readily as the basis for our procedure. This process is that of *extension.* All that we need of it for our purposes is the following. If f is a function defined in V_k with x-coördinate system, and x_0 is a fixed point, we introduce normal coördinates $(y^1, \cdots, y^k)$ with origin at this point. Then if

$$f^*(y^1, \cdots, y^k) \equiv f(x^1, \cdots, x^k) \tag{2.1}$$

we define

$$f_{a_1 \ldots a_n}(x_0) = \left(\frac{\partial^n f^*}{\partial y^{a_1} \cdots \partial y^{a_n}}\right)_{y=0}. \tag{2.2}$$

The functions $f_{a_1 \ldots a_n}$ defined by (2.2) constitute the components of an absolute tensor, covariant of rank n.[4]

Since we shall make extensive use of normal coördinates it will be convenient to assemble for reference the principal facts on which we shall rely. In the neighborhood of any given point of V_k it is possible to introduce normal coördinates y^i with origin at the point in question. If f is a scalar function defined over a portion of V_k we shall use f^* to denote this scalar as a function of normal coördinates about a chosen point. We shall use ψ_{ij} to denote the components, in normal coördinates, of the fundamental metric tensor of the space, and $C_{\alpha\beta}{}^{\gamma}$ to denote the Christoffel symbols in normal coördinates. The normal coördinates may be so altered by a linear transformation that at the origin of the system

$$\psi_{ij}(0) = \psi^{ij}(0) = \delta_j{}^i. \tag{2.3}$$

The equations of a geodesic through the origin are

$$y^i = a^i s \tag{2.4}$$

where s is the distance along the geodesic, measured from the origin, and the a's are constants satisfying the relation

$$\sum_1^k a^{i^2} = 1. \tag{2.5}$$

Along the geodesic the following fundamental relations hold.[5]

[4] T. Y. Thomas, *The Differential Invariants of Generalized Spaces*, Cambridge University Press (1934), p. 97.

[5] T. Y. Thomas, *op. cit.*, p. 85-87.

$$(2.6)\qquad \begin{cases} s^2 = \psi_{ij}(0)y^i y^j = \psi_{ij}(y)y^i y^j \\ \psi_{ij}(0)y^i = \psi_{ij}(y)y^i \end{cases}$$

$$(2.7)\qquad C_{\alpha\beta}{}^{\gamma} y^\alpha y^\beta = 0.$$

With the quantities (2. 2) we define

$$(2.8)\qquad \begin{aligned} D_0(f,x) &= |\, f(x) \,| \\ D_n(f,x) &= \{ g^{\alpha_1\beta_1} \cdots g^{\alpha_n\beta_n} f_{\alpha_1 \ldots \alpha_n} f_{\beta_1 \ldots \beta_n} \}^{1/2}. \end{aligned}$$

Using normal coördinates with origin at x, and taking account of (2. 2), (2. 3) we see that

$$D_n(f,x) = \left\{ \sum \left| \frac{\partial^n f^*}{\partial y^{\alpha_1} \cdots \partial y^{\alpha_n}} \right|^2_{y=0} \right\}^{1/2}.$$

Thus $D_n(f,x)$ is a real, non-negative scalar.

We are now in a position to enunciate a theorem.

THEOREM 3. *Let f be defined in a connected domain R of V_k. Then, x_0 being a point in R and $D_n(f,x)$ being given by* (2. 8), *f is identically zero provided that conditions* a)–c) *of Theorem* 2 *are fulfilled.*

Proof. Consider a geodesic sphere of radius r in R with center at x_0, and a fixed geodesic passing through x_0. Along this geodesic f becomes a function $F(s)$ of the distance s from x_0. Let x_1 be an arbitrary point on the given geodesic, within the geodesic sphere. Then if we choose normal coördinates with origin at x_1 we have equations similar to (2. 4) for the geodesic, s being replaced by $s - s_1$, where s_1 is the value of s at $x = x_1$. Clearly

$$F^{(n)}(s_1) = a^{\alpha_1} \cdots a^{\alpha_n} \left(\frac{\partial^n f^*}{\partial y^{\alpha_1} \cdots \partial y^{\alpha_n}} \right)_{y=0}$$

and therefore, by (2. 5) and the Schwarz inequality

$$|\, F^{(n)}(s_1) \,| \leqq D_n(f, x_1) \leqq m_n.$$

Hence also $F^{(n)}(0) = 0$. The reasoning then follows that of the previous proofs.

THEOREM 4. *The hypothesis and conclusion here are the same as in Theorem 3, except that in the definition of $D_n(f,x)$* [formula (2. 8)], *$f_{\alpha_1 \ldots \alpha_n}$ shall signify the components of the n-th successive covariant derivative of f.*

Proof. Consider as before a geodesic sphere with center at x_0, and the function $F(s)$. It is clear that it is sufficient to prove that for an arbitrary geodesic through x_0 the inequalities

$$(2.9) \qquad | F^{(n)}(s) | \leqq D_n(f, x)$$

are satisfied, where x and s refer to the same point. We shall first turn our attention to a lemma independent of the above considerations.

LEMMA 1. *Let S be a region in V_k which is entirely covered by a system of normal coördinates y^i with origin in the region. Let f be a scalar defined in S. If the geodesics through the origin are defined by (2.4) subject to condition (2.5), so that the variables (s, a) furnish a fixed set of "polar" coördinates covering S, we define*

$$F(s; a) = f^*(y).$$

In the region S with the origin excluded the derivatives

$$F^{(n)}(s; a) \equiv \frac{\partial^n F(s; a)}{\partial s^n}$$

are then seen to be scalars. As such they are given by the formulae

$$(2.10) \qquad F^{(n)}(s; a) = \psi^{i_1 a_1} \cdots \psi^{i_n a_n} \Omega_{,i_1} \cdots \Omega_{,i_n} f^*_{a_1 \ldots a_n}$$

where $\Omega_{,i}$ are the components of the covariant derivative of the scalar

$$(2.11) \qquad \Omega = \{\psi_{ij}(y) y^i y^j\}^{1/2}$$

and $f^*_{a_1 \ldots a_n}$ are the components of the n-th successive covariant derivative of f^*.

Proof of the lemma. We have

$$F'(s; a) = a^a \frac{\partial f^*}{\partial y^a} = \frac{y^a}{s} \frac{\partial f^*}{\partial y^a}.$$

Now

$$\Omega_{,i} = \frac{\partial \Omega}{\partial y^i}, \qquad f^*_a = \frac{\partial f^*}{\partial y^a}$$

and, remembering relations (2.6)

$$\frac{\partial \Omega}{\partial y^i} = \frac{\psi_{ij}(y) y^j}{\Omega} = \frac{\psi_{ij}(0) y^j}{\Omega}$$

from which follows

$$(2.12) \qquad \psi^{ia} \frac{\partial \Omega}{\partial y^i} = \frac{\psi^{ia} \psi_{ij} y^j}{\Omega} = \frac{y^a}{s}.$$

Therefore (2.10) has been proved for $n = 1$.

Replacing the scalar $F(s; a)$ by $F'(s; a)$ and applying the result just obtained we get

$$F^{(2)}(s;a) = \psi^{i\alpha}\Omega_{,i}(\psi^{j\beta}\Omega_{,j}f^*{}_\beta)_{,a}$$
$$= \psi^{i\alpha}\psi^{j\beta}(\Omega_{,i}\Omega_{,j}f^*{}_{\beta a} + \Omega_{,i}\Omega_{,j,a}f^*{}_\beta).$$

But this will agree with (2.10) if

$$\text{(2.13)} \qquad \psi^{i\alpha}\psi^{j\beta}\Omega_{,i}\Omega_{,j,a} = 0 \qquad (\beta = 1, \cdots, k).$$

To prove (2.13) we write the left side in the form

$$\frac{y^\alpha}{s}\left(\frac{y^\beta}{s}\right)_{,a} = \frac{y^\alpha}{s}\left\{\frac{\partial}{\partial y^\alpha}\left(\frac{y^\beta}{s}\right) + \frac{y^\gamma}{s}C_{\gamma\alpha}{}^\beta\right\}$$

which is possible because of (2.12) and the fact that the covariant derivative of the tensor ψ_{ij} vanishes. But the last expression above on the right is zero because of the fundamental relations (2.6), (2.7), as is easily verified.

Proceeding in this way, by repeated use of (2.10) for $n = 1$, and the relation (2.13), we conclude (2.10) in general.

If in (2.10) we fix the a's and allow s to approach zero we find, in view of (2.12) and (2.4), (2.5), that

$$\text{(2.14)} \quad \lim_{s\to 0} | F^{(n)}(s;a) | = | a^{\alpha_1} \cdots a^{\alpha_n} f^*{}_{\alpha_1 \ldots \alpha_n}(0) | \leqq \{\Sigma \, | f^*{}_{\alpha_1 \ldots \alpha_n}(0) |^2\}^{1/2}$$

Now we return to the proof of Theorem 4. We shall prove the validity of (2.9). To do this we fix the geodesic and the point x on it. If now we consider a neighborhood of x, and normal coördinates with this point as origin, we may apply the lemma to the scalar f in this neighborhood. In particular, from (2.14) we infer that

$$| F^{(n)}(s) | \leqq \{\Sigma \, | f^*{}_{\alpha_1 \ldots \alpha_n}(0) |^2\}^{1/2} \qquad (n = 1, 2, \cdots).$$

But

$$D_n(f, x) = \psi^{\alpha_1\beta_1}(0) \cdots \psi^{\alpha_n\beta_n}(0) f^*{}_{\alpha_1 \ldots \alpha_n}(0) f^*{}_{\beta_1 \ldots \beta_n}(0)$$
$$= \{\Sigma \, | f^*{}_{\alpha_1 \ldots \alpha_n}(0) |^2\}^{1/2}$$

because of (2.3). Hence $| F^{(n)}(s) | \leqq D_n(f, x)$. Thus the proof is complete.

Remark. In Theorems 3 and 4 we may replace the assumption that x_0 is an interior point of R by the requirement that it be a boundary point of R which is the vertex of a conical sub-region of R filled out entirely by geodesic arcs issuing from x_0, the condition b) being in this case replaced by the condition

b′) $\lim D_n(f, x) = 0$ as $x \to x_0$ along a path in R.

3. In this section we shall obtain three theorems in which restrictions are placed, not on the quantities $D_n(f, x)$ themselves, but on the integrals of these expressions.

THEOREM 5. *If f is defined in a connected domain R in V_k, and $D_n(f, x)$ has the meaning ascribed to it in either of the two preceding theorems, then f vanishes identically provided that conditions* a)–c) *below are fulfilled.*

a) $\int_R D_{n+1}(f, x)\, dv \leqq m_n \qquad (n = -1, 0, 1, \cdots)$

b) $f = 0$ in a neighborhood of a point x_0 in R

c) $\sum m_n^{-1/n} = \infty$,

(m_{-1} is a positive constant, and $dv = \sqrt{g}\, dx^1 \cdots dx^k$ is the volume element in V_k).

Proof. Consider the non-negative function

$$\Phi(x) = \sum_0^\infty \frac{1}{2^n m_n} D_{n+1}(f, x).$$

From a) we see that

$$\int_R \Phi\, dv \leqq 2.$$

Let S_0 be a geodesic sphere with center at x_0, lying in R, and entirely covered by a system of normal coördinates with origin at x_0. Since in such a sphere $\sqrt{g} \geqq c > 0$ we conclude that the integral

$$\int_{S_0} \Phi\, dy^1 \cdots dy^k$$

is finite, y^i being normal coördinates. Now we may evaluate this integral by polar coördinates,

$$\int_{S_0} \Phi\, dy^1 \cdots dy^k = \int_{W_k} \left(\int_0^{r_0} s^{k-1} \Phi\, ds \right) d\omega$$

where r_0 is the radius of S_0, and W_k, $d\omega$ denote the surface of the k-dimensional unit sphere and its surface element, respectively. By Fubini's theorem we conclude that

$$\int_0^{r_0} s^{k-1} \Phi\, ds$$

is finite for almost all geodesics issuing from x_0. But f, and therefore Φ, vanishes in the neighborhood of x_0. Consequently the integral remains finite with the factor s^{k-1} of the integrand suppressed.

It results from these considerations that if we define $F(s)$ as in the preceding work then since $|F^{(n)}(s)| \leqq D_n(f, x)$ and $F^{(n)}(0) = 0$ we have

$$| F^{(n)}(s) | = | \int_0^s F^{(n+1)}(s)\,ds | \leqq \int_0^{r_0} D_{n+1}(f, x)\,ds$$
$$| F^{(n)}(s) | \leqq C^n m_n \qquad 0 \leqq s \leqq r_0$$

where C is a constant depending on f and the particular geodesic. By Theorem D, $F(s)$ must be zero, so that f vanishes along almost all geodesics through x_0 in S_0. It then vanishes identically, for it is continuous. The conclusion of the theorem is then reached as before.

The stringency of condition b) can be removed if we alter c) somewhat. We shall do this first for Euclidean space.

THEOREM 6. *Let f be defined throughout E_k, and let x_0 be an arbitrary point. Then the conditions*

$$\text{a)} \quad \int_{E_k} D_n(f, x)\,dv \leqq m_n$$
$$\text{b)} \quad D_n(f, x_0) = 0$$
$$\text{c)} \quad \sum (m_n + \cdots + m_{n+k-1})^{-1/n} = \infty$$

imply that $f = 0$.

Proof. Let

$$\Phi(x) = \sum_0^\infty \frac{1}{2^n m_n} D_n(f, x).$$

Then by a)

$$\int_{E_k} \Phi\,dv \leqq 2.$$

If we adopt spherical coördinates about the point x_0 and denote by s the distance measured along a half-line issuing from x_0, then we conclude by Fubini's theorem, in a manner similar to that of the previous proof, that for almost all half-lines emanating from x_0 the integral

$$\int_0^\infty s^{k-1}\Phi\,ds$$

is finite, and therefore, for some constant A depending on f and the particular line

$$\int_0^\infty s^{k-1} D_n(f, x(s))\,ds \leqq A^n m_n.$$

Now, in the notation used in proving Theorem 1, $F(s) = f(x(s))$ and $| F^{(n)}(s) | \leqq D_n(f, x(s))$. Thus

$$\int_0^\infty | F^{(n)}(s) |\, s^{k-1} ds \leqq A^n m_n \tag{3.1}$$

$$(3.2) \qquad F^{(n)}(0) = 0.$$

It will be proved in the lemma which follows that (3.1) and (3.2), together with condition c) imply that $F(s) \equiv 0$ $(0 \leqq s < \infty)$. Since f is continuous in E_k, this implies that $f = 0$, thus completing the proof. We turn now to the lemma.

LEMMA 2. *Let F be a real function of class C^∞ on the infinite interval $(0, \infty)$, and let the relations* (3.1), (3.2) *above, and condition* c) *of Theorem* 6 *be fulfilled for an integer $k \geqq 1$. Then $F \equiv 0$.*

Proof. $F^{(n)}$ is absolutely integrable, by (3.1). Therefore, for complex z with positive real part $(R(z) > 0)$ the integral

$$(3.3) \qquad H(z) = \int_0^\infty e^{-zs} F(s)\, ds$$

exists and defines an analytic function. Integrating by parts repeatedly and using (3.2) we obtain

$$z^n H(z) = \int_0^\infty e^{-zs} F^{(n)}(s)\, ds.$$

Now we write $H_n(z) = z^n H(z)$. Then

$$\frac{d^{k-1} H_n(z)}{dz^{k-1}} = (-1)^{k-1} \int_0^\infty s^{k-1} e^{-zs} F^{(n)}(s)\, ds$$

and if $R(z) > 0$ we infer from (3.1) that

$$(3.4) \qquad \left| \frac{d^{k-1} H_n(z)}{dz^{k-1}} \right| \leqq A^n m_n.$$

For a fixed z, $R(z) > 0$, we define

$$(3.5) \qquad \phi(t) = H_n((1-t)z + t) \qquad 0 \leqq t \leqq 1.$$

Integration by parts, applied $k - 2$ times to the equation

$$\phi(1) - \phi(0) = \int_0^1 \phi'(t)\, dt$$

yields

$$\phi(0) = \sum_{\nu=0}^{k-2} (-1)^\nu \frac{1}{\nu!} \phi^{(\nu)}(1) + \frac{(-1)^{k-1}}{(k-2)!} \int_0^1 t^{k-2} \phi^{(k-1)}(t)\, dt.$$

Hence

$$(3.6) \qquad H_n(z) = P_n(z) + \frac{(-1)^{k-1}}{(k-2)!} \int_0^1 t^{k-2} \phi^{(k-1)}(t)\, dt$$

where

5

$$P_n(z) = \sum_{\nu=0}^{k-2} \frac{(z-1)^\nu}{\nu!} H_n^{(\nu)}(1) \tag{3.7}$$

and

$$\phi^{k-1}(t) = (-1)^{k-1}(z-1)^{k-1} H_n^{(k-1)}((1-t)z+t).$$

Combining (3.4) and (3.6) we find

$$| H_n(z) - P_n(z) | \leqq \frac{| z-1 |^{k-1}}{(k-2)!} A^n m_n. \tag{3.8}$$

$P_n(z)$ as a function of n is a polynomial of degree $\leqq k-2$. Hence, differencing with respect to n,

$$\sum_{\nu=0}^{k-1} (-1)^\nu \binom{k-1}{\nu} P_{n-\nu}(z) = 0 \qquad (n = k-1, k, \cdots). \tag{3.9}$$

Now

$$\sum_{\nu=0}^{k-1} (-1)^\nu \binom{k-1}{\nu} H_{n-\nu}(z) = H(z) \sum_{\nu=0}^{k-1} (-1)^\nu \binom{k-1}{\nu} z^{n-\nu} = H(z) z^{n-k+1} (z-1)^{k-1}. \tag{3.10}$$

Thus, combining (3.8), (3.9), (3.10)

$$| H(z) z^{n-k+1} (z-1)^{k-1} | \leqq \sum_{\nu=0}^{k-1} \binom{k-1}{\nu} \frac{| z-1 |^{k-1}}{(k-2)!} A^{n-\nu} m_{n-\nu}.$$

From this inequality we obtain, making the unimportant assumption that $A \geqq 1$, the inequalities

$$| z^n H(z) | \leqq A^{n+k-1} (m_n + \cdots + m_{n+k-1}). \tag{3.11}$$

It then follows by condition c) and a lemma of Carleman [6] (which he uses for the proof of Theorem D) that the analytic function $H(z)$ is identically zero. Then $F \equiv 0$ also, as was to be proved.

Before going on to the next theorem we shall prove another lemma.

LEMMA 3. *Let $f(x)$ be of class C^∞ on $(0, 1)$, and suppose that*

a) $\int_0^1 | f^{(n)}(x) | x^{k-1} dx \leqq A^n m_n$

b) $f^{(n)}(0) = 0$

c) $m_1^{-1} \geqq m_2^{-1/2} \geqq \cdots \geqq m_n^{-1/n} \geqq \cdots \to 0$

d) $\sum (m_{n+k-1})^{-1/n} = \infty$

[6] T. Carleman, *op. cit.*, p. 20, Corollary. The statement is that if $\Phi(z)$ is analytic in the half plane $R(z) \geqq a \geqq 0$, and if $| \Phi(z) z^{\lambda_n} | \leqq \beta_n^{\lambda_n}$, where $0 < \lambda_0 < \lambda_1 < \cdots < \lambda_n < \cdots \to \infty$ and $\beta_n > 0$, then the divergence of the series

$$\Sigma \frac{\lambda_n - \lambda_{n-1}}{\beta_n}$$

implies $\Phi \equiv 0$.

where k is an integer $\geqq 1$ *and A is a constant* (which we may assume to be $\geqq 1$). *Then* $f \equiv 0$.

Proof. We shall reduce the problem to that of Lemma 2. Putting

$$x = \frac{t}{1+t} \qquad g(t) = f(x) \qquad 0 \leqq t < \infty$$

we calculate the derivatives of $g(t)$:

$$(3.12) \quad g^{(n)}(t) = \frac{a_{n,n}}{(1+t)^{2n}} f^{(n)}(x) + \frac{a_{n,n-1}}{(1+t)^{2n-1}} f^{(n-1)}(x) + \cdots + \frac{a_{n,1}}{(1+t)^{n+1}} f^{(1)}(x)$$

where the $a_{n,\nu}$ are constants, with $a_{n,n} = 1$. Thus, using a)

$$(3.13) \qquad \int_0^\infty \frac{|g^{(n)}(t)|}{(1+t)^{k+1}} t^{k-1} dt \leqq A^n \sum_0^n |a_{n,\nu}| m_\nu.$$

If we differentiate (3.12) we obtain the recursion formula

$$(3.14) \qquad a_{n+1,\nu+1} = a_{n,\nu} - (n+\nu+1) a_{n,\nu+1}, \qquad (\nu = 0, \cdots, n)$$

which is correct in all cases if we define $a_{n,0} = 0$, $a_{n,\nu} = 0$ if $\nu > n$. Therefore

$$(3.15) \qquad |a_{n+1,\nu+1}| \leqq |a_{n,\nu}| + 2n |a_{n,\nu+1}|.$$

Consider now the expression

$$P_n(\xi) = \xi(\xi+2) \cdots (\xi + 2n - 2) = \sum_0^n b_{n,\nu} \xi^\nu.$$

$P_n(\xi)$ is a polynomial of degree n in ξ. Clearly $b_{n,\nu} \geqq 0$ and

$$(3.16) \qquad P_n(\xi) \leqq (\xi + 2n)^n \qquad \xi \geqq 0.$$

It is easily verified that $b_{n,n} = 1$ and that

$$(3.17) \qquad b_{n+1,\nu+1} = b_{n,\nu} + 2n b_{n,\nu+1}$$

where we again adopt the conventions $b_{n,0} = 0$, $b_{n,\nu} = 0$ if $\nu > n$. Since $a_{11} = b_{11} = 1$ a comparison of (3.15) and (3.17) shows that

$$(3.18) \qquad |a_{n,\nu}| \leqq b_{n,\nu}.$$

Now let us write $\alpha_n = m_n^{1/n}$. By c) the α's have the property

$$(3.19) \qquad \alpha_1 \leqq \alpha_2 \leqq \cdots \leqq \alpha_n \leqq \cdots \to \infty.$$

From (3.13), (3.18), (3.19) and (3.16) we conclude

$$(3.20)\qquad \int_0^\infty \frac{|g^{(n)}(t)|}{(1+t)^{k+1}}\, t^{k-1}dt \leqq A^n(\alpha_n+2n)^n.$$

Finally, we set

$$h(t)=\frac{g(t)}{(1+t)^{k+1}}.$$

Then

$$h^{(n)}(t)=\sum_0^n(-1)^\nu\binom{n}{\nu}(k+1)\cdots(k+\nu)\frac{g^{(n-\nu)}(t)}{(1+t)^{k+1+\nu}}$$

and so

$$|h^{(n)}(t)|\leqq\sum_{\nu=0}^n\binom{n}{\nu}(k+n)^\nu\frac{|g^{(n-\nu)}(t)|}{(1+t)^{k+1}}.$$

From this result and (3.19), (3.20) we get

$$\int_0^\infty |h^{(n)}(t)|\, t^{k-1}dt\leqq A^n(\alpha_n+3n+k)^n$$

or, for a suitable constant B (depending only on A and k),

$$(3.21)\qquad \int_0^\infty |h^{(n)}(t)|\, t^{k-1}dt\leqq B^n(\alpha_n+4n)^n.$$

Since $h^{(n)}(0)=0$ the conclusion $h(t)\equiv 0$ will follow from Lemma 2 if we can prove that, writing $\beta_n=\alpha_n+4n$, we have

$$(3.22)\qquad \sum(\beta_n{}^n+\cdots+\beta_{n+k-1}^{n+k-1})^{-1/n}=\infty.$$

Now $\beta_1\leqq\beta_2\leqq\cdots\leqq\beta_n\leqq\cdots\to\infty$. Therefore the series in (3.22) will diverge if

$$(3.23)\qquad \sum(\beta_{n+k-1})^{-1-(k-1)/n}=\infty.$$

In proving this we shall distinguish two cases. Suppose first that for all sufficiently large values of n, $\alpha_n\geqq n$, $\beta_n\leqq 5\alpha_n$. In this case the series (3.23) will diverge if

$$\sum(\alpha_{n+k-1})^{-1-(k-1)/n}=\infty.$$

But this is true because of condition d) of the lemma and the definition of α_n. The alternative situation is that in which there exists a strictly increasing sequence $\{n_\nu\}$ of integers such that $\alpha_{n_\nu}<n_\nu$ $(\nu=1,2,\cdots)$. Thus we have

$$(3.24)\qquad \beta_n\leqq 5n_\nu \quad \text{if } n_{\nu-1}<n\leqq n_\nu.$$

Now with the omission of at most some early terms the series (3.23) may be written

$$\sum (\beta_n)^{-n/(n-k+1)} = \sum_{\nu} \sum_{n=n_{\nu-1}+1} (\beta_n)^{-n/(n-k+1)}$$

and so, by (3.24), it diverges if the series

$$\sum_{\nu} \sum_{n=n_{\nu-1}+1} (n_\nu)^{-n/(n-k+1)} \tag{3.25}$$

is divergent.

In proving the divergence of (3.25) we may obviously assume that n_ν tends to infinity as rapidly as we please, for this reduces the size of the partial sums. In particular if we can show that for some constant $\mathcal{E}_0 > 0$ and each integer $m \geqq k$ it is possible to determine an integer $N = N(m)$ so large that

$$\sum_{n=m}^{N} N^{-n/n-k+1} \geqq \mathcal{E}_0$$

then we see that with $n_{\nu-1}$ chosen, n_ν may be selected so as to contribute at least $\mathcal{E}_0$ to the sum of the series, which must then diverge. Now for fixed $m > k$

$$\lim \sum_{n=k}^{m} N^{-n/n-k+1} = 0 \text{ as } N \to \infty.$$

Hence we need only prove that

$$\liminf \sum_{n=k}^{N} N^{-n/n-k+1} > 0 \text{ as } N \to \infty. \tag{3.26}$$

Now

$$\sum_{n=k}^{N} N^{-n/n-k+1} \geqq \int_k^{N+1} N^{-t/t-k+1} dt.$$

Making the change of variable $u = \log N/t - k + 1$ we obtain, as soon as $\log N > 1$,

$$\begin{aligned}\sum_{n=k}^{N} N^{-n/n-k+1} &\geqq \frac{\log N}{N} \int_{a/N-k+2}^{1} e^{-(k-1)u} \frac{du}{u^2} \qquad (a = \log N) \\ &\geqq \frac{\log N}{N} e^{-(k-1)} [(N-k+2)/\log N - 1]\end{aligned}$$

and so

$$\liminf \sum_{n=k}^{N} N^{-n/n-k+1} \geqq e^{-(k-1)} > 0 \text{ as } N \to \infty.$$

This completes the proof of the lemma. It is to be used in the proof of the following theorem.

THEOREM 7. *Let f be defined in a connected domain R in V_k. Then if x_0 is a point in R and $D_n(f, x)$ has the significance ascribed to it in either Theorem 3 or Theorem 4, the conditions*

a) $\int_R D_n(f,x)\,dv \leqq m_n$

b) $D_n(f,x_0) = 0$

together with conditions c), d) *of Lemma* 3, *imply that* $f \equiv 0$.

Proof. We may show, by considerations similar to those involved in the proof of Theorem 5, that for a geodesic sphere with center at x_0 and radius r_0 the inequalities

$$\int_0^{r_0} | F^{(n)}(s) | \, s^{k-1} ds \leqq C^n m_n$$

are satisfied for almost all geodesics through x_0. By a trivial transformation we may assume that $r_0 = 1$. Then since $F^{(n)}(0) = 0$ the previous lemma assures us that $F(s) \equiv 0$ for almost all geodesics. The remainder of the proof then goes as before.

Part II.

4. In this section we shall present two theorems about functions defined in the whole Euclidean space E_k. They differ from the theorems of Part I in two respects. Instead of the quantities $D_n(f,x)$ defined by (1. 1) we consider differential operators applied to f. These need not involve all the partial derivatives of f. The second difference lies in the fact that we replace the single point x_0 by a certain type of point set. For this we lay down the following definition:

DEFINITION. *A point set U contained in a region R of E_k will be called a set of analytic determination, or simply a set of determination, if a function ϕ which is analytic in R and which has the value zero at all points of U must vanish identically in R.*

For the first theorem the differential operators are defined in the following manner. Let p be a fixed positive integer. For each positive integer n let $P_n(\xi_1,\cdots,\xi_k)$ be a homogeneous polynomial in $\xi_1,\cdots,\xi_k$ of degree np, with real coefficients. We define $P_0(\xi_1,\cdots,\xi_k) = 1$. Then if f and all its derivatives are defined in E_k the differential operators

$$P_n\left(\frac{\partial}{\partial x_1},\cdots,\frac{\partial}{\partial x_k}\right)$$

have an obvious meaning when applied to f, the resulting being a function defined in E_k. We use the notation

$$P_n\left(\frac{\partial}{\partial x_1},\cdots,\frac{\partial}{\partial x_k}\right) f = L_n(f,x). \tag{4. 1}$$

THEOREM 8. *Let f be defined in E_k. With a certain fixed Cartesian coördinate system $(x_1, \cdots, x_k)$ let $f(x) = 0$ except when $x_i > 0$ $(i = 1, \cdots, k)$, and let U be a set of determination contained in the region $x_i > 0$ $(i = 1, \cdots, k)$. Then if*

$$\text{a)} \quad | L_n(f, x) | \leqq m_n, \ x \text{ in } E_k$$
$$\text{b)} \quad \sum | P_n(\xi) |^{1/np} (m_n)^{-1/np} = \infty, \ \xi \text{ in } U$$

the function f vanishes identically.

Proof. We shall first carry out the proof under the additional assumption that all the partial derivatives of f are bounded in E_k. (f itself is bounded, because of a).

We form the function

$$F(z) \equiv F(z_1, \cdots, z_k) = \int_0^\infty \cdots \int_0^\infty e^{-(z_1t_1+ \cdots +z_kt_k)} f(t)\, dt_1 \cdots dt_k.$$

Since f is bounded F is analytic if $R(z_i) > 0$, $i = 1, \cdots, k$. Under our temporary assumption we may integrate by parts; since $f(x) = 0$ if any $x_i = 0$ (and similarly for the derivatives) we obtain, for any non-negative integers $\rho_1, \cdots, \rho_k$

$$z_1^{\rho_1} \cdots z_k^{\rho_k} F(z) = \int_0^\infty \cdots \int_0^\infty e^{-(z_1t_1+ \cdots +z_kt_k)} \frac{\partial^{\rho_1+ \cdots +\rho_k}}{\partial t_1^{\rho_1} \cdots \partial t_k^{\rho_k}} f(t)\, dt_1 \cdots dt_k$$

$$P_n(z)F(z) = \int_0^\infty \cdots \int_0^\infty e^{-(z_1t_1+ \cdots +z_kt_k)} L_n(f, t)\, dt_1 \cdots dt_k$$

and if $R(z_i) = x_i > 0$ it follows that

$$| P_n(z)F(z) | \leqq \frac{m_n}{x_1 \cdots x_k}.$$

Now let $(\xi_1, \cdots, \xi_k)$ be any point in U, and place $z_i = \xi_i w$, where $R(w) > 0$. Then by the homogeneity of P_n we obtain

$$| w^{np} P_n(\xi) F(\xi_1 w, \cdots, \xi_k w) | \leqq \frac{m_n}{[R(w)]^k \xi_1 \cdots \xi_k}.$$

For fixed ξ define $\Phi(w) = F(\xi_1 w, \cdots, \xi_k w) \cdot \Phi(w)$ is analytic if $R(w) > 0$. If

$$R(w) \{\xi_1 \cdots \xi_k\}^{1/k} \geqq 1$$

we have

$$| w^{np} \Phi(w) | \leqq \frac{m_n}{P_n(\xi)}. \tag{4.2}$$

But by condition b) it follows from a result of Carleman already referred to,[7]

[7] Cf. footnote 6.

that $\Phi(w) \equiv 0$. In this way we see that the analytic function F vanishes at all points of U, and therefore $F \equiv 0$. This implies that $f \equiv 0$.

We have now to free ourselves of the assumption of boundedness of the partial derivatives of f. We do this by approximating the function f by a function whose derivatives are bounded.

By a general theorem [8] it is known that if K is an absolutely integrable function defined over E_k, with

$$\int_{E_k} K(x_1,\cdots,x_k)\,dx_1\cdots dx_k = 1 \tag{4.3}$$

and if f is bounded and continuous, then for the function $(\lambda > 0)$

$$f_\lambda(x) = \int_{E_k} f\left(x_1+\frac{t_1}{\lambda},\cdots,x_k+\frac{t_k}{\lambda}\right)K(t_1,\cdots,t_k)\,dt_1\cdots dt_k \tag{4.4a}$$

$$f_\lambda(x) = \lambda^k\int_{E_k} f(t_1,\cdots,t_k)K(\lambda(t_1-x_1),\cdots,\lambda(t_k-x_k))\,dt_1\cdots dt_k \tag{4.4b}$$

we have

$$\lim f_\lambda(x) = f(x) \quad \text{as } \lambda \to \infty. \tag{4.5}$$

We shall require of K that it be non-negative, possess derivatives of all orders, and that it vanish outside a fixed, finite hypercube in E_k. Then it is permissible to differentiate under the integral in (4.4a), and we obtain

$$L_n(f_\lambda, x) = \int_{E_k} L_n\left(f, x+\frac{t}{\lambda}\right)K(t)\,dt \tag{4.6}$$

where we are using an obvious abbreviation in notation. Consequently, by condition a), together with (4.3) and the fact that K is non-negative, we conclude

$$|L_n(f_\lambda, x)| \leqq m_n.$$

Since K vanishes outside a finite region in E_k, and f vanishes except when $x_i > 0$, we easily infer from (4.4) that for any fixed $\lambda > 0$, f_λ vanishes except when $x_i > b_i$, where $b_1,\cdots,b_k$ are constants depending on λ. Finally, the partial derivatives of f_λ are bounded in E_k. To see this we make use of (4.4b):

$$\begin{aligned}\frac{\partial^{p_1+\cdots+p_k}}{\partial x_1^{p_1}\cdots\partial x_k^{p_k}} f_\lambda(x) &= (-\lambda)^{p_1+\cdots+p_k}\lambda^k\int_{E_k} f(t)\left[\frac{\partial^{p_1+\cdots+p_k}}{\partial u_1^{p_1}\cdots\partial u_k^{p_k}}K(u)\right]_{u=\lambda(t-x)} dt \\ &= (-\lambda)^{p_1+\cdots+p_k}\int_{E_k} f\left(x+\frac{t}{\lambda}\right)\frac{\partial^{p_1+\cdots+p_k}}{\partial t_1^{p_1}\cdots\partial t_k^{p_k}}K(t)\,dt.\end{aligned}$$

Since the derivatives of K vanish outside a finite region, and f is bounded, the result follows.

---

[8] S. Bochner, *Vorlesungen über Fouriersche Integrale*, Leipzig (1932), p. 191.

But now we can conclude from the first part of the proof, after a trivial change of coördinates, that $f_\lambda \equiv 0$. By (4.5) it follows that $f \equiv 0$. The proof is thus complete.

For the next theorem we introduce the Laplace operator and its iterates.

$$\Delta f(x) = \frac{\partial^2 f}{\partial x_1{}^2} + \cdots + \frac{\partial^2 f}{\partial x_k{}^2}$$
$$\Delta^n f(x) = \Delta(\Delta^{n-1} f(x)).$$

As usual we adopt the convention $\Delta^0 f = f$.

THEOREM 9. *Let f be defined in E_k, and let U be a set of determination in E_k. Then the conditions*

a) $|\Delta^n f(x)| \leqq m_n$, *$x$ in E_k*

b) $\Delta^n f(\xi) = 0$, *ξ in U*

c) $\sum m_n{}^{-1/n} = \infty$

imply that $f \equiv 0$.

Proof. We again make the temporary assumption that the partial derivatives of f are bounded in E_k. For $\alpha \geqq 0$ we define

(4.7a) $F(x_1, \cdots, x_k, \alpha)$
$$= c_k \int_{E_k} f(x_1 + \sqrt{\alpha}\, t_1, \cdots, x_k + \sqrt{\alpha}\, t_k)\, e^{-(t_1^2 + \cdots + t_k^2)/4} dt_1 \cdots dt_k$$

(4.7b) $F(x_1, \cdots, x_k, \alpha)$
$$= c_k \alpha^{-k/2} \int_{E_k} f(t_1, \cdots, t_k)\, e^{-[(t_1 - x_1)^2 + \cdots + (t_k - x_k)^2]/\alpha} dt_1 \cdots dt_k$$

where $c_k = 2^{-k}\pi^{-k/2}$ is a normalizing factor. $F(x_1, \cdots, x_k, \alpha) \equiv F(x, \alpha)$ may be regarded as a transform of $f(x)$. If we write

$$F(x, \alpha) = T_\alpha(f, x)$$

then because of the formula

$$c_k \int_{E_k} e^{-(t_1^2 + \cdots + t_k^2)/4} dt_1 \cdots dt_k = 1$$

it is clear that $T_0(f, x) = f(x)$ and

(4.8) $$|T_\alpha(f, x)| \leqq \max |f|.$$

It is well known that [9]

$$\lim F(x, \alpha) = f(x) \quad \text{as} \quad \alpha \to \infty \tag{4.9}$$

the convergence being uniform in any finite portion of E_k.

Under our temporary assumption we may calculate the derivative of $F(x, \alpha)$ with respect to α by differentiating under the integral in (4.7a). If we do this, and then integrate the resulting equation by parts we obtain

$$\frac{\partial F(x, \alpha)}{\partial \alpha} = T_a(\Delta f, x).$$

Consequently

$$\frac{\partial^n F(x, \alpha)}{\partial \alpha^n} = T_a(\Delta^n f, x). \tag{4.10}$$

Therefore, by (4.8) and condition a)

$$\left| \frac{\partial^n F(x, \alpha)}{\partial \alpha^n} \right| \leqq | \Delta^n f | \leqq m_n, \quad x \text{ in } E_k, \quad \alpha \geqq 0$$

while by condition b)

$$\left(\frac{\partial^n F(\xi, \alpha)}{\partial \alpha^n} \right)_{\alpha=0} = \Delta^n f(\xi) = 0, \quad \xi \text{ in } U.$$

By Theorem D it follows that $F(\xi, \alpha) = 0$ if $\alpha \geqq 0$ and ξ is in U. But for fixed $\alpha > 0$, $F(x, \alpha)$ is analytic in the entire space E_k. This is proved [10] by making use of (4.7b). Consequently, because of the nature of the set U, $F(x, \alpha) = 0$ for each x in E_k, if $\alpha > 0$. By (4.9) we conclude $f \equiv 0$.

Without the assumption of boundedness of the partial derivatives of f the whole difficulty lies in the establishment of (4.10). In order to meet this difficulty we define $f_\lambda(x)$ by (4.4), where K has the same properties as were required before. Then we know that the partial derivatives of f_λ $(\lambda > 0)$ with respect to the x's are bounded in E_k, and that

$$| \Delta^n f_\lambda(x) | \leqq m_n, \quad x \text{ in } E_k, \quad \lambda > 0. \tag{4.11}$$

From (4.6) it follows that

$$\lim \Delta^n f_\lambda(x) = \Delta^n f(x) \quad \text{as} \quad \lambda \to \infty \tag{4.12}$$

---

[9] This is in fact a special case of the theorem referred to in footnote 8. For the case $k = 1$ see E. Borel, *Leçons sur les Fonctions de Variables Reélles*, Paris (1928), pp. 52-53.

[10] The method used by Borel, *op. cit.*, pp. 53-54 for the case $k = 1$ works equally well in general. It consists in defining the integral for complex values of the variables and showing that the resulting function is analytic.

and it is easily verified that the convergence is uniform over each finite position of E_k. Now by the previous work, if

$$F_\lambda(x, \alpha) = T_\alpha(f_\lambda, x)$$

then

$$\frac{\partial^n F_\lambda(x, \alpha)}{\partial \alpha^n} = T_\alpha(\Delta^n f_\lambda, x). \tag{4.13}$$

It is readily proved, using condition a) and (4. 11), (4. 12), that

$$\lim T_\alpha(\Delta^n f_\lambda, x) = T_\alpha(\Delta^n f, x) \quad \text{as} \quad \lambda \to \infty.$$

This follows from (4. 7b), together with the fact that f, f_λ are bounded, and that $f_\lambda \to f$ uniformly over each finite portion of E_k.

The passage to the limit $\lambda \to \infty$ in (4. 13) then establishes (4. 10) and completes the proof.

5. In this concluding section we are concerned with the sphere

$$V_k\colon \quad x_1^2 + \cdots + x^2_{k+1} = 1 \qquad (k \geqq 2)$$

in E_{k+1}. It is an analytic Riemannian space of constant positive curvature, and as such the Laplace operator Δ is defined for it.[11]

We may define a set of determination U on V_k in the same way as it was done in § 4. A function ϕ is analytic over a portion R of V_k if it is analytic in each of the allowable coördinate systems covering R (any two coördinate systems being connected by analytic transformations).

Points of V_k will be denoted by $P, Q, \cdots$.

THEOREM 10. *Let f be defined in V_k $(k \geqq 2)$, and let U be a set of determination in V_k. Then the conditions*

a) $|\Delta^n f(P)| \leqq m_n$, P *in* V_k

b) $\Delta^n f(P_0) = 0$, P_0 *in* U

c) $\sum m_n^{-1/n} = \infty$

imply that $f \equiv 0$.

The proof depends upon the properties of certain expansions in series of ultra-spherical functions. These expansions reduce to the ordinary Laplace series of spherical harmonies when $k = 2$.

[11] The invariant form of the operator is

$$\Delta f = g^{ij} f_{,i,j}.$$

See O. Veblen, *Invariants of Quadratic Differential Forms*, Cambridge University Press (1927), p. 63.

We shall introduce coördinates in V_k by using spherical coördinates in E_{k+1}. We put

$$
(5.1)\qquad
\begin{aligned}
x_1 &= r\cos\theta_1\\
x_2 &= r\sin\theta_1\cos\theta_2\\
&\cdot\quad\cdot\quad\cdot\quad\cdot\quad\cdot\quad\cdot\quad\cdot\\
x_k &= r\sin\theta_1\cdots\sin\theta_{k-1}\cos\theta_k\\
x_{k+1} &= r\sin\theta_1\cdots\sin\theta_k.
\end{aligned}
$$

Then it may be found that if [12]

[12] See E. Heine, *Handbuch der Kugelfunktionen*, Band I, 2nd ed., Berlin (1878), pp. 460-461.

$$
\begin{aligned}
v &= r^k\sin^{k-1}\theta_1\sin^{k-2}\theta_2\cdots\sin\theta_{k-1}\\
u_1 &= r^2\\
u_2 &= r^2\sin^2\theta_1\\
&\cdot\quad\cdot\quad\cdot\quad\cdot\quad\cdot\quad\cdot\\
u_k &= r^2\sin^2\theta_1\cdots\sin^2\theta_{k-1}
\end{aligned}
$$

and F is defined in E_{k+1}, then

$$\sum_{i=1}^{k+1}\frac{\partial^2 F}{\partial x_i^2}=\frac{1}{v}\left[\frac{\partial}{\partial r}\left(\frac{v}{1}\frac{\partial F}{\partial r}\right)+\frac{\partial}{\partial\theta_1}\left(\frac{v}{u_1}\frac{\partial F}{\partial\theta_1}\right)+\cdots+\frac{\partial}{\partial\theta_k}\left(\frac{v}{u_k}\frac{\partial F}{\partial\theta_k}\right)\right]$$

Hence we have, for a function f defined in V_k

$$(5.2)\qquad \Delta f=\sum_{i=1}^{k}\frac{w_i}{\sin^{k-1}\theta_i}\frac{\partial}{\partial\theta_i}\left(\sin^{k-1}\theta_i\frac{\partial f}{\partial\theta_i}\right)$$

where $w_1=1$, $w_i=(\sin\theta_1\cdots\sin\theta_{i-1})^{-2}$, $i=2,\cdots,k$.

Now we put

$$(5.3)\qquad \nu=(k-1)/2$$

and consider the polynomials $P_n^{(\nu)}(x)$ defined by the expansion [13]

$$(5.4)\qquad (1-2rx+r^2)^{-\nu}=\sum_0^\infty r^nP_n^{(\nu)}(x).$$

For $k=2$, $\nu=1/2$ and $P_n^{(\nu)}(x)$ is the Legendre polynomial of order n. It is known that if P, Q are two points on the sphere V_k and γ is the angle $(0\leqq\gamma\leqq\pi)$ between the lines joining these points to the center of the sphere (so that γ is the geodesic distance from P to Q), then for the function

$$(5.5)\qquad F_n(P,Q)=P_n^{(\nu)}(\cos\gamma)$$

[13] A systematic study of the functions $P_n^{(\nu)}(x)$ may be found in N. Nielsen, *Théorie des Fonctions Metasphériques*, Paris (1911), especially Chapter VII. The expansion (5.4) is given on p. 98. The notation of Heine, *op. cit.*, pp. 451-460, is slightly different from that of Nielsen.

we have [14]

$$\Delta F_n = -n(n+2\nu)F_n. \tag{5.6}$$

Since F_n is symmetric in P and Q it does not matter which point is regarded as variable in evaluating the left member of (5.6).

As a consequence of (5.6) and Green's theorem we also have the important relation

$$\int_{V_k} F_n(P,Q)\Delta f(Q)dQ = -n(n+2\nu)\int_{V_k} F_n(P,Q)f(Q)dQ \tag{5.7}$$

where dQ denotes the volume element in the space V_k.

With a function f defined in V_k we associate the series [15]

$$f(P) \sim \sum_0^\infty a_n(P,f) \tag{5.8}$$

where

$$a_n(P,f) = \frac{(n+\nu)\Gamma(\nu)}{2\pi^{\nu+1}}\int_{V_k} F_n(P,Q)f(Q)dQ. \tag{5.9}$$

Concerning this series we assert:

Lemma 4. *If f is continuous on V_k, and $H_n^{(p)}(P)$ denotes the Cesàro means of order p of the series (5.8), then*

$$\lim_{n\to\infty} H_n^{(p)}(P) = f(P) \quad \text{if } p > \nu$$
$$|H_n^{(p)}(P)| \leqq \max|f| \quad \text{if } p \geqq 2\nu + 1 = k.$$

Proof. In proving the first assertion we may assume that the coördinates (5.1) are chosen so that P is the point $x_1 = 1$, $x_i = 0$, $i > 1$. Then $\gamma = \theta_1$ and

$$\int_{V_k} F_n(P,Q)f(Q)dQ = \int_0^\pi d\theta_1 \cdots \int_0^\pi d\theta_{k-1}\int_\pi^\pi P_n^{(\nu)}(\cos\theta_1)\sin^{k-1}\theta_1 f(Q)\omega d\theta_k$$

where $\omega = \sin^{k-2}\theta_2 \cdots \sin\theta_{k-1}$ and Q has the coördinates $(\theta_1, \cdots, \theta_k)$. From this result and the formula [16]

$$P_n^{(\nu)}(1) = \frac{\Gamma(n+2\nu)}{\Gamma(n+1)\Gamma(2\nu)}$$

[14] Heine, *op. cit.*, p. 461.

[15] This series arises naturally through the expansion of the integrand in Poisson's integral when the latter is used to solve the Dirichlet problem for the sphere V_k.

[16] Nielsen, *op. cit.*, p. 95.

it is seen that if we put

$$(5.10)\quad \Phi(\cos\theta_1)=\frac{\Gamma(\nu+1/2)\Gamma(1/2)}{2\pi^{\nu+1}}\int_0^\pi d\theta_2\cdots\int_0^\pi d\theta_{k-1}\int_\pi^\pi f(Q)\omega d\theta_k$$

then the series (5. 8) becomes the formal expansion of $\Phi(\cos\theta)$ in a series of the form

$$(5.11)\qquad \Phi(\cos\theta)\sim\sum_0^\infty c_n P_n^{(\nu)}(\cos\theta)$$

for the point $\theta=0$, the coefficients c_n being given by the formula [17]

$$(5.12)\quad c_n=\frac{(n+\nu)\Gamma(\nu)}{\Gamma(\nu+1/2)\Gamma(1/2)}\,\frac{\Gamma(n+1)\Gamma(2\nu)}{\Gamma(n+2\nu)}\int_0^\pi\Phi(\cos\theta)P_n^{(\nu)}(\cos\theta)\sin^{2\nu}\theta d\theta.$$

$\Phi(\cos\theta_1)$ is continuous, since f is continuous; also $\Phi(1)=f(P)$, since P has coördinates $\theta_1=0$, θ_i arbitrary if $i\geqq 2$. Now it has been proved by Kogbetlianz [18] that the series (5. 11) is uniformly summable (C,p) to Φ if $p>\nu$ and the function is continuous. This proves the first assertion of the lemma.

In connection with the second part of the lemma we consider the series

$$(5.13)\qquad \sum_0^\infty(n+\nu)P_n^{(\nu)}(\cos\theta).$$

If $s_n^{(p)}(\cos\theta)$ denotes the Cesàro means of order p of this series we see from (5. 5) and (5. 9) that

$$(5.14)\qquad H_n^{(p)}(P)=\frac{\Gamma(\nu)}{2\pi^{\nu+1}}\int_{V_k}s_n^{(p)}(\cos\gamma)f(Q)dQ.$$

In particular, for $f\equiv 1$ we find

$$(5.15)\qquad 1=\frac{\Gamma(\nu)}{2\pi^{\nu+1}}\int_{V_k}s_n^{(p)}(\cos\gamma)dQ.$$

Now Kogbetlianz has proved [19] that $s_n^{(p)}(\cos\gamma)\geqq 0$ if $p\leqq 2\nu+1$ and $0\leqq\gamma\leqq\pi$. The second assertion of the lemma now follows at once from (5. 14) and (5. 15).

We now turn to the proof of Theorem 10. We begin with the series

$$(5.16)\qquad F(P,\lambda)=\sum_0^\infty e^{-\lambda n(n+2\nu)}a_n(P,f)\qquad \lambda>0$$

---

[17] Ervand Kogbetlianz, "Sommabilité des Séries Ultrasphériques," *Journal de Mathématiques*, vol. 3 (1924), p. 110.

[18] Kogbetlianz, *loc. cit.*, p. 118, p. 168.

[19] Kogbetlianz, *loc. cit.*, p. 179.

obtained from (5.8) by introduction of exponential factors. Concerning this function we make the following statements.

LEMMA 5. *For each P in V_k the function $F(P,\lambda)$ defined by (5.16) is continuous, with continuous derivatives of all orders on the open interval $0 < 1 < \infty$. Moreover*

$$\lim \frac{\partial^n F(P,\lambda)}{\partial\lambda^n} = \Delta^n f(P) \text{ as } \lambda \to 0 \tag{5.17}$$

and

$$\left| \frac{\partial^n F(P,\lambda)}{\partial\lambda^n} \right| \leqq A \max |\Delta^n f| \qquad 0 < \lambda \leqq 1 \tag{5.18}$$

where A is a constant independent of λ, n, and P.

Granting for a moment the truth of the lemma we can easily deduce from it a proof of Theorem 10. For from the lemma and the conditions of Theorem 10 we conclude by Theorem D that $F(P_0,\lambda) \equiv 0$ if P_0 is in U. Therefore $a_n(P_0,f) = 0$ if P_0 is in U. But $P_n^{(\nu)}(\cos\gamma)$ is a polynomial [20] in $\cos\gamma$, and $\cos\gamma$ is a polynomial in $\sin\theta_i$, $\cos\theta_i$, $\sin\beta_i$, $\cos\beta_i$, $i = 1,\cdots,k$, where θ_i refers to Q and β_i to P in (5.5). Thus $a_n(P,f)$ is a function of P is analytic on V_k, and since it vanishes on U it vanishes identically. Then by Lemma 4, $f(P) \equiv 0$.

It remains only to prove Lemma 5. The series (5.16) is a power series in $t = e^{-\lambda}$. Since (5.8) is summable (C,p) if $p > \nu$ we know that [21]

$$\lim_{n\to\infty} \frac{a_n(P,f)}{n^p} = 0 \qquad p > \nu$$

and from this it follows that (5.16) is convergent for all positive values of λ (i.e. for $0 \leqq t < 1$). We may obviously differentiate term by term with respect to λ. Doing this we obtain

$$\begin{aligned} \frac{\partial F(P,\lambda)}{\partial\lambda} &= -\lambda \sum_0^\infty n(n+2\nu)e^{-\lambda n(n+2\nu)} a_n(P,f) \\ &= \sum_0^\infty e^{-\lambda n(n+2\nu)} a_n(P,\Delta f) \end{aligned}$$

because of (5.7). Thus in general

$$\frac{\partial^m F(P,\lambda)}{\partial\lambda^m} = \sum_{n=0}^\infty e^{-\lambda n(n+2\nu)} a_n(P,\Delta^m f), \qquad \lambda > 0. \tag{5.19}$$

[20] For the expression of $\cos\gamma$, see Heine, *op. cit.*, p. 458, and the second line on p. 461.

[21] See for example Tomlinson Fort, *Infinite Series*, Oxford (1930), p. 210, Theorem 210.

From this it follows that in the proof of Lemma 5 we may restrict our attention to the case $n = 0$.

If we reorder the series (5.16) according to the rule

$$\sum_0^\infty a_n b_n = \sum_0^\infty (a_0 + \cdots + a_n)(b_n - b_{n+1})$$

applied $p+1$ times, where we put $b_n = e^{-\lambda n(n+2\nu)}$, we obtain

$$(5.20) \qquad F(P,\lambda) = \sum_{n=0}^\infty (-1)^{p+1} \binom{n+p}{p} H_n^{(p)}(P) D_n^{p+1} b_n$$

where $D_n^0 b_n = b_n$, $D_n^1 b_n = b_{n+1} - b_n$, and $D_n^{p+1} b_n = D_n^1(D_n^p b_n)$. The justification of this rearrangement is easily made because of the speed with which b_n and $D_n^{(i)} b_n \to 0$ as $n \to \infty$.

Now we put $p = 2\nu + 1 = k$ and $M = \max |f|$. Then by Lemma 4

$$|F(P,\lambda)| \leqq M \sum_0^\infty \binom{n+k}{k} | D_n^{k+1} b_n$$

and it suffices, to establish (5.18), to show that for some constant A we have

$$(5.21) \qquad K(\lambda) = \sum_0^\infty \binom{n+k}{k} | D_n^{k+1} b_n | \leqq A \qquad 0 < \lambda \leqq 1.$$

We write

$$b_n = e^{\lambda\nu^2} e^{-\lambda(n+\nu)^2} = e^{\lambda\nu^2} \phi(n).$$

Then

$$D_n^{k+1} b_n = e^{\lambda\nu^2} \int_0^1 \cdots \int_0^1 \phi^{(k+1)}(n + t_1 + \cdots + t_{k+1}) dt_1 \cdots dt_{k+1}.$$

Since

$$\binom{n+k}{k} \leqq (n+k)^k \leqq (1+k)^k n^k \quad \text{if} \quad n \geqq 1$$

we see that (5.21) will hold if the series

$$\sum_{n=1}^\infty \int_0^1 \cdots \int_0^1 (n + t_1 + \cdots + t_{k+1})^k \times | \phi^{(k+1)}(n + t_1 + \cdots + t_{k+1}) | dt_1 \cdots dt_{k+1}$$

is bounded for $0 < \lambda \leqq 1$. But because of the equation

$$\sum_{n=1}^\infty \int_0^1 \psi(n + t_1) dt_1 = \sum_1^\infty \int_n^{n+1} \psi(s) ds = \int_1^\infty \psi(s) ds$$

the above series is equal to

$$\int_0^1 \cdots \int_0^1 dt_2 \cdots dt_{k+1} \int_1^\infty t_1{}^k \mid \phi^{k+1}(t_1) \mid dt_1 = \int_1^\infty t^k \mid \phi^{k+1}(t) \mid dt.$$

It is therefore sufficient to show that this integral is bounded for $0 < \lambda \leqq 1$. We set $s = \lambda^{1/2}(t + \nu)$ and write

$$\psi(s) = \phi(t) = e^{-s^2}.$$

Then

$$\phi^{(k+1)}(t) = \lambda^{(k+1)/2}\psi^{(k+1)}(s)$$

and therefore, with $s_1 = \lambda^{1/2}(1 + \nu)$,

$$\int_1^\infty t^k \mid \phi^{(k+1)}(t) \mid dt = \int_{s_1}^\infty (s - \nu\lambda^{1/2})^k \mid \psi^{(k+1)}(s) \mid ds.$$

This integral is less than an integral of the form

$$\int_0^\infty Q(\lambda, s) e^{-s^2} ds$$

where $Q(\lambda, x)$ is a polynomial in s and $\lambda^{1/2}$. This is clearly bounded for $0 \leqq \lambda \leqq 1$. Thus (5.21) and (5.18) are established.

We have finally to prove (5.17). Since if $f \equiv 1$ we have $H_n{}^{(k)}(P) \equiv 1$ and $F(P, \lambda) \equiv 1$, we have from (5.20)

$$1 = \sum_0^\infty (-1)^{k+1} \binom{n+k}{k} D_n{}^{k+1} b_n.$$

Therefore in general

$$F(P, \lambda) - f(P) = \sum_0^\infty (-1)^{k+1} \binom{n+k}{k} \{H_n{}^{(k)}(P) - f(P)\} D_n{}^{k+1} b_n.$$

Now for a fixed P in V_k and a given $\mathcal{E} > 0$ we can choose $N = N(\mathcal{E})$ so that

$$\mid H_n{}^{(k)}(P) - f(P) \mid < \mathcal{E} \quad \text{if} \quad n \geqq N.$$

Since $b_n = e^{-\lambda n(n+2\nu)}$ it is clear that $D_n{}^{k+1} b_n \to 0$ as $\lambda \to 0$ for a fixed n. Therefore, from (5.21) it follows that

$$\overline{\lim} \mid F(P, \lambda) - f(P) \mid \leqq \mathcal{E} A \quad \text{as} \quad \lambda \to 0.$$

This proves the validity of (5.17), since $\mathcal{E}$ was arbitrary. Thus theorem 10 is completely proved.

PRINCETON UNIVERSITY.

Made in United States of America

ANNALS OF MATHEMATICS
Vol. 44, No. 4, October, 1943

ANALYTIC AND MEROMORPHIC CONTINUATION BY MEANS OF GREEN'S FORMULA

S. BOCHNER

(Received April 28, 1943)

Poincaré was the first to apply methods of potential theory to the study of analytic functions in several complex variables.[1] Using his method of "balayage" he proved for the entire space that meromorphic functional elements whose difference is holomorphic where they overlap can be merged into one meromorphic function. Later Cousin extended the result to the more general case of a polycylindrical domain, but he used the method of iterating the classical Cauchy integral in one complex variable.

In the present paper we will use a greatly simplified approach to Poincaré's original method, and our main purpose is to show that his theorem is closely connected with the theorem of Hartogs that every analytic function in several complex variables can be continued from the connected boundary of a domain into its whole interior. The connecting link is Green's formula for general harmonic functions and a related version of Cauchy's formula. The first application of this method to Hartogs' theorem is due to Fueter. However Fueter has a very involved approach to the method from the standpoint of quaternions and other Cayley number systems. Our starting point is the ordinary Green's formula, and the specialized results will follow from the fact that an arbitrary differential operator with constant coefficients, being commutative with the Laplacean, is also commutative with Green's integral which is the operational inverse to the Laplacean. This approach will yield old and new results in very general versions and will also permit us to extend Poincaré's theorem to functions on the torus.

CHAPTER I

Stokes' Theorem

1. We will consider a Euclidean space E_n, $n \geqq 3$. If A is a pointset then $\bar{A}$ denotes its closure and $\tilde{A}$ some neighborhood of $\bar{A}$ (no matter how small).

[1] For references to the theorem of Poincaré-Cousin see the article by H. Behnke and P. Tullen, Theorie der Funktionen mehrerer komplexen Veränderlichen, in Ergebnisse der Mathematik, volume III, pp. 64–67; for references to the theorem of Hartogs see the same article, p. 50, and two more recent papers by R. Fueter in Commentarii Mathematici Helvetici, volume 12 (1939), pp. 75–80, and volume 14 (1942), pp. 394–400.

Added in proof, Sept. 1943. Formula (53) of the present paper and a proof of theorem 5 based on it have just been published by Enzo Martinelli, Sopra una dimonstrazione di R. Fueter per una teorema di Hartogs, Comm. Helvetici, 15 (1942–3), submitted Jan. 1943. The present author may be permitted to state that these results have been presented by him in a Princeton graduate course in Winter 1940/41 and were subsequently incorporated, in a Princeton doctorate thesis (June 1941) by Donald C. May, entitled: An integral formula for analytic functions of k variables with some applications.

Virtually all our pointsets will be simplices or finite unions of disjoint simplices. Whenever a symbol like B, C, D will be introduced to denote one or several simplices then the same symbol will also be used to denote the corresponding pointset. If B (or C, D) is an m-dimensional simplex in E_n, then it will be always implicitly understood that B is a topological map of the closure $\bar{S}$ of the rectilinear fundamental simplex S in E_m and the mapping transformation is defined and of class one (with regard to differentiability) in $\bar{S}$. All functions introduced will be likewise of class one unless more restrictive assumptions are explicitly stated. It will be sufficient for our purposes to express Stokes' theorem as the formula

$$(1)\qquad \int_B d\varphi\, d\psi_1 \cdots d\psi_{m-1} = \int_C \varphi\, d\psi_1 \cdots d\psi_{m-1}.$$

The scalars $\varphi, \psi_1, \cdots, \psi_{m-1}$ are defined in $\bar{B}$ and C is the set of the $(m-1)$-dimensional faces of B. The simplex B is oriented and the orientation of C is related to the orientation of B by the well known combinatorial rule. By algebraic addition the formula extends readily from a simplex to a (finite) simplicial manifold. In particular, if D is a simplicial domain in E_n with boundary $\bar{B}$, and X_α, $\alpha = 1, \cdots, n$, is a vector field in $\bar{D}$, then

$$(2)\qquad \int_D \sum_{\alpha=1}^{n} \frac{\partial X_\alpha}{\partial \xi_\alpha} dv_\xi = \int_B \sum_{\alpha=1}^{n} X_\alpha \sigma^\alpha,$$

where

$$(3)\qquad dv_\xi = d\xi_1 \cdots d\xi_n$$

$$(4)\qquad \sigma^\alpha = (-1)^{\alpha-1} d\xi_1 \cdots d\xi_{\alpha-1} d\xi_{\alpha+1} \cdots d\xi_n;$$

or, omitting the symbol of summation with respect to α,

$$(5)\qquad \int_D \frac{\partial X_\alpha}{\partial \xi_\alpha} dv_\xi = \int_B X_\alpha \sigma^\alpha.$$

The symbols (3), (4) are external differential forms, and the symbol (3) will be the ordinary Euclidean volume element if D is cooriented with E_n; this will be always assumed.

2. We consider m-dimensional simplices $B_1, \cdots, B_s$ and $(m-1)$-dimensional simplices $C_1, \cdots, C_r$, each in a fixed orientation and all disjoint, and we assume that each C_p is a face of one or several B_q and that each $(m-1)$-dimensional face of each B_q occurs among the C_p. We introduce the incidence number ϵ_{pq} which is 0 if C_p is not a face of B_q and otherwise is $+1$ or -1 depending on whether the given orientation of C_p coincides with that of B_q or not.

If we put

$$(6)\qquad B = B_1 + \cdots + B_s, \qquad C = C_1 + \cdots + C_r,$$

if the functions $\psi_1, \cdots, \psi_{m-1}$ are scalars in $\tilde{B}$, if $\varphi_q(\xi)$ is given in $\tilde{B}_q$, $q = 1, \cdots, s$, and if we put

$$\varphi^p(\xi) = \sum_{q=1}^{s} \epsilon_{pq}\varphi_q(\xi), \qquad p = 1, \cdots r, \tag{7}$$

then (1) implies

$$\sum_{p=1}^{r} \int_{C_p} \varphi^p \, d\psi_1 \cdots d\psi_{m-1} = \sum_{q=1}^{s} \int_{B_q} d\varphi_q \, d\psi_1 \cdots d\psi_{m-1} . \tag{8}$$

In order to assimilate relation (8) to the original relation (1) we introduce the concept of a *conglomerate function*. We take any functional elements φ_q, $q = 1, \cdots, s$, each defined separately on B_q (or $\tilde{B}_q$) without any interrelation, and we call the whole set a conglomerate function on B (or $\tilde{B}$); we denote it either by φ or more explicitly by $\{\varphi_q\}$. If $\{\varphi_q\}$ is defined on $\tilde{B}$, then the functional elements (7) give rise to a new conglomerate function, the latter defined on $\tilde{C}$. The new function will be called the *saltus* of φ on C. We can now interpret (1) to mean (8) if φ in $d\varphi d\psi_1 \cdots d\psi_{m-1}$ is any conglomerate function and the other φ in $\varphi d\psi_1 \cdots d\psi_{m-1}$ is its saltus. The concept of conglomerate functions unifies the common notions of "discontinuity" and "boundary value." In fact, if B is a simplicial manifold, then on any "internal" C_q which separates simplices B_a, B_b, the saltus of φ is $\pm(\varphi_a - \varphi_b)$ and on any boundary face C_q which borders on only one B_c, the saltus is $\pm\varphi$ itself.

3. We are now going to state a lemma of a special nature. If B and C are defined as before, and φ is a conglomerate function on $\tilde{B}$ then for $m \leqq n - 1$ we will investigate the change of the integral

$$\int_B \varphi \, d\psi_1 \cdots d\psi_m$$

for a small deformation of B. It will suffice to consider a deformation in which only one vertex (and the faces containing it) are involved. We denote this vertex by A_0 and its deformed image by A_0'. Consider all those among the simplices $B_1, B_2, \cdots$ which contain the vertex A_0 and denote them by $B_1, \cdots, B\mu$. Similarly denote by $C_1, \cdots, C_\lambda$ all faces containing A_0. By B_q', $q = 1, \cdots, \mu$, we denote the deformed image of B_q, by B_p'' we denote the m-dimensional complex which is the combinatorial product of A_0' and C_p, in this order of combination, and finally we denote by S_q the $(m + 1)$-dimensional simplex which is the product of A_0' and B_q. We now claim that the difference

$$\sum_{q=0}^{\mu} \int_{B_q} \varphi_q \, d\psi_1 \cdots d\psi_m - \sum_{q=0}^{\mu} \int_{B_q'} \varphi_q \, d\psi_1 \cdots d\psi_m \tag{9}$$

has the value

$$\sum_{q=0}^{\mu} \int_{S_q} d\varphi_q \, d\psi_1 \cdots d\psi_m + \sum_{p=0}^{\lambda} \int_{B_p''} \varphi^p \, d\psi_1 \cdots d\psi_m , \tag{10}$$

where $\{\varphi^p\}$ on $\tilde{C}$ is the saltus of $\{\varphi_q\}$ on $\tilde{B}$.

The proof consists in a simple combinatorial analysis. Consider any fixed simplex B_1 with vertex A_0 and write it combinatorially as $A_0A_1A_2 \cdots A_m$ where $A_1, \cdots, A_m$ are other vertices. We now introduce the $(m + 1)$-dimensional simplex S_1, namely the simplex

$$A_0'A_0A_1 \cdots A_m. \tag{11}$$

Its m-dimensional faces in natural orientation are the sum of

$$A_0A_1A_2 \cdots A_m - A_0'A_1A_2 \cdots A_m \tag{12}$$

and

$$A_0'A_0A_2 \cdots A_m - A_0'A_0A_1A_3 \cdots A_m + \cdots \pm A_0'A_0A_1 \cdots A_{m-1}. \tag{13}$$

Now, (12) is simply $B_1 - B_1'$. As for (13), its first term is the product of A_0' with $A_0A_2 \cdots A_m$. The latter is the negative of $\epsilon_{1p}C_p$, for some p, where ϵ_{1p} is our previous incidence coefficient. Thus the first term in (13) is $-\epsilon_{1p}B_p''$ for some p. The same holds for any other term in (13). It is always $-\epsilon_{1p}B_p''$ for some p, the sign being always negative whether it was originally positive or negative in (13). The equality of (9) and (10) follows now by applying Stokes' formula to S_1 and its boundary, and to any other S_q and its boundary, and then adding up over those S_q which correspond to all those B_q which contain the given vertex A_0.

Chapter II

Analytic Functions of Real Variables

4. Green's formula. The most familiar general version of Green's formula is

$$\begin{matrix}\gamma_n f(x)\\ 0\end{matrix} = \int_B \left(f\frac{\partial G}{\partial \xi_\alpha} - G\frac{\partial f}{\partial \xi_\alpha}\right)\sigma^\alpha - \int_D (f\cdot\Delta G - G\cdot\Delta f)\, dv_\xi, \tag{14}$$

and we are adding the kindred formula

$$\begin{matrix}\gamma_n f(x)\\ 0\end{matrix} = \int_B f\frac{\partial G}{\partial \xi_\alpha}\sigma^\alpha - \int_D \left(f\cdot\Delta G + \frac{\partial f}{\partial \xi_\alpha}\frac{\partial G}{\partial \xi_\alpha}\right) dv_\xi. \tag{15}$$

The function G is

$$\frac{1}{2-n}\Big(\sum_\alpha(\xi_\alpha - x_\alpha)^2\Big)^{\frac{2-n}{2}} + H(\xi, x), \tag{16}$$

where $H(\xi, x)$ is analytic in $E_n \times E_n$; the function f is a function on $\tilde{D}$, where D is a simplicial domain with boundary $\bar{B}$; the symbol Δ denotes the Laplacean

$$\Delta = \frac{\partial^2}{\partial \xi_1^2} + \cdots + \frac{\partial^2}{\partial \xi_n^2}; \tag{17}$$

the lower value 0 occurs when x is a point of $E_n - \bar{D}$ whereas the upper value $\gamma_n f(x)$ occurs when x is a point of D; and γ_n is the constant

$$\frac{2\pi^{n/2}}{\Gamma\left(\frac{n}{2}\right)}, \tag{18}$$

which constant is the $(n - 1)$-dimensional volume of the sphere $\xi_1^2 + \cdots + \xi_n^2 = 1$. For x outside $\bar{D}$ our formulas are direct consequences of (5) if $X_\alpha = f\frac{\partial G}{\partial \xi_\alpha} - G\frac{\partial f}{\partial \xi_\alpha}$, or $X_\alpha = f\frac{\partial G}{\partial \xi_\alpha}$, and for x in D they are obtained from the same values X_α by first excising a small sphere with center at x and then letting the radius of the sphere tend to 0.

If $H(\xi, x) = 0$ in (16), that is

$$G(\xi, x) \equiv \frac{1}{2-n}\left(\sum_\alpha (\xi_\alpha - x_\alpha)^2\right)^{\frac{2-n}{2}} \tag{19}$$

then G is harmonic, that is a solution of $\Delta G = 0$. If in addition $f(\xi)$ is also harmonic then (14) reduces to

$$\int_B \left(f\frac{\partial G}{\partial \xi_\alpha} - G\frac{\partial f}{\partial \xi_\alpha}\right)\sigma^\alpha = \gamma_n f(x) \quad \textit{or } 0. \tag{20}$$

Until further notice, the symbol G will always denote the Newtonian kernel (19).

5. Green's integral. We now consider a fixed set of simplices (6) and we specify that the dimension of B shall be $n - 1$. We will call a conglomerate function $\{f_q\}$ on $\tilde{B}$ harmonic if each f_q is harmonic on $\tilde{B}_q$. With such a conglomerate harmonic function $f(\xi)$ we now set up the integral

$$F(x) = \frac{1}{\gamma_n}\int_B \left(f\frac{\partial G}{\partial \xi_\alpha} - G\frac{\partial f}{\partial \xi_\alpha}\right)\sigma^\alpha. \tag{21}$$

The complement $E_n - \bar{B}$ decomposes into a finite number of domains $D_1, \cdots D_t$, D_∞. The domain D_∞ is unbounded, the others are bounded. The bounded ones may be absent; this will happen if $\bar{B}$ does not decompose the space E_n. The union $D_1 + \cdots + D_t + D_\infty$ will be denoted by D, and we will treat D as if it were a union of simplices. Integral (21) defines a harmonic function $F_\tau(x)$ in each D_τ, $\tau = 1, \cdots, t, \infty$. In this sense, $\{F_\tau(x)\}$ is a harmonic conglomerate function on D.

THEOREM 1. *If $f(\xi)$ is a harmonic conglomerate function on $\tilde{B}$, then each $F_\tau(x)$ can be continued analytically across each B_q bordering on D_τ, and the saltus of $F(x)$ on B_q is $f_q(x)$.*

For the proof it may be assumed that B is just one simplex. We deform B into another simplex B' with the same boundary as B in such a way that $B' - B$ shall be part of the boundary of a simplicial domain D. The "side" of B on which D is situated will be termed "positive," the other "negative." If x lies on the negative side of B it lies outside $\bar{D}$ and thus by (20) we have

$$F(x) = \frac{1}{\gamma_n} \int_{B'} \left(f \frac{\partial G}{\partial \xi_\alpha} - G \frac{\partial f}{\partial \xi_\alpha} \right) \sigma^\alpha . \tag{22}$$

Integral (22) however can be continued across B into D. Denoting this continuation by $F^+(x)$ we obviously have for x in D the relation

$$F^+(x) - F(x) = \frac{1}{\gamma_n} \int_{B'-B} \left(f \frac{\partial G}{\partial \xi_\alpha} - G \frac{\partial f}{\partial \xi_\alpha} \right) \sigma^\alpha . \tag{23}$$

However, again by (20), this is $f(x)$, and thus the theorem is proven.

THEOREM 2. *If also the saltus of $f(\xi)$ vanishes identically on $\tilde{C}$, then each F_τ can be continued into $\tilde{D}$.*

More generally, if a small deformation of a vertex A_0 of B enlarges a component D_τ into a domain D'_τ, and if the saltus of $f(\xi)$ vanishes on all those $\tilde{C}_p$ which contain the vertex A_0, then $F_\tau(x)$ can be continued from D_τ into D'_τ.

If we introduce the symbols B_q, B'_q, B''_p and S_q as in section 3, then by section 3 the expression

$$\sum_{q=0}^{\mu} \int_{B_q} \left(f_q \frac{\partial G}{\partial \xi_\alpha} - G \frac{\partial f_q}{\partial \xi_\alpha} \right) \sigma^\alpha - \sum_{q=0}^{\mu} \int_{B'_q} \left(f_q \frac{\partial G}{\partial \xi_\alpha} - G \frac{\partial f_q}{\partial \xi_\alpha} \right) \sigma^\alpha \tag{24}$$

is the sum of

$$\sum_{q=0}^{\mu} \int_{S_q} (f_q \cdot \Delta G - G \cdot \Delta f_q)\, dv_\xi \tag{25}$$

and

$$\sum_{p=0}^{\lambda} \int_{B''_p} \left(f^p \frac{\partial G}{\partial \xi_\alpha} - G \frac{\partial f^p}{\partial \xi_\alpha} \right) \sigma^\alpha . \tag{26}$$

Now, (24) is $\gamma_n(F(x) - F'(x))$ where the integral $F'(x)$ is what will become of $F(x)$ if B is replaced by its deformation B'. The sum (25) is naturally 0, and (26) is 0 if $f^p = 0$ on B''_p, that is if the saltus of f vanishes on $\tilde{C}_p$. Thus $F(x) = F'(x)$, but $F'(x)$ exists in D'_τ, and this proves our assertion.

6. Analytic continuation. We are now viewing the expression (21) as an operation

$$F(x) = Lf(\xi) \tag{27}$$

which transform a function in $\tilde{B}$ into a harmonic function in D. We claim that this operation is commutative with partial derivation.

THEOREM 3. *If $f(\xi)$ is harmonic in $\tilde{B}$ then*

$$L \frac{\partial f}{\partial \xi_\beta} = \frac{\partial}{\partial x_\beta} F(x), \tag{28}$$

provided the saltus of $f(\xi)$ and of $\dfrac{\partial f}{\partial \xi_1}, \cdots, \dfrac{\partial f}{\partial \xi_n}$ each vanishes on C.

More generally if the saltus of $f(\xi)$ and its partial derivatives of order $\leqq N$ vanishes on C, then

$$L\Lambda f = \Lambda L f \tag{29}$$

where

$$\Lambda f \equiv \sum_{\nu_1+\cdots+\nu_n} a_{\nu_1 \cdots \nu_n} \frac{\partial^{\nu_1+\cdots+\nu_n} f}{\partial \xi_1^{\nu_1} \cdots \partial \xi_n^{\nu_n}} \tag{30}$$

with constant coefficients a.

REMARK. It should be noted that in Theorem 2, the saltus of f has to vanish identically on the open set $\tilde{C}$, whereas in Theorem 3 the saltuses have only to satisfy the *boundary condition of vanishing on the $(n-2)$-dimensional set C.*

PROOF. Obviously (29) follows from (28) by induction on N. In addition to the symbols σ^α as defined by (4), we now introduce symbols $\sigma^{\alpha\beta}$ in the following way. For $\alpha < \beta$ we put

$$\sigma^{\alpha\beta} = (-1)^{\alpha+\beta-1} d\xi_1 \cdots d\xi_{\alpha-1} d\xi_{\alpha+1} \cdots d\xi_{\beta-1} d\xi_{\beta+1} \cdots d\xi_n;$$

for $\alpha = \beta$ we put $\sigma^{\alpha\beta} = 0$; and for $\alpha > \beta$ we put $\sigma^{\alpha\beta} = -\sigma^{\beta\alpha}$. The basis for our proof is the identity

$$\sum_{\alpha=1}^{n} \left(f \frac{\partial^2 G}{\partial \xi_\beta \partial \xi_\alpha} + \frac{\partial f}{\partial \xi_\beta} \frac{\partial G}{\partial \xi_\alpha} \right) \sigma^\alpha = \left(f \cdot \Delta G + \sum_{\alpha=1}^{n} \frac{\partial f}{\partial \xi_\alpha} \frac{\partial G}{\partial \xi_\alpha} \right) \sigma^\beta - \sum_{\alpha=1}^{n} d\left(f \frac{\partial G}{\partial \xi_\alpha} \right) \sigma^{\beta\alpha}. \tag{31}$$

From this we deduce that the quantity

$$\sum_{\alpha=1}^{n} \left(\frac{\partial f}{\partial \xi_\beta} \frac{\partial G}{\partial \xi_\alpha} - G \frac{\partial^2 f}{\partial \xi_\beta \partial \xi_\alpha} + f \frac{\partial^2 G}{\partial \xi_\beta \partial \xi_\alpha} - \frac{\partial G}{\partial \xi_\beta} \frac{\partial f}{\partial \xi_\alpha} \right) \sigma^\alpha \tag{32}$$

equals the quantity

$$(f \cdot \Delta G - G \cdot \Delta f) \sigma^\beta - \sum_{\alpha=1}^{n} d\varphi_\alpha \cdot \sigma^{\beta\alpha}, \tag{33}$$

where

$$\varphi_\alpha = f \frac{\partial G}{\partial \xi_\alpha} - G \frac{\partial f}{\partial \xi_\alpha}. \tag{34}$$

Now, since $\dfrac{\partial G}{\partial x_\beta} = -\dfrac{\partial G}{\partial \xi_\beta}$, the difference

$$\gamma_n \left(L \frac{\partial f}{\partial \xi_\beta} - \frac{\partial}{\partial x_\beta} F \right)$$

is the integral of (32) over B. By the equality of (32) and (33) this is

$$- \int_B \sum_{\alpha=1}^{n} d\varphi_\alpha \cdot \sigma^{\beta\alpha}$$

and by section 2 this will vanish if the saltus of each φ_α vanishes on C. By (34), it is enough to assume that $f, \dfrac{\partial f}{\partial \xi_1}, \cdots, \dfrac{\partial f}{\partial \xi_n}$ has vanishing saltus each.

We note a natural consequence of Theorem 3.

COROLLARY. *If the conglomerate function* $f(\xi)$ *is harmonic and the solution of* $\Lambda f = 0$ *and if the saltuses of* f *and of its partial derivatives of order* $\leqq N$ *vanish on* C *then* $F(x)$ *is likewise harmonic and a solution of* $\Lambda F = 0$.

We will say that a differential operator ΛF of type (30) has the *uniqueness property* if any analytic solution of $\Lambda F = 0$ which is defined in the complement of a bounded domain and vanishes at infinity is identically 0. Such an operator is for instance

$$\frac{\partial^2}{\partial x_1^2} + \cdots + \frac{\partial^2}{\partial x_m^2} \tag{35}$$

for $m < n$. In fact, if a function $F(x_1, \cdots, x_m, x_{m+1}, \cdots, x_n)$ is defined in the complement of a bounded domain in E_n then there exists an $(n - m)$-dimensional neighborhood of values $(x_{m+1}^0, \cdots, x_n^0)$, such that for each of those values, $F(x_1, \cdots, x_m, x_{m+1}, \cdots, x_n)$ is defined in

$$-\infty < x_\mu < \infty, \qquad \mu = 1, \cdots, m.$$

By our assumption it is a harmonic function in the entire $(x_1, \cdots, x_m)$-space and vanishes at infinity. Hence it is identically 0 in the latter variables; but being analytic in all n variables, it vanishes identically in all variables.

THEOREM 4. *If a conglomerate function* $f(\xi)$ *is harmonic in* $\tilde{B}$ *and the solution of an equation* $\Lambda f = 0$ *with uniqueness property; and if the saltuses of* $f(\xi)$ *and of its partial derivatives of order* $\leqq N$ *vanish on* C *then the following propositions hold.*

(i) *If the component* D_∞ *of* $E_n - \bar{B}$ *borders on a simplex* B_p *from both sides then* $f_p(x) = 0$.

(ii) *In particular, if* B *does not decompose the space then* $f(x) = 0$.

(iii) *If a simplex* B_p *separates* D_∞ *from* D_τ *then* $f_p(x)$ *can be continued into all of* D_τ.

(iv) *In particular, if* $\bar{B}$ *is the connected boundary of a bounded domain* D *then* $f(x)$ *can be continued (uniquely) into all of* D.

PROOF. *Ad* (i) *and* (ii). By the hypotheses of the theorem, $F_\infty(x)$ vanishes. But, by Theorem 3, in crossing over from D_∞ into D_∞ via B_p, $F(x)$ jumps by $\pm f_p(x)$. Hence $f_p(x) = 0$.

Ad (iii) *and* (iv). On B_p, $F(x)$ has the saltus $f_p(x)$. Therefore $F_\tau(x) - F_\infty(x) = \pm f_p(x)$. However, $F_\infty(x) = 0$. Thus, near B_p, $F_\tau(x)$ coincides with $\pm f_p(x)$. However, $F_\tau(x)$ exists in all of $\tilde{D}_\tau$. Therefore $f_p(x)$ exists in all of D_τ.

Now consider the space E_{2k}, $k \geqq 2$, of the real variables $x_1, y_1, \cdots, x_k, y_k$ and put $z_\alpha = x_\alpha + iy_\alpha$. If $f(z_1, \cdots, z_k)$ is an analytic function of the complex variables $z_1, \cdots, z_k$, then in particular its real and imaginary parts each satisfy the system of equations

$$\frac{\partial^2}{\partial x_1^2} + \frac{\partial^2}{\partial y_1^2} = 0, \cdots, \qquad \frac{\partial^2}{\partial x_k^2} + \frac{\partial^2}{\partial y_k^2} = 0. \tag{36}$$

Each of these equations has the uniqueness property. Also a simultaneous solution of them is harmonic in $(x_1, y_1, \cdots, x_k, y_k)$. Hence we obtain the following special case of theorem 4, (iv).

THEOREM 5. *If an analytic function of several complex variables is defined in the connected boundary of a bounded domain, then it can be continued uniquely into all of the domain.*

CHAPTER III

Analytic Functions of Complex Variables

7. Complex space. We will now consider systematically the space E_{2k} of k complex variables $\zeta_\alpha = \xi_\alpha + i\eta_\alpha$. All occurring functions will be likewise complex-valued. In particular a harmonic function will be one whose real and imaginary part are each harmonic. Replacing the pair of real variables ξ, η by the conjugate complex quantities $\zeta = \xi + i\eta$, $\bar{\zeta} = \xi - i\eta$ we have

$$d\bar{\zeta}d\zeta = 2id\xi d\eta \tag{37}$$

and therefore

$$d\bar{\zeta}_1 d\zeta_1 \cdots d\bar{\zeta}_k d\zeta_k = (2i)^k d\xi_1 d\eta_1 \cdots d\xi_k d\eta_k . \tag{38}$$

If any two of the $2k$ real factors on the right side of (38) are interchanged, the symbol changes its algebraic sign. However, if any two blocks of terms $d\bar{\zeta}_\alpha d\zeta_\alpha$, $d\bar{\zeta}_\beta d\zeta_\beta$ are interchanged the sign is not altered. It will be worth while to introduce the differential form

$$\omega = d\bar{\zeta}_1 d\zeta_1 + \cdots + d\bar{\zeta}_k d\zeta_k . \tag{39}$$

It is an "algebraic" area element, and *not* the scalar square of a length element. As for its (external) powers $\omega^2 = \omega \cdot \omega$, $\omega^3 = \omega^2 \cdot \omega$, $\cdots$, it can be easily seen that

$$\omega^k = k! d\bar{\zeta}_1 d\zeta_1 \cdots d\bar{\zeta}_k d\zeta_k \tag{40}$$

and that, for each α,

$$d\bar{\zeta}_\alpha d\zeta_\alpha \omega^{k-1} = \frac{1}{k} \omega^k. \tag{41}$$

As for the theorem of Stokes, we note that (1) remains valid for all occurring functions being complex-valued. In particular, if D is a domain with boundary $\bar{B}$, then

$$\frac{1}{k} \int_D \frac{\partial X_\alpha}{\partial \bar{\zeta}_\alpha} \omega^k = \int_B X_\alpha d\zeta_\alpha \omega^{k-1} \tag{42}$$

and

$$-\frac{1}{k} \int_D \frac{\partial Y_\alpha}{\partial \zeta_\alpha} \omega^k = \int_B Y_\alpha d\bar{\zeta}_\alpha \omega^{k-1} . \tag{43}$$

We now put

$$V = 2G \tag{44}$$

where G is the function (16). Thus in the special case (19) we have

$$V = \frac{1}{1-k}\Big(\sum_\alpha (\zeta_\alpha - z_\alpha)(\bar{\zeta}_\alpha - \bar{z}_\alpha)\Big)^{1-k} \tag{45}$$

$$\frac{\partial V}{\partial \zeta_\alpha} = \frac{\bar{\zeta}_\alpha - z_\alpha}{\big(\sum_\beta |\zeta_\beta - z_\beta|^2\big)^k}. \tag{46}$$

In the general case (16), on putting

$$X_\alpha = f\frac{\partial V}{\partial \zeta_\alpha}, \qquad Y_\alpha = V\frac{\partial f}{\partial \bar{\zeta}_\alpha}$$

we obtain, as analogues to (14) and (15) the formulas

$$\begin{aligned}\begin{matrix}(2\pi i)^k f(z, \bar{z})\\ 0\end{matrix} &= \int_B \Big(f\frac{\partial V}{\partial \zeta_\alpha}\,d\zeta_\alpha + V\frac{\partial f}{\partial \bar{\zeta}_\alpha}\,d\bar{\zeta}_\alpha\Big)\omega^{k-1} \\ &\quad - \frac{1}{k}\int_D \Big(f\frac{\partial^2 V}{\partial \zeta_\alpha \partial \bar{\zeta}_\alpha} - V\frac{\partial^2 f}{\partial \zeta_\alpha \partial \bar{\zeta}_\alpha}\Big)\omega^k\end{aligned} \tag{47}$$

and

$$\begin{matrix}(2\pi i)^k f(z, \bar{z})\\ 0\end{matrix} = \int_B f\frac{\partial V}{\partial \zeta_\alpha}\,d\zeta_\alpha\,\omega^{k-1} - \frac{1}{k}\int_D \Big(f\frac{\partial^2 V}{\partial \zeta_\alpha \partial \bar{\zeta}_\alpha} + \frac{\partial f}{\partial \bar{\zeta}_\alpha}\frac{\partial V}{\partial \zeta_\alpha}\Big)\omega^k. \tag{48}$$

Since

$$4\frac{\partial^2}{\partial\zeta\partial\bar{\zeta}} = \frac{\partial^2}{\partial\xi^2} + \frac{\partial^2}{\partial\eta^2}$$

we have

$$4\sum_{\alpha=1}^{k}\frac{\partial^2}{\partial\zeta_\alpha\partial\bar{\zeta}_\alpha} = \sum_{\alpha=1}^{k}\Big(\frac{\partial^2}{\partial\xi_\alpha^2} + \frac{\partial^2}{\partial\eta_\alpha^2}\Big) \equiv \Delta. \tag{49}$$

Thus, since V is harmonic, if we also assume that f is harmonic, then (46) leads to

$$\int_B \Big(f\frac{\partial V}{\partial \zeta_\alpha}\,d\zeta_\alpha + V\frac{\partial f}{\partial \bar{\zeta}_\alpha}\,d\bar{\zeta}_\alpha\Big)\omega^{k-1} = (2\pi i)^k f(z, \bar{z}) \quad or \quad 0, \tag{50}$$

which is the analogue to (20). It should be observed that (50) is genuinely different from (20). All that can be said is that for real f, relation (50) is "the real part" of relation (20). The difference will be significant for general point-sets B, and will in fact lead to an improvement of Theorem 4.

8. Cauchy's formula. Now, if in particular f is an analytic function in $z_1, \cdots, z_k$ alone, thus

$$\frac{\partial f}{\partial \bar{z}_\alpha} = 0, \qquad \alpha = 1, \cdots, k, \tag{51}$$

(which assumption in particular implies that f is harmonic) then (50) reduces to the very simple formula

$$f(z) \text{ or } 0 = \frac{1}{(2\pi i)^k} \int_B f \frac{\partial V}{\partial \zeta_\alpha} d\zeta_\alpha \omega^{k-1}. \tag{52}$$

If we write the formula in full we obtain

THEOREM 6. *If $f(z)$ is analytic in $\tilde{D}$ then the formula*

$$f(z) = \frac{1}{(2\pi i)^k} \int_B \frac{f(\zeta)(\bar{\zeta}_\alpha - \bar{z}_\alpha)\, d\zeta_\alpha \omega^{k-1}}{\left(\sum_\beta (\zeta_\beta - z_\beta)(\bar{\zeta}_\beta - \bar{z}_\beta)\right)^k} \tag{53}$$

holds for z in D.

This is a generalization of Cauchy's formula in one variable, however with the peculiarity that in the cases $k \geqq 2$ we are considering, the integrand contains $\bar{z}_\beta$ explicitly. Thus to start with it is not at all evident that the integral will be independent of $\bar{z}_\beta$ let alone be $f(z)$.

Now take a point $(z_1, \cdots, z_k)$ in D and replace in the integral (53) the conjugate point $(\bar{z}_1, \cdots, \bar{z}_k)$ by a point $(z_1', \cdots, z_k')$ in the vicinity of the conjugate point. The right side of (53) will then be a function

$$F(z_1, \cdots, z_k; z_1', \cdots, z_k')$$

in $2k$ complex variables. By Theorem 6, the function

$$F(z_1, \cdots, z_k; z_1', \cdots, z_k') - f(z_1, \cdots, z_k) \tag{54}$$

vanishes on the manifold $z_1' = \bar{z}_1, \cdots, z_k' = \bar{z}_k$. This implies that (54) vanishes identically, that is

$$f(z_1, \cdots, z_k) \equiv F(z_1, \cdots, z_k; z_1', \cdots, z_k').$$

Now assume that D contains the origin, and let $z_1', \cdots, z_k'$ be the origin. Then, at least for points z near the origin we have

$$f(z) = \frac{1}{(2\pi i)^k} \int_B \frac{f(\zeta)\bar{\zeta}_\alpha\, d\zeta_\alpha \omega^{k-1}}{[(\zeta_1 - z_1)\bar{\zeta}_1 + \cdots + (\zeta_k - z_k)\bar{\zeta}_k]^k}. \tag{55}$$

This formula is much closer to Cauchy's one-variable formula than (53) is, however the integral (55) does not necessarily converge in all of D. The domain of convergence can be described geometrically, but the description would be of no consequence in the present context.

9. Cauchy's integral. As an analogue to (21) we now introduce the integral

$$F(z, \bar{z}) = \frac{1}{(2\pi i)^k} \int_B f(\zeta) \frac{\partial V}{\partial \zeta_\alpha} d\zeta_\alpha \omega^{k-1} \tag{56}$$

for an analytic conglomerate function $f(\zeta_1, \cdots, \zeta_k)$ in $\tilde{B}$. We first of all have the following analogue to Theorems 1 and 2.

THEOREM 7. *The saltus of the conglomerate function* (56) *on B is $f(z)$.*

Also if the saltus of $f(\zeta)$ vanishes identically on all $\tilde{C}_p$ containing a vertex A_0 of B, then a small deformation of A_0 leads to analytic continuation of $F_\tau(z, \bar{z})$ from D_τ into D'_τ.

The following theorem is an analogue to Theorem 3 with

$$\Lambda f \equiv \frac{\partial f}{\partial \bar{z}_\beta} = \frac{\partial f}{\partial x_\beta} + i \frac{\partial f}{\partial y_\beta}, \tag{57}$$

but it is a great improvement on Theorem 3, insofar as it does not involve a boundary condition on the *derivatives* of f.

THEOREM 8. *If the saltus of $f(\zeta)$ vanishes on C, then F is analytic in $z_1, \cdots, z_k$, that is, $\dfrac{\partial F}{\partial \bar{z}_\beta} = 0$, $\beta = 1, \cdots, k$.*

Also, in D_∞ the functional element $F_\infty(z)$ is identically 0.

For use in the proof, we introduce the symbol $s^{\bar{\alpha}}$ which is the differential form arising from

$$d\bar{\zeta}_1 d\zeta_1 \cdots d\bar{\zeta}_k d\zeta_k \tag{58}$$

by dropping $d\bar{\zeta}_\alpha$ and by $s^{\bar{\alpha}\bar{\beta}}$ we denote the symbol arising from (28) by dropping both $d\bar{\zeta}_\alpha$ and $d\bar{\zeta}_\beta$ if $\alpha \neq \beta$, and the symbol 0 if $\alpha = \beta$. Now

$$\frac{(2\pi i)^k}{(k-1)!} \frac{\partial F}{\partial \bar{z}_\beta} = \int_B - \frac{1}{(k-1)!} \sum_{\alpha=1}^{k} f \frac{\partial^2 V}{\partial \zeta_\alpha \partial \bar{\zeta}_\beta} d\zeta_\alpha \omega^{k-1}, \tag{59}$$

but the integral is, identically in f, V,

$$f \cdot \Delta V \cdot s^{\bar{\beta}} + \sum_{\alpha=1}^{k} d\left(f \frac{\partial V}{\partial \zeta_\alpha}\right) s^{\bar{\alpha}\bar{\beta}} - \sum_{\alpha=1}^{k} \frac{\partial V}{\partial \zeta_\alpha} df s^{\bar{\alpha}\bar{\beta}}. \tag{60}$$

In our case we have $\Delta V = 0$, and we also have $df.\ s^{\bar{\alpha}\bar{\beta}} = 0$ since f is independent of both $\bar{\zeta}_\alpha$ and $\bar{\zeta}_\beta$. Therefore, by Stokes' theorem, (59) is

$$\int_C f \frac{\partial V}{\partial \zeta_\alpha} s^{\bar{\alpha}\bar{\beta}}$$

and thus (59) vanishes if the saltus of f vanishes on C. The second half of the theorem has been proven in section 6.

Finally we prove the following analogue of Theorem 4.

THEOREM 9. *If the conglomerate function $f(\zeta_1, \cdots, \zeta_k)$ is analytic in B and if its saltus vanishes on C, then the following propositions hold:*

(i) *If the component D_∞ borders on a simplex B_p from both sides then $f_p(\zeta) = 0$.*

(ii) *If, in particular, $\bar{B}$ does not decompose E_n, then $f(\zeta) = 0$.*

(iii) *If a simplex B_p separates D_∞ from D_τ then $f_p(\zeta)$ can be continued into all of D_τ.*

As a special conclusion we observe that if the set of all zeroes of an analytic function $f(\zeta_1, \cdots, \zeta_k)$ contains a bounding cycle of dimension $2k - 2$ then the function vanishes identically.

As a counterpart to this property we observe that if $f(\zeta_1, \cdots, \zeta_k)$ is $\neq 0$ everywhere on the boundary $\bar{B}$ of a domain D then $f \neq 0$ everywhere in D. In fact f^{-1} is analytic in a neighborhood of B and therefore it can be continued into all of D.

Chapter IV

Integrability Conditions

10. We are interested in the system of equations

$$
(61) \qquad \begin{aligned}
\frac{\partial f}{\partial \bar{z}_\alpha} &= f_\alpha(z, \bar{z}), & \alpha &= 1, \cdots, k, \\
\frac{\partial f_\alpha}{\partial \bar{z}_\beta} &= \frac{\partial f_\beta}{\partial \bar{z}_\alpha}, & \alpha, \beta &= 1, \cdots, k,
\end{aligned}
$$

in any domain D in E_{2k}. The functions f_α are given and the function f is a desired solution. The system is in some respects similar to the classical system

$$
(62) \qquad \begin{aligned}
\frac{\partial f}{\partial x_\alpha} &= f_\alpha(x), & \alpha &= 1, \cdots, n, \\
\frac{\partial f_\alpha}{\partial x_\beta} &= \frac{\partial f_\beta}{\partial x_\alpha}, & \alpha, \beta &= 1, \cdots, n,
\end{aligned}
$$

in a domain D in E_n. Assuming all functions f_α to be analytic in x or $(z, \bar{z})$ respectively then either system has an analytic solution in a neighborhood of a point. For (62) this follows easily if all $f_\alpha(x)$ are expanded in power series. Also if the functions $f_\alpha(x)$ contains additional analytic parameters then so does the solution f. Hence we may conclude the existence of local solutions of (61) by viewing $\bar{z}_1, \cdots, \bar{z}_k$ as independent variables and $z_1, \cdots, z_k$ as parameters.

As for solutions in the large, we will first prove a theorem which will apply to either system (61) or (62). The theorem will not be needed, but the proof will be based on an application of Green's formula which apparently has not been noticed before. After that we will prove another theorem for the system (61). It will be the exact counterpart within the scope of the general Green's integral to Cousin's method as based on the classical Cauchy's integral.

Theorem 10. *If D is a simplicial domain with boundary $\bar{B}$, if $\{f_\alpha\}$ is given in D and if either system* (61) *or* (62) *has a solution in $\tilde{B}$, then there also exists a solution in D.*

We will treat only system (62) for system (61) the proof is analogous. With the given functions f_α in $\tilde{D}$ and the solution f in $\tilde{B}$ we set up the integral

$$\gamma_n F(x) = \int_B f \frac{\partial G}{\partial \xi_\alpha} \sigma^\alpha - \int_D f_\alpha \frac{\partial G}{\partial \xi_\alpha} dV_\xi .$$

By formula (15), if f exists in all of $\tilde{D}$, then $F(x) = f(x)$. Now take a fixed point x^0 in D and a sufficiently small neighborhood D' with boundary $\bar{B}'$. If it is sufficiently small, then there exists a function $g(x)$ with $f_\alpha = \dfrac{\partial g}{\partial x_\alpha}$ in D' such that, for x in D',

$$\gamma_n g(x) = \int_{B'} f \frac{\partial G}{\partial \xi_\alpha} \sigma^\alpha - \int_{D'} f_\alpha \frac{\partial G}{\partial \xi_\alpha} dV_\xi .$$

The reader will now easily see that our theorem will follow from the following

LEMMA. *If the boundary B of a domain D consists of two disjoint parts, $B = B' + B''$, if f in $\tilde{B}'$ and g in $\tilde{B}''$ are each a solution of* (62), *then for x in $E_n - \bar{D}$, the partial derivatives of the function*

$$\int_{B'} f \frac{\partial G}{\partial \xi_\alpha} \sigma^\alpha + \int_{B''} g \frac{\partial G}{\partial \xi_\alpha} \sigma^\alpha - \int_D f_\alpha \frac{\partial G}{\partial \xi_\alpha} dV_\xi$$

vanish identically.

In order to prove the lemma we note that the partial derivative of the given expression with respect to x_β, for x outside $\bar{D}$, is

$$-\int_{B'} f \frac{\partial^2 G}{\partial \xi_\alpha \partial \xi_\beta} \sigma^\alpha - \int_{B''} g \frac{\partial^2 G}{\partial \xi_\alpha \partial \xi_\beta} \sigma^\alpha + \int_D f_\alpha \frac{\partial^2 G}{\partial \xi_\alpha \partial \xi_\beta} dV_\xi .$$

By (31) this is

$$\int_B f_\beta \frac{\partial G}{\partial \xi_\alpha} \sigma^\alpha - \int_B f_\alpha \frac{\partial G}{\partial \xi_\alpha} \sigma^\beta + \int_D f_\alpha \frac{\partial^2 G}{\partial \xi_\alpha \partial \xi_\beta} dV_\xi .$$

On applying Stokes' theorem to the integrals over B, this turns out to be

$$\int_D \left(\frac{\partial f_\beta}{\partial \xi_\alpha} - \frac{\partial f_\alpha}{\partial \xi_\beta}\right) \frac{\partial G}{\partial \xi_\alpha} dV_\xi$$

and this in its turn is obviously 0.

11. An existence theorem. We will now prove a genuine existence theorem.

THEOREM 11. *The system* (61) *has a solution if the domain D is a general polycylinder, that is the topological product of a domain in each of the variables $z_1, \cdots, z_k$ separately.*

The theorem will follow from a lemma on functions in the plane of one complex variable $z = x + iy$.

If $\varphi(z, \bar{z})$ is analytic in $\tilde{D}$, then the integral

$$F(z, z) = \frac{-1}{2\pi i} \int_D \frac{\varphi(\zeta, \bar{\zeta})\, d\bar{\zeta}\, d\zeta}{\zeta - z} \tag{63}$$

is analytic in z, $\bar{z}$ and $\dfrac{\partial F}{\partial \bar{z}} = \varphi(z, \bar{z})$. Also if φ depends analytically on additional parameters, then so does F.

For z outside $\bar{D}$ the integral (63) is obviously analytic in z and other parameters and independent of $\bar{z}$. Hence, in order to prove that the stated properties of $F(z, z)$ hold in the neighborhood of a point z^0 in D we may assume that D is an arbitrarily small neighborhood of z^0. If the neighborhood is sufficiently small then there exists a function $\Phi(z, \bar{z})$ of which $\varphi(z, \bar{z})$ is the derivative with respect to $\bar{z}$, and thus we have to prove the properties of $F(z, \bar{z})$ only in the specialized case

$$F(z, \bar{z}) = \frac{-1}{2\pi i} \int_D \frac{\partial \Phi}{\partial \bar{\zeta}} \frac{d\bar{\zeta}\, d\zeta}{\zeta - z}. \tag{64}$$

If we introduce the Newtonian kernel

$$V = \log | \zeta - z |^2,$$

then (64) is

$$F(z, \bar{z}) = \frac{-1}{2\pi i} \int_D \frac{\partial \Phi}{\partial \bar{\zeta}} \frac{\partial V}{\partial \zeta} d\bar{\zeta}\, d\zeta. \tag{65}$$

Now, formula (48) which we have claimed only for $k \geqq 2$ also holds for $k = 1$. Thus, for z in D, we have

$$\Phi(z, \bar{z}) = \frac{1}{2\pi i} \int_B \Phi \frac{d\zeta}{\zeta - z} + F(z, \bar{z}). \tag{66}$$

But the boundary integral is independent of $\bar{z}$, and this proves our lemma.

We are now going to prove Theorem 11. Since we are going to apply induction on k, we will have to prove our theorem in the sharper form that some solution of (61) depends analytically on any parameters occurring in f_β. We assume the existence of a solution of

$$\frac{\partial f}{\partial \bar{z}_\beta} = f_\beta(z_1, z_1; z_2, z_2, \cdots, z_k, z_k) \tag{67}$$

in $\beta = 2, \cdots, k$; with $z_1, \bar{z}_1$ as parameters. Denoting the solution of (67) by h and introducing the new function $g = f - h$, then the system (61) reduces to a system

$$\frac{\partial g}{\partial \bar{z}_1} = g_1(\bar{z}_1; z_1, z_2, \cdots, z_k)$$

$$\frac{\partial g}{\partial \bar{z}_2} = \cdots = \frac{\partial g}{\partial \bar{z}_k} = 0,$$

and this system is solvable by our lemma.

CHAPTER V

Meromorphic Functions in Euclidean Space

12. A simplified problem of Mittag-Leffler type. We are now returning to the topic of Chapter II, and the symbols $C_1, \cdots, C_r; B_1, \cdots, B_s; D_1, \cdots, D_t$,

D_∞ shall have the previous meaning. In addition to the incidence number ϵ_{pq} between C_p and B_q we also introduce the incidence number $\eta_{q\tau}$ between B_q and D_τ. An elementary combinatorial statement ("the boundary of a boundary is zero") implies the relation

$$\sum_{q=1}^{s} \epsilon_{pq}\,\eta_{q\tau} = 0 \tag{68}$$

and its validity is not impaired by the fact that the domains D_τ are not necessarily simplices. We furthermore introduce an arbitrary pointset S having the following properties: it is closed and nowhere dense, and if R is any domain then $R - S$ (that is $R - R\cdot S$) is also a domain. The set S will be termed an "exceptional" set. Obviously if a harmonic function φ in $R - S$ has an analytic continuation in all of R then the continuation is unique. If such a continuation exists we will say that "the singularities of φ at S are removable" or that "φ exists in (all of) R," or that "φ is analytic in R."

We now assume the existence of a harmonic function φ_τ in $\tilde{D}_\tau - S$, for each $\tau = 1, \cdots, t, \infty$, and we further assume that for each B_q the saltus

$$f_q = \sum_{\tau=1}^{\infty} \eta_{q\tau}\,\varphi_\tau \tag{69}$$

is analytic not only in $\tilde{B}_q - S$ but in all of $\tilde{B}_q$. We therefore may introduce the conglomerate function

$$F(x) = \frac{1}{\gamma_n} \sum_{q=1}^{s} \int_{B_q} \left(f_q \frac{\partial G}{\partial \xi_\alpha} - G \frac{\partial f_q}{\partial \xi_\alpha} \right) \sigma^\alpha . \tag{70}$$

Also, the saltus of $\{f_q\}$ in $\tilde{C}_p$ is defined as

$$f^p = \sum_q \epsilon_{pq} f_q = \sum_q \sum_\tau \epsilon_{pq}\,\eta_{q\tau}\,\varphi_\tau$$

and owing to relation (68) it turns out to be identically 0. Thus we may apply Theorems 1, 2, 3 and we obtain the following result.

THEOREM 12. *If the conglomerate function $\{\varphi_\tau\}$ is defined and harmonic in $\{\tilde{D}_\tau - S\}$, and if its saltus $\{f_p\}$ has none other than removable singularities in $\tilde{B}$, then the same saltus will be produced by the harmonic conglomerate function* (70) *which has no singularities whatsoever in $\{\tilde{D}_\tau\}$. Also, if $\{\varphi_\tau\}$ is a solution of a system of differential equations with constant coefficients,*

$$\Lambda\varphi = 0, \tag{71}$$

then $\{F_\tau\}$ is also a solution.

Thus, the functional elements

$$F_1 - \varphi_1 = F_2 - \varphi_2 = \cdots = F_\infty - \varphi_\infty$$

define a function Φ with the following properties. It is one functional element in $E_n - S$, it is harmonic and a solution of (71), *and in each $\tilde{D}_\tau$, the difference $\Phi - \varphi_\tau$ is analytic without exception.*

A seemingly striking situation arises, if the system (71) has the uniqueness property, and the set S is bounded. Being analytic outside a bounded domain, the function Φ must be analytic in the entire space and therefore $\varphi_\tau \equiv \Phi - F_\tau$ has only removable singularities in $\tilde{D}_\tau$ each. However, it should be noted that in this case, to start with, φ_∞ itself has to have an analytic continuation in the entire space. Now, let B'_λ be one of our faces lying in the boundary of D_∞. If it separates D_∞ from D_∞ then the saltus of φ is $\varphi_\infty - \varphi_\infty \equiv 0$, and thus naturally has no singularities. If however, B'_λ separates D_∞ from another domain D_c, then the saltus is $\pm(\varphi_\infty - \varphi_c)$ and if it is assumed to be free from singularities then so is φ_c itself. In other words: if we introduce the closet set $\bar{D}_1 + \cdots + \bar{D}_t$, and if we consider its interior D_0 and the frontier B_0; then our assumption requires *explicitly* that in a neighborhood $\tilde{B}_0$ the function $\{\varphi_\tau\}$ itself shall have no singularities, and only for "internal" faces B''_μ is the explicit requirement less stringent. If such a face separates two domains D_a, D_b then the assumption requires only that the saltus $\varphi_a - \varphi_b$ shall have no singularities. The theorem then allows the conclusion that actually the singularities of each φ_τ itself are removable everywhere in $\tilde{D}_\tau$.

13. A simplified problem of Weierstrass type. Everything else being as before, we consider a conglomerate function $\{\psi_\tau(x)\}$ with the following properties. Each $\psi_\tau(x)$ is analytic in $\tilde{D}_\tau$ and different from 0 in $\tilde{D}_\tau - S$. The "multiplicative" saltus

$$\prod_{\tau=1}^{\infty} \psi_\tau^{\eta_{q\tau}}$$

is by assumption analytic everywhere in $\tilde{B}_q$, and in the neighborhood of every point of $\tilde{D}_\tau - S$ the function $\log \psi_\tau(x)$ and its partial derivatives of first order both harmonic and a solution of (71), the latter having uniqueness property.

For fixed α, we introduce the conglomerate function

$$\varphi_\tau(x) = \frac{\partial}{\partial x_\alpha} \log \psi_\tau(x) = \frac{1}{\psi_\tau} \frac{\partial}{\partial x_\alpha} \psi_\tau,$$

and by Theorem 12 it is analytic in $\tilde{D}_\tau$. Thus, for fixed τ there exist analytic functions $F_1(x), \cdots, F_n(x)$ in $\tilde{D}_\tau$ such that

$$\frac{\partial \psi}{\partial x_\alpha} = \psi \cdot F_\alpha, \qquad \alpha = 1, \cdots, n.$$

On expanding ψ in a power series in a neighborhood of any point in $\tilde{D}_\tau$, the reader will easily convince himself that *every $\psi_\tau(x)$ is different from 0 in all of $\tilde{D}_\tau$*.

14. The theorem of Cousin. Returning to the topic of section 12, we now set up the modified integral

$$F^0(x) = \frac{1}{\gamma_n} \sum_{\mu=1}^{m} \int_{B''_\mu} \left(f_\mu \frac{\partial G}{\partial \xi_\alpha} - G \frac{\partial f_\mu}{\partial \xi_\alpha} \right) \sigma^\alpha$$

extended only over the "internal" faces B''_μ. Our present setup implies that we require the saltus function f_q to be analytic throughout $\tilde{B}_q$ only in case of internal faces, with no assumption made about the other faces.

Theorems 1 and 2 will permit the conclusion that for each D_a *whose closure is contained in* D_0 the corresponding functional element F_a^0 has no singularities in $\tilde{D}_a$, and if two such simplices have a face B''_μ in common, then again

$$F_a^0 - \varphi_a = F_b^0 - \varphi_b. \tag{72}$$

Therefore if a number of such simplices $D_a, D_b, \cdots$ are the simplicial subdivision of a subdomain D^0 of D_0, then the functional elements (72) merge into a function Φ^0 in $\tilde{D}^0 - S$. However it must not be assumed that Φ^0 will be a solution of (71) because $\{\varphi_\tau\}$ is. All that can be said about the function (or functions)

$$h^0 = \Lambda\Phi^0 = \Lambda F^0 - \Lambda\varphi = \Lambda F^0$$

is that it is harmonic without singularities in D^0 (and indeed in $\tilde{D}^0$). We may now pose the problem of constructing a harmonic solution of the "unhomogeneous" system

$$\Lambda H = h^0$$

for any harmonic h^0 in any (or some specialized type of) domain D^0. If such a solution exists then $\Phi^0 - H$ will have in D^0 the same properties as were stated in Theorem 12 for the function itself.

A classical case arises if E_n is the space E_{2k} of the variables $z_\alpha = x_\alpha + iy_\alpha$, $\alpha = 1, \cdots, k$. Each φ_τ shall be a mereomorphic element in $\tilde{D}_\tau$, that is a quotient of two analytic functions in $z_1, \cdots, z_k$ alone; with S being the union of the zero-sets of the denominators. The system (71) is

$$\frac{\partial\varphi}{\partial\bar{z}_1} = 0, \cdots, \frac{\partial\varphi}{\partial\bar{z}_k} = 0. \tag{73}$$

Theorem 11 will now give in substance a new proof for the classical theorem of Cousin about merging meromorphic elements in a polycylinder into one meromorphic function.

Chapter VI

Meromorphic Functions on Torus Space

15. Green's formula. We consider the locally Euclidean n-dimensional torus T_n with all periods 1:

$$T_n: \quad 0 \leqq x_\alpha < 1, \qquad \alpha = 1, \cdots, n.$$

Stokes' theorem (1) for ordinary or conglomerate functions remains in force. The scalar φ has to be strictly a function on the torus; that is, as a function in E_n it has to be strictly periodic in each variable x_α with period 1. However, any quantity ψ appearing only in a differential $d\psi$ is permitted to alter by an additive constant if continued along a closed path in T_n. We now consider on the torus a function $G(\xi, x) = G(\xi - x)$ of the type (16). Formulas (14) and (15) are valid again, and if we apply them to the special domain $D = T_n$, $B = 0$, we obtain

$$\gamma_n f(x) = -\int_{T_n} \left(f\cdot\Delta G + \frac{\partial f}{\partial \xi_\alpha}\frac{\partial G}{\partial \xi_\alpha} \right) dV_\xi \tag{74}$$

$$\gamma_n f(x) = -\int_{T_n} (f\cdot\Delta G - G\cdot\Delta f)\, dV_\xi \tag{75}$$

On putting $f(\xi) = 1$ we obtain in particular

$$\gamma_n = -\int_{T_n} \Delta G\cdot dV_\xi,$$

thus ΔG cannot be 0. But it might be a constant, and if so, the constant must of necessity be $-\gamma_n$. That is,

$$\Delta G = -\gamma_n. \tag{76}$$

Furthermore a solution of (76) must of necessity be unique. The actual existence of such a G can be proven in many ways. Suffice it to state here, that except for a constant factor, it is the sum of the series

$$\sum_{h_1,\cdots,h_n=-\infty}^{\infty}{}' \frac{\exp\{2\pi i[h_1(x_1 - \xi_1) + \cdots + h_n(x_n - \xi_n)]\}}{h_1^2 + \cdots + h_n^2}.$$

The series is not absolutely convergent but can be "summed" by general summation processes. From now on, G will be this solution of (76). Therefore, we now have

$$\begin{matrix}\gamma_n f(x)\\ 0\end{matrix} = \int_B \left(f\frac{\partial G}{\partial \xi_\alpha} - G\frac{\partial f}{\partial \xi_\alpha}\right)\sigma^\alpha + \gamma_n \int_D f\, dV_\xi \tag{77}$$

and in particular

$$\begin{matrix}\gamma_n f(x)\\ 0\end{matrix} = \int_B \left(f\frac{\partial G}{\partial \xi_\alpha} - G\frac{\partial f}{\partial \xi_\alpha}\right)\sigma^\alpha + \textit{constant}. \tag{78}$$

16. Green's integral. We again consider simplices $B_1, \cdots, B_s; C_1, \cdots, C_r$. The complement $T_n - \bar{B}$ decomposes into domains $D_1, \cdots, D_t$, with no "D_∞" being present. We again set up the integral

$$F(x) = \frac{1}{\gamma_n}\int_B \left(f\frac{\partial G}{\partial \xi_\alpha} - G\frac{\partial f}{\partial \xi_\alpha}\right)\sigma^\alpha \tag{79}$$

for an arbitrary harmonic conglomerate function $f(\xi)$ in $\tilde{B}$, and we note that Theorems 1 and 2 carry over literally. In the proof to Theorem 1 relation (77) has to be used instead of (20). In the proof to Theorem 2, the value of (25) is not 0, but some other constant, namely

$$-\gamma_n \sum_{q=0}^{\mu} \int_{B_q} f_q \, dV_\xi$$

and thus, in the proof, $\gamma_n(F(x) - F'(x))$ is constant, and hence $F(x)$ continuable.

The remarkable thing about the torus is the way in which Theorems 1 and 2 merge immediately with a part of Theorem 12.

THEOREM 13. *If $f(\xi)$ is a harmonic conglomerate function in $\tilde{B}$ and if its saltus vanishes identically in $\tilde{C}$, then f is in its turn the saltus of the harmonic conglomerate function F in $D \equiv T_n$.*

In particular, if S is an exceptional set, if φ is a conglomerate harmonic function in $\tilde{D} - S$, and if its saltus f has no singularities in $\tilde{B}$, then the function F is a harmonic conglomerate function without singularities having the same saltus as φ.

Finally we will analyze Theorem 3.

17. Meromorphic functions. Denoting again (79) by Lf we see on the basis of (31)–(34) that

$$L \frac{\partial f}{\partial \xi_\beta} - \frac{\partial}{\partial x_\beta} Lf = -\gamma_n \int_B f\sigma^\beta \tag{80}$$

and that in particular we have

$$L \frac{\partial f}{\partial \xi_\beta} - \frac{\partial}{\partial x_\beta} Lf = \textit{constant}; \tag{81}$$

provided again that the saltuses of $f, \dfrac{\partial f}{\partial \xi_1}, \cdots, \dfrac{\partial f}{\partial \xi_n}$ vanish on C. Thus we will have

$$L \frac{\partial f}{\partial \xi_\beta} - \frac{\partial}{\partial x_\beta} Lf = 0$$

if and only if

$$\int_B f\sigma^\beta = 0. \tag{82}$$

It will now be easy for the reader to verify the following analogue to Theorem 3.

THEOREM 14. *If $f(\xi)$ is harmonic in $\tilde{B}$ and if its saltus vanishes identically in $\tilde{C}$, then*

$$L\Lambda f - \Lambda Lf = \textit{constant} \tag{83}$$

If Λ be represented in some (or any) way as

$$\Lambda \equiv \sum_{\beta=1}^{n} \frac{\partial}{\partial x_\beta} \Lambda_\beta \tag{84}$$

where $\Lambda_1, \cdots, \Lambda_n$ are any operators, then

$$L\Lambda f - \Lambda L f = -\gamma_n \int_B \left(\sum_{\beta=1}^{n} \Lambda_\beta f \cdot \sigma^\beta \right). \tag{85}$$

In particular we have

$$L\Lambda f = \Lambda L f$$

if and only if

$$\int_B \left(\sum_{\beta=1}^{n} \Lambda_\beta f \cdot \sigma^\beta \right) = 0. \tag{86}$$

Therefore, if $\Lambda f = 0$ *and if* (86) *holds, then* ΛF *is likewise* 0.

For

$$\Lambda f = \left(\frac{\partial^2}{\partial x_1^2} + \cdots + \frac{\partial^2}{\partial x_m^2} \right) f, \qquad m < n$$

(86) is

$$\int_B \left(\sum_{\beta=1}^{n} \frac{\partial f}{\partial x_\beta} \sigma^\beta \right) = 0$$

Or, if

$$\Lambda f = \frac{\partial f}{\partial x_1} + i \frac{\partial f}{\partial x_2} = 2 \frac{\partial f}{\partial \bar{z}},$$

where

$$z = x_1 + i x_2,$$

then we may put $\Lambda_1 = 1, \Lambda_2 = i, \Lambda_3 = \cdots = \Lambda_n = 0$, and therefore

$$\sum_{\beta=1}^{n} \Lambda_\beta f \cdot \sigma^\beta = f \cdot (d\xi_1 - i d\xi_2) \cdot \sigma^{1,2} \equiv f \, d\bar{\xi} \cdot \sigma^{1,2}.$$

Turning now to the torus space T_{2k} of the complex variables $z_\alpha = x_\alpha + i y_\alpha$, $\alpha = 1, \cdots, k$, we therefore obtain the following analogue to Theorem 12.

THEOREM 15. *If* $\varphi_\tau(z_1, \cdots, z_k)$ *is meromorphic in* $\tilde{D}_\tau$, $\tau = 1, \cdots, t$, *but each* $\varphi_a - \varphi_b$ *is analytic in* $\tilde{B}_q$ *and if the* k *periodicity conditions*

$$\int_B f \, d\bar{\xi}_\alpha \cdot \omega^{k-1} = 0, \qquad \alpha = 1, \cdots, k \tag{87}$$

are fulfilled, then there exists a meromorphic function $F(z_1, \cdots, z_k)$ *on the torus, such that* $F - \varphi_\tau$ *is analytic in* $\tilde{D}_\tau$.

If the conditions (87) are not stipulated, then by (83) we have at any rate

$$\frac{\partial F}{\partial \bar{z}_\alpha} = c_\alpha, \qquad \alpha = 1, \cdots, k$$

where c_α is a constant. In this case the function

$$F(z_1, \cdots, z_k; \bar{z}_1, \cdots, \bar{z}_k) - c_1\bar{z}_1 - \cdots - c_k\bar{z}_k$$

in the plane will be strictly meromorphic, however if continued along a closed path on the torus it alters by an additive constant which need not be 0.

The additive constants do not alter the algebraic singularities, and in this sense we have

THEOREM 16. *If an exceptional set S on T_{2k} is locally the singular set of a meromorphic function, then it has the same property in the large.*

PRINCETON UNIVERSITY

ANNALS OF MATHEMATICS
Vol. 45, No. 4, October, 1944

GROUP INVARIANCE OF CAUCHY'S FORMULA IN SEVERAL VARIABLES

BY S. BOCHNER

(Received February 1, 1944)

1. If we iterate Cauchy's formula

$$f(z) = \frac{1}{2\pi i}\int_C \frac{f(\zeta)\, d\zeta}{\zeta - z} \tag{1}$$

we obtain

$$f(z_1, \cdots, z_k) = \frac{1}{2\pi i}\int_{C_k} f(\zeta_1, \cdots, \zeta_k)\, \frac{d\zeta_1 \cdots d\zeta_k}{(\zeta_1 - z_1) \cdots (\zeta_k - z_k)}. \tag{2}$$

In (2), C_k is a k-dimensional "characteristic manifold" of a special topological type, its prototype being the torus

$$| \zeta_1 | = 1, \cdots, | \zeta_k | = 1. \tag{3}$$

The special nature of this manifold corresponds to the special nature of the kernel

$$G(\zeta; z) = \frac{1}{(\zeta_1 - z_1) \cdots (\zeta_k - z_k)} \tag{4}$$

under the integral, and S. Bergmann[1] and A. Weil[2] have studied alterations of the formula to be brought about by suitably modifying both C_k and $G(\zeta; z)$.

In the present paper we will deal with the same question but from a different approach. In the work of Bergmann and Weil the emphasis is put on obtaining a formula to suit a prescribed domain of validity in z, whereas our emphasis will be on retaining prescribed structural features of the formula as such. We note that in the special case (4), the kernel depends only on the differences $\zeta_1 - z_1, \cdots, \zeta_k - z_k$, and this feature will be retained unaltered. We further note that the torus (3) is invariant and transitive under the group of transformations

$$\zeta_1 \to e^{i\alpha_1}\zeta_1, \cdots, \zeta_k \to e^{i\alpha_k}\zeta_k.$$

This feature will be likewise retained but other groups of affine transformations will be admitted.

Part I will deal with the group of translations, Part II with some cases of the unitary group, and Part III with general groups. The general case will be based

[1] *On the surface integrals of functions of two complex variables*, Amer. Jour. of Math. 63 (1941), 295–318.

[2] *L'integrale de Cauchy et les fonctions de plusieurs variables*, Math. Annalen, 111 (1935), 178–182.

on an operational version of Cauchy's theorem in which no reference is made to "realization" by integrals.

PART I

OCTANT-SHAPED DOMAINS

2. Tubes

Given any k complex variables

$$z_p = x_p + iy_p ; \qquad p = 1, \cdots, k, \tag{5}$$

we consider a k-dimensional domain P in the space of the real variables $x = (x_1, \cdots, x_k)$, and with the point set P we construct in the $(2k)$-dimensional space of the variables (5) the domain

$$\{x \in P; -\infty < y_p < \infty, p = 1, \cdots, k\}. \tag{6}$$

Such a domain has been called a *tube*,[3] and the point set P its *base*. A tube will be usually denoted by the letter T. We will say that the domain P is *radiated* if it is convex, does not contain the origin $x = (0, \cdots, 0)$ but is composed of rays which emanate from the origin. That is, if $x^0 = (x_1^0, \cdots, x_k^0)$ is a point of P, then the segment

$$x = tx^0, \qquad 0 < t < \infty \tag{7}$$

also belongs to P. We now pick another set of k real variables denoting them by $\alpha = (\alpha_1, \cdots, \alpha_k)$ and we consider all those α for which the inequality

$$\alpha_1 x_1 + \cdots + \alpha_k x_k > 0 \tag{8}$$

holds for all $x \in P$. They obviously form a convex set. We make the explicit assumption that the set is not empty and that it actually has an interior thus being the closure of a domain. This (open) domain will be denoted by G and called the conjugate to P. The domain G is again radiated. We now set up the integral

$$\int_G e^{-(\alpha_1 x_1 + \cdots + \alpha_k x_k)} \, dv_\alpha \tag{9}$$

where $dv_\alpha = d\alpha_1 \cdots d\alpha_k$ and we make the final explicit assumption that this integral is boundedly convergent in every compact subset of P. If all these assumptions are fulfilled we will say that the tube T is *octant-shaped*, and we associate with it the "kernel"

$$K(z_1, \cdots, z_k) = \int_G e^{-(z_1\alpha_1 + \cdots + z_k\alpha_k)} \, dv_\alpha \tag{10}$$

which is obviously defined and analytic in all of T.

[3] S. BOCHNER, *Bounded analytic functions in several variables and multiple Laplace integrals*, Amer. Jour. of Math. 59 (1937), 732–738.

The boundary of T is likewise "tubular", and the most important part (for us) of the boundary will be the pointset

$$\{\zeta_p = 0 + i\eta_p;\ -\infty < \eta_p < \infty;\ p = 1, \cdots, k\}. \tag{11}$$

It will be called the *spine* of T. The Euclidean volume element on the spine is, of course, $dv\eta = d\eta_1 \cdots d\eta_k$; however there will be a formal advantage in using not the volume element as such but the "oriented" external differential

$$d\zeta_1 \cdots d\zeta_k \equiv i^k d\eta_1 \cdots d\eta_k \equiv i^k dv\eta, \tag{12}$$

and whenever this differential will occur under the integral, the spine as a manifold will be denoted by iE_k.

3. Functions of integrable square[3]

Functions in tubes can be subjected to Fourier analysis. If $f(z)$ is analytic in T then we say that it is of integrable square, or belongs to L_2, if there exists a finite constant $M = M_f$, such that

$$\frac{1}{(2\pi)^k} \int_{E_k} | f(x + iy) |^2 \, dv_y \leqq M_f \tag{13}$$

for all x in P (and not only for every compact subset of P). If (13) holds then $f(x + iy)$, for each x, can be represented as a Fourier integral in y, and these representations coalesce into an absolutely convergent Laplace-integral

$$f(z) = \int_{E_k} e^{-(z_1\alpha_1 + \cdots + z_k\alpha_k)} \varphi(\alpha_1, \cdots, \alpha_k) \, dv_\alpha. \tag{14}$$

The function $\varphi(\alpha)$ is measurable, and by the equality of Parseval-Plancherel we have

$$\int_{E_k} | \varphi(\alpha) |^2 e^{-2(x_1\alpha_1 + \cdots)} \, dv_\alpha = \frac{1}{(2\pi)^k} \int_{E_k} | f(x + iy) |^2 \, dv_y ; \tag{15}$$

therefore

$$\int_{E_k} | \varphi(\alpha) |^2 e^{-2(x_1\alpha_1 + \cdots)} \, dv_\alpha \leqq \mathrm{M}_f . \tag{16}$$

If we now recall the definition of the conjugate set G, and if we realize that the Lebesgue measure of the boundary of G is zero (since G is convex), then a simple argument will show that the function $\varphi(\alpha)$ must be zero, almost everywhere, outside the conjugate domain G. Thus, for instance, we can rewrite relations (14) and (16) into

$$f(z) = \int_G e^{-(z_1\alpha_1 + \cdots)} \varphi(\alpha) \, dv_\alpha \tag{17}$$

and

(18) $$\int_G |\varphi(\alpha)|^2 e^{-2(x_1\alpha_1+\cdots)}\, dv_\alpha \leqq \mathrm{M}_f .$$

Since $x = (0, \cdots, 0)$ is a boundary point of P and for all points of P relation (8) holds, we may now derive from (18) the limit relations

(19) $$\int_G |\varphi(\alpha)|^2\, dv_\alpha \leqq \mathrm{M}_f$$

and

(20) $$\lim_{x\to 0} \int_G |\varphi(\alpha)|^2 (1 - e^{-2(x_1\alpha_1+\cdots)})\, dv_\alpha = 0.$$

The first implies the existence of a function of integrable square

(21) $$f(i\eta) \sim \int_G e^{-i(\eta_1\alpha_1+\cdots)} \varphi(\alpha)\, dv_\alpha ,$$

and the second establishes the limit relation

(22) $$\lim_{x\to 0} \int_{-\infty}^{\infty} |f(i\eta) - f(x + i\eta)|^2\, dv_\eta = 0.$$

Thus our function $f(z)$ has automatically generalized boundary values (in the strong topology of L_2) almost everywhere on the spine of T; also, these boundary values are identical with the limit-values

$$\lim_{x\to 0} f(x + i\eta)$$

whenever such limit values exist on a measurable set of the spine.

Now, for each z in T, $K(z - i\eta)$ is a function of integrable square in η, namely

(23) $$K(z - i\eta) \sim \int_G e^{i(\eta_1\alpha_1+\cdots)} e^{-(z_1\alpha_1+\cdots)}\, dv_\alpha .$$

By the theory of Fourier convolutions (resultants), relations (21) and (23) imply the relation

(24) $$\frac{1}{(2\pi)^k} \int_{E_k} f(i\eta) K(z - i\eta)\, dv_\eta = \int_G e^{-(z_1\alpha_1+\cdots)} \varphi(\alpha)\, dv_\alpha .$$

If we combine this with (17) and recall (12), we finally obtain the following theorem.

THEOREM 1. *If T is an octant-shaped tube and $K(z)$ is its kernel as defined by* (10), *then for functions $f(z)$ of class L_2 we have the formula*

(25) $$f(z) = \frac{1}{(2\pi i)^k} \int_{iE_k} K(z - \zeta) f(\zeta)\, d\zeta_1 \cdots d\zeta_k$$

for all z in T.

The most obvious case of an octant-shaped domain is the tube

$$x_1 > 0, \cdots, x_k > 0.$$

Its basis is "self-conjugate" and the kernal is $1/z_1 \cdots z_k$. However, a more significant type of domain, also self-conjugate, has come to the fore recently and we will discuss it in some detail.

4. Matrices[4]

If $k = n^2$, then the variables

$$z_1, \cdots, z_k \tag{26}$$

can be re-indexed into a square scheme

$$z_{pq}, \qquad p, q = 1, \cdots, n. \tag{27}$$

The space of these independent complex variables will be called the *non-symmetric (matrix) space*. The letter z will sometimes denote the set of variables in either indexing, but most of the time it will denote the matrix (27) as an algebraic concept.

If the matrix is symmetric, $z_{pq} = z_{qp}$, then we choose the elements

$$\begin{matrix} z_{11}, z_{12}, \cdots, z_{1n} \\ z_{22}, \cdots, z_{2n} \\ \cdot \\ \cdot \\ z_{nn} \end{matrix} \tag{28}$$

as independent complex variables, and the letter z will again denote either the set of these variables or the resulting matrix. In this case $k = n(n + 1)/2$, and the space will be called the symmetric space.

We will require a lemma.

LEMMA 1. *In non-symmetric space, if a and b are arbitrary constant matrices, then the Jacobian*

$$J = \frac{\partial(w_1, \cdots, w_k)}{\partial(z_1, \cdots, z_k)}$$

of the linear transformation

$$w = bza \tag{29}$$

has the value

$$J = (\det | ba |)^n. \tag{30}$$

In symmetric space, if a is an arbitrary matrix and a' is its transposed, then the Jacobian of the transformation

[4] See C. L. SIEGEL, *Symplectic geometry*, Amer. Jour. of Math. 65 (1943), 8–10.

(31) $$w = a'za$$

has the value

(32) $$(\det | a |)^{n+1}.$$

PROOF. Since the Jacobian of successive transformations is multiplicative, it is sufficient to verify (30) for the special cases: $a = 1$ and b arbitrary, or $b = 1$ and a arbitrary, and we will only verify the second case. Thus we have to prove that

(33) $$J(a) = (\det | a |)^n$$

for the transformation $w = za$. In the first place, (33) can be verified by direct computation for a diagonal matrix a. In the second place, we have $J(b^{-1}cb) = J(b^{-1})J(c)J(b) = J(c)$, and thirdly we have $\det | b^{-1}cb | = \det | c |$. Thus, (33) holds for any a of the form $b^{-1}cb$ with c diagonal. The general case follows now by analytic continuation in the elements of the matrix a.

In the symmetric case we have to assume $b = a'$ in order to make sure that a symmetric z will transform into a symmetric w. The proof proceeds along the same lines as before.

5. The symmetric case

We write $z = x + iy$, where $x = \{x_{pq}\}$ is the matrix of the real parts and $y = \{y_{pq}\}$ is the matrix of the imaginary parts of the variables (28). The matrices x and y are real and symmetric, and hence hermitian. We now consider the set of those matrices x whose roots are all strictly positive, in symbols $x > 0$. The corresponding pointset will hence forth be denoted by P_n^0 and the tube with P_n^0 as base will be denoted by T_n^0. It is easily seen that P_n^0 is a radiated set. We now take another real symmetric matrix $\beta = \{\beta_{pq}\}$ in which again $(n(n + 1))/2$ elements are variable, and we set up the familiar expression

(34) $$\operatorname{tr}(x\beta) = \sum_{p,q} x_{pq}\beta_{qp}.$$

In the double sum each non-diagonal term occurs twice, and it can be written as $x_1\alpha_1 + \cdots + x_k\alpha_k$ where each α_j is a β_{pq} or double a β_{pq}; also

(35) $$dv_\alpha = 2^{n(n-1)/2}\, dv_\beta.$$

Relation (8) states that $\operatorname{tr}(x\beta) > 0$; and the latter will hold for all $x > 0$ if and only if $\beta > 0$ (as a matrix). Thus the conjugate domain G consist again of all positive matrices, and we will simply denote it likewise by P_n^0. For the kernel, if any, we now have the equation

(36) $$K(x) = 2^{n(n-1)/2} \int_{P_n^0} e^{-\operatorname{tr}(x\beta)}\, dv_\beta.$$

This integral occurs in the work of Siegel.[5] He actually considers the more general integral

[5] C. L. SIEGEL, *Ueber die analytische Theorie der quadratischen Formen*, Annals of Math. 36 (1935), 585, Hilfsatz 37.

$$\int_{P_n^0} e^{-\operatorname{tr}(x\beta)} (\det |\beta|)^{s-(n+1)/2} \, dv_\beta \tag{37}$$

and he shows that it has the value

$$(\det |x|)^{-s} \prod_{m=1}^{n} \pi^{(m-1)/2} \Gamma\left(s - \frac{m-1}{2}\right). \tag{38}$$

(We will indicate the process of evaluation in the discussion of the non-symmetric case.) If we now put $s = (n+1)/2$, then Theorem 1 implies the following theorem:

THEOREM 2. *In the symmetric case we have the formula*

$$f(z) = c_n^0 \int_{iE_k^0} \frac{f(\zeta) \, d\zeta_1 \cdots d\zeta_k}{(\det |z_{pq} - \zeta_{pq}|)^{(n+1)/2}} \tag{39}$$

where

$$c_n^0 = \frac{\Gamma\left(\frac{2}{2}\right)\Gamma\left(\frac{3}{2}\right) \cdots \Gamma\left(\frac{n+1}{2}\right)}{2^n \pi^{(n^2+3n)/2} i^{n(n+1)/2}} \tag{40}$$

for functions of integrable square in T_n^0.

We remark that for even n, the value of the denominator

$$(\det |z_{pq} - \zeta_{pq}|)^{(n+1)/2}$$

is to be obtained by analytic continuation, which is easily proven to be unique, from its initial value $+1$, for $z = 1$ (matrix), $\zeta = 0$ (matrix).

6. The non-symmetric case

The analysis is now more involved than in the symmetric case, however the final formula will be even simpler in structure. We will employ two systems of variables which we will denote by $\{z_{pq}\}$ and $\{Z_{pq}\}$ respectively. They shall be connected by the affine transformation

$$Z_{pp} = z_{pp}; \qquad p = 1, \cdots, k,$$

$$\left.\begin{aligned} Z_{pq} &= \frac{z_{pq} + z_{qp}}{2}, \qquad & z_{pq} &= Z_{pq} + iZ_{qp} \\ Z_{qp} &= \frac{z_{pq} - z_{qp}}{2i}, \qquad & z_{qp} &= Z_{pq} - iZ_{qp} \end{aligned}\right\}, \qquad p < q.$$

If we introduce the real and imaginary parts of all variables, $z_{pq} = x_{pq} + iy_{pq}$, $Z_{pq} = X_{pq} + iY_{pq}$, then we have the relations

$$\begin{aligned} X_{pp} &= x_{pp} & Y_{pp} &= y_{pp} \\ X_{pq} &= \frac{x_{pq} + x_{qp}}{2} & Y_{pq} &= \frac{y_{pq} + y_{qp}}{2} \\ X_{qp} &= \frac{y_{pq} - y_{qp}}{2} & Y_{qp} &= \frac{x_{qp} - x_{pq}}{2} \end{aligned}$$

where again $p < q$. The coordinates z_{pq} will be called *primary* coordinates, and the coordinates Z_{pq} will be called *secondary* coordinates, and we will employ both systems concurrently.

We introduce a third system of quantities in terms of the primary coordinates. If z is a matrix and if z^* is the conjugate-complex transposed matrix ("adjoint" in Hilbert Space terminology), then we put

$$r = \frac{z + z^*}{2}.$$

Hence

$$r_{pq} = \frac{z_{pq} + \bar{z}_{qp}}{2}$$

or

$$\begin{aligned} r_{pp} &= X_{pp} \\ r_{pq} &= X_{pq} + i\,X_{qp} \\ r_{qp} &= X_{pq} - i\,X_{qp} \end{aligned}\Bigg\}, \qquad p < q.$$

The matrix r is hermitian. Furthermore if

$$\varphi(r_{11}, r_{12}, r_{21}, \cdots) \tag{41}$$

is an analytic expression in the quantities r_{pq}; if we view the quantities $\{r_{pq}\}$ as dependent on $\{X_{pq}\}$; and if we then perform the analytic continuation of replacing each X_{pq} by Z_{pq}; then the function (41) will go over into

$$\varphi(Z_{11}, Z_{12} + iZ_{21}, Z_{12} - iZ_{21}, \cdots)$$

which in its turn is

$$\varphi(z_{11}, z_{12}, z_{21}, \cdots). \tag{42}$$

In other words, if we consider a domain which is a tube in terms of secondary coordinates, then the transition from (41) to (42) is the continuation of the function φ from the base of the tube into the tube proper.

Such a tube will be now defined. It shall consist of those points of the space for which all roots of the matrix r are strictly positive, $r > 0$. Its basis (in terms of the components X_{pq}) will be denoted by P_n, and the resulting tube, which can also be described as

$$r \in P_n; \qquad -\infty < Y_{pq} < \infty,$$

will be denoted by T_n. It is easily seen that P_n is radiated, and the computations to follow will imply that T_n is octant-shaped. In order to describe the conjugate set G we have to consider the expression

$$\operatorname{tr}(X\beta) \equiv \sum_{pq} X_{pq}\beta_{pq} \tag{43}$$

for an arbitrary real matrix β. If we put

$$\alpha_{pp} = \beta_{pp}$$

$$\text{(44)}\qquad \left.\begin{aligned} \alpha_{pq} &= \frac{\beta_{pq} + i\beta_{qp}}{2} \\ \alpha_{qp} &= \frac{\beta_{pq} - i\beta_{qp}}{2} \end{aligned}\right\}, \qquad p < q.$$

then (43) is identical with

$$\text{(45)}\qquad \operatorname{tr}(r\alpha) = \sum_{p,q} r_{pq}\alpha_{qp}\,.$$

For fixed α, if α is diagonal, then $\operatorname{tr}(r\alpha)$ will be positive for all positive r, if and only if, α itself is positive. For α non-diagonal we can put $\alpha = u\alpha^0u^{-1}$ where α^0 is diagonal and u is unitary, and the relations

$$\operatorname{tr}(r\alpha) = \operatorname{tr}(ru\alpha^0u^{-1}) = \operatorname{tr}([u^{-1}ru]\alpha^0)$$

will lead to the same conclusion. Thus we see that G is a domain in the β-variables which in terms of the auxiliary hermitian matrix α is again described by $\alpha > 0$. This domain will be denoted specifically by G_n.

As for the kernel of T_n, its value in P_n is given by the integral

$$\text{(46)}\qquad K(r) = \int_{G_n} e^{-\operatorname{tr}(r\alpha)}\,dv_\beta$$

and, by our previous remark, its value in primary coordinates of T_n will be $K(z)$. Suppose first that r is diagonal, $r_{pp} = \lambda_p$; $r_{pq} = 0$, $p \neq q$. The transformation

$$\alpha_{pq} \to \frac{\alpha_{pq}}{\sqrt{\lambda_p\lambda_q}}, \qquad p \lesseqgtr q,$$

or rather the corresponding transformation in the β-variables, transforms G_n into itself, and replaces dv_β by $(dv_\beta)/((\lambda_1 \cdots \lambda_n)^n.)$ Since $\lambda_1 \cdots \lambda_n = \det|r|$ and $\alpha_{pp} = \beta_{pp}$, we therefore obtain, for r diagonal,

$$\text{(47)}\qquad K(r) = \frac{a_n}{(\det|r|)^n},$$

where

$$\text{(48)}\qquad a_n = \int_{G_n} e^{-\operatorname{tr}(\beta)}\,dv_\beta\,.$$

If $r = ur^0u^{-1}$, where r^0 is diagonal and u is unitary, we replace in the integral, α by $u\alpha u^{-1}$. On the basis of Lemma 1 in connection with the identity $\operatorname{tr}\{ur^0u^{-1}\alpha\} = \operatorname{tr}\{r^0u^{-1}\alpha u\}$ it is now possible to verify that $K(r) = K(r^0)$. Since we also have $\det|r| = \det|r^0|$, we may thus conclude that (47) holds for an arbitrary hermitian matrix r. In order to compute the constant a_n, it is advisable to set up the more general expression

(49) $$A_n(s) = \int_{G_n} e^{-\operatorname{tr}(\beta)} (\det |\alpha_{pq}|)^{s-n}\, dv_\beta ,$$

for $s \geqq n$, and to prove the inductive relation

(50) $$A_n(s) = (4\pi)^{n-1}\Gamma(s-n+1)A_{n-1}(s)$$

for $n \geqq 2$. The coordinates

(51) $$\beta_{\rho\sigma}\,; \qquad \rho, \sigma = 2, 3, \cdots, n$$

are subject to the condition $\det |\alpha_{\rho\sigma}| > 0$, and their totality describes a point-set that may be denoted by G_{n-1}. If the quantities (51) are fixed, then the remaining coordinates

(52) $$\beta_{11}; \qquad \beta_{1\rho},\, \beta_{\sigma 1}; \qquad \rho, \sigma = 2, \cdots, n$$

are variables subject to the restriction

(53) $$\det \begin{vmatrix} \beta_{11} & , & \dfrac{\beta_{1\rho} + i\beta_{\sigma 1}}{2} \\ \dfrac{\beta_{1\rho} - i\beta_{\sigma 1}}{2}, & & \alpha_{\rho\sigma} \end{vmatrix} > 0,$$

and we are going to integrate (49) with respect to the variables (52) first. In order to simplify the computation we will first assume that the matrix $|\alpha_{\rho\sigma}|$ is a diagonal matrix $\begin{vmatrix} \lambda_2 & & \\ & \ddots & \\ & & \lambda_n \end{vmatrix}$ and the value of its determinant will be denoted by D. Since $tr(\beta) = \beta_{11} + \beta_{22} + \cdots$ we thus have to compute the integral

(54) $$D^{s-n} \int e^{-\beta_{11}} \left(\beta_{11} - \sum_{\rho=2}^{n} \frac{\beta_{1\rho}^2 + \beta_{\rho 1}^2}{4\lambda_\rho}\right)^{s-n} d\beta_{11} \prod_{\rho=2}^{n} d\beta_{1\rho}\, d\beta_{\rho 1} ,$$

the integration extending over the point set

(55) $$\beta_{11} > 0, \qquad \beta_{11} - \sum_{\rho=2}^{n} \frac{\beta_{1\rho}^2 + \beta_{\rho 1}^2}{4\lambda_\rho} > 0.$$

In terms of new coordinates

$$\beta = \beta_{11}, \qquad \xi_{\rho-1} = \frac{\beta_{1\rho}}{\sqrt{4\lambda_\rho \beta_{11}}}, \qquad \eta_{\rho-1} = \frac{\beta_{\rho 1}}{\sqrt{4\lambda_\rho \beta_{11}}}$$

the integral is

$$4^{n-1} D^{s-n+1} \int e^{-\beta} \beta^{s-1} \left(1 - \sum_{\rho=1}^{n-1} \xi_\rho^2 + \eta_\rho^2\right)^{s-n} d\beta \prod d\xi_\rho\, d\eta_\rho$$

and the point set is

$$\beta > 0, \qquad 1 - \sum_{\rho=1}^{n-1} \xi_\rho^2 + \eta_\rho^2 > 0.$$

A direct computation gives the value

$$D^{s-n+1}(4\pi)^{n-1}\Gamma(s - n + 1),$$

and a re-check will show that we obtain the same value for non-diagonal matrices $|\alpha_{p\sigma}|$. This completes the proof of (50). Since $A_1(s) = \Gamma(s)$, we finally obtain

$$A_n(s) = (4\pi)^{n(n-1)/2} \prod_{m=1}^{n} \Gamma(s - m + 1) \tag{56}$$

and in particular

$$a_n = (4\pi)^{n(n-1)/2} \prod_{m=1}^{n-1} m! \tag{57}$$

If we insert (47) in (25) we obtain the integral

$$\frac{a_n}{(2\pi i)^{n^2}} \int_{iE_k} \frac{f(\zeta)\, dZ_1^0 \cdots dZ_k^0}{(\det |z_{pq} - \zeta_{pq}|)^n},$$

where ζ_{pq} are primary and Z_p^0 are secondary coordinates on the spine. Finally, if we replace secondary coordinates by primary coordinates even on the spine, we obtain the following theorem.

THEOREM 3. *In the non-symmetric case we have the formula*

$$f(z) = \frac{1!2! \cdots (n-1)!}{(2\pi i)^{n(n+1)/2}} \int \frac{f(\zeta)\, d\zeta_1 \cdots d\zeta_k}{(\det |z_{pq} - \zeta_{pq}|)^n} \tag{58}$$

for functions in T_n which are of integrable square relative to secondary coordinates; the integration extending over the spine of T_n.

In Part II we will transfer formulas (39) and (58) from the tube to a type of unit sphere, thus bringing them much closer to being Cauchy formulas, and in Part III we will make an analogous investigation for the general formula (25).

PART II

SPHERE-SHAPED DOMAINS

7. The Cayley transformation

If z is any matrix and z^* is the adjoint matrix, then the products zz^* and z^*z are non-negative hermitian and, although not necessarily equal, they have the same roots.[6] The set of points z for which $zz^* < 1$, (or $z^*z < 1$) will be called the "unit sphere". It will be denoted by S_n in the non-symmetric, and by S_n^0 in the symmetric case. Thus, explicitly, $\{z_{pq}\}$ belongs to S_n (or S_n^0) if for every vector $(\lambda_1, \cdots, \lambda_n) \neq (0, \cdots, 0)$,

$$z_{pr}\bar{z}_{qr}\lambda_p\bar{\lambda}_q < \lambda_r\bar{\lambda}_r$$

or if

$$z_{rp}\bar{z}_{rq}\lambda_p\bar{\lambda}_q < \lambda_r\bar{\lambda}_r .$$

[6] Compare P. R. HALMOS, *Finite dimensional vector spaces*, Princeton (1942) 138–140.

Unit-matrices ζ are those for which $\zeta\zeta^* = 1$. Thus if U_n (or U_n^0 in the symmetric case) is the set of all unitary matrices, then U_n is a proper part of the boundary of S_n.

We now claim that for $\zeta \in U_n$, $z \in S_n$ we have

$$\det | \zeta - z | \neq 0. \tag{59}$$

Since $\zeta - z = (1 - z\zeta^*)\zeta$ and $z\zeta^*(z\zeta^*)^* = z\zeta^*\zeta z^* < 1$ it is obviously sufficient to show that $\det | 1 - z | \neq 0$ for $z \in S_n$. If this were false, there would exist a vector $\lambda \neq 0$ for which $\lambda_r = z_{rp}\lambda_p$. On multiplying this relation with its conjugate complex, we would obtain $\lambda_r\bar{\lambda}_r = z_{rp}\overline{z_{rq}}\lambda_p\bar{\lambda}_q$ which is impossible for z in S_n. Since $\zeta = -1$ is unitary, we see that $\det | 1 + z | \neq 0$, and thus for $z \in S_n$ we can set up the quotient

$$w = \frac{1 - z}{1 + z} \tag{60}$$

("Cayley transformation"). Obviously

$$\begin{aligned} w + w^* &= (1 + z)^{-1}(1 - z) + (1 - z^*)(1 + z^*)^{-1} \\ &= (1 + z)^{-1}\{(1 - z)(1 + z^*) + (1 + z)(1 - z^*)\}(1 + z^*)^{-1} \\ &= 2(1 + z)^{-1}(1 - zz^*)(1 + z^*)^{-1}. \end{aligned}$$

Now, in general, if p is positive hermitian and q is arbitrary, then qpq^* is again positive hermitian. Thus, $1 - zz^* > 0$ implies $w + w^* > 0$, and the latter inequality characterizes a point of the tube T_n in primary coordinates.

Conversely, let $w \in T_n$. We certainly have $\det | 1 + w | \neq 0$. Otherwise we would have a relation $\lambda_r = - w_{rp}\lambda_p$ which in its turn would lead to

$$2\lambda_r\bar{\lambda}_r = - (w_{pq} + \bar{w}_{qp})\lambda_p\bar{\lambda}_q .$$

This however contradicts $w + w^* > 0$. Thus we can set up the quotient

$$z = \frac{1 - w}{1 + w} \tag{61}$$

which is the formal converse to (60), and we obtain

$$\begin{aligned} 1 - zz^* &= 1 - (1 + w)^{-1}(1 - w)(1 - w^*)(1 + w^*)^{-1} \\ &= (1 + w)^{-1}\{(1 + w)(1 + w^*) - (1 - w)(1 - w^*)\}(1 + w^*)^{-1} \\ &= 2(1 + w)^{-1}\{w + w^*\}(1 + w^*)^{-1}. \end{aligned}$$

This implies $1 - zz^* > 0$. Hence we have proven

LEMMA 2. *The Cayley transformation is a one-to-one map between S_n and T_n; or between S_n^0 and T_n^0.*

Now, if ζ is any point of U_n, and if $\det | 1 + \zeta | \neq 0$, then

$$\omega = \frac{1 - \zeta}{1 + \zeta} \tag{62}$$

is a point of the spine iE_k. Conversely, if ω is any point whatsoever of iE_k, then $\zeta = (1 - \omega)/(1 + \omega)$ is a point of U_n. Thus, if we denote by A_1 the subset of U_n which is characterized by $\det | 1 + \zeta | = 0$, then (62) is a one-to-one map between $V_1 = U_n - A_1$ and iE_k. More generally, take any fixed number ρ of absolute value 1, denote by A_ρ the subset of U_n for which $\det | 1 + \rho\zeta | = 0$, and put $V_\rho = U_n - A_\rho$ then the transformation

$$\rho\zeta = \frac{1 - \omega}{1 + \omega}$$

is a one-to-one map between V_ρ and iE_k. Now, if $\rho_0 = 1, \rho_1, \cdots, \rho_n$ are $n + 1$ different values of ρ then $\{V_{\rho_0}, V_{\rho_1}, \cdots, V_{\rho_n}\}$ is a complete covering of U_n, since the polynomial $\det | 1 + \rho\zeta |$ cannot have more than $n + 1$ roots ρ. Finally, the intersection of A_1 with any V_ρ is the image in U_n of the pointset

$$\det \left| 1 + \frac{1}{\rho} \frac{1 - \omega}{1 + \omega} \right| = 0$$

in iE_k, and this is an analytic manifold of dimension $\leqq k - 1$. Thus, U_n (and similarly U_n^0) is a k-dimensional analytic manifold and we have

LEMMA 3. *The Cayley transformation is a one-to-one map between* $U_n - A$ *and* iE_k, *where* A *is a lower-dimensional analytical subset of* U_n; *similarly, between* $U_n^0 - A^0$ *and* iE_k^0.

Finally, we will state a last pre-requisite.

LEMMA 4. *The Jacobian* $\partial(z)/\partial(w)$ *of the transformation* (61) *is*

$$(-2)^{n^2} \frac{1}{(\det | 1 + w |)^{2n}}$$

in the non-symmetric case, and

$$(-2)^{n(n+1)/2} \frac{1}{(\det | 1 + w |)^{n+1}}$$

in the symmetric case.

In fact, from $z(1 + w) = 1 - w$ we obtain successively

$$\begin{aligned} dz\cdot(1 + w) + z\cdot dw &= - dw \\ dz\cdot(1 + w) &= -(1 + z)\cdot dw \\ dz &= -(1 + z)\cdot dw\cdot(1 + w)^{-1} \\ &= -2(1 + w)^{-1}\cdot dw\cdot(1 + w)^{-1} \end{aligned}$$

and Lemma 4 follows now from Lemma 1.

8. The non-symmetric case

We note that U_n, being a group space, is orientable. Therefore, the differential $d\zeta_1 \cdots d\zeta_k$ can be defined in the large, and its algebraic sign can be changed by a change of orientation.

THEOREM 4. *If $f(z)$ is analytic and bounded in S_n and has boundary values in the ordinary sense at all points of U_n, then the formula*

$$f(z) = \frac{1!2! \cdots (n-1)!}{(2\pi i)^{n(n+1)/2}} \int_{U_n} \frac{f(\zeta)\, d\zeta_1 \cdots d\zeta_k}{\det |\zeta_{pq} - z_{pq}|)^n} \tag{63}$$

is valid for z in S_n.

If we apply the transformations (61) and (62), we have by Lemma 4,

$$d\zeta_1 \cdots d\zeta_k = \pm 2^{n^2} \frac{d\omega_1 \cdots d\omega_k}{(\det |1 + \omega|)^{2n}}$$

and the relation

$$\begin{aligned} \zeta - z &= (1+\omega)^{-1}(1+\omega) - (1-w)(1+w)^{-1} \\ &= 2(1+\omega)^{-1}[w - \omega](1+w)^{-1} \end{aligned}$$

implies the equality

$$(\det |\zeta - z|)^n = 2^{n^2}(\det |w - \omega|)^n(\det |(1+\omega)(1+w)|)^{-n}.$$

Also the exceptional subset A of U_n, which is not affected by the transformation, can be neglected in the integral (63) on the basis of Lemma 3. Thus, formula (63), whether true or false, is equivalent with formula

$$\mathrm{F}(w) = c_n \int_{iE_k} \frac{\mathrm{F}(\omega)\, d\omega_1 \cdots d\omega_k}{(\det |w - \omega|)^n} \tag{64}$$

where

$$F(w) = \frac{f(z(w))}{(\det |1 + w|)^n} \tag{65}$$

and

$$c_n = \frac{1!2! \cdots (n-1)!}{(2\pi i)^{n(n+1)/2}}. \tag{66}$$

By assumption, $f(z(w))$ is bounded; therefore $F(w)$ is of integrable square in T_n, and its boundary values on the spine derive from the values $f(\zeta)$ on U_n. Thus (64) is true by Theorem 3, and the proof of Theorem 4 is completed.

9. Group-invariant volume

We note that in non-symmetric space, by Lemma 1, the differential form

$$c_n \frac{d\zeta_1 \cdots d\zeta_k}{(\det |\zeta_{pq}|)^n} \tag{67}$$

is invariant for arbitrary transformations $\zeta \to a\zeta b$. For σ unitary, $\zeta \to \sigma\zeta$ is an automorphism of U_n as group space and we can now deduce that (67) is the volume element of the group invariant volume on U_n. In fact, we can define the additive set function

$$\mu(B) = c_n \int_B \frac{d\zeta_1 \cdots d\zeta_k}{(\det|\zeta|)^n}$$

for Borel sets on U_n. It is group invariant, and it is absolutely continuous relative to the group volume, therefore by a general theorem[7] it is a constant multiple of the volume. Also $\mu(U_n) = 1$, and thus it is identical with the volume. We will now denote (67) by dv, and since $\zeta - z = (1 - z\zeta^*)\zeta$ we now obtain

THEOREM 5. *Under the assumptions of Theorem 4 we have*

$$f(z) = \int_{U_n} \frac{f(\zeta)\, dv}{(\det|1 - z\zeta^*|)^n} \tag{68}$$

where dv is the group invariant volume element of U_n, with the entire group having volume 1.

If we introduce the kernel

$$H(z; \bar{\zeta}) = (\det|1 - z\zeta^*|)^{-n} \tag{69}$$

then formula (68) reads

$$f(z) = \int H(z; \bar{\zeta}) f(\zeta)\, dv. \tag{70}$$

The kernel (69) is analytic in $z_1, \cdots, z_k; \bar{\zeta}_1, \cdots, \bar{\zeta}_k$ and has a power series expansion

$$H(z; \bar{\zeta}) = \sum_{p=0}^{\infty} H_p(z; \bar{\zeta}) \tag{71}$$

where each $H_p(z; \bar{\zeta})$ is a homogeneous polynomial (with real coefficients) of dimension p in either z or $\bar{\zeta}$. Also $H(z; \bar{\zeta}) = H(\bar{\zeta}; z)$ and therefore

$$H_p(z; \bar{\zeta}) = H_p(\bar{\zeta}; z). \tag{72}$$

It is easily seen that the series (71) is uniformly convergent for ζ varying in the closure of S_n and z varying in some neighborhood of the origin; (actually, z may vary in any compact subset of S_n). On substituting (71) in (70) we therefore obtain

$$f(z) = \sum_{p=0}^{\infty} f_p(z) \tag{73}$$

where

$$f_p(z) = \int H_p(z; \bar{\zeta}) f(\zeta)\, dv. \tag{74}$$

[7] S. BOCHNER, *Additive set functions on groups*, Annals of Math. 40 (1939), 787, Theorem 3.

By (74), $f_p(z)$ must be of dimension p or identically 0, therefore the series (73) is the diagonal form of the power series of $f(z)$. From this we conclude that

$$\int H_p(z; \bar{\zeta}) f_q(\zeta)\, dv = \begin{cases} 0, & p \neq q \\ f_p(z), & p = q. \end{cases} \tag{75}$$

If we multiply both sides by $\overline{g_p(z)}$ and recall (72), and integrate with respect to z, then we obtain the following theorem on unitary groups.

THEOREM 6. *If $g_p(\zeta)$ and $f_q(\zeta)$ are homogeneous polynomials of different dimensions then*

$$\int_{U_n} g_p(\zeta)\, \overline{f_q(\zeta)}\, dv = 0. \tag{76}$$

Also, if $\{\varphi_r(\zeta)\}$ is a complex system of orthonormal polynomials then

$$H(z; \bar{\zeta}) = \sum_{r=0}^{\infty} \varphi_r(z) \overline{\varphi_r(\zeta)}, \tag{77}$$

the latter series being summed diagonally.

As for series (77), since $\{\varphi_r(\zeta)\}$ is a complete system of homogeneous polynomials, there exists an expansion, diagonally convergent, of the form

$$H(z; \bar{\zeta}) = \sum c_r(z) \overline{\varphi_r(\zeta)}$$

with suitable polynomials $c_r(z)$. Their identity with $\varphi_r(z)$ follows now from (70) if applied to $f(z) = \varphi_r(z)$.

10. The symmetric case

This case has a complication. We do not know whether the manifold U_n^0, which is not a group space is orientable, and therefore we are not justified in setting up the differential $d\zeta_1 \cdots d\zeta_k$ for the entire manifold. Nevertheless we can set it up for the manifold $V_n^0 = U_n^0 - A^0$ where A^0 is the exceptional subset occurring in Lemma 3. Thus, in analogy to (63), we can now obtain the formula

$$f(z) = c_n^0 \int_{V_n^0} \frac{f(\zeta)\, d\zeta_1 \cdots d\zeta_k}{(\det |\zeta_{pq} - z_{pq}|)^{(n+1)/2}} \tag{78}$$

where c_n^0 is the constant (40). However there is an exact analogue to formula (68). It is the formula

$$f(z) = \int_{U_n^0} \frac{f(\zeta)\, dv}{(\det |1 - z\zeta^*|)^{(n+1)/2}}, \tag{79}$$

where formally,

$$dv = c_n^0 \frac{d\zeta_1 \cdots d\zeta_k}{(\det |\zeta_{pq}|)^{(n+1)/2}}. \tag{80}$$

In order to justify the analogy, we consider the group of affine transformations

$$\zeta \to u' \zeta u, \tag{81}$$

where u is an arbitrary (non-symmetric) unitary transformation and u' is its transposed. These transformations are a *transitive* group of motions, and the existence of such a group leads to many properties similar to those of a group space proper. In particular there exists a group-invariant volume, and it is again unique, thus leading to formula (79). We will not insist on details of the proof. Obviously Theorem 6 has also a strict analogue in the symmetric case.

PART III

GENERAL GROUP INVARIANT DOMAINS.

11. An abstract theorem

Let G denote a compact group of affine transformation operating on a fixed set of variables $t = (t_1, \cdots, t_k)$. An element of G will be denoted by the letter σ, and the actual transformation by $t \to \sigma t$. The transposed transformation will be denoted by σ'. Thus, if (t, u) is the inner product $t_1u_1 + \cdots + t_ku_k$, then $(\sigma t, u) = (t, \sigma' u)$. We term G a *comprehensive* group, if any polynomial in t which is absolutely invariant for all transposed transformations σ' is of necessity a constant. For instance, the unitary group U_n is a comprehensive group. In fact, if an element of this group is denoted by ζ and if $P(t)$ is a polynomial in the set of variables $t = \{t_{pq}\}$, and if $P(t)$ is group invariant, then

$$f(\zeta; t) = P(\zeta t) - P(t) \tag{82}$$

is identically 0 for $\zeta \in U_n$. We now consider the function

$$f(z; t) = P(zt) - P(t) \tag{83}$$

for an arbitrary (non-unitary) matrix z, and fixed numerical values of t. Being a polynomial in z, we can apply to it Theorem 3, and since its boundary values on U_n are zero, it vanishes identically in z, that is $P(zt) = P(t)$. If we put $t = 1$ (unit matrix) we obtain

$$P(z) = P(1) = \text{constant},$$

thus proving that the unitary group is comprehensive. In the case of the unitary group the property just stated is part of the so-called "unitarian trick,"[8] and our concept of a comprehensive group is a generalization of this property, by explicit definition, to general affine groups. We will now state a theorem in which this concept has a vital part.

THEOREM 7. *If a distributive operation from (all) polynomials to polynomials is commutative with partial differentiation and invariant under the transformations of a comprehensive group, then the operation is a constant multiple of the identity.*

If we denote the operation by

$$F(x) = Lf(\xi) \tag{84}$$

then the assumptions are

[8] H. WEYL, *The classical groups*, Princeton (1943), 177, Lemma (7.1.A).

$$L(af + bg) = aLf + bLg \tag{85}$$

$$\frac{\partial}{\partial x_p} F(x) = L \frac{\partial f}{\partial \xi_p}; \qquad p = 1, \cdots, k, \tag{86}$$

$$F(\sigma x) = Lf(\sigma \xi); \qquad \sigma \epsilon G \tag{87}$$

and the conclusion is

$$F(x) = c_0 f(x), \tag{88}$$

where c_0 is a constant. We denote the polynomial $L(\xi_1^{p_1} \cdots \xi_k^{p_k})$ by

$$F_{p_1 \cdots p_k}(x) \tag{89}$$

and we introduce the formal power series

$$F(x; t) = \sum_p \frac{F_{p_1 \cdots p_k}(x) t_1^{p_1} \cdots t_k^{p_k}}{p_1! \cdots p_k!}. \tag{90}$$

Assumption (86) is equivalent with the system of relations

$$\frac{\partial F}{\partial x_p} = t_p F; \qquad p = 1, \cdots, k. \tag{91}$$

If we introduce the new expression $G(x; t) = e^{-(x,t)}F(x, t)$, then these relations are $\frac{\partial G}{\partial x_p} = 0$. Therefore $G(x; t)$ is constant in x, and thus we obtain for the system of polynomials (89) the generating function

$$F(x; t) = e^{(x,t)} A(t_1, \cdots, t_k) \tag{92}$$

where $A(t)$ is a formal power series, as yet unrestricted. In order to apply the critical assumption (87) we introduce in addition to the series (90) the formal series

$$e^{(\xi,t)} = \sum \frac{\xi_1^{p_1} \cdots \xi_k^{p_k} t_1^{p_1} \cdots t_k^{p_k}}{p_1! \cdots p_k!}, \tag{93}$$

and after extending our operation to formal power series with polynomials as coefficients we obviously have

$$F(x; t) = Le^{(\xi,t)}.$$

By including assumption (87) we obtain the chain of equalities

$$e^{(\sigma x,t)} A(t) = F(\sigma x; t) = Le^{(\sigma \xi,t)} = Le^{(\xi,\sigma' t)} = F(x; \sigma' t) = e^{(x,\sigma' t)} A(\sigma' t),$$

and hence the relation

$$A(t) = A(\sigma' t).$$

Since the group is comprehensive, this relation can hold only if the power series $A(t)$ reduces to a constant c_0, that is $F(x, t) = c_0 e^{(x,t)}$. Therefore,

$$F_{p_1 \cdots p_k}(x) = c_0 x_1^{p_1} \cdots x_k^{p_k},$$

and this is the conclusion (88). We observe that there is nothing in our assumptions to prevent the constant c_0 from having the value zero.

A seemingly very special type of distributive operations having property (86) is given by the expression

$$F(x) = \sum_m f(x + \xi^m)\mu_m ,$$

where $\{\xi^m\}$ are fixed points, and $\{\mu_m\}$ are fixed constants. However a transition from sums to Stieltjes integrals will lead from the special type to what is well-nigh the most general type. If C is any bounded Borel set and $\omega(A)$ is a completely additive set function of its Borel subsets, then the expression

$$F(x) = \int_C f(x + \xi)\, d\omega(\xi) \tag{94}$$

will obviously define a distributive operation having property (86). It will be easily seen that in order to obtain property (87) it is sufficient to assume that the set C and the set function $\omega(A)$ shall be group invariant; that is $\sigma C = C$, and $\omega(\sigma A) \equiv \omega(A)$.

An important type of Stieltjes integrand is generated by an external differential form. We assume that C is an orientable compact m-dimensional differentiable manifold C_m, $0 < m < k$, and we assume that a differential form

$$d\omega(\xi) = \sum K_{p_1 \cdots p_m}(\xi)\, d\xi_{p_1} \cdots d\xi_{p_m} \tag{95}$$

is defined in a neighborhood D of C_m. The set function $\omega(A)$ is then defined as $\int_A d\omega(\xi)$. The set function will be group invariant if the manifold C and the differential form are group invariant. This being so, if the group G is a comprehensive group, then the conclusion of Theorem 7, if written explicitly, is the formula

$$\int_{C_m} f(x + \xi) \sum K_{p_1 \cdots p_m}(\xi)\, d\xi_{p_1} \cdots d\xi_{p_m} = c_0 f(x). \tag{96}$$

We will also write it more succinctly in the form

$$\int_{C_m} f(x + \xi)\, d\omega(\xi) = c_0 f(x). \tag{97}$$

The formula is valid for the class of polynomials and in most cases, for no class of function which is substantially larger. It is easy to state a further condition under which the constant c_0 is $\neq 0$; the condition being that the group G shall be *transitive* on C_m. In fact, if G is transitive, the set function $\omega(A)$ must be a numerical multiple of a group invariant volume, and c_0 is this multiple, thus being different from zero.

12. Cauchy's formula

If, for a fixed point x, we replace the variable ξ in (97) by the variable $\xi - x$ then the manifold C_m goes over into a translated manifold which we will denote by $C_m(x)$, and formula (97) goes over into

$$\int_{C_m(x)} f(\xi)\, d_\xi\, \omega(\xi - x) = c_0 f(x). \tag{98}$$

The integrand in formula (98) has the shape we are aiming at, however the contour $C_m(x)$ is dependent on x; whereas in a "Cauchy formula" as we understand the concept the contour must be independent of x. We are thus naturally led to seeking conditions under which

$$\left(\int_{C_m(x)} - \int_{C_m}\right) f(\xi)\, d_\xi\, \omega(\xi - x) = 0. \tag{99}$$

By assumption, the differential form (95) is defined in some neighborhood D of C_m. Hence there exists a neighborhood N of the origin, such that for $x \in N$, $C_m(x)$ is also situated in D. It is easily seen that $C_m(x) - C_m$ is the oriented boundary of an $(m + 1)$-dimensional complex B in D. If by the theorem of Stokes, we transform the left side of (99) into an integral over B, then the relation (99) will hold provided

$$df \cdot d_\xi \omega(\xi - x) + f \cdot (d_\xi \omega(\xi - x))' = 0. \tag{100}$$

Now, if the latter relation were to hold for all polynomials $f(\xi_1, \cdots, \xi_k)$ in the (real) variables $\xi_1, \cdots, \xi_k$, then it would hold locally for general differentiable functions $f(\xi)$ and this would imply that the differential form (95) is identically 0. In other words there exists no Cauchy formula in real variables.

However the situation is different for complex variables and the reader will have no difficulties in proving the following theorem.

THEOREM 8. *If $\zeta_1, \cdots, \zeta_k$ are complex variables and $\bar\zeta_1, \cdots, \bar\zeta_k$ are conjugate complex variables; if $0 \leqq m < k$; if the tensor*

$$K_{p_1 \cdots p_m}(\zeta; \bar\zeta), \qquad p_1 < p_2 < \cdots p_m$$

is defined and differentiable in the neighborhood D of a compact orientable differentiable manifold C_{k+m} of dimension $k + m$; if the external derivative of the differential form

$$d\omega(\zeta; \bar\zeta) = d\zeta_1 \cdots d\zeta_k \sum_p K_{p_1 \cdots p_m}(\zeta; \bar\zeta)\, d\bar\zeta_{p_1} \cdots d\bar\zeta_{p_m} \tag{101}$$

is identically zero; and if the manifold and the differential form are invariant under the transformations

$$\{\zeta \to \sigma\zeta;\ \bar\zeta \to \bar\sigma\bar\zeta\}$$

of a comprehensive group G; then the relation

$$c_0 f(z) = \int_{C_{k+m}} f(\zeta)\, d_\zeta\, \omega(\zeta - z; \bar\zeta - \bar z)$$

holds for all analytic polynomials $f(\zeta) \equiv f(\zeta_1, \cdots, \zeta_k)$.

If the group G is transitive on C_{k+m}, then the constant c_0 is $\neq 0$.

The case $m = 0$ is the case of complex variables at its purest, whereas the case $m = k - 1$ is nearest to the case of Green's formula for real variables. The latter case has been treated extensively in a previous paper,[9] and will not be commented upon. In the case $m = 0$, the tensor reduces to a scalar $K(\zeta; \bar{\zeta})$ and the differential form (101) will be

$$K(\zeta; \bar{\zeta})\, d\zeta_1 \cdots d\zeta_k .$$

Its derivative is identically zero whenever $K(\zeta; \bar{\zeta})$ is independent of the conjugate quantities $\bar{\zeta}_1, \cdots, \bar{\zeta}_k$ thus being an analytic function $K(\zeta) \equiv K(\zeta_1, \cdots, \zeta_k)$. Thus we obtain formula

$$c_0 f(z) = \int_{C_k} f(\zeta) K(\zeta - z)\, d\zeta_1 \cdots d\zeta_k$$

provided C_k and $K(\zeta)\, d\zeta_1 \cdots d\zeta_k$ are invariant for some comprehensive group in the complex variables given. This formula obviously includes formulas (63) and (68).

Addendum

Another Special Case

We consider in the space of the real variables $u_1, \cdots, u_k$, $k \geqq 2$, the semi-cone

$$(P) \qquad u_1 > (u_2^2 + \cdots + u_k^2)^{1/2}.$$

The square-root is positive and therefore u_1 is automatically positive. The other variables are not algebraically restricted. It is easily seen that the relation

$$\alpha_1 u_1 + \cdots + \alpha_k u_k > 0$$

will hold for all u in P if and only if

$$(G) \qquad \alpha_1 > (\alpha_2^2 + \cdots + \alpha_k^2)^{1/2}.$$

Furthermore, by a known formula[10],

$$\int_G e^{-(\alpha_1 w_1 + \cdots + \alpha_k w_k)}\, dv_\alpha = \frac{2\pi^{(k+1)/2}\Gamma(k-1)}{\Gamma\left(\frac{k+1}{2}\right)\cdot \{w_1^2 - w_2^2 - \cdots - w_k^2\}^{k/2}}.$$

Thus we obtain the formula

$$F(w) = \frac{\Gamma\left(\frac{k}{2}\right)}{2\pi^{(k/2)+1} i^k} \int_{iE_k} \frac{F(\omega)\, d\omega_1 \cdots d\omega_k}{[(w_1 - \omega_1)^2 - (w_2 - \omega_2)^2 - \cdots - (w_k - \omega_k)^2]^{k/2}}$$

for functions of integrable square in the tube with basis P.

---

[9] S. Bochner, *Analytic and meromorphic continuation by means of Green's formula*, Annals of Math. 44 (1943), 652–673.

[10] S. Bochner, Vorlesungen ueber Fouriersche Integrale, p. 189.

Following E. Cartan[11] we now introduce new variables by the relations

$$w_1 = \frac{2z_1}{z_1^2 + \cdots + z_k^2} - 1$$

$$w_p = \frac{2iz_p}{z_1^2 + \cdots + z_k^2}, \qquad p = 2, \cdots, k,$$

and this leads to the formula

$$f(z) = \frac{\Gamma\left(\frac{k}{2}\right)}{2\pi i\, \pi^{k/2}} \int_U \frac{f(\zeta)\, d\zeta_1 \cdots d\zeta_k}{[(z_1 - \zeta_1)^2 + \cdots + (z_k - \zeta_k)^2]^{k/2}}.$$

The characteristic manifold U is described by

$$| \zeta_1^2 + \cdots + \zeta_k^2 - \zeta_1 | = | \zeta_1 |$$

$$\zeta_p(\bar{\zeta}_1^2 + \cdots + \bar{\zeta}_k^2) = \bar{\zeta}_p(\zeta_1^2 + \cdots + \zeta_k^2), \qquad p = 2, \cdots, k$$

and the domain of validity in z is described by

$$| z_1^2 + \cdots + z_k^2 - z_1 | < | z_1 |$$

$$2(| z_1 - 1 |^2 + | z_2 |^2 + \cdots + | z_k |^2) \leqq 1 + | (z_1 - 1)^2 + z_2^2 + \cdots + z_k^2 |^2.$$

This domain can be proven bounded.

PRINCETON UNIVERSITY

---

[11] Domaines borne's homogènes de l'espace de n variables complexes. Hamburgische Abhandlungen, 11 (1936), p. 149.

ANNALS OF MATHEMATICS
Vol. 45, No. 4, October, 1944

BOUNDARY VALUES OF ANALYTIC FUNCTIONS IN SEVERAL VARIABLES AND OF ALMOST PERIODIC FUNCTIONS

BY S. BOCHNER

(Received June 22, 1944)

INTRODUCTION

If an analytic function $f(z)$ in $|z| < 1$ is such that

$$\int_0^{2\pi} |f(re^{i\theta})| \, d\theta$$

is bounded for $0 < r < 1$ then there exists a boundary function $f(e^{i\theta})$ such that

$$\lim_{r \to 1} \int_0^{2\pi} |f(re^{i\theta}) - f(e^{i\theta})| \, d\theta = 0; \tag{1}$$

or if we "eliminate" the boundary function $f(e^{i\theta})$ then relation (1) is equivalent with the relation

$$\lim_{r \to 1, \rho \to 1} \int_0^{2\pi} |f(re^{i\theta}) - f(\rho e^{i\theta})| \, d\theta = 0.$$

This theorem of F. Riesz[1] has the distinction of being true only for analytic functions and not for harmonic functions in general as many similar theorems are. Correspondingly, the proof is very "analytic." It is based on the canonical decomposition

$$f(z) = g(z) \cdot B(z)$$

where $g(z) \neq 0$ and $B(z)$ has boundary values 1 almost everywhere (Blaschke product).

Now, for functions of several complex variables no comparable decomposition exists, and yet we will succeed in generalizing the theorem of Riesz proper. If $f(z_1, \cdots, z_k)$ is analytic for $0 \leqq |z_j| < 1, j = 1, \cdots, k$, and if

$$\int_0^{2\pi} \cdots \int_0^{2\pi} f(r_1 e^{i\theta_1}, \cdots, r_k e^{i\theta_k}) \, dv_\theta$$

is bounded then there exists an integrable boundary function $f(e^{i\theta_1}, \cdots, e^{i\theta_k})$ in

$$0 < \theta_j < 2\pi; \qquad j = 1, \cdots, k \tag{2}$$

such that

$$\int_0^{2\pi} \cdots \int_0^{2\pi} |f(r_1 e^{i\theta_1}, \cdots, r_k e^{i\theta_k}) - f(e^{i\theta_1}, \cdots, e^{i\theta_k})| \, dv_\theta \tag{3}$$

---

[1] Ueber die Raudwerte einer analytischen Funktion, Mathem. Zeitschrift, 18 (1922), 87–95.

tends to 0 as $(r_1, \cdots, r_k) \to (1, \cdots, 1)$. The theorem can be stated entirely in terms of Fourier series. If we consider an additive set function of bounded variation on the Borel sets of the torus (2); if we introduce its Fourier series in the usual manner; and if the series has the special form

$$\sum_{m_1=0}^{\infty} \cdots \sum_{m_k=0}^{\infty} a_{m_1 \cdots m_k} e^{i(m_1\theta_1 + \cdots + m_k\theta_k)}$$

that is if the Fourier coefficient $a_{m_1 \cdots m_k}$ vanishes whenever at least one index m_j is negative; then the set function is of necessity absolutely continuous and therefore the indefinite integral of a Lebesgue point function. Actually we will show that for the conclusion to hold the non-vanishing Fourier coefficients need not be restricted rigidly to the "octant" $0 < m_j < \infty$ but may be contained in any convex cone which is strictly conical.

The theorem of Riesz can be deduced from a sharper theorem of Hardy-Littlewood[2] which states that the inequality

$$\int_0^{2\pi} | f(re^{i\theta}) | \, d\theta < \gamma, \qquad 0 < r < 1$$

implies the sharper inequality

$$\int_0^{2\pi} \sup_{0<r<1} | f(re^{i\theta}) | \, d\theta \leqq \alpha \cdot \gamma$$

where α is a universal constant. Now, if we replace the variable z by e^{-z} then the theorems of Riesz and Hardy-Littlewood are theorems on periodic analytic functions which are bounded in the half-plane $x > 0$. Our procedure will be to give at first a qualified extension of the theorem of Hardy-Littlewood to almost periodic functions of Bohr. This done, it will be quite easy to obtain the stated theorems in several variables by applying a fundamental relation which connects the two categories of functions.

After that we will turn our attention to almost periodic functions in their own right. We will give both an unequalified extension of the theorem of Hardy-Littlewood and a full generalization of the theorem of Riesz. The "boundary function" whose existence is involved in the theorem of Riesz will be a Besicovitch function. Nothing less general can be guaranteed. However the analytic function given in $x > 0$ may be a Besicovitch function to start with, and as a matter of fact we will set it up in a new scope of generality.

I. A Lemma on Mean Values

As a rule we will denote by $M_y f(y)$ the mean-value

$$\lim_{T \to \infty} \frac{1}{2T} \int_{-T}^{T} f(y) \, dy$$

---

[2] A maximal theorem with function-theoretic applications. Acta Mathematica, 54 (1930), 81–116.

provided it exists. We will also write $\bar{M}_y(y)$ for the limit superior and $\underline{M}_y f(y)$ for the limit inferior, whenever $f(y)$ is real-valued.

THEOREM 1.[3] *If $K(y)$ is an even function in $(-\infty, \infty)$; if in $0 \leqq y < \infty$ $K(y)$ is differentiable and monotonely decreasing; and if*

$$\int_0^\infty K(y) < \infty,$$

thus implying that $yK(y) \to 0$ as $y \to \infty$; then for any non-negative function $\varphi(y)$ which is Lebesgue integrable over any finite interval we have the relations

$$\limsup_{\epsilon \to 0} \int_{-\infty}^{\infty} \varphi(y)\, \epsilon\, K(\epsilon y)\, dy \leqq \bar{M}_y\, \varphi(y) \cdot \int_{-\infty}^{\infty} K(y)\, dy \tag{4}$$

and

$$\underline{M}_y\, \varphi(y) \cdot \int_{-\infty}^{\infty} K(y)\, dy \leqq \liminf_{\epsilon \to 0} \int_{-\infty}^{\infty} \varphi(y)\, \epsilon\, K(\epsilon y)\, dy. \tag{5}$$

PROOF. Since $K(y)$ was assumed even, our relations may be proven in the form

$$\limsup_{\epsilon \to 0} \int_0^\infty \varphi(y)\, \epsilon\, K(\epsilon y)\, dy \leqq \bar{M}_y\, \varphi(y) \cdot \int_0^\infty K(y)\, dy \tag{6}$$

$$\underline{M}_y\, \varphi(y) \cdot \int_0^\infty K(y)\, dy \leqq \liminf_{\epsilon \to 0} \int_0^\infty \varphi(y)\, \epsilon\, K(\epsilon y)\, dy; \tag{7}$$

where $\varphi(y)$ is defined only in $0 \leqq y < \infty$ and where *for the remainder of this proof only*, the symbols $\bar{M}$ and $\underline{M}$ are the lim superior and lim inferior of $\frac{1}{T}\int_0^T \varphi(y)\, dy$ as $T \to \infty$.

The reader will next convince himself easily that we may now also assume that $\varphi(y)$ vanishes in a neighborhood of the origin. In particular, if we introduce the function

$$\Phi(y) = \frac{1}{y}\int_0^y \varphi(t)\, dt, \qquad 0 < y < \infty,$$

we may assume that

$$\Phi(y) \leqq \gamma, \qquad 0 < y < \infty.$$

Now

$$\int_0^\infty \varphi(y)\, \epsilon\, K(\epsilon y)\, dy = -\epsilon^2 \int_0^\infty yK'(\epsilon y)\Phi(y)\, dy$$

where $-K'(\epsilon y) \geqq 0$. Corresponding to any $\rho > 0$ there exists an A, such that

$$\Phi(y) \leqq \bar{M}_y \varphi(y) + \rho, \qquad A < y < \infty.$$

[3] Compare S. Bochner, Fouriersche Integrale, p. 30.

Therefore

$$-\epsilon^2 \int_0^\infty yK'(\epsilon y)\Phi(y)\,dy \leqq -\gamma \int_0^A \epsilon^2 yK'(\epsilon y)\,dy - (\bar{M} + \rho)\epsilon^2 \int_A^\infty yK'(\epsilon y)\,dy$$

$$\leqq -\gamma \int_0^{\epsilon A} yK'(y)\,dy + (\bar{M} + \rho) \int_0^\infty K(y)\,dy,$$

and this proves (6). On the other hand, for each $\rho > 0$ there exists a B such that

$$\Phi(y) \geqq \underline{M}_y\varphi(y) - \rho, \qquad B < y < \infty,$$

and (7) follows easily from

$$-\epsilon^2 \int_0^\infty yK'(\epsilon y)\Phi(y)\,dy \geqq -(\underline{M} - \rho)\epsilon^2 \int_B^\infty yK'(\epsilon y)\,dy.$$

II. A Theorem of Hardy-Littlewood

THEOREM 2. *There exists an absolute constant* α *having the following properties.*

(i) *If*

$$\frac{1}{2\pi} \int_{-\pi}^{\pi} |f(re^{i\theta})|\,d\theta \leqq \gamma, \qquad 0 < r < 1,$$

then

$$\frac{1}{2\pi} \int_{-\pi}^{\pi} \sup_{0<r<1} |f(re^{i\theta})|\,d\theta \leqq \alpha\cdot\gamma. \tag{8}$$

(ii) *If*

$$\int_{-\infty}^{\infty} |f(x + iy)|\,dy \leqq \gamma, \qquad 0 < x < \infty \tag{9}$$

then

$$\int_{-\infty}^{\infty} \sup_{0<x<\infty} |f(x + iy)|\,dy \leqq \alpha\cdot\gamma. \tag{10}$$

(iii) *If*

$$\frac{1}{2T} \int_{-T}^{T} |f(x + iy)|\,dy \tag{11}$$

is bounded in $0 < T < \infty$, $\xi < x < \infty$, *for every fixed* $\xi > 0$, *and if*

$$\bar{M}_y |f(x + iy)| \leqq \gamma, \qquad 0 < x < \infty, \tag{12}$$

then for all $x_1 > x_0 > 0$ *we have*

$$\underline{M}\{\max_{x_0\leqq x\leqq x_1} |f(x + iy)|\} \leqq \alpha\cdot\gamma. \tag{13}$$

(iv) *The best constants* α *in each of the relations* (8), (10), (13) *have the same values.*

PROOF. *Ad (i).* This is a renowned theorem of Hardy-Littlewood.[2] The proof will not be repeated. The other parts of the theorem will be derived from (i).

Ad (ii). If $r = r(\theta)$ is a continuous function in $\theta_0 \leqq \theta \leqq \theta_1$, where $-\pi < \theta_0 < \theta_1 < \pi$, and if $0 < r(\theta) < 1$, then (8) implies

$$\frac{1}{2\pi}\int_{\theta_0}^{\theta_1} |f(r(\theta)e^{i\theta})|\, d\theta \leqq \alpha\cdot\gamma. \tag{14}$$

Now, let $f(x + iy)$ be analytic in $x > 0$, and let

$$\{x = x(y);\ r \leqq y \leqq s\} \tag{15}$$

be a fixed rectilinear polygonal Jordan arc situated in a fixed rectangle

$$\{p \leqq x \leqq q;\ r \leqq y \leqq s\} \tag{16}$$

where

$$0 < p < q; \qquad -\infty < r < s < \infty.$$

If $a > 0$ is sufficiently large, then the circle

$$|z - a| = a - \frac{p}{2}$$

will have the following properties. The curve (15) is contained in its interior, and any radius of the circle which meets the curve at some point meets it at one point only. Therefore, relation (14) implies the relation

$$\left(a - \frac{p}{2}\right)\int_r^s |f(x(y) + iy)|\,\frac{x'(y)\cdot y + a - x(y)}{(a - x(y))^2}\,dy$$
$$\leqq \alpha \int_{-\pi(a-p/2)}^{\pi(a-p/2)} \left| f\left(a + \left(a - \frac{p}{2}\right)e^{i\lambda/a-p/2}\right)\right| d\lambda.$$

We now let $a \to \infty$. The expression on the left converges to

$$\int_r^s |f(x(y) + iy)|\, dy.$$

In the integral on the right the path of integration converges to the straight line $x = \frac{p}{2}$, $-\infty < y < \infty$, and the integral will itself converge towards its formal limit which is

$$\int_{-\infty}^{\infty} \left| f\left(\frac{p}{2} + iy\right)\right| dy,$$

provided the function $f(z)$ in the half-plane $x > 0$ satisfies an estimate

$$|f(z)| \leqq \frac{M}{1 + |z|^3}, \qquad x > 0, \tag{17}$$

say. Thus for a function of this description the last inequality implies

$$\int_r^s |f(x(y) + iy)| \, dy \leqq \alpha \cdot \gamma. \tag{18}$$

Next, if $f(z)$ is an arbitrary function for which (9) holds then a sequence of functions $f_0(z)$ can be found which approximate to $f(z)$ in the norm (9) and such that each $f_0(z)$ satisfies an estimate (17) for some M. The sequence converges uniformly in (16), and thus (18) holds for $f(z)$ in general. We now hold the function $f(z)$ and the rectangle (16) fast, and vary the curve. By convergent sequences of curves we can reach any semi-continuous functions $y(x)$, and therefore (18) implies

$$\int_r^s \max_{p \leqq x \leqq q} |f(x + iy)| \, dy \leqq \alpha \cdot \gamma.$$

Letting $p \to 0$, $q \to +\infty$, $r \to -\infty$, $s \to +\infty$ we finally obtain (10).

Ad (iii). For $0 < \epsilon < \epsilon_0$ we introduce the function

$$g(z) = \frac{f(z)}{(1 + \epsilon z)^2}. \tag{19}$$

Since (19) is bounded for $x > \xi (> 0)$ we see that

$$\int_{-\infty}^{\infty} | g(x + iy) | \, dy$$

is bounded for $x > \xi$, and by (ii) we obtain for $x_1 > x_0 > \xi$ the inequality

$$\int_{-\infty}^{\infty} \max_{x_0 \leqq x \leqq x_1} | g(x + iy) | \, dy \leqq \alpha \int_{-\infty}^{\infty} | g(\xi + iy) | \, dy.$$

Therefore,

$$\int_{-\infty}^{\infty} \frac{\max_{x_0 \leqq x \leqq x_1} | f(x + iy) |}{(1 + \epsilon_0 x_0)^2 + \epsilon^2 y^2} \, dy \leqq \alpha \int \frac{| f(\xi + iy) |}{1 + \epsilon^2 y^2} \, dy.$$

We now multiply both sides by ϵ and apply Theorem 1. If we put $K(y) = (1 + y^2)^{-1}$ in (4) we obtain

$$\limsup_{\epsilon \to 0} \epsilon \int_{-\infty}^{\infty} \frac{| f(\xi + iy) |}{1 + \epsilon^2 y^2} \, dy \leqq \bar{M}_y | f(\xi + iy) | \cdot \pi,$$

and if we put $K(y) = [(1 + \epsilon_0 x_0)^2 + y^2]^{-1}$ in (5) we obtain

$$\underline{M}_y \max_{x_0 \leqq x \leqq x_1} | f(x + iy) | \frac{\pi}{1 + \epsilon_0 x_0}$$
$$\leqq \limsup_{\epsilon \to 0} \epsilon \int_{-\infty}^{\infty} \frac{\max_{x_0 \leqq x \leqq x_1} | f(x + iy) | \, dy}{(1 + \epsilon_0 x_0)^2 + \epsilon^2 y^2}.$$

If we put together the last three relations and let $\epsilon_0 \to 0$ we finally obtain (13).

Ad (iv). If α_1, α_2, α_3 are the best constants in (8), (10), (13) respectively, then our argument shows $\alpha_1 \geqq \alpha_2 \geqq \alpha_3$. However $\alpha_1 \leqq \alpha_3$ and thus $\alpha_1 = \alpha_2 = \alpha_3$, as asserted.

THEOREM 3. *If $f(z)$ is an analytic almost periodic function*[4] *in $x > 0$ [that is in each half plane $\xi \leqq x < \infty$, $\xi > 0$], and if*

$$M_y \,|\, f(x + iy) \,|\, \leqq \gamma, \qquad 0 < x < \infty, \tag{20}$$

then for each $x_0 > 0$ we have

$$M_y\{\sup_{x_0 \leqq x \leqq \infty} |\, f(x + iy) \,|\} \leqq \alpha \cdot \gamma. \tag{21}$$

PROOF. In fact, relation (13) is in the present case

$$M_y\{\max_{x_0 \leqq x \leqq x_1} |\, f(x + iy) \,|\} \leqq \alpha \cdot \gamma \tag{22}$$

for $x_1 > x_0 > 0$. Also, (20) implies that the exponents in the Dirichlet series of $f(z)$ are non-negative, and therefore for $x_1 \to \infty$ the integrand in (22) converges to the integrand in (21), uniformly in $-\infty < y < \infty$. This proves (21).

If $f(z)$ is pure-periodic we may pass from (21) to

$$M_y\{\sup_{0 < x < \infty} |\, f(x + iy) \,|\} \leqq \alpha \cdot \gamma \tag{23}$$

by ordinary Lebesgue theory. However, for almost periodic functions such an unqualified passage to the limit is not admissible without a detailed investigation, and the very existence of the mean-value (23) is a matter requiring investigation. We will not investigate this problem in the direct manner just outlined; but we will see in Part IV that there is a possibility of generalizing relation (23) to the almost periodic case.

III. FUNCTIONS IN SEVERAL VARIABLES

We will describe a well known connection between periodic functions in several variables and almost periodic functions in one variable. For fixed k, we consider the class of all continuous periodic functions

$$f(t_1, \cdots, t_k) \sim \sum a_{m_1 \cdots m_k} e^{i(m_1 t_1 + \cdots + m_k t_k)} \tag{24}$$

on the torus

$$0 \leqq t_j < 2\pi, \qquad j = 1, \cdots, k, \tag{25}$$

and we take a fixed set of positive real numbers

$$\mu_1, \cdots, \mu_k \tag{26}$$

which we assume to be linearly independent with regard to rational coefficients. Thus there is no relation of the form

$$g_1\mu_1 + \cdots + g_k\mu_k = 0$$

[4] As defined by H. Bohr, and not in any more general sense, for the present.

for integers $\{g_j\}$ other than $g_1 = \cdots = g_k = 0$. We now make the substitution

$$t_1 = \mu_1 t, \cdots, t_k = \mu_k t,$$

and the resulting function

$$g(t) = f(\mu_1 t, \cdots, \mu_k t) \tag{27}$$

is a Bohr almost periodic function with the expansion

$$\sum a_{m_1 \cdots m_k} e^{i(m_1\mu_1 + \cdots + m_k\mu_k)t}.$$

We will also denote the function $g(t)$ by

$$L(f) \equiv L(f; t).$$

The transformation

$$f \to Lf$$

is functional operation having the following properties. It is distributive, and it also transforms the real and imaginary parts of f separately into the real and imaginary parts of g. It is multiplicative,

$$L(f_1 \cdot f_2) = Lf_1 \cdot Lf_2,$$

and if $f_n \to f$ uniformly, then $Lf_n \to Lf$ uniformly. Therefore if

$$f_1(t), \cdots, f_n(t) \tag{28}$$

are real functions, and if

$$\lambda(u_1, \cdots, u_n)$$

is a continuous function in the real variables $u_1, \cdots, u_n$ which is defined and continuous on the closure of the trajectory (28), then

$$L(\lambda(f_1, \cdots, f_n)) = \lambda(Lf_1, \cdots, Lf_n). \tag{29}$$

Furthermore, the range of values of $g(t)$ is dense in the range of values of $f(t)$, and in particular,

$$\sup_t | g(t) | = \max_t | f(t_1, \cdots, t_k) | .$$

This will be of no consequence; but what will be of consequence is the relation

$$M_t g(t) = \frac{1}{(2\pi)^k} \int_0^{2\pi} \cdots \int_0^{2\pi} f(t_1, \cdots, t_k)\, dt_1 \cdots dt_k, \tag{30}$$

or written differently,

$$M_t L(f; t) = M_t f(t_1, \cdots, t_k).$$

Now, for $p \geqq 1$, and indeed for $p > 0$, the function

$$\lambda(u_1, u_2) = (u_1^2 + u_2^2)^{p/2}$$

is continuous. Therefore if we put $f = f_1 + if_2$ and apply (29) we see that

$$L(|f|^p) = |Lf|^p,$$

and if we substitute this in (31) we obtain

$$M_t |Lf|^p = M_t |f|^p. \tag{31}$$

Furthermore, the function

$$\lambda(u_1, \cdots, u_n) = \max (u_1, \cdots, u_n)$$

is continuous in $u_1 \geqq 0, \cdots, u_n \geqq 0$, therefore if functions

$$f_1, \cdots, f_n$$

are given, and if

$$g_v = Lf_v \qquad \nu = 1, \cdots, n$$

then

$$\max \{|g_1|, \cdots, |g_n|\} = L(\max \{|f_1|, \cdots, |f_n|\}),$$

and thus we obtain also

$$M_t \max \{|g_1|, \cdots, |g_n|\} = M_t \max \{|f_1|, \cdots, |f_n|\}. \tag{32}$$

We are now ready for the main theorem stated in the Introduction.

THEOREM 4. *If a power series*

$$f(s_1, \cdots, s_k) = \sum_{m_1=0}^{\infty} \cdots \sum_{m_k=0}^{\infty} a_{m_1 \cdots m_k} e^{-(m_1 s_1 + \cdots + m_k s_k)}$$

in the complex variables

$$s_j = \sigma_j + it_j, \qquad j = 1, \cdots, k,$$

is absolutely convergent in the domain

$$\sigma_1 > 0, \cdots, \sigma_k > 0 \tag{33}$$

and if for all $(\sigma_1, \cdots, \sigma_k)$ *in* (33) *we have*

$$\frac{1}{(2\pi)^k} \int_0^{2\pi} \cdots \int_0^{2\pi} |f(\sigma_1 + it_1, \cdots, \sigma_k + it_k)| \, dt_1 \cdots dt_k < \gamma \tag{34}$$

then the limiting series

$$\sum a_{m_1 \cdots m_k} e^{-i(m_1 t_1 + \cdots + m_k t_k)} \tag{35}$$

is the Fourier series of a function of class L^1.

More generally, if a Laurent series

$$f(s_1, \cdots, s_k) = \sum_{-\infty}^{\infty} \cdots \sum_{-\infty}^{\infty} a_{m_1 \cdots m_k} e^{-(m_1 s_1 + \cdots + m_k s_k)}$$

is absolutely convergent in a k-dimensional domain P of the $(\sigma_1, \cdots, \sigma_k)$*-space, if P is the union of half-lines issuing from but not containing the origin, and if relation* (34) *holds in P, then the limiting series is the Fourier series of a function in* L^1.

PROOF. Since P is a domain, and thus contains a neighborhood, it contains a point whose coordinates are linearly independent. Denote them by

$$\mu_1, \cdots, \mu_k .$$

By assumption, P contains the half-line

$$\sigma_j = \mu_j u, \qquad j = 1, \cdots, k; \qquad 0 < u < \infty.$$

Therefore we can introduce the function

$$g(w) = f(\mu_1 w, \cdots, \mu_k w), \qquad w = u + iv,$$

of the one complex variable w. This function is almost periodic, and, on the basis of formula (31), our relation (34) implies the kindred relation

$$M_v \mid g(u + iv) \mid \leqq \gamma, \qquad 0 < u < \infty, \tag{36}$$

since in terms of our previous operation we have

$$g(u + it) = Lf(\mu_1 u + it_1, \cdots, \mu_k u + it_k).$$

Now, take a sequence of numbers

$$u_1 > u_2 > u_3 > \cdots \to 0,$$

and consider the function

$$g_n(t) = g(u_n + it).$$

The latter is the transform of

$$f_n(t) = f(\mu_1 u_n + it_1, \cdots, \mu u_n + it_k).$$

By theorem 3, relation (36) implies

$$M_t \max \{ \mid g_1 \mid, \cdots, \mid g_n \mid \} \leqq \alpha\cdot\gamma,$$

and by relation (32) this is equivalent with

$$M_t \max \{ \mid f_1 \mid, \cdots, \mid f_n \mid \} \leqq \alpha\cdot\gamma, \tag{37}$$

$n = 1, 2, \cdots$. By ordinary Lebesgue theory on the torus (25) this implies that the sequence of functions $f_n(t_1, \cdots, t_k)$ is majorized by an integrable function on the torus (25). On the other hand, their Fourier series are convergent term-by-term to the limiting series (35). By a familiar lemma, the limiting series is the Fourier series of a function in L^1, and this completes the proof of theorem 4.

We note that in the first part of theorem 4, the assumption that our Laurent series shall be strictly a power series was redundant. It follows from the assumption that relation (34) shall hold in the entire "octant" (33). Similar restrictions on the power series are implied in relation (34) depending on the type of domain P we choose in lieu of the octant (33). In fact, if P contains a half-line $(\sigma_1 u, \cdots, \sigma_k u)$, $0 < u < \infty$, then relation (34) implies that a coefficient $a_{m_1 \cdots m_k}$ can be different from zero only if

$$(38) \qquad m_1\sigma_1 + \cdots + m_k\sigma_k \geqq 0.$$

This restriction is automatic. However if we are willing to introduce this restriction explicitly, then we may obtain another type of theorem.

THEOREM 5. *If* (35) *is the Fourier series of a set function of bounded variation on the Borel sets of the torus* (25), *and if the Fourier coefficients are* $\neq 0$ *only for such indices* $m_1, \cdots, m_k$ *as satisfy relation* (38) *for all* $(\sigma_1, \cdots, \sigma_k)$ *of some* k-*dimensional neighborhood* N, *no matter how small, then the set-function is absolutely continuous, and thus* (35) *is the Fourier series of a function in* L^1.

PROOF. We denote by P the point set consisting of all half-lines issuing from the origin and each containing at least one point of N. Obviously, (38) is satisfied for all points of P. We also introduce a sequence of exponential polynomials

$$f^{(\nu)}(t_1, \cdots, t_k) = \sum r^{(\nu)}_{m_1 \cdots m_k} a_{m_1 \cdots m_k} e^{-i(m_1 t_1 + \cdots + m_k t_k)}$$

which converge term by term towards (35) such that

$$M_t \mid f^{(\nu)}(t_1, \cdots, t_k) \mid \leqq \gamma,$$

where γ is the variation of our set function. From the restriction on the Fourier coefficients it follows that we also have

$$(39) \qquad M_t \left| \sum r^{(\nu)}_m a_m e^{-(m_1 s_1 + \cdots + m_k s_k)} \right| \leqq \gamma$$

for $(\sigma_1, \cdots, \sigma_k)$ in P. We can now let $\nu \to \infty$ in (39), and then apply theorem 4.

It should be noted that theorem 5 is quite sharp; it is not permissible to let the neighborhood N shrink to a point. In fact, for $k = 2$, if

$$\sum a_{m_1, 0} e^{-i(m_1 t_1 + 0 \cdot t_2)}$$

is the Fourier series of a set function in one variable which is not absolutely continuous, then it is the Fourier series of a set function in two variables which is not absolutely continuous, although

$$\sigma_1 m_1 + \sigma_2 0 \geqq 0$$

for $\sigma_1 = 0$, $\sigma_2 = 1$ and all integers m_1.

REMARK ON L^p, $p > 1$. The latter comment emphasizes the distinction between our theorem concerning L^1, and a theorem of M. Riesz concerning L^p, $p > 1$. The theorem of M. Riesz states that if

$$\sum_{-\infty}^{\infty} a_m e^{-imt}$$

is the Fourier series of a function in L^p, then so is its "portion".

$$\sum_{m=0}^{\infty} a_m e^{-imt}.$$

As we have demonstrated somewhere else,[5] its generalization is that whenever (35) is a series in L^p, then so is its "portion"

$$\sum \sigma_1 m_1 + \cdots + \sigma_k m_k + \sigma_0 > 0 ,$$

this portion arising by dissection of the original series along any single one hyperplane. The fact is that the theorems of M. Riesz on conjugates is a theorem on functions in real variables (some of its proofs notwithstanding), whereas the theorem we are discussing in the present paper is not so.

IV. Almost Periodic Functions

We will consider in $-\infty < y < \infty$ almost periodic functions $f(y)$ of the most general kind, such being the functions of the Besicovitch class B^1. They result from closing the space of Bohr functions by the norm $M_y \mid f(y) \mid$. We will say that the functions $f(y)$, $g(y)$ are *equivalent* if

$$M_y \mid f(y) - g(y) \mid = 0.$$

In many important respects equivalent functions cannot be distinguished from one another. If $f(y)$ and $g(y)$ are both real, then we will say that $g(y)$ *exceeds* $f(y)$ if the difference $g(y) - f(y)$ is equivalent, in the sense just described, with a non-negative function. Investigations of almost periodic functions on general groups[6] have revealed many properties of Besicovitch functions analogous to those of periodic functions. In particular we will require the following two properties which are implied in general results.

Property 1. *If a sequence of real functions in B^1 are monotonely increasing, $g_n(y) \leqq g_{n+1}(y)$, and if their norms are bounded from above,*

$$M_y g_n(y) \leqq \gamma, \qquad n = 1, 2, \cdots,$$

then there exists a function $g(y)$ in B^1 which exceeds all $g_n(y)$ and for which

$$M_y g(y) \leqq \gamma.$$

Property 2. *If $\{f_n(y)\}$ is a sequence of complex functions in B^1, if all $\mid f_n(y) \mid$ are exceeded by some $g(y)$ in B^1, and if, for $n \to \infty$, the Fourier series of the sequence $f_n(y)$ are convergent term-by-term towards a limiting trigonometric series, then the latter series is again the Fourier series of some function in B^1.*

There is no clear-cut definition of functions B^1 of the complex variable $z =$

---

[5] S. Bochner, Additive set functions on groups, Annals of Math. 40 (1939), 769–799.

[6] In addition to the author's paper quoted in 5) see E. van Kampen, Almost periodic functions and compact groups, Annals of Math. 37 (1936), 78–91.

$x + iy$. All that can be required without prejudice is that the function shall have a Dirichlet expansion of the form

$$\sum a_n e^{-\lambda_n z}, \tag{40}$$

in the sense that for all x of some interval it shall be a B^1-function in the y-variable having the Fourier series

$$\sum_n (a_n e^{-\lambda_n x}) e^{-i\lambda_n y}. \tag{41}$$

Now, for our purpose it will be best to require just that and nothing more. Thus we start from a formal series of the type (40), and we assume that for each x in $0 < x < \infty$, the series (41) is the Fourier series of some element in B^1. Denoting this element by $f(x; y)$ we obtain a family of functions, where the variable x is a parameter designating membership in the family. We will call $f(x; y)$ a *formal* function.

THEOREM 6. *If the formal function $f(x; y)$ in $0 < x < \infty$ satisfies the inequality*

$$M_y \,|\, f(x; y) \,| \leqq \gamma, \qquad 0 < x < \infty, \tag{42}$$

then there exists a B^1-function $g(y)$ which exceeds $| f(x; y) |$ for every x, such that

$$M_y g(y) \leqq \alpha \cdot \gamma, \tag{43}$$

where α is the universal constant of theorem 2.

PROOF. We associate with (40) a sequence of approximating exponential polynomials

$$f^{(k)}(x; y) = \sum r_n^{(k)} e^{-\lambda_n z} \tag{44}$$

of the Fejer-type.[7] They have the properties

$$\lim_{k\to\infty} M_y \,|\, f^{(k)}(\xi; y) - f(\xi; y) \,| = 0, \qquad 0 < \xi < \infty \tag{45}$$

$$M_y \,|\, f^{(k)}(\xi; x) \,| \leqq \gamma, \qquad k = 1, 2, \cdots. \tag{46}$$

On account of (42) we may assume that all λ_n are non-negative, and therefore, by a known theorem[8] we have

$$M_y \,|\, f^{(k)}(\xi; y) \,| \geqq M_y \,|\, f^{(k)}(x; y) \,| \tag{47}$$

and

$$M_y \,|\, f^{(l)}(\xi; y) - f^{(k)}(\xi; y) \,| \geqq M_y \,|\, f^{(l)}(x; y) - f^{(k)}(x; y) \,| \tag{48}$$

for

$$\xi < x.$$

Letting $l \to \infty$, this implies

$$M_y \,|\, f(\xi; y) - f^{(k)}(\xi; y) \,| \geqq M_y \,|\, f(x; y) - f^{(k)}(x; y) \,| \tag{49}$$

[7] A. S. Besicovitch, Almost periodic functions, 1932, p. 103.

[8] Hardy, Ingham and Polya, Theorems concerning mean values of analytic functions, Proceedings of Royal Soc. Lond. 113 (1927), 542–569.

for

$$0 < \xi < x.$$

Since the function (44) is uniformly continuous in any strip we hence conclude the continuity property

(50) $$\lim_{x \to \xi} M_y \mid f(\xi; y) - f(x; y) \mid = 0$$

for $\xi > 0$.

We now apply theorem 3 in the half-plane $x > \xi$, instead of $x > 0$. In particular, if $x_1, \cdots, x_\rho$ are real numbers, all $> \xi$, then

(51) $$M_y \max_{1 \leqq \sigma \leqq \rho} \mid f^{(k)}(x_\sigma; y) \mid \leqq \alpha \cdot \gamma.$$

From the expression

$$\max(u_1, u_2) = \frac{u_1 + u_2 + \mid u_1 - u_2 \mid}{2}$$

and from similar expressions for $\max(u_1, \cdots, u_\rho)$ the reader will conclude that we may let $k \to \infty$ in (51) and hence obtain

(52) $$M_y g_\rho(y) \leqq \alpha \cdot \gamma, \qquad \rho = 1, 2, \cdots,$$

where

(53) $$g_\rho(y) = \max_{1 \leqq \sigma \leqq \rho} \mid f(x_\sigma; y) \mid$$

is a B^1-function each. We now take a sequence of numbers $\{x_\rho\}$ which is a dense set in $0 < x < \infty$, and we form the sequence of functions (53). They are increasing, and, by (52), their norms are bounded. By Property 1 of B^1-functions, there exists a limit-function $g(y)$ which exceeds all $g_\rho(y)$ and hence all $\mid f(x_\rho; y) \mid$. Thus all functions

$$g(y) - \mid f(x_\rho; y) \mid$$

exceed the function zero. By (50) it is now possible to conclude that $g(y) - \mid f(\xi; y) \mid$ exceeds zero for all $\xi > 0$, and this completes the proof of our theorem.

THEOREM 7. *If $f(x; y)$ is a formal function with expansion* (40) *and if* (42) *holds, then there exist a B^1-function $f(0; y)$ with expansion*

(54) $$\sum a_n e^{-i\lambda_n y}$$

such that

(55) $$\lim_{x \to 0} M_y \mid f(0; y) - f(x; y) \mid = 0.$$

PROOF. Since the functions $\mid f(x; y) \mid$ are exceeded by a function $g(y)$, the existence of a function with expansion (54) follows from Property 2 of B^1-functions. Relation (55) will then follow, if we approximate to $f(0; y)$ by our exponential polynomials

$$f^{(k)}(0; y) = \sum r_n^{(k)} a_n e^{-i\lambda_n y},$$

and then apply (48) and (49).

Remark on Functions B^p, $p > 1$. The following theorem is a secondary result since it applies to harmonic functions in general.

Theorem 8. *For $p > 1$, if $h(x; y)$ is a function of class B^p in $-\infty < y < \infty$ for $0 < x < \infty$, if its Fourier series has the form*

$$\sum e^{-\lambda_n x}(a_n \cos \lambda_n y + b_n \sin \lambda_n y)$$

and if

$$M_y \mid h(x; y) \mid^p \leqq \gamma, \qquad 0 < x < \infty,$$

then there exists a B^p-function $h(0; y)$ whose Fourier series is

$$\sum (a_n \cos \lambda_n y + b_n \sin \lambda_n y) \tag{56}$$

and

$$\lim_{x\to 0} M_y \mid h(0; y) - h(x; y) \mid^p = 0. \tag{57}$$

Proof. By a property of B^p-functions, the existence of the function $h(0; y)$ with series (56) follows directly from the facts that the family of functions $\{h(x; y)\}$ is bounded in norm and that their Fourier series are convergent term-by-term as $x \to 0$. Relation (57) follows again by applying algorithmic summation.

Finally we will demonstrate that there is no analogue to theorem 5 for almost periodic functions of unrestricted sequences of exponents $\{\lambda_n\}$.

A counter-example. We take a sequence of positive numbers $\mu_1, \mu_2, \mu_3, \cdots$ with the properties that no two general partial sums

$$\mu_{n_1} + \mu_{n_2} + \cdots + \mu_{n_r}$$

for $1 \leqq r < \infty$, $n_1 < n_2 < \cdots < n_r$, have the same numerical value. As for the existence of such sequences we note that any sequence of numbers which decrease to zero contains a subsequence of the given description. If we now form the functions

$$f_n(z) = \prod_{m=1}^{n} (1 + e^{-\mu_m z}),$$

then

$$M_y f_n(x + iy) = 1,$$

for all n and all x; also the expansions of $f_n(z)$ converge term-by-term, as $n \to \infty$, towards a series containing as sub-series the expansion

$$\sum_m e^{-\mu_m z}.$$

However, for $\mu_m \to 0$, the terms $\{e^{-\mu_m x}\}$ do not converge to zero for any x as $m \to \infty$, and thus the entire series cannot be the Fourier series of a B^1-function $f(x; y)$ for any real value x whatsoever.

Princeton University

Reprinted from the Proceedings of the NATIONAL ACADEMY OF SCIENCES,
Vol. 38, No. 8, pp. 726–732. August, 1952

LOCAL TRANSFORMATIONS WITH FIXED POINTS ON COMPLEX SPACES WITH SINGULARITIES

BY S. BOCHNER AND W. T. MARTIN

PRINCETON UNIVERSITY, MASSACHUSETTS INSTITUTE OF TECHNOLOGY AND THE INSTITUTE FOR ADVANCED STUDY

Communicated May 28, 1952

We will supplement a recent paper of ours[1] and we first of all describe the nature of the complex space there introduced. Let S be the unit sphere in Euclidean space of k complex variables $z_1, \ldots, z_k$, and over S we consider a system of polynomials in ζ's,

$$\psi_\nu\,(z_1, \ldots, z_k, \zeta_1, \ldots, \zeta_\nu) = 0,\ \nu = 1, \ldots, n \qquad (1)$$

with coefficients which are analytic in S. This defines for us a topological space which will be denoted by U. If R is any subset of S then $U(R)$ will denote those points of U whose base points are in R. Now there is a function $D(z)$ analytic and not identically zero in S such that for E denoting the points of S where D vanishes the subset $U(E)$ are possible singularities on U. However, $U - U(E) = U(S - E)$ shall be a complex analytic coordinate space.

If m_ν is the degree of ψ_ν in the variable ζ_ν and if $m = m_1 \ldots m_n$ then the system (1) has m sets of solutions

$$\left\{\zeta_1^{(\mu)}(z), \ldots, \zeta_n^{(\mu)}(z)\right\},\ \mu = 1, \ldots, m. \qquad (2)$$

At each point of $S - E$ these sets are assumed distinct, and they shall form n continuous functional elements in a neighborhood of every point of $S - E$.

A function $F(z, \zeta)$ was called *holomorphic* in U if it is defined throughout U, is bounded in $U(R)$ for every compact subset R of S, and is given in $U(S - E)$ by an expression of the form

$$F(z, \zeta) = \frac{1}{\Delta^2(z)} \sum P_{\lambda_1 \ldots \lambda_n}(z)\zeta_1^{\lambda_1} \ldots \zeta_n^{\lambda_n} \qquad (3)$$

where the P's are analytic in S. In (3) the summation is taken over the range

$$\lambda_1 = 0, 1, \ldots, m_1 - 1; \ \ldots; \ \lambda_n = 0, 1, \ldots, m_n - 1 \tag{4}$$

and the function $\Delta^2(z)$ is the square of the determinant of the matrix

$$([\zeta_1^{(\mu)}(t)]^{\lambda_1} \ldots [\zeta_n^{(\mu)}(t)]^{\lambda_n}) \tag{5}$$

where μ ranges over $1, \ldots, m$ and the λ's over (4). The function $\Delta^2(z)$ is analytic in S, and vanishes in S at most at points of E.

Such a space being given, we now say that a mapping T of U into U is a *continuous holomorphic mapping* if it is given by

$$T: \begin{array}{l} w_j = F_j(z, \zeta), j = 1, \ldots, k \\ \zeta_\nu = G_\nu(z, \zeta), \nu = 1, \ldots, n \end{array} \tag{6}$$

where the F_j and G_ν are continuous and holomorphic in U, and the G's are related to the F's in such a manner that the points $(w, \zeta) \in U$. If T and T' are two such mappings, and if TU is not entirely in $U(E)$ then we shall show that the iterated transformation $T'T$ is also a continuous holomorphic mapping of U into U. This result is an immediate consequence of the following theorem which we will establish.

THEOREM 1. *Let (6) be a continuous holomorphic mapping of U into U such that $TU - U(E)$ is not empty. Let $H(z, \zeta)$ be continuous and holomorphic in U. Then*

$$H[F_1(z, \zeta), \ldots, F_k(z, \zeta); \ G_1(z, \zeta), \ldots, G_n(z, \zeta)] \tag{7}$$

is continuous and holomorphic in U.

Proof: Clearly the function (7) is continuous in U. In order to see that it is holomorphic we first develop $H(z, \zeta)$ as in the right-hand side of (3), and make the transformation (6). This leads to a representation of (7) as the quotient of two expressions

$$\sum P_{\lambda_1 \ldots \lambda_n}[F_1(z, \zeta), \ldots, F_k(z, \zeta)][G_1(z, \zeta)]^{\lambda_1} \ldots [G_n(z, \zeta)]^{\lambda_n} \tag{8}$$

and

$$\Delta^2[F_1(z, \zeta), \ldots, F_k(z, \zeta)], \tag{9}$$

valid in that part of U for which the expression (9) is different from zero. Since $TU - U(E)$ is assumed non-empty, and since $\Delta^2(z)$ vanishes in S only over points of E, the function (9) is not identically zero.

The remainder of the proof of Theorem 1 will follow from three lemmas, of which Lemma 2 will be preliminary to Lemma 3.

LEMMA 1. *Let $F_1(z, \zeta), \ldots, F_s(z, \zeta)$ be s functions each continuous and holomorphic in U, and let $\Phi(w_1, \ldots, w_s)$ be analytic in a region B. If the values*

$$w_\sigma = F_\sigma(z, \zeta), \ \sigma = 1, \ldots, s, \ (z, \zeta) \in U$$

lie in B, then the function

$$F(z, \zeta) = \Phi[F_1(z, \zeta), \ldots, F_s(z, \zeta)] \tag{10}$$

is continuous and holomorphic in U.

LEMMA 2. *Let $I(z, \zeta)$ be holomorphic in U, let $\Omega(z)$ be analytic and not identically zero in S, and let $F(z, \zeta)$ be continuous in U. If*

$$F(z, \zeta) = \frac{I(z, \zeta)}{\Omega(z)} \text{ in } U(S - E - E_\Omega) \tag{11}$$

where E_Ω is the subset of S where Ω vanishes, then F is holomorphic in U.

LEMMA 3. *If $H_1(z, \zeta)$ and $H_2(z, \zeta)$ are holomorphic in U, if $H_2(z, \zeta)$ is continuous and $\not\equiv 0$ in U, if $F(z, \zeta)$ is continuous in U and if*

$$F(z, \zeta) = \frac{H_1(z, \zeta)}{H_2(z, \zeta)} \tag{12}$$

in the part of $U(S - E)$ where $H_2 \neq 0$, then F is holomorphic in U.

Proof of Lemma 1: The function (10) is clearly continuous. In order to see that it is holomorphic we first fix $\lambda_1, \ldots, \lambda_n$ in (4) and denote by $\Delta'_{\lambda_1 \ldots \lambda_n}(z)$ the matrix obtained from the matrix (5) by replacing the elements of the column, one element of which is displayed in (5), by $f_1(z), \ldots, f_m(z)$, where $f_\mu(z) = F(z, \zeta_1^{(\mu)}(z), \ldots, \zeta_n^{(\mu)}(z))$. We denote by $\Delta_{\lambda_1 \ldots \lambda_n}(z)$ the determinant of $\Delta'_{\lambda_1 \ldots \lambda_n}(z)$ and define

$$P_{\lambda_1 \ldots \lambda_n}(z) = \Delta_{\lambda_1 \ldots \lambda_n}(z)\Delta(z). \tag{13}$$

The functions $P_{\lambda_1 \ldots \lambda_n}$ so defined arise from symmetric functions in $(\zeta_1^{(\mu)}, \ldots, \zeta_n^{(\mu)})$, $\mu = 1, \ldots, m$, and are continuous in S and analytic in $S - E$. Since the set E is a removable exceptional set, it follows that each of the P's is analytic in all of S.

Next consider the $(m + 1) \times (m + 1)$-matrix

$$\begin{pmatrix} F(z, \zeta) \ldots \{\zeta_1^{\lambda_1} \ldots \zeta_n^{\lambda_n}\} \ldots \\ \text{-------------------------} \\ f_\mu(z) \ldots \{[\zeta_1^{(\mu)}(z)]^{\lambda_1} \ldots [\zeta_n^{(\mu)}(z)]^{\lambda_n}\} \ldots \\ \text{-------------------------} \end{pmatrix}$$

where we have displayed the first and $(\mu + 1)$-st elements of first column, and the same elements of a typical one of the remaining columns, for the λ's in (4). The determinant of this matrix is zero throughout U since it is zero whenever $\zeta_1 = \zeta_1^{(\mu)}(z), \ldots, \zeta_n = \zeta_n^{(\mu)}(z)$ for any $\mu = 1, \ldots, m$. But if we expand the determinant in terms of the elements of the first row, and multiply through by $\Delta(z)$, we obtain

$$\Delta^2(z)F(z, \zeta) = \sum P_{\lambda_1 \ldots \lambda_n}(z)\zeta_1^{\lambda_1} \ldots \zeta_n^{\lambda_n}, \tag{14}$$

the summation being extended over (4). Thus F is holomorphic in U.

Proof of Lemma 2: As in the proof of Lemma 1, form the functions (13). By the continuity of F and symmetry they are continuous in S, and by (11) and symmetry they are analytic in $S - E - E_\Omega$, and hence in S. Thus (14) holds, and F is holomorphic.

Proof of Lemma 3: By one of the basic assumptions of our space (property (v) of Section 1 in "Complex Spaces"), corresponding to any continuous holomorphic function $H_2(z, \zeta) \not\equiv 0$ in U they are a function $G(z, \zeta)$ strictly holomorphic in U and a function $\Omega(z)$ analytic and not identically zero in S such that $\Delta^2(z)H_2(z, \zeta)G(z, \zeta) \equiv \Omega(z)$ in U. It is clear from this that

$$H_2(z, \zeta) \neq 0 \text{ in } U(S - E_\Omega).$$

Hence

$$\frac{1}{H_2(z, \zeta)} = \frac{\Delta^2(z)G(z, \zeta)}{\Omega(z)} \text{ in } U(S - E_\Omega).$$

Thus the function (12) has a representation (11), and by Lemma 2, (12) is holomorphic.

As regards Theorem 1, the first lemma implies that the two functions (8) and (9) are continuous and holomorphic in U, and as we have seen, the function (7) is expressible as their quotient in the part of U where the denominator (9) is not zero; thus Lemma 3 completes the proof.

THEOREM 2. *Let Γ be an arbitrary compact space, let*

$$F(z, \zeta; \alpha) = \frac{1}{\Delta^2(z)} \sum_{\lambda_1=0}^{m_1-1} \cdots \sum_{\lambda_n=0}^{m_n-1} P_{\lambda_1\ldots\lambda_n}(z; \alpha)\zeta_1^{\lambda_1}\ldots\zeta_n^{\lambda_n}, \quad (15)$$

be a family of functions, continuous and holomorphic in U for each α in Γ, and continuous in α for each (z, ζ) in U. If there are positive numbers $\rho_0 \leq 1$ and M such that

$$|F(z, \zeta; \alpha)| \leq M, \text{ for } \alpha \in \Gamma \text{ and } (z, \zeta) \in U(\rho_0), \quad (16)$$

where $U(\rho_0)$ is the subset of U for which $\|z\| < \rho_0$, then the functions $P_\lambda(z; \alpha)$ in (15) have the following properties:

1°. *For every positive number $\rho < \rho_0$, the P's are uniformly bounded and equi-uniformly continuous in z in S_ρ: $\|z\| \leq \rho$;*

2°. *For each λ in (4), $P_\lambda(z; \alpha)$ is continuous in (z, α) in $S_\rho \times \Gamma(\rho < \rho_0)$; and*

3°. *The coefficients $p_{s_1\ldots s_k;\lambda}(\alpha)$ in the power-series expansions*

$$P_\lambda(z; \alpha) = \sum_{s_1=0}^{\infty} \cdots \sum_{s_k=0}^{\infty} p_{s_1\ldots s_k;\lambda}(\alpha)z_1^{s_1}\ldots z_k^{s_k}$$

are continuous in α.

Proof: The first part of Property 1°, namely the fact that there is a positive number B_ρ such that

$$|P_\lambda(z;\ \alpha)| \leq B_\rho \text{ for } \alpha \in \Gamma,\ z \in S_\rho,\ \lambda \text{ in } (4),$$

is a consequence of Lemma 4.1 of "Complex Spaces." The continuity of $P_\lambda(z;\ \alpha)$ in α follows from the relationship

$$P_{\lambda_1\ldots\lambda_n}(z;\ \alpha) = \Delta_{\lambda_1\ldots\lambda_n}(z;\ \alpha)\Delta(z) \text{ in } S,$$

where $\Delta_{\lambda_1\ldots\lambda_n}(z;\ \alpha)$ is defined as in (13), and from the continuity of the functions $f_\mu(z;\ \alpha) = F(z, \zeta_1^{(\mu)}(z), \ldots, \zeta_n^{(\mu)}(z);\ \alpha)$ in α for fixed z in S.

The remaining part of 1° and properties 2° and 3° are consequences of the following lemma.

LEMMA 4. *Let Γ be a compact space and let $r_1, \ldots, r_k$ be positive numbers. If*

$$P(z;\ \alpha) = \sum_{s_1=0}^{\infty} \cdots \sum_{sk=0}^{\infty} p_{s_1\ldots sk}(\alpha) z_1^{s_1} \ldots z_k^{sk}$$

is a family of functions continuous in α for each z in the multi-cylinder

$$C(r):\ |z_1| < r_1, \ldots, |z_k| < r_k$$

and if there is a positive constant B such that

$$|p_{s\ldots sk}(\alpha)| \leq \frac{B}{r_1^{s_1}\ldots r_k^{sk}},\ \alpha \in \Gamma,\ s_1, \ldots, s_k = 0, 1, 2, \ldots$$

then

(i) *for every set of positive numbers $\rho_1, \ldots, \rho_k$ with $\rho_j < r_j$, the functions $P(z;\ \alpha)$ are equi-uniformly continuous in z in the multi-cylinder*

$$\bar{C}(\rho):\ |z_1| \leq \rho_1, \ldots, |z_k| \leq \rho_k,$$

and continuous in (z, α) in $C(r) \times \Gamma$; and

(ii) *the coefficients $p_{s_1\ldots sk}(\alpha)$ are continuous in α.*

This lemma, which is actually a theorem due to Osgood, is taken as known.

We now take Γ to be a compact group, and with each element α in Γ we associate a continuous holomorphic 1-1 mapping $T(\alpha)$ of U onto U,

$$T(\alpha):\quad \begin{aligned} w_j &= F_j(z, \zeta;\ \alpha),\ j = 1, \ldots, k \\ \xi_\nu &= G_\nu(z, \zeta;\ \alpha),\ \nu = 1, \ldots, n \end{aligned} \qquad (17)$$

As in Section 1, for any pair α, β in Γ we can define the iterated transformation $T(\alpha)T(\beta)$. We require that this transformation be identical with the transformation $T(\alpha\beta)$, where $\alpha\beta$ is the group product in Γ.

Under these conditions we will call $\{T(\alpha)\}$ a compact group of continuous holomorphic mappings of U onto U.

THEOREM 3. *If (17) is a compact group of continuous holomorphic mappings of U onto U, if the functions $F_j(z, \zeta; \alpha)$ are continuous in α for each fixed (z, ζ) in U, and if the functions F_j have the special form*

$$F_j(z, \zeta; \alpha) = z_j + O(||z||^{1+\epsilon}), j = 1, \ldots, k; \ \alpha \in \Gamma \qquad (18)$$

for $||z||$ sufficiently small, and for some fixed $\epsilon > 0$, then there exists a transformation of U into the base space S,

$$R_1: \ w_j = H_j(z, \zeta) = z_j + O(||z||^{1+\epsilon}), j = 1, \ldots, k \qquad (19)$$

defined by k continuous holomorphic functions which has the property that

$$R_1T(\alpha) = R_1 \text{ for all } \alpha \text{ in } \Gamma. \qquad (20)$$

Proof: Since $T(\alpha)$ maps U onto U we have

$$\left|\sum_{j=1}^{k} F_j(z, \zeta; \alpha)\right|^2 < 1. \qquad (21)$$

By (21), (18) and Schwarz's lemma for holomorphic functions[2] we have

$$\sum_{j=1}^{k} |F_j(z, \zeta; \alpha)|^2 \leq ||z||^2, \sum_{j=1}^{k} |F_j(z, \zeta; \alpha) - z_j|^2 \leq [2||z||^{1+\epsilon}]^2. \qquad (22)$$

Also by (21) the functions $F_j(z, \zeta; \alpha)$ satisfy the hypotheses of Theorem 2 for every $\rho_0 \leq 1$, and hence the related analytic functions $P_{\lambda_1 \ldots \lambda_n}, j(z; \alpha)$ enjoy properties 1°, 2° and 3° of Theorem 2.

If we denote by $T_1(\alpha)$ the transformation

$$T_1(\alpha): \ w_j = F_j(z, \zeta; \alpha) \qquad j = 1, \ldots, k$$

from U into the base space S, then we may form the transformation[3–5]

$$R_1 = \int_\Gamma T_1(\alpha)\, d\alpha$$

where $d\alpha$ refers to the Haar measure on Γ, with total measure unity The transformation R_1 is given by

$$w_j = H_j(z, \zeta) = \int_\Gamma F_j(z, \zeta; \alpha)\, d\alpha.$$

Each H_j is continuous in U, and on integrating termwise the series developments as in (15) for F_j we see that each H_j is holomorphic in U Also the H's satisfy the same inequalities (22) as the F's.

In order to prove (20) we first note that for any two elements α, γ in Γ

$$T_1(\alpha)T(\gamma) = T_1(\alpha\gamma).$$

Thus

$$R_1T(\gamma) = \int T_1(\alpha)\, d\alpha . T(\gamma) = \int T_1(\alpha)T(\gamma)\, d\alpha = \int T_1(\alpha\gamma)\, d\alpha.$$

By the invariance of Haar measure the latter integral is merely R_1.

If R_1 were a mapping from U into U, rather than from U into S, then as in the classical case (20) would imply that each $T(\alpha)$ is the identity.

[1] "Complex Spaces with Singularities," *Ann. Math.* (to appear). In the present note this paper will be referred to as "Complex Spaces."

[2] Schwarz's lemma as given in "Complex Spaces" was formulated for single holomorphic functions and is now being used for several simultaneously.

[3] Cartan, H., *Actualités sci. et ind.*, 198, Paris, 1935.

[4] Bochner, S., *Ann. Math.*, **46**, 372–381 (1945).

[5] "Several Complex Variables," Princeton, 1948.

ON THE ADDITION THEOREM FOR MULTIPLY PERIODIC FUNCTIONS

S. BOCHNER

We shall comment on the following classical proposition.

THEOREM 1. (i) *The Weierstrass function satisfies a relation*

$$P_0(\wp(z+w), \wp(z), \wp(w)) = 0, \tag{1}$$

identically in independent variables, z, w, where $P_0(\alpha, \beta, \gamma)$ is a polynomial not $\equiv 0$ with constant coefficients.

(ii) *More generally, any doubly periodic meromorphic function $f(z)$ satisfies a polynomial relation*

$$P_{00}(f(z+w), f(z), f(w)) = 0. \tag{2}$$

(iii) *Even more generally, for any three such functions*

$$f(z), \qquad \phi(z), \qquad \psi(z) \tag{3}$$

with the same periods, the last two not $\equiv c$, we have a relation

$$P(f(z+w), \phi(z), \psi(w)) = 0. \tag{4}$$

(iv) *Also, $\wp(z+w)$ is a rational function of $\wp(z)$, $\wp'(z)$, $\wp(w)$, $\wp'(w)$; and more generally if $\phi^1(z)$, $\phi^2(z)$ are rationally independent [as $\wp(z)$, $\wp'(z)$ are] and $\psi^1(w)$, $\psi^2(w)$ are rationally independent, then $f(z+w)$ is a rational function of $\phi^1(z)$, $\phi^2(z)$, $\psi^1(w)$, $\psi^2(w)$.*

On the face of it, part (i) is a particular case of part (ii), and the latter part of (iii), but actually all three parts are algebraically equivalent by the following elementary argument. For given periods, we introduce the closed Riemann surface V_2 of genus $p=1$ on which our meromorphic functions are suitably defined, and we take it as known that on such or any other closed Riemann surface of any genus $p \geqq 0$, any two meromorphic functions are algebraically dependent one on the other. That is to say, if for any $f(z) \neq c$ on V_2 we introduce the function field

$$K[f(z)] \tag{5}$$

over complex constants, then any other $g(z)$ is algebraic over (5), thus satisfying a polynomial relation $P^0(f(z), g(z)) = 0$. In particular, the functions (3) are connected with the Weierstrass function by relations

Received by the editors January 25, 1951.

$$P_1(\wp(z+w), f(z+w)) = 0, \qquad P_2(\wp(z), \phi(z)) = 0, \tag{6}$$
$$P_3(\wp(w), \psi(w)) = 0,$$

and if we start out from (1) then formal algebraic elimination as between (1) and (6) will produce (4). A similar argument will equate the two halves of part (iv). Also a suitable way of expressing (4) is to say that the two-variable function

$$f(z, w) = f(z + w), \tag{7}$$

whose domain of existence is the product $V_2 \times V_2$ of our torus with itself, is algebraic over the function field

$$K[\phi(z); \psi(w)], \tag{8}$$

and the second half of (iv) states that it is contained in

$$K[\phi^1(z), \phi^2(z); \psi^1(w), \psi^2(w)]. \tag{9}$$

An "addition theorem" suggests colorful connections to arithmetic, and such are indeed manifest in statements of many facets. The point we wish to make, however, is this—that to our theorem, as stated, no such connection is as yet relevant and that it is a special case of a vastly more general and entirely colorless proposition in which additivity is in no way referred to.

If we take any doubly periodic periodic meromorphic function $f(z)$, and introduce the two-variable function (7), then for each fixed w (with some possible exceptions) the latter is again such a function in z, and for each fixed z it is such a function in w, and this is all that is needed to validate our conclusions. In fact, if we realize that for any three (or more) meromorphic functions on V_2 there is a point of the space in the neighborhood of which they are all holomorphic (without any polar singularities), then Theorem 1 is obviously contained in the following "local" theorems which we shall immmediately state for multi-dimensional variables as well.

THEOREM 2. *If A is a domain in the space of k complex variables $z=(z^{(1)}, \cdots, z^{(k)})$ and B is a domain in the space of l complex variables $w=(w^{(1)}, \cdots, w^{(l)})$; if we are given a finite number of holomorphic functions*

$$\phi^1(z), \cdots, \phi^r(z) \tag{10}$$

in A, and a finite number of holomorphic functions

$$\psi^1(w), \cdots, \psi^s(w) \tag{11}$$

in B; and if a holomorphic function $f(z, w)$ in $A \times B$ has the property *that for each w in B it is algebraic over*

$$K[\phi^1(z), \cdots, \phi^r(z)] \tag{12}$$

and for each z in A it is algebraic over

$$K[\psi^1(w), \cdots, \psi^s(w)]; \tag{13}$$

then in $A \times B$ it is algebraic over

$$K[\phi^1(z), \cdots, \phi^r(z); \psi^1(w), \cdots, \psi^s(w)]. \tag{14}$$

THEOREM 3. *If the function $f(z, w)$ is for each w in B contained in the field* (12) *and for each z in A in the field* (13), *then in $A \times B$ it is contained in the field* (14).

There is also a mixed theorem as follows.

THEOREM 4. *If $f(z, w)$ is for each w algebraic over* (12), *of a degree not greater than g, where g is independent of w, and if for each z it is contained in* (13), *then in $A \times B$ it is algebraic over* (14), *of degree not greater than g.*

A particular consequence of Theorems 2 and 3 is the following known generalization of Theorem 1 to many dimensions.[1]

THEOREM 5. *If for k complex variables there are given k algebraically independent meromorphic functions*

$$\phi^1(z), \cdots, \phi^k(z) \tag{15}$$

with the same 2k "independent" periods, then for any other such function $f(z)$—which in particular may be one of the $\phi^\rho(z)$—the new function $f(z+w)$ is algebraic over

$$K[\phi^1(z), \cdots, \phi^k(z); \phi^1(w), \cdots, \phi^k(w)]; \tag{16}$$

and if there are given $k+1$ rationally independent functions

$$\phi^0(z), \phi^1(z), \cdots, \phi^k(z), \tag{17}$$

then $f(z+w)$ is rational over

$$K[\phi^0(z), \cdots, \phi^k(z); \phi^0(w), \cdots, \phi^k(w)]. \tag{18}$$

Although syllogistically subordinate to Theorems 2, 3, our Theorem 5 has nevertheless the distinguishing feature that in it the two-variable function occurring is actually a one-variable function in which the variable z has been replaced by the two-fold argument

[1] A. Krazer and W. Wirtinger, *Abelsche Funktionen und allgemeine Thetafunktionen*, Encyclopädie der mathematischen Wissenschaften II B 7, p. 827.

$z+w$. Now this distinguishing feature has a generalization of its own and it is as follows.

Let the domain A of our theorems be a coordinate neighborhood of a complex manifold V_{2k}; let W_{2l} be a (complex) Lie group acting on V_{2k}, and let B be a neighborhood of the identity in W_{2l}. Now, if $f(z)$ is meromorphic on V_{2k} and has no singularities in A, and if w is a point of B, and if we denote by $f(w(z))$ the function resulting from carrying out the homeomorphism "w," then we obviously obtain a two-variable function $f(z, w)$ on $A \times B$ to which our Theorems 2, 3, 4 can be applied, provided, of course, that finite systems of functions $\{\phi^\rho(z); \psi^\sigma(w)\}$ as prescribed in the theorems are available on A, B.

A pertinent though seemingly "trivial" illustration arises if for V_{2k} we take the ordinary Gaussian sphere V_2 and for W_{2l} the group W_6 of projective transformations

$$z' = \frac{w' + w''z}{w'' + w''''z}$$

on it. A meromorphic function on V_2 is any ordinary rational function in z,

$$f(z) = R(z) \equiv \frac{\sum a_\mu z^\mu}{\sum b_\nu z^\nu}, \tag{19}$$

and the quantities w', w'', w''', w'''' are obviously meromorphic functions on the rational surface

$$w'w'''' - w''w''' = 1 \tag{20}$$

constituting W_6. Now, the two-variable function

$$f(z, w) = R\left(\frac{w' + w''z}{w''' + w''''z}\right) \tag{21}$$

has the property that for each w it is meromorphic in z and for each z meromorphic in w and, in conformity with Theorem 3, it is indeed meromorphic in (z, w). Also, if $f(z)$ is algebraic of a degree g, then for each w, $f(z, w)$ is algebraic of a degree not greater than g, and for each z it is rational in w, and, in conformity with Theorem 4, it is indeed algebraic in (z, w) of degree g.

The example obviously falls under the following assertions. (α) If the holomorphic function $f(z, w)$ in $A \times B$ is for each w algebraic in the local parameters z, and for each z algebraic in the parameters w, then it is algebraic in (z, w). (β) If it is rational in z and w separately it is rational in them jointly. (γ) If for each w it is algebraic in z of degree not greater than g, and for each z rational in w, then it is algebraic in (z, w) of degree not greater than g. Now, assertion (β) was first stated by Weierstrass,[2] and then proven by A. Hurwitz;[3] and assertions (α) and (γ) were set up and proven by ourselves[4] when reproducing the proof of Hurwitz in the simplified version given to it by H. Kneser;[5] and the proofs to follow of the general Theorems 2, 3, 4 will be nearly literally the same as those given in the paper cited in footnote 4 for the particular assertions (α), (β), (γ). But before giving the proofs we wish to point out that although the problem of Weierstrass and Hurwitz dealt directly only with meromorphic functions of the ordinary kind, that is to say, with functions on complex projective spaces, yet the context in which it arose was one dealing with multi-periodic functions and Abelian integrals primarily; so that our general versions of the theorems may be said to be related to the original ones not only syllogistically but even genetically as well, although our proof itself would hardly betoken such a development.

The proof of our theorems rests on the following lemma, which by itself does not yet refer to analyticity or even continuity.

LEMMA 1. *If* $F_1(z, w), \cdots, F_N(z, w)$ *are functions in* $A \times B$, *not all* $\equiv 0$, *and if they satisfy a relation*

$$c_1(w)F_1(z, w) + \cdots + c_N(w)F_N(z, w) = 0 \tag{22}$$

with arbitrary coefficients $\{c_n(w)\}$ *for which*

$$|c_1(w)|^2 + \cdots + |c_N(w)|^2 > 0, \tag{23}$$

then they also satisfy such a relation with other coefficients $C_n(w)$, *not all* $\equiv 0$, *from the smallest function ring over complex coefficients containing all functions* $\{F_n(a, w)\}$ *for all special values* $z = a$ *and* $n = 1, \cdots, N$.

In particular if each $F_n(z, w)$ *belongs to* (13), *then there are coefficients* $C_n(w)$ *belonging likewise to* (13).

---

[2] C. Weierstrass, *Untersuchungen über die 2r-fach periodischen Functionen von r Veränderlichen*, J. Reine Angew. Math. vol. 89 (1880) pp. 1–8; or, *Werke*, vol. 2, pp. 125–133.

[3] A. Hurwitz, *Beweis des Satzes, dass eine einwertige Function beliebig vieler Variabeln, welche überall als Quotient zweier Potenzreihen dargestellt werden kann, eine rationale Function ihrer Argumente ist*, J. Reine Angew. Math. vol. 95 (1883) pp. 201–206.

[4] S. Bochner and W. T. Martin, *Several complex variables*, Princeton, 1948, pp. 199–203.

[5] H. Kneser, *Einfacher Beweis eines Satzes über rationale Funktionen Zweier Veränderlschen*, Abh. Math. Sem. Hamburgischen Univ. vol. 9 (1933) pp. 195–196.

PROOF. We write relation (22) for N values $z_1, \cdots, z_N$, thus obtaining a system of N homogeneous equations in the quantities $c_n(w)$. Because of (23), the determinant

$$D(z; w) = \begin{vmatrix} F_1(z_1, w) & \cdots & F_N(z_1, w) \\ \cdot & \cdot & \cdot \\ F_1(z_N, w) & \cdots & F_N(z_N, w) \end{vmatrix} \tag{24}$$

must vanish identically in $w \in B$; $z_1 \in A, \cdots, z_N \in A$. We now replace the letter z_N by z, and develop the determinant in terms of its last row. This leads to a relation

$$\sum_{n=1}^{N} C_n(z_1, \cdots, z_{N-1}; w) F_n(z, w) = 0 \tag{25}$$

in which the functions

$$C_n(z_1, \cdots, z_{N-1}; w) \tag{26}$$

are, but for $\pm$ signs, the $(n-1)$-dimensional determinants of the matrix

$$\left\| \begin{matrix} F_1(z_1, w) & \cdots & F_N(z_1, w) \\ \cdot & \cdot & \cdot \\ F_1(z_{N-1}, w) & \cdots & F_1(z_{N-1}, w) \end{matrix} \right\|, \tag{27}$$

and we first assume that not all functions (26) vanish identically in all variables. Hence there exist numerical values $z_1 = a_1, \cdots, z_{N-1} = a_{N-1}$ such that the functions

$$C_n(w) = C_n(a_1, \cdots, a_{N-1}; w)$$

do not vanish identically in w, and for such numerical values relation (25) reduces to relation (22) with new coefficients $c_n(w)$ having the property stated in the lemma. If, however, all functions (26) vanish identically, then the determinant (24) has a principal minor

$$\Delta = \left\| \begin{matrix} F_{\alpha_1}(z_{\alpha_1}, w) & \cdots & F_{\alpha_m}(z_{\alpha_1}, w) \\ \cdot & \cdot & \cdot \\ F_{\alpha_1}(z_{\alpha_m}, w) & \cdots & F_{\alpha_m}(z_{\alpha_m}, w) \end{matrix} \right\|$$

which vanishes identically, but which has at least one subdeterminant of order $m-1$ which does not vanish identically. By the result just proved there exists a relation

$$\sum_{\mu=1}^{m} C_{\alpha\mu}(w) F_{\alpha\mu}(z, w) = 0,$$

and by inserting vanishing coefficients $C_n(w)$ we obtain the conclusion.

PROOF OF THEOREM 3. We form the products

$$(\phi^1(z))^{n_1} \cdots (\phi^r(z))^{n_r}$$

for all combinations of integers $n_1 \geqq 0, \cdots, n_r \geqq 0$ and denote the resulting sequence of "monomials" in some ordering by $p_0(z)$, $p_1(z), \cdots$. By assumption of the theorem we have for each w a relation

$$\left(\sum_{\mu=0}^{m} a_\mu(w)p_\mu(z)\right)f(z, w) + \sum_{\nu=0}^{n} b_\nu(w)p_\nu(w) = 0 \tag{28}$$

with m, n depending on w, and

$$\sum_{\mu=0}^{m} | a_\mu(w) |^2 + \sum_{\nu=0}^{n} | b_\nu(w) |^2 > 0, \tag{29}$$

and if we multiply it by a factor we may presuppose the normalization

$$\sum_{\mu=0}^{m} | a_\mu(w) |^2 + \sum_{\nu=0}^{n} | b_\nu(w) |^2 = 1. \tag{30}$$

By the same argument as on p. 201 of the paper cited in footnote 4, there exist some combination of degrees (m, n) for which relation (28) holds for a subdomain of B, and if we now denote the subdomain itself by B, then (28) will hold everywhere in $A \times B$ for a given set of degrees (m, n). If now we put $N = m+n$, $F_\mu(z; w) = p_\mu(z)f(z, w)$ for $\mu = 1, \cdots, m$, and $F_{m+\nu}(z; w) = p_\nu(z)$ for $\nu = 1, \cdots, n$, then our lemma becomes applicable, and the conclusion is that there is also a relation (28) with coefficients $a_\mu(w)$, $b_\nu(w)$ belonging to the function ring (13), and not all $\equiv 0$, which is precisely the assertion of the theorem.

PROOF OF THEOREMS 2 AND 4. In this case we have for each w a relation

$$\sum_{\mu=0}^{m} \sum_{\gamma=0}^{g} a_{\mu\gamma}(w)p_\mu(z)f(z, w)^\gamma = 0, \tag{31}$$

and for some subdomain B_0 of B this relation is available for the same (m, g), where in the case of Theorem 4 the integer g is the one specified in the theorem. We now put $N = m(g+1)$, and apply the lemma to the functions

$$F_n(z, w) = p_\mu(z)f(z, w)^\gamma.$$

As a consequence of the lemma we obtain a new relation with the same (m, g) in which the $a_{\mu\gamma}(w)$ belong to the smallest function ring containing all $p_\mu(a)f(a, w)^\gamma$, that is, all $f(a, w)^\gamma$, and in the case of Theorem 4 this terminates the proof immediately. In the case of Theorem 2 we have to add the step that the new coefficients are *algebraic* over the smallest function ring containing all $f(a, w)$, and by an elementary theorem on field extensions a second choice of coefficients $a_{\mu\gamma}(w)$, with other (m, g), will produce a final relation (28) in which the coefficients $a_{\mu\gamma}(w)$ are contained in the function ring (13) itself, as claimed.

REMARK. The two function fields (12) and (13), of which (14) is the "product," have been assumed to have each a finite basis over the field of complex numbers, but the proof of the theorems just completed shows that this requirement of finiteness can be relaxed in the following way. The function field (13) of holomorphic functions on B can have an arbitrary potency, and we shall denote it by K_w. The field on A however—we shall denote it by K_z—cannot be allowed to be arbitrary, but it need not be finite either. Rather, it suffices to assume that it is the limit of an increasing sequence of subfields K_z^q, $q=1, 2, \cdots$, of which each has a finite basis, such that, in the case of Theorem 2, our function $F(z, w)$ is for each w algebraic over some K_z^q, $q=q(w)$, and contained in it in case of Theorem 4. The conclusion then is that $F(z, w)$ is algebraic, or rational, over the product $K_w \times K_z$, and indeed over $K_w \times K_z^{q_0}$, for some suitable index q_0 depending on the function.

PRINCETON UNIVERSITY

ALGEBRAIC AND LINEAR DEPENDENCE OF AUTOMORPHIC FUNCTIONS IN SEVERAL VARIABLES

By S. BOCHNER

[Received June 13, 1951]

THEOREM 1. *Let $S = S_{2k}$ be a (non-compact) complex co-ordinate space of $2k$ real dimensions, and let S^0 be a domain in it whose closure $\overline{S^0}$ can be covered by a finite number of allowable co-ordinate systems.*

If $k+2$ holomorphic functions

$$f_1, \cdots, f_{k+2} \tag{1}$$

on S are such that for every homogeneous polynomial

$$Q_n(w_1, \cdots, w_{k+2}) \not\equiv 0 \tag{2}$$

of any degree n the function

$$\phi(P) = Q_n[f_1(P), \cdots, f_{k+2}(P)] \tag{3}$$

satisfies the relation

$$\sup_{P \in S} |\phi(P)| \leqslant M^n . \sup_{P \in S^0} |\phi(P)| \tag{4}$$

for a suitable constant $M > 1$, then there exists one such polynomial $Q_n^0(w)$ for which

$$Q_n^0(f_1, \cdots, f_{k+2}) \equiv 0. \tag{5}$$

PROOF. It is easily seen that we can select in S a finite number of co-ordinate neighbourhoods

$$U_1, \cdots, U_m, \tag{6}$$

each uniformized by the unit sphere

$$r \equiv (z_1 \bar{z}_1 + \dots + z_k \bar{z}_k)^{1/2} < 1, \tag{7}$$

such that the m closed sub-spheres

$$r \leqslant \tfrac{1}{2} \tag{8}$$

together constitute a covering of the closed set $\overline{S^0}$. On the other hand, if $\phi(z) \equiv \phi(z_1, \ldots, z_k)$ is holomorphic in (7), if its power series at the origin

$$\phi(z) = \sum_{n_1+\ldots+n_k \geqslant p} a_{n_1 \ldots n_k} z_1^{n_1} \ldots z_k^{n_k} \tag{9}$$

has no terms of order $< p$, and if $|\phi(z)| < 1$ in (7), then in the sub-sphere (8) we have

$$|\phi(P)| < \frac{1}{2^p} \cdot \tag{10}$$

This is a trivial extension of Schwarz's lemma from one to several variables [2].

Now, with the given functions (1) we form for fixed n the monomials

$$\phi_A(z) = f_1^{n_1} \ldots f_{k+2}^{n_{k+2}}, \; A = 1, \ldots, N,$$

where $n_1 + \ldots + n_{k+2} = n$ and $N = \binom{n+k+1}{k+1}$; and the assertion is that for n sufficiently large, these N monomials are linearly dependent. If this were not so, then for any $p = p_n$ satisfying

$$m \binom{p+k-1}{k} < \binom{n+k+1}{k+1} \tag{11}$$

we could find a linear combination

$$\phi = \sum_{A=1}^{N} C_A \phi_A(P)$$

such that

$$\sup_{P \in S} |\phi(P)| = 1, \tag{12}$$

and that at the centres of the m neighbourhoods (6) we would have expansions of the form (9). Now, by Schwarz's lemma, we would have inequality (10) everywhere in $\overline{S^0}$, and on the other hand, by (4) and (12), in at least one point of $\overline{S^0}$ we would have

$$|\phi(P)| \geqslant \frac{1}{M^n}; \tag{13}$$

and if we compare (10) and (13) we would obtain

$$p \leqslant \frac{\log M}{\log 2} \cdot n. \tag{14}$$

However, for k fixed, and n and p large, we have

$$\binom{n+k+1}{k+1} \sim \frac{1}{(k+1)!} n^{k+1}, \; \binom{p+k-1}{k} \sim \frac{1}{k!} p^k,$$

and therefore it is compatible with (1) to have

$$p_n \geqslant C_0\, n^{1+1/k},$$

where C_0 is fixed and n is large. But this is not compatible with (14), and the proof of the theorem is completed.

REMARK. It is obviously possible to give an estimate for a degree n satisfying (5) in terms of the constants $\{k, M, m\}$ only.

Now, let $V = V_{2k}$ be a compact complex co-ordinate space, let $\tilde{V}$ be its (non-compact) universal covering space, and let $\Gamma = \tilde{V}/V$ be the fundamental group. Since V is compact, there is no holomorphic function on it, other than a constant, that is, but on a non-compact $\tilde{V}$ there might be some; and we call a function on $\tilde{V}$ automorphic if for every α in Γ we have

$$f(\alpha P) = \eta_\alpha(P) . f(P), \tag{15}$$

for some suitable factor $\eta_\alpha(P)$, itself also holomorphic on $\tilde{V}$. This is a broader notion of "automorphy" than sometimes conceived of; and an assemblage of functions

$$\{\eta_\alpha(P)\}, \quad \alpha \in \Gamma, \tag{16}$$

will qualify for the role of such factors of automorphy if they satisfy the law of composition

$$\eta_\beta(P) \cdot \eta_\alpha(\beta P) = \eta_{\alpha\beta}(P) \tag{17}$$

for all pairs of elements α, $\beta \in \Gamma$, and this is all they need satisfy.

THEOREM 2. *If we are given a compact complex space V, its covering space $\tilde{V}$, the fundamental group Γ, and a fixed system of factors* (16) *which satisfy* (17), *then any $k + 2$ holomorphic functions on $\tilde{V}$ which are automorphic with regard to* (16) *satisfy an equation* (5).

PROOF. We can choose in $\tilde{V}$ a domain S^0 having the following properties : the union R of S^0 with part of its boundary constitutes a "fundamental domain" in $\tilde{V}$ relative to the group of automorphisms Γ; and there is a finite set of elements $\alpha_1, \ldots, \alpha_p$

of Γ such that the interior of the union of R with its "translations" $a_1(R), \ldots, a_p(R)$ constitutes a neighbourhood S of the closure $\overline{S^0}$. Also $\overline{S^0}$ is compact in S and can be covered by a finite number of holomorphic co-ordinate systems there. This being so, if we denote by M the supremum of $|\eta_a(P)|$ for P in R and $a = a_1, \ldots, a_p$, then all assumptions of Theorem 1 are fulfilled and the conclusion of Theorem 2 follows.

If $\tilde{V}$ is a domain in Euclidean space E_{2k}, and thus is covered by one co-ordinate system, then the elements of Γ can be represented "globally" by functions

$$z_j^a = f_j^a(z_1, \ldots, z_k), \quad j = 1, \ldots, k, \tag{18}$$

and in this case any power of the Jacobian, that is to say the quantity

$$\left[\frac{\partial(z_1^a, \ldots, z_k^a)}{\partial(z_1, \ldots, z_k)}\right]^W, \tag{19}$$

for any constant W, is an admissible factor $\eta_a(z)$, and in ordinary parlance the concept of an automorphic function is restricted to such a situation only. However, such a restriction is unsatisfactory even within the frame-work of nineteenth century analysis. In fact, if $\tilde{V}$ is the entire E_{2k} and Γ consists of ordinary translations, so that V is a multi-torus, then (19) has the constant value 1 and is useless as a "factor", and other factors must be introduced. Such a factor was indeed introduced by Jacobi, and it is the oldest one on record, apparently. It has the form $e^{2\pi i P(z,a)}$, where $P(z, a)$ is linear in the space variables z, and quadratic in the group parameters describing "a", and the automorphic functions (in our general sense) pertaining to such are the Jacobi functions so-called ("fonctions intermédiares"). Jacobi's selection of his factor has been fully vindicated by P. Appell [1], who showed that in case of "multi-periodicity" the most general factor (16) with only property (17) to start with is, in substance, reducible to one of Jacobi's description ; and this seems to be the only case known in which the factor (16) in its generality has been "reduced" to a particular case of itself, by which to analyze and control it.

THEOREM 3. *If S, S^0 are as in Theorem 1; if F is a module of holomorphic functions on S, each bounded; and if for all elements $f(P)$ of F we have*

$$\sup_{P \in S}|f(P)| \leqslant M. \sup_{P \in S^0}|f(P)|, \tag{20}$$

for some fixed M; then for some suitable integer N (which, incidentally, depends only on S, S^0, M), any $N+1$ elements of F are linearly dependent. That is to say, the module has a finite basis of at most N elements.

PROOF. As in the proof of Theorem 1, if the theorem were false we would have an element $f(P)$ whose expansion at the centres of (6) would begin with terms of order $\geqslant p$, where p is any fixed number for which

$$m\binom{p+k-1}{k} < N.$$

If we put $\sup_{P \in S}|f(P)| = 1$, then at a suitable point of $\overline{S^0}$ we would have

$$\frac{1}{M} \leqslant |f(P)| \leqslant \frac{1}{2^p},$$

which however for large p, and hence for large N, is contradictory.

This now leads to the following conclusion.

THEOREM 4. *If V, $\tilde{V}$, $\eta_a(P)$ are as in Theorem 2, then the module of automorphic functions has a finite basis.*

In a previous paper [2] we have established theorems on the existence of finite bases in certain modules of scalars, tensors, tensor densities, and such-like objects on compact spaces, but we note that the conclusion of our present Theorem 4 is contained in the latter paper only for the particular case in which $\tilde{V}$ is a domain and the factor is of the special form (19). Thus Theorem 4, and Theorem 3, are new relative to the statements previously made. We note however that the method of the previous paper will also immediately establish the following theorem, which in its turn is vastly more general than our present Theorem 3 itself.

THEOREM 5. *If S_n is a (real) separable co-ordinate space, and S^0 a domain in it with compact closure $\overline{S^0}$, and if F is a module of continuous functions on S_n (over real or complex numbers), then in order that F have a finite basis, it is sufficient to know that the following two assumptions are satisfied :* (i) *for a suitable constant $M > 1$ we have*

$$\sup_{P \in S} |f(P)| \leqslant M . \sup_{P \in S^0} |f(P)|$$

for all $f \in F$, and (ii) *with each point P of S there is associable a neighbourhood U, and a sub-neighbourhood $U' \subset U$, such that any set of elements in F which are equally bounded in U are equally uniformly continuous in U'.*

As in the previous paper, we can also generalize Theorem 5 from scalars f to arbitrary tensor densities of any description, and we are not stating this generalization because we have no fruitful interpretation for it, for the present. But we would like to make the observation that the two theorems on automorphic functions, that sufficiently many of them are linearly dependent and that $k+2$ of them are polynomially dependent,—are very different propositions, first impressions notwithstanding, and that further progress in the theory may be expected to come more from a study of the linear dependences between objects other than scalars, rather than from a further study of the algebraic dependences between the scalars themselves.

REFERENCES

1. P. APPELL : Sur les fonctions périodiques de deux variables, *Journal de Mathematiques Pures et Appliquées*, (4) 7 (1891), 157-219.

2. S. BOCHNER : Tensor fields with finite bases, *Annals of Math.* (to be published).

3. S. BOCHNER and W. T. MARTIN : *Several Complex Variables*, Princeton (1948), 59.

Princeton University,
Princeton N. J., U.S.A.

HARTOGS' THEOREM IN COMPLEX SPACES WITH SINGULARITIES

By S. BOCHNER and W. T. MARTIN

[Received July 11, 1952]

1. Introduction. Hartogs [3] has proved the following theorem distinctive of analytic functions in several complex variables.

THEOREM A. *If a function is defined in a domain of* E_{2k}, *and if it is analytic in each variable separately, then it is analytic in all variables.*

Note that, *ab initio,* absolutely no assumption is made on the function, as a function in all its variables jointly.

A closely related result is the following theorem [3,1]. Consider two sets of complex variables $(z_1,\dots,z_k)$ and $(w_1,\dots,w_l)$, and two domains D and $\tilde{D}$ in (z, w)-space, each given as a union of intersections relative to one set of variables,

$$D: \quad [z \in R, w \in H_z], \tag{1}$$

$$\tilde{D}: \quad [z \in R, w \in \tilde{H}_z], \tag{2}$$

where R is a domain in z-space, and where every connected component of $\tilde{H}_z$ contains a non-empty component of H_z. The theorem to which we are referring is the following.

THEOREM B. *If* $\phi(z,w)$ *is analytic in* (z, w) *in* D *and, for each* z *in* R, *is analytic in* w *in* $\tilde{H}_z$, *then* $\phi(z, w)$ *is analytic in* (z, w) *in* $\tilde{D}$.

This theorem obviously generalizes immediately from pairs of domains $D \subset \tilde{D}$ in the product of $E_{2k}(z)$ with $E_{2l}(w)$, to given pairs in the product of any two complex spaces $S^{2k}(z)$, $S^{2l}(w)$.

Now, in a recent paper [2] we have considered holomorphic functions on complex spaces in whose space structure certain singularities (that is, deviations from strictly Euclidean character) were allowed to occur ; and in the present paper the two theorems stated above will be carried over to products of such spaces as well.

2. Holomorphy in each variable separately. For our first result we consider two topological spaces U and V, each having the character of the space U defined in [2, Sec. 3]. We shall prove the following result.

THEOREM 1. (i) *Let* $F(P,Q)$ *be defined for* $[P\in U, Q\in V]$. *If* F *is holomorphic in each variable separately, then it is holomorphic in* (P, Q) *in* (U, V).

(ii) *The word holomorphic in* (i) *may be replaced throughout by the words strictly holomorphic.*

In the hypothesis of (i), when we say that F is holomorphic in each variable separately we mean that for each point P_0 in U, $F(P_0, Q)$ is holomorphic in V, and for each point Q_0 in V, $F(P, Q_0)$ is holomorphic in U.

Now as in [2], the space U is defined in terms of a base region R in ordinary z-space, and n polynomial relations

$$\psi_\nu(z, \zeta_1,\ldots, \zeta_\nu) = 0, \qquad \nu = 1,\ldots, n, \tag{3}$$

in $\zeta_1,\ldots, \zeta_n$, with coefficients which are analytic in R. In a similar manner, the space V is defined in terms of a base region R' in w-space, and p polynomial relations

$$\psi'_\nu(w, \xi_1,\ldots, \xi_\nu) = 0, \qquad \nu = 1,\ldots, p, \tag{4}$$

in $\xi_1,\ldots, \xi_p$ with coefficients analytic in R'. We denote by m the product $m = m_1\ldots m_n$, where m_ν is the degree of the polynomial ψ_ν in the variable ζ_ν, and similarly we denote by m' the product $m'_1\ldots m'_p$. As in [2] the square $\Delta^2(z)$ of the determinant of the matrix

$$\left.\begin{matrix}\left[\{[\zeta_1{}^{(\mu)}(z)]^{\lambda_1}\cdots[\zeta_n{}^{(\mu)}(z)]^{\lambda_n}\}\right] \\ \mu=1,\ldots, m; \lambda_1=0,1,\ldots, m_1-1;\ \ldots; \lambda_n=0,\ldots, m_n-1\end{matrix}\right\}, \tag{5}$$

is analytic and not identically zero in R. We denote by E the subset of R where $\Delta^2(z)$ vanishes. The function $\Delta'^2(w)$ is defined in a similar manner relative to the space V.

For the proof of (i) we first introduce the functions

$$f_{\mu,\mu'}(z,w) = F(z, \zeta_1^{(\mu)}(z),\ldots,\zeta_n^{(\mu)}(z); w, \xi_1^{(\mu')}(w),\ldots, \xi_p^{(\mu')}(w)) \quad (6a)$$

and then we form the product

$$\prod_{\mu=1}^{m}\prod_{\mu'=1}^{m'}[f-f_{\mu,\mu'}(z,w)] = f^{mm'} + B_1(z,w)\, f^{mm'-1} + \ldots + B_{mm'}(z,w) \quad (6b)$$

Now let w^0 be any point in R'. Since $F(P,Q)$ is holomorphic in P for Q fixed in V, the symmetric functions $B_1(z,w^0),\ldots, B_{mm'}(z,w^0)$ are single-valued and analytic in z in $R-E$, and bounded in every compact set T in R. Since the set E is a removable exceptional set, see [1, p. 173], the functions $B_j(z,w^0)$, $j=1,\ldots,mm'$, are analytic in z in R. Similarly for each point z^0 in R, the functions $B_j(z^0,w)$ are analytic in w in R'. By Hartogs' own theorem, (Theorem A), the B's are analytic in (z,w) in (R,R'). This means that they are bounded in (T,T') for every pair of compact sets

$$T \subset R, \quad T' \subset R'.$$

Hence the mm' functions (6a) are bounded in (T,T') and thus our function $F(P,Q)$ is bounded for

$$P \in U(T), \qquad Q \in V(T). \quad (7)$$

Having secured boundedness, the further reasoning is routine. We define

$$P_{\lambda_1,\ldots,\lambda_n,\lambda_1',\ldots,\lambda_n'}(z,w) = \Delta_{\lambda_1,\ldots,\lambda_n,\lambda_1',\ldots,\lambda_n'}(z,w)\,\Delta(z,w), \quad (8)$$

where $\Delta(z,w)$ is the determinant of the $(mm')\times(mm')$ matrix

$$\left[\{[\zeta_1^{(\mu)}(z)]^{\lambda_1}\ldots[\zeta_n^{(\mu)}(z)]^{\lambda_n}[\xi_1^{(\mu')}(w)]^{\lambda_1'}\ldots[\xi_p^{(\mu')}(w)]^{\lambda_p'}\}\right], \quad (9)$$

for

$$\lambda_1 = 0,1,\ldots,m_1-1;\ldots;\lambda_n = 0,1,\ldots;m_n-1,$$
$$\lambda_1' = 0,1,\ldots,m_1'-1;\ldots;\lambda_p' = 0,1,\ldots;m_p'-1, \quad (10)$$

and

$$\mu = 1, \ldots, m, \qquad \mu' = 1, \ldots, m', \tag{11}$$

and where $\Delta_{\lambda_1,\ldots,\lambda_n,\lambda_1',\ldots,\lambda_n'}(z,w)$ is the determinant of the matrix obtained from the matrix (9) by replacing the elements of the column, one element of which is displayed in (9), by $f_{\mu,\mu'}(z,w)$ for μ, μ' ranging over (11). By symmetry and holomorphy of F in each variable separately we conclude as in the case of the B's that each of the functions (8) is analytic in (z, w) in (R, R').

By (6a) and (8) we see that

$$\Delta^2(z,w)\, F(z, \zeta; w, \xi) = \sum_{\lambda, \lambda'} P_{\lambda_1,\ldots,\lambda_n,\lambda_1',\ldots,\lambda_n'}(z, w)\, \zeta_1^{\lambda_1} \ldots \zeta_n^{\lambda_n}\, \xi_1^{\lambda'_1} \ldots \xi_p^{\lambda'_p} \tag{12}$$

in (U, V) where the sum is taken over (10).

We observe that $\Delta^2(z, w)$ is analytic and not identically zero in (R, R'), being given as the product

$$\Delta^2(z, w) = [\Delta(z)]^{2m'}\, [\Delta'(w)]^{2m}. \tag{13}$$

Thus the relation (12), together with the boundedness of F in (7), yields the desired conclusion of (i).

3. Strict holomorphy in each variable. For the proof of (ii) we will utilise the boundedness of F in $U(T)$, $V(T')$ and the relation (12) already obtained. By the strict holomorphy of $F(P, Q)$ in Q for P fixed we also have

$$F(z, \zeta; w, \xi) = \sum_{\lambda'} A_{\lambda'}(z, \zeta; w)\, \xi_1^{\lambda'_1} \ldots \xi_p^{\lambda'_p}, \tag{14}$$

in (U, V), where the A's are analytic in w in R' for each fixed point (z, ζ) in U. The relation (14) and the boundedness of F in (7) tells us that the functions

$$\Delta'^2(w)\, A_{\lambda'}(z, \zeta; w) \tag{15}$$

are bounded for

$$[(z, \zeta) \in U(T),\ w \in T']. \tag{16}$$

Since $\Delta'^2(w)$ and the A's are analytic in w, by Lemma 4.4 of [2] we conclude that the functions

$$A_{\lambda'}(z, \zeta; w) \tag{17}$$

themselves are bounded in (16) even though $\Delta'(w)$ may have zeros in R'. We will use this fact a little later.

Next, a comparison of (12) and (14) gives

$$\Delta^2(z, w)\, A_{\lambda'}(z, \zeta; w) = \sum_{\lambda} P_{\lambda, \lambda'}(z, w)\, \zeta_1^{\lambda_1} \dots \zeta_n^{\lambda_n}, \tag{18}$$

for $[(z, \zeta) \in U, w \in R']$. By solving for $P_{\lambda, \lambda'}$, we now obtain

$$\Delta^2(z)\, P_{\lambda, \lambda'}(z, w) = \Delta^2(z, w)\, A_{\lambda, \lambda'}(z, w)\, \Delta(z), \tag{19}$$

where $\Delta(z)$ is the determinant of the $m \times m$ matrix (5) and $A_{\lambda, \lambda'}(z, w)$ is the determinant of the matrix obtained from (5) by replacing the elements of the column, one element of which is displayed in (5), by

$$A_{\lambda'}[z, \zeta_1^{(\mu)}(z), \dots, \zeta_n^{(\mu)}(z); w], \tag{20}$$

for $\mu = 1, \dots, m$. By symmetry, boundedness of (17) in (16), and Hartogs' theorem, we conclude that the functions

$$Q_{\lambda, \lambda'}(z, w) = A_{\lambda, \lambda'}(z, w)\, \Delta(z), \tag{21}$$

are analytic in (z, w) in (R, R'). Thus by (19)

$$\Delta^2(z)\, P_{\lambda, \lambda'}(z, w) = \Delta^2(z, w)\, Q_{\lambda, \lambda'}(z, w), \tag{22}$$

where $Q_{\lambda, \lambda'}(z, w)$ is analytic in (R, R'). This means that we can replace the representation (12) by

$$\Delta^2(z)\, F(z, \zeta; w, \xi) = \sum_{\lambda, \lambda'} Q_{\lambda, \lambda'}(z, w)\, \zeta_1^{\lambda_1} \dots \zeta_n^{\lambda_n}\, \xi_1^{\lambda'_1} \dots \xi_p^{\lambda'_p}. \tag{23}$$

Similarly, on interchanging the roles of z and w, we obtain a representation (23′) with $\Delta^2(z)$ replaced by $\Delta'^2(w)$ and $Q_{\lambda, \lambda'}(z, w)$ by $R_{\lambda, \lambda'}(z, w)$, where the R's are analytic in (R, R'). On comparing (23) and (23′) we obtain the equalities

$$\Delta'^2(w)\, Q_{\lambda, \lambda'}(z, w) = \Delta^2(z)\, R_{\lambda, \lambda'}(z, w), \tag{24}$$

throughout (R, R').

We now use the following known lemma on analytic functions of several variables, from the local theory of ideals.

LEMMA A. *If $\Phi(z, w)$ and $\Psi(z, w)$ are analytic for $z \in R$, $w \in R'$, if $\phi(w)$ is analytic and $\not\equiv 0$ in R, if $\psi(z)$ is analytic and $\not\equiv 0$ in R', and if*

$$\phi(w)\, \Phi(z, w) = \psi(z)\, \Psi(z, w) \qquad \textit{in } (R, R'),$$

then there is an analytic function $\Psi_1(z, w)$ *in* (R, R') *such that*

$$\Phi(z, w) = \psi(z)\,\Psi_1(z, w), \qquad \textit{in } (R, R').$$

In other words, if $\psi(z)$ *is a factor of* $\phi(w)\,\Phi(z, w)$, *then it is a factor of* $\Phi(z, w)$ *itself.*

By this lemma, relation (24) tells us that there are functions

$$C_{\lambda,\lambda'}(z, w)$$

analytic in (R,R') such that

$$Q_{\lambda,\lambda'}(z, w) = \Delta^2(z)\,C_{\lambda,\lambda'}(z, w).$$

Thus we may replace the representation (23) by

$$F(z, \zeta; w, \xi) = \sum_{\lambda,\lambda'} C_{\lambda,\lambda'}(z, w)\,\zeta_1^{\lambda_1}\cdots\zeta_n^{\lambda_n}\,\xi_1^{\lambda_1'}\cdots\zeta_p^{\lambda_p'}. \tag{25}$$

This means that F is strictly holomorphic in (R, R'), and thus the proof of part (ii) of Theorem 1 is complete.

4. Holomorphic continuation. We now consider two domains (1) and (2) in (z, w)-space. Let R' be a domain in w-space which contains the union of the sets $\tilde{H}_z$,

$$\sum_{z \in R} \tilde{H}_z \subset R', \tag{26}$$

and as in Section 2 consider the two spaces U and V defined in terms of the polynomial relations (3) and (4) respectively for z in R and w in R'.

For z fixed in R we denote by V_z the subset of V for which $w \in H_z$, and by $\tilde{V}_z$ the subset of V for which $w \in H_z$. If P is any point (z, ζ) in U, we will also denote V_z by V_P and $\tilde{V}_z$ by $\tilde{V}_P$.

We shall prove the following theorem.

THEOREM 2. (i) *If* $F(P, Q)$ *is holomorphic in* (P, Q) *in*

$$[P \in U,\ Q \in V_P], \tag{27}$$

and for each P *in* U *is holomorphic in* Q *in* $\tilde{V}_P$, *then* F *is holomorphic in* (P, Q) *in*

$$[P \in U,\ Q \in \tilde{V}_P]. \tag{28}$$

(ii) *The word holomorphic in* (i) *may be replaced throughout by the words strictly holomorphic.*

We will first prove part (ii), that is the strictly holomorphic version. By hypothesis we have a representation (25) in (27), where the C's are analytic in (1), and we have a representation (14) in (28), where the A's are analytic in w in $\tilde{H}_z$ for each fixed point (z, ζ) in U. A comparison of these two representations yields the relations

$$A_{\lambda'}(z, \zeta; w) = \sum_{\lambda} C_{\lambda,\lambda'}(z, w)\, \zeta_1^{\lambda_1} \dots \zeta_n^{\lambda_n}, \tag{29}$$

valid for $[(z, \zeta) \in U,\ w \in H_z]$. We may solve for the C's in terms of the A's, obtaining

$$\Delta^2(z)\, C_{\lambda,\lambda'}(z, w) = A_{\lambda,\lambda'}(z, w)\, \Delta(z) = Q_{\lambda,\lambda'}(z, w), \tag{30}$$

where the $A_{\lambda,\lambda'}$ and $Q_{\lambda,\lambda'}$ are defined as in (19) and (21).

Now the left-hand side of (30) is analytic in (z, w) in (1), while the right hand side is analytic in w in $\tilde{H}_z$ for each fixed z in R. Thus by Theorem B, the function $Q_{\lambda,\lambda'}(z, w)$ is analytic in (z, w) in (2). The desired result on the strict holomorphy of F in (28) will follow from a lemma on analytic functions which we next state and prove.

Lemma 1. *If $C(z, w)$ is analytic in* (1), *if $Q(z, w)$ is analytic in* (2), *if $\phi(z)$ is analytic and not identically zero in R and if*

$$\phi(z)\, C(z, w) = Q(z, w)$$

holds in (1), *then $C(z, w)$ has an analytic continuation into* (2).

For the proof it is sufficient to establish the conclusion when H_z and $\tilde{H}_z$ are multi-cylinders independent of z,

$$H_z : |w_j| < p_j;\ \tilde{H}_z : |w_j| < P_j,\ j = 1, \dots, l,$$

where $0 < p_j < P_j$.

We develop $C(z, w)$ in a series

$$C(z, w) = \sum_{n_1,\dots,n_l=0}^{\infty} c_{n_1,\dots,n_l}(z) w_1^{n_1} \dots w_l^{n_l}. \tag{31}$$

The coefficients $c_{n_1,\dots,n_l}(z)$ are analytic in R and the series (31) is absolutely-continuously convergent in $[z \in R, |w| < p_j]$. On the other hand the series

$$\sum \phi(z).\ c_{n_1,\dots,n_l}(z) w_1^{n_1} \dots w_l^{n_l} \tag{32}$$

is absolutely-continuously convergent in $[z \in R, |w_j| < P_j]$. From this latter fact, if T is any compact set in R and if $0 < P'_j < P_j$, we see that there is a positive number $M = M(T, P'_1, \dots, P'_l)$ such that

$$|\phi(z) c_{n_1,\dots,n_l}(z) P'^{n_l}_1 \dots P'^{n}| \leqslant M, \text{ for } z \in T, n_1, \dots, n_l = 0, 1, \dots. \quad (33)$$

Since ϕ and the c's are analytic in R, we conclude from (33) and Lemma 4.4 of [2] that there is a constant M' such that

$$c_{n_1,\dots,n_l}(z) P'^{n_l}_1 \dots P'^{n_l}_l| \leqslant M', \text{ for } z \in T, n_1, \dots, n_l = 0, 1, 2, \dots. \quad (34)$$

Thus the series (31) converges absolutely-continuously in $[z \in R, |w_j| < P'_j]$ and since P' is any number less than P_j, the same is true if $[z \in R, |w_j| < P_j]$. This concludes the proof of the lemma.

Applying the lemma to the functions C in (30) we see that the $C_{\lambda, \lambda'}$ are analytic in (2) and hence F is strictly holomorphic in (28). This yields part (ii) of Theorem 2.

For part (i) of the theorem, we first note that since F is holomorphic in (27) the function

$$\Delta^2(z, w)\ F(z, \zeta; w, \xi) \quad (35)$$

is strictly holomorphic there. On the other hand, from the holomorphy of $F(P, Q)$ for Q in V_p for P fixed in U we know that

$$\Delta'^2(w)\ F(z, \zeta; w, \xi)$$

is holomorphic for (w, ξ) in $\tilde{V}_z$ for (z, ζ) fixed in U. By the relation (13) this latter fact tells us that (35) is strictly holomorphic for (w, ξ) in $\tilde{V}_z$ for (z, ζ) fixed in U. Hence part (ii) of Theorem 2 applies to the function (35), and we conclude that (35) is strictly holomorphic in (28). If in addition we can show that F itself is bounded in every compact subset of (28) then we will have F holomorphic in (28).

In order to prove boundedness of F we define the functions (6a) and the symmetric functions $B_1(z, w), \dots, B_{mm'}(z, w)$ in (6b). By the holomorphy of F in (P, Q) in (27) the B's are analytic in (z, w) in (1), and by the holomorphy of F in Q in $\tilde{V}_P$

for P fixed in U, the B's are analytic in w in $\tilde{H}_z$ for z fixed in R. Hence by Theorem B they are analytic in (z, w) in (2). Thus $F(z, \zeta; w, \xi)$ is bounded whenever (z, w) belongs to any compact subset of (2). This, together with the strict holomorphy of (35) in (28) yields the holomorphy of F in (28).

This concludes the proof of Theorem 2.

We next give a global version of Theorem 2.

First we introduce the concept of a *full-part* of a space V of the character considered in Section 2. A subset W of V is called a *full-part* of V if it consists of all points $[w, \xi_1^{(\mu)}(w), \ldots, \xi_p^{(\mu)}(w)]$ for w in an open set of the base region R' and $\mu = 1, \ldots, m'$.

Next, let $S_{2k}^{(w)}$ be a complex space with singularities, defined as in [2, Sec. 10], with a covering by open sets, $\{V_\sigma\}$, each having the character of the space V. We say that $S_{2k}^{(w)}$ is *fully-connected* if to every V_0 in $\{V_\sigma\}$ and every point Q in $S_{2k}^{(w)}$ there exists a finite chain, $V_1, \ldots, V_r$, with $Q \in V_r$, such that the intersection $V_{\tau+1} \cap V_\tau$ contains a full-part of $V_{\tau+1}$, $\tau = 1, \ldots, r-1$.

As a corollary to Theorem 2 we have the following result.

COROLLARY : (i) *Let $S_{2k}^{(w)}$ be a fully-connected complex space with singularities, and let V_0 be any element of the covering $\{V_\sigma\}$. If $F(P, Q)$ is holomorphic in (P, Q) in*

$$P \in U, \; Q \in V_0$$

and holomorphic in Q in $S_{2k}^{(w)}$ for each fixed point P in U, then F is holomorphic in (P, Q) in

$$P \in U, \; Q \in S_{2k}^{(w)}.$$

(ii) *In* (i) *the word holomorphic may be replaced throughout by the words strictly holomorphic.*

Acknowledgement. This paper was written while the second-named author was in residence at the Institute for Advanced Study, Princeton, New Jersey.

REFERENCES

1. S. BOCHNER and W. T. MARTIN: *Several complex variables*, Princeton (1948).

2. S. BOCHNER and W. T. MARTIN: Complex spaces with singularities, *Annals of Math.* (to be published).

3. F. HARTOGS: Zur Theorie der analytischen Funktionen mehrerer unabhangiger Veränderlichen, insbesondere über die Darstellung derselben durch Reihen, welche nach Potenzen einer Veränderlichen fortschreiten, *Math. Annalen*, 62 (1906), 1-88.

Princeton University
Princeton, New Jersey

Massachusetts Institute of Technology
Cambridge, Massachusetts

The theorem of Morera in several variables.

Memoria di SALOMON BOCHNER (a Princeton, N. J., U. S. A.).

Sunto. - *È dato dal seguente capoverso.*

The classical theorem of MORERA has been generalized from one to several variables by VOLTERRA and SEVERI (1), and SEVERI's proof is very brief and elegant. We will re-prove this theorem (our Theorem 7) by a chain of reasoning which, to the contrary, will be lengthy and elaborate, but our actual contributions are the intermediary results and the method of proof, rather than the final proposition itself.

In the Euclidean E_{2k} of k complex variables $z^j = x^j + iy^j$, $j = 1, \dots, k$, we will distinguish between functions which have power series in $z^j - z_0^j$ locally; and such ones which have power series in $x^j - x_0^j$, $y^j - y_0^j$, or, what is the same, in $z^j - z_0^j$, $\bar{z}^j - \bar{z}_0^j$ locally. The first ones we will call « holomorphic »; and the second ones we will term « real-analytic », even if their values themselves will be complex numbers, as a rule. Also, if a function has power series in $\bar{z}^j - \bar{z}_0^j$ only, we will call it « anti-holomorphic ».

1. General Cycles.

In the complex (x, y)-plane we formally introduce the variables

$$z = x + iy, \qquad \bar{z} = x - iy$$

and we denote an arbitrary function $F(x, y)$ by $f(z, \bar{z})$ in accordance with a well-known symbolism (2). The classical theorem of MORERA then states that if such a function is defined and continuous in a domain D and if we then have

$$\int_{B_1} f dz = 0$$

(1) F. SEVERI, *Sur une propriété fondamentale des fonctions analytiques de plusieurs variables complexes*, Comptes Rendus, Paris, 192, (1931).

(2) Compare, for instance, S. BOCHNER and W. T. MARTIN, *Several complex variables*, (1948), p. 37.

over every 1-cycle B_1, then f is holomorphic:

$$\frac{\partial f}{\partial \bar{z}} = 0,$$

and, of course, dually, if we have

$$\int_{B_1} f d\bar{z} = 0$$

then f is anti-holomorphic

$$\frac{\partial f}{\partial z} = 0.$$

Now, perhaps the most natural generalization of this theorem is as follows.

THEOREM 1 [3]. – *In the space E_{2k} of the k complex variables $z^\alpha = x^\alpha + iy^\alpha$, $\alpha = 1, \dots, k$, if*

$$f = f(z^1, \dots, z^k, \bar{z}^1, \dots, \bar{z}^k) \tag{1}$$

is continuous in a domain D, and if we then have

$$\int_{B_{p+q}} f dz^{\alpha_1} \dots dz^{\alpha_p} d\bar{z}^{\beta_1} \dots d\bar{z}^{\beta_q} = 0 \tag{2}$$

over every $(p+q)$-dimensional cycle B_{p+q}, where

$$1 \leq \alpha_1 < \alpha_2 < \dots < \alpha_p \leq k, \qquad 1 \leq \beta_1 < \beta_2 < \dots < \beta_q \leq k \tag{3}$$

are fixed indices with

$$0 \leq p, \qquad 0 \leq q, \qquad 1 \leq p + q \leq 2k - 1, \tag{4}$$

then f is holomorphic in those z^β, which are different from the given ones, and anti-holomorphic in those z^α which are different from the given ones; that is

$$\frac{\partial f}{\partial \bar{z}^\alpha} = 0, \qquad \alpha \neq \alpha_1, \dots, \alpha_p \tag{5}$$

$$\frac{\partial f}{\partial z^\beta} = 0, \qquad \beta \neq \beta_1, \dots, \beta_q. \tag{6}$$

---

(3) For the propositions of which Theorem 1 is the converse see B. SEGRE, *Sull'estensione della formula integrale di Cauchy e sui residui degli integrali n-pli nella teoria delle funzioni di n variabili complesse*, « Atti I Congresso U. M. I. », Bologna, (1937), and E. MARTINELLI, *Formule Integrali e topologia nella teoria delle funzioni di più variabili complesse*, « Acta Pontif Acad. Scient. », IX, p. 235-250.

Proof. We first note that, for instance.

$$\int_{B_k} f dz^1 \dots dz^k = 0 \tag{7}$$

implies that f is holomorphic in all z^α, that is

$$\frac{\partial f}{\partial \bar{z}^\beta} = 0, \qquad \beta = 1, \dots, k;$$

and that, for instance, for $k \geq 2$,

$$\int_{B_2} f dz^1 d\bar{z}^1 = 0$$

implies that f is « arbitrary » continuous in z^1, $\bar{z}^1$, that is in x^1, y^1 but that it is constant in all the other variables, that is

$$\frac{\partial f}{\partial z^\gamma} = 0 = \frac{\partial f}{\partial \bar{z}^\gamma}, \qquad \gamma = 2, \dots, k,$$

and hence

$$\frac{\partial F}{\partial x^\gamma} = 0 = \frac{\partial F}{\partial y^\gamma} = 0, \qquad \gamma = 2, \dots, k.$$

For the proof we note that if f has also continuous first derivatives in all x^γ, y^γ, then the theorem is a direct consequence of STOKES' theorem. In fact, if B_{p+q} bounds a chain B_{p+q+1} then (2) implies

$$\int_{B_{p+q+1}} df dz^{\alpha_1} \dots dz^{\alpha_p} d\bar{z}^{\beta_1} \dots d\bar{z}^{\beta_q} = 0, \tag{8}$$

and since B_{p+q+1} can be an arbitrary simplex, this implies

$$df dz^{\alpha_1} \dots dz^{\alpha_p} d\bar{z}^{\beta_1} \dots d\bar{z}^{\beta_q} = 0, \tag{9}$$

where

$$df = \frac{\partial f}{\partial z^1} dz^1 + \dots + \frac{\partial f}{\partial z^k} dz^k + \frac{\partial f}{\partial \bar{z}_1} d\bar{z}^1 + \dots + \frac{\partial f}{\partial \bar{z}_k} d\bar{z}^k = 0.$$

However, for fixed α_ρ, $\rho = 1, \dots, p$; β_σ, $\sigma = 1, \dots, q$; and variable $\alpha \neq \alpha_\rho$, $\beta \neq \beta_\sigma$ the $(k - p) + (k - q)$ monomials

$$dz^\alpha dz^{\alpha_1} \dots dz^{\alpha_p} d\bar{z}^{\beta_1} \dots d\bar{z}^{\beta_q}, \qquad d\bar{z}^\beta dz^{\alpha} \dots dz^{\alpha_p} d\bar{z}^{\beta_1} \dots d\bar{z}^{\beta_q}$$

are all linearly independent, and this implies relation (5), (6).

Next, if it is only known that f is continuous, we will have to apply an approximation process and the underlying approximation theorem is as follows ([4]).

In (real) Euclidean $E_n: (t^1, \ldots, t^n)$ it is possible to construct a sequence of functions $\{\varphi_\mu(t^1, \ldots, t^n)\}$, $\mu = 1, 2, 3, \ldots$, having the following properties.

(i) Each $\varphi_\mu(t)$ is defined in all of E_n but vanishes outside a sphere

$$K_\mu : (t^1)^2 + \ldots + (t^n)^2 \leq \delta_\mu^2,$$

where

$$\lim_{\mu \to \infty} \delta_\mu = 0.$$

(ii) Each $\varphi_\mu(t) \varepsilon C^\infty$ in E_n, that is, it has continuous partial derivatives of every mixed order.

(iii) $\varphi_\mu(t) \geq 0$, and

(iv)
$$\int_{E_n} \varphi_\mu(t) dv_t = \int_{K_\mu} \varphi_\mu(t) dv_t = 1.$$

Now, if $F(x^1, \ldots, x^n)$ is defined and continuous and also bounded in a domain D, then we supplement it by values 0 outside D, and we set up the expressions

$$F_\mu(x^1, \ldots, x^n) = \int_{E_n} F(x^1 + t^1, \ldots, x^n + t^n) \varphi_\mu(t^1, \ldots, t^n) dv_t$$

$$= \int_{K_\mu} F(x^1 + t^1, \ldots, x^n + t^n) \varphi_\mu(t^1, \ldots, t^n) dv_t.$$

for $x = (x^\nu)$ in D; $\mu = 1, 2, \ldots$. These expressions are approximating functions in the following sense ([5]).

LEMMA 1. – *(i) Each $F_\mu(x) \varepsilon C^\infty$ in E_n, and if $D' \subset D$ is a subdomain whose closure is likewise contained in D, $\overline{D'} \subset D$, then*

$$\lim_{\mu \to \infty} F_\mu(x) = F(x)$$

uniformly in every D'.

(ii) If, moreover, $F \varepsilon C^r$ in D, $r \geq 1$, then we also have

$$\lim_{\mu \to \infty} \frac{\partial^{q_1 + \ldots + q_n} F_\mu}{\partial (x^1)^{q_1} \ldots \partial (x^n)^{q_n}} = \frac{\partial^{q_1 + \ldots + q_n} F}{\partial (x^1)^{q_1} \ldots \partial (x^n)^{q_n}}$$

for $0 \leq q_1 + \ldots + q_n \leq r$, again uniformly in D'.

([4]) S. BOCHNER and W. T. MARTIN, loc. cit., p. 109-13.

([5]) BOCHNER and MARTIN, loc. cit., p. 36.

Returning to our Theorem 1, if (2) holds for « all » cycles B_{p+q}, and if we pick a fixed such cycle, then it also holds for « translated » cycles $B_{p+q} - (t)$ for small $|t|$. However in the relation (2), instead of translating the cycle, by $-t$, we can also translate the argument of the function by $+t$. That is, if we denote the function (1) for the moment by

$$F(x^1, \dots, x^n) \qquad (n = 2k),$$

then (2) also holds for the function

$$F(x^1 + t^1, \dots, x^n + t^n)$$

for sufficiently small t. For given B_{p+q}, we can then multiply the resulting relation by $\varphi_\mu(t)$ and integrate over K_μ, for μ sufficiently large, and thus, we obtain

$$\int_{B_{p+q}} f_\mu(z, \bar{z}) dz^{\alpha_1} \dots dz^{\alpha_p} d\bar{z}^{\beta_1} \dots d\bar{z}^{\beta_q} = 0,$$

for $\mu \geq \mu(B_{p+q})$. But $f_\mu \in C^1$ and thus we have our conclusions

$$\frac{\partial f_\mu}{\partial z^\alpha} = 0, \qquad \alpha \neq \alpha_1, \dots, \alpha_p$$

$$\frac{\partial f_\mu}{\partial \bar{z}^\beta} = 0, \qquad \beta \neq \beta_1, \dots, \beta_q,$$

for $\mu \geq \mu(D')$, for every subdomain D' as before. However, in D', f_μ converges uniformly towards f; and by WEIERSTRASS' theorem on the uniform limit of holomorphic functions, which also holds for anti-holomorphic functions, we now obtain the relations (5), (6) themselves, q. e. d.

2. Rigidly-places products of one-cycles.

In the z^j-plane, $j = 1, \dots, k$, we take a one-dimensional cycle B_1^j which bounds a two-chain B_2^j and we consider the products

(10) $$B_{2k} = B_2^1 \times B_2^2 \times \dots \times B_2^k$$

and

(11) $$B_k = B_1^1 \times B_1^2 \times \dots \times B_1^k.$$

For a given domain D, if (10) is contained in D, then we will say that (11) is a rigidly-placed products of one-cycles.

A « repeated » application of STOKES' theorem gives formally

$$\int_{B_k} f dz^1 \dots dz^k = \int_{B_{2k}} \frac{\partial^k f}{\partial \bar{z}^1 \dots \partial \bar{z}^k} d\bar{z}^1 dz^1 d\bar{z}^2 dz^2 \dots d\bar{z}^k dz^k$$

$$= (2i)^k \int_{B_{2k}} \frac{\partial^k f}{\partial \bar{z}^1 \dots \partial \bar{z}^k} d V_{x,y}$$

where

$$dV_{x,y} = dx^1 dy^1 dx^2 dy^2 \ldots dx^k dy^k$$

is the ordinary volume-element, and, again with the aid of lemma 1, the following conclusion can be drawn.

THEOREM 2. - *If* $f \in C^k$, *and*

$$\frac{\partial^k f}{\partial \bar{z}^1 \ldots \partial \bar{z}^k} = 0, \tag{12}$$

then we have

$$\int_{B_k} f dz^1 \ldots dz^k = 0 \tag{13}$$

for every rigidly-placed product (11).

Conversely, if (13) *holds for all such cycles in D, and if* $f \in C^k$, *then f satisfies the equation* (12), *and if* $f \in C^0$, *then f is a uniform limit in D′ of solutions of* (12).

If a function (1) is real-analytic and if we expand it in a power-series of z^α, $\bar{z}^\alpha$ in the neighborhood of the origin, say, then it is a solution of (12) if and only if the expansion has the form

$$\sum_{j=1}^{k} \sum_{(n)} f_{n_1 \ldots n_{j-1} n_{j+1} \ldots n_k}(z)(\bar{z}^1)^{n_1} \ldots (\bar{z}^k)^{n_k} \tag{14}$$

in which the coefficients f_n are holomorphic in their variables. But we do not know whether any solution of (12) must automatically be real-analytic, and thus no absolute assertion can be formulated.

3. **Affinely-placed products of one-cycles.**

We will say that a k-cycle B_k in D is an affinely-placed product of one-cycles if there is a non-singular holomorphic transformation with constant coefficients

$$z^j = \sum_{l=1}^{k} a_{jl} w^l + a_j, \tag{15}$$

such that the image of B_k in the w-space is a rigidly-placed product of one-cycles as previously defined.

THEOREM 3. - *If a function has the form*

$$f = \sum_{0 \le p_1 + \ldots + p_k \le k-1} a_{p_1 \ldots p_k}(z)(\bar{z}^1)^{p_1} \ldots (\bar{z}^k)^{p_k} \tag{16}$$

with holomorphic coefficients

$$a_p(z) = a_{p_1 \ldots p_k}(z^1, \ldots, z^k), \tag{17}$$

then we obviously have

$$\frac{\partial^k f}{\partial \bar{z}^{\alpha_1} \dots \partial \bar{z}^{\alpha_k}} = 0, \tag{18}$$

for all possible combinations

$$1 \leq \alpha_1 \leq k, \qquad 1 \leq \alpha_2 \leq k, \dots, \qquad 1 \leq \alpha_k \leq k, \tag{19}$$

and we have

$$\int_{B_k} f dz^1 dz^2 \dots dz^k = 0$$

for every affinely placed B_k.

Proof. If we introduce the linear operators

$$L_j = \sum_{l=1}^{k} \bar{a}_{lj} \frac{\partial}{\partial \bar{z}^l} \tag{20}$$

with any constant coefficients a_{lj} then (18) implies

$$L_1 L_2 \dots L_k f = 0, \tag{20'}$$

and this means that we have property (12) even after any affine transformation (15). And the conclusion now follows from Theorem 2.

THEOREM 4. - *If f continuous in D_1 and if we have*

$$\int_{B_k} f dz^1 \dots dz^k = 0 \tag{21}$$

for every affinely-placed B_k, then f is real-analytic, and we have (18), *and it can be represented as an expression* (16).

Proof. Assume first that it is already known that f is real-analytic. It is then in particular C^k, and (21) implies by Theorem 2 that we have (20') for all possible operators (20), and hence (18). But f being also real analytic, we can expand it in the neighborhood of every point z_0^α in a power-series in the quanties $z^\alpha - z_0^\alpha$, $\bar{z}^\alpha - \bar{z}_0^\alpha$, and due to (18) the expansion must be of the form

$$\sum_{0 \leq p_1 + \dots + p_k \leq k-1} a_p(z\,;\, z_0)(\bar{z}^1 - \bar{z}_0^1)^{p_1} \dots (\bar{z}^k - \bar{z}_0^k)^{p_k} \tag{22}$$

where $a_p(z\,;\, z_0)$ is holomorphic in the z^α in the neighborhood of the given point z_0^α. Now, (22) is a polynomial of degree $\leq k-1$ in the $\bar{z}^\alpha - \bar{z}_0^\alpha$, and as such it is also a polynomial in the $\bar{z}^\alpha$ themselves:

$$\sum_{0 \leq p_1 + \dots + p_k \leq k-1} b_p(z\,;\, z_0)(\bar{z}_1^1)^{p_1} \dots (\bar{z}^k)^{p_k}$$

with ceefficients $b_p(z\,;\,z_0)$ which are holomorphic in the z^α, in a neighborhood of z_0^α. But by a known uniqueness theorem these coefficients must be identical with one another in the intersection of two such neighborhoods, and hence there is a representation (16) jointly for the entire domain D.

But now we must eliminate the assumption that f is real-analytic, and we will perform this elimination in the following two steps, where the second step will again be based on our process of approximation as ensuing from lemma 1.

Step 1. If $f \varepsilon C^{k+1}$ and (18) holds for all combinations (19), then f is real-analytic.

Step 2. If we are given a sequence of function f_μ in D, if each f_μ has a representation (16) with holomorphic coefficients (17) and if the sequence is uniformly convergent,

$$\lim_{\mu \to \infty} f_\mu = f,$$

then f itsel can likewise be represented in the form (16).

Now Step 1 obviously follows from the following more general theorem.

THEOREM 5. - *If for some $l \geq 1$, $l \gtreqless k$, we are given a function $f \varepsilon C^{l+1}$ and if the expressions*

$$g_{\alpha_1 \dots \alpha_l}(z, \bar{z}) = \frac{\partial^l f}{\partial \bar{z}^{\alpha_1} \dots \partial \bar{z}^{\alpha_l}} \tag{23}$$

for all possible combinations $\alpha_1, \dots, \alpha_l$ are real-analytic (for instance, if they are all 0), then the function f itself is real-analytic.

Proof. We will need the following lemma.

LEMMA 2. - If we are given formal power series

$$g_\alpha(t) = \sum_{p_j \geq 0} a_{\alpha p_1 \dots p_k} t_1^{p_1} \dots t_k^{p_k}, \quad \alpha = 1, \dots, k, \tag{24}$$

in symbols $t_1, \dots, t_k$ with coefficients from an arbitrary vector field over rational numbers, and if the integrability conditions

$$\frac{\partial g_\alpha}{\partial t_\beta} = \frac{\partial g_\beta}{\partial t_\alpha} \tag{25}$$

are fulfilled formally, then the series without constant term

$$G(t) = \sum_{p_1 + \dots + p_k > 0} a_{p_1 \dots p_k} t_1^{p_1} \dots t_k^{p_k} \tag{26}$$

with the coefficients

$$a_{\ _1 \dots p_k} = \frac{\sum_\alpha a_{\alpha p_1 \dots (p_\alpha - 1) \dots p_k}}{p_1 + \dots + p_k} \tag{27}$$

is such that we have

$$\frac{\partial G}{\partial t_\alpha} = g_\alpha(t), \qquad (\alpha = 1, \dots, k). \tag{28}$$

Proof of Lemma 2 (6). If (26) is to satisfy (28) then on putting

$$H(\lambda_1, \ldots, \lambda_k; s) = G(\lambda_1 s, \ldots, \lambda_k s) \tag{29}$$

we must have

$$\begin{aligned} \frac{\partial H}{\partial s} &= \sum_\alpha \frac{\partial G}{\partial t_\alpha} \lambda_\alpha = \sum_\alpha g_\alpha \lambda_\alpha \\ &= \sum_{p_1+\ldots+p_k>0} \frac{1}{s}(p_1 + \ldots + p_k) a_{p_1 \ldots p_k} (\lambda_1 s)^{p_1} \ldots (\lambda_k s)^{p_k} \\ &= \sum_{p,\alpha} \lambda_\alpha a_{\alpha p_1, \ldots, p_k} (\lambda_1 s)^{p_1} \ldots (\lambda_k s)^{p_k} \end{aligned} \tag{30}$$

so that (27) is the only possible choice of coefficients to that purpose. Conversely, if we now set

$$w_\alpha(\lambda; s) = \frac{\partial H}{\partial \lambda_\alpha} - s h_\alpha$$

where

$$h_\alpha(\lambda; s) = g_\alpha(\lambda_1 s, \ldots, \lambda_k s)$$

then we obtain

$$\frac{\partial w_\alpha}{\partial s} = \frac{\partial}{\partial \lambda_\alpha}\left(\frac{\partial H}{\partial s}\right) - h_\alpha - s \sum_\beta \frac{\partial g_\alpha}{\partial t_\beta} \lambda_\beta$$

and by (30) this is

$$h_\alpha + s \sum_\beta \lambda_\beta \frac{\partial g_\beta}{\partial t_\alpha} - h_\alpha - s \sum_\beta \frac{\partial g_\alpha}{\partial t_\beta} \lambda_\beta = 0,$$

by (25). Therefore, if (25) holds, then $w_\alpha(\lambda; s)$ must be independent of the formal variable s. But $w_\alpha(\lambda; 0) = 0$, and therefore

$$\frac{\partial H}{\partial \lambda_\alpha} = s h_\alpha,$$

or what is the same

$$\frac{\partial G}{\partial t_\alpha} = g_\alpha,$$

as claimed.

Turning now to Theorem 5, it suffices to prove the theorem in the neighborhood of a point, and we will prove it by induction on l. For $l = 1$, we have $f \varepsilon C^2$, and (23) is

$$g_\alpha(z, \bar{z}) = \frac{\partial f}{\partial \bar{z}^\alpha}, \qquad (\alpha = 1, \ldots, k).$$

(6) Compare our paper: *Formal Lie groups*, « Annals of Math. », 47, (1946), 192-201.

Differentiating once more we obtain

$$\frac{\partial g_\alpha}{\partial \bar{z}^\beta} = \frac{\partial g_\beta}{\partial \bar{z}^\alpha},$$

and by lemma 2 there exists locally a real-analytic function $G(z\ \bar{z})$ such tha

$$\frac{\partial G}{\partial \bar{z}^\alpha} = g_\alpha(z, \bar{z}).$$

For the difference $\varphi = f - G$ we obtain

$$\frac{\partial \varphi}{\partial \bar{z}^\alpha} = 0,$$

thus φ is holomorphic, and in particular real-analytic, and $f = \varphi + G$ is likewise so.

For $l > 1$ if we put

$$h = \frac{\partial^{l-1} f}{\partial \bar{z}^{\alpha_1} \dots \partial \bar{z}^{\alpha_{l-1}}}$$

for any fixed combination $\alpha_1, \dots, \alpha_{l-1}$, then by assumption of theorem 5

$$\frac{\partial h}{\partial \bar{z}^\alpha} \equiv g_{\alpha_1 \dots \alpha_{l-1}\alpha}, \qquad (\alpha = 1, \dots, k),$$

is real-analytic, and by what we have just proven, the function h is real-analytic itself. Therefore the expressions

$$h_{\alpha_1 \dots \alpha_{l-1}} = \frac{\partial^{l-} f}{\partial \bar{z}^{\alpha_1} \dots \partial \bar{z}^{\alpha_{l-1}}}$$

are real-analytic and if the theorem was known to hold now for $l - 1$, then f is real-analytic, and thus theorem holds for l itself, as claimed.

We will make a remark to Theorem 5 for its own sake.

THEOREM 6. - *In Theorem 5, instead of assuming $f \in C^{l+1}$, it is also sufficient to assume $f \in C^l$ and to add the explicit assumption that the integrability condition*

$$\frac{\partial g_{\alpha_1 \dots \alpha_{l-1}\alpha}}{\partial \bar{z}^\beta} = \frac{\partial g_{\alpha_1 \dots \alpha_{l-1}\beta}}{\partial \bar{z}^\alpha} \tag{31}$$

shall be valid.

Now, (31) is certainly valid if $g_{\alpha_1 \dots \alpha_{l-1}\alpha_l} = 0$, as is especially the case for equations (18) with $l = k$, so that in step 1 it would have been sufficient to assume $f \in C^k$ instead of $f \in C^{k+1}$. But for our context itself this is of no difference.

Next, for the proof of step 2, it again suffices to deal with a neighborhood of a point, and if we expand each $a_p(z)$ in a power series, then it is not hard to see that the following lemma, of some interest in itself, is sufficient for the purpose. We will again in this lemma, as in lemma 2, place all indices as subscripts, instead of as superscripts.

LEMMA 3. - *In the domain*

$$0 \leq |z_j| \leq 2, \qquad (j = 1, \dots, k), \tag{32}$$

we consider the absolutely convergent expansion

$$f(z, \bar{z}) = \sum_{p,q} a_{p_1 \dots p_k; q_1 \dots q_k} z_1^{p_1} \dots z_k^{p_k} \bar{z}_1^{q_1} \dots \bar{z}_k^{q_k},$$

in which

$$0 \leq p_j < \infty, \qquad 0 \leq q_j \leq l, \tag{33}$$

for some finite l, and we assume that we have

$$|f(z, \bar{z})| < 1 \tag{34}$$

for all points in (32).

There exist a constant $L = L(k, l)$ *such that*

$$\sum_{p,q} |a_{p,q}| \; |z_1|^{p_1+q_1} \dots |z_k|^{p_k+q_k} \leq L$$

in

$$0 \leq |z_j| \leq \frac{1}{4}. \tag{35}$$

Proof. If we put $z_j = \rho_j e^{i\varphi_j}$, $0 \leq \varphi_j < 2\pi$, $\rho_j > 0$, then

$$f = \sum_{p,q} a_{p,q} \rho_1^{p_1+q_1} \dots \rho_k^{p_k+q_k} e^{i[(p_1-q_1)\varphi_1 + \dots + (p_k - q_k)\varphi_k]},$$

and if we put $p_j = q_j + r_j$ then this is

$$\sum_r \left(\sum_q a_{q+r,q} \rho_1^{2q_1+r_1} \dots \rho_k^{2q_n+r_n}\right) e^{i(r_1\varphi_1 + \dots + r_k\varphi_k)}.$$

The latter expansion is a multiple FOURIER series, and in such one each coefficient is maximized by the value of the function, so that, due to (34), we have

$$\left|\sum_q a_{q+r,q} \rho_1^{2q_1} \dots \rho_k^{2q_k}\right| \leq \rho_1^{-r_1} \dots \rho_k^{-r_k}$$

for $0 < \rho_j < 2$. Therefore, for

$$\frac{1}{2} \leq \rho_j \leq 2 \tag{36}$$

we have

$$\left|\sum_q a_{q+r,q} \rho_1^{2q_1} \dots \rho_k^{2q_k}\right| \leq 2^{(r_1 + \dots + r_k)}.$$

Next, there exist a number $L_0 = L_0(k, l)$ such that if we have

$$|\sum_{q_j=0}^{l} b_{q_1 \dots q_k} \rho_1^{2q_1} \dots \rho_k^{2q_k}| \leq 1 \tag{37}$$

in (36), then

$$\sum_{q_j=0}^{l} | b_{q_1 \dots q_k} | \leq L_0 .$$

In fact, if such a constant would not exist, there would be a sequence of systems of coefficients $\{ c^{\mu}_{q_1 \dots q_k} \}$, $\mu = 1, \dots,\ 0 \leq q_j \leq l$, such that

$$\sum_{q_j=0}^{l} | c^{\mu}_{q_1 \dots q_k} | = 1,$$

and yet

$$|\sum_{q_j=0}^{l} c^{\mu}_{q_1 \dots q_k} \rho_1^{2q_1} \dots \rho_k^{2q_k}| \leq \varepsilon_{\mu}$$

for $\rho_1, \dots, \rho_k$ in (36), with $\varepsilon_{\mu} \to 0$. Due to the finiteness of l, we would then have coefficients $c_{q_1 \dots q_k}$ not all 0, for which

$$\sum_{q_j=0}^{l} c_{q_1 \dots q_k} \rho_1^{2q_1} \dots \rho_k^{2q_k} = 0$$

for ρ_j in (36), which for polynomials in ρ_j is impossible.

Therefore, (37) implies

$$| a_{q+r, q} | \leq 2^{(r_1 + \dots + r_k)} L_0 .$$

and therefore for $p_j \geq 0$, $q_j \geq 0$ we have

$$| a_{p, q} | \leq 2^{p_1 + \dots p + _k} L_0$$

and hence

$$\sum_{p, q} | a_{p, q} | 4^{-(p_1 + q_1 + \dots + p_k + q_k)} \leq k^l L_0(k, l) = L_1(k, l),$$

which proves the lemma.

4. Holomorphically-placed products af one-cycles.

A k-cycle B_k in D will be called a holomorphically-placed product if it is contained in some domain D' in D such that a suitable one-one holomorphic mapping

$$z^j = \varphi_j(w^1, \dots, w^k)$$

of D' transforms B_k into a cycle which is rigidly-placed in the w-space.

THEOREM 7. – *If f is continuous and if we have* (21) *for all holomorphically-placed cycles, then f is holomorphic.*

Proof. First of all, by theorem 4, f can be represented by an expression (16) and uniquely so. Now, we claim that (16) must be independent of $\bar{z}^\alpha$. In fact, if it would depend on $\bar{z}^1$, say, then the holomorphic transformation

$$z^1 = w^1 + (w^1)^k, \qquad z^2 = w^2, \dots, z^k = w^k$$

would transform (16) into a polynomial

$$b_{p_1 \dots p_k}(w)(\bar{w}^1)^{p_1} \dots (\bar{w}^k)^{p_k}$$

which is of a degree $\geq k$, but by the uniqueness of (16) in every system of holomorphic variables, this is impossible. Hence the theorem.

Annals of Mathematics
Vol. 57, No. 3, May, 1953
Printed in U.S.A.

COMPLEX SPACES WITH SINGULARITIES

By S. Bochner and W. T. Martin

(April 21, 1952)

§1. Introduction

In the case of one complex variable if a Riemann surface A is spread out over a Riemann surface B and A has (isolated) ramification points of the familiar kind, then by adding such points to A the local Euclidean character is preserved and the analytic structure can be then adjusted so as to absorb the ramification points conformally as well.

For several variables the situation is entirely different. A pertinent analog to a ramification point is a lower-dimensional singular subset as may occur in a complex locus that is defined by a system of equations

$$c_\nu(z_1, \cdots, z_k, w_1, \cdots, w_n) = 0, \qquad \nu = 1, \cdots, n$$

in the (z, w)-space. The entire locus need not be locally Euclidean and this topological possibility is inherent to the set-up. In algebraic geometry, for *compact* algebraic varieties, such singularities are being "removed" by algebraic processes which substitute for them other varieties algebraically closely related to them, even though topologically somewhat different from them. From an analytic approach, however, it is of interest to leave the singularities as they are and to study functions on varieties as they present themselves.

The first task of course is to define such an object. Such definitions have been given before [1, 4, 8, 9, 11, 13] and we will introduce our own formulation both locally and globally. As regards defining what a holomorphic function is to be (compare [1, 15]) we will have occasion to do this in two versions one "stricter" than the other; and the purpose of giving a stricter version will be to include a fundamental theorem on algebraic dependences of automorphic functions which it will thus be possible to retain.

For the non-strict version we will retain the maximum modulus theorem, Schwarz's lemma, other convexity theorems and what is least obvious Vitali's theorem.

Part I

§2. The polynomial relations

We consider a system of n relations

$$\psi_\nu(z_1, \cdots, z_k, \zeta_1, \cdots, \zeta_\nu) = 0, \qquad \nu = 1, \cdots, n \tag{2.1}$$

in which ψ_ν is a polynomial in $\zeta_1, \cdots, \zeta_\nu$ with coefficients which are analytic functions of $z_1, \cdots, z_k$ in a fixed region S of E_{2k}. We denote by m_ν the degree of ψ_ν in the variable ζ_ν, and we assume that the coefficient of $\zeta_\nu^{m_\nu}$ in ψ_ν is unity.

We make the following assumptions on the system (2.1). We ask that for each z in S the system have

$$m = m_1 \cdots m_n \tag{2.2}$$

(not necessarily distinct) sets of solutions (and no others)

$$\zeta_1^{(\mu)}(z), \cdots, \zeta_n^{(\mu)}(z), \qquad \mu = 1, \cdots, m \tag{2.3}$$

with the following properties:

(i) the solutions are continuous in z in S in the sense that if z^0 is any point of S and if $\varepsilon > 0$, then there is a positive number $\delta = \delta(\varepsilon, z^0)$ such that for every z' in $\| z - z^0 \| < \varepsilon$ there is a permutation $(p_1, \cdots, p_m)$ of $(1, \cdots, m)$ for which

$$\textstyle\sum_{\mu=1}^{m} | \zeta_\nu^{(\mu)}(z^0) - \zeta_\nu^{(p_\mu)}(z') |^2 < \varepsilon^2, \qquad \nu = 1, \cdots, n; \tag{2.4}$$

the solutions (2.3) being bounded in every compact set in S;

(ii) there is an analytic function $D(z)$ in S, $D(z) \not\equiv 0$, such that at each point of the set $S - E$, where E is the subset of S at which D vanishes, the m sets (2.3) are distinct; which by (i) implies that in a neighborhood of each point of $S - E$ there are n continuous functional elements so that

(iii) each functional element in (2.3) is analytic in a neighborhood of every point of $S - E$; at each point z^0 in $S - E$ each set in (2.3) is obtainable from any one such set by analytic continuation along suitable paths in $S - E$; and analytic continuation of these m sets around any closed path in $S - E$ merely permutes the sets;

(iv) the square $\Delta^2(z)$ of the determinant of the $m \times m$ matrix

$$([\zeta_1^{(\mu)}(z)]^{\lambda_1} \cdots [\zeta_n^{(\mu)}(z)]^{\lambda_n}) \tag{2.5}$$

for $\mu = 1, \cdots, m$ and for

$$\lambda_1 = 0, 1, \cdots, m_1 - 1; \cdots, \lambda_n = 0, 1, m_n - 1 \tag{2.6}$$

is analytic in S and vanishes in S at most over the points E where $D(z)$ vanishes;

(v) corresponding to any polynomial

$$F(z, \zeta) = \textstyle\sum_{\lambda_1=0}^{m_1-1} \cdots \sum_{\lambda_n=0}^{m_n-1} P_{\lambda_1 \cdots \lambda_n}(z)\zeta_1^{\lambda_1} \cdots \zeta_n^{\lambda_n} \tag{2.7}$$

with coefficients $P_\lambda(z)$ analytic in S and not all identically zero there exists another such polynomial

$$G(z, \zeta) = \textstyle\sum_{\lambda_1=0}^{m_n-1} \cdots \sum_{\lambda_n=0}^{m_n-1} Q_{\lambda_1 \cdots \lambda_n}(z)\zeta_1^{\lambda_1} \cdots \zeta_n^{\lambda_n} \tag{2.8}$$

with coefficients analytic in S and not all identically zero, and a function $\Omega(z)$ which is analytic and $\not\equiv 0$ in S such that

$$F(z, \zeta_1^{(\mu)}(z), \cdots, \zeta_n^{(\mu)}(z)) \cdot G(z, \zeta_1^{(\mu)}(z), \cdots, \zeta_n^{(\mu)}(z)) \equiv \Omega(z) \tag{2.9}$$

throughout S, for $\mu = 1, \cdots, m$.

The latter parts of property (iii) constitute an irreducibility requirement and for example in the case $n = 1$ of a single equation $\psi(z, \zeta) = 0$ they rule out reducible situations such as

$$\psi(z, \zeta) = [\zeta - a(z)][\zeta^2 - z_1 \cdots z_k]$$

where $a(z)$ is analytic in S.

Condition (v) tells us that there are functions G and Ω such that in the part of S where $\Omega \not\equiv 0$ we have $1/F = G/\Omega$ in the sense of (2.9).

We need not be concerned with the question of ascertaining or stating the extent to which these conditions are independent of each other.

Now let $\Phi(z)$ be analytic and $\not\equiv 0$ in S, and consider a function

$$F(z, \zeta) = \frac{1}{\Phi(z)} \sum_{\lambda_1=0}^{m_1-1} \cdots \sum_{\lambda_n=0}^{m_n-1} Q_{\lambda_1 \cdots \lambda_n}(z) \zeta_1^{\lambda_1} \cdots \zeta_n^{\lambda_n} \tag{2.10}$$

where the Q's are analytic in S. The function is analytic in (z, ζ) in $[z \in S - E_\Phi, \mid \zeta_\nu \mid < \infty]$, where E_Φ is the subset of S where Φ vanishes. By property (iii) in the neighborhood of every point of $S - E - E_\Phi$ there are m functional elements

$$f_\mu(z) = F[z, \zeta_1^{(\mu)}(z), \cdots, \zeta_m^{(\mu)}(z)], \qquad \mu = 1, \cdots, m \tag{2.11}$$

analytic there, and if $\phi(t_1, \cdots, t_m)$ is symmetric and analytic for $\mid t_\mu \mid < \infty$, then the various functional elements

$$\phi[f_1(z), \cdots, f_m(z)] \tag{2.12}$$

constitute one analytic function in $S - E - E_\Phi$.

If we assume that for every compact set T in S there is a constant K_T such that

$$\mid f_\mu(z) \mid \leqq K_T, \qquad \mu = 1, \cdots, m \tag{2.13}$$

in $T \cap (S - E_\Phi)$ (or only in $T \cap (S - E - E_\Phi)$), then since $E + E_\Phi$ is a *removable* exceptional set [5, p. 173], this function (2.12) has an analytic continuation into all of S.

As a special case, under condition (2.13), the symmetric polynomials

$$\begin{aligned} B_1(z) &= -[f_1(z) + \cdots + f_m(z)], B_2(z) \\ &= f_1 f_2 + \cdots + f_{m-1} f_m, \cdots, B_m(z) = (-1)^m f_1 \cdots f_m \end{aligned} \tag{2.14}$$

are analytic in S, and hence the m functions (2.11) satisfy in $S - E_\Phi$ a pseudo-algebraic equation

$$f^m + B_1(z) f^{m-1} + \cdots + B_m(z) = 0 \tag{2.15}$$

in which the B's are analytic in S. Since the roots of (2.15) are defined throughout S we may define the $f_\mu(z)$ in E_Φ so as to satisfy (2.15) throughout S. The roots are continuous in S in the same manner as in (i).

As a consequence, if $h(t_1, \cdots, t_m)$ is continuous and symmetric in $|t_\mu| < \infty$, then the function

$$h[f_1(z), \cdots, f_m(z)] \tag{2.16}$$

is continuous in S. In particular the function

$$\hat{f}(z) = \max [| f_1(z) |, \cdots, | f_m(z) |] \tag{2.17}$$

is continuous in S.

We summarize these results as follows:

LEMMA 2.1. *Let F and f_μ be as in* (2.10) *and* (2.11) *and assume that* (2.13) *holds in $T \cap (S - E - E_\Phi)$ for every compact subset T of S. Then the functions $f_\mu(z)$ can be defined in E so as to be the roots of a pseudo-algebraic equation* (2.15) *throughout S, where the coefficients $B_1, \cdots, B_m$ are analytic in S. The roots $f_\mu(z)$ are continuous in S, and the functions* (2.16) *and* (2.17) *are continuous in S, where, in* (2.16), *$h(t_1, \cdots, t_m)$ is any symmetric continuous function in $| t_\mu | < \infty$.*

Consider again the function (2.10) under the hypothesis (2.13). Then as we have seen, the f_μ satisfy an equation (2.15) and hence they are bounded in every compact set in S. Denote by $\Delta_{\lambda_1 \cdots \lambda_n}(z)$ the determinant of the matrix obtained from the matrix (2.5) by replacing the elements of the column, one element of which is displayed in (2.5), by $f_1(z), \cdots, f_m(z)$. The product

$$\Delta_{\lambda_1 \cdots \lambda_n}(z) \cdot \Delta(z) \tag{2.18}$$

is bounded in every compact set in S since the $f_\mu(z)$ and $\zeta_\nu^\mu(z)$ are. Also if we insert into $\Delta_{\lambda_1 \cdots \lambda_n}(z)$ the series for the f_μ obtained from (2.10) then it is easily seen that

$$\Delta_{\lambda_1 \cdots \lambda_n}(z) = (Q_{\lambda_1 \cdots \lambda_n}(z)/\Phi(z)) \cdot \Delta(z) \tag{2.19}$$

in $S - E_\Phi$. Hence the product (2.18) is equal to

$$\frac{Q_{\lambda_1 \cdots \lambda_n}(z)}{\Phi(z)} \Delta^2(z) \tag{2.20}$$

in $S - E_\Phi$. But (2.20) is analytic in $S - E_\Phi$. Since E_Φ is a removable exceptional set there is an analytic function $P_{\lambda_1 \cdots \lambda_n}(z)$ in S which is equal to (2.20) in $S - E_\Phi$.

Thus we have the following result.

LEMMA 2.2 *If $F(z, \zeta)$ has a representation* (2.10) *with* (2.13) *holding, then it also has a representation*

$$F(z, \zeta) = \frac{1}{\Delta^2(z)} \sum_{\lambda_n = 0}^{m_1 - 1} \cdots \sum_{\lambda_1 = 0}^{m_1 - 1} P_{\lambda_1 \cdots \lambda_n}(z) \zeta_1^{\lambda_1} \cdots \zeta_n^{\lambda_n} \tag{2.21}$$

where the P's are analytic in S, and $\Delta^2(z)$ is the square of the determinant of the matrix (2.5).

We shall now prove the following result.

LEMMA 2.3. *Let (2.21) be given and assume that the functions $f_\mu(z)$ are bounded in every compact set in S. If $\hat{f}(z)$ is identically constant in a neighborhood of some point of S, then the function (2.21) is identically constant:*

$$(2.22) \qquad P_{\lambda_1\cdots\lambda_n}(z) \equiv \begin{cases} c\Delta^2(z) \text{ in } S, \text{ if } \lambda_1^2 + \cdots + \lambda_n^2 = 0, \\ 0 \qquad \text{ in } S, \text{ if } \lambda_1^2 + \cdots + \lambda_n^2 > 0. \end{cases}$$

PROOF. Since every open set in S contains points of $S - E$ there are a constant γ, a point z^0 in $S - E$ and a neighborhood R of z^0 such that

$$(2.23) \qquad \hat{f}(z) \equiv \gamma \text{ in } R.$$

Hence $| f_\mu(z) | \leqq \gamma$ in R for $\mu = 1, \cdots, m$, and $| f_{\mu_0}(z^0) | = \gamma$ for some index μ_0 in $1 \leqq \mu_0 \leqq m$. Since $f_{\mu_0}(z)$ is analytic in a neighborhood of z^0 this means that there is a constant c (with $| c | = \gamma$) such that $f_{\mu_0}(z) \equiv c$ in a neighborhood of z^0. By property (iii)

$$(2.24) \qquad f_\mu(z) \equiv c \text{ in } S - E, \qquad \mu = 1, \cdots, m\cdot$$

By Cramer's rule and property (iv) we can solve for the P's in terms of the f's, obtaining

$$(2.25) \qquad P_{\lambda_1\cdots\lambda_n}(z) = \Delta_{\lambda_1\cdots\lambda_n}(z)\cdot\Delta(z)$$

where $\Delta_{\lambda_1\cdots\lambda_n}(z)$ is defined as in (2.18). Using (2.24) we obtain (2.22).

§3. The topological space U

Consider a system of n relations

$$(3.1) \qquad \psi_\nu(z, \zeta_1, \cdots, \zeta_\nu) = 0, \qquad \nu = 1, \cdots, n$$

as in Section 2. Let E_{2k+2n} be the Euclidean space of $k + n$ complex variables $z_1, \cdots, z_k, \zeta_1, \cdots, \zeta_n$, and in E_{2k+2n} consider all points (z, ζ) for which z lies in S and (z, ζ) solves the system. By (2.3) this means that we consider all points (z, ζ) for which

$$(3.2) \qquad z \in S, \qquad \zeta_\nu = \zeta_\nu^{(\mu)}(z), \qquad \nu = 1, \cdots, n,$$

for some $\mu = 1, \cdots, m$. This gives a well defined point-set in E_{2k+2n}; we impart to this set the topology of E_{2k+2n} and we denote the topological space which we obtain in this manner (that is, the point set (3.2) so topologized) by U. We emphasize that this topologization embraces all points of our point-set (3.2) without any exception. After this we take the function $D(z)$ and the set E

$$(3.3) \qquad E{:}D(z) = 0 \qquad z \text{ in } S,$$

defined as in (ii) of Section 2, and we select in U those points whose pre-images z in S satisfy the equation (3.3); we denote this subset of U by M.

If P is any point of the difference set

$$(3.4) \qquad U - M$$

then it arises from a point (z^0, ζ^0) in E_{2k+2n} where

$$z^0 \in S - E, \qquad \zeta_\nu^0 = \zeta_\nu^{(\mu_0)}(z^0), \qquad \nu = 1, \cdots, n \tag{3.5}$$

for some μ_0 in $1 \leqq \mu_0 \leqq m$. And we know by property (iii) of Section 2 that the functions $\zeta_\nu^{(\mu_0)}(z)$, $\nu = 1, \cdots, n$ are analytic in a $2k$-dimensional neighborhood of z^0. Also all points (z, ζ) for z in a sufficiently small neighborhood of z^0 and

$$\zeta_\nu = \zeta_\nu^{(\mu_0)}(z), \qquad \nu = 1, \cdots, n \tag{3.6}$$

belong to $U - M$. By property (iii) of Section 2, this makes the set (3.4) a complex analytic coordinate space, with the z's as allowable coordinates in a neighborhood of every point, and the incident topology of (3.4) is the same as that of being a subspace of U.

REMARK. If $\Phi(z)$ is any function which is analytic and $\not\equiv 0$ and which is such that its zeros include all of those of $D(z)$ in S, and if we denote by $M(\Phi)$ the subset of U consisting of those points of U whose pre-images z in S satisfy the equation $\Phi(z) = 0$, then again the space $U - M(\Phi)$ is a complex analytic coordinate space as just described.

It is convenient to speak of points (z, ζ) in U. If P^0 is a point in U and if we refer to its coordinates as (z^0, ζ^0) we mean that P^0 corresponds to a point in E_{2k+2n} with coordinates (z^0, ζ^0) where (3.5) holds for some index μ_0 in $1 \leqq \mu_0 \leqq m$. If P^0 happens to belong to the subspace $U - M$, then, as we know, $z^0 \in S - E$, there is a single such index μ_0, and the functions (3.6) are analytic in a neighborhood of z^0. But for points of M we may have more than one such index μ_0, and we may not have analyticity of the functions (3.6).

A neighborhood of a point P^0 in U consists of all points (z, ζ) which simultaneously lie in a $(2k + 2n)$-dimensional neighborhood of (z^0, ζ^0) in E_{2k+2n} and satisfy the relations

$$\zeta_1 = \zeta_1^{(\mu)}(z), \cdots, \zeta_n = \zeta_n^{(\mu)}(z) \tag{3.7}$$

for some index $\mu = \mu(z)$ in $1 \leqq \mu \leqq m$.

If S_1 is a subset of S we will denote by $U(S_1)$ the subset of U whose pre-images lie in S_1.

We say that a function $F(P)$ is *holomorphic* in U if it is defined throughout U, is bounded in $U(T)$ for every compact subset T of S, and is given by an expression of the form

$$\frac{1}{\Delta^2(z)} \sum_{\lambda_1=0}^{m_1-1} \cdots \sum_{\lambda_n=0}^{m_n-1} P_{\lambda_1 \cdots \lambda_n}(z) \zeta_1^{\lambda_1} \cdots \zeta_n^{\lambda_n} \tag{3.8}$$

in $M = U(S - E)$, where the P's are analytic in S.

If F is holomorphic in U and if $a(z)$ is analytic in S then by aF we mean the function whose value at any point (z, ζ) of U is $a(z)$ times the value of F there.

Clearly if F and G are holomorphic in U so is $aF + bG$ where a and b are arbitrary analytic functions in S. We can also see that the product FG is holomorphic in U, as follows. First the product is bounded in every $U(T)$ and has a representation

$$F \cdot G = \frac{1}{\Delta^4(z)} \sum \cdots \sum R_{\lambda_1 \cdots \lambda_n}(z) \zeta_1^{\lambda_1} \cdots \zeta_n^{\lambda_n} \tag{3.9}$$

where the R's are analytic in S, and the sum is a polynomial of degree $\leqq 2(m_\nu - 1)$ in ζ_ν. By making use of the polynomial relations (2.1), the polynomial in the numerator of (3.9) can be reduced to one of the same form but of degree $\leqq m_\nu - 1$ in ζ_ν. By Lemma 2.2 there is then a representation of the same form but with denominator $\Delta^2(z)$. Thus FG is holomorphic in U.

Finally if F is holomorphic and $\not\equiv 0$ in U, then by property (v) of Section 2, there are a function (2.8) holomorphic in U, and a function $\Omega(z)$ analytic and $\not\equiv 0$ in S such that

$$\Delta^2 \cdot F \cdot G = \Omega.$$

Now the function $\Delta^2 G/\Omega$ need no longer be holomorphic in U but it so is if it happens to be bounded in every $U(T)$ for T a compact subset of S. Hence if F is holomorphic and if $1/F$ is bounded in every $U(T)$, then $1/F$ is holomorphic.

Along with holomorphic functions we shall on certain occasions treat a subclass, namely *strictly holomorphic* functions. We say that a function is strictly holomorphic in U if it is defined at every point of U by an expression of the form (2.10) with $\Phi(z) \equiv 1$, and the Q's analytic throughout S. Since the function (2.10) (with $\Phi \equiv 1$) is a continuous function of (z, ζ) for $(z \in S, | \zeta_\nu | < \infty)$ it follows that it is a continuous function in U, in the topology of U.

EXAMPLE.

$$n = 1, \qquad k = 1, \qquad s: | z | < 1$$

$$\psi(z, \zeta) \equiv \zeta^2 - 2z^2\zeta + \tfrac{1}{4}z^2, \qquad \zeta = z^2 \pm z\sqrt{z^2 - \tfrac{1}{4}}$$

$$D(z) = \Delta^2(z) = 4z^4 - z^2$$

The function

$$F \equiv \frac{P_0 + P_1\zeta}{\Delta^2(z)}$$

where

$$P_0(z) = z^3 - 4z^5, \qquad P_1(z) = 4z^3 - z$$

is given by

$$F = \pm\sqrt{z^2 - \tfrac{1}{4}}$$

on U and hence is bounded. Hence this F is holomorphic in U. Clearly it has no representation $F = Q_0 + Q_1\zeta$ where $Q_0(z)$ and $Q_1(z)$ are analytic in the entire unit circle $| z | < 1$.

From the definitions given in this section in conjunction with expression

(2.25) we obtain the following "intrinsic" characterization of holomorphic functions on U.

THEOREM 3.1. *If a function is defined on the complex space $U - M$, if it is holomorphic there in the ordinary sense, and if it is bounded there, then it is holomorphic on U according to our present definition.*

§4. Convergence relations

LEMMA 4.1. *Let*

$$F(P;\alpha) = \frac{1}{\Delta^2(z)} \sum_{\lambda_1=0}^{m_1-1} \cdots \sum_{\lambda_n=0}^{m_n-1} P_\lambda(z;\alpha)\zeta_1^{\lambda_1}\cdots\zeta_n^{\lambda_n} \tag{4.1}$$

be a family **A** *of holomorphic functions in U, the functions being equi-uniformly bounded in $U(T)$ for every compact subset T of S:*

$$|F(P;\alpha)| \leqq K_T \text{ in } U(T), \qquad \alpha \in \mathbf{A}. \tag{4.2}$$

Then there is a constant B_T depending only upon T and the original system (2.1) *such that*

$$|P_\lambda(z;\alpha)| \leqq B_T K_T \text{ in } T, \qquad \alpha \in \mathbf{A} \tag{4.3}$$

for (λ) *in* (2.6).

PROOF. As before we write

$$f_\mu(z;\alpha) = F(z, \zeta_1^{(\mu)}(z), \cdots, \zeta_n^{(\mu)}(z);\alpha), \qquad \mu = 1, \cdots, m. \tag{4.4}$$

Cramer's rule applied to (4.4) and (4.1) yields

$$\frac{P_{\lambda_1\cdots\lambda_n}(z;\alpha)}{\Delta^2(z)} = \frac{\Delta_{\lambda_1\cdots\lambda_n}(z;\alpha)}{\Delta(z)} \quad \text{in} \quad S - E, \tag{4.5}$$

where $\Delta_{\lambda_1\cdots\lambda_n}(z;\alpha)$ is defined as in (2.18). Hence

$$P_{\lambda_1\cdots\lambda_n}(z;\alpha) = \Delta_{\lambda_1\cdots\lambda_n}(z;\alpha)\cdot\Delta(z) \text{ in } S - E. \tag{4.6}$$

If we expand the determinant $\Delta_{\lambda_1\cdots\lambda_n}(z;\alpha)$ in terms of the minors of the column containing $f_1(z;\alpha), \cdots, f_m(z;\alpha)$, use (4.2) and the boundedness of the ζ's in T we see that there is a constant B_T such that

$$|\Delta_{\lambda_1\cdots\lambda_n}(z;\alpha)\Delta(z)| \leqq B_T K_T \text{ in } T; \qquad \alpha \in \mathbf{A}, \lambda \text{ in } (2.6).$$

Hence (4.3) holds first in $T - E$, and then in T.

COROLLARY 4.1. *If $\{F_s(P)\}$ is any infinite sequence of functions of A, then there is a subsequence which converges in $S - E$ to a holomorphic function $F(P)$, the convergence being uniform in $U(T^*)$ for every compact subset T^* of $S - E$.*

In order to see this we first note that by (4.3) there is a subsequence $\{F(P;s_\sigma)\}$, $\sigma = 1, 2, \cdots$ such that the sequences $P_{\lambda_1\cdots\lambda_n}(z;s_\sigma)$ converge uniformly in every compact set T of S to analytic functions $P_{\lambda_1\cdots\lambda_n}(z)$ in S. The function $F(P)$ formed as in (3.8) with these P's is defined in $U(S - E)$ and is the limit there of the sequence $\{F(P;s_\sigma)\}$. Moreover the convergence is uniform in $U(T^*)$ for every compact subset T^* of $S - E$. Since the limitands are equi-uniformly bounded in $U(T)$ (see (4.2)) it follows that $| F(P) | \leqq K_T$ in $U(T)$. Hence $F(P)$ may be defined in $U(E)$ so as to be holomorphic in U:

COROLLARY 4.2. *If a sequence $\{F_s(P)\}$ of holomorphic functions in U converges uniformly in $U(T)$ for every compact set T in S, then the limit function is again holomorphic in U.*

We shall next consider similar results for strictly holomorphic functions.

LEMMA 4.2. *Let*

$$F\ (P; \alpha) = \sum_{\lambda_1=0}^{m_1-1} \cdots \sum_{\lambda_n=0}^{m_n-1} P_\lambda(z; \alpha)\zeta_1^{\lambda_1} \cdots \zeta_n^{\lambda_n} \tag{4.7}$$

be a family $\mathbf{A}'$ of functions each strictly holomorphic in U. Let $z^0 \in S$. If there are a neighborhood R of z^0 in S and a positive number K such that

$$| F(P; \alpha) | \leqq K \text{ in } U(R), \qquad \alpha \in \mathbf{A}' \tag{4.8}$$

then there are a neighborhood R_1 of z^0 and a positive number K_1 such that

$$| P_\lambda(z; \alpha) | \leqq K_1 \text{ in } R_1, \qquad \alpha \in \mathbf{A}', \ (\lambda) \text{ in } (2.6) \tag{4.9}$$

PROOF. As before we write

$$f_\mu(z; \alpha) = F(z; \zeta_1^{(\mu)}(z), \cdots, \zeta_n^{(\mu)}(z); \alpha), \qquad \mu = 1, \cdots, m \tag{4.10}$$

$$\hat{f}(z; \alpha) = \max_{\mu=1,\cdots,m} | f_\mu(z; \alpha) |. \tag{4.11}$$

Now Cramer's rule applied to (4.10) and (4.7) yields

$$\Delta^2(z)P_{\lambda_1\cdots\lambda_n}(z; \alpha) = \Delta_{\lambda_1\cdots\lambda_n}(z; \alpha)\Delta(z) \tag{4.12}$$

where $\Delta(z)$ is the determinant of the matrix (2.5), and $\Delta_{\lambda_1\cdots\lambda_n}(z; \alpha)$ is the determinant of the matrix obtained from (2.5) by replacing the elements of the particular column one element of which is displayed in (2.5), by $f_1(z; \alpha), \cdots, f_m(z; \alpha)$.

If R' is any neighborhood of z^0, with $R' \subset R$, $\bar{R}' \subset S$, then in R' the roots (2.3) are bounded and by (4.9) the functions $f_\mu(z; \alpha)$ are uniformly bounded (by K). Hence by Hadamard's theorem on determinants, there is a constant K' such that the right-hand side of (4.12) is bounded by K' in R'. Hence

$$| \Delta^2(z)P_\lambda(z; \alpha) | \leqq K' \text{ in } R' \text{ for all } \alpha \in \mathbf{A}', \qquad (\lambda) \text{ in } (2.6). \tag{4.13}$$

We will next prove the following two results on analytic functions; the second of these results, together with (4.13) will yield Lemma 4.2.

LEMMA 4.3. *If $\Omega(z)$ is analytic and $\neq 0$ in a sphere*

$$\| z - z^0 \| < r \tag{4.14}$$

then there are

(i) *positive numbers δ, ε and ρ with*

$$2\rho < r, \qquad 2\varepsilon < r \tag{4.15}$$

and

(ii) *a point* $(\beta_1, \cdots, \beta_k)$ *on the unit sphere* $\| \beta \| = 1$ *such that*

$$(4.16) \quad | \Omega(z_1 + \beta_1 t, \cdots, z_k + \beta_k t) | \geqq \delta \text{ for } \| z - z^0 \| < \varepsilon \text{ and } | t | = \rho.$$

LEMMA 4.4. *Let* $\Omega(z)$ *be analytic and not identically zero in a neighborhood* R' *of a point* z^0, *and let* $\{h_\alpha(z)\}$ *be a family of functions each analytic in* R'. *If there is a positive number* K' *such that*

$$(4.17) \quad | \Omega(z) h_\alpha(z) | \leqq K' \text{ in } R'$$

for all α *then there are a neighborhood* R'' *of* z^0 *and a positive constant* K'' *such that*

$$(4.18) \quad | h_\alpha(z) | \leqq K'' \text{ in } R''$$

for all α.

For the proof of Lemma 4.3, without loss of generality we take the point z^0 to be the origin. Then the function $\Omega(b_1 t, \cdots, b_k t)$ is an analytic function of the $k + 1$ complex variables $b_1, \cdots, b_k, t$ for $\| b \| = 1$ and $| t | < r$, and it is not identically zero since $\Omega(z) \not\equiv 0$. Hence there is a point $(\beta_1, \cdots, \beta_k)$ on $\| \beta \| = 1$ such that

$$\Omega (\beta_1 t, \cdots, \beta_k t) \not\equiv 0 \text{ in } t.$$

Since $\Omega(\beta_j t)$ is analytic in t for $| t | < r$, it follows that there is a positive number $\rho < r/2$ such that $\Omega(\beta_j t)$ is different from zero on the circle $| t | = \rho$. With this ρ set

$$\delta = \tfrac{1}{2} \min_{|t|=\rho} | \Omega(\beta_j t) |$$

so that

$$(4.19) \quad 0 < 2\delta \leqq | \Omega(\beta_j t) | \text{ for } | t | = \rho.$$

By the uniform continuity of $\Omega(z)$ in any sphere interior to $\| z \| < r$ it follows from (4.19) that there is a positive number $\varepsilon < r/2$ such that

$$(4.20) \quad \delta \leqq | \Omega(z_j + \beta_j t) | \text{ for } \| z \| < \varepsilon \text{ and } | t | = \rho.$$

This concludes the proof of Lemma 4.3.

We now turn to the proof of Lemma 4.4. Let $r > 0$ be such that the sphere $\| z - z^0 \| < r$ is contained in R', and let β, ε, δ and ρ be as in Lemma 4.3. Then by (4.17) and (4.16)

$$(4.21) \quad | h_\alpha(z_j + \beta_j t) | \leqq \frac{K'}{\delta} \quad \text{for} \quad \| z - z^0 \| < \varepsilon, \qquad | t | = \rho, \quad \text{all } \alpha.$$

By the maximum modulus principle for analytic functions of one complex variable t, the inequality (4.21) holds for $t = 0$. This yields the inequality (4.18) with

$$(4.22) \quad K'' = \frac{K'}{\delta}, \qquad R'': \| z - z^0 \| < \varepsilon.$$

Lemma 4.4 applied to the inequality (4.13) yields Lemma 4.2.

If T is a compact subset of S and if there is a neighborhood R_T of T in S such that the functions $F(P, \alpha)$ are equi-uniformly bounded in $U(R_T)$ then with the help of the Heine-Borel theorem Lemma 4.2 tells us that there is a neighborhood R_{1T} of T in which the functions $P_{\lambda_1 \cdots \lambda_n}(z; \alpha)$ are uniformly bounded in R_{1T}. Hence we have the following theorem.

THEOREM 4.1. *If* $\mathbf{A}'$ *is a family of functions* F *each strictly holomorphic in* U *and equi-uniformly bounded in* $U(T)$ *for every compact subset* T *of* S, *then every infinite sequence* $\{F_s(P)\}$ *of functions* $F_s \in \mathbf{A}'$ *contains a subsequence which converges uniformly on every* $U(T)$; *the limit function is strictly holomorphic in* U.

We will close this section with a few remarks on analytic functions of $z_1, \cdots, z_k$ in a neighborhood of the origin.

Let $\phi(z_1, \cdots, z_k)$ be analytic in a neighborhood of the origin. We say that ϕ has a *zero of order* β at the origin, where β is a positive integer, if ϕ has a development of the form

$$\phi(z) = b_\beta(z) + b_{\beta+1}(z) + \cdots, \qquad b_\beta(z) \not\equiv 0 \tag{4.23}$$

valid in a neighborhood of the origin, where each $b_\sigma(z)$ is a homogeneous polynomial of degree σ in $z_1, \cdots, z_k$.

If $\phi(z)$ is analytic for $\| z \| < \rho_0$ and if it has a zero of order $\geqq \beta$ at the origin, then Schwarz's lemma states that for $\rho < \rho_0$

$$| \phi(z) | \leqq M(\rho) \left(\frac{\| z \|}{\rho} \right)^\beta \quad \text{for} \quad \| z \| < \rho \tag{4.24}$$

where

$$M(\rho) = \max_{\|z\| = \rho} | \phi(z) |. \tag{4.25}$$

Now suppose we have s functions $\phi_1(z), \cdots, \phi_s(z)$ each analytic in a neighborhood $\| z \| < \rho_0$ of the origin. Given a positive integer β, we can always determine complex constants $c_1, \cdots, c_s$ not all zero so that the linear combination

$$c_1\phi_1 + \cdots + c_s\phi_s \tag{4.24}$$

has a zero of order $\geqq \beta$ at the origin, provided β and s satisfy the inequality

$$\binom{k + \beta - 1}{k} < s. \tag{4.27}$$

This is evident since we need in general to be able to impose

$$1 + \binom{k}{1} + \binom{k+1}{2} + \cdots + \binom{k + \beta - 2}{\beta - 1} = \binom{k + \beta - 1}{k} \tag{4.28}$$

conditions.

PART II

§5. Algebroid functions

In Section 2 we showed that if $F(P)$ was holomorphic in U,

$$(5.1)\qquad F(P) = \frac{1}{\Delta^2(z)} \sum_{\lambda_1=0}^{m_1-1} \cdots \sum_{\lambda_n=0}^{m_n-1} P_{\lambda_1\cdots\lambda_n}(z)\zeta_1^{\lambda_1}\cdots\zeta_n^{\lambda_n}$$

then the m functions

$$(5.2)\qquad f_\mu(z) = F(z, \zeta_1^{(\mu)}(z), \cdots, \zeta_n^{(\mu)}(z)), \qquad \mu = 1, \cdots, m$$

were roots of a pseudo-algebraic equation

$$(5.3)\qquad f^m + B_1(z)f^{m-1} + \cdots + B_m(z) = 0$$

in which the B's are analytic in S.

In this part we will not be specifically concerned with holomorphic functions (5.1) but rather with algebroid functions

$$(5.4)\qquad f(z) = \{f_1(z), \cdots, f_m(z)\}$$

which are solutions of equations of the form (5.3). We shall obtain results on the maximum principle, Schwarz's lemma and convexity theorems for the function

$$(5.5)\qquad \hat{f}(z) = \max[|f_1(z)|, \cdots, |f_m(z)|]$$

We now introduce the following definition:

DEFINITION 5.1. By a *properly algebroid* (p.a.) function $f(z)$ in a region S of E_{2k} we mean an m-valued function

$$(5.6)\qquad f(z) = \{f_1(z), \cdots, f_m(z)\}$$

where the f_μ are solutions of a pseudo algebraic equation

$$(5.7)\qquad f^m + B_1(z)f^{m-1} + \cdots + B_m(z) = 0,$$

with coefficients $B_\mu(z)$ analytic in S.

If we say that f is p.a. at a point z^0 we will mean that there is a neighborhood of z^0 in which it is p.a.

We mention explicitly that we make no assumption on the discriminant of the equation (5.7).

If $\phi(t_1, \cdots, t_m)$ is defined, continuous and symmetric for $|t_\mu| < \infty$, $\mu = 1, \cdots, m$, then by the continuity of the coefficients $B_\mu(z)$ in S, it follows that $\phi[f_1(z), \cdots, f_m(z)]$ is continuous in S. Thus in particular *the function*

$$(5.8)\qquad \hat{f}(z) = \max[|f_1(z)|, \cdots, |f_m(z)|]$$

is continuous in S.

Along with the function $\hat{f}(z)$ we also introduce the sequence of functions

$$s_\alpha(z) = \frac{[f_1(z)]^\alpha + \cdots + [f_m(z)]^\alpha}{m}, \qquad \alpha = 1, 2, \cdots. \tag{5.9}$$

Since the symmetric functions $s_\alpha(z)$ are expressible as polynomials in the coefficients $B_\mu(z)$, it follows that *the functions $s_\alpha(z)$ are analytic in S.*

We shall later use the following lemma expressing relationships between $\hat{f}$ and the B's, and between $\hat{f}$ and the s's.

LEMMA 5.1.

(i) *The coefficients $B_\mu(z)$ in (5.7) satisfy the inequalities*

$$|B_\mu(z)| \leqq \binom{m}{\mu} [\hat{f}(z)]^\mu \quad \text{in} \quad S, \qquad \mu = 1, \cdots, m; \tag{5.10}$$

(ii) *for every point z in S we have*

$$|s_\alpha(z)|^{1/\alpha} \leqq \hat{f}(z), \qquad \alpha = 1, 2, \cdots, \tag{5.11}$$

and

$$\limsup_{\alpha\to\infty} |s_\alpha(z)|^{1/\alpha} = \hat{f}(z). \tag{5.12}$$

PROOF. The first assertion, (5.10), follows at once from the formulas

$$B_1(z) = -[f_1(z) + \cdots + f_m(z)], \cdots, B_m(z) = (-1)^m f_1 \cdots f_m. \tag{5.13}$$

The inequality (5.11) follows immediately from the definitions (5.9) and (5.8). The final assertion, (5.12), is a consequence of the following lemma which we next prove.

LEMMA 5.2. *Let $g_1, \cdots, g_m$ be complex constants. Then*

$$\limsup_{\alpha\to\infty} |g_1^\alpha + \cdots + g_m^\alpha|^{1/\alpha} = \max[|g_1|, \cdots, |g_m|] \tag{5.14}$$

where α approaches infinity through positive integral values.

For the proof of this lemma it is clearly sufficient to consider the case in which all the g's have the same absolute value, $|g_1| = \cdots = |g_m|$. Without loss in generality we assume that this common absolute value is unity. We write then

$$g_\mu = e^{i\theta_\mu}, \qquad \mu = 1, \cdots, m$$

where $0 \leqq \theta_\mu < 2\pi$. For every $\varepsilon > 0$ there are an infinite number of positive integers

$$\alpha_1, \alpha_2, \alpha_3, \cdots; \qquad \alpha_1 < \alpha_2 < \alpha_3 < \cdots$$

such that

$$|e^{i\theta_1\alpha} + \cdots + e^{i\theta_m\alpha}| > m - \varepsilon \text{ for } \alpha = \alpha_1, \alpha_2, \cdots. \tag{5.15}$$

(This may be seen in various ways. In particular it follows at once from a result due to Harald Bohr [**6**, p. 88] which states that if $p(t)$ is an a.p. function and if

$\varepsilon > 0$, then there is a length $L = L(\varepsilon) > 0$ such that every interval of length L contains at least one *integer* which is an ε-translation number for $p(t)$).

The inequality (5.15) tells us that

$$\limsup_{\alpha\to\infty} \mid e^{i\theta_1\alpha} + \cdots + e^{i\theta_m\alpha} \mid^{1/\alpha} \geqq 1.$$

Since the limitand is $\leqq m$ this yields

$$\limsup_{\alpha\to\infty} \mid e^{i\theta_1\alpha} + \cdots + e^{i\theta_m\alpha} \mid^{1/\alpha} = 1.$$

This concludes the proof of Lemma 5.2, and hence of Lemma 5.1.

§6. A convexity principle

Let a set $\{S, B, z^0, h(z), \sigma\}$ be given where S is a region in E_{2k}, B is a subset of the closure of S, z^0 is a point of S, $h(z)$ is a real non-negative function on B, and σ is a real non-negative constant.

DEFINITION 6.1. The set $\{S, B, z^0, h(z), \sigma\}$ is called an A-set if the two properties

(i) $\phi(z)$ is analytic in $S + B$, and

(ii) $\mid \phi(z) \mid \leqq h^\alpha(z)$ on B for some positive integer α

always imply that

(6.1) $$\mid \phi(z^0) \mid \leqq \sigma^\alpha \qquad \text{(for the positive integer } \alpha \text{ of (ii))}.$$

We shall prove the following result.

LEMMA 6.1. *Let* $\{S, B, z^0, h(z), \sigma\}$ *be an* A-*set and let* $f(z) = \{f_1(z), \cdots, f_m(z)\}$ *be* p.a. *in* $S + B$. *If*

(6.2) $$\hat{f}(z) \leqq h^\alpha(z) \text{ on } B \text{ for some positive integer } \alpha$$

then

(6.3) $$\hat{f}(z^0) \leqq \sigma^\alpha.$$

PROOF. For each positive integer β, the function (5.9) is analytic in $S + B$, and by (5.11) and (6.2)

(6.4) $$\mid s_\beta(z) \mid \leqq h^{\alpha\beta}(z) \qquad \text{in } B.$$

Since we are dealing with an A-set, this implies that

(6.5) $$\mid s_\beta(z^0) \mid \leqq \sigma^{\alpha\beta}$$

By (5.12) this yields (6.3).

REMARK. If in the definition of A-sets we replace the condition that α be a positive integer by the condition that α be a positive number, we can conclude from $\hat{f}(z) \leqq h^\beta(z)$ on B for some $\beta > 0$ that $\hat{f}(z^0) \leqq \sigma^\beta$. This suggests that $\log \hat{f}(z)$ behaves like $\log \mid \phi(z) \mid$. We will make no particular use of this remark, but in Section 8 we shall look further into the similarity of behaviour of $\log \hat{f}(z)$ and $\log \mid \phi(z) \mid$ where $\phi(z)$ is analytic.

§7. Schwarz's lemma for p.a. functions

By using the principle enunciated in the preceding section (Lemma 6.1) we shall prove a form of Schwarz's lemma for p.a. functions.

THEOREM 7.1. *Let* $f(z) = \{f_1(z), \cdots, f_m(z)\}$ *be* p.a. *in the unit sphere*

$$C\colon \| z \| < 1 \tag{7.1}$$

and let

$$\hat{f}(z) \leqq 1 \qquad \text{in } C \tag{7.2}$$

(i) *If*

$$\hat{f}(0) = 0 \tag{7.3}$$

then

$$\hat{f}(z) \leqq \| z \|^{1/m} \qquad \text{in } C \tag{7.4}$$

(ii) *If there are positive constants c and* β *such that*

$$\hat{f}(z) \leqq c \| z \|^{\beta} \qquad \text{in } C \tag{7.5}$$

then

$$\hat{f}(z) \leqq \| z \|^{\beta} \qquad \text{in } C. \tag{7.6}$$

For the proof we will use the following two lemmas.

LEMMA 7.1. *If* $f(z)$ *is* p.a. *in the unit sphere C and if* (7.2) *and* (7.3) *hold, then*

$$\hat{f}(z) \leqq (2^m - 1)^{1/m} \| z \|^{1/m} \qquad \text{in } C. \tag{7.7}$$

LEMMA 7.2. *If* $\phi(z)$ *is analytic in the unit sphere C and if there are positive constants c and* β *such that*

$$| \phi(z) | \leqq \min \{1, c \| z \|^{\beta}\} \text{ in } C, \tag{7.8}$$

then

$$| \phi(z) | \leqq \| z \|^{\beta} \qquad \text{in } C. \tag{7.9}$$

For the proof of Lemma 7.1 we see by (7.2) and part (i) of Lemma 5.1 that the symmetric functions $B(z)$ in (5.7) satisfy the inequalities

$$| B_{\mu}(z) | \leqq \binom{m}{n} \qquad \text{in } C$$

By (7.3) and (5.13), each $B_{\mu}(z)$ vanishes at the origin. Hence by Schwarz's lemma for analytic functions of several complex variables [5, p. 61]

$$| B_{\mu}(z) | \leqq \binom{m}{n} \| z \| \qquad \text{in } C. \tag{7.10}$$

Inserting (7.10) into (5.7), and again using (7.2) we see that

$$\hat{f}^m(z) \leqq \| z \| \sum_{\mu=1}^{m} \binom{m}{\mu} = (2^m - 1) \| z \|$$

This completes the proof of Lemma 7.1.

Lemma 7.2 is a result on analytic functions of several complex variables which can be derived from Schwarz's lemma in one complex variable in the following manner. From (7.8) we conclude not only that ϕ vanishes at the origin, but also that the function

$$\phi(a_1 t, \cdots, a_k t) \tag{7.11}$$

as a function of t, for any fixed point a in C, has a zero at $t = 0$ of order $\geqq \beta'$, where β' is the largest integer less than $\beta + 1$. This follows from the fact that the function (7.11) is analytic for $|t| \leqq 1$, and from the fact that it satisfies the inequality

$$|\phi(a_j t)| \leqq \min\{1, c \| a \|^\beta |t|^\beta\} \text{ for } |t| \leqq 1. \tag{7.12}$$

From here on we will use only the first part of the inequality (7.12),

$$|\phi(a_j t)| \leqq 1, \qquad |t| \leqq 1 \tag{7.13}$$

together with the fact that (7.11) has a zero of order $\geqq \beta'$ at $t = 0$. If we apply the maximum modulus theorem to the function in (7.11) divided by $t^{\beta'}$, then by (7.13) we see that

$$|\varphi(a_j t)| \leqq |t|^{\beta'}, \qquad |t| \leqq 1. \tag{7.14}$$

Now let z be any point in C other than the origin, let θ be a number in the interval $\| z \| < \theta < 1$, and choose

$$a_j = \frac{\theta z_j}{\| z \|}, \qquad t = \frac{\| z \|}{\theta}.$$

Then (7.14) yields

$$|\varphi(z)| \leqq \left(\frac{\| z \|}{\theta}\right)^{\beta'}. \tag{7.15}$$

Since this holds for every θ in $\| z \| < \theta < 1$, it follows that

$$|\phi(z)| \leqq \| z \|^{\beta'} \qquad \text{in } C. \tag{7.16}$$

Since β' was defined to be the largest integer less than $\beta + 1$ we a fortiori have (7.9).

We now use Lemma 7.2 to prove part (ii) of Theorem 7.1. Since Lemma 7.2 is valid for every pair of positive constants c and β it follows that the set $\{S, B, z^0, h(z), \sigma\}$ with

$$S = C, \qquad B = C, \qquad z^0 \text{ any point in } C,$$

$$h(z) = \min\{1, c \| z \|^\beta\}, \qquad \sigma = \| z^0 \|^\beta$$

is an A-set, and hence by the principle expressed in Lemma 6.1, we have part (ii) of Theorem 7.1.

By use of this part of Theorem 7.1, and Lemma 7.1, we have part (i) of Theorem 7.1.

We remark that in the conclusion of Lemma 7.2 the exponent β can be replaced by β'. This, however, does not imply that the exponent β in the conclusion (7.6) can be replaced by β'.

§8. Subharmonic character of $\log \hat{f}(z)$

If $f(z)$ is p.a. in a region S of E_{2k} and if $\hat{f}(z)$ is the maximum function (6.3), then we may use the principle on A-sets (Lemma 7.1) to prove that $\log \hat{f}(z)$ is what Thorin [14] calls a *bisubharmonic* function, or what we have called a *Hartogs* function [5].

DEFINITION 8.1. A real valued function $u(z_1, \cdots, z_k)$ of the $2k$ real variables $x_1, y_1, \cdots, x_k, y_k (z_j = x_j + iy_j)$ in a domain S is called *bisubharmonic* in S if for every point z^0 in S and every point $(a_1, \cdots, a_k)$ on the unit sphere $\| a \| = 1$, the function

$$u(z_1^0 + a_1 t, \cdots, z_k^0 + a_k t)$$

is a subharmonic function of t in a neighborhood of $t = 0$.

Important examples of bisubharmonic functions are

$$| \phi(z) |^{\lambda}, \qquad \log | \phi(z) |$$

where $\phi(z)$ is analytic in S, and λ is a positive constant.

We shall prove the following theorem.

THEOREM 8.1. *Let* $f(z)$ *be* p.a. *in a region* S *of* E_{2k}. *Then*

$$\log \hat{f}(z) \tag{8.1}$$

is bisubharmonic in S.

PROOF. Since $\hat{f}(z) \geqq 0$ and since $\hat{f}(z)$ is continuous in S (see (5.8)), it follows that $\log \hat{f}(z)$ is upper semi-continuous in S.

Now let $z^0 \in S$, $\| a \| = 1$, and let $\rho = \rho(z^0, a)$ be so small that the circle

$$(z_1^0 + a_1 t, \cdots, z_k^0 + a_k t), \qquad | t | < \rho \tag{8.2}$$

lies in S. Let R be any region in the t-plane with boundary B' such that $R + B'$ lies in $| t | < \rho$. Finally let $h(t)$ be any function such that

(8.3) $\log h(t)$ is continuous in $R + B'$ and harmonic in R

and such that

$$\hat{f}(z_j^0 + a_j t) \leqq h(t) \tag{8.4}$$

for t on B'. We need to show that this same inequality holds for t in R. For this purpose we first note that if $\phi(z)$ is analytic in S and if for some $\alpha > 0$

$$| \phi(z_j^0 + a_j t) | \leqq h^{\alpha}(t) \tag{8.5}$$

for t on B', then the same inequality (8.5) holds for t in R since

$$\log | \phi(z_j^0 + a_j t) |$$

is subharmonic in t in $|t| < \rho$ and since $\alpha \log h(t)$ is harmonic in R and continuous in $R + B'$. Hence by Lemma 6.1 we have (8.4) for t in R. This, together with the upper semi-continuity of $\log \hat{f}$ implies that $\log \hat{f}(z)$ is a bisubharmonic function of z in S.

We can also prove Theorem 8.1 directly from (ii) of Lemma 5.1 by making use of the following result on bisubharmonic functions.

If $\{u_n(z_1, \cdots, z_k)\}$ is a sequence of bisubharmonic functions in a region S and if the sequence is uniformly bounded above in every compact set in S, then the function

$$u(z) = \lim_{n\to\infty} \sup u_n(z)$$

is bisubharmonic in S whenever u is upper semi-continuous in S.

This result is an immediate consequence of the definition and the analogous result for subharmonic functions, [12].

Now let $f(z)$ be p.a. in S. Then each of the functions $s_\alpha(z)$ of (5.9) is analytic in S and hence each of the functions

$$\frac{1}{\alpha} \log | s_\alpha(z) |$$

is bisubharmonic in S. Moreover by (5.11)

$$\frac{1}{\alpha} \log | s_\alpha(z) | \leqq \log \hat{f}(z)$$

and we know that $\log \hat{f}(z)$ is bounded above in every compact set in S. Finally by (5.12) we again conclude that $\log \hat{f}(z)$ is bisubharmonic in S.

We now state two further properties of bisubharmonic functions which we will later use.

LEMMA 8.1. *Let $u(z) \geqq 0$ in a domain S of E_{2k}. Then*

(i) $\log u$ *is bisubharmonic in S if and only if u^λ is bisubharmonic there for every $\lambda > 0$;*

(ii) $\log u$ *is bisubharmonic in S if and only if*

$$u(z)e^{\mathrm{Re}(b\cdot z)} \qquad (b\cdot z = b_1z_1 + \cdots + b_kz_k)$$

is bisubharmonic there for every choice of the complex constants $b_1, \cdots, b_k$.

These two properties are immediate consequences of the definition of bisubharmonic functions and the known analogous properties of subharmonic functions, [12].

By use of Theorem 8.1 and Lemma 8.1 we obtain the following result for several p.a. functions.

LEMMA 8.2. *Let*

$$f^{(\sigma)}(z) = \{f_1^{(\sigma)}(z), \cdots, f_{m_\sigma}^{(\sigma)}(z)\}, \qquad \sigma = 1, \cdots, s \tag{8.6}$$

be s functions, each p.a. *in a region S of E_{2k}. Then for every $\lambda > 0$ the function*

$$\log \{[\hat{f}^{(1)}(z)]^\lambda + \cdots + [\hat{f}^{(s)}(z)^\lambda]\} \tag{8.7}$$

is bisubharmonic in S.

PROOF. Since $\log \hat{f}^{(\sigma)}(z)$ is bisubharmonic and since $\text{Re}(b\cdot z)$ is bisubharmonic (even biharmonic) it follows from part (i) of Lemma 8.1 that

$$[\hat{f}^{(\sigma)}(z)\, e^{\text{Re}(b\cdot z)}]^{\lambda}$$

is bisubharmonic in S for every $\lambda > 0$ and for every choice of the complex constants $b_1, \cdots, b_k$. Hence the sum

$$\{[\hat{f}^{(1)}(z)]^{\lambda} + \cdots + [\hat{f}^{(s)}(z)]^{\lambda}\}e^{\lambda \text{Re}(b\cdot z)}$$

is bisubharmonic, and by part (ii) of Lemma 8.1 this yields the desired conclusion of Lemma 8.2.

§9. Convexity theorems

If $f(z) = \{f_1(z), \cdots, f_m(z)\}$ is p.a. in a region S of E_{2k} then as we saw in Theorem 8.1 the function $\log \hat{f}(z)$ is bisubharmonic in S. Thus we have at out disposal the various known results on bisubharmonic functions. Thorin [14] has given several such results which we interpret for p.a. functions.

THEOREM A (*Principle of the maximum*). *Let $f(z)$ be* p.a. *in a region S of E_{2k}. Then $\hat{f}(z)$ has no maximum in S unless $\hat{f}(z)$ is a constant. If S' is any bounded region which with its boundary B' is contained in S, and if $\hat{f}(z) \leqq M$ on B', then $\hat{f}(z) \leqq M$ in S'.*

Before giving the remaining results we will recall two types of domains which occur frequently in the study of analytic functions of several complex variables, namely Reinhardt regions, and tubes.

Let B_0 be a region in real Euclidean k-dimensional space. By a *Reinhardt region R with basis B_0* we will understand the region in $(z_1, \cdots, z_k)$-space which consists of all points z for which

$$(\log |z_1|, \cdots, \log |z_k|) \in B_0 .$$

By a tube T with basis B_0 we will understand the region in $(z_1, \cdots, z_k)$-space consisting of all points z for which

$$(\text{Re}\, z_1, \cdots, \text{Re}\, z_k) \in B_0 .$$

THEOREM B (Three circles theorem and Hardy's theorem).

(i) *Let $f(z)$ be* p.a. *in a ring $\sigma_1 < \| z \| < \sigma_2$. Then*

$$\log \max_{\|z\|=\rho} \hat{f}(z)$$

is a convex function of $\log \rho$ *in* $\sigma_1 < \rho < \sigma_2$.

(ii) *Let $f(z)$ be* p.a. *in a Reinhardt region R with convex basis B_0. Then*

$$\log \left[\max_{\substack{0 \leqq \theta_j \leqq 2\pi \\ j=1,\cdots,k}} \hat{f}(\rho_1 e^{i\theta_1}, \cdots, \rho_k e^{i\theta_k}) \right] \tag{9.1}$$

and

$$\log \int_0^{2\pi} \cdots \int_0^{2\pi} [\hat{f}(\rho_1 e^{i\theta_1}, \cdots, \rho_k e^{i\theta_k})]^{\lambda}\, d\theta_1 \cdots d\theta_k \tag{9.2}$$

are convex functions of $\log \rho_1, \cdots, \log \rho_k$ *in* B_0, *this being true for* (9.2) *for every* $\lambda > 0$.

THEOREM C (Three lines theorem and a related result). *Let* $f(z)$ *be* p.a. *in a tube* T *with convex basis* B_0, *and let* $\hat{f}(z)$ *be bounded in* T. *Then*

$$\log \left[\sup_{\substack{-\infty<y_j<\infty \\ j=1,\cdots,k}} \hat{f}(x_1 + iy_1, \cdots, x_k + iy_k) \right] \tag{9.3}$$

and

$$\log \int_{-\infty}^{\infty} \cdots \int_{-\infty}^{\infty} [\hat{f}(x_1 + iy_1, \cdots, x_k + iy_k)]^{\lambda} \, dy_1 \cdots dy_k \tag{9.4}$$

are convex functions of $x_1, \cdots, x_k$ *in* B_0, *this being true for* (9.4) *for every* $\lambda > 0$ *for which* (9.4) *is bounded above in* B_0.

As we remarked earlier these results are essentially reformulations, for p.a. functions, of results given by Thorin for bisubharmonic functions. For their proof we may refer to Thorin's paper. It might be interesting, however, to give the proof of one of them by use of the notion of A-sets. As an example we prove part (i) of Theorem B by this method.

By Hadamard's three spheres theorem for analytic functions of several complex variables [5], the following set is an A-set:

$S: \rho_1 < \| z \| < \rho_3$; $\quad B$: the union of $\| z \| = \rho_1$ and $\| z \| = \rho_3$;

z^0: an arbitrary point of S;

$$h(z) = \begin{cases} M_1 \text{ on } \| z \| = \rho_1 \\ M_3 \text{ on } \| z \| = \rho_3 \end{cases} \qquad M_1, M_3 \text{ constants} \geqq 0;$$

$$\sigma = (M_1^{b_1} M_3^{b_3})^{1/b_2}$$

where

$$b_1 = \log (\rho_3/\rho_2) \qquad b_2 = \log (\rho_3/\rho_1) \qquad b_3 = \log (\rho_2/\rho_1), \qquad \rho_2 = \| z^0 \|.$$

If $f(z)$ is p.a. in the ring $\sigma_1 < \| z \| < \sigma_2$ and if for $\sigma_1 < \rho_1 < \rho_2 < \rho_3 < \sigma_2$ we set

$$M_\gamma = \max_{\| z \| = \rho_\gamma} \hat{f}(z), \qquad \gamma = 1, 2, 3,$$

then Lemma 6.1 with the set $\{S, B, z^0, h(z), \sigma\}$ as above, tells us that

$$M_2 \leqq (M_1^{b_1} M_3^{b_3})^{1/b_2}$$

which means that $\log \max \hat{f}(z)$ is a convex function of $\log \rho$.

We make the following remarks.

REMARK 1. If $k > 1$, that is if we deal with the case of more than one complex variable, then any function $f(z)$ which is p.a. in a ring $\sigma_1 < \| z \| < \sigma_2$ can be extended so as to be p.a. in the sphere $\| z \| < \sigma_2$.

This follows at once from Hartog's theorem on analytic continuation; the functions $B_1(z), \cdots, B_m(z)$ of (5.7) have analytic continuations from the ring into the sphere, see for example [11, p. 206; 5, p. 82].

REMARK 2. If $f^{(\sigma)}(z)$ are s p.a. functions as in (8.6), then each of the theorems Theorems A, B and C, remain valid if we replace $\hat{f}(z)$ by the expression

$$\{[\hat{f}^{(1)}(z)]^p + \cdots + [\hat{f}^{(s)}(z)]^p\}^{1/p} \tag{9.6}$$

for any $p > 0$.

This follows from Lemma 8.2.

In connection with the maximum principle, Theorem A, the following formulation (which is an immediate consequence of Lemma 6.1) is frequently of use for the case of more than one complex variable.

REMARK 3. If S is any domain in E_{2k}, if B is any subset of the boundary with the property that

$$|\phi(z)| \leqq \sup |\phi(w)|, \qquad z \text{ in } S, w \text{ on } B,$$

always holds for analytic functions ϕ, then the inequality

$$\hat{f}(z) \leqq \sup \hat{f}(w), \qquad z \text{ in } S, w \text{ on } B,$$

always holds for p.a. functions f.

PART III

§10. A complex space S_{2k} with singularities

We now define globally what we will term "a complex space S_{2k} with singularities."

We take a connected topological space with a covering by open sets $\{V_\sigma\}$. We ask that each V_σ be a fixed topological image of a topological space $U = U_\sigma$ of the type defined in the first paragraph of Section 3. We assume that the region $S = S_\sigma$ in E_{2k} associated with U_σ is in each case the unit sphere

$$C: \| z \| < 1.$$

Thus for each σ there are given a topological space U_σ having the character of the space U of Section 3, referred to the region $S = C$, and a topological map t_σ of U_σ onto V_σ,

$$U_\sigma \underset{t_\sigma}{\rightarrow} V_\sigma. \tag{10.1}$$

With each U_σ there are associated $n = n(\sigma)$ functions $\psi_{\nu,\sigma}(z, \zeta_1, \cdots, \zeta_\nu)$, $\nu = 1, \cdots, n$, as in (2.1), functions $D_\sigma(z)$, $\Delta_\sigma^2(z)$, as well as the other functions and properties discussed at the beginning of Sections 2 and 3. With each U_σ we now associate a function $\Phi_\sigma(z)$ which is analytic and not identically zero in C, and such that the set $\Phi_\sigma(z) = 0$ includes the points E_σ where $D_\sigma(z) = 0$. We select in U_σ those points whose pre-images z satisfy the equation $\Phi_\sigma(z) = 0$ and denote this subset of U_σ by $M_\sigma = M(U_\sigma, \Phi_\sigma)$. The functions Φ_σ will be held fixed.

We denote by N_σ the image of M_σ,

$$(10.2) \qquad M_\sigma \underset{t_\sigma}{\longrightarrow} N_\sigma$$

and we assume that the topological map t_σ of the subspace $U_\sigma - M_\sigma$ onto the subspace

$$(10.3) \qquad V_\sigma - N_\sigma$$

is *holomorphic*. In this way we endow the subspace (10.3) with a complex structure of the sort described for $U - M$. Finally we assume that the union

$$(10.4) \qquad W = \textstyle\sum_\sigma \{V_\sigma - N_\sigma\},$$

which is a subspace of S_{2k}, is locally Euclidean and is endowable with a complex structure globally in such a manner that the individual complex structure of the parts (10.3) shall be holomorphically compatible with it.

An object with all these properties is what we term a complex space S_{2k} with singularities.

If $0 < \rho < 1$ we shall denote by $U_\sigma(\rho)$ the subset of those points of U_σ for which $\| z \| < \rho$, and by $V_\sigma(\rho)$ the image of $U_\sigma(\rho)$ under t_σ.

§11. Holomorphic functions in S_{2k}

We say that a function F is *holomorphic* in S_{2k} if it is defined at every point of S_{2k} and is holomorphic in each V_σ in the earlier sense; that is, if it is bounded in each $V_\sigma(\rho)$ $(\rho < 1)$ and if in each $V_\sigma - N_\sigma$ it has a representation of the form

$$(11.1) \qquad \frac{1}{\Delta^2(z)} \sum_{\lambda_1=0}^{m_1-1} \cdots \sum_{\lambda_n=0}^{m_n-1} P_{\lambda_1\cdots\lambda_n}(z)\zeta_1^{\lambda_1} \cdots \zeta_n^{\lambda_n}$$

where the coefficients P are analytic in the entire unit sphere.

In writing (11.1) in $V_\sigma - N_\sigma$ we work for convenience with the coordinates (z, ζ) of the pre-image U_σ. The values of the ζ's in (11.1) are, of course, dependent upon the z's, and the various quantities Δ, m_μ, P_λ and the ζ's depend upon σ.

Since F is holomorphic in each V_σ we may use the work of Parts I and II.

We shall prove the following result.

LEMMA 11.1. *A function which is holomorphic and continuous on a compact S_{2k} is identically constant.*

If F is continuous, so is $| F |$, and since for this lemma S_{2k} is assumed compact, $| F |$ assumes its maximum at some point Q of S_{2k}. But Q must lie in some V_σ. Let (z^0, ζ^0) be the pre-image of Q in U_σ. Since $| F |$ assumes its maximum at Q, the function $\hat{f}(z) = \hat{f}_\sigma(z)$ defined as in (2.17) and (2.10) relative to the series (11.1), has a maximum at z^0. Now $\hat{f}(z)$ is bisubharmonic in C (see Theorem 8.1 and Lemma 8.1), and cannot have a maximum in an inner point of C unless it is constant. By Lemma 2.3 this implies that there is a constant c such that

$$P_{\lambda_1\cdots\lambda_n}(z) = \begin{cases} c\Delta^2(z) \text{ in } C, \text{ if } \lambda_1^2 + \cdots + \lambda_n^2 = 0 \\ 0 \quad\ \text{ in } C, \text{ if } \lambda_1^2 + \cdots + \lambda_n^2 > 0. \end{cases}$$

Hence $F = c$ in $V_\sigma - N_\sigma$ and by continuity in V_σ. By analytic continuation we conclude that $F = c$ in the complex coordinate subspace (10.4), and hence in the total space S_{2k}.

We shall now drop the requirement that S_{2k} be compact but for the next result we shall add the assumption that each V_σ is such that its point-set closure in S_{2k} is compact, that is we will ask that each V have compact closure. With this additional assumption on our space S_{2k} we shall prove the following result.

LEMMA 11.2. *Let S_{2k} be defined as in Section 10 with the added property that each of the open sets V_σ has compact closure. Let $\{F_s\}$ be a sequence of holomorphic functions which converge uniformly in the union*

$$\sum_\sigma V_\sigma(\rho)$$

for every $\rho < 1$. Then the limit function F is holomorphic in S_{2k}.

This result is an immediate consequence of Corollary 4.2.

As before we say that a function $F(P)$ is *strictly holomorphic* in S_{2k} if it is defined at every point of S_{2k} and if in each V_σ it has a representation of the form

$$\sum_{\lambda_1=0}^{m_1-1} \cdots \sum_{\lambda_n=1}^{m_n-1} P_{\lambda_1 \cdots \lambda_n}(z)\, \zeta_1^{\lambda_1} \cdots \zeta_n^{\lambda_n} \tag{11.2}$$

where the coefficients P_λ are analytic in the entire unit sphere C.

A strictly holomorphic function is continuous on S_{2k}, and if S_{2k} happens to be compact, then such a function is identically constant.

§12. Algebraic and linear dependence

THEOREM 12.1. *Let $S = S_{2k}$ be a complex space with singularities, defined as in Section 11, with the added property that there is a positive number $r < 1$ such that the open sets $V_\sigma(r)$ cover S. Let S^0 be a domain in S whose closure $\bar{S}^0$ can be covered by a finite number of the open sets $V_\sigma(r)$.*

If $k + 2$ strictly holomorphic functions

$$F_1, \cdots, F_{k+2} \tag{12.1}$$

on S have the property that there is a constant $M > 1$ such that for every homogeneous polynomial

$$Q_p(t_1, \cdots, t_{k+2}) \not\equiv 0 \tag{12.2}$$

of any degree p, the function

$$G(P) = Q_p[F_1(P), \cdots, F_{k+2}(P)] \tag{12.3}$$

satisfies the relation

$$\sup_{P \epsilon S} |G(P)| \leqq M^p \sup_{P \epsilon S^0} |G(P)|, \tag{12.4}$$

then there exists one such polynomial $Q_p^0(t)$ for which

$$Q_p^0[F_1, \cdots, F_{k+2}] = 0. \tag{12.5}$$

The proof is essentially the same as that given in [**3**]. We repeat it here, with modifications where necessary. The other parts of this section also parallel results given in [**3**].

Denote by $V_1(r), \cdots, V_q(r)$ the (finite number of) open sets which cover $\bar{S}^0$, and denote by m_0 the sum

$$m_0 = m^{(1)} + \cdots + m^{(q)}, \tag{12.6}$$

where $m^{(\tau)}$ is the number of sets of solutions (2.3) associated with the set $U_\tau = t_\tau^{-1} V_\tau$.

With the given functions (12.1) we form for fixed p the monomials

$$G_A(P) = F_1^{p_1} \cdots F_{k+2}^{p_{k+2}}, \qquad A = 1, \cdots, \tilde{p} \tag{12.7}$$

where

$$p_1 + \cdots + p_{k+2} = p \quad \text{and} \quad \tilde{p} = \binom{p+k+1}{k+1} \tag{12.8}$$

and the claim is that for p sufficiently large these $\tilde{p}$ monomials are linearly dependent. In order to see this assume for the moment that for each p these $\tilde{p}$ monomials are linearly independent and, for fixed p let $\beta = \beta_p$ be a positive integer satisfying the inequality

$$m_0 \binom{\beta+k-1}{k} < \binom{p+k+1}{k+1}, \tag{12.9}$$

where m_0 has the value (12.6).

Now with β so defined we determine $\tilde{p} = \binom{p+k+1}{k+1}$ constants $c_1, \cdots, c_{\tilde{p}}$ such that the linear combination

$$G(P) = \textstyle\sum_{A=1}^{\tilde{p}} c_A G_A(P) \tag{12.10}$$

which is strictly holomorphic in S, satisfies the two relations

$$\sup_{P \in S} | G(P) | = 1 \tag{12.11}$$

and

$$| G(P) | \leqq r^\beta \qquad \text{in } \bar{S}_0 . \tag{12.12}$$

In order to achieve (12.11) and (12.12) we first determine the constants c_A (not all zero) so that in each V_τ (or in the pre-image U_τ), $\tau = 1, \cdots, q$, G has a development of the form (11.2) with each $P_{\lambda_1 \cdots \lambda_n}(z)$ having a zero of order $\geqq \beta$ at $z = 0$. As we noted at the end of Section 4 (see (4.28)) this requires

$$m^{(\tau)} \cdot \binom{\beta+k-1}{k}$$

conditions for each V_τ, and with $\tilde{p}$ defined as in (12.8) we can always achieve this because of (12.9).

With the c_A so determined the function (12.10) is bounded in $\bar{S}^0$ since it is bounded in each of the finitely many sets $V_\tau(r)$ which cover $\bar{S}^0$, and by the basic assumption (12.4) it is bounded in S. Since it is not identically zero we can achieve (12.11) without destroying the property that in its development (11.2) in each U_τ, each $P_{\lambda_1 \cdots \lambda_n}$ has a zero of order $\geqq \beta$ at $z = 0$. Having this and (12.11) we use Schwarz's lemma. First we note that each of the P's satisfies an inequailty of the form (4.24), and hence in each V_τ, $\tau = 1, \cdots, q$ we have

$$| G(P) | \leqq \min \{1, c \| z \|^\beta\} \qquad P = (z, \zeta) \in V_\tau,$$

for a suitable constant c. By Schwarz's lemma (Theorem 7.1) this yields (12.12) in $V_\tau(r)$ and hence in $\bar{S}_0$.

It follows from (12.11) and (12.4) that there is a point P' in $\bar{S}^0$ such that

$$(12.13) \qquad | G(P') | \geqq \frac{1}{M^p}.$$

A comparison of (12.12) and (12.13) gives

$$(12.14) \qquad \beta_p \leqq \frac{\log M}{\log 1/r}\, p.$$

Now, for k fixed, and p and β_p large, we have

$$\binom{p + k + 1}{k + 1} \sim \frac{1}{(k+1)!}\, p^{k+1}, \qquad \binom{\beta + k - 1}{k} \sim \frac{1}{k!}\, \beta^k$$

and therefore it is compatible with (12.9) to have

$$\beta_p \geqq C_0 p^{1+(1/k)}$$

where C_0 is fixed and p is large. But this is not compatible with (12.14), and hence our temporary assumption that for every p the $\tilde{p} = \binom{p + k + 1}{k + 1}$ monomials (12.7) are linearly independent leads to a contradiction. This completes the proof of the theorem.

We now take a space as before and for the present purpose we denote it by $\tilde{S}_{2k}$. On it we envisage a group $\Gamma: (\gamma)$ of topological transformations with the following properties.

(i) There is a region S^0 in $\tilde{S}_{2k}$ whose closure $\bar{S}^0$ can be covered by a finite number of the open sets $\{V_\sigma(r)\}$ (r fixed, $0 < r < 1$);

(ii) to every point P of $\tilde{S}_{2k}$ there exists an element γ of Γ such that the image γP lies in $\bar{S}^0$;

(iii) there exist an open neighborhood S of $\bar{S}^0$ and a finite number of elements $\gamma_1, \cdots, \gamma_\alpha$ of Γ such that the union of the images $\gamma_1 \bar{S}^0, \cdots, \gamma_\alpha \bar{S}^0$ cover S.

We call a function $F(P)$ *automorphic* relative to this set-up if $F(P)$ is strictly holomorphic on $\tilde{S}_{2k}$ and if with every γ of Γ there is associated a strictly holomorphic function $\eta_\gamma(P)$ on $\tilde{S}_{2k}$ such that

$$F(\gamma P) = \eta_\gamma(P)F(P) \tag{12.15}$$

on $\tilde{S}_{2k}$. The factors of automorphy, $\{\eta_\gamma(P)\}$ shall satisfy the chain relation

$$\eta_{\gamma\delta}(P) = \eta_\gamma(\delta P)\eta_\delta(P) \tag{12.16}$$

for all pairs of elements $\gamma, \delta \in \Gamma$, so that the equations (12.15) shall be compatible for the various elements of Γ.

We shall now prove

THEOREM 12.2. *If we are given a complex space* $\tilde{S}_{2k}$, *and a system of factors* $\{\eta_\gamma(P)\}$ *of automorphy which satisfy* (12.16), *then any* $k + 2$ *strictly holomorphic functions on* $\tilde{S}_{2k}$ *which are automorphic with regard to* (12.15) *satisfy an equation* (12.5) *as before.*

PROOF. Each of the functions $\eta_\gamma(P)$, $\gamma = \gamma_1, \cdots, \gamma_\alpha$, is bounded in $\bar{S}^0$ since $\bar{S}^0$ is covered by a finite number of the sets $V_\sigma(r)$ with $r < 1$. We denote by M the supremum of $|\eta_\gamma(P)|$ for P in $\bar{S}^0$ and $\gamma = \gamma_1, \cdots, \gamma_\alpha$. Since the union of $\gamma_1 \bar{S}^0, \cdots, \gamma_\alpha \bar{S}^0$ covers S, (12.15) implies (12.4). Thus all assumptions of Theorem 12.1 are fulfilled and we have Theorem 12.2.

COROLLARY 12.1. *If* S, S^0 *are as in Theorem* 12.1; *if* **F** *is a module of strictly holomorphic functions on* S, *each bounded; and if for all elements* $F(P)$ *of* **F** *we have*

$$\sup_{P\in S} |F(P)| \leqq M \sup_{P\in S^0} |F(P)| \tag{12.17}$$

for some fixed M; *then for some suitable integer* N *depending only on* S, S^0, M *any* $N + 1$ *elements of* F *are linearly dependent.*

PROOF. As in the proof of Theorem 12.1; if the corollary were false, we would have an element $F(P)$ of **F** whose expansion in each of the sets V_τ, $\tau = 1, \cdots, q$ would be of the form (11.2) with each $P_\lambda(z)$ having a zero of order $\geqq \beta$ at $z = 0$, where β is any positive integer for which

$$m_0\binom{\beta + k - 1}{k} < N. \tag{12.18}$$

If we put $\sup_{P\in S} |F(P)| = 1$, then at a suitable point P' of $\bar{S}^0$ we would have

$$\frac{1}{M} \leqq |F(P')| \leqq r^\beta$$

which for large N, and hence large β, is contradictory.

This leads to the following result

COROLLARY 12.2. *If* $\tilde{S}_{2k}$, $\{\eta_\gamma(P)\}$ *are as in Theorem* 12.2, *then the module of automorphic functions has a finite basis.*

PRINCETON UNIVERSITY;
THE INSTITUTE FOR ADVANCED STUDY AND
THE MASSACHUSETTS INSTITUTE OF TECHNOLOGY

Bibliography

1. H. Behnke and K. Stein, *Modifikation komplexer Mannigfaltigkeiten und Riemannscher Gebiete*, Math. Ann., vol. 124 (1951) pp. 1–16.
2. S. Bochner, *On compact complex manifolds*, J. Indian Math. Soc., vol. 11 (1947) pp. 1–21.
3. S. Bochner, *Algebraic and linear dependence of automorphic functions in several variables*, J. Indian Math. Soc., vol. 16 (1952) pp. 1–6.
4. S. Bochner, *Structure of complex spaces*, Conference on Riemann Surfaces in Commemoration of the Hundredth Anniversary of Riemann's Inaugural Dissertation, Princeton, 1951, Annals of Math. Studies (to appear).
5. S. Bochner and W. T. Martin, Several complex variables, Princeton, 1948.
6. H. Bohr, *Zur Theorie der fast periodischen Funktionen, I. Eine Verallgemeinerung der Theorie der Fourierreihen*, Acta Math., vol. 45 (1925) pp. 29–127.
7. Newton S. Hawley, *A theorem on compact complex manifolds*, Ann. of Math., vol. 52 (1950) pp. 637–641.
8. H. Hermes, *Analytische Mannigfaltigkeiten in Riemmannschen Bereichen*, Math. Ann., vol. 120 (1949) pp. 539–562.
9. B. O. Koopman and A. B. Brown, *The Riemann multiple space and algebroid functions*, Trans. Amer. Math. Soc., vol. 36 (1934) pp. 618–626.
10. P. Lelong, *Les fonctions plurisousharmoniques*, Ann. L'École Norm., vol. 62 (1943) pp. 301–338.
11. W. F. Osgood, Lehrbuch der Funktionentheorie, vol. II, part I, 1929, second edition, Leipzig.
12. T. Rado, *Subharmonic functions*, Ergebnisse der Math. und ihrer Grenzgebiete, vol. 5, part 1 (1942) pp. 1–56.
13. W. Rückert, *Zum Eliminationsproblem der Potenzreihenidealen*, Math. Ann. vol. 107 (1932) pp. 259–281.
14. G. O. Thorin, *Convexity theorems*, Meddelanden Från Lunds Univ. Mat. Sem., vol. 9 (1948) pp. 1–58.
15. O. Zariski, *Theory and applications of homomorphic functions on algebraic varieties over arbitrary ground fields*, Memoirs of the Amer. Math. Soc., No. 5 (1951) 90 pp

STRUCTURE OF COMPLEX SPACES

S. Bochner

§1. COMPLEX AND REAL COORDINATE SPACES

The natural generalization of the classical Riemann surface from one to several variables is the connected complex coordinate space S_{2k} and we define it as follows, see [14], p. 54. It is a neighborhood space, locally Euclidean, of an even dimension $n = 2k$; and there are given a covering of it by a family of neighborhoods $\{U_\nu\}$, and topological mappings $V_\nu = \chi_\nu (U_\nu)$ into the ordinary space E_{2k} of the variables $z_\alpha = x_\alpha + iy_\alpha$, $\alpha = 1,\ldots,k$, such that if two neighborhoods U_μ, U_ν have a non-empty intersection U, and if we put $V = \chi_\mu(U)$, $V' = \chi_\nu(U)$, then the functions

$$x'_\alpha = \varphi_\alpha(x,y), \quad y'_\alpha = \psi_\alpha(x,y), \quad \alpha = 1,\ldots,k \tag{1}$$

which transmit the map $V' = \chi_\nu \chi_\mu^{-1} (V)$ shall be the real and imaginary parts of certain holomorphic functions

$$z'_\alpha = f_\alpha (z_1,\ldots,z_n) \quad \alpha = 1,\ldots,k. \tag{2}$$

We will call our space a <u>real</u> space if we do not assume that our functions (1) have this holomorphic structure, and we will then assume that they are twice differentiable say. Formally, by putting $\Phi_\alpha = \varphi_\alpha + i\,\psi_\alpha$, we can then always rewrite (1) into

$$z'_\alpha = \Phi_\alpha(z,\bar z), \quad \bar z'_\alpha = \bar\Phi_\alpha , \tag{3}$$

and our real space is also a complex space if and only if we have

$$\frac{\partial \Phi_\alpha}{\partial \bar z_\beta} = 0 = \frac{\partial \bar\Phi_\alpha}{\partial z_\beta}, \quad \alpha,\beta = 1,\ldots,k. \tag{4}$$

Furthermore, for the Jacobian of the transformation (1) we can always write

$$\frac{\partial (z'_1,\ldots,z'_k, \bar z'_1,\ldots,\bar z'_k)}{\partial (z_1,\ldots,z_k, \bar z_1,\ldots,\bar z_k)}, \tag{5}$$

see [14], p. 39, and since for a complex space (4) this is

$$\frac{\partial (z'_1, \ldots, z'_k)}{\partial (z_1,\ldots,z_k)} \cdot \frac{\partial (\bar z'_1, \ldots, \bar z'_k)}{\partial (\bar z_1,\ldots,\bar z_k)}, \tag{6}$$

it follows that a complex space is always orientable, perforce.

§2. SEPARABILITY

For $k = 1$, T. Rado [17] has shown that our (connected) complex space has a covering by countably many complex neighborhoods automatically. It appears from many indications that for $k \geq 2$ complex structure virtually ceases to have influence on "separability" and we will briefly reproduce Rado's reasoning, in somewhat altered form, in order to show when and why it ceases to be applicable.

THEOREM 1. Let S be any arcwise connected neighborhood space and let $\{S_\alpha\}$ be the family of its perfectly separable arcwise connected open subspace containing a given point. Let E be another such space, and let it be possible to associate with every S_α a perfectly separable open subspace D_α of E such that for $S_\alpha > S_\beta$, $S_\alpha \neq S_\beta$ we have $D_\alpha > D_\beta$, $D_\alpha \neq D_\beta$.
Now, if E itself is perfectly separable, the S itself is likewise perfectly separable.

In fact, if we form the point set

$$D_\infty = \Sigma_\alpha D_\alpha \quad ,$$

then due to Lindelof there is a countable subset $\{D_n\}$, $n = 1,2,\ldots,$ such that

$$D_\infty = \sum_{n=1}^{\infty} D_n,$$

and if we take their preimages $\{S_n\}$ and introduce their union

$$S_\infty = \sum_{n=1}^{\infty} S_n$$

then the latter must be maximal among the separable subspaces of S introduced, but a maximal such element must be S itself, q.e.d.

Now, in Rado's case, S is a connected complex space, and E is the Gaussian sphere, and the D_α's are concentric circles, finite or infinite, or E itself. If now with each S_α we associate that D_α which arises by conformal mapping of its universal covering space by holding the common point and a given line-element of it fixed, then Theorem 1 applies and Rado's assertion follows. However, for $k \geq 2$, Theorem 1 ceases to be applicable towards the same end, not because of any delicate features of the uniformization theory failing to hold, but already on the crude grounds that purely topologically, even among algebraic surfaces only, there are too many types to fit as "subspaces" D_α into a single one, as Theorem 1 would require them to do.

§3. WHEN A REAL SPACE IS ALSO A COMPLEX SPACE

For $k = 1$, an orientable real space is also automatically a complex space if and only if it carries a conformal structure, and we are going to generalize this to $k \geq 2$, first apparent obstacles notwithstanding

For $k = 1$, conformal structure is a line element which in every coordinate system V_α has the form

$$ds^2 = \lambda(x,y)\ (dx^2 + dy^2) = g(z,\bar{z})dz\ d\bar{z}, \tag{7}$$

and for $k \geq 2$, even if S_{2k} is compact, it is probably not always possible to have a line element of the form

$$\lambda(x,y) \sum_{\alpha=1}^{k} (dx_\alpha^2 + dy_\alpha^2) = g(z,\bar{z}) \sum_{\alpha=1}^{k} dz_\alpha\ d\bar{z}_\alpha\ . \tag{8}$$

However, as we have recently pointed out in [9], if S_{2k} has a countable covering by complex neighborhoods, we can always introduce a line-element having the Hermitian form

$$ds^2 = \sum_{\alpha,\beta=1}^{k} g_{\alpha\bar{\beta}}\ dz_\alpha d\bar{z}_\beta\ , \qquad g_{\beta\bar{\alpha}} = \overline{g_{\alpha\bar{\beta}}} \tag{9}$$

and this is our generlization of (7).

Now, conversely, for $k = 1$, if a real space is orientable, and carries a metric (7) then it is automatically complex. For $k \geq 2$, even if a real space is orientable and compact, then the presence of a conformal metric (8), and let alone of an Hermitian metric (9), will not make it complex, as already disproved by the ordinary four-sphere. However the addition to (9) of a certain structural requirement which for a complex space is fulfilled anyway, will so make it, always.

THEOREM 2. If a real space is such that for the functions (3) we have

$$\det \left| \frac{\partial \Phi_\alpha}{\partial z_\beta} \right|_{\alpha,\beta = 1,\dots,k} \neq 0 \tag{10}$$

at all points, and if a non-singular Hermitian metric (9) is known to exist (it need not be definite) then the space is automatically a complex one.

In fact, the equality

$$\Sigma\, g_{\alpha\bar\beta}\, dz_\alpha\, d\bar z_{\beta} = \Sigma\, g'_{\gamma\bar\delta}\, dz'_\gamma\, d\bar z'_\delta \tag{11}$$

will imply the identities

$$g_{\alpha\bar\beta} = g'_{\gamma\bar\delta}\, \frac{\partial \Phi_\gamma}{\partial z_\alpha}\, \frac{\partial \bar\Phi_\delta}{\partial \bar z_\beta}$$

$$0 = g'_{\gamma\bar\delta}\, \frac{\partial \bar\Phi_\gamma}{\partial z_\alpha}\, \frac{\partial \Phi_\delta}{\partial \bar z_\beta} \tag{12}$$

and their conjugate complex ones, and for a non-singular matrix $\{g'_{\gamma\bar\delta}\}$ together with (10), relations (12) will imply relations (4), as claimed.

Next, on a real space, an object with coordinates $(\zeta_1, \ldots, \zeta_k)$ will be called a (covariant) vectorfield if we have

$$\sum_{\alpha=1}^{k} \zeta_\alpha\, dz_\alpha = \sum_{\beta=1}^{k} \zeta'_\beta\, dz'_\beta \quad , \tag{13}$$

but unless postulated explicitly the components need not be holomorphic in the given coordinates, in any manner.

THEOREM 3. If on a real space there are given r vectorfields

$$\left\{\zeta_\alpha^{(\rho)}\right\}, \quad \alpha = 1, \ldots, k; \quad \rho = 1, \ldots, r, \tag{14}$$

$r \geq k$, and if the rank of the matrix (14) has always its maximal value k, then the space is also complex.

In fact, (13) implies

$$0 = \sum_{\beta=1}^{k} \zeta'_\beta\, \frac{\partial \Phi_\beta}{\partial \bar z_\alpha}, \quad \alpha = 1, \ldots, k,$$

and if this holds for r vectorfields of rank k then we obtain (4) as claimed.

Note, that our vectorfields generate the Hermitian metric tensor

$$g_{\alpha\bar\beta} = \sum_{\rho=1}^{r} \zeta_\alpha^{(\rho)}\, \overline{\zeta_\beta^{(\rho)}} \tag{15}$$

and we might have attempted to obtain the conclusion of Theorem 3 by way of Theorem 2 already proven. But this would have necessitated adding assumption (10), which in the direct proof of Theorem 3 is not required, however.

§4. RIGIDITY

We will draw a geometric conclusion, for $k \geq 2$, from Hartog's Theorem that if in $E_{2k} : (z_1, \ldots, z_k)$ a function is holomorphic in a connected neighborhood of a connected boundary of a bounded domain, then it can be continued (uniquely) into the entire domain; and we also note that in [1] and [13] we have generalized this to analytic functions in real variables which satisfy an equation

$$\Sigma_{0 \leq p_1 + \ldots + p_n \leq 2h} \, a_{p_1 \ldots p_n} \frac{\partial^{p_1 + \ldots + p_n} f}{\partial x_1^{p_1} \ldots \partial x_n^{p_n}} = 0$$

in all variables, and an additional equation

$$\Sigma_{0 \leq q_1 + \ldots + q_m \leq m} \, b_{q_1 \ldots q_m} \frac{\partial^{q_1 + \ldots + q_m} f}{\partial x_1^{q_1} \ldots \partial x_m^{q_m}} = 0$$

in fewer than all variables,

$$m < n,$$

both equations with constant coefficients, the second otherwise unrestricted, and the first being elliptic semi-homogeneous as defined in [13].

An immediate conclusion from Hartog's own theorem is that if we can map in E_{2k} the neighborhood of one boundary holomorphically into the neighborhood of another boundary then there exists a unique continuation of the map which carries the domain bounded into the domain bounded, and the reader will see that this can be interpreted in the following manner.

THEOREM 4. For $k \geq 2$, if in a complex S_{2k} we remove a "cap" lying all in one complex neighborhood and replace it by another such cap holomorphically, then the new cap is identical with the old one, unless, of course, it is possible to reorganize the complex structure altogether so as to identify points near the surface of excision with points away from it.

However, it is know already from algebraic geometry that it is very well possible to replace the simple cap by a mitre-like cover which arises from the simple cap by a birational transformation in which all points are mapped one-one except for the center of the cap which is being inflated into a hypersurface P_{2k-2} of the type of projective space. But the mitre is no longer a topological cell, and the problem remains of deciding whether it is possible to replace the cap by a holomorphically inequivalent topological cell also.

§5. KAEHLER METRIC GENERATED BY TENSOR FIELDS

A Kaehler metric is an Hermitian metric (9) with the added condition

$$\frac{\partial g_{\alpha\bar{\beta}}}{\partial z_\gamma} = \frac{\partial g_{\gamma\bar{\beta}}}{\partial z_\alpha} \tag{16}$$

and its conjugate complex; for a detailed discussion, compare [3] and [2], [6]. For $k = 1$, (16) is automatically fulfilled, for $k \geq 2$ there are compact complex manifolds on which no Kaehler metric can exist.

A Kaehler metric certainly does exist if we are given r vector-field as in Theorem 3, but this time holomorphic, and with the curl-property

$$\frac{\partial \zeta_\alpha^{(\rho)}}{\partial z_\beta} = \frac{\partial \zeta_\beta^{(\rho)}}{\partial z_\alpha} \tag{17}$$

added. In fact, the metric (15) already introduced will now have the Kaehler property in consequence of (17); and the meaning of (17) is that we can introduce the r simple Abelian integrals of the first kind

$$w_\rho(z) = \int^z \sum_{\alpha=1}^{k} \zeta_\alpha^{(\rho)} dz_\alpha , \tag{18}$$

which although perhaps not uniquely determined globally (they may have additive periods) are nevertheless creating locally a one-one holomorphic immersion of our S_{2k} into the E_{2r} : $(w_1, \ldots, w_r)$. The metric (15) is then induced by the metric

$$dw_1 \, d\bar{w}_1 + \ldots + dw_r \, d\bar{w}_r \tag{19}$$

in E_{2r}, and since the latter is Kaehlerian, (15) is likewise so. Also, the metric (15) has then non-positive Ricci-curvature

$$-R_{\alpha\bar{\beta}} \leq 0, \tag{20}$$

see [3], p.190; and if S_{2k} is also compact, then (18) maps it locally one-one into a certain compact complex multi-torus, and the Euler characteristic of S_{2k} has then algebraic sign $(-1)^k$ or 0, see [7], p. 255/6.

A parallel but different situation arises if the vector fields given are not covariant but contravariant, thus

$$\left\{ \zeta_{(\rho)}^{\alpha} \right\} \tag{21}$$

again with maximum rank k everywhere. The matrix

$$g^{\alpha\bar{\beta}} = \sum_{\rho=1}^{r} \zeta^{\alpha}_{(\rho)} \overline{\zeta^{\beta}_{(\rho)}} \tag{22}$$

is again positive definite Hermitian, but the requirement that its inverse $g_{\alpha\bar{\beta}}$ shall have the Kaehler property (16) is this time a very restrictive one, see [10] and [2]. It more or less demands that we have

$$\zeta^{\beta}_{(\rho)} \frac{\partial \zeta^{\alpha}_{(\sigma)}}{\partial z_{\beta}} - \zeta^{\beta}_{(\sigma)} \frac{\partial \zeta^{\alpha}_{(\rho)}}{\partial z_{\beta}} = 0, \tag{23}$$

$\alpha = 1, \ldots, k$; $\rho, \sigma = 1, \ldots, r$, but it turns out that the metric is then a flat metric, with the space being spread out over the flat E_{2k}, and, if compact, being a complex multitorus directly, see [10].

If we wish to construct a Kaehler metric by immersing S_{2k} holomorphically into a projective P_{2s-2} with homogeneous coordinates

$$w_1 : w_2 \cdots : w_s, \tag{24}$$

$s \geq k + 1$, it is no longer feasible to state conditions, for this to be possible, in terms of the vectorial gradients of the coordinates w_{σ} only, but we will also have to involve these coordinates functions themselves, undifferentiated. Since only the ratios of the coordinates count, the coordinates as objects on S_{2k} need not be holomorphic scalar functions proper, but it suffices that they be holomorphic scalar densities of some joint weight g, no matter what the weight. Such a density is an object having in each coordinate system a holomorphic component $\varphi(z)$ with a law of transformation

$$\varphi'(z') = \left(\frac{\partial(z)}{\partial(z')}\right)^{g} \varphi(z); \tag{25}$$

and if we are given s such densities $\varphi_{\sigma}(z)$ for which the matrix

$$\left| \varphi_{\sigma}, \frac{\partial \varphi_{\sigma}}{\partial z_{\alpha}} \right| \tag{26}$$

of s rows and k+1 columns has everywhere its maximal rank k+1, then we obtain an immersion into P_{2s-2} and thus a Kaehler metric on S_{2k}, compare [10].

And as regards securing sufficiently many densities of the same weight, the weight unlimited, we have noted in [10] that they may be obtained if there are given sufficiently many tensorfields

$$\zeta^{\beta_1 \ldots \beta_q}_{\alpha_1 \ldots \alpha_p} \tag{27}$$

or even tensor densities, of arbitrary type (p,q,g) by forming certain determinants with their components. Thus we obtain the following indeterminate proposition which could be made precise by drawing up specific assumptions laboriously.

THEOREM 5. A complex S_{2k} can be immersed holomorphically one-one into a projective space, and thus carries a Kaehler metric, if there are on it sufficiently many suitably independent holomorphic tensor densities (27) of any given type.

§6. UNIFORMIZATION

We will say that a complex space S_{2k} can be uniformized, if its universal covering space is holomorphically equivalent with a domain D in complex E_{2k}.

THEOREM 6. (Heuristic) If S_{2k} can be uniformized then it has a Kaehler metric for which the Ricci curvature is negative.

THEOREM 7. (conjectural, dubious) If S_{2k} has a Kaehler metric for which the Ricci curvature is negative, it can be uniformized.

THEOREM 8. (conjectural) If a complex space S_{2k} can be uniformized, then a lower dimensional complex space $S_{2\ell}$, $\ell < k$ which can be imbedded in it holomorphically locally one-one can be likewise uniformized.

As in [3] and [4], for the motivation of Theorem 6 we take as our prospective covering space a bounded domain D in E_{2k} on which there is a transitive group G of complex homeomorphisms

$$z_j' = f_j(z, \theta), \tag{28}$$

$\theta \varepsilon G$. With a Haar measure we introduce, if the integrals converge absolutely, the metric form

$$g_{\alpha\bar{\beta}} = \int_G \left(\sum_{j=1}^{k} \frac{\partial f_j(z, \theta)}{\partial z_\alpha} \cdot \frac{\overline{\partial f_j(z, \theta)}}{\partial z_\beta} \right) d\theta, \tag{29}$$

and we note that this metric is group-invariant, Kaehlerian, and semi-definite. If we add the explicit assumption that it is strictly positive definite it follows rigorously that its Ricci curvature is non-positive,

$$-R_{\alpha\bar{\beta}} \leq 0, \tag{30}$$

this following from the fact that the metric (29) can be interpreted as a limit of metrics each of which arises from an imbedding in a Euclidean space with a metric (19). Actually, in most interesting cases the Ricci curvature is strictly negative, and furthermore the metric is an Einstein metric

$$-R_{\alpha\bar{\beta}} = bg_{\alpha\bar{\beta}} \quad , \tag{31}$$

and finally the four-indices curvature itself is likewise negative.

Now, take a discrete subgroup Γ of G and form the space $S_{2k} = D/\Gamma$ of which D is the covering space. Since (29) is invariant with respect to G it gives rise to a Kaehler metric on S_{2k} itself, and (30) holds, as claimed.

As regards Theorem 7 we note that by a general theorem of (Hadamard and) E. Cartan a real space with Riemann metric and non-positive four-indices curvature has a covering space which differentiably is a Euclidean domain, and we are setting down the conjecture that for a complex space with Kaehler metric the weaker assumption (30) will produce the even stronger conclusion that the covering space is even holomorphically a domain.

Theorem 8 is now a combination of 6 and 7. In fact if S_{2k} is uniformizable then (30) holds; for an imbedded $S_{2\ell}$ we have then (30) even more so [this is a rigorous theorem, see [3], p. 190] and thus, conversely, $S_{2\ell}$ is likewise uniformizable.

We further note that a compact S_{2k} whose universal covering space is a <u>bounded</u> domain D has itself as a rule only a finite number of complex homeomorphisms, see [4] and also [16]. Thus, if it is an algebraic variety, as it probably always is, the field of meromorphic functions on it has only a finite number of automorphisms.

Also, see [2], (30) implies that there exists on it no holomorphic tensor field with only upper indices

$$\zeta^{\alpha_1 \ldots \alpha_p}$$

and if in addition it is an Einstein space (31) then there even exists no mixed holomorphic tensor field

$$\zeta^{\alpha_1 \ldots \alpha_p}_{\beta_1 \ldots \beta_q}$$

in which the number of upper indices predominates, $p > q$, see [11].

§7. COMPLEX MANIFOLDS WITH ALGEBRAIC SINGULARITIES

We are now going to give a brief extract from a forthcoming paper [15] jointly with W. T. Martin.

Consider, say over

$$|z_1|^2 + \dots + |z_k|^2 < 1, \tag{32}$$

a system of polynomial equations

$$P_m(\zeta_m, \zeta_{m-1}, \dots, \zeta_1; \quad z_1, \dots, z_k) = 0, \; m = 1,\dots,n, \tag{33}$$

in variables $\zeta_1, \dots, \zeta_n$ with coefficients holomorphic in (32) such that in the m-th polynomial the highest occurring power of ζ_m has total coefficient 1. We also assume that the system (33) is "irreducible" in the sense that there exists in (32) a certain holomorphic function

$$\Delta(z_j) = \Delta_U(z_j),$$

which function we choose and hold fast, such that for z not satisfying

$$\Delta(z_j) = 0 \tag{34}$$

the systems of solutions $(\zeta_1, \dots, \zeta_n)$ of (33) are all different from one another, and constitute themselves into holomorphic functional elements

$$\zeta_1 = \zeta_1(z), \dots, \zeta_n = \zeta_n(z) \tag{35}$$

which are freely continuable along any path for which $\Delta(z_j) \neq 0$ throughout.

If now we consider the system (33) as a system of equations in the space E_{2k+2n} of the complex variables (z_j, ζ_m), for z in (32) and all such ζ's as will solve the equations, then we obtain a well defined point set in E_{2k+2n}, and we may impart to it the topology of the latter space. We do this, and if so topologized, we denote our pointset by U; and we emphasize that this topologization embraces all points of our pointset without any exception. After this, we select in U those points whose z-coordinates satisfy (34) and we denote this subset of U by $N(U,\Delta)$. Now, the difference set

$$U - N(U,\Delta)$$

is obviously such that the functions (35) may be employed towards endowing it with a complex structure as defined in Section 1, and we so endow it.

If now we wish to define a "complex manifold S_{2k} with singularities" globally, we can do this in the following way. It is a topological space together with a covering by open sets U_α each of which is first of all topologically of the kind introduced. With each U_α there is associated a function Δ_α as before, and we are endowing the subspace

$$U_\alpha - N(U_\alpha, \Delta_\alpha) \tag{36}$$

with the complex structure as before. And, what is decisive, we assume that the union of sets

$$\Sigma_\alpha \left\{ U_\alpha - N(U_\alpha, \Delta_\alpha) \right\} \quad , \tag{37}$$

which is a subspace of S_{2k}, is locally Euclidean, and is endowable with a complex structure globally in such a manner that the individual complex structures of the parts (36) shall be holomorphically compatible with it.

We say that a function $f(z)$ is holomorphic on such a S_{2k} if it is defined and continuous everywhere and if on every U_α it has a representation by an expression

$$\Sigma_{(\nu)} \zeta_1^{\nu_1} \cdots \zeta_n^{\nu_n} f_{\nu_1 \ldots \nu_n}(z), \tag{38}$$

$\nu_m \geq 0$, which is polynomial in the ζ's with coefficients which are holomorphic in (32).

A first theorem is that if a function $f(z)$ is defined in (37) only, and is bounded in every (36), and if there is a representation (38) in every (36), then it has a holomorphic continuation into all of S_{2k}. Furthermore, a holomorphic function in a subdomain assumes the maximum of its modulus on the boundary, so that in particular on a compact such S_{2k} it must be a constant. Next, there are satisfactory generalizations of Schwarz' lemma, and of other convexity theorem. And finally, and this is the least obvious generalization, it is again true that the uniform limit of holomorphic functions is again a holomorphic function.

If we wish likewise to extend the notions of holomorphic vector and tensor fields to our new manifolds, the exceptional points

$$\Sigma_\alpha N_\alpha (U_\alpha, \Delta_\alpha)$$

included, then the experience from algebraic geometry shows that in the (z, ζ) variables certain singularities for the components must be envisaged, if the objects are not to be restricted unduly. This can be done adequately and leading theorem may be then generalized. We can for instance retain the theorem, recently proven in [9] that on a compact S_{2k} the linear space of holomorphic tensor fields (27) of any one type (p,q,g) has again a finite basis, always.

§8. AUTOMORPHIC FUNCTIONS

Perhaps the most notable generalization possible, from regular manifolds to manifolds with singularities, is for a basic theorem affecting automorphic functions so-called, and the following very comprehensive version of it, that was recently established by ourselves in [12] goes over directly. If S_{2k} is compact and $\tilde{S}_{2k}$ is its universal covering space, then we call $f(P)$ automorphic if it is defined and holomorphic on $\tilde{S}_{2k}$ and for any element γ of the fundamental group we have

$$f(\gamma P) = \eta_{\gamma}(P)\, f(P),$$

where each factor of automorphy $\eta_{\gamma}(P)$ is itself holomorphic on $\tilde{S}_{2k}$. Now, take such a system of factors $\eta_{\gamma}(P)$ and consider $k + 2$ functions

$$f_o, f_1, \ldots, f_{k+1}$$

each of which is automorphic relative to these given factors, and the theorem is that there exists a non-vanishing polynomial with constant coefficients $Q(w_o, w_1, \ldots, w_{k+1})$ such that

$$Q\Big(f_o(P),\ f_1(P),\ \ldots,\ f_{k+1}(P)\Big) = 0.$$

This is the theorem, and the main application is that for $f_o(P) \neq 0$, the $k+1$ meromorphic quotients

$$g_{\ell}(P) = \frac{f_1\ (P)}{f_o\ (P)}$$

are connected by an (inhomogeneous) polynomial equation, so that, if k among the latter functions are not so connected, our space S_{2k}, singularities and all, may be viewed as the locus of an algebraic surface as defined in algebraic geometry, in a certain sense.

BIBLIOGRAPHY

[1] Analytic and meromorphic continuation by means of Green's formula, Annals of Math., 44 (1943), 652-673.

[2] Vector fields and Ricci curvature, Bull. American Math. Soc., 52 (1946), 776-797.

[3] Curvature in Hermitian metric, Bull. American Math. Soc., 53 (1947), 179-195.

[4] On compact complex manifolds, Jour. Indian Math. Soc., 51 (1947), 1-21.

[5] Curvature and Betti numbers, Annals of Math., 49 (1948), 379-390.

[6] Curvature and Betti numbers II, Annals of Math., 50 (1949), 77-93.

[7] Euler-Poincaré characteristic for locally homogeneous and complex spaces, Annals of Math., 51 (1950), 741-761.

[8] Vector fields on complex and real manifolds, Annals of Math., 52 (1950), 647-649.

[9] Tensor fields with finite basis, Annals of Math., 53 (1951), 400-411.

[10] Complex spaces with transitive commutative groups of transformations, Proc. Nat. Acad. Sci., 37 (1951), 356-359.

[11] Tensor fields and Ricci curvature in Hermitian metric, Proc. Nat. Acad. Sci., 37 (1951), 704-706.

[12] Algebraic and linear dependence of automorphic functions in several variables, Jour. Indian Math. Soc., 16 (1952), 1-6.

[13] Partial differential equations and analytic continuation, Proc. Nat. Acad. Sci., 38 (1952), 227-30

[14] BOCHNER, S., MARTIN, W. T., Several Complex variables, Princeton University Press, 1948. 216 pp.

[15] BOCHNER, S., MARTIN, W. T., Complex spaces with singularities, (to appear).

[16] HAWLEY, N. S., A theorem on compact complex manifolds, Annals of Math., 52 (1950), 637-641.

[17] RADO, T., Über den Begriff der Riemannschen Fläche, Acta Szeged, 2 (1925), 101-121.

I. GREEN'S FORMULA AND ANALYTIC CONTINUATION

S. Bochner

For anyalytic functions in more than one complex variable there is a theorem of Hartog's that if a function is given on the connected boundary of a bounded domain, then it can be continued analytically into all of the domain. The class of functions to which this theorem applies was considerably generalized in our paper [1]: "Analytic and meromorphic continuation by means of Green's formula," Annals of Mathematics 44 (1943), 652-673; and it was further expanded in our recent note [2]: "Partial differential equations and analytic continuation," Proceedings of the National Academy of Sciences 38 (1952), 227-30. Now, in §1 of the present paper the leading theorem of [2] will be given its final version known to us (see Theorem 5) and, furthermore, details of the proof will be modified and added.

The real and imaginary parts of analytic functions of complex variables are solutions of a system of Cauchy-Riemann equations in real variables. In the case of more than one complex variable, this system is quite complicated and, as it turns out, much too restrictive for our theorem. At first in [1], and then more systematically and generally in [2], we introduced instead a system consisting of only two equations, both with constant coefficients: an elliptic one in all variables and some other one in fewer than all variables; the second equation was the one by which the actual continuation was brought about. However, the second equation could only operate if the function was first represented by a certain Green's formula, and it was the sole task of the elliptic equation to secure just such a formula. Now, in further analyzing certain aspects of our theorem, we found it pertinent to try to give up the elliptic equation altogether and to hypothesize directly a Green's formula having the requisite properties. This will be done in the present paper.

In §2 we will be dealing in a similar fashion with another theorem in several complex variables which although closely related to the previous one is different from it nevertheless. Following up a suggestion of Severi's, this theorem was presented more systematically than had been done before in Chapter IV of the book by Bochner-Martin: Several Complex Variables, Princeton, 1948. It will now be given a rather more general version than previously.

§1. EUCLIDEAN SPACES

In Euclidean E_n: $(\xi^1, \ldots, \xi^n)$ we take a p-form, $1 \leq p \leq n-1$

$$\frac{1}{p!} \sum_{(\alpha)} A_{\alpha_1 \ldots \alpha_p} d\xi^{\alpha_1} \ldots d\xi^{\alpha_p} \tag{1}$$

and we assume that each component of the skew tensor is a finite linear partial differential expression involving an unspecified function $f(\xi)$ with coefficients which are functions of the difference $\xi - x = (\xi^\beta - x^\beta)$, where $x = (x^\beta)$ is another variable point of the given space. Thus we have

$$A_{\alpha_1 \ldots \alpha_p} = \sum_{(\nu)} G_{\alpha_1 \ldots \alpha_p}^{\nu_1 \ldots \nu_n} (\xi - x) \wedge_{\nu_1 \ldots \nu_n} f(\xi) \tag{2}$$

where

$$\wedge_{\nu_1 \ldots \nu_n} f(\xi) = \frac{\partial^{\nu_1 + \ldots + \nu_n} f(\xi)}{(\partial \xi^1)^{\nu_1} \ldots (\partial \xi^n)^{\nu_n}} \tag{3}$$

with $\nu_1 \geq 0, \ldots, \nu_n \geq 0$, $\nu_1 + \ldots + \nu_n \leq N - 1$ for some N sufficiently large but finite. In this sense we denote the form (1) by

$$G_p(\xi - x; f(\xi); d\xi) \tag{4}$$

and we stipulate that the individual functions $G_{(\alpha)}^{(\nu)}(t)$, $t = \xi - x$, which occur in (2) shall be defined and real analytic in a certain open set T of the Euclidean E_n: (t^β).

We now introduce the requirement

$$d_\xi G_p \equiv 0 \tag{5}$$

that is

$$\sum_{q=1}^{p+1} (-1)^q \frac{\partial}{\partial \xi^{\alpha_q}} A_{\alpha_1 \ldots \alpha_{q-1} \alpha_{q+1} \ldots \alpha_{p+1}} = 0 \tag{6}$$

and this is a system of equations

$$\sum_{(\nu)} H_{\alpha_1 \ldots \alpha_{p+1}}^{\nu_1 \ldots \nu_n} (\xi - x) \wedge_{\nu_1 \ldots \nu_n} f(\xi) = 0 \tag{7}$$

$$0 \leq \nu_1 + \ldots + \nu_n \leq N .$$

in which the coefficients $H^{(\nu)}_{(\alpha)}$ are linear combinations of the coefficients $G^{(\nu)}_{(\alpha)}$ and their first derivatives.

We now introduce fictitiously an "elliptic operator" $\Delta_\xi f(\xi)$ and we will say that a function $f(\xi)$ which is defined and analytic in a domain U of E_n satisfies there the equation

$$\Delta_\xi f(\xi) = 0 \tag{8}$$

if the relations (7) are fulfilled for ξ in U and $\xi - x$ in T. For every U the space of solutions of (8) is a vector space having the following closure property: Let $f_1(\xi)$, $f_2(\xi)$, ... be a sequence of solutions. If every point ξ^β of U has a complex neighborhood into which all the functions of the sequence can be continued analytically, and if these functions converge uniformly in these neighborhoods, then the limit function is again a solution.

Furthermore, if we put $\xi = t + x$ in (7), we obtain

$$\sum_{(\nu)} H^{(\nu)}_{\alpha}(t) \wedge^t_{\nu_1 \cdots \nu_n} f(t+x) = 0 \tag{9}$$

and therefore, locally, if $f(\xi)$ is a solution of (7) then so is also the translated function $f(\xi+h)$ for a sufficiently small constant displacement $h = (h^\beta)$. This property is crucial to our purpose, and we express it symbolically by saying that our (fictitious) operator Δ has "constant coefficients"; and if we combine all properties enumerated, we obtain the further property that if $f(\xi)$ is a solution, then so are also all first derivatives $\wedge_\beta f(\xi)$, and hence, also all mixed partial derivatives, and hence, also every finite linear combination with constant coefficients

$$\wedge f = \sum_{(\mu)} a_{\mu_1 \cdots \mu_n} \wedge_{\nu_1 \cdots \nu_n} f(\xi) \tag{10}$$

$$0 \le \mu_1 + \cdots + \mu_n \le M$$

this operator being formed literally, and not just only fictitiously or symbolically.

We now take in U a p-dimensional chain B_p and we form the integral

$$g(x) = \int_{B_p} G(\xi - x;\ f(\xi);\ d\xi) \tag{11}$$

for those points x for which it is definable, that is, for the open set X which is such that for x in X and ξ in B_p, $\xi - x$ is in T.

If we hold B_p fixed, then $g(x)$ is analytic in T; and if we vary $f(\xi)$ and then introduce the functional

$$g(x) = L(f(\xi); x) \tag{12}$$

then the latter is distributive, that is,

$$L(c_1f_1 + c_2f_2; x) = c_1L(f_1; x) + c_2L(f_2; x). \tag{13}$$

Next, if two chains B_p, B'_p are homologous relative to $U - (x)$, then by (5) we have

$$\int_{B_p} = \int_{B'_p} \tag{14}$$

and to this extent the integral (11) is "independent of the path" if $f(\xi)$ is a solution of (8). But the decisive property is yet to be stated and it is as follows.

THEOREM 1. If B_p is a cycle then the functional (12) is commutative with translations and partial differentiations. That is, (12) implies

$$g(x+h) = L(f(\xi + h); x) \tag{15}$$

for small h locally, and also

$$\frac{\partial g}{\partial x^\beta} = L(\frac{\partial f}{\partial \xi^\beta}; x) \tag{16}$$

and more generally

$$\Lambda_x\, g(x) = L(\Lambda_\xi f(\xi); x) \tag{17}$$

for any operator (10).

PROOF. We have

$$g(x+h) = \int_{B_p} G(\xi-x-h; f(\xi); d\xi) \tag{18}$$

and if we replace ξ by $\xi + h$ (18) is equal to

$$\int_{B'_p} G(\xi - x;\ f(\xi + h);\ d\xi) \tag{19}$$

where $B'_p = B_p - (h)$ results from B_p by a translation. If x is given then for sufficiently small h, B'_p is homologous to B_p, and by (14) the integral (19) is

$$\int_{B_p} G(\xi - x;\ f(\xi + h);\ d\xi) = L(f(\xi + h);\ x)$$

as claimed in (15).

Due to (13) the result just obtained implies that the integral (11) carries a difference quotient

$$\frac{1}{h}[f(\xi^1, \ldots, \xi^\beta + h, \ldots, \xi^n) - f(\xi^1, \ldots, \xi^\beta, \ldots, \xi^n)]$$

into the corresponding difference quotient for $g(x^\alpha)$. If now we let $h \longrightarrow 0$, then we obtain (16), as can be easily proved, and then also (17), as claimed.

Theorem 1 will suffice for our present purposes. However, for a certain type of conclusion that was attempted in [1] and [2] a partial generalization of Theorem 1 is required for the case in which B_p is not a cycle, and we will state the generalization without proof.

THEOREM 2. If we are given a symbolic equation (8) and an operator (10), then there exists a (p-1)-form

$$Y_{p-1}(\xi - x;\ f(\xi);\ d\xi)$$

in which the partial derivatives of $f(\xi)$ occurring are of order $\leq N - 1 + M - 1$, and which has the following property: If B_p is a p-chain and B_{p-1} is its boundary and if $f(\xi)$ satisfies (8) in a domain containing $B_p + B_{p-1}$, then we have

$$L(\Lambda_\xi f(\xi);\ x) - \Lambda g(x) = \int_{B_{p-1}} Y_{p-1}(\xi - x;\ f(\xi);\ d\xi)$$

and thus we again have (17) provided that the function $f(\xi)$ and its derivatives of order $\leq N - 1 + M - 1$ are zero on the boundary B_{p-1}.

From now on we will consider only the dimension $p = n - 1$, and we will assume that the domain T on which the coefficients $G^{(\nu)}_{(\alpha)}(t)$ are defined is the entire E_n except for the origin, so that $L(f(\xi); x)$ is defined for x in $E_n - B_{n-1}$.

THEOREM 3. If B_{n-1} bounds a small subdomain D^o of U then we have

$$\oint_{B_{n-1}} G(\xi - x; f(\xi); d\xi) = \sum_{(r)} c_{r_1 \ldots r_n} \wedge_{r_1 \ldots r_n} f(x) \tag{20}$$

$$r_1 \geq 0, \ldots, r_n \geq 0$$

where $c_{r_1 \ldots r_n}$ are constants independent of $f(\xi)$.

PROOF. It follows from (14) that for the proof of (20) we may assume that D^o is a coordinate sphere $|\xi| < 2\rho$, and that even the double sphere $|\xi| < 4\rho$ is still contained in U, and that the points x are restricted to the small sphere $|x| < \rho$. Now, on replacing ξ by $\xi + x$ we obtain

$$\begin{aligned} g(x) &= \oint_{B_{n-1} - (x)} G(\xi; f(\xi + x); d\xi) \\ &= \oint_{B_{n-1}} G(\xi; f(\xi + x); d\xi) \end{aligned}$$

and for sufficiently small ρ we may now insert the Taylor expansion

$$f(\xi + x) = \sum_{(s)} \frac{(\xi^1)^{s_1} \ldots (\xi^n)^{s_n}}{s_1! \ldots s_n!} \wedge_{s_1 \ldots s_n} f(x)$$

and integrate term by term. Hence the conclusion.

The theorem is of no practical importance to us, but it puts into a proper perspective the assumption we are now going to make that the right side in (20) shall be identically $f(x)$, that is,

$$\oint_{B_{n-1}} G(\xi - x; f(\xi); d\xi) = f(x) \tag{21}$$

and we express this assumption by saying that our integral (11) is a <u>Green's formula</u>. The first consequence of this assumption is:

THEOREM 4. If B is a simplex (which may be

part of any chain over which we integrate) then the function

$$(22) \qquad g(x) = \int_B G(\xi - x;\ f(\xi);\ d\xi)$$

can be continued across B from either side and the difference of the two values in the vicinity of B is f(x),

$$(23) \qquad g_+(x) - g_-(x) = \pm f(x)$$

the algebraic sign being determined by the orientation of B.

PROOF. (Cf. [1], p. 656-57.) Denoting one side of B as positive, we deform B into a simplex B' on its negative side but keep the edges fixed. By (14) we have for x on the positive side

$$g_+(x) = \int_B = \int_{B'}$$

and thus $g_+(x)$ can be continued a certain distance into the negative side. Denoting the continuation still by $g_+(x)$, we now have for points x in the domain bounded by B' - B

$$g_+(x) - g_-(x) = \int_{B'} - \int_B = \int_{B'-B}$$

and by (21) this is $\pm f(x)$ as claimed.

THEOREM 5. If we are given a Green's formula and if the coefficients $G^{(\nu)}_{(\alpha)}(t)$ in (2) are of slow growth at infinity in the sense that for some $r_o > 0$ we have

$$(24) \qquad \wedge_{r_1 \dots r_n} G^{\nu_1 \dots \nu_n}_{\alpha_1 \dots \alpha_{n-1}}(t) = O(|t|^{-n-1}), \qquad |t| \longrightarrow \infty$$

for

$$(25) \qquad r_1 \geq 0,\ \dots,\ r_n \geq 0, \qquad r_1 + \dots + r_n \geq r_o$$

then, if D is a bounded domain having a connected boundary B_{n-1} and if U is a neighborhood of B_{n-1} and if an analytic function $f(x)$ in U satisfies the associated "elliptic equation"

(26)
$$\Delta_x f(x) = 0$$

in all its n variables and an additional equation

$$\wedge f \equiv \sum_{(q)} a_{q_1 \ldots q_m} \wedge_{q_1 \ldots q_m} f(x) = 0 \qquad (m<n)$$

with constant coefficients not all zero in fewer than all variables, then the function $f(x)$ has an analytic continuation into all of $D + U$.

PROOF. By Theorem 4, if we restrict U sufficiently, we can put

$$f(x) = g_+(x) - g_-(x)$$

where $g_+(x)$ exists and is analytic in $D + U$, $g_-(x)$ is analytic in $(E^n - D) + U$, and the assertion of the theorem will follow if we show that $g_-(x)$ has an analytic continuation into all of E^n.

Let us denote $g_-(x)$ by $\phi(x)$. We know the following about this function: It is defined and analytic in the exterior of a sphere

(27)
$$|x| > R$$

and there exists an $r' > 0$ such that

$$\wedge_{r_1 \ldots r_n} \phi(x) = O(|x|^{-n-1}), \qquad |x| \longrightarrow \infty$$

for $r_1 + \ldots + r_n \geq r'$ and we have

$$\wedge \phi(x) \equiv 0$$

If we now take fixed values $r_1, \ldots, r_n$, and fixed coordinates $x_o^{m+1}, \ldots, x_o^n$ for which

(28)
$$(x_o^{m+1})^2 + \ldots + (x_o^n)^2 > R^2$$

and introduce the function

$$\Psi(x^1, \ldots, x^m) = \Lambda_{r_1 \ldots r_n} \phi(x^1, \ldots, x^m, x_o^{m+1}, \ldots, x_o^n)$$

then the latter has the following properties: It is defined and analytic in E_m: $(x^1, \ldots, x^m)$, we have

(29) $$\Lambda_{q_1 \ldots q_m} \Psi(x) = O(|x|^{-m-1}), \qquad |x| \longrightarrow \infty$$

for any $q_1 \geq 0, \ldots, q_m \geq 0$, and we have

(30) $$\Lambda \Psi(x) \equiv 0$$

We claim that under these circumstances we have

$$\Psi(x) \equiv 0$$

In fact, because of (29) we may introduce the Fourier transform

$$\chi(\alpha_j) = \int_{E^m} e^{i(\alpha_1 x^1 + \ldots + \alpha_m x^m)} \Psi(x) dv_x$$

and it follows easily that the Fourier transform of $\Lambda \Psi(x)$ is then

(31) $$\chi(\alpha_j) \sum_{(q)} a_{q_1 \ldots q_m} (i\alpha_1)^{q_1} \ldots (i\alpha_m)^{q_m}$$

Now, (30) implies the vanishing of (31), and hence the vanishing of $\chi(\alpha)$, and by uniqueness of Fourier transforms the vanishing of $\Psi(x)$. Therefore we have

(32) $$\Lambda_{r_1 \ldots r_n} \phi(x^1, \ldots, x^m, x^{m+1}, \ldots, x^n) = 0$$

for $(x^1, \ldots, x^m)$ in E_m and $(x^{m+1}, \ldots, x^n)$ in (28), but since the left side of (32) is analytic in (27), the relation (32) also holds identically in (27). And since this is true for all combinations $r_1, \ldots, r_n$ with $r_1 + \ldots + r_n = r'$ it follows that $\phi(x)$ is a polynomial of total degree $\leq r' - 1$, and thus is certainly continuable into all E_n.

§2. COORDINATE SPACES

We now take an arbitrary analytic coordinate space X_n: $(\xi^1, \ldots, \xi^n)$ and in it, immediately for $p = n - 1$, a form

$$G(\xi;\ x;\ f(\xi);\ d\xi) = \frac{1}{(n-1)!}\sum_{(\alpha)} A_{\alpha_1\cdots\alpha_{n-1}}\, d\xi^{\alpha_1}\cdots d\xi^{\alpha_{n-1}}$$

where

$$A_{\alpha_1\ldots\alpha_{n-1}} = \sum_{(\nu)} G^{\nu_1\ldots\nu_n}_{\alpha_1\cdots\alpha_{n-1}}(\xi;\ x)\wedge_{\nu_1\cdots\nu_n} f(\xi)$$

for $\xi \neq x$. The point x itself also ranges over X_n, and any two points ξ_o, x_o $(\xi_o \neq x_o)$ have non-overlapping coordinate neighborhoods for which the tensoroid components $G^{(\nu)}_{(\alpha)}(\xi;\ x)$ are defined and analytic in $(\xi;\ x)$.

We again introduce the requirement $d_\xi G = 0$, that is

$$\sum_{q=1}^{n} (-1)^q \frac{\partial}{\partial \xi^{\alpha_q}} A_{\alpha_1\cdots\alpha_{q-1}\alpha_{q+1}\cdots\alpha_n} = 0$$

and it amounts to a tensoroid system of equations

$$\sum_{(\nu)} H^{(\nu)}_{(\alpha)}(\xi;\ x)\wedge_{\nu_1\ldots\nu_n} f(\xi) = 0$$

which we again abbreviate to

(33) $$\Delta_\xi f(\xi) = 0$$

and we again consider analytic functions in open sets U which satisfy this "elliptic equation" there. We wish to point out, however, that we are not retaining in any manner whatsoever the previous assumption that the variables ξ, x shall occur as the difference $\xi - x$ only, and thus the "constancy of coefficients" in the operator Δ is being entirely dispensed with.

We again introduce the integral

(34) $$g(x) = \int_{B_{n-1}} G(\xi;\ x;\ f(\xi);\ d\xi)$$

for a chain in U. It is again analytic and distributive, and we again have (14); and if B_{n-1} is the boundary of a small domain D^o, then for points of D^o we have

$$g(x) = \sum_{(r)} c^{r_1\ldots r_n}(x)\wedge_{r_1\ldots r_n} f(x)$$

where the coefficients $c^{r_1 \dots r_n}(x)$ are now functions of x.

We again call the integral a Green's formula if we have, in particular,

$$\oint_{B_{n-1}} G(\xi;\ x;\ f(\xi);\ d\xi) = g(x) \tag{35}$$

and theorem 4 is valid again.

We now take a second analytic coordinate space Y_m: $(y^1, \dots, y^m)$, $(m \gtreqless n)$ which will be a space of "parameters" and we form the product space

$$Z_{n+m} = X_n \times Y_m \tag{36}$$

in the usual manner. Also, for any point set θ in Z_{n+m}, we will employ the representation by its layers in the X_n-space, thus

$$\theta = \left\{ y \in Y_m;\ x \in D(y) \right\} \tag{37}$$

where for each y, $D(y)$ is a point set in X_n, perhaps empty.

We now take a domain (37) in (36) and another such domain

$$\tilde{\theta} = \left\{ y \in Y_m;\ x \in \tilde{D}(y) \right\} \tag{38}$$

and we introduce the following

DEFINITION. We call the domain $\tilde{\theta}$ an enlargement of the domain θ if

(i) $\tilde{\theta} \supset \theta$ that is $\tilde{D}(y) \supset D(y)$ for every y,

(ii) for each (x,y) in $\tilde{\theta}$ there is in $D(y)$ a cycle $B_{n-1} = B_{n-1}(x,y)$ which bounds an n-complex $K_n(x,y)$ in $\tilde{D}(y)$, $B_{n-1} = \mathrm{bd}(K_n)$, such that $x \in K_n(x,y)$.

(iii) for any (x',y') in $\tilde{\theta}$ there is a neighborhood $N = N(x',y')$ in $\tilde{\theta}$ such that for $(x,y) \in N$ the cycle $B_{n-1}(x',y') - B_{n-1}(x,y)$ bounds a chain $H_n(x,y)$ in $D(y)$, with $x \notin H_n(x,y)$, that is $B_{n-1}(x',y') \approx B_{n-1}(x,y)$ in $D(y) - x$; and finally

(iv) there exists a point (x_0, y_0) in θ and a neighborhood $N_0(x_0,y_0)$ in θ such that for any (x,y) in $N_0(x_0,y_0)$ the complex $K_n(x,y)$ of (ii) lies in $D(y)$ itself, and not only in the larger $\tilde{D}(y)$.

We note that condition (iv) is implied by the simpler though less general condition

(iv)' for (x,y) in $N_o(x_o,y_o)$ we have $\tilde{D}(y) = D(y)$.

Now, the leading statement is as follows:

THEOREM 6. If $f(x,y)$ is defined in a domain θ of Z_{n+m} and is analytic as a function of (x,y) there, and if for each y it is a solution of our elliptic equation

$$\Delta_x f(x, y) = 0 \tag{39}$$

then $f(x, y)$ has an analytic continuation into any given enlargement $\tilde{\theta}$ of θ.

REMARK. There are enlargements which in a certain sense are not enlargeable themselves. If, however, we take any two enlargements $\tilde{\theta}_1$, $\tilde{\theta}_2$ of the kind introduced and if the point sets $\tilde{\theta}_1 - \theta$, $\tilde{\theta}_2 - \theta$ have a non-vacuous intersection which has parts not connected with θ itself, then we do not claim that the two continuations will necessarily coincide there; it is not known to us what the actual situation is.

PROOF. For every (x,y) in $\tilde{\theta}$, we introduce the quantity

$$g(x, y) = \int_{B_{n-1}(x,\, y)} G(\xi;\, x;\, f(\xi,\, y);\, d\xi). \tag{40}$$

Now, in a neighborhood $N(x', y')$ [defined on p. 15, property (iii)] we can replace the variable cycle $B_{n-1}(x, y)$ by the fixed cycle $B_{n-1}(x',y')$ and it now follows that $g(x, y)$ is analytic in $N(x', y')$ and hence everywhere in θ. But by property (iv) it coincides with $f(x, y)$ itself in a certain neighborhood N_o of θ, and by analytic continuation it coincides with $f(x, y)$ everywhere in θ, as claimed.

We now take in Z_{n+m} a closed point set which is the locus of a (finite or infinite) simplicial chain R_{n-1+m} and a neighborhood V_{n+m} of the latter and we also introduce the open set

$$\theta_{n+m} = Z_{n+m} - R_{n-1+m} \tag{41}$$

We are also introducing the representatives

$$R_{n-1+m} = \left\{ y \in Y_n;\quad x \in B_{n-1}(y) \right\}$$

$$V_{n+m} = \left\{ y \in Y_m; \quad x \in U_n(y) \right\}$$
$$\theta_{n+m} = \left\{ y \in Y_m; \quad x \in D_n(y) \right\}$$

and we are making the following assumptions.

(v) each $B_{n-1}(y)$ is the locus of a ("regular" or) "singular" cycle which we denote by the same symbol, and

(vi) for each (x', y') in θ_{n+m} there exists a neighborhood $N(x', y')$ and a cycle $B_{n-1}(x', y')$ in $U_n(y')$ such that for (x,y) in $N(x', y')$ the cycle $B_{n-1}(y) - B_{n-1}(x', y')$ bounds an n-dimensional complex in $U_n(y)$ which does not contain the point x.

THEOREM 7. Under the assumptions just made, if $f(x, y)$ is defined and analytic in V_{n+m} and satisfies (39) there, then the integral

$$g(x, y) = \int_{B_{n-1}(y)} G(\xi; x; f(\xi; y); d\xi) \tag{42}$$

defines an analytic function in θ_{n+m} having the following properties:

If we denote the (connected) components of θ_{n+m} by $\theta^1, \theta^2, \ldots,$ and denote the value of (42) in θ^a by $g^a(x, y)$, and if a "regular" simplex R^o_{n-1+m} of the chain R_{n-1+m} separates two components θ^a, θ^b; then the function $g^a(x, y)$ can be continued a certain distance across R^o into θ^b, and $g^b(x, y)$ can be so continued into θ^a, and in the vicinity of R^o we have the saltus relation

$$g^a(x, y) - g^b(x, y) = \pm f(x, y)$$

For the moment we call a component θ^a <u>non-bounded</u> if there is in Y_m some (small) neighborhood N_m such that for y in N_m the cross section $D^a_n(y)$ of $\theta^a_{n+m}(y)$ is the entire X_n. For these values y, the cycle $B_{n-1}(y)$ can be taken as null and thus $g^a(x, y) \equiv 0$. Hence the following consequences.

THEOREM 8. If in Theorem 7 all components θ^a are non-bounded then $f(x, y) \equiv 0$. If V_{n+m} is connected then the conclusion also holds if there are two non-bounded components θ^a, θ^b

having on their boundaries a joint simplex R^{o}_{n-1+m} separating them.

If there are altogether only two components θ^1, θ^2, and θ^2 is non-bounded, then $f(x, y)$ can be continued from the intersection of V_{n+m} with θ^1 into all of θ^1, and this is the most frequently occurring situation underlying Theorem 6.

Reprinted from the Proceedings of the NATIONAL ACADEMY OF SCIENCES,
Vol. 41, No. 10, pp. 746–752. October, 1955

EXISTENCE OF FUNCTIONALLY INDEPENDENT AUTOMORPHIC FUNCTIONS

BY S. BOCHNER AND R. C. GUNNING

DEPARTMENT OF MATHEMATICS, PRINCETON UNIVERSITY

Communicated July 11, 1955

In the theory of automorphic functions of n complex variables, the existence of n functionally independent meromorphic functions is demonstrated, separately, in the multiperiodic case by means of theta functions, in the case of a bounded domain by means of Poincaré series, and in the modular case by means of Eisenstein series. We have found that these constructions differ but little from one another, and we are going to established the existence theorem in a unified version which will subsume all those known to us.

1. We take a complex analytic manifold D of complex dimension n, and a countable group Γ of analytic homeomorphisms of D onto itself, transformations with fixed points being permitted. A meromorphic function $f(z)$ on D is automorphic if $f(Tz) = f(z)$ for all $T \in \Gamma$.

Definition 1: A set of *factors of automorphy*[1, 2] is a set of *meromorphic* functions $\eta_T(z)$ on D indexed by the elements $T \in \Gamma$ and satisfying

$$\eta_{ST}(z) = \eta_S(Tz)\eta_T(z).$$

Definition 2: A countable collection of sets of factors of automorphy $\{\eta_{\alpha, T}(z)\}$, $\alpha = 1, 2, \ldots$, is called a *complete system of factors* if the following conditions are fulfilled:

i) For each α there is a holomorphic function $h_\alpha(z) \not\equiv 0$ such that the functions $h_\alpha(z)[\eta_{\alpha, T}(z)]^{-1}$ are holomorphic.

ii) For each α there is a subgroup Γ_α of Γ such that (a) $\eta_{\alpha, T}(z) \equiv 1$ whenever $T \in \Gamma_\alpha$; (b), letting Λ_α be a set of representatives of the right cosets of Γ_α in Γ, the series

$$\sum_{T \in \Lambda_\alpha} h_\alpha(z)[\eta_{\alpha, T}(z)]^{-1} \tag{1}$$

is absolutely uniformly convergent on any compact subset of D; (c) the sum function (1) is not identically zero.

iii) There is a fixed set $M \subset D$ (M being a countable union of closed subsets of D each of which is of topological dimension at most $2n - 1$) such that, for any $z^\circ \in D - M$, each function $\eta_{\alpha, T}(z)$ is holomorphic in a neighborhood of z°, and the point set

$$\tilde{V}(z^\circ) = \{z \mid \eta_{\alpha, T}(z) = \eta_{\alpha, T}(z^\circ), \text{all } T \in \Gamma \text{ and all } \alpha\} \tag{2}$$

is zero-dimensional in a neighborhood of z°.

Note that in definition 2 the convergence of the series (1) is independent of the choice of the representatives of the various cosets. The set (2) may also be defined by

$$\tilde{V}(z^\circ) = \{z \mid \eta_{\alpha, T}(z) = \eta_{\alpha, T}(z^\circ), \text{all } T \in \Lambda_\alpha \text{ and all } \alpha\}.$$

Assuming that $\{\eta_{\alpha, T}(z)\}$ form a complete system of factors of automorphy, we may introduce the functions

$$\varphi_{\alpha, k}(z) = \sum_{T \in \Lambda_\alpha} [\eta_{\alpha, T}(z)]^{-k}$$

for all integers $k \geq 1$. These functions are meromorphic on D, and $\varphi_{\alpha, 1}(z) \not\equiv 0$. It is also clear that they are independent of the choice of representatives of the various cosets and that they satisfy the functional equations

$$\varphi_{\alpha, k}(Tz) = [\eta_{\alpha, T}(z)]^k \varphi_{\alpha, k}(z)$$

for all $T \in \Gamma$. Therefore,

$$f_{\alpha, k}(z) = \varphi_{\alpha, k}(z)[\varphi_{\alpha, 1}(z)]^{-k} \tag{3}$$

are well-defined automorphic functions.

Analytic functions $w_1(z), \ldots, w_m(z)$, $m \leq n$, are functionally independent on D if the Jacobian matrix

$$\frac{\partial(w_1, \ldots, w_m)}{\partial(z_1, \ldots, z_n)} \tag{4}$$

has its maximal rank m at some point of D.

THEOREM 1. *If the manifold D and the group Γ admit a complete system of factors of automorphy, then there exist at least n functionally independent automorphic functions.*

Proof: From the collection of automorphic functions (3) we select a maximal set of functionally independent functions

$$w_1(z) = f_{\alpha_1, k_1}(z), \ldots, w_m(z) = f_{\alpha_m, k_m}(z); \tag{5}$$

that is to say, the functions (5) are functionally independent, but, by adjoining to the set (5) any additional functions from (3), we secure a set of functions which are not functionally independent. We shall assume that $m < n$, and from this derive a contradiction.

The set M' of all points of D at which the Jacobian matrix (4) is of rank less than m is a closed subset of D of topological dimension at most $2n - 2$. The sets

$$M_\alpha = \{z \mid \varphi_{\alpha, 1}(z) = 0\}$$

and

$$M_\alpha' = \{z \mid h_\alpha(z) = 0\},$$

where $h_\alpha(z)$ are the functions introduced in part i of definition 2, are also closed subsets of D of topological dimension at most $2n - 2$. Letting M be the set introduced in part iii of definition 2, the set

$$\tilde{N} = M \cup M' \cup (\textstyle\bigcup_\alpha M_\alpha) \cup (\textstyle\bigcup_\alpha M_\alpha')$$

is of topological dimension at most $2n - 1$, so that there is a point $z^\circ \epsilon D - \tilde{N}$. We shall then consider the subvariety

$$V(z^\circ) = \{z \mid w_1(z) = w_1(z^\circ), \ldots, w_m(z) = w_m(z^\circ)\}.$$

Since the matrix (4) is of rank m at z°, we may introduce the new co-ordinates

$$w_1, \ldots, w_m, \qquad z_{m+1}, \ldots, z_n$$

in a neighborhood of z°; then locally $V(z^\circ)$ will be a linear subspace of topological dimension at least 2. For any other function $f_{\alpha, k}(z)$, the Jacobian matrix

$$\frac{\partial(f_{\alpha, k}(z), w_1, \ldots, w_m)}{\partial(w_1, \ldots, w_m, z_{m+1}, \ldots, z_n)}$$

is by hypothesis of rank m, which implies immediately that

$$\frac{\partial f_{\alpha, k}(z)}{\partial z_{m+1}} \equiv \ldots \equiv \frac{f_{\alpha, k}(z)}{\partial z_n} \equiv 0$$

in a neighborhood of z°; thus all the functions (3) are constant on the subvariety $V(z^\circ)$.

Following Siegel,[3] we introduce for each α the function

$$\Psi_\alpha(\zeta, z) = \sum_{T \epsilon \Lambda_\alpha} [\zeta - \varphi_{\alpha, 1}(z)\eta_{\alpha, T}(z)]^{-1}, \tag{6}$$

noting that it is a meromorphic function of the point z on D and of the variable ζ. As a function of ζ, for fixed z, $\Psi_\alpha(\zeta, z)$ has simple poles at the points

$$\zeta = \varphi_{\alpha, 1}(z)\eta_{\alpha, T}(z) \tag{7}$$

for the various $T \epsilon \Lambda_\alpha$. Since, by our construction, the numbers $|\varphi_{\alpha, 1}(z)\eta_{\alpha, T}(z)|$ are bounded from below by some positive number δ for all z in a neighborhood of z°, the function (6) may be expanded as a power series,

$$\Psi_\alpha(\zeta, z) = -\sum_{k=1}^{\infty} f_{\alpha, k}(z)\zeta^{k-1}, \tag{8}$$

for z near z° and $|\zeta| < \delta$.

Restricting the variable z to the subvariety $V(z^\circ)$, we have proved that the coefficients in the power-series expansion (8) become constants, so that the function (6) is independent of $z \epsilon V(z^\circ)$. In terms of the poles (7), this means that

$$\varphi_{\alpha, 1}(z)\eta_{\alpha, T}(z) = \varphi_{\alpha, 1}(z^\circ)\eta_{\alpha, T}(z^\circ)$$

for all $z \epsilon V(z^\circ)$. If $T \epsilon \Gamma_\alpha$, this equation becomes simply

$$\varphi_{\alpha, 1}(z) = \varphi_{\alpha, 1}(z^\circ),$$

so that we must have indeed

$$\eta_{\alpha, T}(z) = \eta_{\alpha, T}(z^\circ)$$

for all $z \epsilon V(z^\circ)$. This contradicts the crucial condition iii of definition 2, and the theorem follows from this contradiction.

2. Our first application shall be to the multiperiodic case. We assume that the

period matrix (the $n \times 2n$ matrix whose columns represent vectors generating the invariant lattice) is a Riemann matrix;[4, 5, 6] by a standard normalization, we may assume that this matrix is of the form

$$(I\Omega),$$

where I is a square unit matrix, $\Omega = (\omega_{jk})$ is a square matrix, and there is a diagonal matrix $\Delta = (\delta_{jj})$ with integer entries such that $\Delta\Omega$ is symmetric and $i\Delta(\bar{\Omega} - \Omega)$ is positive definite. The transformations of the group Γ may then be written explicitly in the form

$$T(m_1, \ldots, m_n;\ m_1', \ldots, m_n'): z_j \rightarrow z_j + m_j' + \sum_{k=1}^{n} \omega_{jk} m_k,$$

and as factors of automorphy we shall take the functions

$$\eta_{T(m_1,\ldots,m_n;\ m_1',\ldots,m_n')}(z) = \exp \frac{\pi}{i} \left\{ \sum_{j=1}^{n} m_j \delta_{jj}(2z_j - \omega_{jj}) + \sum_{j,\,k=1}^{n} \delta_{jj}\omega_{jk} m_j m_k \right\}. \quad (9)$$

THEOREM 2. *If in the multiperiodic case the period matrix is a normalized Riemann matrix, the factors of automorphy* (9) *form a complete system of factors.*

Proof: (i) This is an example of a case in which a single set of holomorphic factors of automorphy is sufficient. The first condition of definition 2 is trivially satisfied by taking $h(z) \equiv 1$.

ii) Let Γ_1 (Γ_α with $\alpha = 1$) be the subgroup of Γ consisting of the transformations

$$T(0, \ldots, 0;\ m_1', \ldots, m_n'),$$

and select as the coset representatives Λ_1 the transformations

$$T_0(m_1, \ldots, m_n) = T(m_1, \ldots, m_n;\ 0, \ldots, 0).$$

On any compact subset of D the series

$$\sum_{T \in \Lambda_1} [\eta_T(z)]^{-1} \quad (10)$$

is eventually dominated by the series

$$\sum_{m_1,\ldots m_n = -\infty}^{+\infty} \exp {}^1/_2\pi i \left\{ \sum_{j,k=1}^{n} \delta_{jj}\omega_{jk} m_j m_k \right\},$$

which converges as a consequence of the hypothesis that the period matrix is a normalized Riemann matrix. Since (10) is the Fourier expansion of the sum function, the latter is not identically zero. This verifies the second condition.

iii) Suppose that for some points z° and z,

$$\eta_{T_0(m_1,\ldots,m_n)}(z) = \eta_{T_0(m_1,\ldots,m_n)}(z^\circ) \quad (11)$$

for all $T_0(m_1, \ldots, m_n)$. Using the explicit form (9) of these factors of automorphy, condition (11) means that

$$\sum_{j=1}^{n} m_j \delta_{jj} z_j = \sum_{=1}^{n} m_j \delta_{jj} z_j^\circ + q(m_1, \ldots, m_n)$$

for all $m_1, \ldots, m_n$ and some integers q depending upon the indices $m_1, \ldots, m_n$; in particular, by setting $m_r = 1$ and $m_1 = \ldots = m_{r-1} = m_{r+1} = \ldots = m_n = 0$, we secure

$$z_r = z_r{}^{\circ} + \frac{q_r}{\delta_{rr}} \tag{12}$$

for some integers q_r. Whatever the choice of z°, the set of all points z satisfying a condition of the form (12) with any arbitrary integers q is surely zero-dimensional at z°. This then completes the proof of the theorem.

3. We next consider the case in which D is a bounded subdomain of the complex affine space and Γ is any properly discontinuous group of analytic homeomorphisms of D onto itself. Letting $J_T(z)$ be the complex Jacobian determinant of the transformation T and $h_\alpha(z)$ be the set of all not identically vanishing polynomials with rational coefficients, we introduce the factors of automorphy

$$\eta_{\alpha,\, R,\, T}(z) = [J_T(z)]^{-2R}[h_\alpha(Tz)]^{-1}h_\alpha(z) \tag{13}$$

for all integers $R \geq 1$.

THEOREM 3. *If D is a bounded subdomain of a complex affine space and Γ is a properly discontinuous group of analytic homeomorphisms of D onto itself, then a subcollection of the sets of factors of automorphy* (13) *form a complete system of factors.*

Proof: (i) As the functions $h_\alpha(z)$ of part i of definition 2, we may take the polynomials $h_\alpha(z)$ appearing in (13).

ii) Taking $\Gamma_\alpha = \mathrm{I}$ and $\Lambda_\alpha = \Gamma$ for each α, it is well known[4, 7] that the series

$$\sum_{T \in \Gamma} [J_T(z)]^2$$

is absolutely uniformly convergent on every compact subset of D; since each polynomial $h_\alpha(z)$ is bounded on D, the series (1) likewise converges in this case.

Now select a point $\tilde{z}^{\circ} \in D$ with rational co-ordinates, and let

$$\Gamma' = \{T \mid T \in \Gamma \text{ and } |J_T(\tilde{z}^{\circ})| < 1\};$$

there will be only finitely many transformations $I, T_1, \ldots, T_p$ in $\Gamma - \Gamma'$, including the identity transformation I. Considering only those polynomials $h_\alpha(z)$ for which

$$h_\alpha(\tilde{z}^{\circ}) = 1, \qquad h_\alpha(T_1\tilde{z}^{\circ}) = \ldots = h_\alpha(T_p\tilde{z}^{\circ}) = 0, \tag{14}$$

we have

$$\varphi_{\alpha,\, R,\, 1}(\tilde{z}^{\circ}) = 1 + \sum_{T \in \Gamma'} h_\alpha(T\tilde{z}^{\circ})[J_T(\tilde{z}^{\circ})]^{2R} \neq 0$$

whenever $R \geq R(\alpha)$ for some suitable index $R(\alpha)$ sufficiently large. We shall consider only those factors of automorphy (13) for which the component polynomials $h_\alpha(z)$ satisfy (14) and for which $R \geq R(\alpha)$; condition ii is thus verified for this subcollection.

iii) Let $N = \{z \mid z = T\tilde{z}^{\circ} \text{ for some } T \in \Gamma\}$, $N_{\alpha,\, T} = \{z \mid h_\alpha(Tz) = 0\}$, N_T be the set of all fixed points of the transformation T, and

$$M = N \cup \bigcup_{T \in \Gamma,\, T \neq I} N_T \cup \bigcup_{T \in \Gamma,\, \alpha} N_{\alpha,\, T}.$$

We may of course always assume that the group Γ is nontrivial. Suppose, then, that for points $z^\circ \in D - M$ and $z^1 \in D$,

$$\eta_{\alpha, R, T}(z^1) = \eta_{\alpha, R, T}(z^\circ) \tag{15}$$

for all factors of automorphy $\eta_{\alpha, R, T}(z)$ admissible under the restrictions imposed in part ii above. Since (15) must hold for all R suitably large, we have

$$h_\alpha(z^1)h_\alpha(Tz^\circ) = h_\alpha(z^\circ)h_\alpha(Tz^1) \tag{16}$$

for all polynomials $h_\alpha(z)$ satisfying (14). If z^1 is not congruent to z° under Γ, select any $T \in \Gamma$ other than the identity transformation; we may then construct a polynomial $h_\alpha(z)$ satisfying (14) and such that $|h_\alpha(z^\circ)| < 1$, $|h_\alpha(Tz^\circ)| > 2$, and $^1/_2 < |h_\alpha(z^1)|, |h_\alpha(Tz^1)| < 1$. This clearly contradicts (16) and is thus sufficient to demonstrate condition iii of definition 2 and complete the proof of the theorem.

4. For our final example we shall consider Siegel's modular group of degree p, where $n = {}^1/_2\, p(p + 1)$. If A is a matrix, then $A > 0$ means that it is positive definite, and ${}^tA$ denotes its transpose. The manifold D is the space of all symmetric $p \times p$ complex matrices Z such that

$$\frac{1}{2i}(Z - \bar{Z}) > 0,$$

and the group Γ is the set of all transformations

$$T: Z \rightarrow (AZ + B)(CZ + D)^{-1},$$

where A, B, C, D, are $p \times p$ integer matrices satisfying

$${}^t\begin{pmatrix} A & B \\ C & D \end{pmatrix} \begin{pmatrix} 0 & I \\ -I & 0 \end{pmatrix} \begin{pmatrix} A & B \\ C & D \end{pmatrix} = \begin{pmatrix} 0 & I \\ -I & 0 \end{pmatrix}. \tag{17}$$

The transformations T thus correspond to the $2p \times 2p$ integer matrices satisfying (17). We then introduce the factors of automorphy

$$\eta_T(z) = \det\,(CZ + D)^{p+2}, \tag{18}$$

where "det" denotes the determinant of the matrix; these functions are holomorphic and nonvanishing on D.[3,8]

THEOREM 4. *For the modular group of degree p, the factors of automorphy* (18) *form a complete system of factors.*

Proof: (i) The first condition is trivial, as in Theorem 2.

ii) Let Γ_1 (Γ_α with $\alpha = 1$) be the subgroup of Γ consisting of all transformations whose corresponding matrices are of the form

$$\begin{pmatrix} {}^tU & SU^{-1} \\ 0 & U^{-1} \end{pmatrix},$$

where S is a $p \times p$ symmetric integer matrix and U is an integer matrix with det $U = 1$. Then, for any system Λ_1 of representatives of the right cosets of Γ_1 in Γ, it is known[8, 9] that the series

$$\sum_{T \in \Lambda_1} [\eta_T(z)]^{-1} = \sum_{T \in \Lambda_1} \det\,(CZ + D)^{-p-2}$$

is absolutely uniformly convergent on any compact subset of D. A Fourier expansion[3] shows that the sum function is not identically zero.

iii) Let us note that for an arbitrary $p \times p$ symmetric matrix A, the matrix

$$\begin{pmatrix} 0 & I \\ -I & A \end{pmatrix} \tag{19}$$

satisfies (17) and therefore corresponds to a transformation $T \epsilon \Gamma$. Suppose, now, that for two points z° and z^1 of D, $\eta_T(z^1) = \eta_T(z^\circ)$ for all $T \epsilon \Gamma$. For the particular transformations (19) this may be written explicitly as

$$\det (A - Z^\circ) = \epsilon \det (A - Z^1), \tag{20}$$

where $\epsilon^{p+2} = 1$, and A is an arbitrary symmetric integer matrix. Since (20) is a polynomial relation in the variables A which is true for all integer values of its arguments, it must be true for all symmetric matrices, and, further, $\epsilon = 1$. This, in turn, is only possible when $z^\circ = z^1$, so that condition iii of definition 2 is verified in this case also, and the proof is complete.

[1] S. Bochner, "On Compact Complex Manifolds," *J. Indian Math. Soc.*, **11,** 1–21, 1947.

[2] R. C. Gunning, "General Factors of Automorphy," these PROCEEDINGS, **41,**496–498, 1955.

[3] C. L. Siegel, "Einführung in die Theorie der Modulfunktionen n-ten Grades," *Math. Ann.*, **116,** 617–657, 1939.

[4] C. L. Siegel, "Analytic Functions of Several Complex Variables" (Institute for Advanced Studies, 1952), mimeographed notes.

[5] F. Conforto, *Funzioni abeliani e matrici di Riemann* (Rome, 1942).

[6] G. Frobenius, "Über die Grundlagen der Theorie der Jacobischen Funktionen," *J. reine u. angew. Math.*, **97,** 16–48, 1884.

[7] A. Borel, "Les Fonctions automorphes de plusieurs variables complexes," *Bull. Soc. Math. France,* **80,** 167–182, 1952.

[8] M. Sugawara, "Über eine allgemeine Theorie der Fuchsschen Gruppen und Theta-Reihen," *Ann. Math.*, **41,** 488–494, 1940.

[9] H. Braun, "Konvergenz verallgemeinter Eisensteinschen Reihen," *Math. Z.*, **44,** 387–397, 1939.

BOCHNER, S.
Math. Zeitschr. Bd. 63, S. 230—242 (1955)

Green-Goursat Theorem.

In memory of ISSAI SCHUR.

By
S. BOCHNER.

In a recent paper [1] we have generalized from complex to real variables GOURSAT's classical proof for the formula

$$\oint f(z)\,dz = 0 \tag{1}$$

in which only the existence of the derivative

$$\lim_{\zeta \to 0} \frac{f(z+\zeta) - f(z)}{\zeta} = f'(z) \tag{2}$$

is being assumed, but nothing further on the behaviour of $f'(z)$ as a function of the variable point z is being stipulated (explicitly).

In the present paper our previous theorem will be rounded off by incorporating two further generalizations, both noteworthy, we think (part I). Then, a pecularity of the one-dimensional case will be commented upon (part II), and finally an effect of the entire generalization on the CAUCHY formula with remainder term,

$$f(z, \bar z) = \frac{1}{2\pi i} \int_C \frac{f(\zeta, \bar\zeta)\, d\zeta}{\zeta - z} - \frac{1}{2\pi i} \int_D \frac{\frac{\partial f(\zeta, \bar\zeta)}{d\bar\zeta}\, d\bar\zeta\, d\zeta}{\zeta - z}, \tag{3}$$

will be discussed (part III).

Most modern papers on GREEN and STOKES theorem are very much more elaborate and refined than is the present one (see for instance [2] and authors there quoted), but we do not think that our theorems 1 and 2 are actually directly contained in any of them or that the direction or aim of the present paper has been anticipated as such.

Part I. Green's Theorem for Two Variables.

Theorem 1. *Let D be a bounded (open) domain in the (x, y)-plane and let C be its boundary. The formula*

$$\int_C A\,dy - B\,dx = \int_D (A_x + B_y)\,dx\,dy \tag{4}$$

holds under the following assumptions 1.—5.

1. *The boundary C consists of one or several simple closed rectifiable curves.*

2. *The functions $A(x, y)$, $B(x, y)$ are defined and continuous in the closure $\bar D = D + C$.*

3. *At each point* x_0, y_0 *of* D *the functions* A, B *have* STOLZ *derivatives, meaning that*

$$(5)\qquad \begin{cases} A(x, y) = A(x_0, y_0) + A_{x_0} \cdot (x - x_0) + A_{y_0} \cdot (y - y_0) + o(|z - z_0|) \\ B(x, y) = B(x_0, y_0) + B_{x_0} \cdot (x - x_0) + B_{y_0} \cdot (y - y_0) + o(|z - z_0|), \end{cases}$$

where $z = x + iy$.

4. *The sum*

$$(6)\qquad \Phi(x, y) \equiv A_x + B_y$$

is LEBESGUE *integrable over* D, *meaning that*

$$(7)\qquad \int_D |A_x + B_y| \, dx \, dy < \infty.$$

5. *At each point* x_0, y_0 *of* D *we have the generalized continuity property*

$$(8)\qquad \lim_{\varepsilon \downarrow 0} \frac{1}{\pi \varepsilon^2} \int_{|z - z_0| < \varepsilon} |\Phi(x, y) - \Phi(x_0, y_0)| \, dx \, dy = 0.$$

Remark 1. In [*1*] we have established this theorem under the more special assumptions that $\Phi(x, y)$ is continuous (in the ordinary sense) and that the curves C are rectilinear polygons, and the present more general version would reflect the natural scope of the theorem, if it were not for the fact that the continuity property 5 can be relaxed to a certain extent, for a sufficiently smooth boundary at any rate (compare Remark 9 in part II). In this connection we would like to point out that according to a theorem of LEBESGUE,—for its one-dimensional version see [*3*], p. 13,—any LEBESGUE integrable function $\Phi(x, y)$ in D possesses the continuity property (8) at the points (x^0, y^0) of D automatically, but with the possible exception of a set of measure 0, whereas in our theorem no *general* set of measure 0 could be excepted, although certain null sets of special structure could.

Remark 2. As already emphasized in [*1*] the novelty of this version of GREEN's theorem lies in the fact that no assumption on the behaviour of any of the functions A_x, B_y and also A_y, B_x *separately* is being made, but only on their combination (6) which is the only one occurring explicitly in the formula.

Remark 3. If we introduce the differential operators

$$(9)\qquad \frac{\partial}{\partial z} = \frac{1}{2} \frac{\partial}{\partial x} + \frac{1}{2i} \frac{\partial}{\partial y}, \qquad \frac{\partial}{\partial \bar{z}} = \frac{1}{2} \frac{\partial}{\partial x} - \frac{1}{2i} \frac{\partial}{\partial y},$$

and if we designate an arbitrary function $F(x, y)$ symbolically by $f(z, \bar{z})$ then our theorem implies

$$(10)\qquad \int_C f(z, \bar{z}) \, dz = \int_D \frac{\partial f}{\partial \bar{z}} \, d\bar{z} \, dz, \qquad d\bar{z} \, dz = 2i \, dx \, dy,$$

under the assumption that f is continuous in $D + C$, and is STOLZ differentiable in D as a function of (x, y), and that the function $\frac{\partial f}{\partial \bar{z}}$ is LEBESGUE integrable over D and has the property (8) at each point of it.

Remark 4. It is easily verified that if $\Phi(x, y)$ has property (8) then for any continuous factor $\lambda(x, y)$ the product $\Psi(x, y) = \Phi(x, y)\, \lambda(x, y)$ will have it likewise. Also if D_ε is a neighborhood of (x_0, y_0) which contains a circle $|z - z_0| < \alpha\, \varepsilon$ and is itself contained in a circle $|z - z_0| < \beta\, \varepsilon$, where α, β are independent of ε, then (8) implies

$$\int_{D_\varepsilon} |\Psi(x, y) - \Psi(x_0, y_0)|\, dx\, dy = o(\varepsilon^2), \qquad \varepsilon \to 0. \tag{11}$$

Remark 5. If $x = x(u, v)$, $y = y(u, v)$ is a continuously differentiable one-one mapping of a domain S of the (u, v)-plane into our domain D, and if

$$J(u, v) \equiv \frac{\partial(x, y)}{\partial(u, v)} > 0, \tag{12}$$

then we have

$$\int_D \Phi(x, y)\, dx\, dy = \int_S \Psi(u, v)\, du\, dv, \tag{13}$$

where $\Psi(u, v) = \Phi(x, y)\, J(u, v)$, in the sense that if either integral (13) exists then so does the other. Furthermore if (u_0, v_0) maps into (x_0, y_0), then for small increments we have

$$\Delta z_0 = O(\Delta w_0) \quad \text{and} \quad \Delta w_0 = O(\Delta z_0),$$

and remark 4 leads to the following conclusion. If $\Phi(x, y)$ possesses property (8) as function in (x, y) then $\Psi(u, v)$ also possesses this property as function in (u, v).

Remark 6. If on a joint interval $0 \le t \le 1$ a sequence of functions $z_n(t) = x_n(t) + i\, y_n(t)$ is convergent to $z(t)$, if all $z_n(t)$ are of bounded variation and their variations are jointly bounded,

$$V_{0,1}\{z_n(\cdot)\} \le M, \tag{14}$$

and if a sequence of continuous functions $\varphi_n(t)$ is uniformly convergent to $\varphi(t)$ on $0 \le t \le 1$, then

$$\lim_{n \to \infty} \int_0^1 \varphi_n(t)\, dz_n(t) = \int_0^1 \varphi(t)\, dz(t) \tag{15}$$

[Riemann-Stieltjes integrals, say]. This implies as follows.

Remark 7. If on a closed pointset $\overline{D}$ in (x, y)-plane there is given a sequence of (continuous) rectifiable curves $C_n : z_n(t)$, $0 \le t \le 1$; if they converge uniformly in $0 \le t \le 1$ to the curve $C : z(t)$; if the lengths of the curves C_n are jointly bounded; and if $A(x, y)$ and $B(x, y)$ are continuous functions on $\overline{D}$; then we have

$$\begin{aligned} &\lim_{n \to \infty} \int_{C_n} A\big(x_n(t), y_n(t)\big)\, dy_n(t) - B\big(x_n(t), y_n(t)\big)\, dx_n(t) \\ &= \int_C A\big(x(t), y(t)\big)\, dy(t) - B\big(x(t), y(t)\big)\, dx(t), \end{aligned}$$

that is

$$\lim_{n \to \infty} \int_{C_n} A\, dy - B\, dx = \int_C A\, dy - B\, dx. \tag{16}$$

Remark 8. In the plane of $w=u+iv$ let S be a bounded domain which is bounded by the (outer) circle

$$R^0 : w = w^0 + r^0 e^{i\vartheta}, \qquad 0 \leq \vartheta < 2\pi$$

and (perhaps) by the additional (inner) circles

$$R^m : w = w^m + r^m e^{i\vartheta}, \qquad m = 1, \ldots, n.$$

If for $0<\varepsilon<1$ we introduce the circles

$$R^0 : w = w^0 + (1-\varepsilon)\, r^0 e^{i\vartheta}$$
$$R^m : w = w^m + (1+\varepsilon)\, r^m e^{i\vartheta}$$

then for ε sufficiently small they bound a domain S_ε interior to S which for $\varepsilon\to 0$ converges to the latter. Now, let $z=f(w)$ be a conformal mapping from the domain S into our previous domain D which also carries the circles R^0, R^m into rectifiable curves C^0, C^m respectively. It then carries the approximating circles R^0_ε, R^m_ε into approximating curves C^0_ε, C^m_ε which are uniformly convergent to the boundary curves C^0, C^m in parameter, and, what is decisive for us, the lengths of the curves C^0_ε, C^m_ε are jointly bounded in ε, see [4], p. 194. Thus, if we put $C_\varepsilon = C^0_\varepsilon + C^1_\varepsilon + \cdots + C^m_\varepsilon$, then by remark 7 we have

$$\lim_{\varepsilon\to 0} \int_{C_\varepsilon} A\,dy - B\,dx = \int_C A\,dy - B\,dx \tag{17}$$

for any functions A, B continuous in $D+C$.

Proof of Theorem 1. Following the classical procedure of GOURSAT we first assume, as we did in [1], that the domain D is a rectilinear triangle, which we will denote by D_0, and that the assumptions 2.—5. of the theorem are fulfilled in an open neighborhood U of the closed set $\bar{D}_0 = D_0 + C_0$. Assuming (4) not to hold, we write,

$$\left| \int_{C_0} A\,dy - B\,dx - \int_{D_0} \Phi(x,y)\,dx\,dy \right| = \alpha > 0, \tag{18}$$

and we subdivide D_0 into 4 rectilinear triangles $D_1^{(\nu)}$, $\nu = 1, \ldots, 4$, any two of which are congruent so that their sides are half as long as the corresponding ones in D_0 itself. Due to the "additivity" of relation (4), assumption (18) implies that we have

$$\left| \int_{C_1} A\,dy - B\,dx - \int_{D_1} \Phi(x,y)\,dx\,dy \right| \geq \frac{\alpha}{4}$$

for at least one of the triangles $D_1^{(\nu)}$, which we have chosen fixed and denoted by D_1. Continuing in this manner we obtain a sequence of triangles $D_0 \supset D_1 \supset D_2 \supset D_3 \ldots$ with the following properties. Firstly we have

$$\left| \int_{C_n} A\,dy - B\,dx - \int_{D_n} \Phi(x,y)\,dx\,dy \right| \geq \frac{\alpha}{4^n}; \tag{19}$$

secondly we have

$$\text{length}\,[C_n] = 0\left(\frac{1}{2^n}\right)$$

and
$$\frac{1}{4^n} = 0\,(\text{area } [D_n])\,;$$

and thirdly we have that the closures $\overline{D}_n$ have a point in common, and this point we denote by x_0, y_0.

If now we put $A(x, y) = A' + A''$, $B = B' + B''$ where

$$A'(x, y) = A(x_0, y_0) + A_{x_0} \cdot (x - x_0) + A_{y_0} \cdot (y - y_0)$$
$$B'(x, y) = B(x_0, y_0) + B_{x_0} \cdot (x - x_0) + B_{y_0} \cdot (y - y_0),$$

and A'', B'' are corresponding "remainder functions", then the parts A', B' are linear in (x, y), and for such one we take our theorem as known, as does also GOURSAT in his case. However, since A, B are STOLZ differentiable at (x_0, y_0), we have

$$|A''| + |B''| = o(|z - z_0|) = o\left(\frac{1}{2^n}\right), \quad \text{for } z \text{ on } C_n;$$

and also

$$\Phi''(x, y) \equiv A''_x + B''_y = \Phi(x, y) - \Phi(x_0, y_0).$$

Therefore,

$$\int_{C_n} A''\,dy - B''\,dx - 0\left(\max_{x \varepsilon C_n} [|A''_n| + |B''_n|]\right) \cdot \text{length } [C_n] = o\left(\frac{1}{4^n}\right),$$

and

(20) $$\int_{D_n} \Phi''\,dx\,dy = o(\text{area } [D_n]) = o\left(\frac{1}{4^n}\right),$$

and this would not be compatible with relation (19) for $A = A''$, $B = B''$; whence the conclusion.

The protective neighborhood U was needed because the point (x_0, y_0) in the proof might be on the boundary C of $D \equiv D_0$, but we can now dispense with it by the following reasoning. Let the origin $(0, 0)$ be a point of D and let $z = z(t)$ be a parametrization of C. If we introduce the triangle $C_\varepsilon : z_\varepsilon = (1 - \varepsilon)\,z(t)$, $0 < \varepsilon < 1$, then it bounds a subdomain D_ε of D of which D is such a neighborhood, so that our conclusion applies to D_ε at first. However, we have

$$\lim_{\varepsilon \to 0} \int_{D_\varepsilon} \Phi(x, y)\,dx\,dy = \int_D \Phi(x, y)\,dx\,dy$$

whenever the LEBESGUE integral on the right side exists, and on the other hand we have

$$\lim_{\varepsilon \to 0} \int_{C_\varepsilon} A\,dy - B\,dx = \int_C A\,dy - B\,dx$$

by remark 7, and thus (4) holds if D is a triangle.

Next, by combinatorial "additivity" it holds whenever the bounding curves $C^0, C^1, \ldots, C^n$ are rectilinear polygons. But then, again by the use of remark 7 ,it can be easily shown that it holds whenever the curves are arbitrarily convex, and especially whenever they are circles.

Turning now to the most general domain D as envisaged in the theorem, we introduce a conformal mapping of it into a domain S with circular boundary curves R, as in remark 8; and if, as in remark 8, we approximate to D by subdomains D_ε whose boundaries C_ε map into circles R_ε of S, then on account of relation (17), our assertion will follow from the following lemma, finally.

Lemma 1. Let U be a neighborhood of $D+C$, and let $x=x(u,v)$, $y=y(u,v)$ be a one-one continuously differentiable mapping, with positive Jacobian, of the neighborhood U into a neighborhood V in (u,v)-plane, and let $S+R$ be the image of $D+C$, and let it be known that our theorem has already been established for this domain S; then it also holds for the domain D.

In fact, if we put

$$A_1(u,v) = A(x,y)\frac{\partial y}{\partial v} - B(x,y)\frac{\partial x}{\partial v}$$

$$B_1(u,v) = B(x,y)\frac{\partial x}{\partial u} - A(x,y)\frac{\partial y}{\partial u}$$

then we have identically

$$A\,dy - B\,dx = A_1\,du - B_1\,du$$

$$\left(\frac{\partial A}{\partial x} + \frac{\partial B}{\partial y}\right)\frac{\partial(x,y)}{\partial(u,v)} = \frac{\partial A_1}{\partial u} + \frac{\partial B_1}{\partial v},$$

and all that is needed for the proof of the lemma has already been anticipated in remark 5.

Part II. Several Variables.

For more than two dimensions our theorem also holds (compare theorem 1) in [*1*]) but the case of "general" boundary surfaces is fraught with difficulties which we do not wish to enter into. However we do wish to make a statement resembling the present two-dimensional one, and we will make it for a certain type of simplex therefore.

For any $k \geqq 1$, in the Euclidean $E_k:(u_j)$ we take a rectilinear k-dimensional simplex S, whose boundary we denote by R, and we take some open neighborhood V of $R+S$ also. We also take an arbitrary continuously differentiable mapping with positive Jacobian,

$$x_j = x_j(u_1,\ldots,u_k), \qquad j=1,\ldots,k$$

of the neighborhood V in $E_k:(u_j)$ into a neighborhood U in $E_k:(x_j)$, and the resulting image of (S,R) shall be denoted by (D,C) then.

Theorem 2. *In $E_k:(x_j)$ let D be any k-dimensional simplex as just described and let C be its boundary. Then the formula*

$$\int_C \sum_{j=1}^{k} A_j(x_k)(-1)^{j-1}dx_1\ldots dx_{j-1}dx_{j+1}\ldots dx_k = \int_D \Phi(x_j)\,dv_x, \tag{21}$$

$$\Phi(x_j) \equiv \frac{\partial A_1}{\partial x_1} + \cdots + \frac{\partial A_k}{\partial x_k}, \tag{22}$$

holds under the following assumptions. Each of the functions

$$A_1(x_j),\ldots,A_k(x_j)$$

is defined and continuous in $D+C$ *and has a* STOLZ *derivative at each point* (x_j^0) *of* D. *The function* $\Phi(x_j)$ *is* LEBESGUE *integrable over* D,

$$\int_D |\Phi(x_j)|\, dv_x < \infty,$$

and at each point (x_j^0) *of* D *we have*

$$\int_{|x-x^0|<\varepsilon} |\Phi(x_j) - \Phi(x_j^0)|\, dv_x = o(\varepsilon^k), \qquad \varepsilon \to 0. \tag{23}$$

Remark 9. Assumption 5 in theorem 1 was needed only for the proof of the inequality (20), and therefore we may state as follows. If in theorem 1 (or 2) the domain D is a rectilinear polyhedron, then assumption (8), or (23), can be relaxed in the following manner. If $\{T\}$ is a family of k-dimensional rectilinear simplices in E_k any two of which are similar or antisimilar to one another and if each T contains the point (x_j^0) on its closure then we have

$$\int_T (\Phi(x_j) - \Phi(x_j^0))\, dv_x = o(\text{volume } [T]), \tag{24}$$

as volume $[T]$ tends to 0, without absolute value signs under the integral.

Now, GOURSAT's method of proof for theorem 2 can also be made to apply to the one-dimensional case $k=1$, in which case D is an interval $a<x<b$, and C is the combinatorial pair of points $x=a$, $x=b$, and the resulting relation is

$$A(b) - A(a) = \int_a^b \frac{dA}{dx}\, dx. \tag{25}$$

Property (24) reads in this case

$$\lim_{h+k\to 0} \frac{1}{h+k} \int_{x^0-h}^{x^0+k} \frac{dA}{dx}\, dx = 0, \qquad h \geq 0,\ k \geq 0,\ h+k>0,$$

and thus the following statement is included in theorem 2.

Theorem 3. *Relation* (25) *holds whenever* $A(x)$ *is continuous in* $a \leq x \leq b$, *differentiable at each point of* $a<x<b$, *and* $\frac{dA}{dx}$ *is* LEBESGUE *integrable over* $a<x<b$, *and the function*

$$B(x) = \int_a^x \frac{dA}{d\xi}\, d\xi$$

has a derivative

$$\frac{dB(x)}{dx} = \frac{dA(x)}{dx} \tag{26}$$

at each point in $a<x<b$.

This is a simple enough theorem, but it is not entirely "trivial" nevertheless. In fact, if we introduce the difference $F(x) = B(x) - A(x)$ then (26) implies

$$\frac{dF(x)}{dx} = 0, \tag{27}$$

and in order to conclude (25) it is necessary to know that (27) implies $F(x) = \text{constant}$, which is ROLLE's theorem, no less. Thus we see that, in the one-dimensional case, GOURSAT's method of proving (25) by subdividing the interval (a, b) into equal halves successively is as good as ROLLE's reasoning, and for $k \geq 2$ it is a rather efficient replacement for it.

We are now returning to the k-dimensional case, and to the following compare [5] and chapter VIII in [6].

We take a domain D and a subdomain D_0 whose closure $\overline{D}_0$ is contained in D, and we denote by ϱ the distance from D_0 to the boundary of D. If for $\sigma < \varrho$ we take in the sphere

$$(28) \qquad S_\sigma : (t_1^2 + \cdots + t_k^2)^{\frac{1}{2}} \leq \sigma$$

an integrable function $K(t) \equiv K(t_1, \ldots, t_k)$, for which

$$(29) \qquad \int_{S_\sigma} K(t)\, dv_t = 1\,,$$

say, and if a function $A(x) \equiv A(x_1, \ldots, x_k)$ is defined and continuous in D, then for x in D_0, we can introduce the "average"

$$(30) \qquad A^K(x) = \int_{S_\sigma} A(x - t)\, K(t)\, dv_t$$

which, incidentally, for $\sigma \to 0$, converges to $A(x)$ there. This function (30) is a "smoothing" of $A(x)$ itself in the following manner (among others). If we replenish the function $K(t)$ in (28) to a function in the entire E_k by putting it $= 0$ for $|t| > \sigma$, and if we also replenish $A(x)$ to a function in E_k by putting it $= 0$ for x in $E_k - D$ then we can write

$$(31) \qquad A^K(x) = \int_{E_k} A(x - t)\, K(t)\, dv_t$$

and also

$$(32) \qquad A^K(x) = \int_{E_k} A(t)\, K(x - t)\, dv_t\,.$$

Now, in the first place, (32) implies that if $K(t)$ as a function in E_k belongs to differentiability class $C^{(1)}$ there, then $A^K(x)$ belongs to $C^{(1)}$ in E_k, and especially in D; and, in the second place, if $A(x)$ itself belongs to $C^{(1)}$ in D, then $A^K(x)$ belongs to $C^{(1)}$ in D_0 and we can also interchange differentiation with integration, thus

$$(33) \qquad \frac{d A^K(x)}{\partial x_j} = \int_{S_\sigma} \frac{\partial A(x - t)}{\partial x_j} K(t)\, dv\,.$$

Now, if $A(x)$ has only a STOLZ derivative at each point then the second property cannot be asserted, but our theorem 2 will permit us to make a certain statement nevertheless.

Before making it, we define as follows. If $K(t)$ is integrable in (28) then we say that it is a $C^{(1)}$-kernel, if for every function $A(x)$ which is continuous in D, the corresponding function (32) is continuous in D_0. As already remarked, if $K(t)$, as a function in E_k is $C^{(1)}$ there, then it is a $C^{(1)}$-kernel in S_σ, but there are important $C^{(1)}$-kernels which are more general than that. For instance for $h \leqq \frac{\sigma}{k}$, the function

$$K(t) = \begin{cases} \frac{1}{(2h)^k}, & |t_j| \leqq h,\ j = 1, \ldots, k \\ 0, & \text{some } |t_j| > 0, \end{cases}$$

is a $C^{(1)}$-kernel, the corresponding function (32) being the familiar average

$$\frac{1}{(2h)^k} \int_{-h}^{h} \cdots \int_{-h}^{h} A(x_1 + t_1, \ldots, x_k + t_k)\, dv_t$$

then.

Theorem 4. *If io a domain D in E_k we are given functions $A_1(x), \ldots, A_k(x)$, $\Phi(x)$ with exactly the same properties as in theorem 2, thus being connected by (22) among others, and if for a $C^{(1)}$-kernel $K(t)$ in S_σ we introduce the averages*

$$A_j^K(x) = \int_{S\sigma} A_j(x - t)\, K(t)\, dv_t \tag{34}$$

$$\Phi^K(x) = \int_{S\sigma} \Phi(x - t)\, K(t)\, dv_t \tag{35}$$

then we have the relation

$$\sum_{j=1}^{k} \frac{\partial A_j^K(x)}{\partial x_j} = \Phi^K(x) \tag{36}$$

for almost all points x in D_0. (If $\Phi^K(x)$ is continuous in D_0, then this holds at all points without exception.)

Proof. For x in D_0 and t in S_σ, the original relation (22) implies

$$\sum_{j=1}^{k} \frac{\partial A_j(x - t)}{\partial x_j} = \Phi(x - t)$$

and therefore

$$\int_{S\sigma} \left(\sum_{j=1}^{k} \frac{\partial A_j(x - t)}{\partial x_j} \right) K(t)\, dv_t = \int_{S\sigma} \Phi(x - t)\, K(t)\, dv_t,$$

and our theorem asserts, that under the assumptions of theorem 2, the left side can be re-arranged thus:

$$\sum_{j=1}^{k} \frac{\partial}{\partial x_j} \int_{S\sigma} A_j(x - t)\, K(t)\, dv_t,$$

this involving two interlocking interchanges between summation, integration and differentiation.

We take in D_0 a function $\lambda(x) = \lambda(x_1, \ldots, x_k)$ which is $C^{(1)}$ there and which vanishes outside a polyhedral neighborhood U in D_0 (a so-called "testing function" in the terminology of [5]). For fixed t in S_σ, the function in x,

(37) $$\bar{A}_j(x) = \lambda(x)\, A_j(x-t),$$

has a STOLZ derivative at each point of D_0, and the "divergence"

$$\bar{\Phi}(x) = \sum_{j=1}^{k} \frac{\partial \bar{A}_j(x)}{\partial x_j} \equiv \lambda(x)\, \Phi(x-t) + \sum_{j=1}^{k} \frac{\partial \lambda(x)}{\partial x_j} A_j(x-t)$$

has the same properties in D_0, as $\Phi(x)$ has in D. However, all functions $\bar{A}_j(x)$ are 0 on the boundary of U, and thus by theorem 2 we have

$$\int_U \bar{\Phi}(x)\, dv_x = 0,$$

that is

$$\int_U \left[\lambda(x)\, \Phi(x-t) + \sum_{j=1}^{k} \frac{\partial \lambda}{\partial x_j} A_j(x-t)\right] dv_x = 0.$$

If we now multiply this by $K(t)$ and integrate with respect to t over S_σ, then by FUBINI's theorem the two integrations can be interchanged, and we have

$$\int_U \left[\lambda(x)\, \Phi^K(x) + \sum_{j=1}^{k} \frac{\partial \lambda}{\partial x_j} A_j^K(x)\right] dv_x = 0.$$

However, the functions $A_j^K(x)$ are $C^{(1)}$ and λ is 0 outside U, so that

$$\int_U \sum \frac{\partial \lambda}{\partial x_j} A_j^K\, dv_x = -\int_U \lambda \sum \frac{\partial A_j^K}{\partial x_j}\, dv_x,$$

and altogether we obtain

$$\int_{D_0} \lambda(x) \left[\Phi^K(x) - \sum_{j=1}^{k} \frac{\partial A_j^K}{\partial x_j}\right] dv_x = 0.$$

But, if this is so for every testing function $\lambda(x)$ as introduced, then we indeed must have relation (36) for almost all points of D_0, as claimed.

Part III. Cauchy Formula with Remainder Term.

In E_2: (ξ, η) if $H(\xi, \eta)$ is LEBESGUE measurable and

(38) $$\int_{E_2} \frac{|H(\xi,\eta)|}{1 + |\xi| + |\eta|}\, d\xi\, d\eta < \infty$$

then for almost all (x, y) the integral

(39) $$G(x, y) = \int_{E_2} \frac{H(\xi,\eta)}{\zeta - z}\, d\xi\, d\eta$$

exists, $\zeta=\xi+i\eta$, $z=x+iy$, and $G(x, y)$ is LEBESGUE integrable over any bounded domain. This follows from FUBINI's theorem by showing that

$$\int\limits_{|z|\leq A} dx\,dy \int\limits_{E_2} \left|\frac{H(\xi,\eta)}{\zeta - z}\right| d\xi\, d\eta < \infty, \tag{40}$$

for any $A<\infty$ (integrate with respect to x, y first), and it is also easily seen that if $H(\xi,\eta)$ is bounded in a neighborhood, then the integral (39) exists at all points of the neighborhood without exception and is likewise bounded there.

Also, if we apply to the integral

$$\int\limits_{|\zeta|\leq 1} \frac{H(\xi,\eta)\,d\xi\,d\eta}{\zeta - z}, \qquad |z|<1,$$

E. E. LEVI's transformation of coordinates

$$\begin{aligned} \xi - x &= \varrho(\cos\vartheta - x), & \eta - y &= \varrho(\sin\vartheta - y) \\ 0 &\leq \vartheta < 2\pi, & 0 &< \varrho \leq 1, \end{aligned}$$

then it goes over into

$$\int\limits_{\varrho=0}^{1} \int\limits_{\vartheta=0}^{2\pi} h\left(x(1-\varrho) + \varrho\cos\vartheta,\ y(1-\varrho)+\varrho\sin\vartheta\right) \frac{1-\cos\vartheta\, x - \sin\vartheta\, y}{e^{i\vartheta} - z}\, d\varrho\, d\vartheta,$$

and we hence conclude that if the function $H(\xi,\eta)$ is $C^{(r)}$, $r\geq 1$, in a neighborhood U, then $G(x, y)$ is also $C^{(r)}$ there, and if $H(\xi,\eta)$ is analytic in the real variables ξ,η there, then $G(x,y)$ is likewise so in x, y. However, $G(x,y)$ cannot be a holomorphic function in z unless $H(\xi,\eta)$ is $\equiv 0$ in U, as the following theorems will imply.

Now, if in (10) we replace the letter z by ζ, and subsequently the function $f(\zeta,\bar\zeta)$ by the function

$$\frac{f(\zeta,\bar\zeta)}{\zeta - z},$$

then by the customary procedure of excising a small neighborhood around z the following theorem ensues.

Theorem 5. *If D and C are as in theorem 1, if $f(\zeta,\bar\zeta)$ is continuous in $D+C$ and* STOLZ *differentiable in D, and if $f_{\bar\zeta}(\zeta,\bar\zeta)$ is* LEBESGUE *integrable over D and possesses the continuity property 5 of theorem 1 there, then we have*

$$f(z,\bar z) = \frac{1}{2\pi i}\int\limits_{C} \frac{f(\zeta,\bar\zeta)\,d\zeta}{\zeta - z} - \frac{1}{2\pi i}\int\limits_{D} \frac{f_{\bar\zeta}(\zeta,\bar\zeta)\,d\bar\zeta\,d\zeta}{\zeta - z} \tag{41}$$

at all points z of D at which the second integral exists (the exceptional points having measure zero at most).

If $f_{\bar\zeta}$ is also locally bounded in D, then the formula is valid at all points, and if $f_{\bar\zeta}$ is $C^{(r)}$ or real-analytic, then $f(\zeta,\bar\zeta)$ itself is likewise so.

Of interest is the following companion theorem.

Theorem 6. *If $h(\zeta, \bar{\zeta})$ is* LEBESGUE *integrable on D so that the function*

$$(42) \qquad g(z, \bar{z}) = -\frac{1}{2\pi i} \int_D \frac{h(\zeta, \bar{\zeta})\, d\bar{\zeta}\, d\zeta}{\zeta - z}$$

exist almost everywhere and is LEBESGUE *integrable over D itself, then for any testing function $\lambda(z, \bar{z})$ in D we have*

$$(43) \qquad \int_D \left(\frac{\partial \lambda}{\partial \bar{z}} \cdot g + \lambda \cdot h\right) d\bar{z}\, dz = 0,$$

and if furthermore it so happens that $g(z, \bar{z})$ has a STOLZ *derivative everywhere and $g_{\bar{z}}$ has the suitable properties, then*

$$(44) \qquad \frac{\partial g}{\partial \bar{z}} = h(z, \bar{z})$$

almost everywhere.

Proof. From (42) we obtain

$$\int_D \frac{\partial \lambda}{\partial \bar{z}} g\, d\bar{z}\, dz = -\int_D h(\zeta, \bar{\zeta}) \left[-\frac{1}{2\pi i} \int_D \frac{\frac{\partial \lambda}{\partial \bar{z}} d\bar{z}\, dz}{z - \zeta} \right] d\bar{\zeta}\, d\zeta.$$

However, if in (41) we replace $f(z, \bar{z})$ by $\lambda(z, \bar{z})$ and recall that λ is 0 on the boundary of D, then we see that the inner integral has the value $\lambda(\zeta, \bar{\zeta})$, which proves (43). Next, as in the proof to theorem 4, if g has a STOLZ derivative everywhere, and $g_{\bar{z}}$ has the suitable properties (for instance if it is continuous) then (43) implies

$$\int_D \lambda \left(h - \frac{\partial g}{\partial \bar{z}}\right) d\bar{z}\, dz = 0$$

and this implies (44) a.e.

Theorem 5 (but not theorem 6) has a natural extension to k complex variables (see [7], p. 660/1). If we put

$$\omega = d\bar{\zeta}_1\, d\zeta_1 + \cdots + d\bar{\zeta}_k\, d\zeta_k \quad \text{(external form)}$$

and

$$V_j = \frac{\bar{\zeta}_j - \bar{z}_j}{\left(\sum_p |\zeta_p - z_p|^2\right)^k}$$

then a generalization of (41) is

$$(45) \quad (2\pi i)^k f(z_1, \bar{z}_1, z_2, \bar{z}_2, \ldots, z_k, \bar{z}_k) = \int_C f \sum_j V_j\, d\zeta_j\, \omega^{k-1} - \frac{1}{k} \int_D \sum_j \frac{\partial f}{\partial \bar{\zeta}_j} V_j\, \omega^k,$$

and an extension of theorem 5 is as follows.

Theorem 7. *If D is a polyhedral domain, and the function f has a* STOLZ *derivative in D, and if each of the k functions*

$$\frac{\partial f}{\partial \bar{\zeta}_1}, \ldots, \frac{\partial f}{\partial \bar{\zeta}_k} \tag{46}$$

is LEBESGUE *integrable and possesses the continuity property of theorem 2, then* (45) *holds at all points at which the second integral in* (45) *exists (the exceptional points having measure 0 at most).*

If the functions (46) *are also locally bounded in D, then the formula is valid at all points, and if they are $C^{(r)}$ as real-analytic then f itself is likewise so.*

References.

[1] BOCHNER, S.: Functions in one complex variable as viewed from the theory of functions in several variables, in "Lectures on Functions of a Complex Variable", p. 315. University of Michigan Press 1955. — [2] KRICKEBERG, K.: Über den Gaußschen und den Stokesschen Integralsatz, I and II. Math. Nachr. **10**, 261—314 (1953); **11**, 35—60 (1954). — [3] LEBESGUE, H.: Séries trigonométriques. 1906. — [4] SEIDEL, W.: Ränderzuordnung bei konformen Abbildungen. Math. Ann. **104**, 182—243 (1931). — [5] BOCHNER, S.: Linear partial differential equations with constant coefficients. Ann. of Math. **47**, 202—212 (1946). — [6] BOCHNER, S., and W. T. MARTIN: Several complex variables. Princeton 1948. — [7] BOCHNER, S.: Analytic and meromorphic continuation by means of Green's formula. Ann. of Math. **44**, 652—674 (1943).

Department of Mathematics, Princeton University, Princeton N.J. (USA.).

(Eingegangen am 2. April 1955.)

Druck der Universitätsdruckerei H. Stürtz AG., Würzburg

FUNCTIONS IN ONE COMPLEX VARIABLE AS VIEWED FROM THE THEORY OF FUNCTIONS IN SEVERAL VARIABLES

SALOMON BOCHNER

The theory of functions in several variables, holomorphic or harmonic or the like, is of course guided by the facts, proofs, and viewpoints of the classical theory of holomorphic (and also harmonic) functions in one complex variable, but it may be said that the ever-growing theory of functions in several variables bids fair to exercise certain retroactions on the classical theory itself; of these I shall present several instances in this article.

1. The theorem of Cauchy-Goursat. If we put $f(z) = u(x, y) + iv(x, y)$, $A_1 = iu - v$, $A_2 = -u - iv$, then the Cauchy-Riemann equations,

$$u_x = v_y, \qquad u_y = -v_x, \tag{1}$$

imply $\partial A_1/\partial x + \partial A_2/\partial y = 0$. Therefore the Cauchy-Goursat theorem

$$\int_B f(z)\, dz = 0 \tag{2}$$

is contained in Green's theorem

$$\int_B A_1\, dy - A_2\, dx = \int_S \left(\frac{\partial A_1}{\partial x} + \frac{\partial A_2}{\partial y}\right) dx\, dy, \tag{3}$$

which is nonholomorphic and comprehensive. However, it is commonly felt that in proving (2) by way of (3) more continuity properties would have to be imposed upon the partial derivatives u_x, u_y, v_x, v_y than are implied in Goursat's requirement that $f(z)$ shall have a limit

$$\lim_{h\to 0} \frac{f(z + h - f(z)}{h} \qquad (= f'(z)) \tag{4}$$

at every point z of a domain (and nothing else); it is also vaguely felt that Goursat's own proof for (2), which requires only this property of $f(z)$, has a peculiar affinity to complex variables.

Now we find that this is not so, and by use of Goursat's method we shall establish a comprehensive version of the general (nonharmonic) formula (3) of which (2) will be a direct special case. In order to underline the nonaffinity of the method to holomorphy, we shall do this for any number of real variables as well. However, we shall deal only with domains having "rectilinear" boundaries, because the subsequent transition from such boundaries to general ones is no longer part of Goursat's own distinctive reasoning.

In a domain D of the real Euclidean $E_k : (x_1, \cdots, x_k)$, $k \geqq 2$, we take a real- or complex-valued function $A(x_j)$—or in fact a Banach-valued function—and we say that it has a total derivative at a point $x^0 = (x_j^0)$ if we can put

$$(5) \qquad A(x_j) = A(x_j^0) + \textstyle\sum_{j=1}^k a_j \cdot (x_j - x_j^0) + \epsilon(x^0; x),$$

where the quantities $a_j = a_j(x^0)$ are independent of x, and

$$(6) \qquad \epsilon(x^0; x) = o(|\, x - x^0 \,|), \qquad \text{as } x \to x^0,$$

in the sense that for any $\eta > 0$ there is a $\delta = \delta(x^0, \eta) > 0$ such that we have

$$(7) \qquad |\, \epsilon(x^0; x)| \leqq \eta$$

whenever

$$|\, x - x^0 \,| \equiv [\textstyle\sum_j (x_j - x_j^0)^2]^{\frac{1}{2}} \leqq \delta.$$

If $A(x)$ has a total derivative at x^0, then the partial derivatives

$$(8) \qquad \frac{\partial A}{\partial x_j}, \qquad j = 1, \cdots, k,$$

exist at x^0, and their values are the coefficients a_j, so that the latter are uniquely determined; but the mere existence of the partial derivatives at a point does not secure the existence of a total derivative as we have defined it. If the partial derivatives (8) exist in D and are themselves continuous functions in D, then the total derivative exists everywhere in D, but we emphasize that the converse is not true, and that the continuity of the derivatives (8) will not be required in our discussion.

If in $E_2 : (x, y)$ we introduce the symbol $z = x + iy$ and, following the usual custom (of which we disapprove), denote the general function $A(x, y)$ by $f(z)$, then $f(z)$ has a Goursat derivative $f'(z)$ at a point z^0, as defined by (4), if and only if

$$f(z) = f(z^0) + f'(z^0)(z - z^0) + o(|z - z^0|). \tag{9}$$

For the function $A(x, y)$ this means that it has a total derivative at (x^0, y^0). If we put

$$A(x, y) = A(x^0, y^0) + a(x^0, y^0)(x - x^0) + b(y^0, y^0)(y - y^0) + \epsilon(x^0, y^0; x, y),$$

then

$$a \cdot (x - x^0) + b \cdot (y - y^0) = f'(z^0)[(x - x^0) + i(y - y^0)],$$

and because of the uniqueness of a, b we have the following lemma:

LEMMA 1. *A function $f(z) = u(x, y) + i\, v(x, y)$ has a Goursat derivative $f'(z^0)$ at a point (x^0, y^0) if and only if the real and imaginary parts u, v have each a total derivative there, and the equations* (1) *hold.*

We return to the general case.

THEOREM 1. *If in a domain D of $E_k : (x_j)$ there are given k functions*

$$A_1(x_j), \cdots, A_k(x_j), \tag{10}$$

if each of them has a total derivative at every point, and if the "divergence"

$$\operatorname{div} A \equiv \frac{\partial A_1}{\partial x_1} + \cdots + \frac{\partial A_k}{\partial x_k} \tag{11}$$

is continuous in D, then the Green-Stokes relation

$$\int_B \sum_{j=1}^{k} A_j(x)(-1)^{j-1}\, dx_1 \cdots dx_{j-1}\, dx_{j+1} \cdots dx_k = \int_S \operatorname{div} A\, dx_1 \cdots dx_k \tag{12}$$

holds for every nonsingular rectilinear simplex S which together with its (oriented) boundary B is contained in D.

COROLLARIES. (i) *In particular if $A_1, \cdots, A_k$ have total derivatives everywhere, and*

$$\frac{\partial A_1}{\partial x_1} + \cdots + \frac{\partial A_k}{\partial x_k} = 0, \tag{13}$$

then we have

$$\int_B \sum_{j=1}^{k} A_j(x)(-1)^{j-1}\, dx_1 \cdots dx_{j-1}\, dx_{j+1} \cdots dx_k = 0.$$

(ii) *By Lemma* 1, *Cauchy's theorem* (2) *holds if $f(z)$ has a Goursat derivative at every point.*

(iii) *If a function $u(x, y)$ in D has continuous partial derivatives u_x, u_y in D, if each of these has a total derivative everywhere in D, so that u_{xx}, u_{yy} are defined, and if $u_{xx} + u_{yy} = 0$, then*

$$\int_B (u_x\, dy - u_y\, dx) = 0$$

for the boundary B of any triangle S in D.

Proof. As in a previous paper [2], the abbreviations

$$\sigma^j = (-1)^{j-1}\, dx_1 \cdots dx_{j-1}\, dx_{j+1} \cdots dx_k$$

and $dv_x = dx_1 \cdots dx_k$ are introduced. If the theorem were false, then for a certain pair (S_1, B_1) we would have

$$\left| \int_{B_1} A_j \cdot \sigma^j - \int_{S_1} \operatorname{div} A \cdot dv_x \right| = \alpha > 0. \tag{14}$$

Following Goursat, we subdivide S_1 into 2^k congruent rectilinear simplices $\{S_2^{(\nu)}\}$ so that their one-dimensional edges are half as long as the corresponding edges of S_1 itself. Because of the "additivity" of the relation (12), assumption (14) implies that we have

$$\left| \int_{B_2} A_j \cdot \sigma^j - \int_{S_2} \operatorname{div} A \cdot dv_x \right| \geqq \frac{\alpha}{2^k}$$

for at least one of the simplices $S_2^{(\nu)}$, which we have chosen fixed and denoted by S_2, with boundary B_2. Now, continuing in this manner, we obtain a sequence of simplices $S_1 \supset S_2 \supset S_3 \supset \cdots$ with the following properties: First, we have

$$\left| \int_{B_n} A_j \cdot \sigma^j - \int_{S_n} \operatorname{div} A \cdot dv_x \right| \geqq \frac{\alpha}{2^{kn}}. \tag{15}$$

Secondly, we have

$$\delta(S_n) = O\left(\frac{1}{2^n}\right), \qquad n \to \infty,$$

where $\delta(S_n)$ is the point-set diameter of S_n ; if we denote the volume of S_n by V_n and the $(k-1)$-dimensional volume of B_n by Ω_n, we have, by congruence,

$$\Omega_n = \frac{\Omega_1}{2^{n(k-1)}}, \qquad V_n = \frac{V_1}{2^{nk}}.$$

Thirdly, the point-set closures of the sets S_n have a point in common, which we denote by (x_j^0).

If now we put $A_j(x) = A_j'(x) + A_j''(x)$, where

$$A_j'(x) = A_j(x^0) + \sum_{l=1}^{k} \left.\frac{\partial A_j}{\partial x_l}\right|_{x=x^0} \cdot (x_l - x_l^0),$$

and where A_j'' is the remainder, then the functions A_j' are linear functions in the variables x_l ; for such functions we take our theorem as known (as Goursat does also in his case). However, for the remainder function A_j'' we have

$$A_j'' = o(|x - x^0|) = o\left(\frac{1}{2^n}\right), \quad \text{for} \quad x \text{ on } B_n, \quad n \to \infty,$$

$$\sum_{j=1}^{k} \frac{\partial A_j''}{\partial x_j} = o(1), \quad \text{for} \quad x \text{ in } S_n, \qquad n \to \infty,$$

the latter by continuity of $\operatorname{div} A'' = \operatorname{div} A - \operatorname{div} A'$. Therefore we have

$$\left| \int_{B_n} A_j'' \cdot \sigma^j \right| = O(\max_{x \in B_n} |A_j''| \cdot \Omega_n) = o\,(1/2^{nk}),$$

$$\int_{S_n} \frac{\partial A_j''}{\partial x_j}\, dv_x = O(\max_{x \in S_n} |\operatorname{div} A''| \cdot V_n) = o(1/2^{nk}),$$

and this would not be compatible with relation (15) for $A_j = A_j''$ and $\alpha > 0$.

2. The theorem of Morera. This is a (local) converse to Cauchy's theorem. It states that if a continuous function $A(x, y)$ is such that

$$\int_B A(x, y)(dx + i\, dy) = 0 \tag{16}$$

for every triangle B, say, then $A(x, y) = f(z)$ has a Goursat derivative everywhere. A natural line of proof would be to utilize Green's theorem:

$$\int_B A(x, y)(dx + i\,dy) = \int_S \left(-\frac{\partial A}{\partial x} + i\frac{\partial A}{\partial y}\right) dx\,dy, \tag{17}$$

in the following manner. By assumption (16), the left side in (17) is zero; therefore, so is the right side, also. But if the right side vanishes for every triangle S, then we must have $-A_x + iA_y = 0$, which means that the Goursat derivative $f'(z)$ exists as claimed. However, on the face of it, this proof has the difficulty of requiring that $A(x, y)$ shall be known to have partial derivatives A_x, A_y in D, whereas the underlying requirement (16) itself can be stated if $A(x, y)$ is merely known to be continuous; the emphasis of the theorem lies in the fact that only continuity of $A(x, y)$, say, shall be presupposed given. In keeping with this desideratum, the traditional proof of the theorem proceeds in another manner, more or less as follows. From (16), the integral

$$g(z) \equiv G(x, y) = \int_{x^0,y^0}^{x,y} A(x, y)(dx + i\,dy)$$

is independent of the path, thus defining, for fixed x^0, y^0, a function of the point x, y, and this function $g(z)$ has a Goursat derivative $g'(z)$ whose value is $A(x, y)$. But at this state various consequences from Cauchy's theorem are already presupposed known, one of them being that such a derivative $g'(z)$ itself has a derivative, which means that $A(x, y)$ has a Goursat derivative, as claimed.

If we turn to several variables, various generalizations and modifications of Morera's theorem suggest themselves rather convincingly. But the last-named traditional proof is all but useless for carrying the syllogistic burden of such generalizations, whereas the natural line of proof, namely the one by way of Stokes's theorem, far from being abortive, turns out to be as pliable and adaptable as one could wish. We shall give a brief outline of its adaptation [7].

In Euclidean $E_k : (t_1, \cdots, t_k)$ it is possible to construct a sequence of functions $\phi^n(t_1, \cdots, t_k)$ having the following properties.

Each $\phi^n(t)$ is defined in all of E_k and is of class C^∞ in E_k; that is, it has continuous partial derivatives of every mixed order, but vanishes outside a sphere $K_n : (t_1^2 + \cdots + t_k^2) \leqq \delta_n^2$ where $\lim_{n\to\infty} \delta_n = 0$. Furthermore, we have $\phi^n(t) \geqq 0$ and

$$\int_{E_k} \phi^n(t)\, dv_t \equiv \int_{K_n} \phi^n(t)\, dv_t = 1.$$

Now let $F(x_1, \cdots, x_k)$ be defined and continuous, and also bounded, in a domain D; we supplement it by values 0 outside D. If we set up the expressions

$$F^n(x_1, \cdots, x_k) = \int_{E_k} F(x_1 + t_1, \cdots, x_k + t_n)\phi^n(t_1, \cdots, t_n)\, dv_t$$

$$\equiv \int_{K_n} F(x_1 + t_1, \cdots, x_n + t_k)\phi^n(t_1, \cdots, t_k)\, dv_t,$$

$n = 1, 2, \cdots$, then they approximate to $F(x_j)$ in D in the following sense:

LEMMA 2. *Each $F^n(x_j) \in C^\infty$ in E_k, and if $D' \subset D$ is a subdomain whose closure is likewise contained in D, $\bar{D}' \subset D$, then*

$$\lim_{n\to\infty} F^n(x_j) = F(x_j) \tag{18}$$

uniformly in every D'.

If, moreover, $F \in C^r$, $r \geqq 1$, then we also have

$$\lim_{n\to\infty} \frac{\partial^{q_1+\cdots+q_k} F^n(x_j)}{\partial x_1^{q_1} \cdots \partial x_k^{q_k}} = \frac{\partial^{q_1+\cdots+q_k} F(x_j)}{\partial x_1^{q_1} \cdots \partial x_k^{q_k}} \tag{19}$$

for $0 \leqq q_1 + \cdots + q_r \leqq r$, again uniformly in D'.

Now take a continuous vector field $A_1(x_1), \cdots, A_k(x_k)$, and assume that we have

$$\int_B A_j(x) \cdot \sigma^j = 0, \tag{20}$$

for every B, as before. For a *given* B, since this also holds for "every other" B, it holds in particular for translated B's; thus

$$\int_{B-(t)} A_j(x) \cdot \sigma^j = 0, \tag{21}$$

at least for (t) sufficiently small. After a change of variables we thus obtain

$$\int_B A_j^n(x) \cdot \sigma^j = 0, \tag{22}$$

where

$$A_j^n(x) = \int_{E_k} A_j(x + t)\phi^n(t)\, dv_t\,, \tag{23}$$

and $A_j(x) = 0$ outside D, by convention.

However, A_j^n has continuous derivatives, by Lemma 2, and (12) applies to it. If (22) holds for every B, then we obtain $\operatorname{div} A^n = 0$. Hence we have the following theorem:

THEOREM 2. *If, in a domain D in E_k, for a continuous vector field $\{A_j(x)\}$ we have the Morera assumption* (20) *for all B as before, then the vector field is a continuous limit in D of a sequence of vector fields $\{A_j^n(x)\}$ each of which is of class C^∞ and satisfies*

$$\sum_{j=1}^{k} \frac{\partial A_j^n(x)}{\partial x_j} = 0. \tag{24}$$

Further, the approximating vector fields can be chosen to be weighted averages, as in (23).

If we apply this to Morera's own original situation, then the function $A(x, y) = f(z)$ is a continuous limit of functions $f^n(z)$, each of which is holomorphic. If we apply the Weierstrass limit theorem, which is another consequence of the Cauchy-Goursat theorem already assumed known, then $f(z)$ itself must be holomorphic, which is the Morera conclusion, differently arrived at.

Now the Weierstrass limit theorem is not restricted to holomorphy, but is more or less coextensive with "ellipticity," and it is at any rate present for the solutions of the Laplacean

$$\frac{\partial^2 \phi}{\partial x_1^2} + \cdots + \frac{\partial^2 \phi}{\partial x_k^2} = 0 \tag{25}$$

for any $k \geqq 2$; hence the following corollary:

THEOREM 3. *If, in a domain D in E_k, a function $\phi(x_j)$ has continuous partial derivatives ϕ_{x_j}, $j = 1, \cdots, k$, and if we have*

$$\int_B \sum_{j=1}^{k} \frac{\partial \phi}{\partial x_j} \cdot \sigma^j = 0 \tag{26}$$

for all B, as before, then $\phi(x_j)$ is analytic in the real variables x_j and is a solution of (25).

If we seek an actual generalization of Morera's own theorem to several complex variables, then the first version provable by our method is as follows:

THEOREM 4. *In the space* E_{2k} *of the* k *complex variables* $z_\alpha = x_\alpha + iy_\alpha$, $\alpha = 1, \cdots, k$, *if a function*

$$f = f(z_1, \cdots, z_k; \bar{z}_1, \cdots, \bar{z}_k)$$

is continuous in a domain D, *and if we have*

$$\int_{B_{p+q}} f\, dz_{\alpha_1} \cdots dz_{\alpha_p}\, d\bar{z}_{\beta_1} \cdots d\bar{z}_{\beta_q} = 0 \tag{27}$$

over every $(p + q)$*-dimensional cycle* B_{p+q}, *where*

$$1 \leqq \alpha_1 < \alpha_2 < \cdots < \alpha_p \leqq k, \qquad 1 \leqq \beta_1 < \beta_2 < \cdots < \beta_q \leqq k$$

are fixed indices with $0 \leqq p$, $0 \leqq q$, $1 < p + q \leqq 2k - 1$, *then we have*

$$\frac{\partial f}{\partial z_\alpha} = 0, \qquad \alpha \neq \alpha_1, \cdots, \alpha_p \quad \textit{and} \quad \frac{\partial f}{\partial \bar{z}_\beta} = 0, \qquad \beta \neq \beta_1, \cdots, \beta_q,$$

these Goursat derivatives existing.

Another version, more or less due to Severi, which is harder to obtain, is as follows:

THEOREM 5. *In the same space* E_{2k}, *if* f *is continuous and if we have*

$$\int_{B_k} f\, dz_1 \cdots dz_k = 0$$

for all holomorphically placed cycles B_k *in* D, *then* f *is holomorphic in* $z_1, \cdots, z_k$; *a cycle is called holomorphically placed in* D *if a suitable one-to-one holomorphic transformation*

$$w_j = \phi_j(z_1, \cdots, z_k), \qquad j = 1, \cdots, k,$$

of D *into a domain* D' *in* $E_{2k} : (w_1, \cdots, w_k)$ *will carry* B_k *into a direct product* $B^1 \times \cdots \times B^k$, *where* B^j *is a one-cycle in the* w^j*-plane,* $j = 1, \cdots, k$.

3. Removable singularities. The classical theorem is that if $f(z)$ is defined and holomorphic in the "punctured" neighborhood

$0 < |z| \leqq \rho$, and if we have $f(z) = O(1)$ as $|z| \to 0$ or, more generally,

$$\int_{C_\epsilon} |f(\zeta)| \cdot |d\zeta| = o(1) \qquad \text{as } \epsilon \to 0, \tag{28}$$

where C_ϵ is the circumference $|\zeta| = \epsilon$, then $f(z)$ has an analytic continuation into the entire neighborhood $|z| \leqq \rho$. The usual proof (even when disguised by the use of the Laurent expansion) amounts to setting up for $\epsilon < |z| < \rho$ the Cauchy formula

$$f(z) = \frac{1}{2\pi i}\int_{C_\rho} \frac{f(\zeta)\, d\zeta}{\zeta - z} + \frac{1}{2\pi i}\int_{C_\epsilon} \frac{f(\zeta)\, d\zeta}{\zeta - z}$$

and then letting $\epsilon \to 0$. The second integral "disappears" as a consequence of (28) and the remaining integral

$$f(z) = \frac{1}{2\pi i}\int_{C_\rho} \frac{f(\zeta)\, d\zeta}{\zeta - z}$$

has the continuation property claimed. The method is obviously quite universal. It applies, for instance, if we replace the single point $z = 0$ by any compact "exceptional" set A in D, and C_ϵ by any curve or system of curves encircling A and "constricting" on A as $\epsilon \to 0$, and if we leave assumption (28) literally as before.

The reasoning applies likewise to harmonic and similar functions in k variables, whenever the functions can be represented by an "associated" Green's formula over $(k - 1)$-dimensional hypersurfaces. The formula may involve not only the function $f(z)$ itself but also certain of its partial derivatives, provided we extend the analogue of assumption (28) to include such derivatives as needed. Furthermore, the Cauchy integral over a single variable, with other variables being parameters, has been used extensively in the theory of functions in several complex variables for related theorems, some of which may not yet have been re-proved differently.

Nevertheless, when studying the problem afresh ([3], [8, Chap. VIII]) we have found that it is possible and rather profitable to replace integrals over hypersurfaces, which encircle and thus avoid the singularities, by certain k-dimensional volume integrals which are taken directly across the singularities themselves. The results ensuing have turned out to be in certain respects more general than those previously known. When proceeding in this manner we may start in E_k, $k \geqq 2$, with a system of linear equations in several functions $f_1(x_j), \cdots, f_r(x_j)$:

$$\sum_{\rho=1}^{r} \sum_{0 \leqq n_1 + \cdots + n_k \leqq N} a^{\sigma\rho}{}_{n_1 \cdots n_k} \frac{\partial^{n_1 + \cdots + n_k} f_\rho(x)}{\partial x_1^{n_1} \cdots \partial x_k^{n_k}} = 0, \tag{29}$$

$$\sigma = 1, \cdots, s,$$

having the following properties: (i) the coefficients $a^{\sigma\rho}{}_{(n)}$ are constants (this probably could be weakened), (ii) any solution $\{f_\rho(x)\}$ of class C^N in a domain D is, say, automatically analytic there, and (what is decisive) (iii) if we are given a sequence of solutions $\{f_\rho^n\}$, $n = 1, 2, 3, \cdots$, and if they have (locally) a limit $\{f_\rho\}$ in the L_1-norm,

$$\lim_{n \to \infty} \int (| f_1 - f_1^n | + \cdots + | f_r - f_r^n |) \, dv_x = 0,$$

then $\{f_\rho\}$ equals an (analytic) solution almost everywhere.

The last property is, of course, the Weierstrass limit theorem, but stated in a stronger form than in the elementary part of the classical theory, although it is well known to so hold. The reason for our requiring the theorem in this stronger form is that it then more cogently expresses the "ellipticity" of the system (29) without a Green's formula being needed to supplement it.

We now take in a domain D an "exceptional" set A which is bounded and relatively closed, but not necessarily compact, so that it may extend to the boundary of D, and which may also decompose D into various connected parts. We shall, however, assume that the k-dimensional Lebesgue measure of A is zero, $v(A) = 0$, although even this could be weakened to an extent. We next denote by A_ϵ the points in $D - A$ having a distance $< \epsilon$ from A, and our general theorem is as follows:

THEOREM 6. *If* $\{f_\rho(x)\}$ *is a solution of* (29) *in* $D - A$ *and if we have*

$$\int_{A_\epsilon} (| f_1 | + \cdots + | f_r |) \, dv_x = o(\epsilon^N), \qquad \epsilon \to 0, \tag{30}$$

where N *is the integer occurring in* (29), *then* $\{f_\rho(x)\}$ *has an analytic continuation into all of* D.

Note that assumption (30) is more restrictive the larger N is, but our assumption does not involve the partial derivatives of order $\leqq N - 1$, as a direct use of Green's integral might be expected to do.

For $k = 2$, holomorphy of $f(z)$ amounts to $f_x - if_y = 0$, so that $N = 1$, and for $f(z)$ in $0 < |z| < \rho$ our condition (30) is

$$\int_{0<|z|<\epsilon} |f(z)| \, dx \, dy = o(\epsilon),$$

which at any rate includes (28). But now take for $z = x + iy$ a holomorphic function $\phi(z)$ in the upper half-plane $y > 0$, and assume that on the segment $0 < x < 1$ of $y = 0$ it has boundary values 0, approached uniformly in x, say. It is known that such a function $\phi(z)$ is $\equiv 0$; the usual interpretation of this statement is that it is a uniqueness theorem, by which a holomorphic function is $\equiv 0$ if it is 0 on a segment only, even if this segment is on the boundary. However, from our approach, this is a theorem on removable singularities; it is deducible from Theorem 6 in the following manner. Denote by D the domain in the z-plane which arises by removing the segment $0 < x < 1$, $y = 0$. But now we consider this segment as an exceptional set A, so that $D - A$ decomposes into the two half-planes $y > 0$, $y < 0$, and in $D - A$ we define a holomorphic function $f(z)$ to be $= \phi(z)$ for $y > 0$ and $= 0$ for $y < 0$. Now A_ϵ is the point set $0 < x < 1$, $0 < |y| < \epsilon$, and if we assume $\phi(z) \to 0$ as $y \downarrow 0$ for $0 < x < 1$, uniformly in x, say, then we obtain $\int_{A_\epsilon} |\phi(z)| \, dx \, dy = o(\epsilon^1)$. By Theorem 6, $\phi(z)$ has an analytic continuation into the (connected) domain D, but since it is $= 0$ for $y < 0$, it is $\equiv 0$, and thus $\phi(z) \equiv 0$, as claimed.

4. Inversion of holomorphic transformations. In real variables, if a function $y = \phi(x)$ is C^1 and if $\phi'(x^0) \neq 0$, then locally there exists an inverse function $x = \psi(y)$ and it is likewise C^1. Now this statement has a converse, but only of the following kind. If a mapping $y = \phi(x)$ is C^1, and if the mapping has an inverse $x = \psi(y)$ which is likewise known to be C^1, then we must have $\phi'(x) \neq 0$ and also $\psi'(y) \neq 0$; this follows from the fact that the identity $\psi(\phi(x)) = x$ may be differentiated thus:

$$\frac{d\psi}{dy} \cdot \frac{d\phi}{dx} = 1.$$

If, however, we know only that the mapping $y = \phi(x)$ is C^1 and that it has an inverse mapping $x = \psi(y)$ which is one-to-one (and thus automatically C^0), then we may well have $\phi'(x^0) = 0$ at a point, say; this is exemplified by the mapping $y = x^3$, which is one-to-one on $(-\infty, \infty)$ and for which $dy/dx = 0$ at $x = 0$.

For a holomorphic transformation $w = \phi(z)$, however, the situation is much more precise. Not only does $\phi'(z^0) \neq 0$ imply that the mapping is locally one-to-one and that the inverse function is likewise holomorphic, but, conversely, if the mapping is one-to-one, then the inverse must be holomorphic, and thus we must have $\phi'(z) \neq 0$. Strangely enough, this converse is usually not derived by itself, as just stated, but is obtained as part of a fuller analysis, as follows. If we have $\phi'(z^0) = 0$, then, since $\phi'(z)$ is itself holomorphic, it must have at z^0 a zero of a specific order $n \geqq 2$; it is then shown—for example, by means of residues (that is, fundamentally, by means of Kronecker's integral)—that the mapping is strictly n to 1 except at the origin, and this in particular is incompatible with the assumption that it be one-to-one.

We turn now to several complex variables. If for a holomorphic transformation

$$w_j = \phi_j(z_1, \cdots, z_k), \qquad j = 1, \cdots, k, \tag{31}$$

the Jacobian

$$\frac{\partial(w_1, \cdots, w_k)}{\partial(z_1, \cdots, z_k)} \tag{32}$$

is $\neq 0$ at the origin, for example, then locally the transformation is one-to-one and its inverse is holomorphic. However, the following remarkable fact is likewise true.

THEOREM 7. *If a holomorphic transformation* (31) *is one-to-one, then, locally, its inverse must be holomorphic, too, so that the Jacobian must be* $\neq 0$.

This proposition is of considerable importance, and it is of special interest, because it is one of the extremely few theorems occurring in the theory of functions of several variables which fits the class of holomorphic functions exactly. Almost any other theorem known for functions in k complex variables can be extended to certain classes of functions in $2k$ real variables considerably larger than the original class; on the other hand, certain facts from the classical theory of conformal mapping can be maintained only by imposing assumptions which would amount to taking a subclass of all holomorphic functions—and a badly circumscribed subclass at that. But Theorem 7 seems to fit the class of holomorphic functions precisely.

For the proof of this theorem one could of course imitate the one-dimensional procedure and set about making a "full" analysis of what happens if the Jacobian (32) has a zero "manifold" through the origin (which is now a $2(k-1)$-dimensional point set, topologically) and again obtain the conclusion, in particular, that (31) cannot then be one-to-one. But for $k \geqq 2$ this is easier said than done, and the task arises of reducing the full analysis to a minimum of information from which the desired conclusion would still follow.

It turns out that the theory of removable singularities may be used to good purpose; for $k = 1$ it applies in the following manner. If $\phi'(z^0) = 0$, $w^0 = \phi(z^0)$, then for $z \neq z^0$ we have, locally, $\phi'(z) \neq 0$, and since the map is one-to-one, it follows that $z = \psi(w)$ is holomorphic for $w \neq w^0$. However, $\psi(w)$ is a bounded function, and by Riemann's original theorem it must be holomorphic also at w^0.

For $k \geqq 2$ this scheme works as follows [8, Chap. VIII]. We take in the (z_j)-space the point set A^z on which the Jacobian (32) is zero, and its image A^w in the (w_j)-space under the transformation (31); the inverse functions $z_j = \psi_j(w)$ are then holomorphic for w not on A^w, and they are bounded as well. If now we wish to apply Theorem 6 toward proving that these functions are holomorphic throughout, then it suffices to know that we have $v(A_\epsilon^w) = o(\epsilon)$ as $\epsilon \to 0$, where A_ϵ^w is the ϵ-neighborhood of A^w and v is its k-dimensional volume. Also, it is fairly easy to see that this would follow from $v(A^z) = o(\epsilon)$, where A^z is the original zero variety of the Jacobian in the z-space. Topologically, A^z is $(2k-2)$-dimensional, and if it is "sufficiently regular" then we ought to have

$$v(A_\epsilon^z) = O(\epsilon^2), \tag{33}$$

which is even better than $o(\epsilon)$, of course. However, first impressions notwithstanding, it is not trivial to prove (33) directly as stated, and for the actual completion of the proof of Theorem 7 a certain induction on the dimension k of the space was resorted to, with (33) not actually being proved directly.

The difficulty in obtaining (33) directly is worth describing. It stems from the following circumstances. Take in E_r: $(t_1, \cdots, t_r)$ an (open) simplex S^t, say rectilinear; denote by $\bar{S}^t$ its point set closure, and by $\tilde{S}^t$ a neighborhood of $\bar{S}^t$. Take a one-to-one mapping

$$u_\sigma = \chi_\sigma(t_1, \cdots, t_r) \qquad \sigma = 1, \cdots, s \tag{34}$$

of $\tilde{S}^t$ into a point set $\tilde{S}^u$ in E_s: $(u_1, \cdots, u_s)$, with $s \geqq r + 1$; denote the images of S^t and $\bar{S}^t$ by S^u and $\bar{S}^u$ accordingly. Now if the functions (34) are in differentiability class C^1 on $\bar{S}^t$, or, what is even better, on $\tilde{S}^t$, then the s-dimensional volume $v(S_\epsilon^u)$ of the ϵ-neighborhood of $\bar{S}^u$ is $O(\epsilon^{s-r})$ and this is $O(\epsilon^2)$ if $s \geqq r + 2$. If, however, on $\bar{S}^t$ we know of the map only that it is continuous, then we cannot assert that we have $v(S_\epsilon^u) = o(\epsilon^{s-r-1})$, even if in the interior S^t of $\bar{S}^t$ the functions (34) are analytic.

Returning to our point set A^z, we note that it can be "triangulated" into a finite number of "curved" simplices S^z of topological dimension $\leqq 2k - 2$ each of which is a map of a certain rectilinear simplex S^t, the mapping being analytic in S^t. However, there is no precise statement in the literature on triangulation of holomorphic varieties from which to deduce enough "regularity" of this mapping on the *closures* $\bar{S}^t$ to serve as a basis for conclusion (33); consequently, we had to complete our proof without using such a triangulation; it would be of some interest to settle this point directly. In general, when employing the triangulation of holomorphic manifolds, the C^1-structure of the "triangles" must not be used, unless such use be justified from case to case.

5. Addition theorems for elliptic functions. The so-called Weierstrass preparation theorem, even when occurring in a strictly "classical" context, is a theorem in several variables. We have recently analyzed another such situation, perhaps even more pronounced, in which a statement presumably on one-variable functions is in reality a statement about several variables: it is the addition theorem for doubly periodic functions. The underlying general theorem is as follows [6]:

THEOREM 8. *Let A be a domain in the space of k complex variables $z = (z_1, \cdots, z_k)$ and let B be a domain in the space of l complex variables $w = (w_1, \cdots, w_l)$; let there be given a finite number of holomorphic functions $\phi^1(z), \cdots, \phi^r(z)$ in A, and a finite number of holomorphic functions $\psi^1(w), \cdots, \psi^s(w)$ in B; let a holomorphic function $f(z, w)$*

in $A \times B$ have the property that for each w in B it is algebraic over the function field

$$K[\phi^1(z), \cdots, \phi^r(z)] \tag{35}$$

with complex coefficients, and for each z in A it is algebraic over

$$K[\psi^1(w), \cdots, \psi^s(w)]. \tag{36}$$

Then in $A \times B$, $f(z, w)$ is algebraic over

$$K[\phi^1(z), \cdots, \phi^r(z); \psi^1(w), \cdots, \psi^s(w)]. \tag{37}$$

If a function $f(z, w)$ is for each w in B contained in the field (35) *and for each z in A contained in the field* (36), *then in $A \times B$ it is contained in the field* (37).

An application to one-variable functions arises, for instance, if we take a holomorphic function $F(z)$, say in a neighborhood of the origin, which has the property that, on putting $F_a(z) = F(a + z)$, the resulting family of functions $\{F_a(z)\}$ has a finite algebraic or rational basis over complex numbers there. If we denote a basis by $\phi^1(z), \cdots, \phi^r(z)$, then the function $F(z + w)$ is algebraic or rational over the field

$$K[\phi^1(z), \cdots, \phi^r(z); \phi^1(w), \cdots, \phi^r(w)],$$

and this is usually the meaning of the statement that $F(z)$ has an "addition theorem," no more and no less.

In the case of elliptic functions with given periods, there is algebraically only one independent element ($r = 1$), say the function $\wp(z)$, and rationally there are only two, say $\wp(z)$ and $\wp'(z)$. We thus obtain the "classical" statement that $\wp(z)$ satisfies an algebraic, nondegenerate, polynomial relation

$$P(\wp(z + w); \wp(z); \wp(w)) = 0,$$

and a fractional relation

$$\wp(z + w) = R(\wp(z), \wp'(z); \wp(w), \wp'(w));$$

but unless arithmetical considerations relative to the periods are being envisaged, there is not much profit from knowing this.

6. Automorphic forms. Other theorems on the existence of finite algebraic bases occur in the theory of automorphic functions. We shall state one of them, the one referring to (holomorphic and not meromorphic) automorphic forms, because we have all but trivialized the proof, and the "trivialization" is of some interest for $k = 1$ as well. The proof is a very brief and direct application of the Schwarz lemma, and nothing else, but in order to obtain the conclusion ready-made, the Schwarz lemma has to be applied to the local coordinates of the underlying compact complex space (closed Riemann "surface" for $k = 1$), and not to the coordinates of the uniformizing ordinary domain, even if such a domain exists. Actually, this very simple version of the traditional line of proof was used because of the wish to get away from the assumption usually made that a uniformizing domain exists; on a general compact space the reasoning could only simplify or fail (which it did not).

Let $V = V_{2k}$ be a compact complex coordinate space, let $\tilde{V}$ be its (noncompact) universal covering space, and let Γ be the fundamental group $\tilde{V}/V$. Since V is compact, there are no holomorphic functions on it (other than a constant), but on a noncompact $\tilde{V}$ there might be some; and we call a function f on $\tilde{V}$ automorphic if for every α in Γ we have

$$f(\alpha P) = \eta_\alpha(P)\cdot f(P) \tag{38}$$

for some suitable factor $\eta_\alpha(P)$, itself also holomorphic on $\tilde{V}$. An assemblage of functions

$$\{\eta_\alpha(P)\}, \qquad \alpha \in \Gamma, \tag{39}$$

will qualify for the role of such factors of automorphy if they satisfy the law of composition

$$\eta_\beta(P)\cdot\eta_\alpha(\beta P) = \eta_{\alpha\beta}(P) \tag{40}$$

for all pairs of elements α, $\beta \in \Gamma$ (and this is all they need satisfy). The theorem is as follows (see [5], [9]):

THEOREM 9. *If we are given V, $\tilde{V}$, Γ as just described and any fixed system of factors* (39) *satisfying* (40), *then any $k + 2$ holomorphic functions on V which are automorphic with regard to* (39) *satisfy an equation*

$$Q_n(f_1, \cdots, f_{k+2}) = 0,$$

where $Q_n(w_1, \cdots, w_{k+2})$ is a homogeneous polynomial, of some degree n, which is not $\equiv 0$. Also for given factors (39) *the degree of the polynomial can be chosen independently of the given function $f_1, \cdots, f_{k+2}$.*

The customary assumption (which is not necessary for our proof) is that V is a domain in Euclidean space E_{2k} which is covered by one coordinate system, and that either V is bounded, the elements of Γ are linear transformations

$$z_j^\alpha = f_j^\alpha(z, , \cdots , z_k), \qquad j = 1, \cdots , k, \tag{41}$$

and $\eta_\alpha(P)$ has the value

$$\left(\frac{\partial(z_1 , \cdots , z_k)}{\partial(z_1^\alpha ,\cdots , z_k^\alpha)}\right)^g \tag{42}$$

for some positive integer g, or that $\tilde{V}$ is the entire E_{2k}, Γ consists of all holomorphic translations so that V is a multitorus, and $\eta_\alpha(P)$ is of the form

$$e^{2\pi i P(z,\alpha)}, \tag{43}$$

where $P(z, \alpha)$ is linear in the space variables z and quadratic in the group parameters describing α.

For the latter case it has been shown by P. Appell (see [1], [10]) that even if we start with the most general automorphy factor $\eta_\alpha(P)$ of the kind we have described, then to within a certain "equivalence" it is nevertheless not different from the factor (43); but for bounded domains $\tilde{V}$ no such investigations have yet been undertaken, and even in the classical case of the unit circle

$$\tilde{V}: | z | < 1 \tag{44}$$

the factor

$$\left(\frac{\partial z}{\partial z^\alpha}\right)^g \tag{45}$$

is used exclusively. But actually there are known in the classical theory in (44) certain automorphy factors other than (45), namely,

$$\eta_\alpha(z) = \frac{\partial z}{\partial z^\alpha} \cdot \eta_\alpha^0,$$

where η_α^0 is a number independent of z and $| \eta_\alpha^0 | = 1$; the functions $f(z)$ which transform by these factors are the "Prym differentials." But we know of no investigation to determine under which assumptions a factor $\eta_\alpha^0(z)$ must be of the form $(\partial z/\partial z^\alpha)^g \cdot \eta_\alpha^0$, where η_α^0 is bounded, or perhaps not, and g is a constant exponent, possibly a complex one. The problem is even more interesting if we set up the factor $\eta_\alpha(P)$ not as a single function but as a matrix function

$$\eta_\alpha^{mn}(P), \quad m, n = 1, \cdots , M, \tag{46}$$

and correspondingly set up $f(P)$ as a "vector" $f_m(P)$, so that the equation (38) now becomes

$$f_m(\alpha P) = \sum_{n=1}^{M} \eta_\alpha{}^{mn}(P) f_n(P), \tag{47}$$

and relations (40) are now to be read as matrix relations in the obvious manner.

In the classical theory, factors of the form (46) were introduced long ago [11], but have never been studied conclusively, although the field seems to be a promising one. As a token of what can be done in this direction, one rather general statement [4] may be made in concluding this paper: If we are given any pair of complex spaces V, $\tilde{V}$ as before in any number of variables, and if we are given any automorphy matrix factors (46), then the sets of solutions $\{f_m(P)\}$ of the relations (47) have always a finite additive basis for complex coefficients.

BIBLIOGRAPHY

1. Appell, P. *Sur les fonctions périodiques de deux variables*, J. Math. Pures Appl. (4) 7 (1891), 157–219.

2. Bochner, S. *Analytic and meromorphic continuation by means of Green's formula*, Ann. of Math. 44 (1943), 652–673.

3. —— *Linear partial differential equations with constant coefficients*, Ann. of Math. 47 (1946), 202–212.

4. —— *Tensor fields with finite bases*, Ann. of Math. 53 (1951), 400–411.

5. —— *Algebraic and linear dependence of automorphic functions in several variables*, J. Indian Math. Soc. 6 (1952), 1–6.

6. —— *On the addition theorem for multiply periodic functions*, Proc. Amer. Math. Soc. 3 (1952), 99–106.

7. —— *The theorem of Morera in several variables*, Ann. Mat. Pura Appl. (4) 34 (1953), 27–39.

8. —— and Martin, W. T. *Several complex variables*. Princeton, 1948.

9. —— —— *Complex spaces with singularities*, Ann. of Math. 57 (1953), 490–516.

10. Conforto, F. *Funzioni abeliane e matrici di Riemann*. Libreria dell'Università di Roma. Rome, 1942.

11. Ritter, E. *Über Riemannsche Formenscharen auf einem beliebigen algebraischen Gebilde*, Math. Ann. 47 (1896), 157–221.

Princeton University

Reprinted from the Proceedings of the NATIONAL ACADEMY OF SCIENCES
Vol. 45, No. 1, pp. 46–47. January, 1959.

*ON IDENTICAL VANISHING OF HOLOMORPHIC FUNCTIONS IN SEVERAL COMPLEX VARIABLES**

BY S. BOCHNER

MATHEMATICS DEPARTMENT, PRINCETON UNIVERSITY

Communicated November 7, 1958

We will make some comment on the following recent result of J. J. Kohn.[1] In the complex space $C_n:(z_1, \ldots, z_n)$ if $f_1, \ldots, f_n$ are holomorphic in the unit ball and have continuous boundary values on the boundary, and if on the boundary the linear combination $\bar{z}_1 f_1 + \ldots + \bar{z}_n f_n$ is 0, then the functions $f_1, \ldots, f_n$ are identically 0.

1. We denote by D any bounded *circular* domain in C_n, by B its boundary, and by B_0 an open neighborhood in B. We do not assume that D contains the origin in C_n, but only that B_0 has a positive distance from it. For any integer m, $m \lesseqgtr n$, we take functions $f_1, \ldots, f_m$ which are holomorphic in D and have continuous boundary values on B_0. We also take m polynomials $P_1, \ldots, P_m$ in n symbols $\zeta_1, \ldots, \zeta_n$, homogeneous and of a common degree and we denote the degree by r. Finally, we introduce the function

$$\Phi(\zeta; z) = \sum_{\mu=1}^{m} P_\mu(\zeta) f_\mu(z) \tag{1}$$

in the $2n$ variables $\zeta_1, \ldots, \zeta_n; z_1, \ldots, z_n$. If for $z \in D$ we make the specialization $\zeta_\nu = \bar{z}_\nu$, $\nu = 1, \ldots, n$, then we obtain a function $\Phi(\bar{z}; z)$ which in D is analytic in the $2n$ real variables $x_1, \ldots, x_n; y_1, \ldots, y_n$, $z_\nu = x_\nu + iy_\nu$, and has continuous boundary values on B_0.

THEOREM 1. (i) *If the function $\Phi(\bar{z}; z)$ is 0 on B_0, then the function $\Phi(\zeta; z)$ is identically 0 in $(\zeta; z)$.*

(ii) *Hence, if the polynomials $P_1, \ldots, P_m$ are linearly independent over constants, then $f_1, \ldots, f_m$ are all identically 0.*

Proof: In the case considered by Kohn, $P_1 = \zeta_1, \ldots, P_n = \zeta_n$. These polynomials are independent and his conclusion follows.

In our general case, let $a = (a_1, \ldots, a_n)$ be a fixed point of D, not the origin in C_n. If we put $z_\nu = ta_\nu$, $\nu = 1, \ldots, n$, then we have

$$\Phi(\bar{t}\bar{a}; ta) = \bar{t}^r \sum_{\mu=1}^{m} P_\mu(\bar{a}) f_\mu(ta), \tag{2}$$

and in the complex t-plane the latter sum is defined and holomorphic in a certain annulus with center at $t = 0$. Because of our assumptions, there is a neighborhood D_0 in D such that for each fixed $a \in D$, the sum in equation (2) is 0 on an arc of the t-annulus. Hence the sum is identically 0 in t, and hence for $z \in D_0$, $\Phi(\bar{z}; z)$ is 0. By a uniqueness theorem of E. Cartan, the function $\Phi(\zeta; z)$ is identically 0 as claimed.

2. In the case of the unit ball a statement can be made even if the polynomials $P_1, \ldots, P_n$ are arbitrary non-homogeneous, and the reader will readily verify that this statement subsumes Theorem 1, for such a D.

THEOREM 2. *If D is the unit ball, and $P_1, \ldots, P_m$ are any polynomials in $\zeta_1, \ldots, \zeta_n$, then the vanishing of $\Phi(z; z)$ on a neighborhood B_0 of the boundary of D implies a representation*

$$\Phi(\zeta; z) = (1 - \zeta_1 z_1 - \ldots - \zeta_n z_n)\psi(\zeta; z) \tag{3}$$

in which $\psi(\zeta; z)$ is a polynomial in (ζ_ν) with coefficients which are holomorphic in D.

Proof. If we put

$$u = 1 - \bar{z}_1 z - \ldots - \bar{z}_n z, \tag{4}$$

then D is defined by $u > 0$ and B by $u = 0$. If now we substitute

$$\bar{z}_1 = \frac{1}{z}(1 - u - \bar{z}_2 z_2 - \ldots - \bar{z}_n z_n)$$

in $\Phi(\bar{z}; z)$ we obtain an expression $A(u, \bar{z}_2, \ldots, \bar{z}_n; z_1, \ldots, z_n)$ which is a polynomial in $u, \bar{z}_2, \ldots, \bar{z}_n$ with coefficients in $z_1, \ldots, z_n$ which are holomorphic in a neighborhood D_0 in D bordering on an open part of B_0. For $u = 0$ the expression has value 0. Hence we can put

$$A(u, \bar{z}_2, \ldots, \bar{z}_n; z) = uB(u, \bar{z}_2, \ldots, \bar{z}_n; z).$$

and if herein we replace u by the sum (4), we obtain a representation (3) to the following extent. The object (3) is a polynomial in (ζ_ν) with coefficients which are holomorphic in D_0, and the equality (3) is valid for $z \in D_0$ and $\zeta_\nu = \bar{z}_\nu$. However, we can form the quotient

$$\psi(\bar{z}; z) = \frac{\Phi(\bar{z}; z)}{1 - \bar{z}_1 z_1 - \ldots - \bar{z}_n z_n}$$

in the entire unit ball D, and it is analytic in the real variables $(x_\nu; y_\nu)$. Therefore, $\psi(\zeta; z)$ has a holomorphic continuation from D_0 into D, which completes the proof of the theorem.

* This research was supported by the United States Air Force through the Office of Scientific Research of the Air Force Research and Development Command.

[1] J. J. Kohn, "A Boundary Condition for the Vanishing of n Holomorphic Functions in Complex n-Space," *Proc. Am. Math. Soc.*, **9**, 175–177, 1958.

*LINEAR AND ALGEBRAIC DEPENDENCE OF FUNCTIONS ON COMPACT COMPLEX SPACES WITH SINGULARITIES**

By S. Bochner

MATHEMATICS DEPARTMENT, PRINCETON UNIVERSITY

Communicated November 20, 1958

We will axiomatize the following situation. Take a (non-compact) complex space W^m, entirely regular, and form the quotient space $V = W^m/R$ relative to some equivalence relation R, say relative to the action of a group of homeomorphisms. It is the space V we are interested in. We assume that it is compact, but about its "regularity" (or what may be left of it) we only assume as follows. For some complex dimension n, $n \leq m$, there is in W^m a finite number of separate complex n-cells, each holomorphically immersed, such that the union of their projections covers V. In the axiomatization we eliminate V altogether, and only retain the separate n-cells and a "weak" structure to bind them, as suggested by R.

In the space of n complex variables $z = (z^1, \ldots, z^n)$ we take a finite number of neighborhoods U_a, $a = 1, \ldots, A$, each represented by the unit ball $\|z_a\| < 1$, and, for every a, some subneighborhood T_a, as represented by $\|z_a\| < r_a$, $r_a < 1$. With each z_a in each U_a, we associate permanently some point z_b of some T_b ($b \neq a$ or $b = a$), in symbols

$$z_b = \phi(z_a). \tag{1}$$

That is all. This is our (compact) pseudo-space, and we denote it by S.

1. A *scalar* function on S is a set $f = (f_1, \ldots, f_A)$ such that $f_a(z_a)$ is an ordinary point function on V_a and that equation (1) implies

$$f_a(z_a) = f_b(\phi(z_a)). \tag{2}$$

Our first statement is as follows. *If each $f_a(z_a)$ is holomorphic, then each is a constant C_a* (maximum modulus theorem).

2. More generally, an *automorphic* (or density) function arises if for every $z_a \in U_a$, $a = 1, \ldots, A$, there is given a *fixed* number $\lambda(z_a)$, with a joint bound, $|\lambda(z_a)| < M$, and if we replace equation (2) by

$$f_a(z_a) = \lambda(z_a) . f_b(\phi(z_a)). \tag{3}$$

The automorphic functions $f = (f_1, \ldots, f_A)$ constitute, in an obvious manner, a vector field over complex constants and the next statement is as follows:[1] *For holomorphic automorphic functions this vector space has a finite basis* (linear dependence of automorphic functions).

3. Even more generally we may assume that the "automorphy factor" $\lambda(z_a)$ is not a number but a N by N matrix of numbers, and that each $f_a(z_a)$ has N components, so that equation (3) is an affine transformation between such components. The object $(f_1, \ldots, f_A)$ is then representative of a general tensor (linear vector bundle over S), and the following statement still remains:[2] *Any vector field of holomorphic tensors of a given type has a finite basis.*

4. Finally, we consider meromorphic functions $f = (f_a)$. Each $f_a(z_a)$ is a quotient $p_a(z_a) : q_a(z_a)$ and there is a system of factors $\lambda(z_a)$ which depends on the given function, again with $|\lambda(z_a)| < M$, such that

$$p_a(z_a) = \lambda(z_a) p_b(\phi(z_a)), \quad q_a(z_a) = \lambda(z_a) q_b(\phi(z_a)).$$

Then the following theorem holds:[3] *To any $n + 1$ meromorphic functions $f^1, \ldots, f^{n+1}$ there is a polynomial $P(\zeta^1, \ldots, \zeta^{n+1}) \not\equiv 0$ such that $P(f^1, \ldots, f^{n+1}) \equiv 0$* (algebraic dependence of meromorphic functions).

There are versions of this theorem on other types of complex spaces with singularities.[4]

5. Finally, we note that in all preceding statements the ordinary regular neighborhoods U_a can be replaced by more general local varieties which are themselves already singular. All that is needed is that, for holomorphic functions suitably defined on them, a suitable version of the classical Schwarz's Lemma shall continue to hold. Such generalizations are indeed available.[5]

* This work was supported by the United States Air Force through the Office of Scientific Research of the Air Research and Development Command.

[1] S. Bochner, "Algebraic and Linear Dependence of Automorphic Functions in Several Variables," *J. Indian Math. Soc.*, **16**, 1–6, 1952.

[2] S. Bochner, "Tensor Fields with Finite Basis," *Ann. Math.*, **53**, 402–411, 1951.

[3] J. P. Serre, "Fonctions automorphes," *Séminaire École Norm. Sup.*, **2**, (Paris 1953/4); C. L. Siegel, "Meromorphe Funktionen auf kompakten analytischen Mannigfaltigkeiten," *Nachr, Akad. Wiss. Göttingen*, **2a**, 71–77, 1955.

[4] R. Remmert, "Meromorphe Funktionen in kompakten komplexen Räumen," *Math. Ann.*, **132**, 277–288, 1956.

[5] S. Bochner, and W. T. Martin, "Complex Spaces with Singularities," *Ann. Math.*, **57**, 490–519 1953.

Reprinted from the Proceedings of the NATIONAL ACADEMY OF SCIENCES
Vol. 46, No. 5, pp. 721–723. May, 1960.

CLASSES OF HOLOMORPHIC FUNCTIONS OF SEVERAL VARIABLES IN CIRCULAR DOMAINS*

BY S. BOCHNER

PRINCETON UNIVERSITY

Communicated March 24, 1960

We will extend a theorem of (F. and M. Riesz and) Hardy-Littlewood[1] from holomorphic functions $f(z)$ in the unit disk $|z| < 1$ to holomorphic functions of severa- variables $f(z_1, \ldots, z_k)$ in general circular domains.

THEOREM 1.[1] *If $f(z)$ is holomorphic in $|z| < 1$ and, for some $\lambda > 0$,*

$$\sup_{0 < r < 1} \int_0^1 |f(re^{2\pi i\theta})|^\lambda \, d\theta \equiv C^\lambda < \infty,$$

then

$$\int_0^1 \sup_{0 < r < 1} |f(re^{2\pi i\theta})|^\lambda \, d\theta \leq \alpha_\lambda C^\lambda,$$

where α_λ depends only on λ and nothing else.

Also, for $r_1, r_2 < 1$,

$$\int_0^1 |f(r_1e^{2\pi i\theta}) - f(r_2e^{2\pi i\theta})|^\lambda \, d\theta \to 0 \text{ as } (r_1, r_2) \to (1, 1),$$

and there exists a measurable function $F(\theta)$ on the boundary $\zeta = e^{2\pi i\theta}$ such that, for $0 < r < 1$,

$$\int_0^1 |f(re^{2\pi i\theta}) - F(\theta)|^\lambda \, d\theta \to 0 \text{ as } r \to 1.$$

The quantity

$$M(r) = \int_0^1 |f(re^{2\pi i\theta})|^\lambda \, d\theta$$

is monotonely increasing in r. Using this, it is not hard to verify that Theorem 1 gives rise to the following generalization of itself.

THEOREM 2. *If our holomorphic function depends on a parameter ξ, $f(z) = \varphi(z; \xi)$; if ξ is the point of a Lebesgue measure space $\{X: (\xi), d\mu(\xi)\}$ and $\varphi(z; \xi)$ is measurable in $(z; \xi)$; if for almost all ξ, $\varphi(z; \xi)$ is holomorphic in $|z| < 1$; and if, for $\lambda > 0$,*

$$\sup_{0 < r < 1} \int_X d\xi \int_0^1 |\varphi(re^{2\pi i\theta}; \xi)|^\lambda \, d\theta \equiv C^\lambda < \infty,$$

then, for the same α_λ,

$$\int_X d\xi \int_0^1 \sup_{0 < r < 1} |\varphi(re^{2\pi i\theta}; \xi)|^\lambda \, d\theta \leq \alpha_\lambda C^\lambda;$$

also,

$$\int_X d\xi \int_0^1 \left| \varphi(r_1 e^{2\pi i\theta};\ \xi) - \varphi(r_2 e^{2\pi i\theta};\ \xi)\right|^\lambda d\theta \to 0 \text{ as } (r_1, r_2) \to (1, 1)$$

and there exists a function $\Phi(\theta;\ \xi)$ *measurable in* $(\theta;\ \xi)$ *such that*

$$_X\, d\xi \int_0^1 \left| \varphi(re^{2\pi i\theta};\ \xi) - \Phi(\theta;\ \xi)\right|^\lambda d\theta \to 0.$$

Now, in the space of k complex variables $(z_1, \ldots, z_k)$ let D be a (bounded) circular domain containing the origin. Thus, if $(a_1, \ldots, a_k) \in D$, then also $(za_1, \ldots, za_k) \in D$ for $|z| \leq 1$. Let B denote the boundary of B with relative topology. We take on B any finite Borel measure $\omega(B_0)$, $B_0 \subset B$, subject only to the condition that it be circularly invariant. By this we mean that if we replace all points $(\zeta_1, \ldots, \zeta_k)$ of B_0 by $(e^{2\pi i\theta}\, \zeta_1, \ldots, e^{2\pi i\theta}\, \zeta_k)$ and denote the new set by $B_0{}^\theta$ then $\omega(B_0{}^\theta) = \omega(B_0)$. The leading feature of such a measure is the following. If $\Psi(\zeta_1, \ldots, \zeta_k)$ is a (non-holomorphic) function on B summable with respect to $\omega(B_0)$ then

$$\int_B \Psi(\zeta_1, \ldots, \zeta_k) d\omega = \int_0^1 d\theta \int_B \Psi(e^{2\pi i\theta}\, \zeta_1, \ldots, e^{2\pi i\theta}\, \zeta_k) d\omega.$$

Because of this, Theorem 2 implies as follows.

THEOREM 3. *If* $f(z_1, \ldots, z_k)$ *is holomorphic in a bounded circular domain* D *with origin, and if*

$$\sup_{0<r<1} \int_B \left|f(r\zeta_1, \ldots, r\zeta_k)\right|^\lambda d\omega(\zeta) \equiv C^\lambda < \infty,$$

$\lambda > 0$, *relative to a finite circularly invariant Borel measure* $\omega(B_0)$ *on the boundary* B *of* D*; then*

$$\int_B \sup_{0<r<1} \left|f(r\zeta_1, \ldots, r\zeta_k)\right|^\lambda d\omega(\zeta) \leq a_\lambda C^\lambda.$$

Also

$$\int_B \left|f(r_1\zeta_j) - f(r_2\zeta_j)\right|^\lambda d\omega(\zeta) \to 0 \text{ as } (r_1, r_2) \to (1, 1)$$

and there exists a function $F(\zeta_j) \equiv F(\zeta_1, \ldots, \zeta_k)$ *on* B, *measurable relative to* $\omega(B_0)$ *such that*

$$\int_B \left|f(r\zeta_j) - F(\zeta_j)\right|^\lambda d\omega(\zeta) \to 0 \text{ as } r \to 1.$$

Theorem 3 applies in particular[2] to the sphere

$$S_k\colon\ |z_1|^2 + \ldots + |z_k|^2 < 1$$

with "ordinary" (2k-1)-dimensional surface area $\omega(B_0)$; and it also applies to the polycylinder

$$P_k\colon\ |z_1| < 1, \ldots, |z_k| < 1,$$

if the measure $\omega(B_0)$ on its boundary B is defined in such a manner that

$$\int_B \Psi(\zeta_1, \ldots, \zeta_k) d\omega(\zeta) = \int_0^1 \ldots \int_0^1 \Psi(e^{2\pi i\theta_1}, \ldots, e^{2\pi i\theta_k}) d\theta_1 \ldots d\theta_k.$$

This measure is concentrated entirely on the k-dimensional subset $|\zeta_1| = 1, \ldots, |\zeta_k| = 1$ of the total (2k-1)-dimensional boundary, but it is a finite Borel measure and circularly invariant, and this was all we have required of it.

We note the corollary ($\lambda = 1$) that if $f(z_1, \ldots, z_k)$ is holomorphic in P_k and

$$\sup_{0<r<1} \int_0^1 \ldots \int_0^1 \left|f(re^{2\pi i\theta_1}, \ldots, re^{2\pi i\theta_k})\right| d\theta_1 \ldots d\theta_k < \infty$$

then there exists a multiperiodic function $F(\theta_j) \equiv F(\theta_1, \ldots, \theta_k)$ of class L_1 such that

$$\int_0^1 \ldots \int_0^1 |f(r\theta_j) - F(\theta_j)| d\theta_1 \ldots d\theta_k \to 0, r \to 1.$$

This non-trivial proposition was originally stated and proved by this author,[3] and the proof employed the radial maximum of Hardy-Littlewood, but in a much more complicated manner than the present approach suggests. Also an entirely different proof of this "corollary" was recently given by Helson and Lowdenslager,[4] and we wish to point out that our original proof and their recent method both yield a somewhat stronger result which cannot be easily stated within the present set-up.

Finally we observe that for $\lambda > 1$ we have the following theorem again obtainable from its one variable prototype.[1]

THEOREM 4. *For $\lambda > 1$ all parts of Theorem 3 (and also of Theorem 2) are valid if instead of a holomorphic function in D we consider a function $u(z_1, \ldots, z_k)$ which is the real part of a holomorphic function.*

Also, for $\lambda = 1$ there remains the statement that

$$\sup_{0 < r < 1} \int_B |u(r\zeta_j)| \log^+ |u(r\zeta_j)| d\omega(\zeta) \leq C < \infty$$

implies the relation

$$\int_B \sup_{0 < r < 1} |u(r\zeta_j)| d\omega(\zeta) \leq \alpha C + \beta$$

and the existence of a function $U(\zeta_j)$ for which

$$\int_B |u(r\zeta_j) - U(\zeta_j)| d\omega(\zeta) \to 0, r \to 1.$$

Remark: Note that the log factor in the assumption appears in first power for all complex dimensions k.

One can also extend a theorem of M. Riesz[5] on conjugate functions and the statement is as follows.

THEOREM 5. *If $f(z_j) = u(z_j) + iv(z_j)$ is holomorphic in D and $v(0) = 0$, then, for $p > 1$,*

$$\sup_r \int_B |v(r\zeta_j)|^p d\omega \leq \gamma_p \sup_r \int_B |u(r\zeta_j)|^p d\omega$$

and, for $p = 1$,

$$\sup_r \int_B |v(r\zeta_j)| d\omega \leq \gamma \sup_r \int_B |u(r\zeta_j)| \log^+ |u| d\omega + \delta,$$

where γ_p, γ, δ are fixed constants.

* This research was supported by the United States Air Force through the Office of Scientific Research of the Air Force Research and Development Command.

[1] Hardy, G. H., and J. E. Littlewood, "A maximal theorem with function-theoretic applications," *Acta Mathematica*, **54**, 81–116 (1930).

[2] Rauch, H. E., "Harmonic and analytic functions of several variables and the maximal theorem of Hardy and Littlewood," *Canadian Journal of Mathematics*, **8**, 171–183 (1956); see also Smith, K. T., "A generalization of an inequality of Hardy and Littlewood," *ibid.*, 157–170 (1956).

[3] Bochner, S., "Boundary values of analytic functions in several variables and of almost periodic functions," *Annals of Mathematics*, **45**, 708–722 (1944).

[4] Helson, H., and D. Lowdenslager, "Prediction theory and Fourier Series in several variables," *Acta Mathematica*, **99**, 165–202 (1958).

[5] Riesz, M., "Sur les fonctions conjugueés," *Mathematische Zeitschrift*, **27**, 218–244 (1928).

HARTOGS' THEOREM IN EUCLIDEAN SPACE AND A RELATED COUSIN THEOREM ON THE TORUS†

By S. BOCHNER‡

1. Introduction. The title of this paper refers to the following theorem of Hartogs (which Hartogs himself proved but very faultily). In the space of 2 or more complex variables if a bounded domain has a connected exterior, then any function which is defined and holomorphic in a neighborhood of the boundary of the domain can be continued analytically into the domain itself.

Our own proof of the theorem, as given in [1], applied immediately to a more comprehensive proposition set in real Euclidean E_n, and we generalized it further in [2] and [3], and then also in [4].

A version in [2], with which we would like to resume the topic reads as follows. Take a suitable elliptic operator with constant coefficients and of arbitrary order :

$$\Delta f = \sum_{p_1+\dots+p_n \leqslant 2h} a_{p_1\dots p_n} \frac{\partial^{p_1+\dots+p_n} f}{\partial x_1^{p_1} \dots \partial x_n^{p_n}}, \tag{1.1}$$

and some second (non-elliptic) operator with constant coefficients

$$\Lambda f = \sum_{q_1+\dots+q_m \leqslant m} b_{q_1\dots q_m} \frac{\partial^{q_1+\dots+q_m} f}{\partial x_1^{q_1} \dots \partial x_m^{q_m}}, \tag{1.2}$$

which acts only on a proper part of the variables, that is

$$m < n. \tag{1.3}$$

Now, if $f(x)$ is defined real-analytic in a neighborhood of the boundary, and if $\Delta f = 0$, $\Lambda f = 0$ then f can be continued analytically into the domain.

At present we generalize this as follows. We form the functions

$$\rho(x) = \Delta f, \quad \sigma(x) = \Lambda f \tag{1.4}$$

† This research was supported by the United States Air Force through the Office of Scientific Research of the Airforce Research and Development Command.

‡ Did not attend the Colloquium.

and we do not at all demand that $\rho(x)$ and $\sigma(x)$ shall be zero. All we demand is that the functions $\rho(x)$ and $\sigma(x)$ shall be continuable into the domain. This is enough to conclude that $f(x)$ itself is also so continuable.

We also have theorems along these lines if our domain is bounded by an outer hypersurface and several interior ones.

If one goes over from Euclidean space onto torus then there are no " outer " and " inner " domains, and Hartogs' theorem goes over into a Riemann-Cousin theorem. For instance, we may decompose the torus into two domains D_1, D_2 which are separated by a hypersurface B. Let $f(x)$ be an analytic function with certain properties which is given in a neighborhood of the boundary. The problem becomes one of putting $f(x)$ in the form $f(x) = F_1(x) - F_2(x)$ where $F_1(x)$ is continuable into D_1 and $F_2(x)$ into D_2, and such that F_1, F_2 retain certain desirable properties of $f(x)$ itself. Now, for an elliptic operator on the torus, the Green's function no longer satisfies $\Delta G(\xi) = 0$ but only $\Delta G(\xi) = \text{constant}$ and this introduces certain " residues " whose vanishing is sufficient, and frequently also necessary for an affirmative answer to the existence question connected with the problem. We previously introduced in [1] a residue for the case corresponding to $\rho(x) = 0$, $\sigma(x) = 0$, of the present setting. This residue in the identical form will reappear for non-vanishing $\rho(x)$, $\sigma(x)$, and it will be a primary one, but a further residue, of more secondary importance, will be present as well.

" Inhomogeneous " systems of equations of the type (1.4) with non-vanishing given functions $\rho(x)$, $\sigma(x)$ were introduced in and formed the subject matter of our last paper [4], but the results and outlooks of the present paper are in a different direction. Also, we expect to extend some of the statements of the present paper from the torus to some other homogeneous compact manifolds, but this is being deferred to another occasion.

2. One closed hypersurface in Euclidean space. In real Euclidean E_n, any $n \geqslant 2$, we take a bounded domain D whose exterior is connected. In our leading statements the domain can be deformed a little, and therefore we assume immediately that its boundary B is a combinatorially closed $(n-1)$-dimensional hypersurface in simplicial form, and that each simplex together with its pointset closure is part of a larger " open " $(n-1)$-dimensional simplex, of differentiability class $C^{(1)}$ say. We take some n-dimensional neighborhood of B and denote it always by $\tilde{B}$. In any one case the neighborhood can be replaced by a subneighborhood. Therefore we may think of $\tilde{B}$ as being an ϵ-neighborhood of B for some $\epsilon > 0$, say.

If $n = 2k$, and E_n is viewed as the space of complex variables $z_\alpha = x_\alpha + iy_\alpha$, $\alpha = 1,\ldots, k$, and if $k \geqslant 2$, then Hartogs' own theorem states as follows. If $f(z)$ is given holomorphic in $\tilde{B}$ then it can be continued into D. That is, in the domain

$$\tilde{D} \equiv D \cup \tilde{B} \tag{2.1}$$

there exists a holomorphic function whose restrictions to $\tilde{B}$ is the given function $f(z)$.

A holomorphic function is a simultaneous solution of the system of equations

$$\frac{\partial f}{\partial \bar{z}_1} = 0,\ldots, \frac{\partial f}{\partial \bar{z}_k} = 0. \tag{2.2}$$

Each of these k equations is of the form $\Lambda f = 0$, where in the real space coordinates (x_α, y_α), Λf is a linear partial differential operator with complex constant coefficients, and for $k \geqslant 2$ it is *defective* in the sense that it " effectively " operates on fewer than all coordinates. Our generalization of it is as follows.

I. Usually, the symbol Λf will denote, for $n \geqslant 2$, any fixed operator of the form (1.2) subject to the condition (1.3). The coefficients $b_{q_1 \ldots q_m}$ are complex constants and at least one coefficient is $\neq 0$.

Returning to the holomorphic case for guidance we note that the system (2.2) also implies

$$\frac{1}{4}\left(\frac{\partial^2 f}{\partial \bar{z}_1 \partial \bar{z}_1} + \ldots + \frac{\partial^2 f}{\partial z_k \partial \bar{z}_k}\right) = 0, \tag{2.3}$$

that is $\Delta f = 0$, where Δf is the classical Laplacean. Our generalization of this is any elliptic operator Δf with constant coefficients and " customary " properties. Rather than derive and analyze these properties we prefer to re-state them " axiomatically " from our previous papers [2], [3], [4], in the manner needed.

II. The symbol Δf will always denote an operator of the form (1.1) with constant coefficients having the following properties among others.

III. It gives rise to a *Green's formula*

$$\left.\begin{matrix} f(x) \\ 0 \end{matrix}\right\} = \int_B \Omega(f(\xi), \xi - x, d\xi) - \int_D \Delta f(\xi). \, G(\xi - x) \, dv_\xi \tag{2.4}$$

of the following description.

IV. The function $G(\xi)$ is defined and (real-) analytic in E_n after deletion of the origin $\xi = 0$, and it is Lebesgue integrable in the neighborhood of the origin,

$$\int_{0<|\xi|<1} | G(\xi) | \, dv_\xi < \infty. \tag{2.5}$$

V. As regards its behavior at infinity, there is an integer $r_0 \geqslant 0$ such that for every system of integers

$$r_1 \geqslant 0, \ldots, r_n \geqslant 0, \quad r_1 + \ldots + r_n \geqslant r_0, \tag{2.6}$$

we have

$$\frac{\partial^{r_1+\ldots+r_n} G}{\partial \xi_1^{r_1} \ldots \partial \xi_n^{r_n}} = O(| \xi |^{-n-1}), \quad | \xi | \to \infty. \tag{2.7}$$

VI. The expression $\Omega(f(\xi), \ \xi - x, d\xi)$ which is the integrand under the surface integral in (2.4) is a certain external differential form

$$\Sigma_{(\alpha)} \Omega_{\alpha_1 \ldots \alpha_{n-1}} d\xi_{\alpha_1} \wedge \ldots \wedge d\xi_{\alpha_{n-1}}, \tag{2.8}$$

in which each component $\Omega_{\alpha_1 \ldots \alpha_{n-1}}$ is a certain finite expression

$$\Sigma_{(p)} G_{p_1 \ldots p_n}(\xi - x) \frac{\partial^{p_1+\ldots+p_n} f(\xi)}{\partial \xi_1^{p_1} \ldots \partial \xi_n^{p_n}} \tag{2.9}$$

with a generic function $f(\xi)$. The coefficients $G_{p_1 \dots p_n}(\xi)$ have "axiomatically" the same properties as $G(\xi)$ itself. They are defined and analytic in $E_n - (0)$ and there is an $r_0 \geqslant 0$, which we may assume to be common to all, so that for (2.6) we have

$$\frac{\partial^{r_1+\dots+r_n} G_{p_1 \dots p_n}}{\partial \xi_1^{p_1} \dots \partial \xi_n^{p_n}} = O(|\xi|^{-n-1}), \quad |\xi| \to \infty. \tag{2.10}$$

In actual fact these coefficients are each a finite combination with constant coefficients of partial derivatives of the Green's function $G(\xi)$, but we will have no opportunity to use this explicitly.

VII. Identically in the symbol $f(\xi)$ and the "parameter" x we have

$$d_\xi \Omega(f(\xi), \xi - x, d\xi) = \Delta f(\xi). G(\xi - x) dv_\xi, \tag{2.11}$$

where $d_\xi \Omega$ is the external derivative relative to the space coordinates (ξ), and dv_ξ is the external form

$$d\xi_1 \wedge \dots \wedge d\xi_n. \tag{2.12}$$

Also, $G(\xi)$ is the Green's function of the operator Δ so that in particular

$$\Delta G(\xi) = 0 \tag{2.13}$$

but this again will not be used expressly.

VIII. Relation (2.11) automatically implies the *lower* half of (2.4) in the following sense. If D is a bounded domain whose boundary B consists of one or several disjoint combinatorially closed hypersurfaces, and if x is a point of D_∞ which is the complement of $D \cup B$ in E_n, then the right side of (2.4) has the value 0. Now, the *upper* half of (2.4) refers to points x in D, and in this case the fact is only this, that the right side is a point functional of $f(\xi)$ which is commutative with translations, see [3] and [4]. We could, as we did in [4], operate to an extent with this functional unrestricted, but we prefer, in order to avoid complications, to restrict ourselves to the case in which this functional is the identity. Thus we have as follows. If D is a bounded domain whose boundary B consists of one or several closed hypersurfaces, all suitably oriented, and if $f(\xi)$ is defined in a neighborhood of $D \cup B$ and belongs there to differentiability class $C^{(\infty)}$, (or only to differentiability class $C^{(N)}$, for N sufficiently large); then for x in D the right side in (2.4) has value $f(x)$, and for x in D_∞ it has value 0.

Our last "axiomatic" property is the following known basic result.

IX. If $\rho(x)$ is (real-) analytic in a neighborhood and $f(x)$ is a $C^{(\infty)}$ solution of

$$\rho(x) = \Delta f(x), \tag{2.14}$$

then $f(x)$ itself is analytic.

We will now have our first theorem which is nearest to Hartogs' classical version (and subsumes it), but in actual trend is already rather different from it.

THEOREM 1. *If D is a bounded domain whose boundary is one closed hypersurface B, if $\tilde{B}$ is a neighborhood of B, and Δf, Λf are operators as described, then*

(i) *If $f(x)$ is given (real-) analytic in $\tilde{B}$, and if the functions $\rho(x)$, $\sigma(x)$ as defined by*

$$\rho(x) = \Delta f(x), \quad \sigma(x) = \Lambda f(x) \tag{2.15}$$

in $\tilde{B}$ have analytic continuations into $\tilde{D}$, (see (2.1)), then $f(x)$ itself has an analytic continuation into $\tilde{D}$.

(ii) *Especially if $f(x)$ in $\tilde{B}$ is an (analytic) solution of*

$$\Delta f = 0, \quad \Lambda f = 0, \tag{2.16}$$

then it has an analytic continuation into $\tilde{D}$.

(iii) *If $f(x)$ in $\tilde{B}$ belongs only to class $C^{(\infty)}$, and if the $C^{(\infty)}$ functions $\rho(x)$, $\sigma(x)$ as defined by (2.15) have $C^{(\infty)}$ continuations into $\tilde{D}$ and if these satisfy in $\tilde{D}$ the integrability condition*

$$\Lambda\rho(x) = \Delta\, \sigma(x), \tag{2.17}$$

then $f(x)$ as a solution of (2.15) has a $C^{(\infty)}$ continuation into $\tilde{D}$.

REMARK 1. In $\tilde{B}$ we have

$$\Lambda\rho - \Delta\sigma = (\Lambda\Delta - \Delta\Lambda) f.$$

However operators with constant coefficients are commutative and thus this is 0. Now in the analytic case the relation (2.17) which was automatic in $\widetilde{B}$ continues by itself into $\widetilde{D}$, but in the $C^{(\infty)}$ case it is necessary to stipulate the latter expressly. This necessity is emphasized by the fact that $C^{(\infty)}$ continuations of $\rho(x)$, $\sigma(x)$, even if subject to (2.17), if at all possible are *never* unique.

PROOF. It will suffice to deal with part (iii). With the given functions $f(\xi)$ in $\widetilde{B}$ and $\rho(\xi)$ in $\widetilde{D}$ we set up the expression,

$$\int_B \Omega(f(\xi), \xi - x, d\xi) - \int_D \rho(\xi).\ G(\xi - x)\, dv_\xi, \tag{2.19}$$

in which Ω and G are the data of formula (2.4) and we denote it now and later more briefly by

$$[B, D;\ f(\xi), \rho(\xi)]\,(x). \tag{2.20}$$

It defines two separate functions,

$$\left.\begin{matrix} F(x) \\ F_\infty(x) \end{matrix}\right\} = [B, D;\ f(\xi), \rho(\xi)]\,(x), \tag{2.21}$$

where the upper function $F(x)$ is intended for (and restricted to) x in D and the lower function $F_\infty(x)$ for x in D_∞. The previous axioms IV, V, VI imply immediately as follows.

LEMMA 1. *The function $F_\infty(x)$ is analytic and for* (2.6) *we have*

$$\frac{\partial^{r_1+\cdots+r_n} F_\infty}{\partial x_1^{r_1} \cdots \partial x_n^{r_n}} = O(|x|^{-n-1}), \quad |x| \to \infty. \tag{2.22}$$

The next step is less obvious.

LEMMA 2. *The function $F(x)$ is $C^{(\infty)}$ and for any linear operator with constant coefficients*

$$Lf = \sum_{(s)} l_{s_1 \cdots s_n} \frac{\partial^{s_1 + \cdots + s_n} f}{\partial x_1^{s_1} \cdots \partial x_n^{s_n}}\ , \tag{2.23}$$

we have

$$\left.\begin{matrix} L_x\ F(x) \\ L_x\ F_\infty(x) \end{matrix}\right\} = [B, D;\ L_\xi f(\xi),\ L_\xi \rho(\xi)]\ (x). \tag{2.24}$$

REMARK 2. In the paper [4] the symbol Lf had a specialized meaning which it does not have in the present paper.

PROOF. Firstly, due to (2.5), $F(x)$ is certainly continuous. Next, if in

$$F(x) = [B,\ D;\ f(\xi),\ \rho(\xi)]\ (x) \tag{2.25}$$

we hold x fixed in D, then for a sufficiently small constant vector

$$h = (h_1,\dots,h_n), \tag{2.26}$$

we can envisage the translated function

$$F(x+h) = [B, D;\ f(\xi),\ \rho(\xi)\ ]\ (x+h), \tag{2.27}$$

and since the right side of (2.25) is the expression (2.19) we have

$$F(x+h) = [\ B(h),\ D(h)\ ;\ f(\xi+h),\ \rho(\xi+h)\ ]\ (x), \tag{2.28}$$

where $[\ B(h),\ D(h)\ ]$ arise from $[\ B,\ D]$ if we translate the latter by $-h$, that is,

$$B(h) = E\{\xi \mid \xi + h \in B\},\ \ D(h) = E\{\xi \mid \xi\ + h \in D\}. \tag{2.29}$$

Now $B - B(h)$ is the combinatorial boundary of the combinatorial complex $D - D(h)$, and for x fixed in D it follows from the lower half of (2.4) that for h sufficiently small we have

$$[B - B(h),\ D - D(h)\,;\ f(\xi{+}h),\ \rho(\xi+h)]\ (x) = 0. \tag{2.30}$$

Thus we can replace (2.28) by

$$F(x+h) = [B,\ D;\ f\,(\xi+h),\ \rho(\xi+h)]\,(x). \tag{2.31}$$

We now introduce the special vector

$$h = (h,\ 0,\ 0,\ \dots,\ 0), \tag{2.32}$$

and the difference quotients

$$F_h(x) = \frac{1}{h}\,(F(x+h) - F(x)), \tag{2.33}$$

$$f_h(x) = \frac{1}{h}\,(f\,(x+h) - f\,(x)),\ \rho_h(x) = \frac{1}{h}\,(\rho(x{+}h) - \rho(x)). \tag{2.34}$$

From (2.25) and (2.31) we then obtain

$$F_h(x) = [B,\ D\,;\ f_h\,(\xi),\ \rho_h(\xi)](x). \tag{2.35}$$

Due to (2.5) we can let $h \to 0$ on the right side. Therefore we can also let $h \to 0$ in (2.33), and we obtain

$$\frac{\partial F}{\partial x_1} = \left[B, D; \frac{\partial f}{\partial \xi_1}, \frac{\partial \rho}{\partial \xi_1} \right](x), \tag{2.36}$$

and we also find that $\dfrac{\partial F}{\partial x_1}$ is even continuous. The reasoning also applies to other coordinates, and the outcome is that $F(x)$ belongs at any rate to $C^{(1)}$ and that (2.24) holds for the first order operators

$$Lf \equiv \frac{\partial f}{\partial x_m}, \quad m = 1,\dots,n. \tag{2.37}$$

But, we can iterate the procedure arbitrarily often and the lemma follows for $F(x)$. For $F_\infty(x)$ the proof is quite similar and in fact much easier.

LEMMA 3. *We have*

$$\Delta F(x) = \rho(x), \ \Lambda F(x) = \sigma(x), \tag{2.38}$$

$$\Delta F_\infty(x) = 0, \ \Lambda F_\infty(x) = 0. \tag{2.39}$$

PROOF. For $L = \Delta$ in (2.24) we obtain, because of (2.15),

$$\left.\begin{matrix}\Delta F\\ \Delta F_\infty\end{matrix}\right\} = [B, D; \rho(\xi), \Delta\, \rho(\xi)]\,(x), \tag{2.40}$$

and for $L = \Lambda$ we obtain, because of (2.15) and (2.17),

$$\left.\begin{matrix}\Lambda F\\ \Lambda F_\infty\end{matrix}\right\} = [B, D; \sigma(\xi), \Delta\sigma(\xi)]\,(x). \tag{2.41}$$

Now, $\rho(\xi)$ and $\sigma(\xi)$ are defined in $\tilde{D}$, and the conclusions of the lemma follow from (2.40) and (2.41) if we apply both parts of (2.4) for $f(x) = \rho(x)$ and $f(x) = \sigma(x)$.

LEMMA 4. *The function $F_\infty(x)$ has an analytic continuation into all of E_n; in fact it is a polynomial in $(x_1, \dots, x_n)$ as is every function $\phi(x_1, \dots, x_n)$ of the following kind. It is defined and analytic in a spherical exterior $|x| > R$; there is an $r_0 \geqslant 0$ such that for* (2.6)

$$\frac{\partial^{r_1+\dots+r_n}\phi}{\partial x_1^{r_1}\dots\partial x_n^{r_n}} = O\,(\,|x|^{-n-1}), \quad |x| \to \infty,$$

and there is an operator Λf satisfying our axiom I *for which*

$$\Lambda\phi = 0.$$

PROOF. See [3], pages 8-9.

REMARK 3. Any function of which we prove that it can be continued beyond its original domain of definition will after continuation be denoted by the same symbol as before.

LEMMA 5. *In the intersection*

$$D \cap \widetilde{B}, \tag{2.42}$$

which is an " interior boundary layer " of D we have

$$F(x) - F_{\infty}(x) = f(x). \tag{2.43}$$

PROOF. We deform B into a closed hypersurface B' within (2.42), so that B' decomposes E_n into domains D', D'_{∞} with $D' \subset D$, $D'_{\infty} \supset D_{\infty}$. The pointset difference

$$D'_{\infty} - D_{\infty} \tag{2.44}$$

is then contained in $\widetilde{B}$. Now in D'_{∞} we can introduce the function

$$F'_{\infty}(x) = [B', D'; f(\xi), \rho(\xi)]\,(x). \tag{2.45}$$

At first it only "corresponds" to $F_{\infty}(x)$, but in D_{∞}, which is a subdomain of D'_{∞}, the two are equal. In fact, for x in D_{∞},

$$F_{\infty}(x) - F'_{\infty}(x) = [B - B', D - D'; f(\xi), \rho(\xi)]\,(x), \tag{2.46}$$

and by lower (2.4) this is 0. Thus (2.45) may be viewed as a continuation of F_{∞} from D_{∞} into D'_{∞}. Hence, for x in the domain (2.44), we have for $F(x) - F_{\infty}(x)$ the same expression

$$[B - B', D - D'; f(\xi), \rho(\xi)]\,(x), \tag{2.47}$$

but now x being a point of $D - D'$. By the upper (2.4) this is indeed $f(x)$, as claimed. Finally we note that any point of (2.42) can be made to lie in a domain (2.44) by proper choice of B', which completes the proof of the lemma.

We can now finally deal with Theorem 1 itself by proving its part (iii) as announced. Since $F(x)$ and $F_{\infty}(x)$ both exist in D, formula (2.43) gives a $C^{(\infty)}$ continuation of $f(x)$ from (2.42) into D and hence also from $\tilde{B}$ into $\tilde{D}$. Also relations (2.38) and (2.39) imply for the difference (2.43) in D the relations $\Delta f = \rho$, $\Lambda f = \sigma$, which completes the proof of our assertions.

REMARK 4. Relation (2.43) implies that if $f(x)$ and $\rho(x)$ are analytic in $\tilde{B}$, $F(x)$ can be continued " outwardly " from D into $\tilde{D}$. This by itself can also be obtained " directly ", that is without first proving the continuability of F_∞, as follows. We deform B into a hypersurface B'' which is located in the " exterior " boundary layer $D \cap \tilde{D}_\infty$, and we set up the function

$$F''(x) = [B'', D''; f(\xi), \rho(\xi)]\,(x) \tag{2.48}$$

in the bounded domain $D'' \supset D$ bounded by B''. It follows by the lower (2.4) that, for x in D, the expression (2.48) is equal to (2.25) itself, so that in D, $F''(x) = F(x)$. Next, by Lemma 3, $\Delta F'' = \rho(x)$ in D'', so that $F''(x)$ is analytic in D'', just as $F(x)$ is analytic in D. Hence $F''(x)$ is an analytic continuation of $F(x)$ outwardly.

3. Several closed hypersurfaces in Euclidean space. In the case of Theorem 1 we have introduced and even emphasized its part (iii) which refers to non-analytic data, mainly in order to delineate the " true " range and scope of the entire theory. From now on however, in order to avoid certain inconveniences of formulation and reasoning we will be dealing with the analytic case exclusively.

We are hereby postulating that henceforth all *given* functions are to be analytic (in the real components of the underlying space). For *constructed* functions it will not be immediately obvious whether they are analytic and it will be expressly stated if they are. Also the word " continuation " by itself will mean analytic continuation, both in assumptions and conclusions.

We take in E_n a finite number of strictly disjoint combinatorially closed simplicial hypersurfaces, say in their " natural " orientation and denote them by

$$B_1, B_2, \dots, B_s. \tag{3.1}$$

They decompose E_n into a finite number of domains which in their natural orientation we denote by

$$D_1, \dots, D_{t-1}, D_t. \tag{3.2}$$

The last domain, that is D_t, shall be the one which is unbounded (and which corresponds to D_∞ of the previous section). We introduce the incidence matrix

$$\eta_{q\tau} = 0, +1, -1;\ 1 \leqslant q \leqslant s,\ 1 \leqslant \tau \leqslant t, \tag{3.3}$$

where $\eta_{q\tau} = 0$ if B_q is not on the boundary of D_τ; $= 1$ if B_q is on the boundary of D_τ and is co-oriented with it; and $= -1$ if it is counter-oriented with it. Each B_q "separates" two of our domains (3.2), and we can denote these by D_u, D_v in such a way that

$$\eta_{qu} = +1;\ \eta_{qv} = -1;\ \eta_{q\tau} = 0 \text{ for } \tau \neq u, v. \tag{3.4}$$

With each B_q we associate a pointset neighbhorhood $\tilde{B}_q$, and these neighborhoods shall be so small that any two are disjoint. Also, we denote by $\tilde{D}_\tau$ a neighbhorhood of the closure of D_τ, and it is convenient to take

$$\tilde{D}_\tau = D_\tau \cup \sum_{\eta_{q\tau} \neq 0} \tilde{B}_q. \tag{3.5}$$

We now assume that there are given (analytic) functions

$$\rho_1(x), \rho_2(x), \dots, \rho_{t-1}(x), \rho_t(x) \tag{3.6}$$

and

$$f_1(x), f_2(x), \dots, f_s(x) \tag{3.7}$$

of the following interlocking descriptions. Each ρ_τ is given in $\tilde{D}_\tau$ and f_q in $\tilde{B}_q$, and for each q we have

$$\Delta f_q(x) = \sum_{\tau=1}^{t} \eta_{q\tau}\, \rho_\tau(x) = \rho_u(x) - \rho_v(x), \tag{3.8}$$

in $\tilde{B}_q$. Finally for the present we are imposing the rigorous restriction, which in the next section will be relaxed, that the $\rho(x)$ in the infinite domain D_t is strictly 0, that is,

$$\rho_t(x) = 0,\ x \in \tilde{D}_t. \tag{3.9}$$

In the set-up of Theorem 1 this corresponds to the fact that no $\rho_\infty(x)$ was assigned to $\widetilde{D}_\infty$.

We are introducing again the expression (2.20) and now we are defining it as follows :

$$[B,\, D;\; f(\xi),\, \rho(\xi)]\,(x) = \sum_{q=1}^{s} \int_{Bq} \Omega(f_q(\xi),\; \xi - x,\, d\xi) - \\ - \sum_{\tau=1}^{t} \int_{D_\tau} \rho_\tau(\xi) . G(\xi - x) . dv_\xi. \qquad (3.10)$$

The function

$$F(x) = [\,B,\, D\,;\, f(\xi),\, \rho(\xi)\,]\,(x) \qquad (3.11)$$

is now defined in the union $\bigcup_\tau D_\tau$, and separately we have a function

$$F_\tau(x) \text{ in } D_\tau,\ 1 \leqslant \tau \leqslant t. \qquad (3.12)$$

LEMMA 6. *For given functions ρ_τ, f_q with* (3.8) *if there is also given a function $\phi_\tau(x)$ in $\widetilde{D}_\tau$, $1 \leqslant \tau \leqslant t$, such that*

$$\rho_\tau(x) = \Delta\, \phi_\tau(x), \quad \phi_t(x) = 0, \qquad (3.13)$$

$$f_q(x) = \sum_{\tau=1}^{t} \eta_{q\tau}\, \phi_\tau(x) = \phi_u(x) - \phi_v(x), \qquad (3.14)$$

then the " operator " (3.11) *reproduces $\{\phi(x)\}$, that is*

$$F_\tau(x) = \phi_\tau(x), \quad \tau = 1, \ldots, t. \qquad (3.15)$$

PROOF. Follows by a judicious application of (2.4).

LEMMA 7. *Let x be given fixed in one of the D_{τ_0}. If B'_q is a sufficiently small deformation of B_q, $1 \leqslant q \leqslant s$, and $\{D'_\tau\}$ are the resulting deformations of $\{D_\tau\}$, then*

$$[B,\, D\,;\, f(\xi),\, \rho(\xi)\,]\,(x) = [B',\, D'\,;\, f(\xi),\, \rho(\xi)]\,(x). \qquad (3.16)$$

PROOF. Any small deformation can be built up " additively " from deformations in neighborhoods of single points. For instance, let B_q separate D_u, D_v as in (3.4). Let $\bar{B}_q$ be a small piece of B_q and let the deformation consist of replacing it by a small piece of $\bar{B}'_q$ which lies in D_u, and let R denote the piece of D_u bounded by $\bar{B}_q - \bar{B}'_q$. In this case we have on the one hand, by (2.4) and (3.8),

$$\sum_q \Big[\int_{B_q} \Omega(f_q(\xi), \xi - x, d\xi) - \int_{B'_q} \Omega(f_q(\xi), \xi - x, d\xi) \Big]$$

$$= \int_R (\rho_u - \rho_v) \,.\, G(\xi - x)\, dv_\xi.$$

But, on the other hand, also

$$\sum_\tau \Big[\int_{D_\tau} \rho_\tau \,.\, G \,.\, dv_\xi - \int_{D'_\tau} \rho_\tau \,.\, G\, dv_\xi \Big]$$

$$= \int_{D_u} \rho_u(\xi) \,.\, G(\xi - x)\, dv_\xi - \int_{D_u - R} \rho_u(\xi) \,.\, G(\xi - x)\, dv_\xi$$

$$+ \int_{D_v} \rho_v(\xi) \,.\, G(\xi - x)\, dv_\xi - \int_{D_v + R} \rho_v(\xi) \,.\, G(\xi - x)\, dv$$

$$= \int_R (\rho_u - \rho_v) \,.\, G(\xi - x) \,.\, dv_\xi,$$

and this verifies (3.16) for our deformation.

Lemmas 6 and 7 lead by the same steps as in Section 2 to the following proposition.

THEOREM 2. (i) *The functions* $F_\tau(x)$, (see (3.12)), *are immediately seen to be continuous, and the function* $F_t(x)$ *to be analytic.*

(ii) *But in fact each* $F_\tau(x)$ *is* $C^{(\infty)}$ *and for any linear operator* (2.23) *we have*

$$L_x F(x) = [B, D; L_\xi f(\xi), L_\xi \rho(\xi)]\,(x). \tag{3.17}$$

(iii) *Especially for* $L \equiv \Delta$ *we have*

$$\Delta F_\tau(x) = \rho_\tau(x), \tag{3.18}$$

so that every $F_\tau(x)$ *is analytic.*

(iv) *Furthermore, each* $F_\tau(x)$ *has an analytic continuation into* $\tilde{D}_\tau$, *and also*

(v) *For each* q, *we have in* $\tilde{B}_q$

$$f_q(x) = \sum_{\tau=1} \eta_{q\tau} F_\tau(x) = F_u(x) - F_v(x), \tag{3.19}$$

(see (3.4)).

(vi) *Finally, if there is given some operator* Λf, *at first not necessarily defective as in axiom* I, *and if there are given functions*

$$\sigma_1(x), \dots, \sigma_t(x), \tag{3.20}$$

σ_τ *in* $\widetilde{D}_\tau$, *such that*

$$\Lambda \rho_\tau = \Delta \sigma_\tau, \quad \tau = 1, \dots, t,$$

and

$$\Lambda f_q = \sum_{\tau=1}^{t} \eta_{q\tau} \sigma_\tau, \quad q = 1, \dots, s,$$

then

$$\Lambda F_\tau = \sigma_\tau, \quad \tau = 1, \dots, t.$$

The defectiveness of Λf comes in only at the very end to give the following conclusion à la Hartogs.

THEOREM 3. *If* Λf *is indeed as prescribed in axiom* I, *and if* $\sigma_t = 0$ (*only for* $\tau = t$) ; *then* $F_t = 0$ (*only for* $\tau = t$), *and for any* q *for which* $\eta_{qt} \neq 0$, *the function* $f_q(x)$ *can be continued from* $\widetilde{B}_q$ *into the domain* $\widetilde{D}_\tau$ *for which* $\eta_{q\tau} = -\eta_{qt}$.

REMARK 5. We have imposed the rigorous restriction (3.9) only in order to avoid discussing the integrals

$$\int_{D_t} \rho_t(\xi - x)\, dv_\xi, \quad \int_{D_t} L_\xi\, \rho_t(\xi) \cdot G(\xi - x) dv_\xi$$

over the unbounded domain D_t, and such integrals for $\phi_t(x)$ in the place of $\rho_t(x)$ in Lemma 6. However, due to (2.5) and (2.10), all such integrals, as are actually required in Theorem 2 would become manageable and admissible provided the function $\rho_t(x)$ and certain of its partial derivatives are sufficiently small at infinity.

4. A Cousin type theorem in Euclidean space. The present section consists of two distinct parts. In the first part we will state for Theorem 2, without proof, a generalization of the topological set-up which we have already introduced and featured in our first paper [1]. In the second part we will give an interpretation of Theorem 2 which is fairly meaningful for this theorem itself but becomes appreciably more so after the topological generalizations of the first part has been superimposed.

We are retaining the previous decomposition of E_n into finitely many domains with simplicial boundaries,

$$D_1, \dots, D_{t-1}, D_t, \tag{4.1}$$

with D_t being unbounded and the other domains bounded. However, the bounding $(n-1)$-dimensional simplices need no longer combine into finitely many distinct hypersurfaces each with combinatorial boundary 0. We only assume at present, for better visualization, that each such simplex separates two different domains, although in [1] even this was not required. These $(n-1)$-dimensional *simplices* themselves, *individually*, and each occurring only once, shall now be denoted by

$$B_1, \dots, B_s. \tag{4.2}$$

The pointset closures of any two B_q shall be either disjoint or their intersection shall be the pointset closure of an entire $(n-2)$ dimensional face common to the two. The set of all these faces, each counted only once, shall be denoted by

$$C_1, \dots, C_r. \tag{4.3}$$

The domains (4.1) have "natural" orientations and we adopt those, so that the "external" monomial (2.12) represents the positive volume element dv_ξ. However, for the simplices (4.2) and (4.3) no "natural" orientations are available now. We impart some fixed orientations to them and we introduce two incidence matrices : firstly the matrix

$$\eta_{q\tau} = 0, +1, -1, \quad 1 \leqslant q \leqslant s, \tag{4.4}$$

between B_q, D_τ as before, and secondly the corresponding matrix

$$\epsilon_{pq} = 0, +1, -1, \quad 1 \leqslant p \leqslant r, \quad 1 \leqslant q \leqslant s \tag{4.5}$$

between C_p, B_q. We recall the basic elementary relation

$$\sum_{q=1}^{s} \epsilon_{pq}\, \eta_{q\tau} = 0. \tag{4.6}$$

We again introduce ϵ-neighborhoods $\tilde{B}_q$ of B_q, then neighborhoods $\tilde{D}_\tau$ of D_τ, say as defined by (3.5), and also the neighborhoods

$$\tilde{C}_p \equiv \bigcap_{|\epsilon_{pq}| \neq 0} \tilde{B}_q \tag{4.7}$$

of C_p.

The incidence matrix (4.5), which we did not need in Section 3, will now enter by the following opening. If in each $\tilde{D}_\tau$ there is defined a function $\phi_\tau(x)$, $1 \leqslant \tau \leqslant t$, and if we introduce the "saltuses"

$$f_q(x) = \sum_{\tau=1}^{t} \eta_{q\tau}\, \phi_\tau(x) = \phi_u(x) - \phi_v(x) \tag{4.8}$$

in the $\tilde{B}_q$, then (4.6) implies that the latter satisfy the relations

$$\sum_{q=1}^{s} \epsilon_{pq}\, f_q(x) = 0 \tag{4.9}$$

in the $\tilde{C}_p$. These relations, which were introduced and featured for the first time (apparently) in our paper [1], were not stated in Section 3 because they were there "vacuous", but now they will have to be stipulated expressly.

Definition 1. *We again introduce and define the functions*

$$F(x) = [B, D; f(\xi), \rho(\xi)]\,(x) \tag{4.10}$$

by the same expressions (3.10) *as before, for any data of the following description. Each function $f_q(x)$ is defined in its $\tilde{B}_q$, and, what is important, the relations* (4.9) *holds. Each function $\rho_\tau(x)$ is defined in its $\tilde{D}_\tau$, and again we impose the rigorous restriction $\rho_t = 0$. And the functions $\{f_q\}$ and $\{\rho_\tau\}$ are again connected by the relations* (3.8).

This leads to the following generalizations.

Lemma 6′. *Lemma* 6 *holds literally as before.*

The proof is the same.

Lemma 7′. *Lemma* 7 *holds literally as before.*

This lemma is a serious claim, but we will not go into details. We are referring to pages 654-5 of [1] for a basic lemma needed for the generalization.

THEOREM 2′. *Theorem 2 holds literally as before.*

THEOREM 3′. *Theorem 3 holds literally as before.*

REMARK 5′. Remark 5 applies again.

These things stated, we are now turning to an interpretation of Theorem 2 (and Theorem 2′) which will be the second half of this section. We take in each $\tilde{D}_\tau$ a function $\phi_\tau(x)$ which, as always, is real-analytic, however we are permitting, as an exception, that each $\phi_\tau(x)$ have " singularities." We are not concerned with the actual nature of the singularities, so that singularities are for us points at which the functions are not defined. Specifically we take a closed set E in E_n, and we assume that each ϕ_τ is defined and analytic in the pointset difference

$$\tilde{D}_\tau - E = \tilde{D}_\tau - E_\tau, \tag{4.11}$$

where

$$E_\tau = \tilde{D}_\tau \cap E. \tag{4.12}$$

It is an open set, and we do not have to assume that it is connected, that is, a subdomain of $\tilde{D}_\tau$. With such functions we form the " saltuses "

$$f_q = \sum_{\tau=1}^{t} \eta_{q\tau}\, \phi_\tau, \tag{4.13}$$

each of which is defined at first not in $\tilde{B}_q$ but in

$$\tilde{B}_q - E = \tilde{B}_q - \tilde{B}_q \cap E. \tag{4.14}$$

Now, we expressly assume that the singularities of f_q are " removable ", meaning that there exists an (analytic) function which we also denote by f_q, in all of $\tilde{B}_q$ which in (4.14) coincides with (4.13). If (4.14) is connected, this " larger " $f_q(x)$ is unique, otherwise perhaps not. We need not analyze the possibilities, but for the general topological set-up we do have to make expressly the indispensable assumptions

$$\Sigma\, \epsilon_{pq} f_q = 0, \quad 1 \leqslant p \leqslant r, \tag{4.15}$$

for the enlarged functions $\{f_q\}$. If the functions ϕ_τ have no singularities, (4.15) is a formal consequence of (4.14), as previously noted, but with our very general notion of singularities it is safest to stipulate (4.15) expressly.

Next we form the functions

$$\rho_\tau(x) = \Delta\, \phi_\tau(x). \tag{4.16}$$

To begin with, they are defined only in (4.11), however we assume that each $\rho_\tau(x)$ shall have some analytic continuation into all of $\widetilde{D}_\tau$, and that (in all of $\widetilde{B}_q$) the relation

$$\Delta\, f_q = \sum_{\tau=1}^{t} \eta_{q\tau}\, \rho_\tau \tag{4.17}$$

holds. Finally we are also imposing the rigorous restriction $\phi_t = 0$, and hence also $\rho_t = 0$.

The conclusion from Theorem 2 (2′) is now as follows.

INTERPRETATION. *There exist fixed functions $F_\tau(x)$, $1 \leqslant \tau \leqslant t$, each analytic in all of $\widetilde{D}_\tau$, such that*

$$\Delta F_\tau(x) = \rho_\tau(x) \equiv \Delta\, \phi_\tau(x), \tag{4.18}$$

and also

$$\Sigma\, \eta_{q\tau} F_\tau(x) = f_q(x) \equiv \Sigma\, \eta_{q\tau} \phi_\tau(x). \tag{4.19}$$

If for some further operator Λf the functions

$$\sigma_\tau = \Lambda\, \phi_\tau \tag{4.20}$$

are also continuable into all of $\widetilde{D}_\tau$ and

$$\Delta\, \sigma_\tau = \Lambda\, \rho_\tau, \quad 1 \leqslant \tau \leqslant t, \tag{4.21}$$

then also

$$\Lambda F_\tau(x) = \sigma_\tau(x) \equiv \Lambda\, \phi_\tau(x). \tag{4.22}$$

5. A lemma from the theory of integral invariants. We now take an n-dimensional oriented real torus of total Euclidean volume 1,

$$\det |g_{ij}| = 1, \tag{5.1}$$

and we represent it in E_n by the fundamental domain

$$-\tfrac{1}{2} < x_m \leqslant \tfrac{1}{2}, \quad m = 1, 2, \ldots, n, \tag{5.2}$$

or any other pointset which arises from this by a homogeneous affine transformation with determinant $+1$. We will denote by S_n either the torus as a compact space, or one of these fundamental domains, depending on the context.

Given any constants

$$a^1, a^2, \ldots, a^n, \tag{5.3}$$

we can interpret them on S_n as a contravariant vector. With each such set of constants we will associate on S_n two objects which will be denoted by the use of the letter A; firstly the linear partial differential operator of first order

$$AF(x) \equiv a^1 \frac{\partial F}{\partial x_1} + \ldots + a^n \frac{\partial F}{\partial x_n}, \tag{5.4}$$

and secondly the $(n-1)$-dimensional external form

$$A(d\xi) = (\det |g_{ij}|)^{1/2} \sum_{\alpha=1}^{n} a^\alpha \, d\xi_1 \wedge \ldots \wedge d\xi_{\alpha-1} \wedge d\xi_{\alpha+1} \wedge \ldots \wedge d\xi_n. \tag{5.5}$$

The factor in front, $\det |g_{ij}|$, has numerical value $+1$ by (5.1), but it is technically needed in order to " transform " the vector component a^α into a tensor component $b_{1\ldots\alpha-1,\,\alpha+1\ldots n}$ of a covariant skewtensor of dimension $n-1$.

Finally we associate with our constants (5.3)—if they are real valued—the one parametric group of translations

$$x'_\alpha = x_\alpha + a^\alpha h, \ \alpha = 1, \ldots, n; \ -\infty < h < \infty \tag{5.6}$$

on S_n. If B is a closed oriented hypersurface on S_n, then (5.6) " deforms " it into another such hypersurface, which we denote by B^h, and we introduce the n-dimensional combinatorial complex whose boundary is $B^h - B$. The complex will be denoted by R^h.

If $\tilde{B}$ is a neighborhood of B then for sufficiently small h the pointset underlying R^h is contained in a compact subset of $\tilde{B}$.

LEMMA 8. *If $f(\xi)$ is defined and continuous in $\tilde{B}$, then*

$$\lim_{h\to 0}\frac{1}{h}\int_{R^h} f(\xi)\, d\xi_1\wedge\cdots\wedge d\xi_n = \int_B f(\xi)\; A(d\xi), \tag{5.7}$$

the limit existing.

PROOF. For the purpose of proof one can replace the neighborhood $\tilde{B}$ by a smaller neighborhood and denote the smaller by the same symbol. We take some oriented domain D whose oriented boundary consists of B and some other hypersurface B_0 which is disjoint from the closure of $\tilde{B}$. We can find in all of S_n a continuous function $g(x)$ which in $\tilde{B}$ equals $f(x)$ and in a neighborhood of B_0 has value 0. Now, in (5.7) itself, we may replace $f(x)$ by $g(x)$ and for the proof of (5.7) utilize properties of the latter. Now for $g(x)$, if D^h is the image of D under (5.6) we have identically

$$\frac{1}{h}\int_{R^h} g(\xi)d\xi_1\wedge\cdots\wedge d\xi_n = \frac{1}{h}\left(\int_{D^h}-\int_D\right) g(\xi)\, d\xi_1\wedge\cdots\wedge d\xi_n. \tag{5.8}$$

However, by a theorem from the theory of invariants, see [5], the right side in (5.8) has indeed a limit, as $h\to 0$, and its value is

$$\int_{B^h} g(\xi).\; A(d\xi) - \int_{B_0^h} g(\xi).\; A(d\xi). \tag{5.9}$$

But the second integral is 0, and in the first we may replace the symbol $g(\xi)$ by $f(\xi)$, which proves the lemma.

6. Residues on toruses. Any differential operator with constant coefficients can be interpreted as acting on S_n for any fundamental domain. With the operator (1.1) we associate for a brief moment the ordinary polynominal in the real variables $t_1, \ldots, t_n$,

$$T(t_1, \ldots, t_n) = \sum_{(p)} a_{p_1\ldots p_n}(2\pi\, it_1)^{p_1}\ldots(2\pi\, it_n)^{p_n}, \tag{6.1}$$

and we decompose it in the form

$$T(t) = T_{2h}(t) + T^0(t), \tag{6.2}$$

where $T_{2h}(t)$ consists of the highest terms with

$$p_1 + \dots + p_n = 2h,$$

and $T^0(t)$ of the additional term with

$$p_1 + \dots + p_n < 2h.$$

For our purposes the operator Δf is elliptic on every S_n, if firstly the principal part is positive definite, that is

$$T_{2h}(t_1, \dots, t_n) > 0 \text{ for } t_1{}^2 + \dots + t_n{}^2 > 0, \tag{6.3}$$

and also, (if T^0 is not $\equiv 0$,)

$$(T(t) \equiv)\ T_{2h}(t) + T^0(t) \neq 0 \text{ for } t_1{}^2 + \dots + t_n^2 > 0. \tag{6.4}$$

We have shown ourselves in [6], that for instance for the fundamental domain (5.2) the limit

$$G(\xi) = \lim_{\epsilon \downarrow 0} \sum_{(r)}{}' \frac{e^{-\epsilon(r_1{}^2 + \dots + r_n{}^2)}\, e^{2\pi i(r_1 x_1 + \dots + r_n x_n)}}{T(r_1, \dots, r_n)} \tag{6.5}$$

exists for $(\xi_1 \dots, \xi_n) \neq (0, \dots, 0)$ and constitutes a Green's function with which a Green's formula can be erected. However, due to the compactness of S_n the Green's function does not satisfy relation (2.13) but instead the relation

$$\Delta\, G(\xi) = -1, \tag{6.6}$$

and this introduces a certain " complication " into the set-up. The Green's formula becomes

$$\left.\begin{matrix} f(x) \\ 0 \end{matrix}\right\} = [B, D; f(\xi), \Delta f(\xi)]\,(x) + \int_D f(\xi)\, dv_\xi, \tag{6.7}$$

where, as before, the first term on the right is again the difference of integrals (2.19). However there is the additional term

$$\int_D f(\xi)\, dv_\xi \tag{6.8}$$

which is the same for the upper and lower half and depends on the configuration, that is on D, and on the given function $f(\xi)$, and linearly on the latter. However it does not depend on the variable point x, and in this sense it is a constant. This kind of " additional " quantity we term a *residue*. In the Green's formula, the residue is a volume integral. In the subsequent theorems it will engender other residues, and these will be integrals over $(n-1)$-dimensional hypersurfaces (or simplices). We will not be led to introducing residues which are integrals over lower dimensional simplices but it would be possible to undertake elaboration of the theory that would lead to such.

Proceeding as in Section 3 we are introducing on S_n a decomposition into a finite number of domains

$$D_1, \ldots, D_{t-1}, D \tag{6.9}$$

with separating closed surfaces

$$B_1, \ldots, B_s, \tag{6.10}$$

and with the incidence matrix

$$\eta_{q\tau}, 1 \leqslant q \leqslant s, 1 \leqslant \tau \leqslant t. \tag{6.11}$$

The domains (6.9) have natural orientations and we endow them with those, for simplicity's sake. The hypersurfaces, if in general positions on the torus, do no longer have " natural " orientations, but this causes no syllogistic inconvenience whatsoever. We impart to each B_q some fixed orientation, and we introduce the incidence matrix (6.11) accordingly.

REMARK 6. On the torus we do *not* single out a " last " domain D_t in any manner whatsoever, and there will be no analog whatsoever to the " rigorous restriction " (3.9).

We again introduce neighborhoods $\tilde{B}_q$, $\tilde{D}_\tau$ and, with function f_q in $\tilde{B}_q$, ρ_τ in $\tilde{D}_\tau$, which are linked by

$$\Delta f_q(\xi) = \sum_\tau \eta_{q\tau}\, \rho_\tau(\xi), \tag{6.12}$$

we again set up the functions

$$F_\tau(x) \text{ in } D_\tau, \tag{6.13}$$

defined collectively by

$$F(x) = [B, D; f(\xi), \rho(\xi)](x). \tag{6.14}$$

The right side is identically the same formula as in Section 3, namely

$$\sum_{q=1}^{s} \int_{B_q} \Omega(f_q(\xi), \xi - x, d\xi) - \sum_{\tau=1}^{t} \int_{D_\tau} \rho_\tau(\xi)\, G(\xi - x)\, dv_\xi. \tag{6.15}$$

The functions (6.13) are at first sight certainly continuous. Our first formal statement is an analog to Lemma 6. It will differ from Lemma 6 by the occurrence of a residue, and because of this novelty we will state it as a full theorem. But the proof is quite simple and will be left to the reader.

THEOREM 4. *If there are given functions $\phi_\tau(x)$ in the $\tilde{D}_\tau$ (without any singularities), and we again introduce the functions*

$$\rho_\tau(\xi) = \Delta \phi_\tau(\xi), \tag{6.16}$$

$$f_q(\xi) = \sum_\tau \eta_{q\tau}\, \phi_\tau(\xi), \tag{6.17}$$

which satisfy (6.12), *then*

$$F_\tau(x) = \phi_\tau(x) - \text{residue}, \tag{6.18}$$

where, at present,

$$\text{residue} = \sum_{\tau=1}^{t} \int_{D_\tau} \phi_\tau(\xi)\, dv_\xi. \tag{6.19}$$

The next theorem is the analog to Lemma 7. The proof will also be left to the reader although it requires a little more attention.

THEOREM 5. *If x is held fixed in one of the domains, if B'_q is a sufficiently small deformation of B_q, $1 \leqslant q \leqslant s$, and D'_τ is the resulting deformation of D_τ, $1 \leqslant \tau \leqslant t$, and if we denote by R_q the n-dimensional chain whose boundary is $B'_q - B_q$, that is,*

$$\text{boundary of } R_q = B'_q - B_q, \tag{6.20}$$

then

$$[B', D'; f(\xi), \rho(\xi)](x) - [B, D; f(\xi), \rho(\xi)](x) = -\text{residue}, \tag{9.21}$$

where at present,

$$\text{residue} = \sum_{q} \int_{R_q} f_q(\xi)\, d\xi_1 \wedge \dots \wedge d\xi_n. \tag{6.22}$$

Herein $\{f_q\}$, $\{\rho_\tau\}$ *are general functions satisfying* (6.12).

The next two theorems taken together are the analog to part (ii) of Theorem 3.

THEOREM 6. *Each function* $F_\tau(x)$ *has continuous first derivatives, and if for any constants* (5.3) *we introduce the differential operator*

$$A_x F(x) = a^1 \frac{\partial F}{\partial x_1} + \dots + a^n \frac{\partial F}{\partial x_n}, \tag{5.4}$$

then

$$A_x F(x) = [B,\ D;\ A_\xi f(\xi),\ A_\xi \rho(\xi)\,](x) - \text{residue}, \tag{6.23}$$

where at present

$$\text{residue} = \sum_{q=1}^{s} \int_{B_q} f(\xi).A(d\,\xi), \tag{6.24}$$

and $A(d\,\xi)$ *is defined by* (5.5).

PROOF. We first take the particular constants

$$a^1 = 1, a^2 = 0, \dots, a^n = 0, \tag{6.25}$$

so that (5.4) is

$$A_x\, F = \frac{\partial F}{\partial x}. \tag{6.26}$$

As in Section 2 we introduce the vector $h = (h, 0, \dots, 0)$, the shifted set $B_q(h)$, $D_\tau(h)$, and the difference quotients $F_{\tau h}(x)$, $f_{qh}(x)$, $\rho_{\tau h}(x)$ as defined by (2.33) and (2.34). We then obtain

$$F_h(x) = \frac{1}{h}\{[\,B(h),\ D(h);\ f(\xi+h),\ \rho(\xi+h)]\,(x) - [B,\ D; f(\xi),\ \rho(\xi)]\,(x)\}. \tag{6.27}$$

If now $R_q(h)$ is the chain bounded by $B_q - B_q(h)$, then by Theorem 5 we obtain

$$F_h(x) = [\,B,\ D;\ f_h(\xi),\ \rho_h(\xi)]\,(x) - \text{residue}, \tag{2.28}$$

where

$$\text{residue} = \frac{1}{h}\sum_q \int_{R_q(h)} f_q(\xi)\, d\xi_1 \wedge \dots \wedge d\xi_n. \tag{6.29}$$

We now wish to let $h \to 0$ in (6.28). Obviously

$$\lim_{h\to 0}\, [B, D;\, f_h(\xi),\, \rho_h(\xi)]\,(x) = \left[B, D;\, \frac{\partial f}{\partial \xi_1},\, \frac{\partial \rho}{\partial \xi_1}\right](x),$$

the limit existing. Now, in consequence of Lemma 8, we have

$$\lim_{h\to 0}\, (\text{residue}) = \sum_q \int_{B_q} f_q(\xi).\, d\xi_2 \wedge \dots \wedge d\xi_n, \tag{6.30}$$

this limit also existing. Therefore the limit

$$\lim_{h\to 0} F_h(x) = \frac{\partial F}{\partial x_1} \tag{6.31}$$

exists too, and the limit of (6.28) is the relation (6.23) corresponding to the particular constants (6.25). The same reasoning gives formula (6.23) for the constants $a^m = 1$, $a^\nu = 0$, $\nu \neq m$, any m, and since (6.23) is additive in the vector $(a^1, \dots, a^2)$, the theorem follows.

THEOREM 7. *Each function $F_\tau(x)$ is $C^{(\infty)}$ and for any linear operator Lf, see* (2.23), *we have*

$$L_x F(x) = [B, D;\, L_\xi\, f(\xi),\, L_\xi\, \rho(\xi)]\,(x) - \text{residue}. \tag{6.32}$$

REMARK 7. This "residue" is a quantity which is independent of x and τ. A formula for it will be given in the next section.

PROOF. We apply induction on the order of Lf. If the order is 1, Theorem 7 is a consequence of Theorem 6. Now an operator of an order $s + 1$ can be represented as a sum

$$L = \frac{\partial}{\partial x_1} L^1 + \dots + \frac{\partial}{\partial x_n} L^n + L^0 \tag{6.33}$$

in which $L^0, L^s, \dots, L^n$ is each of order $\leqslant s$. If we assume that the theorem holds for order $\leqslant s$, then

$$L^m F(x) = [B, D;\, L^m_\xi\, f(\xi),\, L^m_\xi \rho(\xi)]\,(x) - \text{constant},$$
$$m = 0, 1, \dots n, \tag{6.34}$$

and by differentiation we obtain

$$\left(\frac{\partial}{\partial x_m} L^m\right) F = \frac{\partial}{\partial x_m} \{[B, D; L^m f(\xi), L^m \rho(\xi)] (x)\}, \quad 1 \leqslant m \leqslant n. \tag{6.35}$$

But if we apply Theorem 6 to the functions $\{L^m f(\xi), L^m \rho(\xi)\}$ in the place of the functions $\{f(\xi), \rho(\xi)\}$ with $A = \dfrac{\partial}{\partial x_m}$, then we finally obtain

$$\left(\frac{\partial}{\partial x_m} L^m\right) F = \left[B, D; \left(\frac{\partial}{\partial \xi_m} L^m\right) f(\xi), \left(\frac{\partial}{\partial \xi_m} L^m\right) \rho(\xi)\right](x) - \\ - \text{residue}. \tag{6.36}$$

Relation (6.32) itself follows now if we add up relations (6.36) for $m = 1, \ldots, n$ and relation (6.34) for $m = 0$.

THEOREM 8. *In particular*

$$\Delta F_\tau(x) = \rho_\tau(x) - \text{residue}, \quad 1 \leqslant \tau \leqslant t \tag{6.37}$$

so that each $F_\tau(x)$ *is even analytic.*

REMARK 8. The residue in (6.37) is a constant in x which does not even depend on τ.

PROOF. If in Theorem 7 we put $L = \Delta$ we obtain

$$\Delta F = [B, D; \Delta f(\xi), \Delta \rho(\xi)] (x) - (\text{res})_1, \tag{6.38}$$

and, by Theorem 4 in connection with (6.12),

$$[B, D; \Delta f(\xi), \Delta \rho(\xi)] = \rho(\xi) - (\text{res})_2, \tag{6.39}$$

where, as it happens,

$$(\text{res})_2 = \sum_\tau \int_{D_\tau} \rho_\tau(\xi) \, dv_\xi. \tag{6.40}$$

This proves (6.37) with

$$\text{residue} = (\text{res})_1 + (\text{res})_2. \tag{6.41}$$

THEOREM 9. (i) *Each* $F_\tau(x)$ *has a continuation from* D_τ *into* $\tilde{D}_\tau$ *and*

(ii) *we have*

$$\sum_{\tau=1}^{t} \eta_{q\tau} F_\tau(x) = f_q(x), \tag{6.42}$$

in $\tilde{B}_q$.

Proof. We single out one of the hypersurfaces. We assume that it is B_1 and that it separates the domain D_u with which it is co-oriented, from the domain D_v with which it is counteroriented. We replace a small piece of B_1 by a small piece of hypersurface located in D_v and let this be the total deformation of our system of hypersurfaces. Let R denote the small piece of D_v bounded by $B'_1 - B_1$. For x in D_u we introduce our original function

$$F_u(x) = [B, D; f(\xi), \rho(\xi)]\,(x), \tag{6.43}$$

and the second function

$$F'_u(x) = [B', D'; f(\xi), \rho(\xi)]\,(x). \tag{6.44}$$

By Theorem 5 we have

$$F'_u(x) = F_u(x) - c, \tag{6.45}$$

where, as it happens, the constant c has the value

$$c = \int_R f_1(\xi)\, dv_\xi. \tag{6.46}$$

Now, by Theorem 8, F'_u is defined and analytic in D'_u which is D_u plus R, and thus, by (6.45), $F_u(x)$ itself has a continuation into D'_u. This proves part (i).

Let the symbol $F_u(x)$ denote the continued function in R, and for x in R we wish to determine the value of the difference

$$F_u(x) - F_v(x). \tag{6.47}$$

It is

$$F'_u(x) - F_v(x) + c = c + [B', D'; f, \rho]\,(x) - [B, D; f, \rho]\,(x)$$

$$= c + \int_{B'_1 - B_1} \Omega(f_1, \xi - x, d\xi) - \int_{D_u + R} \rho_u \, dv -$$

$$- \int_{D_v - R} \rho_v \, dv + \int_{D_u} \rho_u \, dv - \int_{D_v} \rho_u \, dv$$

$$= c + \int_{B'_1 - B_1} \Omega(f_1, \xi - x, d\xi) - \int_R (\rho_u(\xi) - \rho_v(\xi)) \, dv_\xi.$$

Now, by the upper part of (6.7) and (6.46) this is

$$c + f_1(x) - c = f_1(x),$$

which proves part (ii) of the theorem.

THEOREM 9. *If furthermore for some operator Λf and functions $\sigma_\tau(x)$ in the $\tilde{D}_\tau(x)$ we have*

$$\Lambda \rho_\tau(x) = \Delta \, \sigma_\tau(x), \quad 1 \leqslant \tau \leqslant t, \tag{6.48}$$

and

$$\Lambda f_q(x) = \sum_{\tau=1}^{t} \eta_{q\tau} \, \sigma_\tau(x), \quad 1 \leqslant q \leqslant s, \tag{6.49}$$

then

$$\Lambda F_\tau(x) = \sigma_\tau(x) - \text{residue} \tag{6.50}$$

for $1 \leqslant \tau \leqslant t$.

PROOF. By Theorem 7 we have

$$\begin{aligned} \Lambda F_\tau &= [B, D; \Lambda f, \Lambda \rho] - (\text{res})_1 \\ &= [B, D; \Lambda f, \Delta \sigma] - (\text{res})_1. \end{aligned} \tag{6.51}$$

However, by Theorem 4, in connection with (6.49),

$$[B, D; \Lambda f, \Delta \sigma] = \sigma_\tau(x) - (\text{res})_2, \tag{6.52}$$

where

$$(\text{res})_2 = \sum_{\tau=1}^{t} \int_{D_\tau} \sigma_\tau(\xi) \, dv_\xi. \tag{6.53}$$

This proves (6.50) with

$$\text{residue} = (\text{res})_1 + (\text{res})_2. \tag{6.54}$$

7. A Cousin type theorem on the torus. In the expression (6.54) and also in the earlier expression (6.41) which can be subsumed under it, the total residue appears as a sum of two addends. The second addend, which we denoted by $(\mathrm{res})_2$, is a volume integral. It was easy to give immediately a formula for it, as we did, and it is of secondary importance in as much as it is easy to state when its value is zero.

THEOREM 10. *We have, corresponding to the operator* Λf,

$$(\mathrm{res})_2 = \sum_\tau \int_{D_\tau} \sigma_\tau \, dv_\xi \tag{7.1}$$

so that

$$(\mathrm{res})_2 = 0 \tag{7.2}$$

if all $\sigma_\tau = 0$. *In particular,* (7.2) *holds for the operator* Δf *itself, if all* $\rho_\tau = 0$.

Thus, this part of the residue does not occur effectively whenever the functions $f_q(x)$ are solutions of the " homogeneous " equations

$$\Lambda f = 0 \tag{7.3}$$

and

$$\Delta f = 0 \tag{7.4}$$

respectively. And these " homogeneous " forms of the equations are the ones which are basic to holomorphy and to such generalizations of holomorphy as our approach suggests and implies.

But the first addend in the sum (6.54), which we have denoted by $(\mathrm{res})_1$, is the really *primary* part of it, and it is an evolution of Cauchy's own concept of a residue.

If $\Lambda_x F$ is an isobaric operator of the first order, see (5.4), then $(\mathrm{res})_1$ is given again "uniquely " by a simple expression, namely by (6.24). Building on this, one can give an expression for the residue in the general case, but it is no longer unique in its form or in its implications.

THEOREM 11. *If one takes any representation*

$$\Lambda f = \sum_{\mu=1}^{\nu} A_\mu(L^\mu f) + b_0 f \qquad (7.5)$$

in which $A_1,\ldots,A_\nu$ *are a finite number of isobaric operators of first order, if their factors* L^μ *are any other linear operators, and if* b_0 *is a constant, then this implies the formula*

$$(\mathrm{res})_1 = \sum_{\mu=1}^{\nu} \sum_{q=1}^{s} \int_{B_q} L^\mu f(\xi).A_\mu(d\xi). \qquad (7.6)$$

PROOF. "Usually" the constant b_0 in (7.5) will be 0, but even if it is $\neq 0$ it obviously does not contribute anything to the value of $(\mathrm{res})_1$ in (6.51), therefore it does not enter (7.6).

Formula (6.35) implies

$$\Lambda_x F = \sum_{\mu=1}^{\nu} A_\mu\{[B, D; L^\mu f, L^\mu \rho]\,(x)\}, \qquad (7.7)$$

but by Theorem 6 we have for each μ

$$A_\mu\{[B, D; L^\mu f, L^\mu \rho]\,(x)\} = [B, D; A_\mu L^\mu f, A_\mu L^\mu \rho]\,(x) - \\ - \sum_{q=1}^{s} \int_{B_q} L^\mu f_q(\xi)\cdot A_\mu(d\xi). \qquad (7.8)$$

If we insert this in (7.7) it follows that (7.6) is the correct value with which to validate (6.51).

DEFINITION 2. *For given functions* $\{f_q\}$ *if* Λf *(or* Δf*) is some operator, and if for this operator we choose some representation of the form* (7.5), *then the corresponding value* (7.6) *shall be called a principal residue of the operator.*

We now take a set-up as in the second half of Section 4. We take a closed set E in S_n and we assume that there is given a function $\phi_\tau(x)$ in each $\tilde{D} - E$, such that the functions

$$\rho_\tau = \Delta\phi_\tau \qquad (7.9)$$

are continuable into all of $\tilde{D}_\tau$ and the saltuses

$$f_q = \sum_{\tau=1}^{t} \eta_{q\tau}\, \phi_\tau \tag{7.10}$$

into all of $\tilde{B}_q$. If we introduce the functions $F_\tau(x)$ as defined by (6.15) then they have the following properties.

THEOREM 12. *Each $F_\tau(x)$ is defined and analytic in all of $\tilde{D}_\tau$ and the functions $\{F_\tau\}$ have the same saltuses as the functions with singularities $\{\phi_\tau\}$,*

$$\sum_{\tau} \eta_{q\tau}\, F_\tau = f_q = \Sigma \eta_{q\tau}\, \phi_\tau. \tag{7.11}$$

However,

$$\Delta F_\tau = \rho = \Delta \phi_\tau, \tag{7.12}$$

whenever (i) *a principal residue of the operator (relative to the given functions* (7.10)) *is zero and* (ii) *also*

$$\sum_{\tau=1}^{t} \int_{D_\tau} \rho_\tau(\xi)\, dv_\xi = 0. \tag{7.13}$$

If for some other operator Λf the functions

$$\sigma_\tau = \Lambda \phi_\tau \tag{7.14}$$

are also continuable into all of D_τ so that also

$$\Delta \sigma_\tau = \Lambda \rho_\tau, \tag{7.15}$$

then the same functions $\{F_\tau\}$ also satisfy

$$\Lambda F_\tau = \sigma_\tau = \Lambda \phi_\tau, \tag{7.16}$$

provided a principal residue of this operator Λ relative to the functions f_q is zero and also

$$\sum_{\tau=1}^{t} \int_{D_\tau} \sigma_\tau(\xi)\, dv_\xi = 0. \tag{7.17}$$

A situation of considerable interest arises if in the second half of this theorem there is given not just one operator Λf but several operators simultaneously, which we distinguish by an upper index, thus,

$$\Lambda^1, \ldots, \Lambda^k, \tag{7.18}$$

and if furthermore this set of operators "generate" the elliptic operator Δ in the sense that for some linear operators $M_1, \ldots, M_k$ we have

$$\Delta = M_1 \Lambda^1 + \ldots + M_k \Lambda^k. \tag{7.19}$$

Now, instead of starting out with an elliptic operator Δ, and then adding operators (7.18) by which Δ is presentable in a form (7.19), we may also adopt the approach that we start out initially with a system of operators (7.18) which are the only ones given. But the *system* shall be *elliptic* in the sense that among all linear combinations (7.19) for " variable " $M_1,\ldots,M_k$ there is at least one for which Δ is elliptic in our previous sense. We can now state as a corollary to Theorem 12 an "existence" theorem in which the operator Δ no longer enters *explicitly* at all.

THEOREM 13. *If there is given a system of functions with singularities* $\{\phi_\tau\}$, *whose saltuses* $\{f_q\}$ *are without singularities, and an elliptic system of operators* (7.18) *such that the functions*

$$\sigma_\tau^j = \Lambda^j \phi_\tau, \quad 1 \leqslant j \leqslant k, \quad 1 \leqslant \tau \leqslant t,$$

are also without singularities; *and if the principal and secondary residues for these operators are zero*; *then there exist functions* F_τ *without singularities which have the same saltuses* f_q *as the* ϕ_τ *and which satisfy the same equations*

$$\sigma_\tau^j = \Lambda^j F_\tau$$

for the given operators (7.18).

8. Meromorphic functions on the torus. We will now apply Theorem 13 to obtain a statement which, except for an accidental mistake in a formula, was already made in [1].

We consider a complex torus. Thus the space dimension is an even number $n = 2k$, and there are complex coordinates $z_j = x_j + \surd(-1)\, y$; $j = 1,\ldots,k$. Our operators (7.18) shall be

$$\Lambda^j f \equiv \frac{\partial f}{\partial \bar{z}_j}, \; j = 1, \ldots, k, \tag{8.1}$$

and they constitute an elliptic system since

$$\frac{1}{4}\sum_{j=1}^{k}\frac{\partial}{\partial z_j}\cdot\frac{\partial}{\partial \bar{z}_j} \tag{8.2}$$

is the classical Laplacean.

We are only interested in holomorphic functions with singularities and thus we consider a system of functions $\{\phi_\tau(z)\}$ with singularities whose saltuses $f_q(z)$ have holomorphic continuations into the entire sets $\tilde{B}_q$.

Now, it can be computed that the equations for the vanishing of the principal residues are

$$\sum_{q=1}^{s}\int_{B_q} f_q(\xi).\,d\zeta_j \wedge \underbrace{\omega\wedge\ldots\wedge\omega}_{k-1} = 0,\ \ j=1,\ldots,k, \tag{8.3}$$

where $\zeta_j = \xi_j + \sqrt{(-1)}\,\eta_j$ and

$$\omega = d\,\zeta_1 \wedge d\,\bar{\zeta}_1 + \ldots + d\,\zeta_k \wedge d\,\bar{\zeta}_k. \tag{8.4}$$

Thus, if the residues (8.3) are all zero, there are holomorphic functions without singularities having the same saltuses $\{f_q\}$. That is the differences $\phi_\tau - F_\tau$ have saltuses 0, and thus constitute a function on $S_{2k} - E$, and this function has the same singularities as were given by the functional elements $\phi_\tau(z)$.

In the case of one complex variable the result is as follows. If on a complex torus in one variable one wishes to have a function, which is holomorphic except for a finite number of isolated singularities say (which may be essential singularities), and if the principal parts of the singularities are prescribed, then such a singular function exists provided only that the sum of the residues of the principal parts is 0. This theorem in one variable is easily proven, although rarely stated in books on elliptic functions. Our work above for the torus may be therefore viewed as an elaboration of this simple theorem, as it were.

REFERENCES

1. S. Bochner : Analytic and meromorphic continuation by means of Green's formula, *Annals of Math.* 44 (1943), 652-673.

2. S. Bochner : Partial differential equations and analytic continuation, *Proc. Nat. Acad. Sci. U.S.A.* 38 (1952), 227-30.

3. S. Bochner : Green's formula and partial differential equations, in *Contributions to the Theory of Partial Differential Equations, Annals of Math. Studies, Number* 33, Princeton, (1954).

4. S. Bochner : Systèms des équations aux dérivées partielles qui se rattachent aux formules de Green, *Colloques Internationaux du Centre National de la Recherche Scientifique*, LXXI, *La théorie des équations au dérivées partielles, Nancy*, 9-15 *Avril* 1956.

5. E. Goursat : *Lecons sur le problème de Pfaff*, Hermann & Cie, Paris, (1922), p. 221-222.

6. S. Bochner : Zeta Functions and Green's Functions for linear partial differential operators of elliptic type with constant coefficients, *Annals of Math.* 57 (1953), 32-56.

Princeton University
Princeton N.J., U.S.A.

ENTIRE FUNCTIONS IN SEVERAL VARIABLES WITH CONSTANT ABSOLUTE VALUES ON A CIRCULAR UNIQUENESS SET[1]

S. BOCHNER

If for a function $\phi(z)=\sum_0^\infty a_p z^p$ in the entire z-plane we have $|\phi(z)|=1$ on $|z|=1$, then the product

$$\sum_0^\infty a_p z^p \cdot \sum_0^\infty \frac{\bar{a}_q}{z^q}$$

which is analytic in $z\neq 0$ has value 1 on $|z|=1$. Therefore it is identically 1, and thus $\phi(z)\neq 0$ for $z\neq 0$. Therefore $\phi(z)=z^p h(z)$ where $h(z)\neq 0$ everywhere. But again $|h(z)|=1$ on $|z|=1$, and for such an $h(z)$ we have $h(z)=c$, so that $\phi(z)=cz^p$. By the use of the same method, Bojanic and Stoll [1] have recently given the following generalization to functions which are holomorphic in the entire $\mathbf{C}^n$, for any n.

THEOREM 1. *If for an entire function* $f(z)\equiv f(z_1,\cdots,z_n)$ *we have*

$$|f(\zeta_1,\cdots,\zeta_n)| = 1$$

on the set

$$(1) \qquad |\zeta_1| = 1,\cdots, |\zeta_n| = 1$$

then

$$f(z_1,\cdots,z_n) = c\, z_1^{p_1}\cdots z_n^{p_n}.$$

The authors invoke the lemma that if an entire function is $\neq 0$ for $z_1\cdots z_n\neq 0$ then it can be represented as a product

$$c\, z_1^{p_1}\cdots z_n^{p_n} h(z)$$

in which $h(z)\neq 0$ everywhere. We propose to avoid taking recourse to this lemma and to obtain a more systematic theorem in the process.

We replace the point set (1) by a general point set S having the following properties

(i) S is circular, that is, if $(\zeta_1,\cdots,\zeta_n)\in S$, then also $(\zeta_1 t,\cdots,\zeta_n t)\in S$ for any $|t|=1$.

(ii) S is connected.

(iii) S is a uniqueness set for entire functions, in the sense that if

Received by the editors February 27, 1961.

[1] This research was supported by the United States Air Force through the Office of Scientific Research of the Air Force Research and Development Command.

$f(z)$ is 0 on S it is $\equiv 0$. We note that S is of this kind if it is part of the boundary of a domain R such that, for z in R, $f(z)$ can be represented by a suitable Cauchy integral in which integration extends over S only.

THEOREM 2. *If for an entire function we have* $|f(\zeta)| = 1$ *on a set* S *then* $f(z)$ *is a homogeneous polynomial*

$$A(z_1, \cdots, z_n) = \sum_{p_1+\cdots+p_n=p} a_{p_1\cdots p_n} z_1^{p_1} \cdots z_n^{p_n} \tag{2}$$

of some finite degree $p \geqq 0$.

PROOF. For any fixed $(\zeta) \in S$ we form the function

$$\phi(z) = f(\zeta_1 z, \cdots, \zeta_n z)$$

in $\mathbf{C}^1$. By property (i) of S we have $|\phi(z)| = 1$ on $|z| = 1$, and hence by our introductory statement we have $\phi(z) = c\, z^p$, that is

$$f(\zeta_1 z, \cdots, \zeta_n z) = z^p f(\zeta_1, \cdots, \zeta_n).$$

Up to here the exponent p is a function of ζ. However it follows from

$$|z|^p = |f(\zeta_1 z, \cdots, \zeta_n z)|$$

(for $z = 2$, say) that p is a continuous function on S. Furthermore it is integer-valued, and S is connected by property (ii). Therefore p is a constant number.

We can now form the difference

$$f(w_1 z, \cdots, w_n z) - z^p f(w_1, \cdots, w_n).$$

It is an entire function in $w_1, \cdots, w_n$ and z, and it is 0 for $(w) \in S$. By property (iii) it is $\equiv 0$. If we now introduce the power series for $f(z)$, the conclusion of Theorem 2 follows.

We next add one further property of S which is much more structural than the preceding ones.

(iv) On S we have

$$\bar{\zeta}_\nu = \frac{\lambda_\nu(\zeta)}{D(\zeta)}, \qquad \nu = 1, \cdots, n,$$

where $\lambda_1(z), \cdots, \lambda_n(z)$, $D(z)$ are (homogeneous) polynomials in $z_1, \cdots, z_n$.

If now we introduce the factorization into irreducible polynomials

$$D(z) = D_1(z)^{q_1} \cdots D_m(z)^{q_m} \tag{3}$$

then the following conclusion results.

THEOREM 3. *Furthermore, we have*

$$f(z_1, \cdots, z_n) = c\, D_1(z)^{p_1} \cdots D_m(z)^{p_m} \tag{4}$$

for some exponents $p_1 \geqq 0, \cdots, p_m \geqq 0$.

PROOF. If with the coefficients of (2) we form the polynomial

$$B(z_1, \cdots, z_n) = \sum_{p_1+\cdots+p_n=p} \overline{a_{p_1\cdots p_n}\lambda_1^{p_1} \cdots \lambda_n^{p_n}}$$

then our assumption

$$|f(\zeta)|^2 = 1 = f(\zeta)\overline{f(\zeta)}$$

implies that we have

$$A(z)\cdot B(z) = D(z)^p \tag{5}$$

on S. Using again (iii) we conclude that this holds identically in z. But all factors in (5) are polynomials and therefore (5) implies (4) by simple algebra.

REMARK. If also $|D_\nu(\zeta)| = 1$, $\nu = 1, \cdots, n$, then conversely every function (4) has constant absolute value on S.

Theorem 1 subsumes under Theorem 3 if we put

$$D(z) = z_1 \cdots z_n; \qquad \lambda_\nu(z) = \frac{D(z)}{z_\nu}, \qquad \nu = 1, \cdots, n.$$

But we also obtain interesting statements for some types of symmetric domains. Assume for instance that $n = k^2$ and that our variables $z_1, \cdots, z_n$ constitute a square array $\{z_{pq}\}$, $p, q = 1, \cdots, k$. The associated "natural" uniqueness set is formed by the unitary matrices

$$\sum_{r=1}^{k} \zeta_{pr}\bar{\zeta}_{qr} = \delta_{pq}, \tag{6}$$

see [2].

This gives

$$\bar{\zeta}_{\mu\nu} = \frac{\lambda_{\mu\nu}(\zeta)}{\det|\zeta|}$$

where $\lambda_{\mu\nu}$ are certain minors of the matrix $\{\zeta_{pq}\}$ and

$$D(\zeta) = \det|\zeta_{pq}|$$

is its determinant. Now, the determinant is an irreducible polynomial and $|D(\zeta)| = 1$. Hence the following theorem.

THEOREM 4. *If $f(z_{rs})$ is defined holomorphic over the entire matrix space, then it has absolute value* 1 *on the unitary set* (6) *if and only if*

$$f(z) = e^{i\alpha}(\det | z_{rs} |)^p$$

for some integer $p \geqq 0$.

BIBLIOGRAPHY

1. R. Bojanic and W. Stoll, *A characterization of monomials*, Proc. Amer. Math. Soc. this issue, [pp. 115–116].

2. S. Bochner, *Group invariance of Cauchy's formula in several variables*, Ann. of Math. (2) 45 (1944), 686–707.

PRINCETON UNIVERSITY

Reprinted from the PROCEEDINGS OF THE NATIONAL ACADEMY OF SCIENCES
Vol. 53, No. 5, pp. 904–907. May, 1965.

INTRINSIC ANALYTIC CONTINUATION AND ENVELOPES OF HOLOMORPHY*

BY S. BOCHNER

PRINCETON UNIVERSITY

Communicated March 11, 1965

We are returning to an early topic of ours.[1] We will establish a general proposition which will elucidate the nature and construction of envelopes of holomorphy, and also of Weierstrass (-Poincaré) analytic continuations of single analytic functions; especially, we will explain the following difference between the two. On the one hand, as we will redemonstrate in our subsequent Theorem 3, if, in the ordinary space of several complex variables, a domain is transformed holomorphically into another, then the Cartan-Thullen envelopes of holomorphy of the two domains correspond to each other.[2] On the other hand, this is not at all the case for Weierstrass continuation of a single function, as exemplified by the following instance.

Counterexample: If D_z is the half plane $\text{Re } z > 0$, and D_w is the strip $|\text{Im } w| < \pi/2$, then $z = e^w$ is a conformal map from D_w to D_z; so that, if $f(z)$ is a function on D_z, then $g(w) = f(e^w)$ on D_w is "intrinsically" the same function. However, for $f(z) = z$ the Weierstrass continuations are: $\tilde{f}(z) = z$ in C^1, and $\tilde{g}(w) = e^w$ in C^1; and these continuations are not the same because $\tilde{f}(z)$ assumes the value 0 and $\tilde{g}(w)$ does not.

Thus, the *analytisches Gebilde* of H. Weyl[3] is not at all an "intrinsic" construction.

For our comprehensive statements we will not require holomorphy, and only real-analyticity will be assumed; even a suitably defined quasi-analyticity would suffice. We are introducing a system

$$\Sigma = \{S; f^\alpha\}. \tag{1}$$

in which S is what we will call a *base space;* it is a *connected* real-analytic manifold of a *fixed* dimension n; $\{f^\alpha(P)\}$, $P \in S$, is a family of analytic functions on S, and α is an enumerative index which traverses an arbitrary but *fixed* index set A. It is not necessary that f^α be an "ordinary" function; it may be a many-to-one analytic function from S to some analytic manifold W^α of some dimension n_α; or, f^α may be a tensorial object of the classical kind, that is, a cross section in a *tangentially* constructed vector bundle for which S is the topological base space. In fact, all we require of our "functions" is the validity of the following assertion (; remember that S is connected).

LEMMA 1. *If, on S, two functions of the same kind coincide in a neighborhood, then they are identical.*[1]

Definition 1: In what follows, a homeomorphism φ from one base space into another is tacitly assumed to be such that φ and φ^{-1} are both analytic.

Since homeomorphisms (and automorphisms) are special cases of "functions," Lemma 1 implies the following.

LEMMA 2. *Given three spaces S, S_1, S_2, if φ_1 is a homeomorphism of S into a part S_1' of S_1, and φ_2 of S into a part S_2' of S_2, and if the homeomorphism*

$$\varphi_2 \circ \varphi_1^{-1} \tag{2}$$

of S_1' into S_2' can be extended to an analytic mapping of all of S_1 into a part of S_2 which is not all of S_2; then the inverse homeomorphism from S_2' to S_1' cannot be continued to a mapping of S_2 into S_1.

We now hold the system (1) fast, and we consider any ensemble

$$(\Sigma_1, \varphi_1) = \{S_1; f_1^\alpha; \varphi_1\}, \tag{3}$$

in which $\Sigma_1 = \{S_1; f_1^\alpha\}$ is a system like (1) and φ_1 is a homeomorphism from S into a part S_1' of S such that

$$f_1^\alpha(\varphi_1 P) = f^\alpha(P); \qquad P \in S, \ \alpha \in A.$$

Such an ensemble will be called an (analytic) *continuation* of a fixed system (1). Given a second continuation

$$(\Sigma_2, \varphi_2) = \{S_2; f_2^\alpha; \varphi_2\}, \tag{4}$$

we put

$$(\Sigma_1, \varphi_1) < (\Sigma_2, \varphi_2) \tag{5}$$

if the homeomorphism (2) can be continued to a homeomorphism from S_1 to a proper part of S_2.

By Lemmas 2 and 1, relation (5) gives rise to a partial ordering in the family of all continuations of (1). Furthermore, any linearly ordered subfamily

$$(\Sigma_\lambda, \varphi_\lambda) = \{S_\lambda; f_\lambda^\alpha; \varphi_\lambda\} \tag{6}$$

has a least upper bound. In fact, for $\lambda < \mu$, $\varphi_\mu \circ \varphi_\lambda^{-1}$ is a mapping from S_λ to S_μ and this gives rise to a projective limit S_∞ which is the base space in a continuation

$$\{S_\infty; f_\infty^\alpha; \varphi_\infty\}$$

which is the least upper bound of (6).

By a familiar reasoning,[4] we obtain the following conclusion.

THEOREM 1. *Any system (1) has at least one continuation which is maximal in our partial ordering.*

The maximum of Theorem 1 is not unique; in fact, the two Weierstrass continuations of the above *Counterexample* are both "intrinsically" maximal in the sense of Theorem 1. In order to obtain an assertion with uniqueness, we are introducing a restriction.

Definition 2: A system (1) has the *separation property* if the following conditions (*i*) and (*ii*) are fulfilled.

(*i*) If $P_1 \in S$, $P_2 \in S$, $P_1 \neq P_2$; and if U_1 and U_2 are neighborhoods of P_1 and P_2, respectively; then there does not exist a homeomorphism φ from P_1 to P_2 such that

$$\varphi(P_1) = P_2 \tag{7}$$

and

$$f^\alpha(\varphi(P)) = f^\alpha(P); \qquad P \in U_1,\ \alpha \in A.$$

(*ii*) If $P_1 = P_2$, then the only such homeomorphisms are identity mappings of U_1.

THEOREM 2. *If the system (1) has the separation property, then the family of continuations with separation property is not only a partially ordered set, but even a directed set; and, again, any linearly ordered subset has a least upper bound.*

Thus, within the family of continuations with separation property, the system (1) has a maximal continuation, and the maximum is unique.

Proof: We will only comment very briefly on the assertion that corresponding to any two continuations (3) and (4) with separation property, there exists a third

$$(\Sigma_3, \varphi_3) = \{S_3; f^\alpha{}_3; \varphi_3\}, \tag{8}$$

of the same kind, for which

$$(\Sigma_1, \varphi_1) \leq (\Sigma_3, \varphi_3),\ (\Sigma_2, \varphi_2) \leq (\Sigma_3, \varphi_3). \tag{9}$$

There is even a smallest ensemble (8) of this kind, and its base S_3 is obtained as follows. We form the disjoint logical union of S_1 and S_2, and then perform the following identification. We identify a point $P_1 \in S_1$ with a point $P_2 \in S_2$ if there is an isomorphism φ from a neighborhood U_1 of P_1 to a neighborhood U_2 of P_2 such that we have (7) and

$$f_1{}^\alpha(P) = f_2{}^\alpha(\varphi(P)); \qquad P \in U_1,\ \alpha \in A.$$

The reader will verify that our Definition 2 was designed to make this S_3 into a base space, and that an ensemble (8) with the property (9) is created in this way.

Definition 3: The maximal continuation of Theorem 2 will be called the *natural completion* of the system (1).

THEOREM 3. *The envelope of holomorphy of a (one-sheeted) domain D of complex C^n is the natural completion of the system (1) in which $S = D$, and $\{f^\alpha\}$ is composed of all one-component holomorphic functions, scalars, or densities; the latter being one-component functions which under a local change of coordinates are multiplied by a (positive or negative) integer power of the Jacobian.*

In particular, a one-one holomorphic map of the domain D into another domain of *C^n, also maps the corresponding envelopes of holomorphy, even if the latter are many-sheeted.*

Proof: Since D is covered by one coordinate system, there is no "practical" distinction between scalars and densities, and this continues to be so for the envelope of holomorphy $\tilde{D}$ of D, even if $\tilde{D}$ is many-sheeted. Next, since D is one-sheeted, the n coordinate functions

$$q^1 = z^1, \ldots, q^n = z^n \tag{10}$$

guarantee the separation property for the system (1) of Theorem 3. If $\tilde{D}$ is multi-sheeted, the continuation of the functions (10) guarantees only part (*ii*) of Definition 2, but part (*i*) is supplied precisely by the requirement, à la Weierstrass-

Poincaré, that if two points of $\tilde{D}$ have the same projection in C^n, at least one function f^α must have different functional elements at the two points. Therefore $\tilde{D}$ gives rise to a continuation in the abstract sense of Theorem 2. Conversely, on the natural completion of (1) there are continuations of the functions (10). But their Jacobian

$$\frac{\partial(q^1, \ldots, q^n)}{\partial(t^1, \ldots, t^n)} \tag{11}$$

is a density, and so is its reciprocal. Therefore the reciprocal exists as a holomorphic function, which means that on the natural completion the continuations of the functions (10) continue to be *locally* functionally independent. Therefore the base space of the natural completion can be spread out "regularly" over C^n, and is thus a part of $\tilde{D}$, which proves the theorem.

Remark 1: If D itself is many-sheeted to begin with, then the coordinates (10) guarantee at any rate part (*ii*) of Definition 1. If now there are enough holomorphic functions on D to supply part (*i*), then Theorem 3 remains in force.

Remark 2: If, however, there are not enough holomorphic functions on D to do that, then we may first create a new domain D° by identifying points of D in which all holomorphic functions, including (10), have the same functional elements, and now[5] Theorem 3 applies to D°.

Remark 3: The original Weierstrass continuation of a holomorphic function f, say in a domain D of C^m, becomes a natural completion if we introduce the system (1), in which $S = D$, and $\{f^\alpha\}$ consists not only of the given function f, but also of the functions (10), and of the Jacobian (11) and its reciprocal. It is thus highly "nonintrinsic" as far as the function f itself is concerned.

Remark 4: Since meromorphic functions, as generalizations of holomorphic ones, are also subject to Lemma 1, Theorem 3 also applies, literally, to envelopes of meromorphy.

* This research was supported by the U. S. Air Force Office of Scientific Research.

[1] Bochner, S., "Fortsetzung Riemannscher Flächen," *Math. Ann.*, **98**, 406–421 (1927).

[2] Behnke, H., and P. Thullen, "Theorie der Funktionen mehrerer Komplexer Veränderlichen," in *Ergebnisse der Mathematik und ihrer Grenzgebiete* (Berlin: Springer, 1934), vol. 3, especially pp. 70–76.

[3] Weyl, H., *Die Idee der Riemennschen Fläche*, many editions.

[4] This is the so-called "Zorn's lemma." It began to be thus designated in the mid-1930's, but it had been formulated, very expressly, in our ref. 1.

[5] See the article of Wightman, A. S., "Analytic functions of several complex variables," in *Relations de Dispersion et Particules Élémentaires*, Université de Grenoble, Ecole d'été de physique théorique, Les Houches (Paris: Hermann, 1960), pp. 249–250.

SINGULARITIES AND DISCONTINUITIES*

by Salomon Bochner

Singularities of analytic spaces, and of objects on them, are becoming ever more the vital subject matter of the theory of analytic spaces and functions in several complex and also real variables. Almost anything presently under active research in complex analysis falls, or can be subsumed, under the study of singularities, in one form or another.

Most large-scale topics of present-day mathematics were engendered in the nineteenth century, and so was also the topic of singularities. But in the nineteenth century, except in the uncanny vision of a Riemann, singularities were generally felt to be oddities and curiosities, even abnormalities, and their habitat was in marginal areas of mathematics. In the course of the twentieth century however, singularities have become a mainspring of vital normality and the staff of mathematical life, and they now dwell, ever more dominantly, in the very heartland of analysis. As we will later point out in succinct detail, there were analogous changes of outlook and attitude in other areas of knowledge too.

In the Late Victorian and Edwardian eras, complex analysis prided itself on being "beautiful" and "elegant," and singularities, being rarely either, did not quite fit into the tableau. But in the course of the twentieth century, mathematics has been reorienting itself from "beauty" and "elegance" towards relevance and cogency, and the study of singularities has profited immensely thereby. To judge by nascent future-oriented developments of today, the mathematics of the twenty-first century bids fair to be austere and purposeful rather than beautiful and elegant. On the one hand, in avant garde cosmology there are rumblings against presumptions of homogeneity, symmetry, and isotropy,[1] all naive concomitants of "beauty" and "elegance." On the other hand, "hard-nosed" analysts are turning ever more to the study of non-linear partial differential equations, although there is very little about the equations that an analyst not habituated to them can presently relish, syllogistically or esthetically. And please note that very few singularities are generated by homogeneities, symmetries, and isotropies,

* Research supported by NSF-GP-18961.

but that in solutions of non-linear partial differential equations singularities are likely to come up at every turn. Also, there may even be interesting solutions of non-linear partial differential equations that are dotted all over with singularities. Such would have to be solutions of the Navier-Stokes equations that describe phenomena of "turbulence," assuming that there are indeed such phenomena which can be so described.

Removable singularities

Although the nineteenth century did not yet make a proper assessment of singularities, it did knowingly introduce them, and in most contexts in which they occur. Bernhard Riemann even devoted his famed doctoral thesis[2] to them, for objects in one complex variable, that is. The thesis features the theorem on removable singularities named after Riemann, and also the construction of covering surfaces, also named after him, whose purpose it is to "regularize" algebraic singularities rather than "remove" them.

In an early article of the German encyclopedia of mathematics, W. F. Osgood credits the theorem on removable singularities not to Riemann but to Weierstrass,[3] and he does so for the following reason. The theorem can be obtained by showing that if the sum of a Laurent series in a punctured disk is bounded, then the coefficients of the negative powers must be zero; but this statement about Laurent series occurs in a memoir of Weierstrass[4] which, although published after Riemann's thesis, is reputed to have been composed before. Now, this may have indeed been so, but it nevertheless does not deprive Riemann of the credit for the theorem named after him. The fact is that the statement of Weierstrass does not at all have the "feel" of a theorem on "singularities," removable or other. Nor does it presage a future importance of this kind of theorem, as does the thesis of Riemann, in which Riemann immediately starts extending the theorem to solutions of partial differential equations. In the twentieth century this extension has become the "true" version of the theorem on removable singularities,[5] and the spirit of Riemann is hovering over the developments.

Cayley and after

Riemann's thesis, which does not speak of "singularities" but of "discontinuities," appeared as a pamphlet in 1851. A year later the English noun "singularity" appeared, as a technical mathematical term in the title of a note of Arthur Cayley.[6] The author states that, in the case of a curve, a singularity may, for instance, be a double point, or it may be a cusp. He proposes to call a double point a "node" (without, however, making it clear that he is the first to do so), and the second a "spinode." He also suggests the use of several other terms which have not become generally accepted; for instance, he suggests "flecnode" for a double point which is a point of inflection.

Our reference to Cayley's note is taken from the *Oxford English Dictionary*. There seem to be no similar references in large-scale dictionaries to first prominent mathematical occurrences of *singularité* (French), *singolarità* (Italian), and so on. However, it would probably be possible to ascertain when *point singulier* came into use in French differential geometry.

In algebraic geometry of one or several complex variables the general notion of a singularity has not changed materially since Cayley's note. Singularities are still being classified, and they are being "removed" or "simplified" by the use of birational transformations which need not be topological.

In complex analysis today, the basic attitude towards singularities is a different one. Externally, by immediate structure, at least in the "local" case, it is hard to distinguish between singularities in the general holomorphic case and the particular algebraic case, whether it be singularities of functions, or, what is more significant, of spaces. However, the motivation is a different one. General complex analysis is not anxious to "remove" singularities but to "live with" them. It wants to do analysis with them and on them, and it wants to extend properties of smooth spaces to spaces with singularities as far as possible.

Cosmology

In the future, say in the twenty-first century, these differences of approach may be of consequence. A future cosmology may want to view the spatial substratum of the universe together with force and potential fields on it as a "high-dimensional" fiber bundle with an analytic structure, and galaxies or clusters of galaxies as singularities in the bundle. Also, following a precedent set by Albert Einstein himself, even the injection of some features of complex structure might suggest itself.[7] Now, such a cosmology would undoubtedly hesitate to start out with general birational transformations that interfere with the size and distribution of the singularities, that is, the galaxies, but would want to apply statements about global analytic behavior in which the galaxies remain as they are; although, eventually, speculations on nontopological birational "modifications" of singularities (= galaxies) might have an audience too.

Recent developments in relativistic geometrodynamics are perhaps harbingers of such future events.[8] Singularities of space (or space-time) have been occurring in cosmological models since the inception of General Relativity. From the first, there have been points at which the space ceases to have a Euclidean structure or at least the line-element ceases to be regular; and such singular points might vary with (cosmic) time, due to explosions or implosions. Still, until recently, such points were mathematically of the classical differential-geometric variety. But the geometrodynamical "worm holes," and their cognates the "black holes,"[9] are of a more future-oriented kind.

Physics

Actually, in order to highlight the role of mathematical singularities in recent science it even suffices to stay within physics proper without going into cosmology. Already the nineteenth century had occasion to view the essential sources of matter as a kind of singularity. But these singularities manifested themselves in the structure of matter, and not, or not yet, in the structure of the space in which matter was located. But nowadays there is a tendency to envisage the singularities in the very structure of the underlying space itself, which is usually an analytic space, on the whole. On this developing view, the elementary particles in the universe, of which there are enormously many, are as many point singularities in the spatial substratum; and whenever a particle is in motion, the singularity is a moving one.

Very intriguing is the Yukawa potential

$$M \frac{e^{-mr}}{r},$$

which describes the meson field about a nucleon. Near the origin $r = 0$ it behaves like the Newtonian potential in R^3, but for large r it falls off faster than $1/r^{n-2}$ for any n, that is faster than the Newtonian potential in R^n, for any n. This suggests that the nucleon "creates" for and by itself a space which near the nucleon is "approximately" three-dimensional but at distant points acts as if it had a rapidly rising dimension. A singular variety in an infinite dimensional space might conceivably behave this way.

I have just used the word "approximately" advisedly. As Heinz Hopf once observed to me, in a casual exchange of small talk, it is remarkable how "everybody" would agree that our Earth is "approximately" a solid ball, although due to tunnels, bridges, air-pockets and suchlike it is, in fact, homologically very complicated indeed.

Incidentally, these occurrences of singularities in cosmology and physics suggest that mathematically singularities will continue to be hard and difficult for some time to come. If singularities of analytic spaces were easily resolvable, then so might be the structure of elementary matter — in its mathematical aspects at any rate — but of this there is no indication. The nuclear structure of protons, electrons, photons, gravitons, etc., is as mysterious and mystifying as ever.

Condensation of singularities

In 1870, almost two decades after Cayley's abovementioned note, a different mathematical meaning of "singularity," much less specific but quite emphatic, was introduced by Hermann Hankel in a memoir about infinitely oscillating and discontinuous functions.[10] A memorable feature of the memoir is the proclamation of the Principle of *Condensation of Singularities*, literally so named by the author himself.[11] Hankel credits the inspiration for it to a construction in "art. 6" of Riemann's unique memoir on trigonometric series,[12] and he obviously means Riemann's construction of a (Riemann) integrable function which has a dense set of discontinuities. The idea of Hankel's principle is indeed there, but in a setting which would be somewhat unfamiliar to a reader of today, the reason being that nowadays Hankel's principle is usually combined, more or less intimately, with the ternary set construction of Georg Cantor, whereas in Riemann's work it is not, or not yet. In fact, Cantor's ternary set, Hankel's principle, and occasionally the Hamel basis are the building blocks for most counterexamples in C. Carathéodory's leading treatise on functions of a real variable.[13]

Singularities as boundary points

Several years after Hankel a third connotation of the mathematical meaning of "singularity" was introduced knowingly by K. Weierstrass, this one into "classical" complex analysis.[14] His formulation is gauche and limited, and nothing would be gained for our purposes by reproducing it literally. Rather we will represent it immediately in a modernized form and comprehensive version. We will deal immediately with the case of many variables, and with any analytic structure, complex or real.

Let D be an analytic space in any number of dimensions and let it be a domain in a larger analytic space $\tilde{D}$. Let $f(P)$ be an analytic function or some more general function-like object on D, and let P_0 be a boundary point of D *relative to* $\tilde{D}$. Then, *relative to* $\tilde{D}$, $f(P)$ is deemed to be regular at P_0 if there exists an inbetween space D',

$$D \subset D' \subset \tilde{D},$$

which contains the point P_0 such that $f(P)$ has an analytic continuation into D'. If no such D' exists, then P_0 is deemed to be a singular point of f(P) *always relative to* $\tilde{D}$. Also, in such a setting, the singularity at P_0 is *inessential*, if there exists a neighborhood N of P_0 in $\tilde{D}$ and a pair of analytic functions $\phi(P)$ and $\psi(P)$ in N such that, in $N \cap D$, $\psi(P)f(P) = \phi(P)$.

In the case of one complex variable there is one situation in which the choice of $\tilde{D}$ is totally immaterial. Let D' be some (abstract) Riemann surface, P_0 a point of it, and let

$$D \overset{\text{def}}{=} D' - \{P_0\}.$$

Obviously a holomorphic function in D can be viewed as a function in D' which is holomorphic everywhere except for a possible isolated singularity P_0. Such a function has a Laurent expansion at P_0, and if there is a singularity then it is inessential iff there are only a finite number of negative terms in the expansion. This is no longer so if one replaces the holomorphic complex variable $z = x + iy$ by the analytic pair of real variables (x,y) because, for instance, the transformation

$$u = \frac{x}{\sqrt{x^2+y^2}} + x, \qquad v = \frac{y}{\sqrt{x^2+y^2}} + y$$

is an inversely analytic one-one map of the punctured disk $0 < x^2 + y^2 < 1$ into the annulus $1 < u^2 + v^2 < 4$.

A topic in one complex variable in which, on the contrary, the choice of $\tilde{D}$ is all-important is the theory of Taylor expansions, or, what is the same, of functional elements. In this case, D is a disk $|z| < r_0$ in a fixed coordinate system, that is in C^1, and $\tilde{D}$ is C^1 itself. If the disk D is replaced by some other domain D^0 in C^1 which is bounded by a closed Jordan curve, and if a function $f(z)$ on D is transferred into D^0 by a one-one holomorphic map of D into D^0, then the distribution of regular and singular points of the function on the boundary may change radically and unrecognizably, although the mapping is continuous on the boundary too.[15] Some theorems, as for instance those that refer to star-shapedness, may be affected even if the disk is mapped into itself by a linear transformation. It is this all-pervading feature of "noninvariance" which probably accounts for the fact that the theory of Taylor expansions, which was in vogue for nearly a century, has become rather quiescent lately.

In the case of one complex variable, another subject bearing on singularities which has become rather neglected, almost defunct, lately, is the problem of finding a proper conception and definition of a singularity of a function on the boundary of a domain, if the domain is not a single-sheeted part of C^1 but, more generally, a multisheeted spread over it. L. Bieberbach in particular made furious efforts to find the "true" definition of a singularity in this more general situation, but the outcome of his labors has been neither felicitous nor of any consequence.[16]

Holomorphic extension

A striking difference between one and several complex variables is the total absence of a two-dimensional analog to the one-dimensional phenomenon of conformal mapping. But this difference is more striking than meaningful. The difference is too obvious, too abrupt, too nonnegotiable. In analysis there are usually transitions, generalizations, and reinterpretations. It is a curious fact that Riemann himself made no profound study of conformal mapping. He simply credited Dirichlet with having "solved" the boundary value problem of uniformization theory, and he himself kept his attention on matters beyond this fact.

Much more illuminating are the differences in the nature of singularities. It is true that singularities are harsher and more difficult for two and more variables than for one variable, but they also become more diversified and interesting, and there is a gradation of difficulties and of interest with rising dimension. For instance, the two-dimensional situation is still "exceptional" in a sense, inasmuch as, in this case, subspaces of codimension one — which are frequently of some importance — have still dimension one themselves. Furthermore, as if to compensate for novel difficulties in the field of singularities, the multidimensional situation also revealed a novel phenomenon in this field, namely the phenomenon of "holomorphic extension," which is a phenomenon of *non*-singularities, as it were. Also, not only was this phenomenon unknown to classical complex analysis in one variable, in which it does not occur, but also it was not observed in algebraic geometry of several variables either. The phenomenon was revealed in a massive memoir of Fritz Hartogs,[17] and some follow-up papers. Hartogs not only was not an algebraic geometer, but was in fact a loyal disciple of a staunch real-variables man (Alfred Pringsheim).

Now for a brief description of holomorphic extension. If D is a domain in C^1, then corresponding to any boundary point P of D there is a function in D which is singular at P. By a judicious application of Hankel's principle of condensation of singularities there exists a function in D which is singular at every boundary point of it. Not so, for two or more complex dimensions.

"Most" bounded domains D in $\mathbf{C}^n$, $n \geqq 2$, are such that all holomorphic functions $f(z)$ on D can be continued simultaneously into some joint domain D' larger than D. For given D there is even a largest such domain, $\tilde{D}$, and it is the so-called envelope of holomorphy of D. This domain $\tilde{D}$ need not be single-sheeted in $\mathbf{C}^n$. But it has a certain feature of convexity which is retained even when many-sheeted.

The most eye-catching instance of this situation is the so-called Theorem of Hartogs. It states that if a bounded domain D in $\mathbf{C}^n$, $n \geqq 2$, has a connected boundary B, and if N is a neighborhood of B, then any $f(z)$ which is defined and holomorphic in N can be continued into all of D: meaning that the envelope of holomorphy of N includes D. In term of singularities it means, surprisingly, that the attempted continuation of $f(z)$ along any path leading from N into D can never hit a singularity and is thus always successful.

A certain phenomenon involving holomorphic extension, the so-called "Edge of the Wedge" phenomenon, has already been used in quantum field theory.[18] Also, an envelope of holomorphy may be defined and shown to exist, even if D is a bounded set of a certain kind, not necessarily a domain itself.[19] Thus, in a cosmological universe with complex structure, a galaxy, or a cluster of galaxies, even if deemed to be lower-dimensional, might exercise a "holomorphic" control over an open domain enveloping it.

The concept of function

I already noted above that, in my interpretation, Riemann's creation of Riemann surfaces was responsive to the needs of singularities. If this was indeed so then, indirectly, singularities also had a very creative influence on the formation of our general concept of a function.

Before Riemann, in the thinking of Abel when creating the Abelian integrals, and in the thinking of Gauss when creating the functions defined by hypergeometric series, a function, especially a holomorphic function, was somehow given in the entire complex plane and precisely there, even if there were singularities at some points. But Riemann pioneered in the recognition that in order to create or define a function there first of all must be given a domain space, *the* domain space, on which it is to exist. It is this recognition that made Riemann create, at the very outset of his career, his multi-sheeted surfaces. And he more or less created them for the functions which Gauss and Abel had just introduced and not for some pathological functions which Gauss and Abel would have shrugged off.

Nobody sensed such things, in depth, before Riemann. Newton, for instance, constantly had functions in his thinking, and it is this which separates Newton from Archimedes, who had no functions in his thinking at all. But Newton took it for granted that a gravitational function field exists "everywhere" in absolute space, except at "singular" points, of course, although in his undivulged thoughts he may have wondered whether the actual physical universe does indeed extend so far.

Riemann's pioneering insight into the nature of function revealed itself not only in the case of a complex variable but also in the case of a real variable. In his one great excursion into "pure" real variables, namely in his memoir on trigonometric series,[20] Riemann created, perhaps for the first time, a sharply defined class of functions to fit precisely a given purpose, namely his Riemann integrable functions. And, in his own idiom, he established the fact that a bounded function on an interval is Riemann integrable if and only if the set of its discontinuities (alias singularities) is a set of Lebesgue measure 0.

"Discontinuity" in the nineteenth century

As noted before, Riemann does not speak of "Singularität" (= singularity) but of "Unstetigkeit," which is the native German word for "discontinuity." The German language also uses very freely the "borrowed" word "Diskontinuität," and Riemann uses it (at least once) in his fundamental paper on hyperbolic differential equations which was also a seminal work on shock waves.[21] Our expression "removable singularity" is a (would-be) translation of Riemann's expression "hebbare Unstetigkeit." The verbal form "hebbar" (= "liftable"), which comes from the verb "heben," was apparently Riemann's coinage. Its meaning in German is immediately and unmistakably understood, but the form is not in general use, and non-specialized German dictionaries, however large, do not carry it.

Nowadays "Unstetigkeit" may sound softer than "Singularität," but one must not assume that such was also the case in the mid-nineteenth century when Riemann was active. At that time the prevalent dictionary meaning of "singularity" was: "odd," "eccentric," "unusual," mostly in a rather "benign" sense; whereas "Unstetigkeit" had the meaning of "disruption," "abnormality," and suggested something jarring, discomfiting, disturbing, even calamitous. This was so in spite of the fact, or perhaps because of the fact that in the nineteenth century continuity was, philosophically, at any rate, the normal state, the state to be expected, and discontinuity was never felt to be a state entirely in its own right but somehow a deviation and aberration from a state of continuity. A state of discontinuity, however remarkable, was usually contrapuntally submerged in a state of continuity which somehow was the dominant one.

Continuity in nineteenth-century knowledge

Karl Marx's doctrine of a violent revolution to come, which was proclaimed in the middle of the nineteenth century, was a socioeconomic discontinuity on a grand scale, perhaps the greatest such discontinuity ever conceived philosophically. Yet this seemingly unyielding discontinuity was installed in a frame of continuity, inasmuch as the revolution was historically rationalized and dialectically "Hegelianized" to be sociologically predetermined and politically inescapable.

Altogether, in the nineteenth century, all of historiography was dominated by continuity of one kind or another. Continuity was a framework for, or at least a leading strand in, any history, ancient or modern, general or special, political or economical, Biblical or philosophical. One of the leading concerted efforts of the century was a great rehabilitation of the Middle Ages. It had been begun in the eighteenth century, but then by individual euphoric Romantics, whereas in the nineteenth century it was continued by generations of sober professionals; and a leading motivation was to bring out, by similarities and contrasts, the maximum continuity of development between Ancient and Modern Times across the "intercalated" Middle Ages. The effects of these presumptions of continuity were mixed. Some presumptions were overt, some covert; some good; some not so good; some crude, some subtle; some refined, some highly refined; but the presumptions were always there.

Continuity was also triumphant in nineteenth-century biology and geology. In biology, evolution was first Lamarckian and then Darwinian, and both were emphatic avowals of continuity. In a parallel development, geology adopted, somewhat less enthusiastically but rather tenaciously, so-called "uniformitarianism," which is a form of continuity. It is an obstreperous mode of continuity, very hard to define consistently and convincingly. Yet the negation of it seems alarming and is ominously named "catastrophism," as if to stigmatize it and warn the unwary against flirting with it.

In basic physical science, the nineteenth century well knew that any kind of matter has a discrete constitution, being composed, ultimately, of "indivisibles." Chemistry based itself on atoms and molecules, heat was interpreted to be kinetic energy of large assemblages of discrete units of matter, and even electrodynamicists like Wilhelm Weber knew, however inarticulately, that there are ultimate discrete particles of electricity. Yet macrocosmic features of continuity prevailed, both in phenomena and explanations, so that in leading statements about elasticity, chemistry, and thermodynamics, matter was, in the end, spoken of as if it were distributed continuously. Continuity was especially victorious in electrodynamics. Maxwell's own theory — which was so impressive that it became and remained a hallmark of nineteenth-century physics in its entirety — was a pure field and wave theory, with nothing noncontinuous anywhere in sight. The theory was especially applauded for the boldness with which it theoretically subordinated the ancient doctrine of optics under itself, thus establishing that light is undulatory, that is continuous. The long-drawn uncertainty as to whether light is corpuscular or undulatory thus seemed finally settled, with continuity triumphant, of course.

Continuity in nineteenth-century mathematics

In mathematics, the nineteenth century finally came to grips with the phenomena and conception of continuity, frontally, directly, effectively. Greek mathematics, the much-vaunted, dared not do so. Greek rational thought as such was able to bring forth the conception of continuity, the name for it being *synechés*: This term already occurs in Homer; and Aristotle in his *Physica* uses it freely in our present-day sense, more or less. But, unbelievably, Greek mathematics could or would not use it. There is not a trace of a hint of a *direct frontal* use of either the term or even the conception anywhere in Euclid, Archimedes, Apollonius, and others. An explanation of this is a leading problem in the History of Ideas.

The Middle Ages nibbled away at the conception, just enough to intrigue historians but not more; and in the seventeenth and eighteenth centuries, mathematics was much too busy plumbing the depth of the conception *im*plicitly to find time for analyzing it *ex*plicitly. But the nineteenth-century finally went to work.

In the first half of the century Cauchy and others, at long last, gave a workable definition for the continuity of an "ordinary" numerical function. And in the second half of the century, Dedekind and Georg Cantor, at long last, gave a workable characterization of the linear continuum, a characterization which had been gestating for almost 24 centuries, certainly since Aristotle's *Physica* (ca. 350 B.C.), and perhaps already since Zeno of Elea's indestructible "puzzles" (ca. 450 B.C.). In between these achievements in the two halves of the century, the developing theory of functions of a real variable illumined many facets of continuity and discontinuity. Finally, towards the end of the century, the emergent systematic topology began to study continuity in a novel way, in its basic structure and pristine purity, as an appropriately circumscribed mapping from one topological space into another; and systematic topology spent the first half-century of its existence doing nothing other than exploring, exploiting, and diversifying this approach to continuity. After that, in the 1930's topology bethought itself of the fact that analysis does not live on continuity alone, but that it also *must* perform certain operations, notably differentiation and integration. Topology's response to this challenge was the conception of a differentiable structure and the creation of cohomology, and after that topology again became the tool and center of analysis, as it had already been in the thinking of Riemann and of some other great complex-variable analysts of the nineteenth century.

Duality in twentieth-century physics and mathematics

In the twentieth century, continuity's claim to exclusiveness has been challenged in virtually all areas of knowledge and insight, and, in outcome, continuity has been shorn of its exclusiveness though not in the least reduced in its importance. In most academic fields, scientific and humanistic, continuity has been compelled to accept an accommodation with certain features of discontinuity, and, on the whole, a salutary balance between continuity and discontinuity has come about.

The century's most challenging event of this kind was the development of quantum theory in basic physics. A pioneering quantum hypothesis was proposed first by Max Planck. He introduced it rather suddenly, at the very threshold of the century, in November 1900, into the process of black body radiation. At the time, this was a somewhat isolated topic in physics, not part of a standard curriculum. But from there the hypothesis gradually spread by way of more central parts of physics into its very roots, threatening to undermine the entire edifice of exact science and its metaphysical foundation. The metaphysical issue involved seemed formidable. However, unlike the general experience with metaphysical issues, this one was resolved, felicitously and clearly, almost simplistically, by a "decree" of Louis de Broglie. The decree promulgated a comprehensive "duality principle" by dint of which every elementary particle whatsoever has — Janus-like fashion — two aspects, an undulatory (continuous) and a corpuscular (discrete). The decree does not specify whether or when the two aspects are to be contemplated alternately or simultaneously, in coexistensive or complementary realizations, but practicing physicists seem to have no difficulties at all deciding this for themselves.

This startling duality in physics has a paradigmatic counterpart in mathematics, a very un-startling one. But the duality in mathematics is sprawling many-faceted, and wholly un-unified. And since the mathematical paradigm is considerably older than the insight in physics, it suggests itself that perhaps the de Broglie version is only the first in a series of future ones. At any rate, in mathematics there is no "central" pronouncement like de Broglie's, and philosophers seem not to be aware of it at all.

In mathematics, the most "typical" version seems to be the duality between a periodic function and its set of Fourier coefficients. In a Sturm-Liouville setup it is the duality between the set of eigenfunctions and the set of eigennumbers. It is the duality between homology and cohomology, contragradience and cogradience, inverse mapping and forward mapping, a topological vector space and the space of its linear functionals. It is the duality between the two sides in any version of the Poisson summation formula, between the two power series in any Jacobi theta relation, and the two Dirichlet series in any Riemann-Hecke functional equation. Also some of the versions just enumerated are subsumable under Cauchy's residue formula

$$\oint_C f(z)dz = \text{sum of residues of } f(z) \text{ inside } C.$$

This formula itself is a duality relation, even emphatically so, since the sum on the right side is unequivocally discrete, and thus indubitably "discontinuous," but the "sum" on the left, being an integral, is indubitably continuous. Also, the link between the two sides is established by singularities, namely the singularities of the function f.

It is noteworthy that there is no genuine generalization of the residue formula from one to several variables, and what is missing in the setting of several variables is not so much the analog to the formula itself as a certain syllogistic apparatus in which the formula is involved. Consequents of this "defect" manifest themselves in seemingly distant areas of real variables and we are going to adduce one such case.

The "harmonic" system of functions

$$e^{inx}, \qquad n \in \mathbf{Z}$$

is a complete base in the L^2-space of square-integrable functions on $0 \leqq x < 2\pi$. Paley and Wiener have shown[23] that the "modified" nonharmonic system

$$e^{i\lambda_n x}, \qquad n \in \mathbf{Z}$$

remains complete on that interval provided

$$|\lambda_n - n| \leqq D < \frac{1}{\pi^2}, \quad n \in \mathbf{Z}.$$

N. Levinson showed afterwards[24] that the constant $1/\pi^2$ can be replaced by the smaller constant $\frac{1}{4}$; that the statement also holds for L^p, $1 < p \leqq 2$, the constant being $\dfrac{p-1}{2p}$; and that the latter constant is "precise." Now it suggests itself that for square-integrable functions $f(x,y)$ in $0 \leqq x < 2\pi$, $0 \leqq y < 2\pi$ it ought to be admissible to "modify" the basis

$$e^{i(mx+ny)}, \quad (m,n) \in \mathbf{Z}^2$$

into

$$e^{i(mx+ny)}, \quad (m,n) \in \mathbf{Z}^2,$$

provided

$$(\lambda_m - m)^2 + (\mu_n - n)^2 \leqq D^2$$

for some constant $D > 0$. But this most intriguing problem has apparently not been attacked thus far. At any rate, the proof for one variable relies heavily on complex methods which fall into the area of attraction of the residue theorem and does not carry over to the case of two variables.

Synthesis of continuity and discontinuity in other areas of knowledge

In no other area of knowledge has the compromise between continuity and discontinuity been as spectacularly successful as in physics, but efforts to find one have been made almost everywhere.

In biology, the outgoing nineteenth century knew that Darwinian evolution would have to come to terms with discontinuity, as outwardly evidenced by the fact that for instance, a serious contemporary work in the area had the word "discontinuity" in its subtitle.[25] The efforts lasted for decades. Thus, in the words of a dedicated follower of Darwin, "during the first twenty years of the twentieth century, evolutionary studies and theories were in a state of chaos and confusion."[26] The difficulty was that, on the face of it, the basic trait of continuity that is inherent to evolution is in discord with the traits of discontinuity that are inherent to chromosomic mutation and Mendelian inheritance. But in the end, a compromise of interpretation was worked out by which, on the contrary, chromosomic mutation and Mendelian inheritance are the very instrumentalities by which "continuous" evolution is rationalized and enabled to function.[27]

In geology developments have been less felicitous, or at least, less harmonious. There are two contrasting hypotheses, Catastrophism and Uniformitarianism, the first allowing for violent changes, the second insisting on dominant traits of continuity, The nineteenth century, after some early avowals of catastrophism, became ever more uniformitarian. In the twentieth century geologists have been clinging to the hypothesis of uniformitarianism in the face of great difficulties of maintaining it or even of defining it.[28] In midcentury the public became aware of such difficulties when Immanuel Velikovsky, with his sensational book *Worlds in Collision* (1950), began to flout certain academic theories of cosmology and geology, all attuned to uniformitarianism more or less. The result has been that most geologists are more intransigent in their attitude, perhaps because it would be even harder to agree on alternatives to uniformitarianism than on uniformitarianism itself.

From general history I will mention two major injections of discontinuity into continuity, one very "public" and one very "scholarly." The public one is a compound of the theses of Oswald Spengler (*The Decline of the West*) and of Arnold Toynbee (*A Study of History*), both of whom maintain that history should not be viewed as an overall flow in one all-embracing continuous stream, but as a pattern of separate, or at least separable, large-scale units of civilization, each with a theme or leitmotif of its own, even if they may overlap and partially merge or fuse. The scholarly one is a gradual rehabilitation of Herodotus in comparison with Thucydides. In the Late Victorian and Edwardian eras some classicists were vying with each other in "reducing" Herodotus to the role of a mere "story-teller," that is, a mere narrator of episodically disconnected — or at least not properly connected — contexts of events; while Thucydides was glorified as really the first Greek historian by virtue of his presenting everything in a maximally integrated design which is unfolding inexorably. But there are signs nowadays that this imbalance of judgment is being redressed[29] and that Herodotus is being restored to his place as "father of history," even if with some qualifications that cannot be gainsaid. Except for Old Testament historians before him, Herodotus was indeed the father of history, general and social, even if Thucydides was in a sense pioneering in *political* history.

I will also cite an instance from specialized histories. In the Victorian and Edwardian eras, historians of science inclined towards *inductivism*, which presumes that "science advances gradually from observation to theoretization, and that it does so univalently, forcibly, and unerringly," and that "an experimenting or observing scientist can report on facts 'faithfully' without, at all, rendering an 'opinion' on them."[30] Present-day history and philosophy of science cannot accept such a gradualism of approach, and for several decades now they have been groping, as yet inconclusively, for more complex explanations by which spontaneity of cognition and suddenness of insight can be accounted for.

A "discontinuity" in the rise of modern mathematics

By modern mathematics I mean mathematics since around 1600. As I have stated before, throughout the duration of the eighteenth century "there was a near-perfect, and richly yielding, fusion between mathematics and mechanics,"[31] but in the early nineteenth century the two unmistakably separated from each other; there is no explanation for this at all, especially since many of the leading analysts of the nineteenth century continued to contribute a great deal to mechanics. The period of transition, during which the separation took place, falls into an era which I have termed the "age of eclosion" which for *general* purposes I extended from 1776 to 1825.[32] For our *present* purposes we may delimit it as the shorter era 1785–1820, from the time when Lagrange wrote his *Mécanique Analytique* to the time when Cauchy wrote his *Cours d'Analyse*.[33]

Lagrange and Cauchy were both analysts, both within a common French development, and Cauchy seemingly continued where Lagrange had left off. Yet there is an enormous difference of mathematical style between them. However much Lagrange may be doing analysis it is somehow always mechanics, or at the very least he is doing it with mechanics in mind; a function is for Lagrange a mechanical path or orbit, and not the other way round. Cauchy however is always doing analysis, even if the setting is from general mechanics of continuous media, elasticity, or optics; many analysts are probably not even aware of the fact that a considerable number of Cauchy's papers have such settings. Between Lagrange and Cauchy, there was quite a line-up of great analysts: Laplace, Monge,[34] Fourier, Poisson, and others; but it does not seem easy to interpret their activity, by methodology, as a bridge between the late eighteenth and early nineteenth century. At any rate, there are no major studies bearing on this problem, and the extant works on history of mathematics are of no help.

If this problem is viewed as a problem in French mathematics, then the key figure in the period of transition is probably Poisson, whose mathematical interests are very broad. But inexplicably, there are no "Oeuvres Completes de Poisson" as there are for all other French mathematicians of comparable stature. Felix Klein in his history of nineteenth-century mathematics condemns Poisson for prolixity ("Weitschweifigkeit") and, as if in retribution, devotes to him less than one page.[35]

If the problem is viewed as a problem of mathematics at large however, then the key figure in the period of transition becomes C. F. Gauss. His achievements and works, unlike those of Poisson, have been receiving the widest publicity from the first. But as regards his intellectual origin and evolutionary spontaneity Gauss is nonetheless something of a mystery figure. Mathematically he somehow comes from nowhere, and there certainly is no evolutionary link between him and Leibniz, for instance and the reverence for Gauss is such that few would dare to "confine" him to the role of a figure of transition between two eras, however large the eras — although he is perhaps precisely that. Also, if one raised the question who of the two was the greater mathematician, Gauss or Riemann, then the instantaneous answer that it was Gauss, of course, need not be the right one.

Finale. Exotic spheres and singularities

After this digression into the past I dare to offer, as a finale, a fantasy about the present and perhaps the future.

On Weierstrass's view, any obstacle to the continuation of an analytic function is caused by a singularity. When we broaden this view, we may conceive of a singularity or of singularities whenever a function or function-like object of some kind, when given on a subset of a certain space, cannot be continued into all of the space, although *prima facie* it would be justified to expect that such a continuation would be possible. Solutions of differential and other equations are of this kind, and have more or less been known to be so for a long time. More recently topology has created situations of such a kind, except that it frequently speaks not of "singularities" but of "obstructions." In particular, John Milnor's "exotic" or "twisted" spheres have given rise to an occurrence of this kind, and this is the one on which I will comment.

Let S^n be a Euclidean n-sphere with its "natural" differentiable structure and let B^{n+1} be the $(n + 1)$-dimensional Euclidean ball bounded by S^n. An "exotic" or "twisted" structure on S^n — if it exists — is any C^1-structure on it which is C^0-compatible with the natural structure but not C^1-compatible with it. It has been shown by Milnor that for $n = 7$, for instance, exotic structures indeed exist. On the other hand it follows from a rather general proposition that on any S^n any exotic structure differs from the natural one by the fact that the exotic structure cannot be C^1-continued into all of B^{n+1}. In the parlance of singularities this means that any attempt to continue an exotic structure of S^n into all of B^{n+1} must encounter a "singularity" somewhere. Cosmologically we can spin these two facts out into the following fantasy.

Assume that the space (-time) framework of our universe together with some other physical data on it would be presentable as an S^7, and assume further that on this S^7 there physically exist some further force fields — the really decisive ones — and that these depend on the C^1 structure of the S^7 and are the ones by which "signals" in the universe can be created. By Milnor's result, one could have on the same S^7 two different force fields of the latter kind, belonging to different C^1-structures, and we assume that such is indeed the case. Two possibilities now arise, depending on whether man's perception is equally receptive to both sets of phenomena or prefers one over the other. If it is equally receptive to both, then man may perceive and experience two separate sets of phenomena which he can rationally fit into two separate theories, but not into one common one. In the second case he may experience two different kinds of phenomena. The phenomena of the first kind are clearly perceived and can be fitted into a comprehensive theory; but the phenomena of the second kind are only dimly perceived, and their comprehension is small and circumscribed.

Alternately, if the S^7 were not the universe itself but only the "rim of the universe," the universe itself being the B^8 bounded by it, then the inhabitants of B^8 might be able to receive two types of signals emanating from the rim of the universe, one "regular" and orderly corresponding to the natural structure on S^7, the other "singular" and chaotic corresponding to an exotic structure on S^7.

NOTES

1. G. de Vaucouleurs, "The case for a Hierarchical Cosmology," *Science* **167.2** (February 27, 1970), 1203–13.
2. "Grundlagen für eine allgemeine Theorie der Funktionen einer veränderlichen complexen Grösse," *Collected works* (Leipzig, 1892), p. 3.
3. "Allgemeine Theorie der analytischen Funktionen . . .," *Encyclopädie der Mathematischen Wissenschaften*, IIB1, p. 18.
4. Karl Weierstrass, *Collected Works* (Berlin, 1894), I, 63.
5. See S. Bochner and W. T. Martin, *Several complex variables* (Princeton, New Jersey, 1948), Ch. VIII; S. Bochner, "Weak solutions of partial differential equations," *Journal de Mathématique* **35** (1956), 193–202; F. Reese Harvey and John Polking, "Removable singularities of solutions of linear partial differential equations," *Acta Mathematica* **123** (1970), 39–56.
6. "On the singularities of surfaces," *Cambridge and Dublin Mathematical Journal VII* (1852), 166–171; also Cayley, *Collected Papers* (Cambridge, 1889), II, 28.
7. A. Einstein, "A generalization of the relativistic theory of gravitation," *Annals of Mathematics* **46** (1945), 578–584.
8. J. A. Wheeler, *Geometrodynamics* (New York: Academic Press, 1962).
9. See Roger Penrose, "Black holes," *The Scientific American* **226** (May 1972), 38–46.

10. "Untersuchungen über die unendlich oft oscillierenden und unstetigen Funktionen," *Gratulationsprogramm der Tübinger Universität vom* 6. *Marz* 1870 (Leipzig, 1893). This medium of publication made the memoir almost inaccessible even in Germany, but it was rescued from threatening oblivion by being reprinted, after Hankel's early death, in *Mathematische Annalen* **20** (1882), 164–212.

11. The author had a penchant for such principles. A few years earlier he had proclaimed the "Principle of Permanency of Mathematical Laws."

12. "Ueber die Darstellbarkeit einer Funktion durch eine trigonometrische Reihe," Riemann's *Collected Works*, pp. 241–271.

13. *Vorlesungen über reelle Funktionen* (Berlin, 1918).

14. "Zur theorie der eindeutigen analytischen Funktionen," *Mathematische Abhandlungen der Akademie der Wissenschaften zu Berlin*, 1876, 11–60; *Collected Works* (Berlin, 1895), II, 77–124.

15. Salomon Bochner, "The Rise of Functions," Proceedings of the Conference on Complex Analysis, Rice University, 26–29 March 1969, *Rice University Studies* **56**, No. 2, pp. 3–21, especially pp. 15–19.

16. "Neuere Untersuchungen über Funktionen von komplexen Variablen," *Encyclopädie der Mathematischen Wissenschaften*, IIC4, Section 8, pp. 401–403. See also L. Bieberbach, *Lehrbuch der Funktionentheorie*, 4th edition (Leipzig, 1934), I, 208–209.

17. "Zur Theorie der analytischen Funktionen mehrerer unabhängigen Veränderlichen, insbesondere über die Darstellung derselben durch Reihen, welche nach Potenzen einer Veränderlichen fortschreiten," *Mathematische Annalen* **62** (1906), 1–88.

The followup papers are listed in the bibliography of H. Behnke and P. Thullen, *Theorie der Funktionen mehrerer komplexen Veränderlichen* (Berlin, 1934).

18. V. S. Vladimirov, *Methods of the Theory of Functions of Several Complex Variables*, ed. L. Ehrenpreis (Combridge, Mass.: M. I. T. Press, 1966), especially Ch. 5.

19. R. O. Wells, "Holomorphic Hulls and Holomorphic Convexity," Proceedings of the Conference on Complex Analysis, Rice University, 13–15 April 1967, *Rice University Studies* **54**, No. 4 (1968).

20. See Note 12.

21. "Ueber die Fortpflanzung ebener Luftwellen von endlichen Schwingungsweite," *Collected works*, p. 164.

22. To this section and to parts of the preceding one, see my article "Continuity and Discontinuity in Nature and Knowledge," in *Dictionary of the History of Ideas* (New York: Scribner's, in press).

23. R. E. A. C. Paley and N. Wiener, *Fourier Transforms in the Complex Domain* (New York, 1934), Ch. VII.

24. Norman Levinson, *Gap and density theorems* (New York, 1940), Ch. IV.

25. William Bateson, *Materials for the Study of Variation, with special regard to discontinuity in the Origin of Species* (London, 1894). This work is mentioned on page 82 of the book of Alister Hardy cited in Note 27.

26. Sir Gavin de Beer, *Charles Darwin; Evolution by Natural Selection* (London, 1963), p. 183.

27. See for instance, Sir Alister Hardy, *The Living Stream* (London, 1965), Lecture 3.

28. Claude C. Albritton, editor, *The Fabric of Geology* (Reading, Mass., 1963).

29. Kurt von Fritz, *Die griechische Geschichtsschreiburg* (Berlin, 1967).

30. Salomon Bochner, *The Role of Mathematics in the Rise of Science* (Princeton, 1966), p. 62.

31. Ibid., p. 7.

32. Salomon Bochner, *Eclosion and Synthesis* (New York, 1970).

33. Lagrange's *Fonctions analytiques* were written much later than the *Mécanique analytique*, but is is the earlier treatise that counts for our purposes.

34. Gaspard Monge was an analyst doing differential geometry, just as in the twentieth century Marston Morse is an analyst doing topology.

35. Felix Klein, *Vorlesungen über die Entwicklung der Mathematik im* 19. *Jahrhundert* (Berlin, 1926), I, 67–68.

RICE UNIVERSITY

Section G

Differential Geometry

Bochner wrote in the 1930s on the imbedding of Riemannian manifolds and turned in the 1940s and 1950s to a consideration of the topological implications of curvature conditions. In an influential sequence of papers, he introduced criteria for the vanishing of harmonic forms and demonstrated the topological significance of this vanishing. Some of the latter work was surveyed in the book *Curvature and Betti Numbers* (Princeton, 1953) with K. Yano; a Russian translation of this book appeared in 1962. He was particularly interested in differential geometry on complex manifolds.

Reprinted from Duke Mathematical Journal
Vol. 3, No. 2, June, 1937

REMARK ON THE THEOREM OF GREEN

By S. Bochner

The theorem of Green is an identity between an integral over a compact region R of an orientable locally Euclidean separable space S of class two[1] whose dimension will be denoted by n, and an integral over the boundary B of R, in case B is formed by "hypersurfaces":

$$\int_R \operatorname{div} \lambda dv = \int_B \lambda^i \Phi_i \, d\omega. \tag{1}$$

If R is closed, that is, compact and equal to S, thus having "no boundary", the integral over the boundary is to be put equal to zero:

$$\int_S \operatorname{div} \lambda dv = 0. \tag{2}$$

If R is contained in one coördinate neighborhood, the proof of (1) is comparatively simple, provided the boundary B is sufficiently smooth with respect to the coördinate system.[2] But the passage from the local case to a domain R in the large is rather laborious. It requires a cellular subdivision of R into sufficiently small subregions whose boundary is sufficiently smooth, an application of the local theorem to each subregion, and finally a justification of the mutual cancellation of the boundary terms arising from the artificial cellular partitions. Now this procedure is much too complicated and heavy in the case of formula (2) or in case of formula

$$\int_R \operatorname{div} \lambda dv = 0 \qquad (\lambda^i \equiv 0 \text{ on } B). \tag{3}$$

We want to show that these two formulas can be deduced in a much simpler fashion, even avoiding any complication that might be inherent to the local theorem itself. In particular we shall eliminate from the proof the concept of [$(n - 1)$-dimensional] volume on the boundary B.

Our space S being of class two, we can consider on it tensors of class one and tensor densities of class one. In particular, we assume the existence of a positive (non-vanishing) scalar density which, as in the special case of Riemann spaces, will be denoted by $\sqrt{g}$; in going over from coördinates $(x_1, \cdots, x_n)$ to coördinates $(y_1, \cdots, y_n)$, the quantity $\sqrt{g}$ is to be multiplied by the jacobian

Received December 19, 1936.

[1] That is, a space allowing local coördinate transformations with continuous first and second partial derivatives and a positive (non-vanishing) Jacobian.

[2] A. Duschek and W. Mayer, *Differentialgeometrie*, 1930, vol. II, p. 237

$| \partial x/\partial y |$.[3] With such a scalar density we can form an invariant volume element

$$dv \equiv dv_x = \sqrt{g}\, dx_1 \cdots dx_n. \tag{4}$$

It gives rise to a regular Lebesgue measure and a Lebesgue integral as in the Euclidean case $\sqrt{g} = 1$; the only material difference being that in the case of every closed space S the total measure of S is finite.

If λ^i is a contravariant vector, the quantity

$$\operatorname{div} \lambda = \frac{1}{\sqrt{g}} \frac{\partial(\sqrt{g}\, \lambda^i)}{\partial x_i} \tag{5}$$

is an absolute scalar. This follows easily from the fact that the $(n - 1)$-dimensional minors A_{ij} (with appropriate signs) of the matrix

$$\left\| \frac{\partial x_i}{\partial y_j} \right\| \tag{6}$$

satisfy the relation

$$\frac{\partial A_{ij}}{\partial y_j} = 0 \qquad (i = 1, \cdots, n). \tag{7}$$

Hence we can form the integral of div λ over any bounded measurable set A. If A is contained in a coördinate neighborhood, then

$$\int_A \operatorname{div} \lambda dv = \int_A \frac{\partial(\sqrt{g}\, \lambda^i)}{\partial x_i} dx_1 \cdots dx_n. \tag{8}$$

Suppose now that A is a rectangle $a_\nu < x_\nu < b_\nu$, $\nu = 1, \cdots, n$, and that λ^i vanishes on the boundary of A. In this case, for each i,

$$\int_{a_i}^{b_i} \frac{\partial(\sqrt{g}\, \lambda^i)}{\partial x_i} dx_i = 0,$$

and therefore

$$\int_A \operatorname{div} \lambda dv = 0. \tag{9}$$

Also, the integral of div λ is 0 over any open set on which λ^i vanishes. Hence we obtain the first result that (2) is true if λ^i vanishes outside some "rectangle" A. If S is closed, and hence bicompact, we can cover it by a finite number of neighborhoods $U_1, \cdots, U_m$, whose closures are contained in "rectangles" $A_1, \cdots, A_m$ respectively. Corresponding to each μ, $\mu = 1, \cdots, m$, we can easily find a neighborhood V_μ between U_μ and A_μ and a non-negative scalar function φ_μ of class one in A_μ which is $\geqq 1$ in U_μ and 0 outside V_μ. Completing φ_μ by values 0 outside A_μ we have, throughout S, $\varphi_1 + \cdots + \varphi_m \geqq 1$. Hence the functions of class one

[3] O. Veblen, *Invariants of Quadratic Differential Forms*, 1927, pp. 19–25.

$$\psi_\mu = \frac{\varphi_\mu}{\varphi_1 + \cdots + \varphi_m} \tag{10}$$

have the following properties: each vanishes outside some rectangle and their sum is 1. For fixed μ we multiply the vector λ^i by the scalar ψ_μ. The resulting vector $\lambda^i_{(\mu)}$ vanishes outside a rectangle, and hence satisfies (2). But div $\lambda = \sum_{\mu=1}^{m}$ div $\lambda_{(\mu)}$, and this completes the proof of (2).

Formula (3) can be proved in a similar way provided λ^i is *strictly regular up to the boundary*, this expression meaning that, if completed by values 0 outside R, it is a vector of class one throughout S, so that (3) may be written

$$\int_S \operatorname{div} \lambda dv = 0 \qquad (\lambda^i \equiv 0 \text{ outside } R + B). \tag{11}$$

In fact, we can construct non-negative functions of class one $\varphi_1, \cdots, \varphi_m$, each vanishing outside some rectangle and whose sum is $\geqq 1$ for all points of $R + B$. The functions (10) will be of class one in some neighborhood U of $R + B$, although perhaps not throughout S. But λ^i vanishes outside $R + B$, and therefore the vectors $\lambda^i_{(\mu)}$ are again of class one, and we can again prove (3) from

$$\int_S \operatorname{div} \lambda dv = \sum_{\mu=1}^{m} \int_S \operatorname{div} \lambda_\mu \, dv. \tag{12}$$

We can make further statements if S is a Riemann space with a positive-definite fundamental tensor g_{ij} having the following property. *Corresponding to any point x^0 of S there exists a neighborhood $U(x^0)$ such that for any points x, y of $U(x^0)$ the function $\Omega(x, y)$, which is the square of the geodesic distance between x and y, has continuous first partial derivatives in $(x_1, \cdots, x_n; y_1, \cdots, y_n)$ and that, for some constant $C = C(x_0)$,*

$$\| \operatorname{grad}_x \Omega \|^2 \equiv g^{ij}(x) \frac{\partial \Omega}{\partial x_i} \frac{\partial \Omega}{\partial x_j} \leqq C\Omega(x, y). \tag{13}$$

The neighborhood $U(X^0)$ entering our condition can be assumed to be a sphere with center at x^0: $\Omega(x, x_0) < \rho$, $\rho > 0$. If x^0 transverses a set A whose closure is compact, then by the Heine-Borel theorem, this radius ρ and the constant $C(x^0)$ may be chosen uniformly with respect to A: $\rho = \rho(A)$, $C = C(A)$. If $0 < r < r_0 = \frac{1}{2}\rho(A)$, then A_r will denote the open set consisting of all points x for which $\Omega(x, x^0) < r$ for some $x^0 \subset A$. Let $\varphi(t)$ be any non-negative function of the real variable t which has a continuous first derivative in $0 \leqq t \leqq 3$, is positive for $0 \leqq t \leqq \frac{1}{2}$ and 0 for $1 \leqq t \leqq 3$. If $3r < r_0$, the expression

$$P_r(x) = \int_{A_r} \varphi\left(\frac{\Omega(x, y)}{r^2}\right) dv_y \Big/ \int_{A_{3r}} \varphi\left(\frac{\Omega(x, y)}{r^2}\right) dv_y \tag{14}$$

defines a non-negative function of class one in A_{3r} which has the value 1 on A and the value 0 outside A_{2r}. Completing $P_r(x)$ by the values 0 outside A_{3r}, we obtain a non-negative function of class one throughout S which is $\leqq 1$ everywhere, has the value 1 on A and the value 0 outside A_{2r}. Owing to the nature of our function $\varphi(t)$, the denominator is bounded from below by a constant multiple of r^n uniformly for all points x in A_r, and both the numerator and denominator are bounded from above by a constant multiple of r^n, for the given function $\varphi(t)$, and, likewise, if this function is replaced by $\varphi'(t)$. Forming now the partial derivatives of $P_r(x)$ and using (13), we find that

$$\text{(15)} \qquad || \operatorname{grad}_x P_r(x) || \leqq \frac{C_0}{r}, \text{ uniformly for } x \subset A_r.$$

We are now in a position to drop completely in formula (3) the requirement that λ^i be strictly regular up to the boundary and to interpret the boundary condition "$\lambda^i \equiv 0$ on B" as a very light condition requiring that $\lambda^i(x)$ have limit values 0 as x approaches the boundary B from within the region A.

THEOREM. *If R is an open region whose closure $R + B$ is compact and λ^i is a contravariant vector of class one in R, then*

$$\text{(16)} \qquad \int_R \operatorname{div} \lambda dv = 0$$

holds, provided

$$\text{(17)} \qquad \int_R | \operatorname{div} \lambda | \, dv < \infty$$

and

$$\text{(18)} \qquad \lim_{r \to 0} \frac{1}{r} \int_{B_r} \sqrt{g_{ij} \lambda^i \lambda^j} \, dv = 0.$$

Proof. Condition (17) simply states that our integral shall also be absolutely convergent and condition (18) requires that the "length" of $\lambda^i(x)$ shall "in average" tend to zero as x tends to the boundary, in the sense that the integral of $|| \lambda^i(x) ||$ over the set of points having a distance less than r from the boundary shall, as r tends to 0, be small in relation to the quantity r.

Now we shall give the proof itself. We consider the region R, the boundary B, and the exterior $\bar{R}$. In $B + \bar{R}$ we complete λ^i by values 0, and we multiply $\lambda^i(x)$ by the function

$$\text{(19)} \qquad Q_r(x) = 1 - P_r(x),$$

the function $P_r(x)$ being formed with respect to the compact set B. It is easily seen that $\lambda^i_{(r)} = Q_r \cdot \lambda^i$ is strictly regular up to the boundary and vanishes on B. Hence (16) holds for $\lambda^i_{(r)}$. All we have to prove now is the limit relation

$$\text{(20)} \qquad \int_R \operatorname{div} (\lambda - \lambda_{(r)}) dv \equiv \int_{R \cdot B_{2r}} \operatorname{div} (P_r \lambda) dv \to 0 \qquad \text{as } r \to 0.$$

But this follows from

$$\operatorname{div}(P_r \cdot \lambda) = P_r \cdot \operatorname{div} \lambda + \operatorname{grad} P_r \cdot \lambda$$

in connection with (15), (17), (18) and $| P_r(x) | \leqq 1$.

In the theorem we have just proved, the boundary may in part or totally consist of points which are limit points of R and not of $\bar{R}$; in other words, the theorem also takes care of the case in which the vector has singularities along lower dimensional manifolds. In particular it can be concluded that in Green's theorems (1), (2), (3) the vector λ^i may cease to be of class one along sufficiently smooth manifolds of dimension $\leqq n - 2$ as long as its length stays bounded in the neighborhood of these manifolds.

We can also make a statement regarding equation (1) itself. Let S be closed, and suppose that on some given boundary B it is possible to define an area element $d\omega$ and a vector Φ_i such that (1) holds for every vector λ^i which is defined and of class one in a neighborhood of $R + B$. Then the same is true for the exterior $\bar{R}$ of $R + B$, which means that

$$\int_{\bar{R}} \operatorname{div} \lambda \, dv = \int_B \lambda^i(-\Phi_i) \, d\omega \tag{21}$$

for every function λ^i which is defined and of class one in a neighborhood of $\bar{R} + B$, provided that the (n-dimensional) volume of the set B is 0. In fact, (21) follows immediately frcm (1) and (2) whenever λ^i is defined and of class one throughout S. But the values of the two integrals in (21) depend only on the values of λ^i on $\bar{R} + B$. Hence (21) will hold for a vector λ^i which, without altering its values on $\bar{R} + B$, can be modified into a vector which is of class one throughout S. Now if λ^i is of class one in a neighborhood of $\bar{R} + B$, this modification can be obtained by multiplying λ^i with a function $P_r(x)$ that belongs to the compact set $\bar{R} + B$ and by putting it 0 in other points.

PRINCETON UNIVERSITY.

Reprinted from DUKE MATHEMATICAL JOURNAL
Vol. 3, No. 2, June, 1937

ANALYTIC MAPPING OF COMPACT RIEMANN SPACES INTO EUCLIDEAN SPACE

BY S. BOCHNER

Recently H. Whitney[1] has proved that an n-dimensional separable coördinate space S of class C_q $(1 \leqq q \leqq \infty)$ can be mapped topologically onto the Euclidean E_{2n+1} in such a manner that the mapping functions

$$t_\nu = t_\nu\,(x_1, \cdots, x_n) \qquad (\nu = 1, \cdots, 2n+1) \tag{1}$$

belong to class C_q on S and have a Jacobian of rank n throughout S; the quantities $x_1, \cdots, x_n$ in (1) are local coördinates on S varying with the neighborhood.

It is not known whether for an analytic space S the mapping functions (1) can be chosen analytic. It is the purpose of the present paper to point out that they can be so chosen provided S is compact and has an analytic Riemann metric.

The line of reasoning is very simple. With the fundamental tensor $g_{ij}(x)$, we form the Laplacian

$$\Delta\varphi = \frac{1}{g^{\frac{1}{2}}}\frac{\partial}{\partial x_i}\left(g^{\frac{1}{2}}\,g^{ij}\,\frac{\partial\varphi}{\partial x_j}\right) = g^{ij}\,\varphi_{,ij}\,. \tag{2}$$

In the Hilbert space H of all square integrable functions on S, the Laplacian is essentially the inverse of a completely continuous operator. Therefore the solutions φ of the equation

$$\Delta\varphi = \lambda\varphi \tag{3}$$

form a complete basis in H. But the solutions of (3) are analytic if the coefficients g_{ij} are so. Hence every function $t(x)$ on S is the limit in square mean of analytic functions. For differentiable functions on S we shall prove more; if $t(x)$ belongs to C_∞, then corresponding to any $\epsilon > 0$ there exists an analytic function $\psi^\epsilon(x)$ such that *the function* $t(x) - \psi^\epsilon(x)$ *and its gradient* differ by less than ϵ throughout S.[2] We now apply this approximation to the functions (1). It is easy to see that, for ϵ sufficiently small, the approximating transformation

$$t_\nu = \psi_\nu^\epsilon(x_1, \cdots, x_n) \tag{4}$$

will have the same mapping properties as the original transformation (1), which proves our assertion.

Received February 16, 1937.

[1] *Differentiable Manifolds*, Annals of Math., vol. 37 (1936), pp. 645–680; p. 654, Theorem 1.

[2] In order to prove this conclusion it would be sufficient to assume that $t(x)$ belongs to C_2. But the proof would become more elaborate.

The Laplacian $\Delta\varphi$ on a compact space S was treated by Hilbert (for $n = 2$)[3] and by G. Giraud.[4] In order to avoid too many references to these and other papers, which would have to be supplemented in minor points in any case, we shall reproduce a number of known details in the first half of this paper.

Finally, in the last part we shall have occasion to generalize the properties of the Laplacian from scalar functions to tensors of arbitrary rank.

Part I. Definitions

1. We shall consider a space S which is a compact orientable coördinate space of an arbitrary dimension n with an analytic positive-definite fundamental tensor $g_{ij}(x)$. In order to avoid a known trivial complication in writing, we shall exclude $n = 2$, and thus assume $n \geqq 3$.

The family of analytic functions and tensors on S, or on a specified open set in S will be denoted by C_a, and C_q, $0 \leqq q \leqq \infty$, will denote the family of functions or tensors of class q.

2. A point on S will be denoted by $x, y, z, \bar{x}, \xi, \eta, \cdots$, its coördinates by a corresponding subscript. The invariant volume element on S gives rise to a theory of Lebesgue measure and Lebesgue integration with the customary properties, except for the fact that the total measure of S is finite. The measure element will be denoted by dv, or more specifically, by dv_x, dv_y, dv_ξ, etc. Thus

$$\int_A f\, dv \quad \text{or} \quad \int_A f(x)\, dv_x$$

will denote the integral of the function f over the set A. In case $A = S$, we shall simply write

$$\int f\, dv \quad \text{or} \quad \int f(x)\, dv_x .$$

3. The class of functions $f(x)$ which are bounded and measurable on S, or on a specified subset of S, will be denoted by B. Furthermore we shall consider, on the total space S, the Hilbert space H of all (real) measurable functions of integrable square, with the inner product

$$(f, g) = \int fg\, dv.$$

We shall also use the symbol $||f||^2$ for (f, f). It follows easily, as in the case of a Euclidean region S, that every subspace C_q of H is dense in H. It is our main objective to prove that C_a is likewise dense in H.

[3] D. Hilbert, *Grundzüge einer allgemeinen Theorie der linearen Integralgleichungen*, 1912, chapter 18.

[4] *Problèmes mixtes et problèmes sur des variétés closes*, etc., Annales de la Société Polonaise de Mathématique, vol. 12 (1933), pp. 35–54; p. 43.

4. Functions of two variables, $f(x, y)$, $f(x, \xi)$, etc., will be defined either on the product space $S^2 = S \times S$, or on a specified part of it. It is clear what will be meant by the symbols C_a^2, C_q^2, B^2.

5. Corresponding to any point x^0 there exists a neighborhood $U = U(x^0)$ such that any two points $x, y \subset U$ have a unique geodesic distance $R \equiv R(x, y)$. The *square* of the geodesic distance will be denoted by $\Omega \equiv \Omega(x, y)$. If U is sufficiently small, then $\Omega(x, y) \subset C_a^2$. Owing to the bicompactness of S and S^2 we can find a number $\rho_0 > 0$ such that $\Omega(x, y) \subset C_a^2$ if $R(x, y) < \rho_0$.

The local analyticity of $\Omega(x, y)$ becomes obvious if $\Omega(x, y)$ is expressed in normal coördinates. The normal coördinates z_i originating at the point x_i are analytic functions

$$z_i = z_i(x, y) = y_i - x_i + \text{(higher terms in } x \text{ and } y)$$

defined for $x, y \subset U(x^0)$. They are characterized by the relations

$$\bar{g}_{ij}(z)z_j = g_{ij}(x)z_j, \tag{5}$$

in which $\bar{g}_{ij}(z)$ is the fundamental tensor expressed in the z-coördinates originating at the variable point x, with the initial values $\bar{g}_{ij}(x) = g_{ij}(x)$; the tensor $g_{ij}(x)$ being given in a fixed coördinate system of which x and y are arbitrary points. Important properties of the functions $z_i(x, y)$ are

$$\frac{\partial z_i}{\partial x_j} = -\delta_{ij}, \qquad \frac{\partial z_i}{\partial y_j} = \delta_{ij}, \qquad \text{if } z = 0, \tag{6}$$

$$\frac{\partial \bar{g}_{ij}(z)}{\partial z_k} = 0, \qquad \text{if } z = 0, \tag{7}$$

$$\Omega(x, y) = \bar{g}_{ij}(z)z_i z_j \equiv g_{ij}(x)z_i z_j.\text{[5]} \tag{8}$$

6. The family $C_{q,r}^2$, $0 \leqq r < n$, consists of functions $Q(x, y)$ with the following properties. $Q(x, y)$ is not defined for $x = y$. Everywhere else on S^2 it belongs to class C_q^2. Inside some belt $R(x, y) < \rho$, $Q(x, y)$ and each of its derivatives is representable as a quotient

$$\frac{A(x, y)}{R(x, y)^{\mu}}; \tag{9}$$

also, within this belt, $Q(x, y)$ has the order of magnitude $O[R(x, y)^{-r}]$, and each derivative of order $q_0(\leqq q)$ has the order of magnitude $O[R(x, y)^{-r-q_0}]$.

We shall also consider a class of functions B_r^2, $0 \leqq r < n$. The function $Q(x, y)$ belongs to B_r^2 if it is measurable on S^2, bounded outside some belt $R(x, y)$

[5] For the existence of normal coördinates with these properties it is not necessary to assume that our space is analytic. They exist if S is a space of class $p \geqq 2$, and they are themselves of class $p - 2$. Checking all steps of our argument, we could easily see that our Theorem I holds if $p = 6$, but normal coördinates can be avoided altogether and the class of the space reduced by several units. Compare G. Giraud, loc. cit.

$< \rho$, and $O[R(x, y)^{-r}]$ within this belt. For such functions we shall consider the process of convolution. For instance

$$Q' * Q'' = \int Q'(x, \xi) Q''(\xi, y) dv_\xi . \tag{10}$$

7. The symbol $\Phi_\rho(t)$ will denote any function in $0 \leqq t < \infty$ with the following properties. The function has derivatives of all orders, is non-negative throughout, has the value 1 in $0 \leqq t \leqq \frac{1}{2}\rho$, values between 0 and 1 in $\frac{1}{2}\rho < t \leqq \rho$, and the value 0 in $\rho \leqq t < \infty$.

8. The symbol $P(x, y)$ will denote any fixed function with the following properties:

$$P(x, y) = P(y, x), \tag{11}$$

$$P(x, y) \subset C_{4, n-2} , \tag{12}$$

and there exists a $\rho > 0$ such that

$$P(x, y) = \frac{\gamma}{R(x, y)^{n-2}}, \qquad \text{if } R(x, y) < \rho \tag{13}$$

for some prescribed constant γ. We shall prescribe

$$\gamma = \frac{\Gamma(\frac{1}{2}n - 1)}{4\pi^{\frac{1}{2}n}}. \tag{14}$$

The existence of such a function is trivial. For example, the function

$$P(x, y) = \gamma \frac{\Phi_\rho[R(x, y)]}{R(x, y)^{n-2}} \tag{15}$$

will do.

Any function $P(x, y)$ with these properties has the further important property

$$\Delta_x P(x, y) \subset C^2_{2, n-2} . \tag{16}$$

This property states essentially that the Laplacian, if applied to our function $P(x, y)$, does not raise the exponent of its singularity on the diagonal $x = y$; whereas, as a rule, if applied to a function from $C^2_{p,r}$, it increases the exponent r by 2. The proof of (16) follows easily if $R(x, y)$ is expressed in normal coördinates issuing from x; in such coördinates z_i we have[6]

$$\Delta_z R = \left(1 - \frac{n}{2}\right) \frac{\partial \log g(z)}{\partial z_i} z_i \cdot R^{-n},$$

and, by (7), the factor of R^{-n} on the right is $O(R^2)$.

[6] W. Feller, *Lösungen der linearen partiellen Differentialgleichungen zweiter Ordnung vom elliptischen Typus*, Mathematische Annalen, vol. 102 (1930), pp. 633–649; p. 639, footnote.

It will be important for us to consider not the Laplacian $\Delta\varphi$ itself, but the operator

$$(17) \qquad \Delta^c\varphi = \Delta\varphi - c\varphi$$

for some positive constant c. The constant c being fixed, we shall write

$$(18) \qquad K(x, y) = \Delta_x^c P(x, y), \qquad \bar{K}(x, y) = K(y, x) = \Delta_y^c P(y, x).$$

These functions again have the property (16).

9. The distance $R(x, y)$ is comparable in size to the Euclidean distance

$$(19) \qquad E(x, y) = \{(y_i - x_i)^2\}^{\frac{1}{2}}.$$

Hence, the integral

$$(20) \qquad \psi(y) = \int \varphi(x)Q(x, y)\, dv_x$$

exists for every y if $\varphi(x) \subset B$, $Q(x, y) \subset C_{0,r}$, $r < n$. For fixed $Q(x, y)$ and variable $\varphi(x)$ the integral operator (20) will be written

$$(21) \qquad \psi = Q\varphi.$$

We shall be mainly interested in the operators $P\varphi$ and $K\varphi$ resulting from the functions $P(x, y)$, $K(x, y)$ that were introduced before.

Part II. Preliminaries

1. We shall state several properties of the expression (20), all of which are trivially true if $Q(x, y)$ happens to vanish inside some belt $R(x, y) < \rho_0$.

LEMMA 1. *If $\varphi(x) \subset B$, $Q(x, y) \subset C_{0,r}^2$, $r < n$, then the function (20) belongs to C_0; more precisely, the modulus of uniform continuity of $\psi(y)$ depends only on* $M = \underset{x \subset S}{\text{l.u.b.}} |\varphi(x)|$.

Proof. With the function $\Phi_\rho(t)$ of Part I, §7, we form the expression $\psi_\sigma(y) = \int \varphi(x)Q(x, y)[1 - \Phi_\sigma\{R(x, y)\}]dv_x$. For σ fixed, our lemma holds for $\psi_\sigma(y)$ in the place of $\psi(y)$. But $\psi_\sigma(y)$ tends to $\psi(y)$ as $\sigma \to 0$, uniformly in y and M.

LEMMA 2. *If $\varphi(x) \subset B$, $Q(x, y) \subset C_{\nu,r}^2$, $r < n - \nu$, then the function (20) belongs to C_ν, again uniformly in M, and the partial derivatives of $\psi(y)$ may be obtained by partial differentiation under the integral. For example,*

$$(22) \qquad \frac{\partial\psi}{\partial y_1} = \int \varphi(x) \frac{\partial Q}{\partial y_1}\, dv_x.$$

Proof. By using induction, it is sufficient to consider the case $\nu = 1$. Since $\frac{\partial Q}{\partial y_1} \subset C_{0,r+1}$, $r + 1 < n$, the integral

$$(23) \qquad g(y) = \int \varphi(x) \frac{\partial Q}{\partial y_1}\, dv_x$$

represents a function in C_0. We restrict ourselves to a coördinate neighborhood $(y_1, \cdots, y_n)$ and integrate the function $g(y)$ with respect to the variable y_1 between the fixed limits α and β. Since the integrand on the right side is a Lebesgue integrable function on S^2 we can exchange, on the right side, the two processes of integration for almost all $(y_2, \cdots, y_n)$, and hence

$$\int_\alpha^\beta g(y_1, \cdots, y_n)\,dy_1 = \psi(\beta, y_2, \cdots, y_n) - \psi(\alpha, y_2, \cdots, y_n)$$

for almost all $(y_2, \cdots, y_n)$. But both sides of the last relation are continuous and, therefore, it holds throughout. Making now β variable, we find that $\psi(y) \subset C_1$, and that $g(y) = \partial\psi/\partial y_1$.

LEMMA 3. *If $\varphi(x) \subset C_1$ and $Q(x, y) \subset C^2_{2,n-2}$, the function* (20) *belongs to C_2.*

Proof. It will be sufficient to show that the function (23) belongs to C_1. Relation (6) implies

$$\frac{\partial z_i}{\partial x_j} + \frac{\partial z_i}{\partial y_j} = O[R(x, y)].$$

Hence, remembering that $Q(x, y) = A(x, y)\,R(x, y)^{-\mu}$ for $R(x, y) < \rho_0$, and using expression (8) for $\Omega(x, y)$, we easily find that

$$\frac{\partial Q}{\partial y_1} + \frac{\partial Q}{\partial x_1} \tag{24}$$

belongs to $C_{1,n-2}$. Hence, on account of Lemma 2, the non-trivial part of function (23) is the function

$$h(y) = \int \varphi(x)\,\frac{\partial Q(x, y)}{\partial x_1}\,dv_x, \tag{25}$$

and it is sufficient to show that the latter function belongs to C_1. We restrict the point y to a sufficiently small neighborhood $U(y^0)$ whose closure is contained within a sphere $R(x, y^0) < \rho$. Forming with our previous function $\Phi_\rho(t)$ the decomposition

$$\varphi(x) = \{\varphi(x)\Phi_\rho[R(x, y^0)]\} + \{\varphi(x) - \varphi(x)\Phi_\rho[R(x, y^0)]\},$$

we see immediately that the second component of $\varphi(x)$ contributes to the function (25) a component of class C_1. Hence we may assume that $\varphi(x)$ and its derivatives vanish outside some coördinate neighborhood and that y is a point of this neighborhood. Denoting the product of $\varphi(x)$ and the volume density $g^{\frac{1}{2}}$ by $\tilde{\varphi}(x)$, we can therefore write

$$h(y) = \int_A \tilde{\varphi}(x)\,\frac{\partial Q(x, y)}{\partial x_1}\,dx_1 \cdots dx_n,$$

where A is an n-dimensional Euclidean domain and $\tilde{\varphi}(x)$ is a function of C_1 vanishing on the boundary of A. Let y be a fixed point within A, A_ρ the difference of A and a small sphere of radius ρ around y, and B_ρ the inner boundary of A. By Green's formula,

$$\int_{A_\rho} \bar{\varphi}(x) \frac{\partial Q}{\partial x_1} dx_1 \cdots dx_n = -\int_{A_\rho} \frac{\partial \bar{\varphi}}{\partial x_1} Q \, dx_1 \cdots dx_n + \int_{B_\rho} \cos(\xi, \nu)\bar{\varphi}(\xi)Q(\xi, y)\, d\omega_\xi .$$

Since $Q(\xi, y) = O(\rho^{-n+2})$, the surface integral tends to 0 as $\rho \to 0$, and therefore

$$h(y) = -\int_A \frac{\partial \varphi}{\partial x_1} Q(x, y)\, dx_1 \cdots dx_n .$$

By Lemma 2, $h(y)$ belongs to C_1.

LEMMA 4. *If $Q(x, y) \subset B_r^2$ and $L(x, y) \subset B_s^2$, then $Q * L$ belongs to B_{r+s-n}^2 if $r + s - n > 0$; to B_ϵ^2, for any $\epsilon > 0$, if $r + s - n = 0$; and to B_0^2 if $r + s - n < 0$.*

Proof. The lemma is known in case S is a bounded set in Euclidean space and $R(x, y)$ has the value (19). For our present case it follows from the fact that in every coördinate neighborhood the quantity $R(x, y)$ is majorized by (19) from above and below.

LEMMA 5. *If $Q(x, y) \subset C_{0,r}^2$, $r < n$, then, from some positive ν onward, the ν-th iterated kernel*

$$Q_\nu = Q * Q * \cdots * Q \qquad (\nu \text{ factors}) \tag{26}$$

belongs to C_0^2.

Proof. By Lemma 4 there exists a $Q_{\nu-1}(x, y)$ which is bounded. But

$$Q_\nu(x, y) = \int Q(x, \xi)Q_{\nu-1}(\xi, y)d\nu_\xi = \int Q_{\nu-1}(x, \xi)Q(\xi, y)d\nu_\xi .$$

Therefore, by Lemma 1, $Q_\nu(x, y)$ is continuous in x uniformly in y, and continuous in y uniformly in x. Hence $Q_\nu(x, y) \subset C_0^2$.

2. We shall now discuss the operator (21) for $\varphi \subset H$. To start with, the expression (20) is not defined for arbitrary elements φ in H. But it is defined for φ in B, and the subset B of H is dense in H. This suggests the possibility of extending the operator formally into a wider set of H. In fact, the operator (21) can be extended uniquely over the whole of H.

LEMMA 6. *For any $Q(x, y) \subset C_{0,r}^2$, $r < n$, the expression* (20) *defines a unique operator* (21) *which is bounded and completely continuous. Also, for any two such kernels $Q'(x, y)$, $Q''(x, y)$, the relation*

$$Q'[Q''(\varphi)] = (Q' * Q'')(\varphi) \tag{27}$$

holds.

Proof. We first assume that $Q(x, y)$ is continuous throughout. In this case it is known that $Q\varphi$ is bounded and completely continuous for φ in B. B being dense in H, the operator $Q\varphi$ has a unique extension into the whole space H, and this extension is again completely continuous.[7] The value of $Q\varphi$ for an arbitrary element $\varphi \subset H$ can be computed in the following manner. Form any sequence of elements $\{\varphi_m\}$ from B of which φ is the limit in the strong topology of H: $\lim_{m\to\infty} \| \varphi - \varphi_m \| = 0$. The values $\psi_m = Q\varphi_m$ can be computed from (20) and they are in their turn convergent to an element ψ of H. This element is the desired value (21). Finally, relation (27) is obvious for $\varphi \subset B$ by Fubini's theorem, and for general φ it follows by the limiting process we have just described.

In the case of a general kernel $Q(x, y)$ we shall apply an important criterion.[8] An operator $Q\varphi$ is bounded and completely continuous in H, if corresponding to any $\epsilon > 0$ there exists a decomposition

$$Q\varphi = Q^\sigma\varphi + R^\sigma\varphi \tag{28}$$

of the following nature. $R^\sigma\varphi$ is bounded and completely continuous, and $Q^\sigma\varphi$ is a bounded operator whose bound is $\leqq \epsilon$. We put

$$Q^\sigma = Q(x, y)\Phi_\sigma[R(x, y)], \qquad R^\sigma = Q - Q^\sigma.$$

R^σ is continuous, and hence the operator $R^\sigma\varphi$ is bounded and completely continuous. As for the operator $\psi^\sigma = Q^\sigma\varphi$, we have, for $\varphi \subset B$,

$$\| \psi^\sigma \|^2 = \int | \psi^\sigma(y) |^2 \, dv_y = \int\int \varphi(\xi)\varphi(\eta)M^\sigma(\xi, \eta)dv_\xi dv_\eta, \tag{29}$$

where

$$M^\sigma(\xi, \eta) = \int Q^\sigma(\xi, y)Q^\sigma(y, \eta)\, dv_y. \tag{30}$$

The kernel $M^\sigma(\xi, \eta)$ has the following two properties:

$$M^\sigma(\xi, \eta) = 0, \text{ if } R(\xi, \eta) > 2\sigma; \tag{31}$$

there exists an exponent s, $0 \leqq s < n$, and a constant A such that

$$| M^\sigma(\xi, \eta) | \leqq \frac{A}{R(\xi, \eta)^s}, \qquad \text{if } R(\xi, \eta) \leqq 2\sigma. \tag{32}$$

Property (31) is an immediate consequence of the fact that $Q^\sigma(x, y)$ vanishes for $R(x, y) \geqq \sigma$; whereas property (32) is an easy consequence of Lemma 4. Putting

$$| \varphi(\xi)\varphi(\eta) | \leqq \frac{| \varphi(\xi) |^2 + | \varphi(\eta) |^2}{2}$$

in (29), we obtain

$$2\| \psi^\sigma \|^2 \leqq \int \varphi(\xi)^2\left[\int | M^\sigma(\xi, \eta) | \, dv_\eta\right] d_\xi + \int \varphi(\eta)^2\left[\int | M^\sigma(\xi, \eta) | \, dv_\xi\right] dv_\eta. \tag{33}$$

[7] S. Banach, *Opérations linéaires*, 1932, p. 99.

[8] Banach, loc. cit., p. 96, Théorème 2.

But, owing to properties (31) and (32), there exists a positive quantity $\epsilon(\sigma)$, tending to 0 as σ tends to 0, such that

$$\int | M^{\sigma}(\xi, \eta) | \, dv_{\eta} \leqq \epsilon(\sigma), \qquad \int | M^{\sigma}(\xi, \eta) | \, dv_{\xi} \leqq \epsilon(\sigma).$$

Hence each of the integrals in (33) is $\leqq \epsilon(\sigma) \cdot || \varphi ||^2$, and therefore ψ^{σ} is a bounded operator on B whose bound is $\leqq \epsilon(\sigma)$. The extension of ψ^{σ} from B onto the total space H does not increase the bound,[9] and, therefore, the assumptions of the above mentioned criterion are fulfilled. This proves the first half of the lemma, whereas relation (27) can be proved as in the case of a continuous kernel $Q(x, y)$.

Part III. The Laplacian

1. If in Green's formula

$$\int \frac{1}{g^{\frac{1}{2}}} \frac{\partial}{\partial x_i} (g^{\frac{1}{2}} \lambda^i) \, dv_x = 0 \tag{34}$$

[λ^i is a contravariant vector belonging to C_1], we put

$$\lambda^i = \varphi \frac{\partial \varphi}{\partial x_i}, \qquad \text{or } \lambda^i = \varphi \frac{\partial \psi}{\partial x_i} - \psi \frac{\partial \varphi}{\partial x_i},$$

we obtain the relations

$$-\int \varphi \Delta^c \varphi \, dv = \int \left[g^{ij} \frac{\partial \varphi}{\partial x_i} \frac{\partial \varphi}{\partial x_j} + c\varphi^2 \right] dv_x \tag{35}$$

$$\int \varphi \Delta^c \psi \, dv = \int \psi \Delta^c \varphi \, dv. \tag{36}$$

From (35) we conclude

$$(\varphi, \Delta^c \varphi) \leqq -c(\varphi, \varphi), \tag{37}$$

and relation (36) reads

$$(\varphi, \Delta^c \psi) = (\Delta^c \varphi, \psi). \tag{38}$$

Hence, on the dense subspace C_2 of H, the operator $-\Delta^c \varphi$ is positive definite with a lower bound $\geqq c$ and symmetric. By a general theorem on operators in Hilbert space there exists a closed extension of the operator $\Delta^c \varphi$ for which the properties (37) and (38) remain in force.[10] It will be this closure of the Laplacian that will forthwith be denoted by $\Delta^c \varphi$; in fact, in order to simplify the writing, *we shall omit the upper index c, and thus write $\Delta \varphi$ instead of $\Delta^c \varphi$ until further notice.* The domain of $\Delta \varphi$ is a well defined set D of H, $C_2 \subset D \subset H$.

[9] Banach, loc. cit., p. 58.
[10] M. H. Stone, *Linear Transformations in Hilbert Space*, 1932, p. 49, Theorem 2.12.

LEMMA 7. *The operator* $\Delta\varphi$ *satisfies the relations*

$$K\varphi - \varphi = P\Delta\varphi, \qquad \textit{if } \varphi \subset D. \tag{39}$$

$$\bar{K}\varphi - \varphi = \Delta P\varphi, \qquad \textit{if } P\varphi \subset D. \tag{40}$$

Proof. By the definitions given in Part I, §§8 and 9, relation (39) is equivalent to

$$\int [\varphi(x)\Delta_x P(x, y) - \Delta_x\varphi(x)P(x, y)]\, dv_x = \varphi(y). \tag{41}$$

It is sufficient to prove it for $\varphi \subset C_2$. For other φ it follows from the continuity of the operators $K\varphi$ and $P\varphi$. The left side of (41) is the limit, as $\sigma \to 0$, of

$$\int_{S_\sigma} \frac{1}{g^{\frac{1}{2}}} \frac{\partial}{\partial x_i} (g^{\frac{1}{2}}\lambda^i)\, dv_x, \tag{42}$$

where S_σ is the total space S minus the geodesic sphere of radius σ around the fixed point y, and

$$\lambda^i(x) = \varphi(x) \frac{\partial P(x, y)}{\partial x_i} - \frac{\partial\varphi}{\partial x_i} P(x, y).$$

By Green's formula, (42) has the value

$$-\int_{B_\sigma} \lambda^i(\xi) \frac{\partial R(\xi, y)}{\partial \xi_i}\, d\omega_\xi, \tag{43}$$

where B_σ is the boundary of S_σ and $d\omega_\xi$ is the invariant surface element on B_σ.[11] In order to estimate (43) we introduce normal coördinates $z_i = z_i(\xi, y)$ originating at the fixed point y. We also make a linear affine transformation by which $\bar{g}_{ij}(y) = \delta_{ij}$. This gives for the quantity $R(\xi, y)$ the value $(z_i z_i)^{\frac{1}{2}}$. Hence we obtain, by a trivial calculation,

$$\lambda^i(\xi) \frac{\partial R(\xi, y)}{\partial \xi_i} = (2 - n)\gamma\cdot\varphi(\xi) \frac{1}{R^{n-1}} - \gamma \frac{\partial\varphi}{\partial z_i} \frac{z_i}{R^{n-1}}$$

$$= (2 - n)\gamma\cdot\varphi(\xi) \frac{1}{\sigma^{n-1}} + O\left(\frac{1}{\sigma^{n-2}}\right).$$

Therefore, (43) has the value

$$(n - 2)\gamma\omega_{n-1}\varphi(y) + o\left(\frac{\sigma^{n-1}}{\sigma^{n-1}}\right), \tag{44}$$

where ω_{n-1} is the $(n - 1)$-dimensional Euclidean volume of the unit-sphere $z_i z_i = 1$. Substituting from (14), we see that (44) tends to $\varphi(y)$ as $\sigma \to 0$, and this completes the proof of (39).

[11] W. Feller, loc. cit., §2; S. Bochner, *Remark on the theorem of Green*, this Journal, pp. 334–338.

Relation (40) will be proved if we show that for every element u of C_2,

$$(u, \bar{K}\varphi - \varphi - \Delta P\varphi) = 0. \tag{45}$$

Obviously

$$(u, \bar{K}\varphi) = \int u(x)\varphi(y)K(x, y)dv_x dv_y = (\varphi, Ku)$$

and

$$(u, \Delta P\varphi) = (\Delta u, P\varphi) = \int \Delta_x u(x)\cdot\varphi(y)P(y, x)dv_x dv_y$$
$$= (\varphi, P\Delta u).$$

Therefore the left side of (45) has the value

$$(\varphi, Ku - u - P\Delta u),$$

and this vanishes by (39).

2. LEMMA 8. *If* $Q(x, y) \subset C^2_{2,n-2}$, $f \subset C_2$, $\varphi \subset H$, *the relation*

$$\varphi = Q\varphi + f \tag{46}$$

implies that $\varphi \subset C_2$.

Proof. Iterating (46), we obtain, using relation (27),

$$\varphi = Q_\nu\varphi + Q_{\nu-1}f + Q_{\nu-2}f + \cdots + f. \tag{47}$$

Obviously, the functions Qf, $Q_2f = Q(Qf)$, $\cdots$, $Q_{\nu-1}f = Q(Q_{\nu-2}f)$ are all bounded. From a sufficiently high ν onward, $Q_\nu(x, y) \subset B^2$ by Lemma 5, and hence by

$$\left|\int Q_\nu(x, y)\varphi(x)dv_x\right|^2 \leqq \int Q_\nu(x, y)^2 dv_x \int \varphi(x)^2 dv_x, \tag{48}$$

$Q_\nu\varphi$ is bounded. Thus, by (47), φ is bounded. Consequently by Lemma 2, $Q\varphi \subset C_1$, and hence by (46), $\varphi \subset C_1$. But then, by Lemma 3, $Q\varphi \subset C_2$, and therefore, again by (46), $\varphi \subset C_2$.

3. LEMMA 9. *The equation*

$$(\Delta^c\varphi \equiv)\,\Delta\varphi = f \tag{49}$$

has a solution for any $f \subset H$.

Proof. A solution, if any, is unique, by (37). It is sufficient to show that (49) has a solution for $f \subset C_1$; in fact, if we show this, we prove that the operator $f = \Delta\varphi$ has an inverse

$$\varphi = Jf$$

on a dense set in H. By (37) this inverse is a bounded operator and has therefore an extension into the whole of H. If we re-invert the fully extended operator Jf we obviously obtain $\Delta\varphi$; in other words, the counter-domain of $\Delta\varphi$ is the whole space H.

Substituting (49) in (39), we are led to investigate the integral equation

$$K\varphi - \varphi = Pf. \tag{50}$$

The operator $K\varphi$ being completely continuous by Lemma 6, we can apply the generalized Fredholm theory.[12] According to this theory, the equations

$$K\varphi - \varphi = 0, \tag{51}$$

$$\bar{K}\psi - \psi = 0 \tag{52}$$

have the same finite number of independent solutions. If we denote such basic solutions by

$$\varphi_1, \cdots, \varphi_m; \qquad \psi_1, \cdots, \psi_m,$$

for a given element f the equation (50) has a solution, if, and only if,

$$(\psi_\mu, Pf) = 0 \qquad (\mu = 1, \cdots, m). \tag{53}$$

We remark once for all a fact which we shall tacitly use several times, namely, that any solution φ or ψ of (50), (51), (52) is automatically a function of C_2, and therefore a function of D. This follows from Lemma 8 [in conjunction with Lemma 3 for equation (50)]. Let φ_μ be any solution of (51). Combining (51) and (39), we obtain

$$P\Delta\varphi_\mu = 0. \tag{54}$$

Hence for $\psi_\mu = \Delta\varphi_\mu$, we obtain from (40), $\bar{K}\psi_\mu - \psi_\mu = 0$. Also, if $\varphi_1, \cdots, \varphi_m$ are linearly independent, then $\Delta\varphi_1, \cdots, \Delta\varphi_m$ are likewise linearly independent. Hence, the basic solutions of (52) can be assumed to have the form $\psi_\mu = \Delta\varphi_\mu$. Now, $(\psi_\mu, Pf) = (f, P\psi_\mu)$, and vanishes by (54). Thus equation (50) has a solution φ^* for an arbitrary element $f \subset C_1$. Comparing now relations (50) and (39), we obtain for the function $\psi^* = f - \Delta\varphi^*$ the equation $P\psi^* = 0$. By (40), ψ^* is a solution of (52) and may be written in the form $c_1\psi_1 + \cdots + c_m\psi_m$. In other words, the function $\varphi = \varphi^* + c_1\varphi_1 + \cdots + c_m\varphi_m$ is a solution of (49).

Theorem I. *The operator $f = \Delta\varphi - c\varphi$ is for every fixed $c > 0$ the inverse of a completely continuous operator $\varphi = Jf$ in H. The solutions $\varphi \subset H$ of the equation*

$$\Delta\varphi = \lambda\varphi \tag{55}$$

(λ being any constant for which a solution exists) form a complete set of functions in H. They are all, automatically, contained in C_2.

Proof. From (50) we obtain

$$\varphi = K(Jf) - Pf \equiv Jf.$$

[12] Banach, loc. cit., pp. 159–161.

By Lemma 6, Pf is completely continuous, and so is $K(Jf)$, since Jf is at any rate bounded. Therefore, Jf is completely continuous too. Hence, Jf has a pure point spectrum and so has its inverse $\Delta^c\varphi$. Therefore, the solutions of the equation $\Delta\varphi - c\varphi = \lambda\varphi$, or, what amounts to the same, those of equation (55) form a complete basis in H. Substituting (55) in (39), we obtain $(K - \lambda P)(\varphi) = \varphi$, and therefore $\varphi \subset C_2$, by Lemma 8.

Part IV. The mapping theorem

1. The following theorem makes full use of our assumption that the underlying space S is analytic.

THEOREM II. *The solutions φ of* (55) *are all analytic. Thus there exists in the function space H a complete basis $\{\varphi_m\}$ whose elements φ_m are analytic throughout S.*

Proof. The analytic character of the functions φ_m is a purely local property and follows from the following theorem which for $n \geqq 3$ was first proved by J. Hadamard.[13]

LEMMA 10. *If the coefficients $g_{ij}(x)$ (the tensor g_{ij} being symmetric and positive-definite), $b_i(x)$, $c(x)$, $f(x)$ are all analytic in the neighborhood of a point $x^0 = (x_1^0, \cdots, x_n^0)$, then every solution $\varphi(x)$ of the equation*

$$g_{ij}\frac{\partial^2\varphi}{\partial x_i\,\partial x_j} + b_i\frac{\partial\varphi}{\partial x_i} + c\varphi = f \tag{56}$$

which belongs to C_2 is likewise analytic.

Proof. Since Hadamard does not carry out the last steps of his argument we shall give a brief summary of it. By a fundamental theorem of S. Kowalewsky equation (56) has always analytic solutions (locally). Hence it is sufficient to prove our theorem for $f(x) = 0$. We denote for the moment the operator (56) by $F\varphi$ and its formal adjoint by $F^*\varphi$, and we consider Hadamard's elementary solution $G(x, y)$ belonging to $F^*\varphi$. $G(x, y)$ is a function of the form

$$\gamma\frac{U(x, y)}{R(x, y)^{n-2}} + \log R(x, y)\cdot V(x, y); \qquad U(x, x) = 1, \tag{57}$$

the coefficients $U(x, y)$, $V(x, y)$ being analytic in all $2n$ variables $x_1, \cdots, x_n, y_1, \cdots, y_n$; and $F_x^*G(x, y) = 0$. For a sufficiently small spherical surface B our solution $\varphi(x)$ of $F_x\varphi = 0$ satisfies a relation

$$\varphi(y) = \int\left(\rho(\xi)\frac{\partial G(\xi, y)}{\partial\nu} + \sigma(\xi)G(\xi, y)\right)d\omega_\xi \tag{58}$$

for every point y interior to B; the coefficients $\rho(\xi)$ and $\sigma(\xi)$ are continuous, and we shall not need their precise value. This is a consequence of Green's formula, in which the volume integrals vanish because $\varphi(x)$ and $G(x, y)$ are solutions of $F = 0$ and $F^* = 0$, respectively. The term $\varphi(y)$ emanates from the

[13] *Lectures on Cauchy's Problem*, 1923, p. 102.

singularity (57) in just the same way as in the case of relation (39). In fact, the symmetry of the function $P(x, y)$ has not been used at all for the derivation of (39), and a logarithmic term of the form $\log R(x, y) \cdot V(x, y)$ would not alter the estimate (44) for $n \geqq 3$.

Now let A be any closed point set interior to B. For $y \subset A$, $\xi \subset B$, $R(\xi, y)$ has a positive bound from below; hence we can find a region $\bar{A}$ in the space of the *complex* variables $y_1, \cdots, y_n$ over which $R(\xi, y)$ is still analytic and has a real part which has a positive bound from below. Since $U(\xi, y)$ and $V(\xi, y)$ can also be continued into a complex y-neighborhood of the origin, it follows immediately that the function (58) is analytic in some neighborhood of the origin.

2. We again choose a fixed constant $c > 0$, and denote by $\varphi = Jf$ the inverse to $f = \Delta^c\varphi$ and by $J_\nu f$ the ν-th iterate of Jf.

LEMMA 11. *There exists an integer* $\nu \geqq 1$ *and a constant* A *such that for every* $f \subset H$ *the relations*

$$| J_\nu f(x) | \leqq A \cdot || f || \tag{59}$$

$$| \operatorname{grad} J_\nu f(x) | \leqq A \cdot || f || \tag{60}$$

hold.

Proof. We write $f_0 = f$, $f_{\mu+1} = Jf_\mu$, $\mu \geqq 0$. By (50) we have the relations

$$f_1 = Kf_1 - Pf_0, \qquad f_2 = Kf_2 - Pf_1, \qquad f_3 = Kf_3 - Pf_2,$$

etc., from which we deduce the relations

$$f_1 = Kf_1 - Pf_0, \qquad f_2 = KKf_2 - (KP + PK)f_1 + PPf_0,$$

$$f_3 = KKKf_3 - (KKP + KPK + PKK)f_2 + (KPP + PKP + PPK)f_1 - PPPf_0,$$

etc. In general,

$$f_\nu = Q_{\nu 0}f_\nu + Q_{\nu 1}f_{\nu-1} + \cdots + Q_{\nu\nu}f_0, \tag{61}$$

the functions $Q_{\nu\mu}(x, y)$ being "polynomials" of ν-th degree in the functions $P(x, y)$, $K(x, y)$. Since P and K both belong to B^2_{n-2}, there exists by Lemma 4 a $\nu \geqq 1$ such that the functions $Q_{\nu\mu}$ belong to B^2. We choose a ν with this property and hold it fast. Since Jf is a bounded operator, there exists an $\alpha > 0$ such that for every $f \subset H$

$$|| f_\mu(x) || \leqq \alpha \cdot || f || \qquad (\mu = 0, 1, \cdots, \nu).$$

Therefore, by relations (61) and (48), there exists an $A > 0$ such that

$$| f_\nu(x) | \leqq A \cdot || f || \cdot$$

This proves (59). Raising the exponent ν by 1, we can also assume (for some other $A > 0$)

$$| f_{\nu-1}(x) | \leqq A \cdot || f || \, .$$

Relation (60) follows now from

$$f_\nu = Kf_\nu - Pf_{\nu-1}$$

by Lemma 2.

THEOREM III. *If $t(x)$ is any function on S belonging to C_∞, corresponding to any $\epsilon > 0$ there exists an analytic function $\psi^\epsilon(x)$ on S such that*

$$(62) \qquad | t(x) - \psi^\epsilon(x) | \leqq \epsilon, \qquad | \text{grad } [t(x) - \psi^\epsilon(x)] | \leqq \epsilon.$$

Proof. Applying the operator Δ^c ν times to the function $t(x)$, we obtain a function $\varphi(x)$ for which $t(x) = J_\nu\varphi(x)$. By Theorem II there exists an analytic function $\varphi^\epsilon(x)$ such that $\| \varphi - \varphi^\epsilon \| \leqq \epsilon/A$. The function $\psi^\epsilon(x) = J_\nu\varphi^\epsilon(x)$ is again analytic by Lemma 10, and relations (62) hold by Lemma 11, if applied to the function $f(x) = \varphi(x) - \varphi^\epsilon(x)$.

Theorem III completes the proof of the

MAPPING THEOREM. *Any compact analytic Riemann space S_n can be mapped topologically-analytically onto the Euclidean E_{2n+1}.*

Part V. Tensors

If we use the mapping theorem it can be easily shown that not only the Hilbert space of scalars on S has an analytic basis but that the same is true for the Hilbert space H of tensors of any given rank. As an illustration we shall consider tensors of rank 3, $\varphi \equiv \varphi_{\alpha\beta\gamma}$. The inner product is, of course, defined by

$$(63) \qquad (\varphi, \psi) = \int \varphi_{\alpha\beta\gamma}\psi^{\alpha\beta\gamma}\,dv \equiv \int \varphi^{\alpha\beta\gamma}\psi_{\alpha\beta\gamma}\,dv.$$

Given a tensor $\varphi_{\alpha\beta\gamma}$, we form with the functions (1) the scalars

$$A(\rho, \sigma, \tau; x) \equiv \varphi^{\alpha\beta\gamma}(x) \frac{\partial t_\rho}{\partial x_\alpha} \frac{\partial t_\sigma}{\partial x_\beta} \frac{\partial t_\tau}{\partial x_\gamma}$$

$(\rho, \sigma, \tau = 1, 2, \cdots, 2n + 1)$. Conversely, in every coördinate neighborhood on S the tensor components $\varphi^{\alpha\beta\gamma}$ can be expressed as linear combinations of the functions $A(\rho, \sigma, \tau; x)$, the coefficients being rational functions of the partial derivatives $\dfrac{\partial t_\rho}{\partial x_\alpha}$. Since the functions $A(\rho, \sigma, \tau; x)$ are limits in square mean of analytic functions, the tensor $\varphi_{\alpha\beta\gamma}$ is likewise a limit, in square mean, of analytic tensors.

Another more illuminating way of treating tensors is to set up a Laplacian for them and to investigate it as in the case of scalars. The definition of the Laplacian is

$$\Delta\varphi = g^{pq}\varphi_{\alpha\beta\gamma,p,q}.$$

Putting

$$\psi\Delta\varphi \equiv \psi^{\alpha\beta\gamma} g^{pq} \varphi_{\alpha\beta\gamma,p,q},$$

$$\psi \text{ grad } \varphi \equiv \psi^{\alpha\beta\gamma} g^{pq} \varphi_{\alpha\beta\gamma,p},$$

$$\nabla(\psi, \varphi) \equiv g^{pq} \psi^{\alpha\beta\gamma,p} \varphi_{\alpha\beta\gamma,q},$$

we have the formal identity

$$\psi\Delta\varphi = \text{div } (\psi \text{ grad } \varphi) - \nabla(\psi, \varphi)$$

which leads again to (36), (37), (38). In order to have an analogue to (39) and (40), the quantity $P(x, y)$ must be a tensor of rank 3 in both the variables x and y separately. Everything goes through strictly analogously if we replace definition (15) by

$$P(x, y) = P_{\alpha\beta\gamma;\lambda\mu\nu}(x, y) = \frac{\gamma\Phi_p[R(x, y)]}{R(x, y)^{n-2}} \frac{\partial^2\Omega}{\partial x_\alpha\, \partial y_\lambda} \frac{\partial^2\Omega}{\partial x_\beta\, \partial y_\mu} \frac{\partial^2\Omega}{\partial x_\gamma\, \partial y_\nu}.$$

The analogy is immediate up to Lemma 10. This lemma is to be replaced by the following generalization.

LEMMA 12. *If in a system of equations*

$$g^{ij} \frac{\partial^2 \varphi_\mu}{\partial x_i\, \partial x_j} + b_{\mu\lambda i} \frac{\partial \varphi_\lambda}{\partial x_i} + c_{\mu\lambda}\varphi_\lambda = f_\mu \tag{64}$$

($i, j = 1, \cdots, n$; $\lambda, \mu = 1, \cdots, m$; n and m are arbitrary integers) all given functions are analytic, then every solution $\varphi_1, \cdots, \varphi_m$ which belongs to C_2 is also analytic.

Proof. We introduce auxiliary variables $t_1, \cdots, t_m$. Putting $\Phi(x, t) = t_\mu \varphi_\mu(x)$ and multiplying (64) by t_μ, and summing with respect to μ, we obtain

$$g^{ij}(x) \frac{\partial^2 \Phi}{\partial x_i\, \partial x_j} + t_\mu b_{\mu\lambda i} \frac{\partial^2 \Phi}{\partial x_i\, \partial t_\lambda} + t_\mu c_{\mu\lambda} \frac{\partial \Phi}{\partial t_\lambda} = t_\mu f_\mu. \tag{65}$$

Since $\dfrac{\partial^2 \Phi}{\partial t_\lambda\, \partial t_\mu} = 0$, we can add to the left side of (65) the term $\delta_{\lambda\mu} \dfrac{\partial^2 \Phi}{\partial t_\lambda\, \partial t_\mu}$. If we consider $\Phi(x, t)$ as a function in the $n + m$ variables $x_1, \cdots, x_n; t_1, \cdots, t_m$, relation (65) has the form (56), and Lemma 10 applies for sufficiently small values of t, since for small values of t the matrix of the coefficients of the second derivatives is again positive definite. Therefore $\Phi(x, t)$ is analytic, and so are the functions

$$\varphi_\mu(x) = \frac{\partial \Phi}{\partial t_\mu}.$$

PRINCETON UNIVERSITY.

Reprinted from DUKE MATHEMATICAL JOURNAL
Vol. 4, No. 1, March, 1938

DIFFERENTIABLE AND RIEMANN METRIC

BY S. BOCHNER

In a previous paper[1] the author has proved that an n-dimensional compact analytic space S_n can be mapped topologically-analytically onto the Euclidean E_{2n+1} provided S_n has an analytic Riemann metric.

At first sight the notion of Riemann metric appears to be a very special and "arbitrary" case of the general concept of metric as introduced into topology by Fréchet and Hausdorff, and thus, one might think, our theorem makes a topological conclusion depend on an assumption falling well outside the domain of topology. In the present note we shall discuss the mutual relation of the two metrics for coördinate spaces in general. The discussion will be very simple indeed, but it will show a possibility of characterizing the Riemann metric by properties as closely topological as the situation permits.

1. A new characterization of Riemann metric. In a (sufficiently small) n-dimensional coördinate neighborhood S_n of class C_2 with a positive definite or semi-definite tensor $g_{ij}(x)$ of class C_1, the length function

$$L(C) = \int_C \sqrt{g_{ij}(x)\,\dot{x}_i\dot{x}_j}\,dt \tag{1}$$

(for curves of class C_1) gives rise to a geodesic distance $R(x, y)$ having the following properties of a distance function:

$$1.\ D(x, y) \geqq 0,$$

$$2.\ D(x, x) = 0,$$

$$3.\ D(x, y) + D(y, z) \geqq D(x, z),$$

$$4.\ D(x, y) = D(y, x).$$

If g_{ij} is definite throughout or becomes semi-definite in isolated points only, we have the further property

$$5.\ D(x, y) > 0 \text{ if } x \neq y.$$

Also, if S_n belongs to C_p, $p \geqq 2$, and g_{ij} to C_{p-1}, then the square of the geodesic distance

$$\Omega(x, y) = R(x, y)^2 \tag{2}$$

Received September 24, 1937.

[1] *Analytic mapping of compact Riemann spaces into Euclidean space*, this Journal, vol. 3(1937), pp. 339–354.

belongs to C_{p-2}, as a function of the $2n$ variables on the space $S_n \times S_n$. If S_n and g_{ij} are analytic, $\Omega(x, y)$ is also analytic.

Conversely, *if S_n belongs to C_p, $p \geqq 2$, and if a distance function $D(x, y)$ on S_n satisfies properties* 1, 2, 3 *and has the further property that its square*

$$\Omega(x, y) = D(x, y)^2 \tag{3}$$

belongs to C_p on $S_n \times S_n$, then the length function $L(C)$ corresponding to $D(x, y)$ can be represented in the form (1), *the generating tensor g_{ij} belonging to class C_{p-2}. If S_n and $\Omega(x, y)$ are analytic, g_{ij} is also analytic.* In this statement, $L(C)$ is defined by the relation

$$L(C) = \text{l.u.b.} \left\{ \sum_{\kappa=1}^{k-1} D(x^\kappa, x^{\kappa+1}) \right\}, \tag{4}$$

where $x^1, \cdots, x^k$ is any set of successive points on C.

The proof is quite easy. If we replace the variables y_i by new variables $\eta_i = y_i - x_i$, $\Omega(x, y)$ goes over into a new function $W(x, \eta)$ and by Taylor's formula,

$$W(x, \eta) = W(x, 0) + g_i(x)\eta_i + g_{ij}(x)\eta_i\eta_j + \epsilon_{ij}(x, \eta)\eta_i\eta_j,$$

where

$$g_i(x) = \frac{\partial W(x, 0)}{\partial \eta_i}, \qquad g_{ij}(x) = \frac{\partial^2 W(x, 0)}{\partial \eta_i \partial \eta_j} \tag{5}$$

and

$$\epsilon_{ij}(x, \eta) \to 0 \textit{ as } \eta \to 0, \textit{ uniformly in } x. \tag{6}$$

By property 2, $W(x, 0) = 0$, and by property 1, $g_i(x) = 0$, hence

$$W(x, \eta) = g_{ij}(x)\eta_i\eta_j + \epsilon_{ij}(x, \eta)\eta_i\eta_j. \tag{7}$$

By (5), g_{ij} belongs to C_{p-2}, and it is analytic, if $\Omega(x, y)$ is analytic. The tensor character of $g_{ij}(x)$ follows easily from the scalar character of $W(x, \eta)$. The integral representation (1) follows from (7) and (6) in much the same way as in the Euclidean case, the only complication arising from the fact that the norm $N(x)$ of the quadratic form

$$g_{ij}(x)\zeta_i\zeta_j \tag{8}$$

may vanish in points of C. The complication can be easily handled by choosing a finite number of intervals on C along which $N(x) \leqq 2\epsilon$, and outside of which $N(x) \geqq \epsilon$, and by showing that the contribution arising from these intervals to both the expressions (1) and (4) becomes arbitrarily small with ϵ.

If a length function $L(C)$ with the properties

(α) $L(C) \geqq 0$,

(β) $L(0) = 0$,

(γ) $L(C_1) + L(C_2) = L(C_1 + C_2)$

can be generated from a distance function $D(x, y)$ by relation (4), then the generating function is in general not unique. One of the distance functions is uniquely determined, namely, the geodesic distance $R(x, y)$ which is defined as the greatest lower bound of $L(C)$ for all curves C joining x, y. Obviously $R(x, y)$ is characterized by the property that for any other generating $D(x, y)$,

$$D(x, y) \leqq R(x, y).$$

The existence of distance functions other than $R(x, y)$ is well known. For instance, we may choose

$$D(x, y) = R(x, y)[1 - R(x, y)^2]^{\frac{1}{2}}. \tag{9}$$

It is obvious that properties 1, 2, 4, 5 carry over from $R(x, y)$ to $D(x, y)$. As for property 3, all we have to prove is that for sufficiently small positive numbers A, B, C, the relation $A + B \geqq C$ implies the relation

$$A(1 - A^2)^{\frac{1}{2}} + B(1 - B^2)^{\frac{1}{2}} \geqq C(1 - C^2)^{\frac{1}{2}}$$

and this can be verified by a simple calculation.

If we put

$$R(x, y)^2 \equiv W(x, \eta) = g_{ij}(x)\eta_i\eta_j + g_{ijk}(x)\eta_i\eta_j\eta_k + g_{ijkl}(x)\eta_i\eta_j\eta_k\eta_l + \cdots, \tag{10}$$

the square of (9) differs from (10) in the values of the coefficients g_{ijkl}. Thus two distance functions generating the same length (1) may differ in the coefficients of the fourth powers in η of the corresponding functions $W(x, \eta)$; whereas the coefficients g_{ij}, if symmetric in their indices, must be the same. Now, the coefficients g_{ijk} are also uniquely determined, as a consequence of property 4. In fact, this property implies the relation

$$W(x, \eta) = W(x + \eta, - \eta),$$

or

$$g_{ij}(x)\eta_i\eta_j + g_{ijk}(x)\eta_i\eta_j\eta_k + \cdots = \left(g_{ij}(x) + \frac{\partial g_{ij}}{\partial x_k}\eta_k + \cdots\right)\eta_i\eta_j - (g_{ijk}(x) + \cdots)\eta_i\eta_j\eta_k + \cdots,$$

and therefore

$$\left(\frac{1}{6}\frac{\partial^3 W(x, 0)}{\partial\eta_i\,\partial\eta_j\,\partial\eta_k}\right) \equiv g_{ijk} = \frac{1}{2}\frac{\partial g_{ij}}{\partial x_k}.$$

2. **Points of contraction on analytic Riemann spaces.** The points x at which (8) is semi-definite are those at which the determinant

$$|\, g_{ij}(x) \,| \tag{11}$$

has the value 0. They are characterized by the property that there are small curves C_1 issuing from x whose Riemann length becomes small of higher order than the Euclidean length. We shall term them "points of contraction". On non-analytic Riemann spaces property 5 is compatible with the existence of points of contraction having points of accumulation in the interior of the space. For instance, if $n = 1$, property 5 holds for $D(x, y) = | \varphi(x) - \varphi(y) |$, provided $\varphi(x)$ is a strictly monotone function of the given class of differentiability, without extra restrictions on the set of points at which its derivative vanishes.

But for an analytic space, if $D(x, y)^2$ is an analytic function on $S_n \times S_n$, the corresponding points of contraction are isolated. In fact, the function (11) being analytic, the set of points on which it vanishes is, in every closed neighborhood, a finite sum of analytic cells of different dimensions.[2] If there were a cell of dimension ≥ 1, there would exist an analytic curve C whose length (1) had the value 0. The geodesic distance between the endpoints of this curve would also be 0, in contradiction to property 5.

If S_n is compact, the points of contraction, if any, are finite in number. If there are none, our mapping theorem holds. One should expect the theorem to remain true in the general case, possibly with the qualification that the mapping functions need not be analytic at the points of contraction themselves. The author was unable to settle the question. We only mention, without giving details, that a direct generalization of the original method allows one to establish the theorem under additional restrictions on the behavior of $D(x, y)$ in the neighborhood of the critical points. The restrictions are rather severe. They are elaborations of the requirement that, the origin being a point of contraction, the quantity $D(x, y)$ dominate and be dominated by functions of the form

$$[\max (x, y)]^{\lambda} \cdot [(y_1 - x_1)^2 + \cdots + (y_n - x_n)^2]^{\frac{1}{2}},$$

where λ is a positive integer.

PRINCETON UNIVERSITY.

[2] S. Lefschetz, *Topology*, 1930, Chapter VIII.

HARMONIC SURFACES IN RIEMANN METRIC*

BY

S. BOCHNER

If x^i $(i=1, \cdots, n)$ is a vector representing a map from a domain Q in euclidean (u, v)-space into the euclidean x^i-space, the Euler equations of the problem

$$(1)\qquad \iint (x_u^i x_u^i + x_v^i x_v^i)\,du\,dv = \min, \qquad \text{summation over } i,$$

have the simple form

$$(2)\qquad \Delta x^i \equiv x_{uu}^i + x_{vv}^i = 0.$$

The boundary value problem reduces trivially to the classical Dirichlet problem in its simplest form. If the x^i-space is a Riemannian space S with a positive definite metric, the generalization of (1) is

$$\iint g_{pq}(x)(x_u^p x_u^q + x_v^p x_v^q)\,du\,dv = \min,$$

and the Euler equations are

$$\frac{\partial g_{pq}}{\partial x^i}(x_u^p x_u^q + x_v^p x_v^q) - 2\frac{\partial}{\partial u}(g_{iq}x_u^q) - 2\frac{\partial}{\partial v}(g_{iq}x_v^q) = 0.$$

Putting

$$(3)\qquad \frac{\partial g_{pq}}{\partial x^i} = g_{p\alpha}\Gamma_{iq}^{\alpha} + g_{\alpha q}\Gamma_{pi}^{\alpha},$$

we obtain the equivalent equations

$$(4)\qquad \Delta x^i + \Gamma_{pq}^i(x_u^p x_u^q + x_v^p x_v^q) = 0.$$

Any solution of (4) will be called a *harmonic surface.*

Equations (4) are nonlinear in the first derivatives of x^i and the boundary value problem is now decidedly more difficult than in the classical case (2). There is no direct generalization of Poisson's integral and the results obtained directly by Picard's approximation method are geometrically very restricted.†

* Presented to the Society, October 28, 1939; received by the editors June 2, 1939.

† G. Giraud, *Sur le problème de Dirichlet généralisé*, II, Annales de l'École Normale Supérieure, vol. 46 (1929), pp. 131–245.

However S. Bernstein's elaborate method of majorization* can be shown to be effective in our case, and in the present note we shall offer some first results in this direction. In §1 we shall state some elementary properties of harmonic surfaces and in §2 a uniqueness theorem relative to the boundary value problem; and in §3 we shall prove a theorem on compactness of families of harmonic surfaces.

1. **Subharmonic character of harmonic surfaces.** Equations (4) are an extension, from one to two parameters, of the equations for geodesics:

$$\frac{d^2x^i}{ds^2} + \Gamma^i_{pq}\frac{dx^p}{ds}\frac{dx^q}{ds} = 0. \tag{5}$$

As in the latter case, equations (4) can be easily seen to be formally invariant under transformations of coordinates in the x^i-space, the expressions on the left side of (4) being the components of a vector relative to such transformations. Other quantities which are formally invariant are the coefficients

$$E = g_{pq}x^p_u x^q_u, \qquad F = g_{pq}x^p_u x^q_v, \qquad G = g_{pq}x^p_v x^q_v \tag{6}$$

of the fundamental form $Edu^2+2Fdudv+Gdv^2$. These coefficients are scalars.

As in the euclidean case, the expression

$$E - G - 2F(-1)^{1/2} = g_{pq}(x)(x^p_u - (-1)^{1/2}x^p_v)(x^q_u - (-1)^{1/2}x^q_v)$$

is an analytic function of the complex variable $u+iv$. In other words we have the Cauchy-Riemann equations

$$E_u - G_u = -2F_v, \qquad E_v - G_v = 2F_u, \tag{7}$$

$$\Delta E = \Delta G, \qquad \Delta F = 0. \tag{8}$$

Relations (7) are formal identities. They hold for expressions (6), subject to relations (3) and (4). The differentiations involved can be reduced to a minimum by the use of geodesic coordinates. The first derivatives of E, F, G with respect to u and v involve the coefficients g_{pq}, Γ^i_{pq} but no derivatives of Γ^i_{pq}. In order to prove (7) for a fixed value (u_0, v_0) we may select geodesic coordinates originating at the point $x^i(u_0, v_0)$, for which, at the origin,

$$\Gamma^i_{pq} = 0, \qquad g_{pq} = \delta_{pq}. \tag{9}$$

Thus the proof of (7) consists of proving the same trivial identities as in the euclidean case.

* S. Bernstein, *Sur la généralisation du problème de Dirichlet*, Mathematische Annalen, vol. 69 (1910), pp. 82–136, §1.

If S is a space whose Riemannian curvature is nonpositive, then relations (8) can be supplemented by the further relation

$$\Delta E \geqq 0, \qquad \Delta G \geqq 0. \tag{10}$$

We shall derive a more comprehensive result relating to harmonic "surfaces" in several dimensions. If we replace the two parameters u, v by parameters u^α, $\alpha=1, \cdots, m$, and put

$$\Delta x = \frac{\partial^2 x}{\partial u^\alpha \partial u^\alpha} \equiv \sum_{\alpha=1}^{m} \frac{\partial^2 x}{\partial u^\alpha \partial u^\alpha}, \qquad E_{\beta\gamma} = g_{pq}\frac{\partial x^p}{\partial u^\beta}\frac{\partial x^q}{\partial u^\gamma},$$

then the equations

$$\Delta x^i + \Gamma^i_{pq}\frac{\partial x^p}{\partial u^\alpha}\frac{\partial x^q}{\partial u^\alpha} = 0$$

are the Euler equations of the problem

$$\int\int E_{\alpha\alpha}\, du_1 du_2 \cdots du_m \equiv \min.$$

If for given values u^α, we select a coordinate system which is geodesic at the point $x^i(u^\alpha)$, then we obtain

$$\frac{\partial E_{\beta\gamma}}{\partial u^\alpha} = g_{pq}\left(\frac{\partial^2 x^p}{\partial u^\beta \partial u^\alpha}\frac{\partial x^q}{\partial u^\gamma} + \frac{\partial x^p}{\partial u^\beta}\frac{\partial^2 x^q}{\partial u^\gamma \partial u^\alpha}\right) + g_{pa}\Gamma^a_{qr}\left(\frac{\partial x^p}{\partial u^\beta}\frac{\partial x^q}{\partial u^\gamma} + \frac{\partial x^p}{\partial u^\gamma}\frac{\partial x^q}{\partial u^\beta}\right)\frac{\partial x^r}{\partial u^\alpha}$$

and hence, putting $\Gamma^i_{pq}=0$,

$$\Delta E_{\beta\gamma} = g_{pq}\left(\Delta\frac{\partial x^p}{\partial u^\beta}\cdot\frac{\partial x^q}{\partial u^\gamma} + \frac{\partial x^p}{\partial u^\beta}\cdot\Delta\frac{\partial x^q}{\partial u^\gamma}\right) + 2g_{pq}\frac{\partial^2 x^p}{\partial u^\beta \partial u^\alpha}\frac{\partial^2 x^q}{\partial u^\gamma \partial u^\alpha}$$
$$+ g_{pa}\frac{\partial \Gamma^a_{qr}}{\partial x^s}\frac{\partial x^s}{\partial u^\alpha}\left(\frac{\partial x^p}{\partial u^\beta}\frac{\partial x^q}{\partial u^\gamma} + \frac{\partial x^p}{\partial u^\gamma}\frac{\partial x^q}{\partial u^\beta}\right)\frac{\partial x^r}{\partial u^\alpha}.$$

Now, using the formula

$$\frac{\partial}{\partial u^\beta}\Delta x^p = -\frac{\partial \Gamma^p_{ab}}{\partial x^q}\frac{\partial x^q}{\partial u^\beta}\frac{\partial x^a}{\partial u^\alpha}\frac{\partial x^b}{\partial u^\alpha}$$

and a similar formula for $\partial\Delta x^q/\partial u^\gamma$ and putting $R_{pqrs}=g_{pa}(\partial\Gamma^a_{rs}/\partial x^q - \partial\Gamma^a_{rq}/\partial x^s)$, we finally obtain

$$\Delta E_{\beta\gamma} = 2g_{pq}\frac{\partial^2 x^p}{\partial u^\beta \partial u^\alpha}\frac{\partial^2 x^q}{\partial u^\gamma \partial u^\alpha} - R_{pqrs}\left(\frac{\partial x^p}{\partial u^\beta}\frac{\partial x^q}{\partial u^\gamma} + \frac{\partial x^q}{\partial u^\beta}\frac{\partial x^p}{\partial u^\gamma}\right)\frac{\partial x^r}{\partial u^\alpha}\frac{\partial x^s}{\partial u^\alpha}.$$

For any quantities λ^α, putting $\mu_\alpha{}^p = \lambda^\beta(\partial^2 x^p/\partial u^\beta \partial u^\alpha)$, $\nu^p = \lambda^\beta(\partial x^p/\partial u^\beta)$, we hence obtain

$$\Delta E_{\beta\gamma}\overset{\beta}{\lambda}\overset{\gamma}{\lambda} = 2g_{pq}\overset{p}{\mu_\alpha}\overset{q}{\mu_\alpha} - 2R_{pqrs}\overset{p}{\nu}\overset{q}{\nu}\frac{\partial x^r}{\partial u^\alpha}\frac{\partial x^s}{\partial u^\alpha}.$$

Thus, for a space S of nonpositive curvature, $\Delta E_{\beta\gamma}\lambda^\beta\lambda^\gamma \geqq 0$, and this implies relations (10).

Finally we shall indicate a proof for the assertion that in a space S of nonpositive curvature, the curvature $K(u, v)$ of any harmonic surface is also nonpositive. By a general theorem of Ricci,* K is the sum of two quantities K_1 and K_2; K_1 is the so-called relative curvature of the surface and K_2 is the curvature of S at the point $x^i(u, v)$. Since $K_2 \leqq 0$ by assumption, we have only to show that $K_1 \leqq 0$. Now, if (9) holds, the quantity K_1 can be computed by a formula which is formally the same as in the euclidean case, namely

$$(EG - F^2)K_1 = |\overset{i}{x}_{uu}, \overset{i}{x}_u, \overset{i}{x}_v||\overset{i}{x}_{vv}, \overset{i}{x}_u, \overset{i}{x}_v| - |\overset{i}{x}_{uv}, \overset{i}{x}_u, \overset{i}{x}_v||\overset{i}{x}_{uv}, \overset{i}{x}_u, \overset{i}{x}_v|.$$

Each term on the right side is the (numerical) product of two matrices of 3 rows and n columns each. Now, (9) implies $\overset{t}{x}_{vv} = -\overset{t}{x}_{uu}$ and therefore $K_1 \leqq 0$.

2. **The distance function.** If two solutions of (2) over the same region Q have the same limit values on the boundary B of Q, then they are identical throughout Q. We must not expect a similar theorem to hold without restriction for solutions of (4) since it does not even hold for solutions of the simpler equations (5). However, the restrictions required for equations (4) are exactly the same as those needed for equations (5). It is sufficient to assume that the solutions lie in a region U of the space S in which there are no conjugate points. More precisely we assume that any two points x^i and y^i of U can be joined in S by a geodesic arc whose closure contains no conjugate points and whose geodesic length is the shortest geodesic distance between the points x^i and y^i. Denoting the latter distance by $H(x^i, y^i)$, if the space S belongs to class C_4, then $H(x^i, y^i)$ belongs to class C_2 on the domain $U \times U$ minus the diagonal manifold $x^i = y^i$.

For an arbitrary function $W = W(x^i, y^i)$ and arbitrary numbers ξ^i, η^i we shall consider the expression

$$(W_{x^p x^q} - \overset{i}{\Gamma}_{pq}(x)W_{x^i})\overset{p}{\xi}\overset{q}{\xi} + 2W_{x^p y^q}\overset{p}{\xi}\overset{q}{\eta} + (W_{y^p y^q} - \overset{i}{\Gamma}_{pq}(y)W_{y^i})\overset{p}{\eta}\overset{q}{\eta}$$

(the subscripts of W indicating partial derivations) and we shall denote this expression by

$$P(W) \equiv P(W; \xi, \eta).$$

* L. P. Eisenhart, *Riemannian Geometry*, 1926, p. 174.

We shall say that *a function $W(x^i, y^j)$ has positive character, if $P(W; \xi, \eta) \geqq 0$ identically in $x^i, y^j; \xi^i, \eta^j$ (assuming $x^i \neq y^i$, if necessary).*

We shall now prove

LEMMA 1. *The distance function $H(x^i, y^j)$ has positive character.*

We take a geodesic arc issuing from x^i, whose direction cosines at the point of issuance have the given values ξ^i, and on it a point of geodesic distance ϵ. Denoting the coordinates of the latter point by $x^i + \xi^i(\epsilon)$ we will have by familiar formulas from the theory of normal coordinates,*

$$\xi^i(\epsilon) = \epsilon\xi^i - \frac{\epsilon^2}{2!}\Gamma^i_{pq}(x)\xi^p\xi^q + o(\epsilon^2) \tag{11}$$

and

$$H(x^i, x^j + \xi^j(\epsilon)) = \epsilon\{g_{pq}(x)\xi^p\xi^q\}^{1/2}.$$

In particular

$$\frac{d}{d\epsilon}H(x^i, x^j + \xi^j(c)) - 0. \tag{12}$$

Similarly we put

$$\eta^i(\epsilon) = \epsilon\eta^i - \frac{\epsilon^2}{2!}\Gamma^i_{pq}(y)\eta^p\eta^q + o(\epsilon^2) \tag{13}$$

and obtain

$$\frac{d}{d\epsilon}H(y^i, y^j + \eta^j(\epsilon)) = 0. \tag{14}$$

Since there are no conjugate points on the geodesic arc joining x^i and y^j, the second derivative with respect to ϵ of

$$H(x^i, x^j + \xi^j(\epsilon)) + H(x^i + \xi^i(\epsilon), y^j + \eta^j(\epsilon)) + H(y^i, y^j + \eta^j(\epsilon))$$

is greater than or equal to zero for $\epsilon = 0$. On account of (12) and (14), we therefore have

$$\lim_{\epsilon\to 0}\frac{d^2}{d\epsilon^2}H(x^i + \xi^i(\epsilon), y^j + \eta^j(\epsilon)) \geqq 0.$$

Substituting (11) and (13) we hence obtain $P(H; \xi, \eta) \geqq 0$, which is the assertion of our lemma.

LEMMA 2. *If $\chi(t)$ is differentiable twice and $\chi'(t) \geqq 0$ and $\chi''(t) \geqq 0$, and if $W(x^i, y^j)$ has positive character, then $\chi(W(x^i, y^j))$ has positive character too.*

* T. Y. Thomas, *Differential Invariants of Generalized Spaces*, 1934, p. 7.

In fact,

$$P(\chi(W); \xi, \eta) = \chi'(W)\cdot P(\chi(W); \xi, \eta) + \chi''(W)\cdot(W_{x^p}\xi^p + W_{y^p}\eta^p)^2.$$

In particular, we see that the function $(H(x^i, y^j))^2$ has positive character, and this function has the advantage of belonging to class C_2 throughout $U\times U$, the manifold $x^i=y^j$ not being excepted.

LEMMA 3. *If $W(x^i, y^j)$ has positive character and $x^i(u, v)$, $y^j(u, v)$ are two harmonic surfaces, then the function $f(u, v)=W(x^i(u, v), y^j(u, v))$ is subharmonic, that is $\Delta f\geqq 0$.*

In fact, carrying out the differentiations and replacing Δx^i and Δy^i by $-\Gamma^i_{pq}(x)(x_u{}^p x_u{}^q+x_v{}^p x_v{}^q)$ and $-\Gamma^i_{pq}(y)(y_u{}^p y_u{}^q+y_v{}^p y_v{}^q)$ respectively, we obtain

$$\Delta f = P(W; x_u, y_u) + P(W; x_v, y_v),$$

and therefore $\Delta f\geqq 0$.

Since a subharmonic function attains its maximum on the boundary and since the distance function $H(x^i, y^i)$ vanishes only for $x^i=y^i$, we may conclude the following theorem.

THEOREM 1. *If U is a domain of S on which the distance function $H(x^i, y^j)$ has positive character, then for any two harmonic surfaces $x^i(u, v)$, $y^j(u, v)$, the function $H(x^i(u, v), y^j(u, v))$ attains its maximum on the boundary B of Q. If $x^i(u, v)=y^i(u, v)$ on B, then this equality holds throughout $Q+B$.*

If $K(a^i, \sigma)$ is a geodesic sphere $H(a^i, x^j)<\sigma$ in U, then any harmonic surface in U whose boundary lies in $K(a^i, \sigma)$ lies there in its entirety.

The last part of the theorem is a consequence of the first if we put $y^i(u, v)\equiv a^i$.

3. **The Dirichlet integrand on the boundary.** Using polar coordinates $u=\rho\cos\theta$, $v=\rho\sin\theta$, we assume now that Q is the unit circle $\rho<1$, $0\leqq\theta<2\pi$, and we shall denote the same function in Q by $x(u, v)$ and $x(\rho, \theta)$ interchangeably. If a periodic function $\phi(\theta)$ is continuous, and has m continuous derivatives, $m=0, 1, \cdots$, we shall write

$$\|\phi(\theta)\|_m = \max_{0\leqq\theta<2\pi} | d^m\phi(\theta)/d\theta^m |.$$

In case of a vector $\phi^i(\theta)$ we shall write

$$\|\phi(\theta)\|_m = \sum_{i=1}^{n}\|\phi^i(\theta)\|_m.$$

This symbolism will refer exclusively to the variable denoted by θ.

The theorem to follow will apply to a sufficiently small neighborhood U

of a fixed point on S in which the distance function has positive character. Making U sufficiently small and choosing the coordinate system appropriately, we may first assume that U is the entire cell $-\infty < x^i < \infty, i=1, \cdots, n$, the fixed point of U being the origin $x^i=0$. U contains some geodesic sphere $K(0, \sigma')$ whose radius will be also denoted by $3\sigma^0$. Obviously, for $0<\sigma<3\sigma^0$ there exist two functions $r'(\sigma)$ and $r''(\sigma)$, $0<r'<r''$, such that the geodesic sphere $K(0, \sigma)$ contains the euclidean sphere $x^i x^i \leqq r'^2$ and is in its turn contained in the euclidean sphere $x^i x^i \leqq r''^2$. We further observe that for $x^i(u, v)$ in $K(0, 2\sigma^0)$, the quantity

$$E + G = g_{pq}(x)(\overset{p}{x}_u \overset{q}{x}_u + \overset{p}{x}_v \overset{q}{x}_v) \tag{15}$$

is majorized above and below by

$$\overset{i}{x}_u \overset{i}{x}_u + \overset{i}{x}_v \overset{i}{x}_v . \tag{16}$$

Now, for $a^i \subset K(0, 2\sigma^0)$ and $x^i \subset K(0, \sigma^0)$ we put $\xi^i = x^i - a^i$, and we consider the function $W(a^i, \xi^i) - (H(a^i, x^i))^2/2$, and the quantities

$$\eta^i = \eta^i(a^p, \xi^q) = W_{\xi^i}. \tag{17}$$

Since, for $\xi^i=0$, $\partial\eta^p/d\xi^q = g_{pq}(a)$, we may consider the inverse functions

$$\xi^i = \xi^i(a^p, \eta^q). \tag{18}$$

They are defined, and they are inverse to the functions (17), for

$$a^i \subset K(0, 2\sigma^0), \qquad \eta^i\eta^i \leqq r_0^2 \qquad (>0), \tag{19}$$

the number r_0 being independent of a^i.

We now consider a solution $x^i(u, v)$ of (4) in Q which lies in $K(0, \sigma^0)$ and whose partial derivatives are uniformly continuous in Q. The functions $x^i(u, v)$ and their partial derivatives of first and second order have continuous boundary values on B. The function $f(u, v) = f(\rho, \theta) = W(a^i, x^i(u, v) - a^i)$ has partial derivatives of second order which are uniformly continuous in Q, and satisfies

$$\|f(1, \theta)\|_m \leqq M \|x(1, \theta)\|_m, \qquad m = 0, 1, 2,$$

the constant M being independent of $x^i(u, v)$ but not necessarily the same in all relations in which it will occur. The important property of $f(u, v)$ is that it is subharmonic, $\Delta f \geqq 0$ (see Theorem 1). We now put

$$f(\rho, \theta) = g(\rho, \theta) + h(\rho, \theta)$$

where $g(\rho, \theta)$ is the harmonic function with the boundary values $f(1, \theta)$ and

$h(\rho, \theta)$ is a subharmonic function with boundary values 0. For $g(\rho, \theta)$ we have*

$$\left|\frac{\partial g(\rho, \theta)}{\partial \rho}\right| + \left|\frac{\partial g(\rho, \theta)}{\partial \theta}\right| \leqq M\|x(1, \theta)\|_2.$$

For the function $h(\rho, \theta)$ we immediately obtain the relation

$$\lim_{\rho\to 1}\left|\frac{\partial h(\rho, \theta)}{\partial \theta}\right| \equiv \left|\frac{\partial h(1, \theta)}{\partial \theta}\right| = 0,$$

which is an immediate consequence of the assumption that $h(\rho, \theta)$ shall vanish on the boundary. For the radial derivative we obtain the one-sided relation

$$\lim_{\rho\to 1}\frac{\partial h(\rho, \theta)}{\partial \rho} \equiv \frac{\partial h(1, \theta)}{\partial \rho} \geqq 0$$

with no absolute values to start with. It follows from the fact that $h(\rho, \theta)$ is zero on the boundary and, being subharmonic, is nonpositive in Q. Altogether we have the estimates

$$\left|\frac{\partial f(1, \theta)}{\partial \theta}\right| \leqq M\|x(1, \theta)\|_2, \qquad -\frac{\partial f(1, \theta)}{\partial \rho} \leqq M\|x(1, \theta)\|_2,$$

and this can also be written in the form

$$\max\left|\eta^i(1, \theta)\frac{\partial x^i(1, \theta)}{\partial \theta}\right| - \eta^i(1, \theta)\frac{\partial x^i(1, \theta)}{\partial \rho} \leqq M\|x(1, \theta)\|_2.$$

The constant M is independent of θ and a^i, and all combinations (19) are admissible. Hence we obtain the estimate

$$\left|\frac{\partial x^i(1, \theta)}{\partial \theta}\right| + \left|\frac{\partial x^i(1, \theta)}{\partial \rho}\right| \leqq M\|x(1, \theta)\|_2,$$

the second quantity on the left side being also an absolute value. Since

$$\left(\frac{\partial x^i}{\partial u}\right)^2 + \left(\frac{\partial x^i}{\partial v}\right)^2 = \left(\frac{\partial x^i}{\partial \rho}\right)^2 + \frac{1}{\rho^2}\left(\frac{\partial x^i}{\partial \theta}\right)^2,$$

we finally obtain the following theorem.

THEOREM 2. *If S belongs to class C_4, if U is a sufficiently small coordinate neighborhood of the origin, and if $x^i(u, v) = x^i(\rho, \theta)$ is a harmonic surface in U*

* H. Whitney, *Analytic coordinate systems in a manifold*, Annals of Mathematics, (2), vol. 38 (1937), p. 809, Lemma 4.

whose partial derivatives of the second order are uniformly continuous in the unit circle Q, then the inequality

$$\{E(1, \theta) + G(1, \theta)\}^{1/2} \leqq M \sum_{i=1}^{n} \max \left| \frac{\partial^2 x^i(1, \theta)}{\partial \theta^2} \right|$$

holds for a constant M which is independent of the given surface.

If S has nonpositive curvature, then the sharper inequality

$$\{E(u, v) + G(u, v)\}^{1/2} \leqq M \sum_{i=1}^{n} \max \left| \frac{\partial^2 x^i(1, \theta)}{\partial \theta^2} \right| \tag{20}$$

holds for the same constant M.

The second half of the theorem follows from the first since for spaces of nonpositive curvature, $E+G$ attains its maximum on the boundary of Q. But the inequality (20) probably holds for spaces S of any curvature.

From Theorem 2 we could easily deduce the existence theorem* that any vector $\phi^i(\theta)$ in U which has two continuous derivatives is the boundary of a harmonic surface in U.

---

* J. Léray et J. Schauder, *Topologie et équations fonctionelles*, Annales de l'École Normale Supérieure, vol. 51 (1934), pp. 45–78.

PRINCETON UNIVERSITY,
PRINCETON, N. J.

BLOCH'S THEOREM FOR REAL VARIABLES

S. BOCHNER

The famous theorem of Bloch states that every function

$$F(z) = z + a_2 z^2 + \cdots \tag{1}$$

of the complex variable z in the unit-circle $|z| \leqq 1$ maps some subdomain of the unit-circle univalently onto a circle of a fixed radius whose size is independent of $f(z)$. The purpose of the present paper is to point out the following generalization to n real variables, $n \geqq 2$.

THEOREM 1. *Corresponding to any integer n, $n=1, 2, 3, \cdots$, and any positive constant K, $K>0$, there exists a positive radius*

$$R_0 = R_0(n, K), \qquad R_0 > 0,$$

having the following property:

If the real functions

$$u_i = f_i(x_1, \cdots, x_n), \qquad i = 1, 2, \cdots, n, \tag{2}$$

are defined and solutions of the Laplace equation

$$\frac{\partial^2 f}{\partial x_1^2} + \cdots + \frac{\partial^2 f}{\partial x_n^2} = 0 \tag{3}$$

in a neighborhood of the sphere

$$x_1^2 + \cdots + x_n^2 \leqq 1, \tag{4}$$

if their Jacobian

$$J(x_1, \cdots, x_n) = \frac{\partial(f_1, \cdots, f_n)}{\partial(x_1, \cdots, x_n)} \tag{5}$$

satisfies the decisive relation

$$\sum_{i,j=1}^{n} \left(\frac{\partial f_i}{\partial x_j}\right)^2 \leqq K \cdot |J(x_1, \cdots, x_n)|^{2/n} \tag{6}$$

in (4), *and if*

$$J(0, \cdots, 0) = 1, \tag{7}$$

then there exists in (4) *an open set S such that the functions* (2) *are a one-*

Received by the editors May 6, 1946.

to-one transformation of S onto a sphere of radius R_0 in the Euclidean space of the variables $(u_1, \cdots, u_n)$.

The decisive restriction (6) can be easily interpreted. If we introduce the differential form

$$ds^2 = g_{ij}(x)dx_i dx_j,$$

where

$$g_{ij}(x) = \sum_m \frac{\partial f_m}{\partial x_i} \cdot \frac{\partial f_m}{\partial x_j},$$

then in the ellipsoid $g_{ij}(x)\xi_i\xi_j = 1$ the quotient of the longest over the shortest axis shall have a finite bound which is independent of x; if, however, the determinant of g_{ij} vanishes at a point, then all numbers $\partial f_m/\partial x_i$ shall vanish there separately. This is a generalization of conformal transformations, for which all axes have the same length at any point.

For $n = 2$ the significance of this generalization has been brought to light in the work of Ahlfors[1] when he proved that Picard's theorem on exceptional values will remain in force for this type of transformation. For $n \geqq 3$, our condition may turn out to be of even greater significance. On the one hand there are no strictly conformal transformations other than trivial ones (inversions in spheres). But on the other hand, general analytic transformations in several *complex* variables of the form

$$w_l = F_l(z_1, \cdots, z_k), \qquad l = 1, \cdots, k; n = 2k, \tag{8}$$

do not obey our restriction for $k \geqq 2$, which fact may be said to be responsible for the complete breakdown of Picard's theorem in that case.

As for our additional requirement that each component $f_i(x)$ shall be a solution of (3), it will be evident from the course of the proof that it could be considerably toned down. However, we are drawing up the condition in this rigid form in order to point up an analogy to the case of complex variables. For one complex variable, (3) requires as in the classical case that each component of (1) separately shall be a solution of the Laplace equation. However, in lieu of the rigorous rela-

[1] Compare the account in R. Nevanlinna, *Eindeutige analytische Funktiones*, p. 343. There were also earlier efforts in that direction. See, for instance, H. Groetzsch, *Ueber möglichst konforme Abbildungen von schlichten Bereichen*, Berichte über der Verhandlungen der K. Sächsischen Akademie der Wissenschaften zu Leipzig, Mathematische-Physische Klasse, vol. 84 (1932) pp. 114–120.

tion between the two components as implied in the Cauchy-Riemann equation we now have the much milder condition (6). For several complex variables, analyticity implies in particular the relations

$$\frac{\partial^2 f_i}{\partial x_l^2} + \frac{\partial^2 f_i}{\partial y_l^2} = 0, \qquad l = 1, \cdots, k,$$

for each pair (x_l, y_l) separately, whereas our present condition (3) replaces this by the summary requirement

$$\sum_{l=1}^{k} \left(\frac{\partial^2 f_i}{\partial x_l^2} + \frac{\partial^2 f_i}{\partial y_l^2} \right) = 0,$$

thus creating a situation which is even more general than the case of one complex variable would indicate.

For the proof of Theorem 1 we introduce the functions

$$M(r) = \max_{x_1^2+\cdots+x_n^2 \leq r^2} \left| J(x_1, \cdots, x_n) \right|^{1/n}$$

and

$$\phi(r) = (1 - r)M(r).$$

Since $\phi(0) = 1$ and $\phi(1) = 0$, there exists a number r^0 such that

$$\phi(r^0) = 1, \qquad \phi(r) < 1 \text{ for } r^0 < r \leq 1.$$

Also, there exists a point

(9) $$x^0 = (\overset{0}{x}_1, \cdots, \overset{0}{x}_n)$$

for which $(x_1^0)^2 + \cdots + (x_n^0)^2$ is precisely equal to $(r^0)^2$ and such that

$$J(\overset{0}{x}_1, \cdots, \overset{0}{x}_n)^{1/n} = M(\overset{0}{r}) = \frac{1}{1 - r^0}.$$

Now introduce the functions

(10) $$g_i(\xi_1, \cdots, \xi_n) = \frac{f_i(\overset{0}{x}_m + \rho\xi_m) - f_i(\overset{0}{x}_m)}{\rho \left| J(x_m^0) \right|^{1/n}}, \qquad i = 1, \cdots, n,$$

where

$$\rho = \frac{1 - r^0}{2},$$

and since for

(11) $$\xi_1^2 + \cdots + \xi_n^2 \leq 1$$

we have

$$\sum_{m=1}^{n} (x_m^0 + \rho\xi_m)^2 \leqq \left(\frac{1+r^0}{2}\right)^2,$$

they will exist in the unit sphere (11). We also introduce their Jacobian

$$G(\xi_1, \cdots, \xi_n) = \frac{\partial(g_1, \cdots, g_n)}{\partial(\xi_1, \cdots, \xi_n)},$$

and we list the following properties:

(i) All functions $g_i(\xi)$ are again solutions of the Laplacean $\partial^2 g/\partial\xi_1^2 + \cdots + \partial^2 g/\partial\xi_n^2 = 0$.

(ii) $G(0, \cdots, 0) = 1$.

(iii) $G(\xi_1, \cdots, \xi_n) \leqq 2^n$ for ξ in (11).

(iv) $\sum_{i,j=1}^{n}(\partial g_i/\partial\xi_j)^2 \leqq K|G(\xi_1, \cdots, \xi_n)|^{2/n} \leqq 4K$.

We note that property (iii) follows from

$$\left|\frac{\partial(g_1, \cdots, g_n)}{\partial(\xi_1, \cdots, \xi_n)}\right| = \frac{|J(x_1, \cdots, x_n)|}{|J(x_1^0, \cdots, x_n^0)|} \leqq \left(\frac{M((1+r^0)/2)}{M(r^0)}\right)^n$$
$$\leqq \left(\frac{2}{1-r^0}(1-r^0)\right)^n = 2^n.$$

We are now going to make use of property (i). Since the first partial derivatives $\partial g_i/\partial\xi_j$ are themselves solutions of the Laplacean, it follows from Poisson's integral for harmonic functions that corresponding to n and K there exists a constant $L = L(n, K)$ such that the validity of property (iv) in (11) will imply the further property

$$\text{(v)} \qquad \sum_{i,j,k=1}^{n} \left(\frac{\partial^2 g_i}{\partial\xi_j\partial\xi_k}\right)^2 \leqq L$$

in

$$\text{(12)} \qquad \xi_1^2 + \cdots + \xi_n^2 \leqq 1/2.$$

However, properties (ii), (iv) and (v), together with

$$g_i(0, \cdots, 0) = 0, \qquad i = 1, \cdots, n,$$

lead to the following conclusion. There exists a radius $R_1 = R_1(n, K, L) \equiv R_1(n, K)$ such that on some subset B of (12)—the subset depending on the given functions $g_i(\xi)$—the functions

$$\eta_i = g_i(\xi_1, \cdots, \xi_n)$$

are a one-to-one transformation of B into the sphere

$$\eta_1^2 + \cdots + \eta_n^2 \leqq R_1^2.$$

This is a "uniform" version of the theorem on implicit functions and can be derived from the Kronecker integral or in some other fashion.[2]

Returning now to the original functions (2), we see from (10) that these functions will transform a subdomain B of (4) into a sphere of radius

$$R_0 = R_1 \cdot \rho \left| J(x_m^0) \right|^{1/n}.$$

However, $\rho |J(x_m{}^0)|^{1/n} = ((1-r_0)/2)\cdot(1/(1-r_0)) = 1/2$, and this completes the proof of Theorem 1.

THEOREM 2. *If all assumptions of Theorem* 1 *are fulfilled in the sphere* $x_1{}^2 + \cdots + x_n{}^2 \leqq \lambda^2$, *then the map contains a sphere of radius* $R = \lambda \cdot R_0(n, K)$.

This follows from Theorem 1, if we replace $f_i(x)$ by $\lambda f_i(x/\lambda)$.

Hence we conclude:

THEOREM 3. *If all assumptions of Theorem* 1 *are fulfilled in the entire space, then the map contains spheres of arbitrarily large radius.*

PRINCETON UNIVERSITY

[2] See, for instance, the first chapter in Carathéodory's *Variationsrechnung*.

VECTOR FIELDS AND RICCI CURVATURE

S. BOCHNER

We shall prove theorems on nonexistence of certain types of vector fields on a compact manifold with a positive definite Riemannian metric whose Ricci curvature[1] is either everywhere positive or everywhere negative. Actually we shall have some relaxations of the requirements both as to curvature and as to compactness. We shall deal with real spaces with a customary metric and with complex analytic spaces with an Hermitian metric. In the latter case we shall impose on the metric a certain restriction, first explicitly stated by E. Kaehler,[2] which will be quite indispensable to our argument. In order to elucidate the rôle of this restriction we shall include a systematic introduction to the theory of Hermitian metric.

For positive curvature we shall have the theorem that on a compact space there exists no vector field for which the *divergence* and *curl* both vanish. In the complex case there exists no vector field whatsoever whose covariant components are analytic functions in the complex parameters. If we only assume that the curvature is nonnegative, then there are some "exceptional" vector fields in directions of spatial flatness. A principal result will be the following theorem on meromorphic functions. If a complex space with positive curvature is covered by a finite number of neighborhoods, if a meromorphic functional element is defined in each neighborhood, and if the difference of any two meromorphic elements is holomorphic wherever the elements overlap, then there exists *one* meromorphic function on the space which differs by a holomorphic function from each meromorphic element given. In a previous paper[3] this conclusion was drawn in the

---

Received by the editors June 28, 1946; published with the invited addresses for reasons of space and editorial convenience.

[1] Also called mean curvature; the definition will be restated later in the text. Interesting facts relating Ricci curvature to Riemann curvature have been given by T. Y. Thomas, *On the variation of curvature in Riemann spaces of constant mean curvature*, Annali di Matematica Pura ed Applicata (4) vol. 13 (1935) pp. 227–238, and *New theorems on Riemann-Einstein spaces*, Rec. Math. (Mat. Sbornik) N.S. vol. 3 (1938) pp. 331–340. Also, for the application of the Laplacean on compact Riemannian spaces see T. Y. Thomas, *Some applications of Green's theorem for compact Riemann spaces*, Tôhoku Math. J. vol. 46 (1940) pp. 261–266.

[2] *Ueber eine bemerkenswerte Hermitische Metrik*, Abh. Math. Sem. Hamburgischen Univ. vol. 9 (1933) pp. 173–186; see also S. Chern, *Characteristic classes of Hermitian manifolds*, Ann. of Math. vol. 47 (1946) pp. 85–121, especially pp. 109–112.

[3] S. Bochner, *Analytic and meromorphic continuation by means of Green's formula*, Ann. of Math. vol. 44 (1943) pp. 652–673, especially p. 672, Theorem 15.

case of a multi-torus; however, in that case, due to the flatness of the space, some necessary and sufficient conditions on the meromorphic elements had to be added. These conditions correspond to the "exceptional" vector fields mentioned before, and they do not occur if the curvature is strictly positive.

For negative curvature our theorem will in the main be a generalization of a theorem of Schwarz and M. Noether[4] from one to several complex variables. It states that on an algebraic curve of genus $p>1$ there is no one-parametric group of rational transformations. In other words, on such a compact two-dimensional surface in one complex variable there exists no continuous group of analytic homeomorphisms. Now, the universal covering surface of such a surface can be identified with the unit circle $|z|<1$ and thus endowed with a hyperbolic metric of (constant) negative curvature. Thus we can express the theorem of Schwarz and Noether in the following second version. If a "real" compact two-dimensional space with a Riemannian metric has (constant) negative curvature then there exists no continuous group of conformal (that is, angle-preserving) transformations into itself. Corresponding to these two versions we shall have two generalizations to several variables. The complex version will be that if a compact complex space in n variables with a suitable Hermitian metric has negative Ricci curvature (not necessarily constant), then there exists no continuous group of complex homeomorphisms. The "real" version will be less satisfactory but rather illuminating. We shall first show that for negative curvature in n variables there are no groups of motions. A motion preserves not only angles but also the lengths. In order to obtain a theorem on transformations which distort the length we shall have to add an assumption on the nature of the distortion. However, this assumption will be automatically fulfilled for $n=2$, and thus technically a proper generalization will be obtained. We shall also have "exceptional" groups in case the curvature is only nonpositive, and we shall have a further statement in case the given compact space is a minimal variety in some higher-dimensional space of nonpositive curvature.

Finally we shall have some converse theorems, but only locally. A converse theorem states that if there are sufficiently many vector fields of one kind or another then the curvature is negative or positive respectively. Locally, a gradient field is a scalar function from which it is derived, and several gradient fields is therefore a mapping into

[4] H. A. Schwarz, *Gesammelte Abhandlungen*, vol. 2, pp. 285–291; M. Noether, Math. Ann. vol. 20 (1882) pp. 59–62 and vol. 21 (1883) pp. 138–140. See also A. Hurwitz, *Mathematische Werke*, vol. 1, pp. 241–259.

Euclidean space. Correspondingly we shall have the following theorems on *complex* space. Any piece of m-dimensional surface in n-dimensional Euclidean space has negative curvature, but the presence of sufficiently many contravariant vector fields in commutation gives rise to a fundamental tensor whose curvature is positive.

The most interesting problem in this connection is the problem of constructing a negative (hyperbolic) curvature from the existence of a noncommutative transitive group of analytic homeomorphisms on a piece of Euclidean or other space. This problem will not be approached at all in the present context.

I. Real spaces

1. **Lemmas.** Let S be an n-dimensional coordinate space with a fixed positive definite Riemannian metric

$$ds^2 = g_{ij}dx_i dx_j. \tag{1}$$

The space is of differentiability class C^3, and all scalars, vectors and tensors belong to class C^2. For any scalar ϕ we can form the Laplacean

$$\Delta\phi \equiv g^{ij}\phi_{,i,j} \equiv \frac{1}{g^{1/2}}\frac{\partial}{\partial x_i}\left(g^{1/2}g^{ij}\frac{\partial\phi}{\partial x_j}\right) \tag{2}$$

and we shall apply it so the square length

$$\phi = \xi^i\xi_i \equiv g^{ij}\xi_i\xi_j \equiv g_{ij}\xi^i\xi^j, \tag{3}$$

where $\{\xi_i\}$ is an arbitrary vector field on all of S. We shall next have a definition.

DEFINITION. A vector ξ_i on S is called a *restrained* vector, if either

$$\Delta\phi(P_0) < 0$$

at some point P_0 in S, or

$$\phi(P) \equiv \text{constant} \tag{4}$$

throughout S.

This definition will be justified by the following theorem.

LEMMA 1. *On a compact space S every vector is restrained relative to every metric.*

On a general S a vector is restrained if its length in some metric attains a (relative) maximum in S, and thus in particular if the length is less than ϵ outside some compact subset depending on ϵ.

PROOF. The compact case can be dealt with by a "global" argument, namely Stokes' formula

$$\int (\phi \cdot \Delta\phi + g^{ij}\phi_{,i}\phi_{,j})dv = 0. \tag{5}$$

In fact, if $\phi \geqq 0$, $\Delta\phi \geqq 0$, then (5) can hold only if $\phi_{,i} \equiv 0$, that is, $\phi \equiv$ constant, as claimed. But there is also a "local" argument which implies the lemma in its entirety. This is the following theorem of E. Hopf.[5] If $\Delta\phi \geqq 0$ in a neighborhood T, and if there exists a point P_0 in T such that $\phi(P) \leqq \phi(P_0)$ for all P in T, then $\phi(P) \equiv \phi(P_0)$ in T.

We now start from the formula

$$\xi_{i,j,k} - \xi_{i,k,j} - \xi_l R^l_{ijk} = 0 \tag{6}$$

where R^l_{ijk} is the curvature tensor. The Ricci tensor arises by contraction,

$$R_{ij} = R^k_{ijk} = g^{lk} R_{lijk} = R^l_{ijl}, \tag{7}$$

and it has the following interpretation.[6] If for a given point we consider n orthogonal unit vectors $\eta^i_{(m)}$, $m = 1, \cdots, n$, which are cooriented with $x_1, \cdots, x_n$, and if we denote by κ_{pq} the curvature of the geodesic surface element which is determined by $\eta^i_{(p)}$ and $\eta^i_{(q)}$, then the sum

$$\kappa_{11} + \kappa_{12} + \cdots + \kappa_{1n}$$

is given by

$$- R_{ij}\eta^i_{(1)}\eta^j_{(1)},$$

thus being dependent only on $\eta^i_{(1)}$ itself. We shall call it the Ricci curvature in the direction $\eta^i_{(1)}$. The curvature is called positive if it is strictly positive in all directions.

LEMMA 2. *The quantity*

$$2^{-1}\Delta\phi = 2^{-1}g^{rs}(g^{ab}\xi_a\xi_b)_{,r,s} = g^{rs}g^{ab}\xi_{a,r}\xi_{b,s} + g^{rs}g^{ab}\xi_{a,r,s}\xi_b \tag{8}$$

has the value

$$g^{rs}g^{ab}\xi_{a,r}\xi_{b,s} - R^{lm}\xi_l\xi_m + A \tag{9}$$

where the remainder term is

[5] *Elementare Bemerkungen ueber die Loesungen partieller Differentialgleichungen zweiter Ordnung vom elliptischen Typus*, Preuss. Akad. Wiss. Sitzungsber. vol. 19 (1927) pp. 147–152.

[6] See L. P. Eisenhart, *Riemannian geometry*, p. 113. We have adopted Eisenhart's normalization throughout, and this explains why positive Ricci curvature corresponds to a negative Ricci quadratic form.

$$A = g^{rs}g^{ab}(\xi_{a,r,s} - \xi_{r,a,s} + \xi_{r,s,a})\xi_b; \tag{10}$$

it also has the value

$$g^{rs}g^{ab}\xi_{a,r}\xi_{b,s} + R^{lm}\xi_l\xi_m + B \tag{11}$$

where

$$B = g^{rs}g^{ab}(\xi_{a,r,s} + \xi_{r,a,s} - \xi_{r,s,a})\xi_b. \tag{12}$$

In fact, from (6) we obtain

$$g^{rs}g^{ab}(\xi_{r,s,a} - \xi_{r,a,s})\xi_b - g^{ab}R^l_a\xi_l\xi_b = 0.$$

If we add this to (8) we obtain (9), and if we subtract it from (8) we obtain (11).

2. **Positive curvature.** Since the term

$$g^{rs}g^{ab}\xi_{a,r}\xi_{b,s} \tag{13}$$

is positive for arbitrary quantities $\xi_{i,j}$ except in the case where the latter are all 0, we first obtain the following conclusion.

THEOREM 1. *The remainder term A is obviously* 0 *whenever*

$$\xi_{i,j} = \xi_{j,i} \quad \textit{and} \quad g^{ij}\xi_{i,j} = 0, \tag{14}$$

that is,

$$\operatorname{curl} \xi = 0 \quad \textit{and} \quad \operatorname{div} \xi = 0, \tag{15}$$

or more generally, whenever

$$(\xi_{i,j} - \xi_{j,i})_{,k} = 0 \quad \textit{and} \quad (g^{ij}\xi_{i,j})_{,k} = 0. \tag{16}$$

Therefore, there exists no restrained vector field which satisfies relation (16) *and*

$$- R^{ij}\xi_i\xi_j \geqq 0 \tag{17}$$

unless we have

$$- R^{ij}\xi_i\xi_j = 0, \qquad \xi_{i,j} = 0. \tag{18}$$

Thus, the only exceptions are fields of parallel vectors in directions of Ricci-flatness, and they cannot arise, for instance, if the space as such has positive Ricci curvature throughout.

A vector field for which $\xi_{i,j} = \xi_{j,i}$ is a local gradient field and can be described in the following way. There exists a covering of the space by a system of neighborhoods $\{U^p\}$, which for a compact space is finite, and a scalar $f^{(p)}(x)$ on each U^p, such that

(19) $$f^{(p)}(x) - f^{(q)}(x) = \text{constant}$$

whenever two neighborhoods U^p, U^q intersect. And ξ_i is defined as

(20) $$\xi_i = \frac{\partial}{\partial x_i} f^{(p)}(x)$$

on U^p. The second part of (14) amounts to assuming that each $f^{(p)}(x)$ shall in its neighborhood be a solution of the Laplacean

(21) $$\Delta f = 0,$$

and our theorem states in particular that on a space of positive Ricci curvature every $f^{(p)}(x)$ is itself constant. It should be noted that this conclusion is trivial if S is simply connected, or if some simply connected covering space of S is compact, since in this case the system of equations (20) can be solved by one function $f(x)$ in S; and since its Laplacean vanishes, it is a constant throughout.

3. **Negative curvature.** We now turn to an analysis of the expression (11) and we shall first have a detailed lemma on the influence of the remainder term B.

LEMMA 3. *The remainder term B vanishes if*

(22) $$\xi_{i,j} + \xi_{j,i} = 0,$$

or more generally if

(23) $$(\xi_{i,j} + \xi_{j,i})_{,k} = 0.$$

Thus it vanishes if the contravariant components ξ^i are an infinitesimal generator of a one-parametric group of motions, that is, metric-preserving homeomorphisms, or, more generally, of affine collineations.

We have $B \geqq 0$, if there exists a scalar function $\lambda(x)$ such that

(24) $$\xi_{i,j} + \xi_{j,i} = \lambda g_{ij},$$

that is, if ξ^i represents a one-parametric group of angle-preserving (conformal) homeomorphisms, provided we also have

(25) $$(1 - n/2) g^{ij} \lambda_i \xi_j \geqq 0,$$

the latter condition being automatically fulfilled for $n=2$.

PROOF. If we multiply (23) by g^{ij}, then due to the symmetry of g^{ij} we obtain

$$g^{rs} \xi_{r,s,a} = 0,$$

and thus (23) implies $B=0$. On any differentiable manifold, with or without a Riemannian metric, a vector ξ^i describes a local group of differentiable homeomorphisms. If a metric is given we can form the covariant components ξ_i, (22) is the necessary and sufficient condition for the transformations to be motions,[7] and (23) for affine collineations.

Condition (24) expresses preservation of angles. If we differentiate it we obtain

$$\xi_{i,j,k} + \xi_{j,i,k} = \lambda_{,k} g_{ij} \tag{26}$$

and hence

$$g^{rs} g^{ab} (\xi_{a,r,s} + \xi_{r,a,s}) \xi_b = g^{sb} \lambda_{,s} \xi_b.$$

If, however, we multiply (26) by $g^{ij} g^{kb} \xi_b$ we obtain

$$- g^{rs} g^{ab} \xi_{r,s,a} \xi_b = - (n/2) g^{sb} \lambda_{,s} \xi_b$$

and thus (25) is equivalent with $B \geqq 0$.

THEOREM 2. *There exists no restrained group of motions* $\{\xi^i\}$ *for which*

$$- R_{ij} \xi^i \xi^j \leqq 0, \tag{27}$$

unless we have

$$\xi_{i,j} = 0, \qquad R_{ij} \xi^i \xi^j = 0, \tag{28}$$

that is, unless it is a group of translations along Ricci-flat geodesics.

For $n = 2$, *the same conclusion holds for conformal mappings* $\{\xi^i\}$, *and for* $n \geqq 3$ *it will hold under the additional assumption* (25).

4. Almost periodic vector fields. Take a compact space S and its universal covering space T. If S has negative curvature, T most likely will not be compact. Introduce on T the group Γ of automorphisms for which S is the set of co-sets, and denote by S_0 a fundamental domain in T. It can be chosen compact. Every scalar or tensor function on S gives rise to such a function on T which is "periodic," that is, invariant under Γ. In particular the tensor g_{ij} can be periodically extended. Now take a vector field ξ_i on T which, though not necessarily periodic, is together with its second derivatives almost periodic with respect to Γ. This means that if $\{\gamma_p\}$ is a sequence in Γ then there exists a sub-sequence $\{\gamma_q\}$ such that the sequence of vector fields $\xi_i(\gamma_q P)$ and their second derivatives is uniformly convergent in T.

[7] Eisenhart, p. 234.

THEOREM 3. *Any almost periodic vector field ξ_i on the covering space T of S is restrained in T, and thus Theorems* 1 *and* 2, *and similar theorems, apply to such vector fields in T.*

PROOF. Due to its almost periodicity, the square length $\phi(P)$ of ξ_i is bounded and uniformly continuous, and there exists a sequence of points P_p such that

$$\phi(P_p) \to \sup_P \phi(P).$$

Each P_p lies in some image $\gamma_p(S_0)$ of S_0, and owing to the compactness of S_0 there exists a subsequence γ_q such that the sequence of translated functions $\phi(\gamma_q P)$ and their second derivatives will converge uniformly in T towards a function $\Phi(P)$ which assumes its maximum in an interior point of T. Now, due to the periodicity of g_{ij}, we have

$$\Delta\phi(P) = \Delta\phi(\gamma_q P) = \Delta\Phi(P),$$

and thus $\Delta\phi \geqq 0$ implies $\Delta\Phi \geqq 0$. By Lemma 1, we therefore have $\Phi = \text{const}$. However this implies $\phi = \text{const.}$, and thus ξ_i is restrained.

5. **Minimal varieties.** If V_n is a subspace of any space V_{n+1} whose Riemann-Christoffel tensor vanishes, then for the Ricci tensor of V_n we have $R_{jk} = g^{il}(b_{ik}b_{jl} - b_{il}b_{jk})$ where b_{ij} are the coefficients of the second differential form.[8] If V_n is a minimal hypersurface of V_{n+1}, then $g^{ij}b_{ij} = 0$,[9] and therefore

$$R_{jk}\xi^j\xi^k = g^{il}\eta_i\eta_l \tag{29}$$

where

$$\eta_i = b_{ik}\xi^k. \tag{30}$$

Thus the vanishing of (29) implies the vanishing of (30), and the vanishing of the latter for a group $\{\xi^k\}$ means that the orbits are asymptotic lines on V_n.[10] However, any geodesic line of V_n which is asymptotic on V_n is a geodesic of V_{n+1} itself. The same reasoning applies to a minimal variety in a space V_m of any higher dimension $m \geqq n+1$, since we have in general

$$R_{jk}\xi^j\xi^k = \sum_{\sigma=1}^{m-n} g^{ij} b_{ik}^{(\sigma)} \xi^k b_{jl}^{(\sigma)} \xi^l.$$

[8] Eisenhart, p. 190.
[9] Eisenhart, p. 178.
[10] Eisenhart, p. 167.

Thus we obtain the following theorem.

THEOREM 4. *If the space V_n is a minimal variety in a space V_m, $m \geqq n+1$, of vanishing Riemann curvature, and if there exists a one-parametric group of motions on V_n, then the orbits are geodesics in the enveloping V_m which are also asymptotic lines on V_n.*

II. HERMITIAN METRIC

6. **Power series.** We take independent *complex* variables

$$t_1, \cdots, t_m \tag{31}$$

and we consider power series in these variables with complex coefficients, each series converging in some (not fixed) neighborhood of the origin. Every "function," that is scalar, scalar density or any component of a tensor shall be such a series.

A "transformation"

$$t_i' = \phi_i(t_1, \cdots, t_m), \qquad i = 1, \cdots, m, \tag{32}$$

shall be given by functions ϕ_i which vanish at the origin, in order to allow unrestricted formation of the product of two transformations, and the Jacobian

$$\frac{\partial(t_1', \cdots, t_m')}{\partial(t_1, \cdots, t_m)}$$

shall be not equal to 0 at the origin. The totality of such transformations forms a group, and we consider an *arbitrary* fixed subgroup Γ of the group. The elements of Γ shall be termed allowable transformations and the resulting quantities t_i' allowable variables. We can now define as usual vectors and tensors

$$\xi_i, \ \xi^i, \ \xi_i{}^k{}_{lm\cdots},$$

with upper and lower indices; also, contraction of an upper and lower index is again permitted. A symmetric affine connection Γ_{jk} is again a system of functions with the customary rules of transformation. It gives rise to a mixed tensor

$$R^l_{ijk} = \frac{\partial}{\partial t_j}\Gamma^l_{ik} - \frac{\partial}{\partial t_k}\Gamma^l_{ij} + \Gamma^m_{ik}\Gamma^l_{mj} - \Gamma^m_{ij}\Gamma^l_{mk} \tag{33}$$

with which we can form the Ricci tensor

$$R_{ij} = R^k_{ijk}; \tag{34}$$

it also gives rise to covariant differentiation of arbitrary tensors and tensor densities as usual. For instance

$$(35)\qquad \xi_{i,j} = \frac{\partial \xi_i}{\partial t_j} - \xi_k \Gamma^k_{ij},$$

$$(36)\qquad \xi^i_{,j} = \frac{\partial \xi^i}{\partial t_j} + \xi^k \Gamma^i_{kj}.$$

Also derivatives of sums, products and contractions are to be formed as usual.

We next introduce a fundamental tensor g_{ij}. For the present we only assume that it is symmetric

$$(37)\qquad g_{ij} = g_{ji},$$

and that the determinant

$$g = \| g_{ij} \|_{i,j=1,\cdots,m},$$

is not equal to 0 at the origin. We emphasize that each component g_{ij} is a complex power series in the complex variables (31) and that the symmetry condition (37) is not at all an "Hermitian" property. There exist functions g^{ij} which are solutions of $g^{ij}g_{ik} = \delta^i_k$, and they are a symmetric tensor. The expressions

$$(38)\qquad \Gamma^i_{jk} = 2^{-1} g^{il}\left(\frac{\partial g_{lk}}{\partial t_j} + \frac{\partial g_{jl}}{\partial t_k} - \frac{\partial g_{jk}}{\partial t_l}\right)$$

are an affine connection, and the tensors g_{ij}, g^{ij} can be used for pulling indices up and down in a manner compatible with covariant differentiation. Also

$$(39)\qquad g_{ij,k} = 0, \qquad g^{ij}{}_{,k} = 0,$$

and[11]

$$(40)\quad R_{ij} = 2^{-1}\frac{\partial^2}{\partial t_i \partial t_j}\log\frac{g}{C_0} - \frac{\partial}{\partial t_k}\Gamma^k_{ij} + \Gamma^m_{ik}\Gamma^k_{mj} - 2^{-1}\Gamma^m_{ij}\frac{\partial}{\partial t_m}\log\frac{g}{C_0},$$

where C_0 is the value of the determinant g at the origin. Also

$$(41)\qquad R^i_{jkl} + R^i_{klj} + R^i_{ljk} = 0$$

and

$$(42)\qquad R_{ijkl} = - R_{jikl} = - R_{klji}.$$

[11] Eisenhart, p. 21.

7. Self-adjointness. We now put $m=2n$ and we replace the variables (31) by the symbols

$$z_1, \cdots, z_n; \bar{z}_1, \cdots, \bar{z}_n. \tag{43}$$

Allowable transformations shall be those having the special form

$$z_\alpha' = \phi_\alpha(z_1, \cdots, z_n), \qquad \bar{z}_\alpha' = \psi_\alpha(\bar{z}_1, \cdots, \bar{z}_n), \tag{44}$$

where for each $\alpha=1, \cdots, n$ the power series $\psi_\alpha(t_1, \cdots, t_n)$ arises from the power series $\phi_\alpha(t_1, \cdots, t_n)$ by replacing each coefficient by its conjugate complex value. Therefore, if each $\bar{z}_\alpha$ is conjugate to z_α, then so is $\bar{z}_\alpha'$ to z_α', and therefore it will be consistent to assume henceforth that z_α are complex parameters and $\bar{z}_\alpha$ their conjugate values. However a function in general will continue to be a power series in all $2n$ variables (43), or rather it will be a power series with complex coefficients in the $2n$ real components x_α, y_α; $z_\alpha=x_\alpha+iy_\alpha$. Special scalar or tensor functions whose power series depend only on z_α will be termed "complex analytic" or "analytic" for short.

From now on *italic* indices $i, j, k, \cdots$ will run from 1 to $2n$; Greek indices $\alpha, \beta, \cdots$ from 1 to n, and starred Greek indices will be n units more, $\alpha^*=\alpha+n$. We shall also star italic indices to indicate a change by n, thus $i^*=i\pm n$, $j^*=j\pm n$.

Let ζ_i be a covariant vector. For a transformation (44) we have

$$\zeta_\alpha' = \zeta_\beta \frac{\partial z_\beta'}{\partial z_\alpha}, \qquad \zeta_{\alpha^*}' = \zeta_{\beta^*} \frac{\partial \bar{z}_\beta'}{\partial \bar{z}_\alpha}.$$

Thus the $2n$ components of ζ_i separate completely into the blocks ζ_α and ζ_{α^*}. For all tensors this separation applies to each upper and lower index independently of the others. For instance, g_{ij} separates into the blocks $g_{\alpha\beta}$, $g_{\beta\alpha}$, $g_{\alpha\beta^*}$ $g_{\alpha^*\beta}$, of n^2 components each.

We call a tensor *self-adjoint* if starring all indices simultaneously changes the value of a component into its conjugate:

$$\zeta^{j^*}_{i^*k^*l^*\cdots} = \overline{\zeta^{j}_{ikl\cdots}}.$$

For a vector it means

$$\zeta_{\alpha^*} = \overline{\zeta_\alpha}, \qquad \zeta^{\alpha^*} = \overline{\zeta^\alpha},$$

and for a *symmetric* tensor g_{ij} it means

$$g_{\alpha\beta} = g_{\beta\alpha} = \overline{g_{\alpha^*\beta^*}} = \overline{g_{\beta^*\alpha^*}}, \tag{45}$$

$$g_{\alpha\beta^*} = g_{\beta^*\alpha} = \overline{g_{\alpha^*\beta}} = \overline{g_{\beta\alpha^*}}. \tag{46}$$

A scalar shall be self-adjoint if it is real-valued. Self-adjointness is preserved under addition, multiplication and contraction. If g_{ij} is self-adjoint, then the inverse g^{ij}, and Γ^i_{jk} based on g_{ij}, are self-adjoint, and so are the Riemann and Ricci tensors. Furthermore it is preserved under covariant differentiation.

From now on all functions are assumed to be self-adjoint.

8. **Restrictions on the fundamental tensor.** The first substantial restriction on the fundamental tensor shall be

$$g_{\alpha\beta} = 0 = g_{\alpha^*\beta^*}. \tag{47}$$

Thus only the "interacting" components $g_{\alpha\beta^*} = g_{\beta^*\alpha}$ can be not equal to 0, and $g_{\alpha\beta^*} = \overline{g_{\alpha^*\beta}}$. We automatically also have

$$g^{\alpha\beta} = 0$$

and (38) gives

$$\Gamma^{\alpha}_{\beta\gamma} = 2^{-1} g^{\alpha\epsilon^*}\left(\frac{\partial g_{\epsilon^*\gamma}}{\partial z_\beta} + \frac{\partial g_{\beta\epsilon^*}}{\partial z_\gamma}\right), \tag{48}$$

$$\Gamma^{\alpha}_{\beta\gamma^*} = 2^{-1} g^{\alpha\epsilon^*}\left(\frac{\partial g_{\beta\epsilon^*}}{\partial \bar{z}_\gamma} - \frac{\partial g_{\beta\gamma^*}}{\partial \bar{z}_\epsilon}\right), \tag{49}$$

$$\Gamma^{\alpha}_{\beta^*\gamma^*} = 0, \tag{50}$$

and 5 other relations obtained by symmetry and self-adjointness. From (35) we now have

$$\zeta_{\alpha,\beta} = \frac{\partial \zeta_\alpha}{\partial z_\beta} - \zeta_\epsilon \Gamma^{\epsilon}_{\alpha\beta}, \tag{51}$$

$$\zeta_{\alpha^*,\beta} = \frac{\partial \zeta_\alpha}{\partial \bar{z}_\beta} - \zeta_\epsilon \Gamma^{\epsilon}_{\alpha\beta^*}. \tag{52}$$

However we also want to have $\Gamma^{\alpha}{}_{\beta\gamma^*} = 0$, and this will be obtained by Kaehler's restriction, which is:

$$\frac{\partial g_{\alpha\gamma^*}}{\partial z_\beta} = \frac{\partial g_{\beta\gamma^*}}{\partial z_\alpha}, \tag{53}$$

and by adjointness also implies

$$\frac{\partial g_{\gamma\alpha^*}}{\partial \bar{z}_\beta} = \frac{\partial g_{\gamma\beta^*}}{\partial \bar{z}_\alpha}. \tag{54}$$

This is equivalent with assuming that there exists (locally) a scalar function $\phi(z, \bar{z})$ such that

$$g_{\alpha\beta^*} = \frac{\partial^2\Phi}{\partial z_\alpha \partial \bar{z}_\beta}. \tag{55}$$

Self-adjointness demands that Φ shall be real-valued, and all these assumptions are indeed invariant under transformations of the type (43). We shall now list consequences of our restrictions.

LEMMA 4. *We have*

$$\zeta_{\alpha,\beta^*} = \frac{\partial \zeta_\alpha}{\partial \bar{z}_\beta}, \qquad \zeta^\alpha{}_{,\beta^*} = \frac{\partial \zeta^\alpha}{\partial \bar{z}_\beta}.$$

In particular if the component ζ_α is complex analytic, we have $\zeta_{\alpha,\beta^*}=0$, and if ζ^α is complex analytic we have $\zeta^\alpha{}_{,\beta^*}=0$.

More generally, if the indices $\alpha_1, \cdots, \alpha_p$ are unstarred, we have

$$\zeta_{\alpha_1,\cdots,\alpha_p,\beta^*} = \frac{\partial \zeta_{\alpha_1,\cdots,\alpha_p}}{\partial \bar{z}_\beta}; \qquad \zeta^{\alpha_1,\cdots,\alpha_p}{}_{,\beta^*} = \frac{\partial \zeta^{\alpha_1,\cdots,\alpha_p}}{\partial \bar{z}_\beta}.$$

For the appreciation of the lemma it should be noted that if the n components ζ_α are complex analytic the contravariant components

$$\zeta^\alpha = g^{\alpha\beta^*}\bar{\zeta}_\beta, \qquad \zeta^{\alpha^*} = g^{\alpha^*\beta}\zeta_\beta$$

will depend on all $2n$ variables (42) since the components $g^{\alpha\beta^*}$ so depend.

LEMMA 5. *For the Riemann tensor we have*

$$R_{\alpha\beta jk} = 0. \tag{56}$$

More generally, due to (42), *only those components of R_{lijk} can be different from* 0 *for which, in both pairs of indices* (l, i) *and* (j, k) *simultaneously, one element in the pair is a Greek unstarred and the other a Greek starred index.*

PROOF. It suffices to prove that $R^{\alpha^*}_{\beta jk}$ vanishes. From (33) we obtain for this component the value

$$\frac{\partial}{\partial t_j}\Gamma^{\alpha^*}_{\beta k} - \frac{\partial}{\partial t_k}\Gamma^{\alpha^*}_{\beta j} + \Gamma^{m}_{\beta k}\Gamma^{\alpha^*}_{mj} - \Gamma^{m}_{\beta j}\Gamma^{\alpha^*}_{mk},$$

and this vanishes since only the pure components $\Gamma^\alpha_{\beta\gamma}$ and $\overline{\Gamma^\alpha_{\beta\gamma}}$ can be not equal to 0.

LEMMA 6. *We also have*

$$R^\alpha_{\beta\gamma^*\delta} = -R^\alpha_{\delta\beta\gamma^*}.$$

PROOF. In the Bianchi identity

$$R_{\beta\gamma^*\delta}^{\alpha} + R_{\gamma^*\beta\delta}^{\alpha} + R_{\delta\beta\gamma^*}^{\alpha} = 0$$

the middle term is 0 by Lemma 5.

LEMMA 7. *For the Ricci tensor we have*

$$R_{\alpha\beta} = 0, \tag{57}$$

$$R_{\alpha\beta^*} = \frac{\partial^2 \log |G|}{\partial z_\alpha \partial \bar{z}_\beta} \tag{58}$$

where G is the determinant

$$G = \|g_{\alpha\beta}\|_{\alpha,\beta=1,\cdots,n}.$$

PROOF. (57) follows from

$$R_{\alpha\beta^*} = g^{lk} R_{l\alpha\beta k}$$

by Lemma 5, and (58) follows from (40).

9. **Positive fundamental tensor.** Finally, we make the assumption that the matrix $g_{\alpha\beta^*}$ is positive-definite. Consequently, for any vector ζ_i the "square length"

$$\phi = g^{ij}\zeta_i\zeta_j = 2g^{\alpha\beta^*}\zeta_\alpha\overline{\zeta_\beta} \tag{59}$$

is non-negative. For the Laplacean we have

$$\Delta\phi = g^{ij}\phi_{,i,j} = g^{\alpha\beta^*}\phi_{,\alpha,\beta^*} + g^{\alpha^*\beta}\phi_{,\alpha^*,\beta}, \tag{60}$$

and by Lemma 4 this is

$$2g^{\alpha\beta^*}\frac{\partial^2\phi}{\partial z_\alpha \partial \bar{z}_\beta}.$$

Now, if we herein put $z_\alpha = x_\alpha + iy_\alpha$, then the resulting expression is a type of Laplacean in the variables (x_α, y_α) to which Lemma 1 will apply.

If Φ is the function occurring in (55), the positiveness of the tensor is expressed by the relation

$$\frac{\partial^2\Phi}{\partial z_\alpha \partial \bar{z}_\beta}\zeta^\alpha\overline{\zeta^\beta} \geqq 0, \tag{61}$$

which is supposed to hold for every vector ζ^α. Also, we shall say that the Ricci curvature is nonpositive if $R_{\alpha\beta^*}\zeta^\alpha\zeta^{\beta^*} \geqq 0$, and on account of (58) this is

$$\frac{\partial^2 \log G}{\partial z_\alpha \partial \bar{z}_\beta} \zeta^\alpha \overline{\zeta^\beta} \geqq 0, \tag{62}$$

which is the same type of relation as (61) is. It is not hard to see that relations (61) and (62) must hold for arbitrary complex numbers $\zeta^1, \cdots, \zeta^n$. In particular, if we introduce a complex curve $z_\alpha = z_\alpha(\omega)$, $\omega = u + iv$, and $\zeta^\alpha = dz_\alpha/d\omega$, then we must have $\partial^2\phi/\partial\omega\partial\bar{\omega} \geqq 0$, that is, $\partial^2\phi/\partial u^2 + \partial\phi^2/\partial v^2 \geqq 0$. In other words the function Φ must be a subharmonic function on any analytic curve. Similarly in the case (62), the function $\log G$ must be subharmonic in the manner stated.

III. Complex spaces

10. Vector fields. A complex space is a coordinate space in $2n$ real variables. Every basic neighborhood is the topological image of a neighborhood of the Euclidean space in n complex parameters. If two basic neighborhoods intersect, then the two images of the intersection are mapped into each other by analytic functions

$$z_\alpha' = \phi_\alpha(z_1, \cdots, z_n), \qquad \alpha = 1, \cdots, n. \tag{63}$$

It is easily seen that all concepts and results of Chapter II can be extended to a complex space in the large. In particular we assume the existence of a tensor g_{ij} with all properties and restrictions as enumerated before. We define a *restrained* vector field as before, and on a compact space every vector field is again restrained.

LEMMA 8. *If the covariant components ζ_α are complex analytic then for the scalar*

$$\phi = 2g^{\alpha\beta^*}\zeta_\alpha\zeta_{\beta^*} \tag{64}$$

we have

$$4^{-1}\Delta\phi = g^{\rho\sigma^*}g^{\alpha\beta^*}\zeta_{\alpha,\rho}\zeta_{\beta^*,\sigma^*} - R^{\alpha\beta^*}\zeta_\alpha\zeta_{\beta^*}.$$

If however the contravariant components ζ^α are analytic then we have

$$4^{-1}\Delta\phi = g^{\rho\sigma^*}g_{\alpha\beta^*}\zeta^\alpha{}_{,\rho}\zeta^{\beta^*}{}_{,\sigma^*} + R_{\alpha\beta^*}\zeta^\alpha\zeta_{\beta^*}.$$

PROOF. In the first case we have on basis of Lemmas 4 and 6

$$4^{-1}\Delta\phi = g^{\rho\sigma^*}(g^{\alpha\beta^*}\zeta_\alpha\zeta_{\beta^*})_{,\rho,\sigma^*} = g^{\rho\sigma^*}g^{\alpha\beta^*}\zeta_{\alpha,\rho}\zeta_{\beta^*,\sigma^*} + g^{\rho\sigma^*}g^{\alpha\beta^*}\zeta_{\alpha,\rho,\sigma^*}\zeta_{\beta^*}.$$

Now

$$\zeta_{\alpha,\rho,\sigma^*} = \zeta_{\alpha,\rho,\sigma^*} - \zeta_{\alpha,\sigma^*,\rho} = -\zeta_\lambda R^{\lambda}_{\alpha,\sigma^*,\rho},$$

and by Lemma 6 this is

$$+ \zeta_\lambda R^\lambda_{\rho\alpha\sigma^*} = - \zeta_\lambda R^\lambda_{\rho\sigma^*\alpha}.$$

Therefore

$$g^{\rho\sigma^*} g^{\alpha\beta^*} \zeta_{\alpha,\rho,\sigma^*} \zeta_\beta = - \zeta_\lambda g^{\alpha\beta^*} R^\lambda_\alpha \zeta_{\beta^*} = - \zeta_\lambda \zeta_{\beta^*} R^{\lambda\beta^*}.$$

In the second case we have, as claimed,

$$g^{\rho\sigma^*} g_{\alpha\beta^*} \zeta^\alpha_{,\rho,\sigma^*} \zeta^{\beta^*} = - g^{\rho\sigma^*} g_{\alpha\beta} \zeta^\lambda R^\alpha_{\lambda\rho\sigma^*} \zeta^{\beta^*} = - g^{\rho\sigma^*} R_{\beta^*\lambda\rho\sigma^*} \zeta^{\beta^*} \zeta^\lambda$$
$$= - g^{\rho\sigma^*} R_{\beta^*\sigma^*\lambda\rho} \zeta^{\beta^*} \zeta^\lambda = R_{\beta\lambda} \zeta^{\beta^*} \zeta^\lambda.$$

From Lemma 8 we obtain directly the following theorem, the first half of which is an unqualified extension of the theorem of Schwarz and Noether from one to several complex variables.

THEOREM 5. *On a compact space of negative Ricci curvature there exists no one-parametric group of analytic homeomorphisms whatsoever. More specifically, if on any space in a restricted vector field the contravariant components ζ^α are complex analytic and $-R_{\alpha\beta^*}\zeta^\alpha\zeta^{\beta^*} \leqq 0$ holds, then we of necessity have $\zeta^\alpha{}_{,\beta} = 0$ and $R_{\alpha\beta^*}\zeta^\alpha\zeta^{\beta^*} = 0$.*

If however the covariant components ζ_α are analytic then the inequality $-R^{\alpha\beta^*}\zeta_\alpha\zeta_{\beta_*} \geqq 0$ must be a strict equality and it implies $\zeta_{\alpha,\beta} = 0$.

It should be noted that the second half of the theorem is *not* restricted to local gradient fields as was Theorem 1.

It is very easy to extend Theorem 5 from vectors to tensors.

THEOREM 6. *If a compact space has positive Ricci curvature then there exists no analytic covariant tensor $\zeta_{\alpha_1 \ldots \alpha_p}$ whatsoever, and for negative curvature there exists no analytic contravariant tensor whatsoever.*

PROOF. For instance, in the case $p = 2$, if we put

$$\phi = 4 g^{\alpha_1\beta^*_1} g^{\alpha_2\beta^*_2} \zeta_{\alpha_1\alpha_2} \zeta_{\beta^*_1\beta^*_2}$$

we obtain

$$8^{-1}\Delta\phi = g^{\rho\sigma^*} g^{\alpha_1\beta_2} g^{\alpha_2\beta_1} \zeta_{\alpha_1\alpha_2,\rho} \zeta_{\beta^*_1\beta^*_2,\sigma^*} + A$$

where

$$A = g^{\rho\sigma^*} g^{\alpha_1\beta_1} g^{\alpha_2\beta_2} \zeta_{\alpha_1\alpha_2,\rho,\sigma^*} \zeta_{\beta^*_1\beta^*_2}.$$

However,

$$\zeta_{\alpha_1\alpha_2,\rho,\sigma^*} = \zeta_{\alpha_1\alpha_2,\rho,\sigma^*} - \zeta_{\alpha_1\alpha_2,\sigma^*,\rho} = - \zeta_{\lambda\alpha_2} R^\lambda{}_{\alpha_1\sigma^*\rho} - \zeta_{\alpha_1\lambda} R^\lambda{}_{\alpha_2\sigma^*\rho}$$
$$= - \zeta_{\lambda\alpha_2} R^\lambda{}_{\rho\alpha_1\sigma^*} - \zeta_{\alpha_1\lambda} R^\lambda_\rho{}_{\alpha_2\sigma^*}.$$

Therefore

$$A = - \zeta_{\lambda\alpha_2} \zeta_{\beta^*_1\beta^*_2} R^{\lambda\beta^*_1} g^{\alpha_2\beta^*_2} - \zeta_{\alpha_1\lambda} \zeta_{\beta^*_1\beta^*_2} R^{\lambda\beta^*_2} g^{\alpha_1\beta^*_1},$$

and the conclusion follows.

11. **Meromorphic functions.** We consider a (self-adjoint) vector field which in the neighborhood of every point can be represented in the form

$$\zeta_\alpha = \frac{\partial f}{\partial z_\alpha}, \qquad \zeta_{\alpha^*} = \overline{\zeta_\alpha} = \frac{\partial \bar{f}}{\partial \bar{z}_\alpha}. \tag{65}$$

It is *not* a local gradient field in the proper sense unless $f=\bar{f}$; that is, if f is real-valued, and in our application it will definitely not be so. We introduce the associate vector

$$\eta_\alpha = \frac{\partial \bar{f}}{\partial z_\alpha}, \qquad \eta_{\alpha^*} = \frac{\partial f}{\partial \bar{z}_\alpha}.$$

These two vectors are not subject to Lemma 4; however they have the following alternate properties

$$\zeta_{\alpha,\beta} = \zeta_{\beta,\alpha}, \qquad \zeta_{\alpha^*,\beta^*} = \zeta_{\beta^*,\alpha^*},$$

$$\eta_{\alpha,\beta} = \eta_{\beta,\alpha}, \qquad \eta_{\alpha^*,\beta^*} = \eta_{\beta^*,\alpha^*},$$

$$\zeta_{\alpha,\beta^*} = \eta_{\beta^*,\alpha}, \qquad \zeta_{\alpha,\beta} = \eta_{\beta,\alpha^*}.$$

For $\phi = 2g^{\alpha\beta^*}\zeta_\alpha\zeta_{\beta^*}$ we obtain

$$8^{-1}\Delta\phi = A + B + C$$

where

$$A = g^{\rho\sigma^*}g^{\alpha\beta^*}(\zeta_{\alpha,\rho}\zeta_{\beta^*,\sigma^*} + \zeta_{\alpha,\sigma^*}\zeta_{\beta^*,\rho}),$$

$$B = g^{\rho\sigma^*}g^{\alpha\beta^*}\zeta_{\alpha,\rho,\sigma^*}\zeta_{\beta^*},$$

$$C = g^{\rho\sigma^*}g^{\alpha\beta^*}\zeta_\alpha\zeta_{\beta^*,\rho,\sigma^*}.$$

If we substitute $\zeta_{\alpha,\rho,\sigma^*} = \zeta_{\rho,\alpha,\sigma^*} = (\zeta_{\rho,\alpha,\sigma^*} - \zeta_{\rho,\sigma^*,\alpha}) + \zeta_{\rho,\sigma^*,\alpha} = \zeta_{\rho,\sigma^*,\alpha} - \zeta_\lambda R^\lambda{}_{\rho\alpha\sigma^*}$ we obtain

$$B = g^{\alpha\beta^*}(g^{\rho\sigma^*}\zeta_{\rho,\sigma^*})_{,\alpha}\zeta_{\beta^*} - R^{\lambda\beta^*}\zeta_\lambda\zeta_{\beta^*},$$

and if we put

$$\zeta_{\beta^*,\rho,\sigma^*} = \eta_{\rho,\beta^*,\sigma^*} = \eta_{\rho,\sigma^*,\beta^*} + (\eta_{\rho,\beta^*,\sigma^*} - \eta_{\rho,\sigma^*,\beta^*}) = \eta_{\rho,\sigma^*,\beta} + \zeta_\lambda \overset{\lambda}{R}{}_{\rho\beta^*\sigma^*}$$

we obtain by Lemma 5

$$C = g^{\alpha\beta^*}(g^{\rho\sigma^*}\eta_{\rho,\sigma^*})_{,\beta}\zeta_{\alpha}.$$

Finally we introduce the assumption

$$g^{\rho\sigma^*}\zeta_{\rho,\sigma^*} = 0, \tag{66}$$

that is,

$$\Delta f = 0. \tag{67}$$

This will also imply $g^{\rho\sigma^*}\eta_{\rho\sigma^*}=0$, and if for the sake of a later application we interchange the variables (z_α) and $(\bar{z}_\alpha)$ we obtain the following lemma.

LEMMA 9. *If on a compact space with positive Ricci curvature a (self-adjoint) vector field ζ_i has the property that in the neighborhood of every point the components ζ_{α^*} can be represented in the form*

$$\zeta_{\alpha^*} = \frac{\partial f}{\partial \bar{z}_\alpha} \quad \text{with} \quad \Delta f = 0 \tag{68}$$

then $\zeta_{\alpha^*}\equiv 0$, that is, f is complex analytic.

If the curvature is only non-negative then $\zeta_{\alpha^*,i}=0$, that is, the derivatives $\partial f/\partial \bar{z}_\alpha$ are not necessarily 0 but have covariant derivative 0.

Finally we shall have the following theorem.

THEOREM 6. *If a compact space with positive Ricci curvature is subdivided into simplices $S_1, \cdots, S_t$, if for each r, $1\leqq r\leqq t$, there is defined a meromorphic functional element $\phi_r(z_1, \cdots, z_n)$ in some neighborhood U_r of the closure of S_r, and if each difference $\phi_p-\phi_q$ is holomorphic in the intersection of U_p and U_q (whenever not empty), then there exists one meromorphic function $\Phi(z_1 \cdots z_n)$ in S such that, for each U_r, the difference $\Phi-\phi_r$ is holomorphic.*

PROOF. The proof proceeds as follows. We replace each ϕ_r in U_r by another function $f_{(r)}$ which has no singularities but which on the other hand may also depend on the conjugate values $\bar{z}_\alpha$. In other words, $\partial f_{(r)}/\partial \bar{z}_\alpha$ need not be not equal to 0. However, each $f_{(r)}$ shall be a solution of the Laplacean $\Delta f=0$, and the difference $f_{(p)}-f_{(q)}$ shall be equal to $\phi_p-\phi_q$ for all (p, q). If such functions $f_{(r)}$ can be found then the components

$$\zeta_{\alpha^*} = \frac{\partial f_{(r)}}{\partial \bar{z}_\alpha}, \qquad \alpha = 1, \cdots, n,$$

are uniquely defined in S, and Lemma 9 applies immediately. Thus, for positive curvature, the components all vanish. In other words, the functions $f_{(r)}$ are automatically analytic, and the expression

$$\Phi = \phi_p - f_{(p)} = \phi_q - f_{(q)}$$

will define a meromorphic function in S. A similar proof holds for non-negative curvature. Therefore the only remaining task is the construction of the functions $f_{(r)}$. This is a problem in potential theory and has a solution in either real or Hermitian space. We shall only indicate its solution since an elaboration of details would lead us far afield.

Suppose we are dealing with a compact m-dimensional space with a real positive fundamental form. We set up the Laplacean, and it is known[12] that there always exists a Green's function $G(x;\xi)$. It has the proper type of singularity for $x=\xi$, and it satisfies the relation

$$\Delta_x G(x;\xi) = -1.$$

Let S_r and U_r have the same meaning as in the theorem and let B_{pq} denote the $(n-1)$-dimensional face separating S_p and S_q, if such a face exists. In the integral to follow each such face will appear twice with opposite orientations. Now, for each r, let $\phi_{(r)}$ be an "arbitrary" function in U_r with "singularities." By this we mean that it need be defined and differentiable only in some subset of U_r. However, the differences

$$\phi_{pq} = \phi_p - \phi_q \tag{69}$$

shall be each defined and differentiable in some neighborhood of the closure of B_{pq}. We now set up the Cauchy-Green integral

$$\frac{1}{\omega}\sum_{p,q=1}^{t}\int_{B_{pq}}\sum_{\mu=1}^{m}\left(\phi_{pq}(\xi)\frac{\partial G(x,\xi)}{\partial\xi_\mu} - G(x,\xi)\frac{\partial\phi_{pq}(\xi)}{\partial\xi_\mu}\right)\sigma^\mu.$$

The symbol σ^μ denotes a certain invariant $(m-1)$-dimensional external differential form, which in the Euclidean case is

$$(-1)^\mu d\xi_1\cdots d\xi_{\mu-1}d\xi_{\mu+1}\cdots d\xi_m,$$

whereas ω is a numerical constant. For variable x the integral as it stands defines a function in the interior of each S_r. Denoting this function by $f_{(r)}(x)$, we can show that it can be continued beyond S_r through the interior of each $(m-1)$-dimensional face, that $f_{(p)}(x) - f_{(q)}(x)$ has the value $\phi_{pq}(x)$ on B_{pq}, and that it is a solution of the Laplacean.[13] All this would be true for arbitrary functions $\phi_{pq}(\xi)$ on B_{pq}, but then $f_{(r)}(x)$ need not be continuable through the interior of the boundary faces of dimension not greater than $m-2$. If however the prescribed saltus function has the special form (69) then $f_{(r)}(x)$ can be continued into an entire neighborhood of the closure of S_r, as needed.

In our complex space with a Laplacean of Hermitian type the

[12] Compare, for instance, S. Bochner, *Analytic mapping of compact Riemann spaces into Euclidean space*, Duke Math. J. vol. 3 (1937) pp. 339–354.

[13] See the paper quoted in footnote 3, p. 657.

existence of the Green's function could be proven by an adaptation of the proof for the real case. However the duplication can be avoided, and the complex case can be directly reduced to the real case, if we make the additional assumption on S that it is basically a real space which in some of its allowable coordinates is also Hermitian. By this we mean that there exists a $2n$-dimensional Riemannian space S_0 which is in topological correspondence with S. Each allowable coordinate system in S is in correspondence with some coordinate system of S_0, and the Laplacean on S is a transcription into complex parameters of the real Laplacean as defined on S_0.

12. **Subspaces of Euclidean space.** The following statements are of a local character. Suppose we are given several analytic covariant vector fields in a neighborhood of n complex parameters. We denote them by $\zeta_{k/\alpha}$, where the subscript k identifies the vector, and we attempt to define a fundamental tensor by the formula

$$g_{\alpha\beta^*} = \sum_{k=1}^{l} \zeta_{k/\alpha}\zeta_{k/\beta^*}.$$

Our first concern must be to secure restriction (53). The obvious way of securing it is to assume that we have

$$\zeta_{k/\alpha} = \frac{\partial \phi_k}{\partial z_\alpha} \tag{70}$$

where $\phi_k(z)$ is an analytic scalar, because in this case

$$g_{\alpha\beta^*} = \frac{\partial^2}{\partial z_\alpha \partial \bar{z}_\beta}\left(\sum_k |\phi_k|^2 \right).$$

This is a crude way of securing (53) and there probably is some more elaborate condition; we shall deal with it nevertheless, since geometrically it corresponds to a mapping of our z-neighborhood into a W-neighborhood by the functions $W_k = \phi_k(f)$.

THEOREM 7. *If*

$$W_k = \phi_k(z_1 \cdots z_n), \qquad k = 1, \cdots, l,$$

is a parametrization of a subspace of Euclidean space, then its Hermitian tensor

$$g_{\alpha\beta^*} = \sum_{k=1}^{l} \frac{\partial \phi_k}{\partial z_\alpha} \frac{\partial \bar{\phi}_k}{\partial \bar{z}_{\alpha^*}}$$

generates a negative Ricci curvature.

PROOF. If we evaluate the determinant G of $g_{\alpha\beta^*}$ then we find for its value the expression

$$\sum_{p=1}^{C_{1,k}} D_p \bar{D}_p, \tag{71}$$

where each D_p is the value of some n-dimensional determinant of the n by l matrix

$$\left| \frac{\partial \phi_k}{\partial z_{\alpha^*}} \right|, \qquad \alpha = 1, \cdots, n; \; k = 1, \cdots, l. \tag{72}$$

We now put

$$R_{\alpha\beta^*} = \frac{\partial^2 \log G}{\partial z_\alpha \partial \bar{z}_\beta}$$

and form the sum

$$-R_{\alpha\beta^*} \lambda^\alpha \overline{\lambda^\beta} \tag{73}$$

for an arbitrary complex number λ^α. Now, if we carry out the differentiations and if we put

$$\eta^p = \frac{\partial D^p}{\partial z_\alpha} \lambda^\alpha,$$

then it turns out that the sum (71) has the value

$$G^{-2} \sum_{p,q} | \eta^p D^q - \eta^q D^p |^2$$

and this is non-negative. Also, if the matrix (71) has rank n, this must be strictly positive for some λ^α, as claimed.

For the positiveness of (71) it is not necessary to assume that our vectors have the special form (70); however in general the matrix (73) is not the Ricci tensor and has no immediate geometrical interpretation.

13. **Commutative analytic groups.** We now assume the existence of contravariant analytic vectors ζ_k^α and we set up the corresponding tensor

$$g^{\alpha\beta^*} = \sum_{k=1}^{l} \zeta_k^\alpha \zeta_{k^*}^{\beta^*}.$$

If we denote its determinant by H, then by our last remark the matrix

$$-\frac{\partial^2 \log H}{\partial z_\alpha \partial \bar{z}_\beta}$$

is positive definite. If we introduce the inverse tensor $g_{\alpha\beta^*}$, then $G = H^{-1}$, and this would prove that the Ricci curvature is positive, provided $g_{\alpha\beta^*}$ satisfies restriction (53). Now the identity

$$g_{\beta\epsilon^*} g^{\epsilon^*\alpha} = \delta_\beta^\alpha$$

implies

$$\frac{\partial g_{\beta\epsilon^*}}{\partial z_\gamma} g^{\epsilon^*\alpha} + g_{\beta\epsilon^*} \frac{\partial g^{\epsilon^*\alpha}}{\partial z_\gamma} = 0$$

or

$$\frac{\partial g_{\beta\delta^*}}{\partial z_\gamma} = - g_{\beta\epsilon^*} \frac{\partial g^{\epsilon^*\alpha}}{\partial z_\gamma} g_{\alpha\delta^*}.$$

In applying (53) we hence obtain

$$\frac{\partial g^{\epsilon^*\alpha}}{\partial z_\beta} g^{\beta\delta^*} = \frac{\partial g^{\delta^*\alpha}}{\partial z_\beta} g^{\beta\epsilon^*}$$

which is equivalent with

$$\sum_{p,q} \frac{\partial \zeta_p^\alpha}{\partial z_\beta} \overline{\zeta_p^\epsilon} \zeta_q^\beta \overline{\zeta_q^\delta} = \sum_{pq} \frac{\partial \zeta_q^\alpha}{\partial z_\beta} \overline{\zeta_q^\delta} \zeta_p^\beta \overline{\zeta_p^\epsilon}.$$

The only obvious way of securing this relation is to assume that

$$\frac{\partial \zeta_p^\alpha}{\partial z_\beta} \zeta_q^\beta - \frac{\partial \zeta_q^\alpha}{\partial z_\beta} \zeta_p^\beta = 0$$

for all α, p, q. But this is the classical condition for commutative groups. Thus if the rank of the matrix $|\zeta_p^\alpha|$ is n, and the group operators are commutative, they give rise to an Hermitian metric with positive curvature.

PRINCETON UNIVERSITY

CURVATURE IN HERMITIAN METRIC

S. BOCHNER

Hermitian metric has the peculiarity of favoring negative curvature over positive curvature. We shall explain this phenomenon by pointing out that in the case of an isometric analytic imbedding the relative curvature is on the whole negative; also, by reduction to a *limiting case* of imbedding we shall explain why an invariant metric in the theory of Fuchsian groups is likely to be hyperbolic; see Hua [6].[1]

However, on the other hand, an Hermitian metric is very rigid, and the possibility of imbedding into a *finitely dimensional* enveloping space is very remote. The classical conjectures about the possibility of Euclidean imbedding are rendered entirely false, but as a compensation, there are more and better theorems about equivalence and uniqueness.

1. **Hermitian metric.** There are many places in the literature where an introduction to the theory of Hermitian metric can be found. We shall refer to our own summary as given in Bochner [2, chap. 2]. We quote from there that in discussing an Hermitian metric in a space V_n of n complex variables $z_1, \cdots, z_n$, the basic variables are the $2n$ conjugate complex quantities

$$z_1, \cdots, z_n; \bar{z}_1, \cdots, \bar{z}_n \tag{1}$$

which we shall also denote indifferently by

$$t_1, t_2, \cdots, t_{2n}. \tag{2}$$

Italic indices run from 1 to $2n$, Greek indices from 1 to n, and starring an index will add to it the value n if it is not greater than n, and subtract n from it if it is not less than $n+1$. All scalars and components of tensors are power series in (1), and they are always *self-adjoint*, meaning that starring all indices in a component of a tensor will change its value into its conjugate complex. Scalars are real-valued.

The fundamental tensor g_{ij} has the properties

$$g_{ij} = g_{ji}, \qquad g_{\alpha\beta} = g_{\alpha^*\beta^*} = 0$$

in addition to the self-adjointness property $g_{\alpha\beta^*} = \bar{g}_{\beta\alpha^*}$; also $g_{\alpha\beta^*}$ is positive-definite. In particular, we have

Received by the editors August 3, 1946.

[1] Numbers in brackets refer to the bibliography at the end of the paper.

$$ds^2 = g_{ij}dt_idt_j = 2g_{\alpha\beta^*}dz_\alpha d\bar{z}_\beta. \tag{3}$$

Finally, we have Kaehler's restriction (non-torsion)

$$\frac{\partial g_{\alpha\gamma^*}}{\partial z_\beta} = \frac{\partial g_{\beta\gamma^*}}{\partial z_\alpha} \tag{4}$$

which is equivalent with

$$g_{\alpha\beta^*} = \frac{\partial^2 \Phi}{\partial z_\alpha \partial \bar{z}_\beta} \tag{5}$$

for some scalar function $\Phi(z, \bar{z})$, locally. On replacing Φ by $(\Phi+\overline{\Phi})/2$ we can always make it real-valued. If we form Γ^i_{jk} by the usual formula, then only the unmixed components $\Gamma^\alpha_{\beta\gamma}$ and their conjugates can be not equal to 0, and

$$\Gamma^\alpha_{\beta\gamma} = g^{\alpha\sigma^*} \frac{\partial g_{\sigma^*\beta}}{\partial z_\gamma}. \tag{6}$$

If we form the curvature tensor R_{ijkl} by the usual formula, then it satisfies all classical relations. Furthermore, only the components of the form $R_{\alpha\beta^*\gamma\delta^*}$, $R_{\alpha\beta^*\gamma^*\delta}$, $R_{\alpha^*\beta\gamma\delta^*}$, $R_{\alpha^*\beta\gamma^*\delta}$ can be different from 0. Finally, we have

$$R^\alpha_{\beta\gamma^*\delta} = \frac{\partial}{\partial \bar{z}_\gamma} \Gamma^\alpha_{\beta\delta}$$

and therefore

$$\begin{aligned} R_{\alpha^*\beta\gamma^*\delta} &= g_{\alpha^*\rho} \frac{\partial}{\partial \bar{z}_\gamma}\left(g^{\rho\sigma^*} \frac{\partial g_{\sigma^*\beta}}{\partial z_\delta}\right) \\ &= \frac{\partial}{\partial \bar{z}_\gamma}\left(g_{\alpha^*\rho} g^{\rho\sigma^*} \frac{\partial g_{\sigma^*\beta}}{\partial z_\delta}\right) - g^{\rho\sigma^*} \frac{\partial g_{\sigma^*\beta}}{\partial z_\delta} \frac{\partial g_{\alpha^*\rho}}{\partial \bar{z}_\gamma}. \end{aligned}$$

Hence we have finally

$$R_{\alpha\beta^*\gamma\delta^*} = \frac{\partial^2 g_{\alpha\beta^*}}{\partial z_\gamma \partial \bar{z}_\delta} - g^{\rho\sigma^*} \frac{\partial g_{\rho\beta^*}}{\partial \bar{z}_\delta} \frac{\partial g_{\sigma^*\alpha}}{\partial z_\gamma}. \tag{7}$$

In particular we have the symmetry relations

$$R_{\alpha\beta^*\gamma\delta^*} = R_{\gamma\beta^*\alpha\delta^*} = R_{\alpha\delta^*\gamma\beta^*} \tag{8}$$

which are unmatched in the case of real variables.

THEOREM 1. *If two tensors g_{ij} of our description in the same coordinates have the same curvature tensor $R_{\alpha\beta^*\gamma\delta^*}$, then they are equal to within an allowable transformation.*

The conclusion also holds if the components of the type $R_\alpha{}^\beta{}_{\gamma\delta^*}$ *or of some other type are assumed to be the same.*

PROOF. We take one of the tensors g_{ij} and we introduce the power series

$$\Phi(z_1, \cdots, z_n; \bar{z}_1, \cdots, \bar{z}_n) \tag{9}$$

for which (5) holds. We can obviously omit from (9) all monomials

$$a z_1^{p_1} \cdots z_n^{p_n} \bar{z}_1^{q_1} \cdots \bar{z}_n^{q_n} \tag{10}$$

for which either $p_1 + \cdots + p_n = 0$ or $q_1 + \cdots + q_n = 0$, and thus assume (9) in the form

$$a_{\alpha\beta^*} z_\alpha \bar{z}_\beta + \Phi_3(z, \bar{z}) \tag{11}$$

where $a_{\alpha\beta^*} = g_{\alpha\beta^*}(0)$, and Φ_3 contains only monomials with

$$p_1 + \cdots + p_n + q_1 + \cdots + q_n \geqq 3.$$

An allowable transformation will carry (11) into

$$\sum_{\alpha=1}^{n} z_\alpha \bar{z}_\alpha + \Phi_3(z, \bar{z}), \tag{12}$$

and on making it real, whenever (12) contains a monomial (10) it also contains $\bar{a}\bar{z}_1^{p_1} \cdots \bar{z}_n^{p_n} z_1^{q_1} \cdots z_n^{q_n}$. Therefore, we can put

$$\Phi_3 = \sum_{\alpha=1}^{n} (z_\alpha \overline{f_\alpha(z)} + \bar{z}_\alpha f_\alpha(z)) + \Phi_{2,2}(z, \bar{z}),$$

where each function $f_1, \cdots, f_n(z)$ is a power series in $z_1, \cdots, z_n$ with terms of total degree not less than 2, and $\Phi_{2,2}$ contains only terms with

$$p_1 + \cdots + p_n \geqq 2 \quad \text{and} \quad q_1 + \cdots + q_n \geqq 2.$$

After the allowable transformation $z_\alpha' = z_\alpha + f_\alpha(z)$, our function has the form

$$\Phi = \sum_{\alpha=1}^{n} z_\alpha \bar{z}_\alpha + \psi_{2,2}(z, \bar{z}). \tag{13}$$

If now we introduce the derivatives

$$\frac{\partial^{p_1 + \cdots + p_n + q_1 + \cdots + q_n} \Phi}{\partial z_1^{p_1} \cdots \partial z_n^{p_n} \partial \bar{z}_1^{q_1} \cdots \partial \bar{z}_n^{q_n}}$$

at the origin, then our normalization has fixed their values in case the total order is 1, 2 or 3; and for the total order 4, 5, 6, $\cdots$ if either $p_1+\cdots+p_n\leqq 1$ or $q_1+\cdots+q_n\leqq 1$. Now, if we write equation (7) in the form

$$\frac{\partial^4\Phi}{\partial z_\alpha\partial\bar z_\beta\partial z_\gamma\partial\bar z_\delta} = g^{\rho\sigma^*}\frac{\partial g_{\rho\beta^*}}{\partial\bar z_\delta}\frac{\partial g_{\sigma^*\alpha}}{\partial z_\gamma} + R_{\alpha\beta^*\ \delta^*}, \tag{14}$$

and if we remember that each component $g^{\rho\sigma^*}$ is a rational function of the components $g_{\alpha\beta^*}$, then the missing derivatives of order 4 can be computed directly from (14); furthermore, the missing derivatives of higher order can be successively computed by taking all possible successive derivatives of (14). Also these computations are unique, which proves our contention. If the components $R_\alpha{}^\beta{}_{\gamma\delta^*}$ are given, the same conclusion will follow from the equations

$$\frac{\partial^2 g_{\alpha\beta^*}}{\partial z_\gamma\partial\bar z_\delta} = g^{\rho\sigma^*}\frac{\partial g_{\rho\beta^*}}{\partial\bar z_\delta}\frac{\partial g_{\sigma^*\alpha}}{\partial z_\gamma} + g_{\beta^*\epsilon}R^{\epsilon}_{\alpha\gamma\delta^*},$$

and this completes the proof of Theorem 1.

THEOREM 2. *If the metric tensors g_{ij} and h_{ij} have the same geodesics, then their affine connections Γ^l_{ij} and Δ^l_{ij} are the same. Thus a projective collineation is automatically an affine collineation.*

PROOF. As in the real case we have, see [**3**, p. 132],

$$\Delta^l_{ij} - \Gamma^l_{ij} = \delta^l_i\psi_j + \delta^l_j\psi_i.$$

Putting $l=\alpha$, $i=\alpha$, $j=\gamma^*$, we obtain $\psi_{\gamma^*}=0$, and similarly $\psi_\gamma=0$.

2. **Sectional curvature.** A two-dimensional surface element through the origin is given in the form

$$t_i = \lambda^i x + \mu^i y, \tag{15}$$

that is

$$z_\alpha = \lambda^\alpha x + \mu^\alpha y, \qquad \bar z_\alpha = \overline{\lambda^\alpha}x + \overline{\mu^\alpha}y, \tag{16}$$

where x and y are real parameters, and the systems of complex numbers $\{\lambda^\alpha\}$ and $\{\mu^\alpha\}$ are nonproportional. If we make the decomposition

$$z_\alpha = x_\alpha + (-1)^{1/2}y_\alpha,\quad \lambda^\alpha = \lambda_0^\alpha + (-1)^{1/2}\lambda_1^\alpha,\quad \mu^\alpha = \mu_0^\alpha + (-1)^{1/2}\mu_1^\alpha,$$

then we can also write

$$x_\alpha = \lambda_0^\alpha x + \mu_0^\alpha y, \qquad y_\alpha = \lambda_1^\alpha x + \mu_1^\alpha y. \tag{17}$$

We now set up the number

$$K = \frac{R_{hijk}\lambda^h\mu^i\lambda^j\mu^k}{(g_{hj}g_{ik} - g_{hk}g_{ij})\lambda^h\mu^i\lambda^j\mu^k} \tag{18}$$

and we claim that it is the real-valued curvature for the surface element (17) and the line element (3). In fact, if we denote the real component y_α by $x_{\alpha+n}$, if we write the line element (3) in the real form

$$h_{ij}(x)dx_idx_j,$$

if we introduce the ordinary curvature tensor corresponding to the latter line element, if we form with that curvature tensor the sectional curvature for the section (17), and if finally we carry out purely formally the transformation of coordinates

$$z_\alpha = x_\alpha + (-1)^{1/2}x_{\alpha+n}, \qquad \bar{z}_\alpha = x_\alpha - (-1)^{1/2}x_{\alpha+n}$$

as if it were an allowable transformation; then by formal properties of invariance the curvature will appear in the form (18).

THEOREM 3. *For $n>1$, if at every point the sectional curvature K is the same for all possible two-dimensional sections, then the curvature tensor is identically zero.*

PROOF. Relation

$$[R_{hijk} - K(g_{hj}g_{ik} - g_{hk}g_{ij})]\lambda^h\mu^i\lambda^j\mu^k = 0 \tag{19}$$

is fulfilled for $2n$ independent variables λ^α, μ^α and their conjugate values $\overline{\lambda^\alpha}$, $\overline{\mu^\alpha}$. We are now applying a fundamental lemma to the effect that whenever a power series

$$\Phi(u_1, \cdots, u_m; v_1, \cdots, v_m)$$

is zero identically in $u_1, \cdots, u_m$ for $v_k = \bar{u}_k$, $k=1, \cdots, n$, then it is identically 0 in the independent variables u_k, v_k. If we apply this to (19) then for K independent of λ and μ, we obtain by a classical formal procedure the relation

$$R_{hijk} = K(g_{hj}g_{ik} - g_{hk}g_{ij}). \tag{20}$$

In particular, we obtain

$$R_{\alpha\beta^*\gamma\delta^*} = -Kg_{\alpha\delta^*}g_{\beta^*\gamma},$$

and on applying the important relation (8) we deduce

$$Kg_{\alpha\delta^*}g_{\beta^*\gamma} = Kg_{\gamma\delta^*}g_{\beta^*\alpha}.$$

If we multiply both sides by $g^{\alpha\delta^*}g^{\beta^*\gamma}$ and contract we obtain $n^2K = K$, and hence $K=0$ for $n>1$.

DEFINITION. We call a section *holomorphic* if it is tangent to an analytically imbedded complex "curve." It is not hard to see that (16) is holomorphic if an only if there exists a *non-real* number ϕ such that $\mu^\alpha=\phi\lambda^\alpha$. If we transform the two real parameters x, y by a suitable nonsingular affine transformation $x'=\phi_1x+\phi_2y$, $y'=\phi_3x+\phi_4y$ with real coefficients ϕ_1, ϕ_2, ϕ_3, ϕ_4, then we can obtain the normalization

$$\mu^\alpha = (-1)^{1/2}\lambda^\alpha; \tag{21}$$

that is, the section can be written in the form

$$z_\alpha = \lambda^\alpha z; \qquad z = x + (-1)^{1/2}y. \tag{22}$$

THEOREM 4. *For a holomorphic section* (21) *we have*

$$\begin{aligned} K &= -\frac{R_{\alpha\beta^*\gamma\delta^*}\lambda^\alpha\lambda^{\beta^*}\lambda^\gamma\lambda^{\delta^*}}{g_{\alpha\beta^*}g_{\gamma\delta^*}\lambda^\alpha\lambda^{\beta^*}\lambda^\gamma\lambda^{\delta^*}} \\ &= -2\,\frac{R_{\alpha\beta^*\gamma\delta^*}\lambda^\alpha\lambda^{\beta^*}\lambda^\gamma\lambda^{\delta^*}}{(g_{\alpha\beta^*}g_{\gamma\delta^*}+g_{\gamma\beta^*}g_{\alpha\delta^*})\lambda^\alpha\lambda^{\beta^*}\lambda^\gamma\lambda^{\delta^*}}\,. \end{aligned} \tag{23}$$

PROOF. This follows from the fact that for an arbitrary section (16) we have

$$R_{hijk}\lambda^h\mu^i\lambda^j\mu^k = R_{\alpha\beta^*\gamma\delta^*}(\lambda^\alpha\mu^{\beta^*}-\lambda^{\beta^*}\mu^\alpha)(\lambda^\gamma\mu^{\delta^*}-\lambda^{\delta^*}\mu^\gamma) \tag{24}$$

and

$$\begin{aligned} &(g_{hj}g_{ik}-g_{hk}g_{ij})\lambda^h\mu^i\lambda^j\mu^k \\ &= g_{hj}\lambda^h\lambda^j g_{ik}\mu^i\mu^k - g_{hk}\lambda^h\mu^k g_{ij}\mu^i\lambda^j \\ &= g_{\alpha\beta^*}g_{\gamma\delta^*}[(\lambda^\alpha\mu^\gamma-\lambda^\gamma\mu^\alpha)(\lambda^{\beta^*}\mu^{\delta^*}-\lambda^{\delta^*}\mu^{\beta^*}) \\ &\quad + (\lambda^\alpha\mu^{\delta^*}-\lambda^{\delta^*}\mu^\alpha)(\lambda^{\beta^*}\mu^\gamma-\lambda^\gamma\mu^{\beta^*})]. \end{aligned} \tag{25}$$

We shall now draw an interesting conclusion.

THEOREM 5. *If at a point all holomorphic sections have the same curvature $K=b$, then at that point we have*

$$R_{\alpha\beta^*\gamma\delta^*} = -\frac{b}{2}(g_{\alpha\beta^*}g_{\gamma\delta^*}+g_{\gamma\beta^*}g_{\alpha\delta^*}). \tag{26}$$

Also, if (26) *holds at every point, then b is a constant.*

PROOF. By assumption we have

$$\left[R_{\alpha\beta^*\gamma\delta^*}+\frac{b}{2}(g_{\alpha\beta^*}g_{\gamma\delta^*}+g_{\beta^*\gamma}g_{\alpha\delta^*})\right]\lambda^\alpha\lambda^{\beta^*}\lambda^\gamma\lambda^{\delta^*} = 0.$$

At first the relation holds whenever λ^{α^*} is conjugate complex to λ^α. By our previous argument it holds for independent variables $(\lambda^\alpha, \lambda^{\beta^*})$, and relation (26) follows now from the fact that both sides of it are symmetric in the pairs of indices (α, γ) and (β^*, δ^*).

Next, Bianchi's relation specializes to

$$R_{\alpha\beta^*\gamma\delta^*,\epsilon} = R_{\alpha\beta^*\epsilon\delta^*,\gamma}, \tag{27}$$

and if (26) holds at all points we obtain

$$b_{,\epsilon}(g_{\alpha\beta^*}g_{\gamma\delta^*} + g_{\gamma\beta^*}g_{\alpha\delta^*}) = b_{,\gamma}(g_{\alpha\beta^*}g_{\epsilon\delta^*} + g_{\epsilon\beta^*}g_{\alpha\delta^*}).$$

This shows that $b(z, \bar{z})$ is independent of the variables z_α. The independence of $\bar{z}_\alpha$ follows from $R_{\alpha\beta^*\gamma\delta^*,\epsilon^*} = R_{\alpha\beta^*\gamma\epsilon^*,\delta^*}$.

3. **Fubini spaces.** For arbitrary real b we put

$$\Phi = \frac{2}{b}\log\left(1 + \frac{b}{2}\sum_{\alpha=1}^{n} z_\alpha\bar{z}_\alpha\right) \equiv \frac{2}{b}\log S. \tag{28}$$

Therefore,

$$g_{\alpha\beta^*} = \frac{\partial^2\Phi}{\partial z_\alpha\partial\bar{z}_\beta} = \frac{\delta_{\alpha\beta^*}}{S} - \frac{b}{2}\frac{\bar{z}_\alpha z_\beta}{S^2}, \tag{29}$$

$$\frac{ds^2}{2} = \frac{\sum_\alpha |dz_\alpha|^2 + \frac{b}{2}\left(\sum_\alpha |z_\alpha|^2 \sum_\beta |dz_\beta|^2 - \left|\sum_\alpha \bar{z}_\alpha dz_\alpha\right|^2\right)}{\left(1 + \frac{b}{2}\sum_\alpha |z_\alpha|^2\right)^2}. \tag{30}$$

At the origin we have $g_{\alpha\beta^*} = \delta_{\alpha\beta^*}$ and on computing (7) we obtain

$$R_{\alpha\beta^*\gamma\delta^*} = -\frac{b}{2}(\delta_{\alpha\beta^*}\delta_{\gamma\delta^*} + \delta_{\alpha\delta^*}\delta_{\beta\gamma^*}).$$

Thus (26) holds at the origin. The validity at other points follows from the existence of a transitive group of analytic homeomorphisms which leave the line element (3) invariant, see Fubini [**4**].

THEOREM 6. *For given b all spaces of constant holomorphic curvature b are equivalent.*

PROOF. The conclusion follows from

$$\frac{\partial^2 g_{\alpha\beta^*}}{\partial z_\gamma\partial\bar{z}_\delta} = g^{\rho\sigma^*}\frac{\partial g_{\rho\beta^*}}{\partial\bar{z}_\delta}\frac{\partial g_{\sigma^*\alpha}}{\partial z_\gamma} - \frac{b}{2}(g_{\alpha\beta^*}g_{\gamma\delta^*} + g_{\alpha\delta^*}g_{\gamma\beta^*})$$

as in the case of Theorem 1.

For the function

$$\psi(z, \bar{z}) = \Phi - g_{\alpha\beta^*}(0)z_\alpha\bar{z}_\beta$$

we then obtain

$$\frac{\partial^2\psi}{\partial z_\alpha\partial\bar{z}_\beta}\bar{z}_\beta = 0, \qquad \alpha = 1, \cdots, n,$$

or

$$\frac{\partial\psi}{\partial\bar{z}_\beta}\bar{z}_\beta = F(\bar{z}).$$

On putting $\bar{z}_\beta = \bar{a}_\beta\zeta$, $H(z, \bar{a}, \zeta) \equiv \psi(z, \bar{a}\ \zeta)$, we then obtain $\zeta\partial H/\partial\zeta = F(\bar{a}\zeta)$; and since $F(\bar{z})$ has no constant term we obtain

$$H(z, \bar{a}, \zeta) = R(z, \bar{a}) + S(\bar{a}\zeta).$$

Therefore

$$\psi(z, \bar{z}) = R(z, \bar{z}/\zeta) + S(\bar{z}).$$

But the left side is independent of ζ, and so we have finally

$$\psi(z, \bar{z}) = R(z) + S(\bar{z})$$

and thus $\partial\psi/\partial z_\alpha\partial\bar{z}_\beta = 0$. Thus $g_{\alpha\beta^*}(z, \bar{z}) \equiv g_{\alpha\beta^*}(0)$, and our line element is flat.

However, as was shown in the proof of Theorem 1, there always exists a coordinate system which is *geodesic* at a prescribed point; meaning that the derivatives $\partial g_{\alpha\beta^*}/\partial z_\gamma$ and therefore also the coefficients $\Gamma^\alpha_{\beta\gamma}$ vanish at the point.

4. **Imbedding.** In addition to our space V_n with the metric (3) we consider a space V_m, $m > n$, in the complex variables $w_1, \cdots, w_m$, $m > n$, with a metric

$$2h_{\xi\eta^*}(w, \bar{w})dw_\xi d\overline{w_\eta} \tag{36}$$

where $\xi, \eta, \zeta, \chi = 1, 2, \cdots, m$; and we assume that there exists a complex-analytic transformation

$$w_\xi = f^\xi(z_1, \cdots, z_n), \qquad \xi = 1, \cdots, m, \tag{37}$$

which is an *isometric* map of V_n into V_m. Thus

$$g_{\alpha\beta^*} = h_{\xi\eta^*}f^\xi{}_{,\alpha}\overline{f^\eta{}_{,\beta}} \tag{38}$$

where $f^\xi{}_{,\alpha} = \partial f^\xi/\partial z_\alpha$.

THEOREM 11. *If we denote the curvature tensor of V_n by $R_{\alpha\beta^*\gamma\delta^*}$ and of V_m by $S_{\xi\eta^*\zeta\chi^*}$, then we have*

$$R_{\alpha\beta^*\gamma\delta^*} = S_{\xi\eta^*\zeta\chi^*} f^\xi{}_{,\alpha}\overline{f^\eta{}_{,\beta}} f^\zeta{}_{,\gamma}\overline{f^\chi{}_{,\delta}} + h_{\xi\eta^*} f^\xi{}_{,\alpha,\gamma}\overline{f^\eta{}_{,\beta,\delta}} \tag{39}$$

where $f^\xi{}_{,\alpha,\gamma}$ denotes the second covariant derivative of $f^\xi(z)$ relative to the metric (3).

PROOF. We write (7) in the form

$$R_{\alpha\beta^*\gamma\delta^*} = \frac{\partial^2 g_{\alpha\beta^*}}{\partial z_\gamma \partial \bar{z}_\delta} - \Gamma^\rho_{\alpha\gamma}\frac{\partial g_{\rho\beta^*}}{\partial \bar{z}_\delta} \tag{40}$$

and similarly

$$S_{\xi\eta^*\zeta\chi^*} = \frac{\partial^2 h_{\xi\eta^*}}{\partial w_\zeta \partial \bar{w}_\chi} - \Delta^\phi_{\xi\zeta}\frac{\partial h_{\phi\eta^*}}{\partial \bar{w}_\chi}, \tag{41}$$

where $\Delta^\phi_{\xi\zeta}$ is the affine connection pertaining to (36). Take a point z^0 and its image w^0. It can easily be seen that it will suffice to prove (39) under the assumption that the z-coordinates are geodesic at z^0 and the w-coordinates are geodesic at w^0. In this case all quantities

$$\frac{\partial g_{\alpha\beta^*}}{\partial z_\gamma}, \qquad \Gamma^\alpha_{\beta\gamma}, \qquad \frac{\partial h_{\xi\eta^*}}{\partial w_\zeta}, \qquad \Delta^\xi_{\eta\zeta}$$

and their conjugates are 0, and we easily obtain (39) by substituting (38) in the right side of (40) and carrying out the differentiation.

We now take a fixed point z^0 in V_n, and we introduce an arbitrary surface element

$$z_\alpha - \overset{0}{z}_\alpha = \overset{\alpha}{\lambda} x + \overset{\alpha}{\mu} y \tag{42}$$

and its image

$$w_\xi - \overset{0}{w}_\xi = \overset{\xi}{L} x + \overset{\eta}{M} y, \tag{43}$$

where

$$L^\xi = f^\xi{}_{,\alpha}\lambda^\alpha, \qquad M^\xi = f^\xi{}_{,\alpha}\mu^\alpha. \tag{44}$$

If we multiply both sides of (39) by

$$(\lambda^\alpha\mu^{\beta^*} - \lambda^{\beta^*}\mu^\alpha)(\lambda^\gamma\mu^{\delta^*} - \lambda^{\delta^*}\mu^\gamma) \tag{45}$$

and add up over $\alpha, \beta, \gamma, \delta$ we obtain the following theorem.

THEOREM 12. *The sectional curvature*

$$\frac{R_{abcd}\lambda^a\mu^b\lambda^c\mu^d}{(g_{ac}g_{bd} - g_{ad}g_{bc})\lambda^a\mu^b\lambda^c\mu^d} \tag{46}$$

in V_n is the sum of the corresponding sectional curvature

$$\frac{S_{ijkl}L^iM^jL^kM^l}{(h_{ik}h_{jl} - h_{il}h_{jk})L^iM^jL^kM^l} \tag{47}$$

in V_m and of the relative curvature

$$\frac{h_{\xi\eta^*}f^\xi{}_{,\alpha,\gamma}\overline{f^\eta{}_{,\beta,\delta}}(\lambda^\alpha\mu^{\beta^*} - \lambda^{\beta^*}\mu^\alpha)(\lambda^\gamma\mu^{\delta^*} - \lambda^{\delta^*}\mu^\gamma)}{(g_{ac}g_{bd} - g_{ad}g_{bc})\lambda^a\mu^b\lambda^c\mu^d}. \tag{48}$$

For a holomorphic section (22) *the relative curvature has the value*

$$-\frac{h_{\xi\eta^*}f^\xi{}_{,\alpha,\gamma}\lambda^\alpha\lambda^\gamma\overline{f^\eta{}_{,\beta,\delta}}\lambda^{\beta^*}\lambda^{\delta^*}}{(g_{\alpha\beta^*}\lambda^\alpha\lambda^{\beta^*})^2}. \tag{49}$$

It is therefore negative or 0, *and it has the value* 0 *only if the section is such that*

$$f^\xi{}_{,\alpha,\gamma}\lambda^\alpha\lambda^\gamma = 0, \qquad \xi = 1, \cdots, m. \tag{50}$$

Condition (50) *formally corresponds to directions of asymptotic lines in the real case.*

For the Ricci tensor we have

$$R_{\alpha\delta^*} = S_{\xi\eta^*}f^\xi{}_{,\alpha}\overline{f^\eta{}_{,\delta}} + h_{\xi\eta^*}g^{\beta^*\gamma}f^\xi{}_{,\alpha,\gamma}\overline{f^\eta{}_{,\beta,\delta}} \tag{51}$$

and the Ricci curvature $-R_{ad}\lambda^a\lambda^d/g_{ad}\lambda^a\lambda^d$ is the sum of $-S_{ij}L^iL^j/h_{ij}L^iL^j$ and of the relative Ricci curvature

$$-\frac{h_{\xi\eta^*}g^{\beta^*\gamma}f^\xi{}_{,\alpha,\gamma}\lambda^\alpha\overline{f^\eta{}_{,\beta,\delta}}\lambda^{\delta^*}}{g_{\alpha\delta^*}\lambda^\alpha\lambda^{\delta^*}}. \tag{52}$$

The last term is negative or 0, *and it is* 0 *if and only if*

$$f^\xi{}_{,\alpha,\gamma}\lambda^\alpha = 0, \qquad \xi = 1, \cdots, m; \gamma = 1, \cdots, n, \tag{53}$$

which is a stronger condition than (50).

If V_m has the Euclidean metric

$$dw_1d\bar{w}_1 + \cdots + dw_m\overline{dw_m}, \tag{54}$$

then (48) is the entire amount of the sectional curvature of V_n. Now, put $m=3$, $n=2$, and

$$\overset{1}{f}(z) = z_1, \qquad \overset{2}{f}(z) = z_2, \qquad \overset{3}{f}(z) = 2^{-1}(\overset{2}{z_1} + \overset{2}{z_2}).$$

The metric of V_2 is then $dz_1d\bar{z}_1+dz_2d\bar{z}_2+(z_1dz_1+z_2dz_2)(\bar{z}_1d\bar{z}_1+\bar{z}_2d\bar{z}_2)$. At the origin $z_1=z_2=0$, the z-system is geodesic, and the numerator in (48) has at that point the value

$$\sum_{\alpha,\beta=1}^{3} (\lambda^\alpha \mu^{\beta^*} - \lambda^{\beta^*}\mu^\alpha)^2. \tag{55}$$

Since the denominator of (48) is always positive, see (25), the algebraic sign of (48) is that of (55). If, for instance, all components λ^α, μ^β are real, that is if our section is totally real, then (55) is positive, and thus we see that the relative sectional curvature may be positive for a nonholomorphic section.

THEOREM 13. *For $n=1$, if a line element*

$$ds^2 = g(z, \bar{z})dzd\bar{z} \tag{56}$$

of the special form

$$g(z, \bar{z}) = \sum_{p=0}^{\infty} a_p z^p \bar{z}^p \tag{57}$$

can be isometrically imbedded in V_m, with the Euclidean line element (54), for some finite m, then the power series (57) is a finite polynomial in z, $\bar{z}$. In particular, the line element

$$ds^2 = \frac{dzd\bar{z}}{(1 - z\bar{z})^2} = \sum_{p=0}^{\infty} (p+1) z^p \bar{z}^p dzd\bar{z}, \tag{58}$$

although strictly hyperbolic, cannot be so imbedded.

For $n \geqq 1$, if in the line element

$$g_{\alpha\beta^*}(z, \bar{z}) dz_\alpha d\bar{z}_\beta \tag{59}$$

in the power series of all $g_{\alpha\beta^*}$ only those monomials

$$z_1^{p_1} \cdots z_n^{p_n} \bar{z}_1^{q_1} \cdots \bar{z}_n^{q_n} \tag{60}$$

occur for which

$$p_1 + \cdots + p_n = q_1 + \cdots + q_n, \tag{61}$$

and if (59) can be isometrically imbedded in a V_m, then all $g_{\alpha\beta^*}(z, \bar{z})$ are finite polynomials. In particular, the line elements arising in all types of matrix spaces[2] *cannot be so imbedded.*

PROOF. For $n=1$, our imbedding means the existence of functions

$$w_k = f_k(z) = \sum_{p=1}^{\infty} c_{k,p} z^p, \qquad k = 1, \cdots, m, \tag{62}$$

such that

[2] See Siegel [7] and Hua [5] and [6].

$$\sum_k \left| \frac{df_k(z)}{dz} \right|^2 = \sum_p a_p z^p \bar{z}^p.$$

Therefore

$$pq \sum_{k=1}^{m} c_{k,p}\overline{c_{k,q}} = \delta_{pq} a_{p-1}. \tag{63}$$

For each p, the system of numbers

$$C_p = \{c_{k,p}\}, \qquad k = 1, \cdots, m,$$

is a vector in complex m-space. By (63) any two of the vectors are orthogonal, and therefore all but at most m of them must be identically 0. Therefore at most m^2 among the numbers $c_{k,p}$ can be not equal to 0. Each $f_k(z)$ is a polynomial, and so is $g(z, \bar{z})$.

For $n \geqq 1$, if there exists an imbedding, then we write the mapping functions in the form

$$f_k(z_\alpha) = \sum_{p=1}^{\infty} c_{k,p}(z_\alpha) \tag{64}$$

where $c_{k,p}$ is a homogeneous polynomial in the variables z_α of degree p, or identically zero. Now, put

$$z_\alpha = \zeta_\alpha t, \qquad \bar{z}_\alpha = \bar{\zeta}_\alpha \bar{t}, \tag{65}$$

where $\{\zeta_\alpha\}$ are arbitrary parameters in the neighborhood of the origin and t is the complex variable proper. Thus, (65) is a family of complex curves. For each one we derive from (59) a line element

$$g(\zeta, \bar{\zeta}; t, \bar{t}) dt d\bar{t},$$

where by our assumptions on (59),

$$g(\zeta, \bar{\zeta}; t, \bar{t}) = \sum_p a_p(\zeta, \bar{\zeta}) t^p \bar{t}^p.$$

On substituting (65) into (64), we obtain

$$f_k(\zeta_\alpha t) = \sum_p c_{k,p}(\zeta_\alpha) t^p,$$

and by our previous argument, corresponding to each special point $\{\zeta_\alpha^0\}$, only m^2 among the numbers $c_{k,p}(\zeta_\alpha^0)$ can be not equal to 0. Since each $c_{k,p}(\zeta)$ is a polynomial, it follows easily that all but m^2 among them must vanish, and thus $f_k(z_\alpha)$ is a polynomial. This proves the theorem.

Thus far we have been discussing imbedding into a V_m, for *finite m*.

The situation is radically different if we allow the imbedding space to become infinitely-dimensional. One possibility is to view certain constructions as limiting processes of ordinary imbedding; this will be our method of interpreting groups of automorphisms. However, another possibility is to actually perform an imbedding into a space of countably many variables $w_1, w_2, w_3, \cdots$ with a suitable metric

$$h_{\xi\eta^*}(w, \bar{w})dw_\xi \overline{dw_\eta}, \qquad \xi, \eta = 1, 2, 3, \cdots .$$

The most immediate line element would be the Hilbert metric

$$dw_1 d\bar{w}_1 + dw_2 d\bar{w}_2 + \cdots .$$

In this case the hyperbolic line element (58) can very well be imbedded, namely by the transformation

$$w_k = z^k/(k+1)^{1/2}, \qquad k = 1, 2, \cdots .$$

Furthermore, our previous theorems remain valid, and thus for instance an elliptic line element cannot be isometrically imbedded in flat Hilbert space. Conceivably every finite-dimensional Hermitian line element without torsion could be imbedded into some universal countably-dimensional space with a fixed elliptic metric. It is interesting to note in this connection that the method of S. Bergman [1] for constructing an Hermitian metric on a domain of several complex variables consists precisely in constructing an analytic imbedding of the domain into such a universal enveloping space, the enveloping domain of Bergman being the countably-dimensional Fubini space with positive curvature $b=1$.

We shall not take up this problem in any way; however, we shall point out the following curious little theorem, which apparently is not true for finite m.

THEOREM 14. *Whenever a V_n can be isometrically imbedded in flat Hilbert space it can also be isometrically imbedded in infinitely-dimensional Fubini space with $b=1$.*

PROOF. By assumption we have relation (5) with

$$\Phi(z, \bar{z}) = \sum_{k=1}^{\infty} | f_k(z_\alpha) |^2.$$

On putting $\Phi = \log \psi$, that is, $\psi = e^\Phi$, we obtain

$$\psi = 1 + \sum_{k=1}^{\infty} | g_k(z_\alpha) |^2$$

for suitable functions $g_k(z_\alpha)$, and this is the substance of the theorem.

5. **Group invariance.** We consider a domain D in the space of the complex variables $z_1, \cdots, z_n$, and an arbitrary topological set Γ, and we consider a continuous function

$$f(\theta; z_\alpha) \tag{66}$$

in $\Gamma \times D$ which is analytic in D. We further assume that there exists some kind of integral in Γ such that

$$\Phi(z, \bar{z}) = \int_\Gamma f(\theta; z)\overline{f(\theta; z)}d\theta \tag{67}$$

exists in D, and that in the neighborhood of any point z^0 it is a uniform limit of an approximating sum

$$\sum_{r=1}^{s} \Delta_r f(\theta_r; z)\overline{f(\theta_r; z)}, \qquad \Delta_r > 0.$$

If we now form

$$g_{\alpha\beta^*} = \frac{\partial^2\Phi}{\partial z_\alpha \partial \bar{z}_\beta}, \tag{68}$$

then our restriction (4) is met; also $g_{\alpha\beta^*}$ is non-negative, and we add the *explicit* assumption that it is strictly positive definite. The approximating tensor

$$h_{\alpha\beta^*} = \frac{\partial^2}{\partial z_\alpha \partial \bar{z}_\beta} \sum_{r=1}^{s} \left| \Delta_r^{1/2} f(\theta_r; z) \right|^2 \tag{69}$$

may be also assumed positive definite, and may be looked upon as resulting from an imbedding of D in Euclidean V_s. If we apply Theorem 12, and make the limiting passage from (69) to (68), we see that the resulting line-element (59) on D has nonpositive holomorphic and Ricci curvature.

We shall now vary the set-up slightly. We consider not a family of functions, but a family of analytic homeomorphisms $w_k = f_k(\theta; z_\alpha)$, $k = 1, \cdots, n$, of D into itself, and we assume that Γ is a locally compact group of such homeomorphisms. Correspondingly, we put

$$\Phi(z, \bar{z}) = \int_\Gamma \left(\sum_{k=1}^{n} \left| f_k(\theta; z) \right|^2 \right) d\theta \tag{70}$$

where $d\theta$ represents integration with respect to a Haar-measure, say a left-invariant one. Thus we have

$$g_{\alpha\beta^*}dz_\alpha d\bar{z}_\beta = \int_\Gamma \left(\sum_{k=1}^{n} \frac{\partial f_k(\theta, z)}{\partial z_\alpha} \overline{\frac{\partial f_k(\theta, z)}{\partial z_\beta}} \right) d\theta dz_\alpha d\bar{z}_\beta \tag{71}$$
$$= \int_\Gamma \left(\sum_{k=1}^{n} | dw_k(\theta, z) |^2 \right) d\theta.$$

Now, if θ' is a fixed element of Γ, and if we put $z'_\alpha = f_\alpha(\theta'; z) = \theta' z_\alpha$, then we have $w_k(\theta\theta'; z) = w_k(\theta; z')$, and we find that the line element is group invariant. We can now state the following very conditional result.

THEOREM 15. *If Γ is a transitive group of homeomorphisms of a domain D, if the general homeomorphism can be written in the form*

$$w_k = f_k(\lambda; a, \bar{a}; z_\alpha) \tag{72}$$

where $a = (a_1, \cdots, a_n)$ is an arbitrary point of D and λ is a set of parameters which describe the stability subgroup Γ_0 of Γ, if $d\lambda$ designates Haar-measure on Γ_0, if there is given in D a volume element

$$dv_a \equiv \rho(a, \bar{a}) \prod_{\alpha=1}^{n} da_\alpha d\bar{a}_\alpha$$

which is invariant under all transformations of Γ, and if the integrals

$$g_{\alpha\beta^*} = \int_{\Gamma_0 \times D} \left(\sum_{k=1}^{n} \frac{\partial f_k(\lambda; a, \bar{a}; z)}{\partial z_\alpha} \overline{\frac{\partial f_k(\lambda; a, \bar{a}; z)}{\partial z_\beta}} \right) dv_a d\lambda$$

converge uniformly in a neighborhood of every point z of D, then the corresponding Hermitian metric has no torsion and is group-invariant, and the curvature of holomorphic sections and the Ricci curvature are all nonpositive.

BIBLIOGRAPHY

1. S. Bergman, Über die Kernfunktion eines Bereiches und ihr Verhalten am Rande. I, J. Reine Angew. Math. vol. 169 (1933) pp. 1–42.

2. S. Bochner, Vector fields and Ricci curvature, Bull. Amer. Math. Soc. vol. 52 (1946) pp. 776–797.

3. L. P. Eisenhart, *Riemannian geometry*, p. 132.

4. G. Fubini, *Teoria dei gruppi discontinui*, 1908.

5. L. K. Hua, *On the theory of automorphic functions of a matrix variable*, Amer. J. Math. vol. 66 (1944) pp. 470–488 and 531–563.

6. ———, *On the theory of Fuchsian groups of several variables*, Ann. of Math. vol. 47 (1946) pp. 167–191.

7. C. L. Siegel, *Symplectic geometry*, Amer. J. Math. vol. 65 (1943) pp. 1–86.

PRINCETON UNIVERSITY

ON COMPACT COMPLEX MANIFOLDS.

BY

S. BOCHNER, *Princeton University.*

[Received 6 September, 1947.]

In the present paper we will generalize a classical theorem (Theorem 1) from one to several complex variables and we are justifying this undertaking on two counts. Firstly, the mere question of finding a suitable wording for the desired generalization will emphasize the discrepancy between one and several variables, and secondly our proof for the k-dimensional statements will be new even for the classical case $k = 1$.

A fundamental difference between one and several variables is as follows. In the classical case the whole geometric theory is based on the comprehensive theory of uniformization, but for several variables no such theory of uniformization exists or is possible. This deficiency is responsible for many difficulties, but it is also productive of new view points and situations which the classical theory cannot evolve.

In some contexts a method for overcoming this deficiency is to introduce a so-called Hermitian metric, and our context will offer an opportunity for summarizing and correlating some recent results of the author's bearing on this method.

1. A Theorem for Curves of Genus $p > 1$.

The classical theorem which will be the subject of this paper is as follows.

Theorem 1. *On a closed Riemann surface of genus $p > 1$ there are only a finite number of conformal one-to-one transformations of the surface into itself.**

* A 'conformal one-to-one transformation' is the same as a '(complex) analytic homeomorphism'. We will not adhere to any rigid terminology.

A complete geometrical proof of this theorem, based on an idea of H. Poincare, will be found in the last chapter of H. Weyl's book, *Die Idee der Riemannschen Flæche.* Our proof will be a variant on that proof, but with considerable alterations. The theorem was originally due to H. A. Schwarz and M. Noether, and further illuminating comment, mainly in an algebraic direction, is due to Weierstrass, A. Hurwitz and Fatou.*

In the statement of the theorem the restriction $p > 1$ is very essential. Indeed for $p = 0$ a closed Riemann surface is a Gaussian sphere, and this surface admits infinitely many complex homeomorphisms. Each projective transformation

$$Z' = \frac{aZ+b}{cZ+d}$$

is one, and their totality is a continuous group of 3 complex or 6 real parameters. For $p = 1$, the surface is a double-torus of the kind encountered in the theory of elliptic functions and every such torus admits a two-dimensional group of translations. But for $p > 1$, the situation is radically different. Not only does there not exist a family of homeomorphisms which depends on a continuous parameter, but their totality cannot be infinite in number. The second half of this statement is by no means a formal consequence of the first half, but compare the remark following Theorem 3.

The only geometric property common to all surfaces with $p > 1$ and not present for $p = 0$ and $p = 1$ is the fact that the universal covering space is conformally equivalent with the unit circle

$$|t| < 1. \tag{1}$$

Every geometric proof of Theorem 1 must therefore make

* In addition to the quotations in H. Weyl, see Appell et Goursat, *Fonctions Algebriques*, (Fatou's Second Edition), Chapters XI and XVI.

use of this fact and must obviously be somehow concerned with the group of transformations

$$t' = e^{i\vartheta} \frac{t-a}{1-\bar{a}t}, \qquad |a| < 1, \qquad (2)$$

which carries (1) into itself. This cannot be avoided. However in Weyl's chain of arguments there are many details concerning the specific nature of such transformations, valid only in one complex variable, and such details must be avoided if there is to be a generalization. We will indeed avoid them by appealing to some general results of recent date, valid for any number of variables, and in order to expound these we will next discuss the concept of a complex space.

2. Complex Manifolds. A space in k complex variables is a Hausdorff space of $2k$ real variables having the following property. Each point is contained in a neighbourhood which can be topologically mapped into a neighbourhood of the Euclidean space of k complex variables, and if two such neighbourhoods intersect then the specified Euclidean images of their intersection are being mapped into each other by a system of functions

$$z'_j = f_j(z_1, \ldots, z_k), j = 1, \ldots, k, \qquad (3)$$

which are complex-analytic in the variables given. Any such space is orientable.

We call S a compact (complex) manifold, if it is connected and is coverable by a finite number of its neighbourhoods. We are defining a manifold in this and no other way, although there is a strong heuristic tendency towards unifying this concept of compact manifold with that of an algebraic variety. An algebraic variety is the locus of a system of homogeneous polynomials

$$F_\lambda(z_1, \ldots, z_n) = 0, \lambda = 1, \ldots, l, \qquad (4)$$

in complex projective space of the variables $z_1 : z_2 : \cdots : z_n$. It has dimension $2(n-r-1)$, where r is the rank of the system. It is locally a manifold, except for "singularities". If there are no singularities, that is, if they are all locally "uniformized", the variety is obviously a compact manifold. There is a classical conjecture that every algebraic variety is algebraically equivalent to another one on which no singularities are present, and on the other hand there are large classes of compact manifolds which can be algebraically generated as a variety. Thus, in all likelihood, there is no substantial difference between a compact manifold and a variety. But the possibility of such an identification will play no role in our reasoning, and for our part, we will be dealing exclusively with a manifold as previously defined.

3. Lie Groups of Homeomorphisms. Let

$$Q = f(P) \tag{5}$$

denote a complex homeomorphism of S, and let Γ denote the family of such homeomorphisms. If we take all topological (non-analytic) homeomorphisms, then they obviously form a group, since the succession of two and the inverse of any one, again belong to the family. In the case of complex-analytic transformations the succession of any two is again obviously complex-analytic. But this is not at all obvious for the inverse of a transformation, and it is actually not true for *real* analytic transformations. Thus, for instance, for the analytic homeomorphism $y = x^3$ of the real line into itself the inverse $x = y^{\frac{1}{3}}$ is not again analytic. But in the complex case the peculiarity just exhibited cannot arise* and an inverse does exist. This is related to the fact, with which it is virtually equivalent, that in the case of a complex one-to-one transformation, the Jacobian must be everywhere $\neq 0$. Thus Γ is a group. Due to the compactness of S this group can be easily topologized. To that purpose we take any distance $\rho(P_1, P_2)$ on S. Some such (non-analytic) metric can

* This point will be discussed at great length in Chapter VIII of the forthcoming book: S. Bochner and W. T. Martin, *Several Complex Variables*, Princeton University Press.

always be introduced. If $Q=f(P)$ and $Q=g(P)$ are any two elements of Γ, then the distance

$$\delta(f, g) = \max_{P \varepsilon S} \rho(f(P), g(P)) \tag{6}$$

obviously defines a topology in Γ, and the topology is equivalent for all distances $\rho(P_1, P_2)$. Now the first general theorem to which we will refer is as follows.

THEOREM 2. *The group* Γ *is a Lie group.*

We will not repeat the proof by which it was obtained* but only note the steps by which it was obtained. It was first concluded that Γ is locally compact. This conclusion was predicated on the two assumptions that S is compact, and that it is complex. But once it is known that Γ is locally compact, both assumptions can be discarded and Theorem 2 follows from the following theorem. *If a locally compact group* Γ *acts on a differentiable (locally Euclidean) space, then the group* Γ *is a Lie group.*

In this connection the term 'Lie group' designates a group which is not necessarily connected and possibly is zero-dimensional. If zero-dimensional it is countable, finite or infinite, and "totally discrete", meaning that for all pairs of distinct elements the distance is uniformly bounded away from zero.

If the group is not zero-dimensional, it has group generators, and such a generator is nothing else but a *contravariant* complex-analytic vector-field

$$\zeta^\alpha \equiv \zeta^\alpha(z_1, \ldots, z_k), \quad \alpha = 1, \ldots, k, \tag{7}$$

which is defined and regular everywhere in S. We point out that for us a vector does not cease to be regular if its components vanish at a point. Being "contravariant" means that for a change of coordinates

$$z'_\alpha = z'_\alpha(z_1, \ldots, z_k), \tag{8}$$

* *See*, S. Bochner and D. Montgomery, Locally compact groups of differentiable transformations, *Annals of Mathematics*, 47 (1946), 653.

the new components are connected with the old ones by the relations

$$\overset{\backsim}{\zeta}^{\alpha} = \zeta^{\beta} \frac{\partial z'_{\alpha}}{\partial z_{\beta}} \tag{9}$$

(summation over β). Thus altogether we can draw from Theorem 2 the following conclusion.

THEOREM 3. *If S is any compact complex manifold then (i) either there exists on S a contravariant complex analytic vectorfield ζ^{α} which is not identically* 0, *or (ii) the totality of all complex homeomorphisms is totally discrete.*

Before proceeding with the subject matter proper we will state a few

CONJECTURES. (α) We will prove under special assumptions that there exists no vector field ζ^{α} and then, again using assumptions, prove that Γ is finite. The second step is probably unnecessary. It is probably true, that whenever on a compact complex S the entire group Γ is discrete, it is automatically finite.

(β) It is possible to represent S as a complex analytic product $S_1 \times S_2$, where S_1 is a complex multi-torus (in several variables) and the group of complex homeomorphisms of S_2 is semi-simple.

(γ) If a separable complex space S (not necessarily compact) is fibred without any singularities by a compact complex manifold S_1, then S is the direct product of S_1 with the base space.

4. COVERING SPACES. Let S be our compact manifold. We introduce its universal covering space denoting it by $\overset{\backsim}{S}$. The fundamental group will be denoted by Δ and its elements by

$$u^0, u^1, u^2, \ldots, \tag{10}$$

the sequence is finite or infinite, and each u^n is a complex homeomorphism of $\overset{\backsim}{S}$ without a fixed point. With each point $\overset{\backsim}{P}$ of $\overset{\backsim}{S}$ we can associate the sequence of points

$\overset{\curvearrowright}{P}{}^n = U^n(\overset{\curvearrowright}{P})$, $n = 0, 1, 2, \ldots$ Any two points of the sequence will be called equivalent. The original manifold S can be identified with the space of all sequences

$$P = \{\overset{\curvearrowright}{P}{}^0, \overset{\curvearrowright}{P}{}^1, \overset{\curvearrowright}{P}{}^2, \ldots\} \tag{11}$$

in a suitable manner. For any given sequence (11) we will say that each $\overset{\curvearrowright}{P}{}^n$ "covers" P or "lies over" P, and conversely we will say that P is a "projection" of $\overset{\curvearrowright}{P}{}^n$. It is possible to choose a subset S^0 of $\overset{\curvearrowright}{S}$ having the following property. The point-sets

$$S^n \equiv U^n\,(S^0) \quad , n = 0, 1, 2, \ldots \tag{12}$$

are mutually disjoint, and

$$\overset{\curvearrowright}{S} = S^0 + S^1 + S^2 + \ldots. \tag{13}$$

The set S^0 is then called a fundamental set. It can be chosen in many special ways. We will only require that its closure in $\overset{\curvearrowright}{S}$ shall be compact in $\overset{\curvearrowright}{S}$, and such a choice is certainly possible.

An analytic homeomorphism

$$\overset{\curvearrowright}{Q} = \overset{\curvearrowright}{f}\,(\overset{\curvearrowright}{P}) \tag{14}$$

of $\overset{\curvearrowright}{S}$ will be called *periodic* if it carries equivalent points $\overset{\curvearrowright}{P}$ into equivalent points $\overset{\curvearrowright}{Q}$. By projection on to S it gives rise to a transformation

$$Q = f(P) \tag{15}$$

and it is easy to see that the latter is a homeomorphism of S. If u^r and u^s are any two elements of (10) then

$$u^r \overset{\curvearrowright}{f} u^s \tag{16}$$

has the same projection as (14) and it is not hard to see that homeomorphisms (16) are the only ones having this property. But the important statement is as follows.

Lemma 1. *Conversely every homeomorphism* (15) *has a*

periodic extension (14), *and if we take any two points* $\overset{\hookrightarrow}{A}$, $\overset{\hookrightarrow}{B}$ *with projections* A, B *such that*

$$B = f\,(A) \tag{17}$$

then there exists a periodic extension which satisfies the initial condition

$$\overset{\hookrightarrow}{B} = \overset{\hookrightarrow}{f}\,(\overset{\hookrightarrow}{A}). \tag{18}$$

Proof. Introduce again the set (11), and another set

$$Q = (\overset{\hookrightarrow}{Q}{}^0, \overset{\hookrightarrow}{Q}{}^1, \overset{\hookrightarrow}{Q}{}^2, \ldots). \tag{19}$$

If (14) is to cover (15), then for every point $\overset{\hookrightarrow}{P}{}^m$ with projection P, the image point $\overset{\hookrightarrow}{Q} = \overset{\hookrightarrow}{f}\,(\overset{\hookrightarrow}{P}{}^m)$ must be suitably chosen among the points (19) in such a manner that for near-by points $\overset{\hookrightarrow}{P}$ the images $\overset{\hookrightarrow}{Q}$ shall also be near-by. From this it follows that if the desired transformation (14) has already been established at a point $\overset{\hookrightarrow}{P}$, it is constructible and uniquely determined in a neighbourhood of $\overset{\hookrightarrow}{P}$. Thus starting with the initial data $\overset{\hookrightarrow}{A}$, $\overset{\hookrightarrow}{B}$ we can construct (14) in some neighbourhood $\overset{\hookrightarrow}{N}\,(\overset{\hookrightarrow}{A})$ of $\overset{\hookrightarrow}{A}$. Also, if another covering of (15) with the same initial condition (18) has been constructed on a larger neighbourhood $\overset{\hookrightarrow}{N}_1 \supset \overset{\hookrightarrow}{N}$, then for the smaller neighbourhood $\overset{\hookrightarrow}{N}\,(\overset{\hookrightarrow}{A})$ the two must be identical. Therefore starting with the initial neighbourhood $\overset{\hookrightarrow}{N}\,(\overset{\hookrightarrow}{A})$ we can construct (14) in a maximal connected open subset $\overset{\hookrightarrow}{N}_1$ of $\overset{\hookrightarrow}{S}$. Since $\overset{\hookrightarrow}{S}$ is connected, if $\overset{\hookrightarrow}{N}_1$ were not all of $\overset{\hookrightarrow}{S}$, it would have a boundary point $\overset{\hookrightarrow}{P}$ in $\overset{\hookrightarrow}{S}$. However by continuity it would then be possible to continue (14) uniquely into a neighbourhood of $\overset{\hookrightarrow}{P}$ thus contradicting the assumption that $\overset{\hookrightarrow}{N}_1$ was maximal.

Thus (14) exists everywhere. Similarly, the inverse $P = f^{-1}(Q)$ with the inverse initial condition $\overset{\curvearrowleft}{A} = \overset{\curvearrowleft}{f}{}^{-1}(\overset{\curvearrowleft}{B})$ has a periodic extension

$$\overset{\curvearrowleft}{P} = \overset{\curvearrowleft}{f}{}^{-1}(\overset{\curvearrowleft}{Q}) \tag{20}$$

and the succession of (14) and (20) projects on the identity. From this it follows that (14) is a homeomorphism of $\overset{\curvearrowleft}{S}$.

Also all extensions (14) for all initial conditions form again a group.

5. Proof of Theorem 1.

We will base it on Theorem 3, first ruling out possibility (i) for $p > 1$. The non-existence of vector-fields was recently proved from curvature assumptions* and the argument will be restated afterwards, but for the present, we will give a much simpler reasoning.

Any scalar function $z = \phi(P)$ on S gives rise to a periodic function on $\overset{\curvearrowleft}{S}$ by putting

$$\phi(\overset{\curvearrowleft}{P^n}) = \phi(P), \; n = 0, 1, 2, \ldots \tag{21}$$

In a similar fashion, all vector fields, tensor fields etc., on S can be periodically extended onto $\overset{\curvearrowleft}{S}$. Now consider a periodic analytic contravariant vectorfield on $\overset{\curvearrowleft}{S}$. On the one hand, we have at present $k = 1$ and thus only one component of the vector exists, and on the other hand, we can cover $\overset{\curvearrowleft}{S}$ by the one coordinate system (1). Thus our vectorfield will be described by *one* complex analytic function in (1). We denote it by $\zeta(t)$. If the element u^n of Δ has the form

$$u^n t = e^{i\vartheta_n} \frac{t - a_n}{1 - \bar{a}_n t}, \; |a_n| < 1, \tag{22}$$

then for a *scalar* $\phi(t)$, periodicity would mean

$$\phi(u^n t) = \phi(t).$$

* S. Bochner, Vector Fields and Ricci Curvature, *Bull. Amer. Math. Soc.*, 52 (1946), 776-97.

2

However, for our vector component $\zeta(t)$ a judicious application of relation (9) leads to the functional equation

$$\zeta(u^n t) = \zeta(t)\frac{d(u^n t)}{dt} \tag{23}$$

$$= \frac{\zeta(t) . e^{i\vartheta_n} (1-a_n \bar{a}_n)}{(1-\bar{a}_n t)^2}.$$

Since for the origin $t=0$, the images $u^n 0 = -e^{i\vartheta_n} a_n$ disperse towards the boundary $|t| = 1$, as $n \to \infty$, we have

$$|a_n| \to 1. \tag{24}$$

Since the closure of S^0 is compact in (1), the value of the factor

$$\frac{e^{i\vartheta_n} (1-\bar{a}_n a_n)}{(1-\bar{a}_n t)^2} \tag{25}$$

tends to 0 as $n \to \infty$, uniformly for t in S^0. From this it is easy to conclude that $\zeta(t)$ is bounded in (1) with boundary values 0. It must therefore be identically 0.

Thus, by Theorem 3, the group Γ on S is countable. Suppose it were infinite,

$$\Gamma = (T^0, T^1, T^2, ..). \tag{26}$$

By Lemma 1, each T^n is the projection of a homeomorphism $\overset{\frown}{T}{}^n$ of (1). Denote the latter by

$$t' = \phi_n(t). \tag{27}$$

Since the closure of S^0 is contained in a circle $|t| \leqslant c < 1$ we may assume by Lemma 1 that

$$|\phi_n(0)| \leqslant c < 1.$$

A subsequence $\{\phi_{n_p}(t)\}$ would then converge in such a manner that the projections T^{n_p} would converge uniformly in S. But then the group Γ would contain arbitrary small elements, contrary to the assumption of discreteness. Thus, Γ must be finite, as asserted in Theorem 1.

6. A Generalization to Several Variables.

Theorem 4. Assumptions. I. *Let $\overset{\frown}{S}$ be a bounded domain in the complex variables $t_1, \ldots, t_k$, and let* (10) *be a group of its homeomorphisms which generates a compact manifold S.*

II. *If we denote the homeomorphism u^n, $n = 0, 1, 2 \ldots$ by*

$$t'_\alpha = \psi_{n\alpha}(t_1, \ldots, t_k), \; \alpha = 1, \ldots, k \tag{28}$$

then uniformly for $(t_1, \ldots, t_k)$ in any closed subset S' of $\overset{\frown}{S}$ the partial derivatives

$$\frac{dt'_\alpha}{dt_\beta} = \frac{d\psi_{n\alpha}(t)}{dt_\beta}, \alpha, \beta = 1, \ldots, k, \tag{29}$$

tend to 0, *uniformly as the image point $(t'_1, \ldots, t'_k)$ approaches the boundary of $\overset{\frown}{S}$.*

Conclusion. *The group of complex homeomorphisms of S is finite.*

Proof. A contravariant vector on $\overset{\frown}{S}$ is described by k analytic functions in $\overset{\frown}{S}$ with the periodicity properties

$$\zeta^\alpha(u^n t) = \zeta^\beta(t) \frac{\partial \psi_{n\alpha}(t)}{\partial \psi_\beta(t)}, \alpha = 1, \ldots, k. \tag{30}$$

By Assumption II, each $\zeta^\alpha(t)$ is bounded with boundary values 0 in $\overset{\frown}{S}$. A function in several complex variables assumes the maximum of its absolute values on the boundary, and thus $\zeta^\alpha(t) \equiv 0$.

Thus by Theorem 3, Γ is discrete. Each element T^n of Γ is the projection of a homeomorphism

$$t'_\alpha = \phi_{n\alpha}(t_1, \ldots, t_k) \quad \alpha = 1, \ldots, k \tag{31}$$

of S and we may assume that for some point of $\overset{\frown}{S}$ the image points $\{t'_\alpha\}$ lie in a compact subset of $\overset{\frown}{S}$. If Γ were infinite, a subsequence of (31) would converge uniformly in the neighbourhood of every point of $\overline{S}_1$ and Γ would have arbitrarily small elements. Thus Γ is finite.

The most immediate special case is that of the sphere

$$\breve{S}: |t_1|^2+\ldots+|t_k|^2 < 1. \tag{32}$$

Its homeomorphisms can be written in the form

$$t'_\alpha = \frac{a_{\alpha 0}+\Sigma_\beta\, a_{\alpha\beta}\, z_\beta}{1+\Sigma_\beta\, a_\beta\, z_\beta}, \alpha = 1,\ldots,k, \tag{33}$$

where

$$\Sigma_\alpha |a_\alpha|^2 < 1 \tag{34}$$

$$\Sigma_\alpha |a_{\alpha 0}|^2 < 1 \tag{35}$$

$$\Sigma_\alpha\, a_{\alpha\beta}\, \bar{a}_{\alpha 0} = a_\beta,\ \beta = 1,\ldots,k \tag{36}$$

$$\Sigma_\alpha\, a_{\alpha\beta}\, \bar{a}_{\alpha\gamma} = a_\beta\, \bar{a}_\gamma,\ \beta \neq \gamma \tag{37}$$

$$\Sigma_\alpha |a_{\alpha\beta}|^2 - |a_\beta|^2 = 1 - \Sigma_\alpha |a_{\alpha 0}|^2,\ \beta = 1,\ldots,k. \tag{38}$$

If we introduce the $k+1$ vectors

$$A_{\alpha\rho} = (a_{1\rho}, a_{2\rho},\ldots,a_{k\rho}),\ \rho = 0,\ 1,\ldots,k$$

and the inner product (A, B) of two vectors, then (35)-(38) can also be written as

$$(A_0, A_0) < 1 \tag{39}$$

$$(A_\beta, A_0) = a_\beta.\ \beta = 1,\ldots,k \tag{40}$$

$$(A_\beta, A_\gamma) = a_\beta\, \bar{a}_\gamma,\ \beta \neq \gamma \tag{41}$$

$$(A_\beta, A_\beta) - |a_\beta|^2 = 1 - (A_0, A_0). \tag{42}$$

It is easily seen that every such fractional transformation is a homeomorphism of (32), that they form a group, and that they are transitive on (32). Furthermore the classical Schwarz's lemma can then be suitably extended to show that there are no other homeomorphisms of (32) except these.

Since (33) transforms the point $(0,\ldots,0)$ into the point A_0 whenever a variable transformation (33) moves the origin towards the boundary of (32), we have

$$(A_0, A_0) \rightarrow 1. \tag{43}$$

Now, at the origin we have

$$\frac{\partial t'_\alpha}{\partial t_\beta} = a_{\alpha\beta} - a_{\alpha 0}\, a_\beta,$$

and hence

$$\sum_\alpha \left| \frac{\partial t'_\alpha}{\partial t_\beta} \right|^2 = (A_\beta, A_\beta) - a_\beta (A_0, A_\beta) - \bar{a}_\beta (A_\beta, 0) + |a_\beta|^2 (A_\beta, A_0).$$

By (40), this is

$$\begin{aligned}&(A_\beta, A_\beta) - 2\,|a_\beta|^2 + |a_\beta|^2\ (A_0, A_0)\\ &= (A_\beta, A_\beta) - |a_\beta|^2 - |a_\beta|^2 [1 - (A_0, A_0)]\end{aligned}$$

and (42) finally leads to

$$(1 - |a_\beta|^2)\,[1 - (A_0, A_0)].$$

Thus (43) implies

$$\sum_\alpha \left| \frac{\partial t'_\alpha}{\partial t_\beta} \right|^2 \to 0,\ \beta = 1, \ldots, k,$$

and this verifies the validity of Assumption II of Theorem 4, at least for the point $t = 0$. Its validity for other points uniformly in a compact set can be obtained by a more complicated estimate; or by a change of coordinates, of the form (33), which moves a variable point into the origin.

7. On assumption II in Theorem 4.

We will somewhat relax this assumption to suit a certain family of domains. The most typical representative of the family is the multicircle

$$|t_1| < 1, \ldots, |t_k| < 1. \tag{44}$$

We denote its entire boundary by B. A boundary point arises if at least one inequality $|t_\alpha| < 1$ is replaced by an equality $|t_\alpha| = 1$. In particular, B contains the points

$$|t_1| = 1, \ldots, |t_k| = 1; \tag{45}$$

the latter point-set will be denoted by B_0. For $k > 1$ it is a proper subset of B, but it has the following important property. If $f(t_1, \ldots, t_k)$ is analytic in (44) and has boundary values 0 at the points of B_0 with no assumptions made for the points of $B - B_0$, it is nevertheless identically 0. This follows immediately from the iterated Cauchy formula

$$f(t_1, \ldots, t_k) = \frac{1}{(2\pi)^k} \int_{C_1} \cdots \int_{C_k} \frac{f(\zeta_1, \ldots, \zeta_k)\, d\zeta_1 \ldots d\zeta_k}{(\zeta_1 - t_1) \ldots (\zeta_k - t_k)}. \tag{46}$$

Now the homeomorphisms* of (44) are

$$t'_\alpha = e^{i\vartheta_\alpha} \frac{t_{\rho_\alpha} - a_\alpha}{1 - \bar{a}_\alpha t_{\rho_\alpha}}, \; |a_\alpha| < 1, \tag{47}$$

where $\alpha = 1, \ldots, k$ and $(\rho_1, \ldots, \rho_k)$ is a permutation of $(1, \ldots, k)$. Now, if $(t_1, \ldots, t_k)$ lies in a closed subset of (44), then for variable transformations (47) the partial derivatives

$$\frac{dt'_\alpha}{dt_\beta} \tag{48}$$

do not necessarily converge to 0 as the point $(t_1', \ldots, t'_k)$ approaches the boundary B, but they do so converge if it approaches the subset B_0. However owing to the property of B_0 just stated, this suffices to establish the conclusion of Theorem 4.

In general, let B denote the boundary of any bounded domain $\tilde{S}$. A subset B_0 of B shall be called *significant* if any analytic function in $\tilde{S}$ with boundary values 0 on B_0 is identically 0. Then in Assumption II of Theorem 4 it suffices to assume that the derivatives (29) converge to 0 as the variable point t' approaches the significant part only.

The domain (44) is a product of circles and the significant subset (46) is the product of their boundaries, which is a much smaller set than the boundary of the products.

More generally, whenever a domain $\tilde{S}$ is a product of domains

$$S_1 \times S_2 \times \ldots \times S_l \tag{49}$$

in lower dimensional factors, then the formal product of their boundaries

$$B_1 \times B_2 \times \ldots \times B_l \tag{50}$$

* Caratheodory, Ueber das Schwarzsche Lemma bei analytischen funktionen von zwei komplexen Veraenderlichen, *Math. Annalen*, 97 (1927), 98.

is a significant subset of the boundary of (49). This follows from the fact that there exists a Cauchy formula* for any domain S_λ, and in the case of a product (49), a formula can hence be obtained by iteration, roughly analogous to (46), which determines the values in (49) from those on (50).

However it must not be assumed that the factorization (49) is the main source for significant sets. They may arise from algebraic sources on domains which on the face of it are very much "irreducible". The best illustration for this are the so-called matrix spaces, and we will briefly describe one case. Assume that the dimension number k is a square, $k = n^2$. The variables

$$z_1, \ldots, z_k \tag{51}$$

can then be arranged as a square matrix

$$z = \{ z_{pq} \}, p, q = 1, \ldots, n. \tag{52}$$

The symbol z shall denote the entire matrix. If z^* denotes the adjoint matrix then the matrix product $z z^*$ is positive Hermitian. The domain S shall be the set of points in the space (50) for which all eigen-values of $z z^*$ are < 1. The boundary B consists of those points for which all eigen-values are $\leqslant 1$ with at least one eigen value strictly $= 1$, and the subset B_0 consists of those points for which all eigen-values are $= 1$. A matrix (51) belongs to (51) if and only if it is unitary. There exists a Cauchy formula†

* Due to R. Fueter and Enzo Martinelli. *See*, Martinelli, Alcumi teoremi integrali per le funzioni analitiche di pui variabili complesse, *Reale Acca Reale* (*Mat.*), 1938 XVI, p. 272. A thorough discussion of the formula with many implications was given in S. Bochner, Analytic and meromorphic continuation by means of Green's formula, *Annals of Mathematics*, 44 (1943), 652-73. However this author's claim in *loc. cit.* footnote (1), that he might have been familiar with the general shape of the formula before Martinelli, was wholly unjustified and is hereby being retracted.

† S. Bochner, Group invariance of Cauchy's formula in several variables, *Annals of Mathematics*, 45 (1944), 699.

$$f(z) = \frac{1!\,2!\ldots(n-1)!}{(2\pi i)^{\frac{n(n+1)}{2}}} \int_{B_0} \frac{f(\zeta)\, d\zeta_1 \ldots d\zeta_k}{(\det|\zeta_{pq} - z_{pq}|)^n} \tag{53}$$

from which it is obvious that B_0 is indeed a significant set.

8. Negative Curvature*

If we start from a manifold S, then hypothesis I in Theorem 4 implies the hypothesis that the covering space $\overset{\frown}{S}$ shall be complex-analytically equivalent with a domain in ordinary space. For $k \geqslant 2$ this hypothesis is implacably severe and wholly unverifiable from the definition of a manifold. In this direction nothing is known, not even in the simplest case, whether real analytic or complex analytic. For instance, if $\overset{\frown}{S}$ is topologically a simplex, and if analytically it is a union of two analytic neighbourhoods, each topologically a simplex and if their intersection is a simplex, then it is not known whether $\overset{\frown}{S}$ can be analytically covered by one coordinate system, that is, whether it can be topologically-analytically mapped into Euclidean space. Such a mapping requires the existence of certain analytic functions on all of $\overset{\frown}{S}$. Now except in the classical case, there is no means known of constructing an analytic function in the large, out of nothing as it were. Some global information must be supplied to start with. The most natural information of this kind is an analytic distance, and this implies the existence† of an analytic positive definite differential form $ds^2 = g_{ij}\, dx_i\, dx_j$ on the entire space. In the real case, this author was the first to prove, as a supplement to

* In this section we will partly presuppose, and partly summarize and draw inferences from the main facts in the author's paper mentioned in sec. 5, page 9, and the further paper, S. Bochner, Curvature in Hermitian metric, *Bulletin of American Math. Soc.*, 53 (1947), 179-95.

† *See* S. Bochner, Differentiable and Riemann Metric, *Duke Journal*, 4 (1938), 51-4.

H. Whitney's work on differentiable manifolds, that if on a compact analytic manifold such an analytic metric exists, the manifold can be topologically analytically imbedded into Euclidean space*.

In the complex case a natural restriction on the metric is to assume that it has the Hermitian form

$$ds^2 = g_{\alpha\bar{\beta}} dz_\alpha d\bar{z}_\beta. \tag{53}$$

A further restriction, very far reaching but very fruitful, is the further assumption

$$\frac{\partial g_{\alpha\bar{\beta}}}{\partial z_\gamma} = \frac{\partial g_{\gamma\bar{\beta}}}{\partial z_\alpha} \tag{54}$$

which automatically implies

$$\frac{\partial g_{\alpha\bar{\beta}}}{\partial \bar{z}_\gamma} = \frac{\partial g_{\alpha\bar{\gamma}}}{\partial \bar{z}_\beta}. \tag{55}$$

This amounts to assuming that locally there exists a scalar function $\Omega(z, \bar{z})$ such that

$$g_{\alpha\bar{\beta}} = \frac{\partial^2 \Omega}{\partial z_\alpha \partial \bar{z}_\beta}. \tag{56}$$

In this case the Riemann-Christoffel tensor has components only of the form

$$R_{\alpha\bar{\beta}\gamma\bar{\delta}} \tag{57}$$

and for the contracted Ricci-tensor $R_{\alpha\bar{\beta}}$, we then have the beautiful relation

$$R_{\alpha\bar{\beta}} = \frac{\partial^2 G}{\partial z_\alpha \partial \bar{z}_\beta}, \tag{58}$$

where G is the determinant of the matrix.

Whenever on a Riemannian space, the two-dimensional sectional Gaussian curvature is positive for all sections through a point, then of necessity the Ricci curvature is positive at that point, meaning that†

$$-R_{\alpha\bar{\beta}} \xi^\alpha \bar{\xi}^\beta > 0 \tag{59}$$

for any vector ξ^α of non-vanishing length. But not conversely. Thus to draw a conclusion from the behaviour

* S. Bochner, Analytic mapping of compact Riemann spaces into Euclidean space, *Duke Journal*, 3 (1937), 339-54.

† The negative sign is due to a conventional normalization.

3

of Ricci curvature is better than to conclude it from the behaviour of the Riemann curvature. We now quote the following theorem.

THEOREM 5. *If on a compact complex manifold the Ricci curvature is positive or zero (but not identically zero), then there exists no covariant complex analytic tensor other than zero.*

If however the Ricci curvature is negative or zero (but not identically zero) then there exists no contravariant analytic tensor other than zero.

It will also be useful to introduce an *ad-hoc* definition. We say that a space $\overset{\frown}{S}$ has the *Montel structure* if every sequence of its complex homeomorphisms for which the images of one special point are contained in a compact subset of $\overset{\frown}{S}$ contains a subsequence which converges uniformly in the neighbourhood of every point of $\overset{\frown}{S}$.

By the use of Theorem 3 and the second half of the proof of Theorem 4, we now obtain on the basis of Theorem 5 the following conclusion.

THEOREM 6. *If a compact complex manifold S has negative Ricci curvature and if its universal covering space has the Montel structure, then there are only a finite number of complex homeomorphisms on S.*

9. COMPARISON BETWEEN THEOREMS 4 AND 6.

We are now going to demonstrate that in substance (although by no means in detail) the assumptions in these two theorems are very much alike.

Suppose first the assumptions of Theorem 4 are fulfilled in detail. Then $\overset{\frown}{S}$, being an ordinary bounded domain, has Montel structure, and we still have to argue that S is likely to possess a metric with negative curvature; or, what would be even better, that $\overset{\frown}{S}$ itself is likely to possess a hyperbolic geometry (metric with negative curvature) which is invariant with respect to all homeomorphisms of $\overset{\frown}{S}$ and not only those of any one fundamental group.

Now the theory of automorphic functions is full of indications that such is indeed the case. First of all, for $k = 1$, we have the classical hyperbolic geometry with

$$ds^2 = \frac{2dz\,d\bar{z}}{(1-z\bar{z})^2}$$

in the unit circle $|z| < 1$, the curvature being $K = -1$. The more general line element

$$ds^2 = \frac{2dz\,d\bar{z}}{(1+\frac{1}{2}b\,z\bar{z})^2} \tag{60}$$

has curvature $\frac{1}{2}b$ and for positive b it is an elliptic geometry on the one-dimensional complex projective space (Gaussian sphere). The generalization of (60) to k complex variables is

$$ds^2/2 = \frac{\Sigma_\alpha |dz_\alpha|^2 + b/2\,(\Sigma_\alpha |z_\alpha|^2 \Sigma_\beta |dz_\beta|^2 - |\Sigma_\alpha \bar{z}_\alpha dz_\alpha|^2)}{(1+b/2\,\Sigma_\alpha |z_\alpha|^2)^2} \tag{61}$$

due largely to Fubini.

The sectional curvature is not constant on all sections, but only on holomorphic sections; but on all sections we have

$$\tfrac{1}{4} \leqslant k/b \leqslant 1, \tag{62}$$

and thus for negative b the sectional curvature is at any rate everywhere negative. The constancy on holomorphic sections expresses itself in the relation

$$R_{\alpha\bar{\beta}\gamma\bar{\delta}} = -b/2\,(g_{\alpha\bar{\beta}}g_{\gamma\bar{\delta}} + g_{\gamma\bar{\beta}}g_{\alpha\bar{\delta}}). \tag{63}$$

In recent years, C. L. Siegel and L. K. Hua* have considered, for several variables, other 'natural' types of domains $\overset{\frown}{S}$ in many matrix spaces, and in all cases proved the existence of a hyperbolic geometry. Finally this author has given a general argument, partly heuristic, for the existence of a hyperbolic geometry in the general case. The argument runs as follows. Start with the original metric

* Quotations in the papers of footnote on page 16, Sec. 8.

$$ds^{(o)2} = \Sigma_\alpha dt_\alpha d\bar{t}_\alpha.$$

If we desire invariance only with respect to the elements u^n of a fundamental group $\triangle$, then with each element u^n we associate the transformed line-element

$$ds^{(n)2} = \Sigma_\alpha dt_\alpha^{(n)} \overline{dt_\alpha^{(n)}} = \Sigma_{\alpha,\beta} g_{\alpha\bar{\beta}}^{(n)} dt_\alpha d\bar{t}_\beta$$

and we then form the sum

$$ds^2 = \Sigma_n ds^{(n)2} \tag{64}$$

provided it is finite. If then (64) exists, it is obviously periodic; but we were also able to show that its sectional curvature on holomorphic sections and its Ricci curvature must indeed be always negative (non-positive). This was obtained from a seemingly different theorem, namely from the fact that in the case of a complex imbedding of one space into a higher dimensional one, the so-called relative parts of these curvatures are negative. In particular, if the higher dimensional space is flat, it contributes no local curvature of its own, and the relative parts of these curvatures constitute their entire amounts for the imbedded space. Finally forming the line element (64) as a sum can be interpreted, in the limit, as if resulting from imbedding of $\overset{\frown}{S}$ into a higher dimensional space, and the negative character of the curvatures is the result.

Thus far we have discussed the relation of Theorems 4 and 6 in one direction. In the opposite direction, suppose that the assumptions of Theorem 6 are fulfilled in detail. Assume, more strongly, that the sectional curvatures themselves, are negative. Then by a known theorem on differentiable manifolds*, the covering space $\overset{\frown}{S}$ is *topologically* homeomorphic to a simplex. The vital question remaining is whether $\overset{\frown}{S}$ has the Montel structure. This seems to be a difficult question indeed, but perhaps within reach of the present day methods and we still

* E. Cartan, *Lecons sur la geometrie des spaces de Riemann*, Chapter VI.

venture to advance a tenuous argument in favour of an affirmative answer.

In the real case, all spaces of constant curvature (positive or negative) are, locally, isometrically equivalent with each other, by a classical theorem of F. Schur. From this it follows, by an argument using geodesics* that their universal covering spaces are the same. In the Hermitian case we have proved, in direct analogy to the theorem of F. Schur, that relation (63) implies the existence of an analytic coordinate system in which the line element has the form (61). Starting from this, it ought to be possible to imitate the argument for the real case with the view to obtaining the following theorem.

THEOREM 7. *If a complex manifold S has an Hermitian metric with property* (54) *and if* (63) *holds, and if the space is complete (meaning that its geodesics are continuable indefinitely) in particular if it is compact, then its universal covering space is analytically isometrically equivalent with the unit sphere* $|z_1|^2+\ldots+|z_k|^2<1$ *for* $b=-2$, *and with the complex projective space for* $b>0$.

In particular, for $b=-2$, $\tilde{S}$ would have the Montel structure. Now, it is our further surmise that the same conclusion ought then to be obtainable for any compact complex manifold S of negative sectional curvature. If the negative value of the sectional curvature lies between k_0 and k_1, $k_0<k_1$, then on the basis of (62) one ought to be able geometrically to establish the Montel structure of $\tilde{S}$ by comparing $\tilde{S}$ with two Fubini spaces, with the constants

$$|b_1|=k_0/4,\quad |b_2|=4\,|k|,$$

which, as it were, envelop $\tilde{S}$ from above and below, and have Montel structure themselves.

* *See* Heinz Hopf, Zum Clifford-Kleinschen Raumproblem, *Math. Annalen*, 95 (1925), 317-8.

ANNALS OF MATHEMATICS
Vol. 49, No. 2, April, 1948

CURVATURE AND BETTI NUMBERS

S. BOCHNER

(Received September 9, 1947)

1. The first theorem

If S_n is a compact differentiable space of dimension n, $n \geqq 2$, and if S'_n is a covering space which is likewise compact, and if B_p and B'_p are the Betti numbers of S_n and S'_n then we have

$$B_p \leqq B'_p, \qquad 1 \leqq p \leqq n - 1. \tag{1}$$

In fact, if ω_p is any external form on S_n, then it has a periodic extension ω'_p onto S'_n, and if two forms ω_p are non-homologous on S_n, then their extensions ω'_p are non-homologous on S'_n, which proves (1). On the other hand, if S_n has a positive-definite Riemannian metric

$$ds^2 = g_{ij}\,dx_i\,dx_j, \tag{2}$$

and if its sectional curvature is constant and positive, that is if

$$R_{hijk} = b(g_{hj}g_{ik} - g_{hk}g_{ij}) \tag{3}$$

with $b > 0$, see [4], p. 83, then its universal covering space S'_p is the ordinary sphere, see [6]. Hence $B'_p = 0$, and thus (1) implies

$$B_p = 0, \qquad 1 \leqq p \leqq n - 1, \tag{4}$$

for any compact space of positive constant curvature.

Constant curvature is a very specialized requirement. Our main objective will be to retain conclusion (4), for one or all values of p, under considerably more liberal assumptions involving positiveness of curvature. But we will also have assertions on non-vanishing of Betti numbers, to suit the structure of complex manifolds, and some of these results will be divested of their complex analytic setting altogether.

For $p = 1$, the conclusion

$$B_1 = 0 \tag{5}$$

was discussed in a previous paper, [1], and it was obtained for a very general condition, namely, the positiveness of the Ricci (mean) curvature of any compact space. Our method of dealing with $p > 1$ will be the same, but more stringent conditions will be required, in that some restraint on the amount of deviation from constant curvature will be imposed.

Actually we will measure the deviation, not from constancy of sectional curvature, but from conformal flatness and our conclusion will be thus more liberal. A space S_n is conformally flat if in the neighborhood of every point there exists some coordinate system for which the line element has the form

$$ds^2 = \rho(y_1, \cdots, y_n)(dy_1^2 + \cdots + dy_n^2).$$

A space of constant sectional curvature is conformally flat, but not conversely, and we will obtain conclusion (4) for all p under the assumption that the Ricci curvature of S_n is everywhere positive and that the deviation of the space from conformal flatness is a sufficiently small fraction of the Ricci curvature; the size of the fraction depending only on n (and p), but not on S_n.

If R_{hijk} is the Riemann-Christoffel tensor, then

$$R_{hk} = g^{ij}R_{hijk} \tag{6}$$

is the (contracted) Ricci tensor, and its further contraction

$$R = g^{hk}R_{hk} = g^{hk}g^{ij}R_{hijk} \tag{7}$$

is the so called scalar curvature. We will also employ the tensor

$$\begin{aligned} C_{hijk} = R_{hijk} - \frac{1}{n-2}(g_{ij}R_{hk} - g_{ik}R_{hj} + g_{hk}R_{ij} - g_{hj}R_{ik}) \\ - \frac{R}{(n-1)(n-2)}(g_{hj}g_{ik} - g_{hk}g_{ij}), \end{aligned} \tag{8}$$

which is H. Weyl's conformal curvature tensor, see [4], p. 89. The space S_n is conformally flat, if and only if

$$C_{hijk} \equiv 0. \tag{9}$$

Now, take a fixed point of the space, all skew symmetric tensors ξ^{ij}, $\xi^{ji} = -\xi^{ij}$, and form the number

$$| C | = \sup_\xi \frac{| \xi^{hk} \xi^{ij} C_{hijk} |}{\xi_{hk} \xi^{hk}} \tag{10}$$

as a measure of conformal un-flatness at the point. In forming (10) we do not restrict ourselves to bi-vectors $\xi^{ij} = \eta^i\zeta^j - \eta^j\zeta^i$, but introduce all possible skew tensors. On the other hand, even if (10) were formed only for bi-vectors, its vanishing would imply relation (9), as in the case of the tensor R_{hijk} itself.

The Ricci tensor (6) is symmetric. Corresponding to any vector ξ^i, the number $-\xi^i\xi^jR_{ij}$ is a certain sum of $(n - 1)$ sectional curvatures. Thus, if all sectional curvatures are positive, we have

$$-\xi^i\xi^jR_{ij} > 0 \tag{11}$$

for $\xi_i\xi^i > 0$, but not conversely. The number

$$L = \inf_\xi \frac{| \xi^i \xi^j R_{ij} |}{\xi_i \xi^i} \tag{12}$$

is the smallest absolute value of an eigenvalue. Finally, if we form (7), then (11) implies

$$-R > 0, \tag{13}$$

but not conversely. Thus, (13) is the "weakest" measure for positiveness of curvature at a point. Also, if (11) holds, we have

$$(14) \qquad -R \geqq nL,$$

since in a geodesic coordinate system we have $g^{hk} = \delta^{hk}$, and in this specialization the number R is the trace of the matrix (6). We also note that the symbol $|R|$ will be the ordinary absolute value of the number R, and after these preliminaries we can state our first theorem.

THEOREM 1. *If on a compact S_n, with definite metric, the Ricci curvature is positive, that is, if* (11) *holds everywhere, and if for some p, we have at all points the relation*

$$(15) \qquad (p-1)|C| < \left(1 - \frac{2(p-1)}{n-2}\right)L + \frac{(p-1)|R|}{(n-1)(n-2)}$$

then for this p we have

$$(16) \qquad B_p = 0.$$

In particular, the mere assumption (11) *implies*

$$(17) \qquad B_1 = 0.$$

Furthermore, if n is even, then assumptions (13) *and*

$$(18) \qquad |C| < \frac{|R|}{2(n-1)}$$

imply

$$(19) \qquad B_{n/2} = 0.$$

On the face of it, conclusion (19) is slightly better than (16), since (13) is less than (11). However, de facto, for a compact space, assumption (13) together with (18) might imply (11), for all that is known.

For $p = 1$, conclusion (17) uses only assumption (11), but as such it can also be deduced from a strong result of Myers [7]. Myers proved by a discussion of geodesics that if on a compact manifold the Ricci curvature is strictly positive, the universal covering space is likewise compact. Therefore, $B_1' = 0$, and (17) follows from (1). However, it should be noted that our proof would allow certain immediate generalizations which would not be easily accessible to a proof based on a discussion of geodesics. Instead of demanding that the Ricci-curvature be strictly positive at all points without exception, we might admit that it be semi-positive in a nowhere dense subset, in particular on a lower dimensional analytical variety. The inequalities (15) and (18) can be relaxed in the same way, and analogous relaxations are admissible in theorem 2. Furthermore, as already stated in [1], there is the possibility of extending our results to some types of non-compact spaces, but this kind of extension will be systematically described at a later date.

2. Proof of Theorem 1

As in [1], we will apply the following criterion.

LEMMA. *If on our compact S_n we set up for a scalar function the Laplace-Beltrami operator*

$$\Delta\varphi = g^{rs}\varphi_{,r,s} \tag{20}$$

and if for a given scalar φ we have

$$\Delta\varphi \geqq 0 \tag{21}$$

everywhere, then

$$\Delta\varphi \equiv 0 \tag{22}$$

everywhere.

We will use the lemma in the following way. We will take a tensor $\xi_{a_1\cdots a_p}$ and then set up as our scalar the square length

$$\varphi = \xi_{a_1\cdots a_p}\,\xi^{a_1\cdots a_p} = g^{a_1b_1}\cdots g^{a_pb_p}\xi_{a_1\cdots a_p}\xi_{b_1\cdots b_p}\,. \tag{23}$$

We will then assume that the tensor is harmonic, and from the curvature assumptions we will deduce from the lemma that such a tensor must vanish identically. However, by the theory of harmonic integrals, the non-existence of such a tensor is equivalent with the vanishing of the Betti number B_p. A tensor is harmonic if first of all it is skew symmetric, that is

$$\xi_{a_1\cdots a_{\rho-1}a_\sigma a_{\rho+1}\cdots a_{\sigma-1}a_\rho a_{\sigma+1}\cdots a_p} = -\xi_{a_1\cdots a_p} \tag{24}$$

and if it satisfies the differential relations

$$\xi_{a_1\cdots a_p,r} = \sum_{q=1}^{p}\xi_{a_1\cdots a_{q-1}ra_{q+1}\cdots a_p,a_q} \tag{25}$$

$$g^{bc}\xi_{a_1\cdots a_{p-1}b,c} = 0. \tag{26}$$

For (23) we have

$$\tfrac{1}{2}\Delta\varphi = \xi_{a_1\cdots a_p,r}\xi^{a_1\cdots a_p,r} + g^{rs}\xi_{a_1\cdots a_p,r,s}\xi^{a_1\cdots a_p} \tag{27}$$

and since the first term on the right is non-negative we only have to concentrate on the expression

$$\Phi = g^{rs}\xi_{a_1\cdots a_p,r,s}\xi^{a_1\cdots a_p} \tag{28}$$

and if we can verify that

$$\Phi \geqq 0 \tag{29}$$

then by the lemma we will have

$$\Phi \equiv 0. \tag{30}$$

By covariant differentiation of (25) we obtain

$$\xi_{a_1\cdots a_p,r,s} = \sum_{q=1}^{p}\xi_{a_1\cdots a_{q-1}ra_{q+1}\cdots a_p,a_q,s}\,. \tag{31}$$

If we substitute this in the right side of (28), and permute certain indices on the basis of (24), we obtain

$$\frac{1}{p}\Phi = g^{rs}\xi_{ra_2\cdots a_p,a_1,s}\xi^{a_1\cdots a_p}. \tag{32}$$

We next interchange the covariant differentiations with respect to a_1 and s by the formula

$$\xi_{ra_2\cdots a_p,a_1,s} - \xi_{ra_2\cdots a_p,s,a_1} = \xi_{ta_2\cdots a_p}R^t{}_{ra_1s} + \sum_{j=2}^{p}\xi_{ra_2\cdots a_{j-1}ua_{j+1}\cdots a_p}R^u_{a_ja_1,s}. \tag{33}$$

Now

$$g^{rs}\xi_{ra_2\cdots a_p,s,a_1} = (g^{rs}\xi_{ra_2\cdots a_p,s}),\, a_1 \tag{34}$$

vanishes by (26) in conjunction with (24), and therefore

$$\frac{1}{p}\Phi = g^{rs}\xi_{ta_2\cdots a_p}R^t{}_{ra_1s}\xi^{a_1\cdots a_p} + \sum_{j=2}^{p}\xi_{ra_2\cdots a_{j-1}\,ua_{j+1}\cdots a_p}R^u_{a_ja_1s}\,\xi^{a_1\cdots a_p}. \tag{35}$$

On the one hand we have

$$g^{rs}R^t{}_{ra_1s} = -R^t{}_{a_1}$$

and on the other hand all terms in the sum are equal to one another by (24), and thus we have

$$\frac{1}{p}\Phi = (p-1)\xi_{tub_3\cdots b_p}\xi^{a_1a_2b_3\cdots b_p}R^u{}_{a_2a_1}{}^t - \xi_{tb_2\cdots b_p}\xi^{ub_2\cdots b_p}R^t{}_u\,. \tag{36}$$

We now write $R^u{}_{a_2a_1}{}^t$ as a difference of $C^u{}_{a_2a_1}{}^t$ and some additional terms as given by formula (8), and after a simple calculation we obtain, again using (24),

$$\begin{aligned}\frac{1}{p}\Phi &= (p-1)\xi_{tub_3\cdots b_p}\,\xi^{a_1a_2b_3\cdots b_p}\,C^u{}_{a_2a_1}{}^t\\ &-\left(1-\frac{2(p-1)}{n-1}\right)\xi_{tb_2\cdots b_p}\,\xi^{ub_2\cdots b_p}R^t{}_u - \frac{R(p-1)}{(n-1)(n-2)}\,\xi_{b_1\cdots b_p}\,\xi^{b_1\cdots b_p}.\end{aligned} \tag{37}$$

We have only treated the case $p \geqq 2$. In the simplest case, $p = 1$, see [1], one obtains

$$\Phi = -\xi_t\xi^uR^t{}_u \tag{38}$$

which is formally contained in (37) for $p = 1$. Also, under the assumption (11) we would have (29) everywhere, and thus of necessity (30), and thus a fortiori $\xi^i = 0$ as claimed in Theorem 1.

For $p = 2$ we have

$$|\,\xi_{tu}\,\xi^{a_1a_2}C^u{}_{a_2a_1}{}^t\,| \leqq |\,C\,|\,\xi_{b_1b_2}\xi^{b_1b_2} \tag{39}$$

by definition (10). Furthermore we have

$$-\xi_{tb}\,\xi^{ub}R^t{}_u \geqq L\xi_{tb}\,\xi^{tb}. \tag{40}$$

In order to see the latter, we introduce at the given space point a geodesic coordinate system, which produces the equalities $\xi_{ub} = \xi^{ub} = \xi_u{}^b$ for any indices. Thus, for an individual b we obtain

$$-\xi_{tb}\,\xi^{ub}R^t{}_u = -\xi_{tb}\,\xi_{ub}\,R^t{}_u \geqq L\xi_{tb}\,\xi_{ub} \tag{41}$$

by definition (12), and (40) follows afterwards by summation over b. Now, if at some point of S_n the components ξ_{tb} would not be all zero, then formulas (15), (37), (39), (40) would give $\Phi > 0$ which is contradictory. Hence $\xi_{tb} = 0$ and $B_2 = 0$.

For $p > 2$, we take again at a fixed point a geodesic coördinate system so that, by numerical accident we have $\xi^{a_1a_2b_2\cdots b_p} = \xi^{a_1a_2}{}_{b_3\cdots b_p}$. We now take a fixed combination of indices $b_3, \cdots, b_p$. All our assumptions conjointly would then give

$$\xi_{tub_3\cdots b_p}\,\xi^{a_1a_2}{}_{b_3\cdots b_p}\,C^u{}_{a_2a_1}{}^t - \left(\frac{1}{p-1} - \frac{2}{n-1}\right)\xi_{tb_2\cdots b_p}\,\xi^{ub_2}{}_{b_3\cdots b_p}\,R^t{}_u$$
$$- \frac{R}{(n-1)(n-2)}\,\xi_{b_1b_2b_3\cdots b_p}\,\xi^{b_1b_2}{}_{b_3\cdots b_p} \geqq 0.$$

Also, if some tensor component were $\neq 0$, then for at least one combination we would obtain a strict inequality, and after adding up over all combinations $(b_3, \cdots, b_p)$ we would obtain $\Phi > 0$, which is contradictory. Thus our tensor must vanish identically, and the proof of Theorem 1 is completed.

3. Skew symmetry

We take a tensor $\xi_{a_1\cdots a_p}$. We again assume that it is skew symmetric as in (24), but instead of the first differential relation (25) we assume that its covariant derivative continues to be skew symmetric in all $p + 1$ indices. This amounts to adding the assumption

$$\xi_{a_1\cdots a_{p-1}\,a_p,r} = -\xi_{a_1\cdots a_{p-1}r,a_p}. \tag{42}$$

Due to the symmetry of g^{ij} this automatically implies the second differential relation (26). Due to (42) we now obtain

$$\tfrac{1}{2}\Delta\varphi = \xi_{a_1\cdots a_p,r}\,\xi^{a_1\cdots a_p,r} - g^{rs}\xi_{a_1\cdots a_{p-1}r,a_p,s}\,\xi^{a_1\cdots a_p} \tag{43}$$

(notice the "minus" sign), but from now on the analysis proceeds as before on account of (26). Thus we obtain the following "obverse" conclusion.

THEOREM 2. *If on a compact S_n the Ricci-curvature is everywhere negative and if again restriction* (15) *holds, then there exists no p-dimensional skew symmetric tensor whose covariant derivative is likewise skew symmetric.*

In particular, for $p = 1$, there is only the one condition

$$\xi_{a,r} = -\xi_{r,a}, \tag{44}$$

which is Killing's equation. This means, then, that on a compact space of negative Ricci curvature there exists no one-parametric group of motions. For further results in this direction see [1] and [3].

For $p \geqq 2$ we will make yet another statement. We consider a skew tensor whose covariant derivative vanishes identically

$$\xi_{a_1 \cdots a_p, r} = 0. \tag{45}$$

It then satisfies both sets of differential relations, and we obtain the following peculiar conclusion.

THEOREM 3. *On a compact S_n, there cannot exist a skew symmetric tensor with vanishing covariant derivative, whenever the Ricci curvature is everywhere positive or everywhere negative and the conformal curvature is restricted by relation* (15).

However, a good deal can be concluded directly from local considerations, and we will state a theorem in this direction.

THEOREM 4. *In a Euclidean neighborhood with a line element* (2) *and constant sectional curvature there cannot exist, for $p < n$, a skew symmetric tensor $\xi_{a_1 \cdots a_p}$ for which the difference*

$$\xi_{a_1 \cdots a_p, r, s} - \xi_{a_1 \cdots a_p, s, r} \tag{46}$$

vanishes identically, and for $p = 1$ it cannot even exist in the case of an Einstein space

$$R_{hk} = b g_{hk} \tag{47}$$

with $b \neq 0$.

PROOF. For $p = 1$ the vanishing of (46) implies

$$\xi_\lambda R^\lambda_{ars} = 0.$$

If we multiply by g^{as} and contract we obtain

$$\xi_\lambda R^\lambda{}_r = 0,$$

and this means that the matrix $R^\lambda{}_r$ cannot be definite, but must be semi-definite or indefinite. Especially we cannot have (47). The latter can also be seen very directly since the substitution of (47) will give

$$b \xi_\lambda g^\lambda{}_r = b \xi_r = 0,$$

and hence $\xi_r = 0$.

For $p = 2$ the vanishing of (46) would lead to

$$\xi_{\lambda a_2} R^\lambda{}_{a_1 rs} + \xi_{a_1 \lambda} R^\lambda{}_{a_2 rs} = 0,$$

and for

$$R^h_{ijk} = g^h_j g_{ik} - g^h_k g_{ij}$$

this would be

$$\xi_{ra_2} g_{a_1 s} + \xi_{a_1 r} g_{a_2 s} = \xi_{sa_2} g_{a_1 r} + \xi_{a_1 s} g_{a_2 r}$$

for all a_1, a_2, r, s. If we now multiply by $g^{a_1 s}$ and contract for s and a_1 we obtain

$$n \xi_{ra_2} + \xi_{a_2 r} = \xi_{ra_2} + (g^{a_1 s} \xi_{a_1 s}) g_{a_2 r},$$

and hence

$$(n - 2)\xi_{ra_2} = (g^{a_1 s}\xi_{a_1 s})g_{a_2 r} = 0.$$

Thus for $n > 2$ we must have $\xi_{ra} = 0$. The case $p > 2$ follows in a similar way.

A *symmetric* tensor with $\xi_{ij,k} = 0$ does always exist, since g_{ik} itself is of this kind. It is the skewness that introduces the spatial peculiarities. This can also be seen from a different approach. Take a skew tensor in any dimension, and consider the external form

$$\omega_p = \xi_{\alpha_1 \cdots \alpha_p} dx_{\alpha_1} \cdots dx_{\alpha_p}. \tag{48}$$

If the tensor satisfies (45) then ω_p is certainly a harmonic form and thus $B_p \geqq 1$. Now, consider any two such forms ω_p and ω_q, $p \gtrless q$, and form their external product

$$\omega_p \times \omega_q .$$

In the newly formed skew tensor every coefficient is a polynomial in factor coefficients, and thus its covariant derivative again vanishes. In particular, it is again harmonic if not identically zero. Now, assume that the dimension of the space is even, $n = 2k$, and consider a skew tensor ξ_{ij} such that the determinant of its square matrix does not vanish identically,

$$\Omega = | \xi_{ij} |_{i,j=1,\cdots,n} \not\equiv 0. \tag{49}$$

Furthermore, introduce the external form

$$\omega = \xi_{ij}\, dx_i\, dx_j \tag{50}$$

and its powers

$$\omega^l = \omega \times \omega \times \cdots \times \omega, \qquad 1 \leqq l \leqq k. \tag{51}$$

Obviously

$$\omega^k = 2^k \Omega\, dx_1 \cdots dx_k , \tag{52}$$

and thus (49) amounts to assuming that

$$\omega^l \not\equiv 0, \qquad 1 \leqq l \leqq k.$$

Hence the following conclusion.

THEOREM 5. *If on a compact S_n, $n = 2k$, there exists a skew tensor ξ_{ij} with*

$$\xi_{pq,r} = 0, \tag{53}$$

and if (49) *holds, then*

$$B_{2l} \geqq 1, \qquad 1 \leqq l \leqq k. \tag{54}$$

Perhaps the following statement is worth making.

THEOREM 6. *If in a compact S_{2k}, with a given metric, there exist a skew tensor with properties* (53) *and* (49), *then there exists no change of metric for which the space will have constant curvature.*

We wish to point out that in all our statements a "metric" is a positive definite metric. But in Theorem 5 the metric is explicitly needed only for the statement of assumption (53). To that end, only the presence of an affine connection Γ^p_{qr} is required, and it would be very desirable to investigate whether at least an indefinite metric would be admissible in the hypotheses of Theorem 5.

4. Hermitian metric

We are going to show that our Theorem 5 is an extension from complex to real variables of an important property of complex manifolds. Take a compact complex space in k complex parameters ($n = 2k$), with a positive definite Hermitian metric

$$ds^2 = 2g_{\alpha\beta^*}\, dz_\alpha\, d\bar{z}_\beta\,. \tag{55}$$

For our own introduction to such metrics see [1] and [2].

We can always set up the external form

$$\omega = 2g_{\alpha\beta^*}\, dz_\alpha\, d\bar{z}_\beta \tag{56}$$

in which the ordering of the differentials dz_α and $d\bar{z}_\beta$ is thus fixed. If we replace the quantities

$$z_1, \cdots, z_k; \bar{z}_1, \bar{z}_2, \cdots, \bar{z}_k \tag{57}$$

by the letters

$$t_1, \cdots, t_k, t_{k+1}, \cdots, t_{2k} \tag{58}$$

then we can write (56) as the external form

$$\omega = h_{ij}\, dt_i\, dt_j \tag{59}$$

in which the skew tensor h_{ij} is defined in the following way:

$$h_{\alpha\beta} = h_{\alpha^*\beta^*} = 0 = g_{\alpha\beta} = g_{\alpha^*\beta^*} \tag{60}$$

$$h_{\alpha\beta^*} = g_{\alpha\beta^*} = g_{\beta^*\alpha} \tag{61}$$

$$h_{\alpha^*\beta} = -\, g_{\alpha^*\beta}\,. \tag{62}$$

The first differential relation (25), that is,

$$h_{pq,r} + h_{rp,q} + h_{qr,p} = 0, \tag{63}$$

is nothing else but Kaehler's condition

$$\frac{\partial g_{\alpha\beta^*}}{\partial z_\gamma} = \frac{\partial g_{\gamma\beta^*}}{\partial z_\alpha}. \tag{64}$$

It is known, see [5] and [8], that it automatically implies the harmonic character of all powers

$$\omega^l, \qquad 1 \leqq l \leqq k. \tag{65}$$

Now, we are going to see that actually even more is true.

THEOREM 7. *In the case of a metric* (55) *with property* (64), *the skew tensor* h_{pq} *has covariant derivative zero,*

$$h_{pq,r} = 0, \tag{66}$$

and obviously

$$| h_{pq} |_{p,q=1,\cdots,2k} = - | g_{\alpha\beta^*} |^2 \neq 0. \tag{67}$$

Therefore, if V_{2k} *is compact we have*

$$B_{2l} \geqq 1, \qquad 1 \leqq l \leqq k. \tag{68}$$

PROOF. In verifying (66) it suffices to prove: (i) $h_{\alpha\beta,\gamma} = 0$, (ii) $h_{\alpha\beta,\gamma^*} = 0$, (iii) $h_{\alpha\beta^*,\gamma^*} = 0$. Throughout the process of verification, constant use has to be made of the fact that in consequence of assumption (64) the affine connection Γ^p_{qr} can be different from zero only if all three indices are starred or unstarred simultaneously. Now, take case (i). If we first write out all terms in full, we obtain

$$h_{\alpha\beta,\gamma} = \frac{\partial h_{\alpha\beta}}{\partial z_\gamma} - h_{\lambda\beta}\, \Gamma^\lambda_{\alpha\gamma} - h_{\lambda^*\beta}\, \Gamma^{\lambda^*}_{\alpha\gamma} - h_{\alpha\lambda}\, \Gamma^\lambda_{\beta\gamma} - h_{\alpha\lambda^*}\, \Gamma^{\lambda^*}_{\beta\gamma},$$

and then in the right sum the first, second and fourth term vanish because of (60); but the third and fifth term vanish only in consequence of the vanishing of the affine connections as resulting from (64). Similarly, in case (ii), all terms vanish identically. But in case (iii), if we omit all terms which vanish identically, we obtain

$$h_{\alpha\beta^*,\gamma} = \frac{\partial h_{\alpha\beta^*}}{\partial z_\gamma} - 0.$$

But the latter, due to (61), is $\frac{\partial g_{\alpha\beta^*}}{\partial z_\gamma}$, which in its turn is $g_{\alpha\beta^*,\gamma}$, and thus vanishes also.

As a consequence of Theorem 7 a compact V_{2k}, on which a Hermite-Kaehler metric is definable, cannot have constant curvature in the original or any other (non-Hermitian) metric. This "consequence" is worth drawing only as far as it applies to non-Hermitian alterations of the original metric; because for the original metric the impossibility of a constant curvature, other than zero, has been shown in [2] to be a purely local property, unrelated to any global outlooks, and follows also from our present Theorem 4. It was shown that at best it is possible to prescribe constancy of curvature on the holomorphic sections only, and this leads uniquely to Fubini's metric. But on non-holomorphic sections the curvature may then differ from the constant value up to a fraction 1/4.

5. Einstein Spaces

For an Einstein space of the form (47) we have

$$R_{ij} = \frac{R}{n} g_{ij} .$$

If

$$R > 0, \tag{69}$$

then $L = nR$, and thus, for $p > 1$, conclusion (15) is

$$| C | < \frac{R}{n}\left(\frac{1}{p-1} - \frac{1}{n-1}\right). \tag{70}$$

From this we first of all conclude

THEOREM 8. *For $n \geqq 6$, in a compact semi-simple group space with a group invariant positive-definite metric we have*

$$| C | \geqq \frac{R(n-2)}{2n(n-1)} \tag{71}$$

at all points. Thus the space deviates from conformal flatness by an amount depending only on the dimension.

In fact, in such a metric the space is an Einstein space with (69). On the other hand it is known that $B_3 \geqq 1$. Thus, (70) must be violated for $p = 3$ in at least one point; and due to the group invariance of the metric it does so at all points.

A compact group space behaves in some respects as a product of spheres, although it is not homeomorphic to such a product except for very special cases. However, from our standpoint, we can make a statement concerning such products.

Take an n-sphere S_n with the line element

$$ds_n^2 = \frac{dx_1^2 + \cdots + dx_n^2}{(n-1)(1 - 1/4(x_1^2 + \cdots + x_n^2))^2}.$$

In this normalization its Ricci tensor has all eigenvalues $+ 1$. Now take a product of such spheres

$$S_{n_1} \times S_{n_2} \times \cdots \times S_{n_r}, \qquad n_1 \leqq n_2 \leqq \cdots \leqq n_r ,$$

and give it the metric

$$ds^2 = ds_{n_1}^2 + ds_{n_2}^2 + \cdots + ds_{n_r}^2 .$$

It is then an Einstein space and since $B_{n_1} \geqq 1$ we see that in at least one point we must have

$$| C | \geqq \frac{R}{n_1 + \cdots + n_r}\left(\frac{1}{n_1 - 1} - \frac{1}{n_1 + \cdots + n_r - 1}\right).$$

For a compact complex V_{2k}, we have $B_2 \geqq 1$. Thus, if it is an Einstein space, we obtain, on putting $n = 2k$, $p = 2$,

$$|C| \geqq \frac{R(k-1)}{k(2k-1)} \tag{72}$$

in at least one point. This is in contrast with the classical case $k = 1$, in which case every Riemann surface can be endowed with an Hermite-Kaehler metric in the large for which it is conformally flat. For $k \geqq 2$ we have just demonstrated this to be impossible for certain spaces, and then with the precise estimate (72). However, without such an estimate, the impossibility is universal.

THEOREM 9. *For $k \geqq 2$, an Hermite-Kaehler line-element cannot be conformally flat unless its Riemann-Christoffel tensor vanishes.*

PROOF. This is generalization of the previously mentioned theorem from [2] which stated the conclusion only for spaces of constant sectional curvature.

In the value (8) for the conformal curvature tensor we replace the indices h, i, j, k by $\alpha, \beta^*, \gamma, \delta^*$ respectively. Since

$$g_{\alpha\beta} = g_{\alpha^*\beta^*} = R_{\alpha\beta} = R_{\alpha^*\beta^*} = 0$$

the vanishing of (8) implies

$$R_{\alpha\beta^*\gamma\delta^*} = \frac{1}{2k-2}(g_{\beta^*\gamma}R_{\alpha\delta^*} + g_{\alpha\delta^*}R_{\beta^*\gamma}) - \frac{Rg_{\alpha\delta^*}g_{\beta^*\gamma}}{(2k-2)(2k-1)}.$$

If we multiply this by $g^{\alpha\beta^*}$ and contract we obtain

$$R_{\gamma\delta^*} = \frac{1}{k-1}R_{\gamma\delta^*} - \frac{R}{(2k-2)(2k-1)}g_{\gamma\delta^*}$$

and thus our line element is an Einstein space. But if a conformally flat space is an Einstein space, it has constant sectional curvature, and for such spaces the conclusion was already drawn in [2].

PRINCETON UNIVERSITY

REFERENCES

[1] BOCHNER, *Vector fields and Ricci Curvature*, Bull., Amer. Math. Soc. 52 (1946), 776–797.
[2] BOCHNER, *Curvature in Hermitian Metric*, Bull., Amer. Math. Soc. 53 (1946), 179–195.
[3] BOCHNER, *On compact complex manifolds*, Indian Math. Soc., Jour., vol. 11 (1947) 1–21.
[4] L. P. EISENHART, Riemannian Geometry, 1926.
[5] W. V. D. HODGE, Harmonic Integrals, 1941.
[6] H. HOPF, *Zum Clifford-Kleinschen Raumproblem*, Mathem. Annalen, 95 (1925), 313–339.
[7] S. B. MYERS, *Riemannian Manifolds with Positive Mean Curvature*, Duke Math. Journal 8 (1941), 401–404.
[8] A. WEIL, *Sur la theorie des formes differentielles attachees a une variete analytique complexe*. Commentarii Math. Helvet. 20 (1947), 110–116.

ANNALS OF MATHEMATICS
Vol. 50, No. 1, January, 1949

CURVATURE AND BETTI NUMBERS. II

S. BOCHNER

(Received June 14, 1948)

Previously, in [3], we have described families of Riemannian manifolds S_n, each including the ordinary sphere and retaining its Betti numbers

$$B_p = 0, \qquad 1 \leqq p \leqq n - 1, \tag{1}$$

for some or all dimensions p. We will now have a similar discussion for the complex projective space P_{2k}. We will describe families of Hermitian manifolds, each including P_{2k} and retaining the Betti-numbers

$$B_{2l+1} = 0, \qquad B_{2l} = 1, \qquad\qquad l = 0, 1, 2, \cdots, \tag{2}$$

for some or all dimensions. The method will be the same but with a certain underlying refinement, due to the necessity of letting the even Betti-numbers possess a unit each. Also, in the actual transition from harmonic integrals to Betti-numbers we will have to rely heavily on a Theorem of Hodge's concerning Hermitian manifolds.

In [3] we have found in H. Weyl's conformal curvature tensor a ready made tool by which to gauge metric deviation from an ordinary sphere. In the complex case we have to draw on a certain analogy of formal structure, and we are defining several such tensors ad-hoc without giving any intrinsic geometric interpretation for their meaning or origin. Such an interpretation would be very desirable. It would shed light on the manner in which the global behaviour of P_{2k} is rooted in its local structure, and it might also aid in extending our present results to general Grassmann varieties and similar spaces. We also note that the relations (2) for P_{2k} itself will not be presupposed known, but to the contrary will be proved anew. This renewed discussion of P_{2k} itself may not be of great consequence, but a similar discussion of complex Grassmann varieties, along analogous lines, would be of great value indeed.

We will be bound to re-state our previous results on real metrics, but we will not reiterate them unchanged. Rather, we will elaborate on them by making more statements than in [3], and we will also state them, more generally, for metric spaces with general (non-symmetric) affine connections to fit. We will have certain numbers in lieu of the Betti-numbers, but we are emphasizing that we are unable to identify them with Betti numbers proper, except in the ordinary case of symmetric connections, although most likely they are equal to them in all cases.

Finally we will make some obvious remarks regarding harmonic tensors on non-compact spaces, but the question of interpreting them in terms of cohomology classes is a major problem in potential theory that will not be attempted for the present.

1. Non-symmetric affine connections

We consider a differentiable manifold S_n with a general affine connection E^i_{jk}. Thus we may have

(3) $$E^i_{jk} \neq E^i_{kj}.$$

Covariant differentiation with respect to it will be denoted by a solidus, thus, e.g.,

(4) $$g_{ij|k} = \frac{\partial g_{ij}}{\partial x_k} - g_{rj} E^r_{ik} - g_{ir} E^r_{jk}.$$

We will also introduce its curvature tensor

(5) $$E^i_{jkl} = \frac{\partial E^i_{jl}}{\partial x_k} - \frac{\partial E^i_{jk}}{\partial x_l} + E^h_{jl} E^i_{hk} - E^h_{jk} E^i_{hl}.$$

We next introduce a symmetric positive definite tensor g_{ij}, with an inverse g^{jk} for which $g_{ij}g^{jk} = \delta_{ik}$. As usual we can and we will use this tensor for pulling indices up and down. Furthermore we assume for this special tensor that

(6) $$g_{ij|k} = 0.$$

If and only if we are dealing with the symmetric case

(7) $$E^i_{jk} = E^i_{kj}$$

is this the familiar Christoffel symbol derived from g_{ij}, see [5]. In general, the customary Christoffel symbol has the value

(8) $$\Gamma^i_{jk} = \frac{1}{2}(E^i_{jk} + E^i_{kj}),$$

and if we introduce its Riemann-Christoffel tensor

(9) $$R^i{}_{jkl},$$

then the two tensors (5) and (9) may behave very differently indeed. Thus, in the case of a compact semi-simple group space we have on the one hand

(10) $$E^i{}_{jkl} = 0$$

and on the other hand

(11) $$R_{jk} = \frac{1}{4} g_{jk},$$

where as usual

(12) $$R_{jk} = R^i{}_{jki} = g^{pq} R_{jpqk}.$$

Also the customary tensor (9) has the properties

(13) $$\begin{aligned} R^i{}_{jkl} + R^i{}_{ljk} + R^i{}_{klj} &= 0 \\ R_{ijkl} &= R_{klij} \\ R_{ijkl} &= -R_{jikl} \\ R_{ijkl} &= -R_{ijlk}, \end{aligned}$$

but for (5) we can only claim

$$E_{ijkl} = -E_{jikl}$$
$$E_{ijkl} = -E_{ijlk} . \tag{14}$$

As for the Ricci contraction, the customary concept is

$$E^i_{jki} , \tag{15}$$

but we will form it completely differently, namely as

$$E^i{}_j = \frac{1}{2} g^{kl}(E^i{}_{klj} + E^i{}_{lkj}) \tag{16}$$

(pseudo-Ricci tensor), and we will then form the contraction

$$E = g^{ij}E_{ij} \tag{17}$$

(pseudo-scalar-curvature). Finally we will also form the pseudo-conformal tensor

$$\begin{aligned} D_{hijk} = E_{hijk} - \frac{1}{n-2}(g_{ij}E_{hk} - g_{ik}E_{hj} + g_{hk}E_{ij} - g_{hj}E_{ik}) \\ - \frac{E}{(n-1)(n-2)}(g_{hj}g_{ik} - g_{hk}g_{ik}). \end{aligned} \tag{18}$$

For (7), all these "pseudo"-quantities are the ordinary ones.

2. Pseudo-harmonic tensors

We take a tensor

$$\xi_{a_1 \ldots a_p} \tag{19}$$

which is skew-symmetric (alternating), that is

$$\xi_{a_1 \ldots a_r \ldots a_s \ldots a_p} = -\xi_{a_1 \ldots a_s \ldots a_r \ldots a_p} , \tag{20}$$

and we call it pseudo-harmonic if

$$\xi_{a_1 \cdots a_p | r} = -\sum_{q=1}^{r} \xi_{a_1 \cdots \underset{(q)}{r} \cdots a_p | a_q} \tag{21}$$

(pseudo-curl $\equiv 0$) and

$$g^{rs}\xi_{a_1 \ldots a_p | r | s} = 0 \tag{22}$$

(pseudo-divergence $\equiv 0$). We also assume that our coordinate space belongs to a differentiability class $C^{(\lambda+2)}$, $\lambda \geqq 2$, and we consider skew-tensors belonging to $C^{(\lambda+2)}$, and no others. For any $1 \leqq p \leqq n - 1$ we shall denote by B_p the maximal number of linearly independent ones (with constant coefficients) among all pseudo-harmonic tensors of dimension p. If the affine connection is symmetric and only then it is known for a (compact) manifold S_n that the number B_p is the customary Betti number.

For any scalar φ on S_n we introduce the Laplacean

$$(23) \qquad \Delta\varphi \equiv g^{rs}\varphi_{|r|s},$$

and as in [1, 3] we have

LEMMA 1. *On a compact S_n, if for a scalar φ we have*

$$(24) \qquad \Delta\varphi \geqq 0$$

(everywhere on S_n) then we have

$$(25) \qquad \Delta\varphi \equiv 0.$$

Next, by direct calculation we obtain

LEMMA 2. *For any tensor* (19) *whatsoever if we put*

$$(26) \qquad \Delta\varphi = \xi_{a_1\ldots a_p}\xi^{a_1\cdots a_p} \equiv g^{a_1b_1}\cdots g^{a_pb_p}\xi_{a_1\ldots a_p}\xi_{b_1\ldots b_p}$$

then we have

$$(27) \qquad \tfrac{1}{2}\Delta\varphi = T_1 + T_2$$

where

$$(28) \qquad T_1 = g^{rs}g^{a_1b_1}\cdots g^{a_rb_r}\xi_{a_1\ldots a_p|r}\xi_{b_1\ldots b_p|s}$$

and

$$(29) \qquad T_2 = g^{rs}\xi_{a_1\ldots a_p|r|s}\xi^{a_1\cdots a_p}.$$

Next, due to the positive-definite nature of g^{ij} we have

LEMMA 3. *For any tensor* (19) *we have*

$$(30) \qquad T_1 \geqq 0;$$

and $T_1 \equiv 0$ if and only if

$$(31) \qquad \xi_{a_1\ldots a_p|r} = 0,$$

that is, if and only if (19) *satisfies in any (and hence in every) coordinate system the system of homogenous differential equations*

$$(32) \qquad \frac{\xi_{a_1\cdots a_p}}{\partial x_r} = -\sum_{q=1}^{r}\xi_{a_1\cdots s\cdots a_p}E^s_{a_qr}.$$

Also, by the nature of such equations, every tensor (19) which satisfies it in a neighborhood of a point satisfies is identically; if a solution has components 0 at a point it vanishes identically; and there are at most $\binom{n}{p}$ linearly independent solutions possible.

Next, as in [3], formula (36), on putting

$$T_3 = \frac{1}{p}T_2$$

we have

LEMMA 4. *For a pseudo-harmonic tensor we have*

$$T_3 = (p-1)\xi_{tub_3\cdots b_p}\,\xi^{a_1a_2b_3\cdots b_p}\,E^{u}{}_{a_2a_1}{}^{t} - \xi_{tb_2\cdots b_p}\,\xi^{ub_2\cdots b_p}\,E^{t}{}_{u}. \tag{33}$$

On the other hand from Lemmas 1, 2, 3 we obtain

LEMMA 5. *If on a compact S_n, for a pseudo-harmonic tensor we have*

$$T_3 \geqq 0 \tag{34}$$

then for this tensor we have

$$T_3 \equiv 0 \tag{35}$$

and it must be a solution of the system (32).

By combining Lemmas 4 and 5 and then using Lemma 3 we obtain first of all

THEOREM I. If *for a compact S_n we have*

$$E_{ijkl} \equiv 0 \tag{36}$$

then every pseudo-harmonic tensor must satisfy (31) *and*

$$B_p \leqq \binom{n}{p}. \tag{37}$$

DEFINITION. The square matrix $-E_{ij}$ will be called *non-negative* if all its eigenvalues are non-negative, and it will be called *positive* if all its eigenvalues are positive.

THEOREM 2. *On a compact S_n, if the pseudo-Ricci tensor $-E_{ij}$ is everywhere non-negative, and positive somewhere, then*

$$B_1 = 0. \tag{38}$$

More generally, if it is non-negative everywhere and of rank $\geqq q$ somewhere, $0 \leqq q \leqq n$, then

$$B_1 \leqq n - q. \tag{39}$$

PROOF. For $p = 1$, (33) reduces to

$$T_3 = -\xi_t \xi^u E^t{}_u. \tag{40}$$

Now our assumption implies (34). Hence, by Lemma 5, ξ_a is a common solution of

$$-\xi_t \xi^u E^t_u = 0 \tag{41}$$

and

$$\frac{\partial \xi_a}{\partial x_r} = \xi_s E^s_{ar}. \tag{42}$$

By (41), at a certain point, any $n - q + 1$ of such vectors must be linearly dependent (with suitable constant coefficients) and by (42) this dependence, with the same coefficients, continues all over S_n.

We next introduce the tensor (18), and as in [3], formula (37), we obtain the identity

$$T_3 = (p-1)\xi_{tub_3\cdots b_p}\xi^{a_1a_2b_3\cdots b_p}D^u{}_{a_2a_1}{}^t \tag{43}$$
$$-\left(1-\frac{2(p-1)}{n-1}\right)\xi_{tb_2\cdots b_p}\xi^{ub_2\cdots b_p}E^t{}_u - \frac{(p-1)\xi_{b_1\cdots b_p}\xi^{b_1\cdots b_p}}{(n-1)(n-2)}E.$$

From this we will obtain

THEOREM 3. *If on a compact S_n, $n > 2$, the tensor* (18) *vanishes,*

$$D_{ijkl} \equiv 0, \tag{44}$$

and if $-E_{ij}$ is non-negative, without being identically 0, *then*

$$B_p = 0 \tag{45}$$

for

$$2 \leqq p \leqq \frac{n}{2}. \tag{46}$$

PROOF. Our assumptions imply that the first term in (43) vanishes and the other two are non-negative. Hence $T_3 \geqq 0$, and hence each of the remaining two terms vanishes, in particular

$$\xi_{b_1\cdots b_p}\xi^{b_1\cdots b_p}E = 0.$$

But we also assumed that in at least one point $-E > 0$, hence the tensor (19) vanishes at that point. But it is a solution of (32) and hence it vanishes identically.

REMARK TO THEOREM 3. By Poincaré duality, on a compact S_n, if (45) holds in (46) it holds in

$$2 \leqq p \leqq n-2. \tag{47}$$

But when extending our theorem to non-compact S_n in Section 6, the restriction (46) will have to stay.

Restriction (44) can be relaxed. If for every point of S_n we introduce the quantities

$$D = \sup_\xi \left|\frac{\xi^{hk}\xi^{ij}D_{hijk}}{\xi^{hk}\xi_{hk}}\right|, \qquad L = \inf_\xi \left|\frac{\xi^i\xi^j E_{ij}}{\xi^i\xi_i}\right| \tag{48}$$

then we obtain the following full generalization of Theorem 3.

THEOREM 4. *If on a compact S_n with non-negative $-E_{ij}$ we have*

$$(p-1)D \leqq \left(1-\frac{2(p-1)}{n-2}\right)L - \frac{(p-1)E}{(n-1)(n-2)} \tag{49}$$

everywhere, and if at a point P_0 there is strict inequality, then

$$B_p = 0, \qquad 1 \leqq p \leqq \frac{n}{2}.$$

PROOF. As in [3], pages 383–384, for a quasi-harmonic tensor, (49) implies (34) and hence (32) and (35). But (19) vanishes at the point P_0. Otherwise, since (49) is there a strict inequality we would obtain $T_3 > 0$ at that point which contradicts (35).

Finally, if we denote by N_p the number of linearly independent skew-tensors, not necessarily pseudo-harmonic, whose covariant derivative is likewise alternating,

$$\xi_{a_1\ldots a_p|r} = -\xi_{a_1\ldots r\ldots a_p|a_s}, \tag{50}$$

then as in [1], 781–782 and in [3] we obtain

THEOREM 5. *On a compact S_n with non-positive $-E_{ij}$*, (36) *implies*

$$N_p = 0, \qquad 1 \leqq p \leqq n-1. \tag{51}$$

Without (36), *if at some point the rank of $-E_{ij}$ is $\geqq q$, then*

$$N_1 \leqq n - q, \qquad 0 \leqq q \leqq n.$$

Finally, if (49) *holds, and is a strict inequality somewhere then we have*

$$N_p = 0, \qquad 1 \leqq p \leqq \frac{n}{2}.$$

3. Modified Curvature Tensors

Henceforth we will deal with symmetric connections only and we will replace the previous symbol "E" with indices by the symbol "R" with the same indices. The tensor (18) has the following properties. (i) For a space with constant curvature we have (44). (ii) The Ricci-contraction of (18) vanishes, that is $g^{jk}D_{ijkl} = 0$. — Now, if we are satisfied to have only property (i), then we may employ the somewhat simpler tensor

$$F_{hijk} = R_{hijk} - \frac{g_{hj}R_{ik} - g_{hk}R_{ij}}{1-n}. \tag{52}$$

If we substitute this in (33) we obtain

$$T_3 = (p-1)\xi_{tub_3\cdots b_p}\xi^{a_1a_2b_3\cdots b_p}F^u{}_{a_2a_1}{}^t - \left(1 - \frac{p-1}{n-1}\right)\xi_{tb_2\cdots b_p}\xi^{ub_2\cdots b_p}R^t_u.$$

Thus on introducing the scalar

$$F = \sup_\xi \left| \frac{\xi^{hk}\xi^{ij}F_{hijk}}{\xi^{hk}\xi_{hk}} \right|$$

we see that Theorem 3 holds with

$$F_{hijk} = 0 \tag{53}$$

in lieu of (44), and Theorem 4 holds with

$$(p-1)F \leqq \left(1 - \frac{p-1}{n-1}\right)L \tag{54}$$

in lieu of (49). Now, (53) is more restrictive than (44), and we are losing rather than gaining. But (54) is simpler than (49); and it is also better than (49) since it is directly valid in the interval (47) without intervention of Poincaré duality. This latter advantage will show up in a more trenchant manner in the case of Hermitian metrics to which we are now turning.

As in [1, 2, 3] we deal with a Hermitian metric

$$ds^2 = 2g_{\alpha\beta^*}\, dz_\alpha\, d\bar{z}_\beta \tag{55}$$

which satisfies Kaehler's assumption

$$\frac{\partial g_{\alpha\gamma^*}}{\partial z_\beta} = \frac{\partial g_{\beta\gamma^*}}{\partial z_\alpha}. \tag{56}$$

We showed in [3] that on a compact V_{2k} with such a metric we have

$$B_{2l} \geqq 1, \qquad 0 \leqq l \leqq k, \tag{57}$$

and that therefore it can never be a homology sphere, say. We recall how (57) followed. The line element (55) can be "externalized" into the two-dimensional form

$$\omega = h_{ij}\, dt_i\, dt_j = 2g_{\alpha\beta^*}\, dz_\alpha\, d\bar{z}_\beta \tag{58}$$

with the skew-tensor h_{kj} being defined by

$$\begin{aligned} h_{\alpha\beta^*} &= g_{\alpha\beta^*} = -h_{\beta^*\alpha} \\ h_{\alpha\beta} &= h_{\alpha^*\beta^*} = g_{\alpha\beta} = g_{\alpha^*\beta^*} = 0. \end{aligned} \tag{59}$$

This can be always done. But, if, as we permanently assume, (56) holds, then we have the important relation

$$h_{ij,k} = 0. \tag{60}$$

Therefore, ω is a harmonic form of dimension 2, and all powers

$$\omega^l = \omega \times \omega \times \cdots \times \omega, \qquad 1 \leqq l \leqq k,$$

are (non-vanishing) harmonic forms of dimension $2l$ which proves (57).

Thus for a Hermite-Kaehler metric there does not exist a "homology sphere" in the ordinary sense, and our previous theorems are vacuous. However, this very fact suggests the problem of selecting other "typical" spaces around which to build analogues of our previous theorem. If turns out that this is possible, and in this spirit we will analyze complex projective space P_{2k} whose Betti-numbers are (2). Most likely similar investigations could be undertaken for more elaborate spaces, as general Grassmann varieties, certain types of quadrics etc. but, for the present, nothing beyond P_{2k} will be envisaged.

We recall from [2] that P_{2k} carries a metric of our kind (given by Fubini), by which it is uniquely determined, and for which the Riemann Christoffel tensor (9) has the components

(61) $$R_{\alpha\beta^*\gamma\delta^*} = -\frac{b}{2}\,(g_{\alpha\beta^*}g_{\gamma\delta^*} + g_{\alpha\delta^*}g_{\beta\gamma^*})$$

with $b > 0$. This immediately verifies the properties

(62) $$R_{\alpha\beta^*\gamma\delta^*} = R_{\gamma\beta^*\alpha\delta^*} = R_{\gamma\delta^*\alpha\beta^*} = R_{\alpha\delta^*\gamma\beta^*}\,,$$

which however are true in all cases. The values of other components of (9) can be obtained from (61) by the following rules, on putting $T_{hijk} \equiv R_{hijk}$:

(63) $$T_{\alpha\beta jk} = T_{hi\gamma\delta} = 0; \quad T_{\beta^*\alpha jk} = -T_{\alpha\beta^* jk}\,, \quad T_{hi\delta^*\gamma} = -T_{hi\gamma\delta^*}\,.$$

If we multiply (61) by $g^{\alpha\delta^*}$ and contract we obtain

(64) $$R_{\beta^*\gamma} = -\frac{b}{2}\,(k+1)\,g_{\beta^*\gamma}\,,$$

and from

(65) $$R = g^{ij}R_{ij} = 2g^{\beta^*\gamma}R_{\beta^*\gamma}$$

follows

(66) $$R = -k(k+1)b.$$

Therefore (61) implies

(67) $$R_{\alpha\beta^*\gamma\delta^*} = \frac{1}{k+1}\,(g_{\alpha\beta^*}R_{\gamma\delta^*} + g_{\alpha\delta^*}R_{\beta\gamma^*}).$$

In formal analogy to (52) we now introduce a tensor G_{hijk} in the following way. For $(\alpha, \beta^*, \gamma, \delta^*)$ it shall have the components

(68) $$G_{\alpha\beta^*\gamma\delta^*} = R_{\alpha\beta^*\gamma\delta^*} - \frac{1}{k+1}\,(g_{\alpha\beta^*}R_{\gamma\delta^*} + g_{\alpha\delta^*}R_{\beta^*\gamma}),$$

and for other combinations of indices it shall hence be obtained by the rules (63). If is not hard to see that this actually defines a tensor.

As already stated, for P_{2k} we have

(69) $$G_{hijk} = 0.$$

But the converse is also true. Indeed, (67) leads to

$$R_{\alpha\delta^*} = g^{\beta^*\gamma}R_{\alpha\beta^*\gamma\delta^*} = \frac{1}{k+1}\left(R_{\alpha\delta^*} + \frac{1}{2}g_{\alpha\delta^*}R\right),$$

that is

$$R_{\alpha\delta^*} = \frac{R}{2k}\,g_{\alpha\delta^*}\,,$$

and if we again substitute this in (67) we obtain (61) with (66). Thus for $-R > 0$ this is P_{2k} by our uniqueness theorem previously mentioned. However, we note that our Theorem 6 will prove directly that (69) implies (2), without the intervention of the aforementioned uniqueness theorem.

The tensor (68) does not retain all properties (62). This will not matter, but, nevertheless, it is worth knowing that these properties can be secured by a process of "symmetrization." We are introducing a second tensor H_{hijk} whose principal components are defined by

$$(70)\quad H_{\alpha\beta^*\gamma\delta^*} = R_{\alpha\beta^*\gamma\delta^*} - \frac{1}{2(k+1)}(g_{\alpha\beta^*}R_{\gamma\delta^*} + g_{\alpha\delta^*}R_{\gamma\beta^*} + g_{\gamma\beta^*}R_{\alpha\delta^*} + g_{\gamma\delta^*}R_{\alpha\beta^*}).$$

The two tensors (68) and (70) will be interchangeable in our assertions. But we will also employ a formal analogue to (18). We call it K_{hijk}, and we define it by

$$(71)\quad \begin{aligned} K_{\alpha\beta^*\gamma\delta^*} = R_{\alpha\beta^*\gamma\delta^*} &- \frac{1}{k+2}(g_{\alpha\beta^*}R_{\gamma\delta^*} + g_{\alpha\delta^*}R_{\gamma\beta^*} + g_{\gamma\beta^*}R_{\alpha\delta^*} + g_{\gamma\delta^*}R_{\alpha\beta^*}) \\ &+ \frac{R}{2(k+2)(k+1)}(g_{\alpha\beta^*}g_{\gamma\delta^*} + g_{\alpha\delta^*}g_{\beta^*\gamma}), \end{aligned}$$

and the other coordinates by (63). It recommends itself for the property

$$(72)\quad g^{\beta^*\gamma}K_{\alpha\beta^*\gamma\delta^*} \equiv 0,$$

in addition to being 0 for (61). But we have no geometric interpretation of (72).

4. Identities for Hermite-Kaehler metric

LEMMA 6. *If $p \geqq 2$, then for any skew-tensor* (19) *the sum*

$$(73)\quad T_4 = \xi^{hkb_3\cdots b_p}\xi^{ij}{}_{b_3\ldots b_p}R_{hijk}$$

has the value

$$(74)\quad 2\xi^{\alpha\delta^*b_3\cdots b_p}\xi^{\beta\gamma^*}{}_{b3\ldots bp}R_{\alpha\beta^*\gamma\delta^*}.$$

It is admissible to replace R_{hijk} by any of the three tensors

$$(75)\quad G_{hijk},\ H_{hijk},\ K_{hijk}$$

simultaneously in (73) *and* (74).

PROOF. In the proof, now and later, we will usually make the simplifying assumption $p = 2$, thus ignoring the presence of the dummy indices $b_3, \cdots, b_p$, and in each case the reader will be able to see for himself how such indices might be inserted step by step.

We decompose the sum (73) into 4^2 sums depending on all possible combinations for starred and unstarred indices for (h, i, j, k). Suppose first (h, k) are both unstarred, $h = \alpha$, $k = \delta$. Then by (63) we need only consider (i, j) both starred, $i = \beta^*$, $j = \delta^*$. This produces the sum

$$\xi^{\alpha\delta}\xi^{\beta^*\gamma^*}R_{\alpha\beta^*\gamma^*\delta} = -\xi^{\alpha\delta}\xi^{\beta^*\gamma^*}R_{\alpha\beta^*\delta\gamma^*},$$

and this vanishes since $\xi^{\beta^*\gamma^*}$ is skew-symmetric and $R_{\alpha\beta^*\delta\gamma^*}$ is symmetric in (β^*, γ^*). Thus we need only consider the case (i) $h = \alpha$, $k = \delta^*$ and (ii) $h = \beta^*$, $k = \gamma$. In case (i), due to (63), it suffices to assume $i = \beta^*$, $j = \gamma$, which gives the terms

$$\xi^{\alpha\delta^*}\xi^{\beta^*\gamma}R_{\alpha\beta^*\gamma\delta^*}\,. \tag{76}$$

In case (ii) we obtain the same terms, hence (74). Also all steps of the reasoning will likewise apply to any of the tensors (75) in lieu of (9), hence the lemma.

LEMMA 7. *For any two-dimensional skew-tensor, the expression*

$$\xi^{hk}\xi^{ij}R_{hijk} \tag{77}$$

is the sum of three addends: of

$$\xi^{hk}\xi^{ij}G_{hijk} \tag{78}$$

or alternately

$$\xi^{hk}\xi^{ij}H_{hijk}\,, \tag{79}$$

of

$$\frac{2}{k+1}\,\xi^{\alpha\delta^*}\,\xi_{\alpha}{}^{\gamma}\,R_{\gamma\delta^*}\,, \tag{80}$$

and of

$$\frac{2}{k+1}\,(g_{\alpha\delta^*}\,\xi^{\alpha\delta^*})(\xi^{\beta^*\gamma}\,R_{\beta^*\gamma}). \tag{81}$$

It also is the sum of

$$\xi^{hk}\xi^{ij}K_{hijk}\,, \tag{82}$$

of

$$\frac{4}{k+2}\,\xi^{\alpha\delta^*}\,\xi_{\alpha}{}^{\gamma}\,R_{\delta^*\gamma} - \frac{1}{(k+2)(k+1)}\,\xi^{\alpha\delta^*}\,\xi_{\alpha\delta^*}\,R, \tag{83}$$

and of

$$\frac{4}{k+1}\,(g_{\alpha\delta^*}\,\xi^{\alpha\delta^*})(\xi^{\beta^*\gamma}\,R_{\beta^*\gamma}) + \frac{1}{(k+2)(k+1)}\,(g_{\alpha\delta^*}\,\xi^{\alpha\delta^*})^2\,R. \tag{84}$$

PROOF. Since (77) is twice (76), we substitute from (68) into (76), and we then obtain the sum of (78) and of

$$\frac{2}{k+1}\,\xi^{\alpha\delta^*}\,\xi^{\beta^*\gamma}(g_{\alpha\beta^*}\,R_{\gamma\delta^*} + g_{\alpha\delta^*}\,R_{\gamma\beta^*}). \tag{85}$$

But this splits into (80) and (81). In the case of H_{hijk} we have to symmetrize the bracket of (85), and due to the skew-ness of ξ^{ij} the value of the product (85) is not altered by this. Finally in the case of the tensor (71) the same straightforward calculation leads to an expression which we can split into (83) and (84) for later convenience.

The entire Lemma 7 can be stated for $p \geqq 2$ by insertion of dummy indices. However we will do this under a special assumption which will be of decisive importance.

DEFINITION. For $p \geqq 2$, a (harmonic) skew-tensor (19) shall be called an *effective* tensor if

$$h^{ij}\xi_{ijb_3\ldots b_p} = 0, \tag{86}$$

where h_{ij} is our tensor as defined by (59).

Due to (59) we have

$$h^{\alpha\beta^*} = -g^{\alpha\beta^*} = -h^{\beta^*\alpha}, \qquad h^{\alpha\alpha} = h^{\beta^*\beta^*} = 0,$$

and thus the left side of (86) is

$$g^{\alpha\beta^*}\,\xi_{\alpha\beta^*b_3\ldots b_p} + g^{\beta^*\alpha}\xi_{\beta^*\alpha b_3\ldots b_p} \equiv 2h^{\alpha\beta^*}\xi_{\alpha\beta^*b_3\ldots b_p}\,.$$

Thus, (19) is an effective tensor if and only if

$$g^{\alpha\beta^*}\xi_{\alpha\beta^*b_3\ldots b_p} \equiv 0 \equiv g_{\alpha^*\beta}\xi^{\alpha^*\beta}{}_{b_3\ldots b_p}\,. \tag{87}$$

Hence, if (19) is effective, the terms (81) and (84) are 0 and thus we have

LEMMA 8. *For $p \geqq 2$, if (19) is an effective skew-tensor then the expression*

$$T_5 = \xi^{hkb_3\cdots b_p}\xi^{ij}{}_{b_3\ldots b_p}R_{hijk} \tag{88}$$

is the sum of

$$\xi^{hkb_3\cdots b_p}\xi^{ij}{}_{b_3\ldots b_p}G_{hijk} \tag{89}$$

or of

$$\xi^{hkb_3\cdots b_p}\xi^{ij}{}_{b_3\ldots b_p}H_{hijk} \tag{90}$$

and of

$$T_6 = \frac{2}{k+1}\,\xi^{\alpha\delta^*b_3\cdots b_p}\,\xi_\alpha{}^\gamma{}_{b_3\cdots b_p}\,R_{\gamma\delta^*}\,.$$

It is also the sum of

$$T_7 = \xi^{hkb_3\cdots b_p}\xi^{ij}{}_{b_3\ldots b_p}K_{hijk}$$

and of

$$\frac{4}{k+2}\,\xi^{\alpha\delta^*b_3\cdots b_p}\,\xi_\alpha{}^\gamma{}_{b_3\cdots b_p}\,R_{\gamma\delta^*} - \frac{1}{(k+1)(k+2)}\,\xi^{\alpha\delta^*b_3\cdots b_p}\,\xi_{\alpha\delta^*b_3\cdots b_p}\,R. \tag{91}$$

5. Effective harmonic tensors

We assume that our skew-tensor is "self-adjoint," see [1], p. 786. This means that if for any one component we replace each starred index by numerically the same unstarred index and vice versa then the numerical value of the component turns into its conjugate complex. For a skew-tensor this means that if we write out the external differential form

$$\xi_{a_1 \ldots a_p} \, dt_{a_1} \cdots dt_{a_p}$$

in the real and imaginary parts of the differentials dt; then the corresponding components are real-valued.

For a self-adjoint tensor the scalar quantity (26) is real-valued; and all statements and implications involving "real-valuedness" and "positiveness" which we are going to make can be justified.—We again form the Laplacean, and again introduce the quantities T_1, T_2 and T_3; and Lemma 1–5 remain in force. We now write

$$T_3 = (p-1)T_4 + T_8$$

where T_4 is defined by (73) and

$$T_8 = -\xi_{tb_2 \ldots b_p} \xi^{ub_2 \cdots b_p} R^t{}_u = T_9 + T_{10}$$

with

$$T_9 = -2\xi_{\alpha\beta^* b_3 \ldots b_p} \xi^{\gamma\beta^* b_3 \cdots b_p} R^\alpha{}_\gamma$$

and

$$T_{10} = -2\xi_{\alpha\beta b_3 \ldots b_p} \xi^{\gamma\beta b_3 \cdots b_p} R^\alpha{}_\gamma .$$

Hence we obtain

LEMMA 9. *For an effective harmonic tensor we have*

$$T_3 = (p-1)T_{11} + T_{12} + T_{10} \tag{92}$$

where T_{11} *is* (89) *or* (90) *and*

$$T_{12} = -2\left(1 - \frac{p-1}{k+2}\right) \xi^{\alpha\delta^* b_3 \cdots b_p} \xi_\alpha{}^\gamma{}_{b_3 \ldots b_p} R_{\gamma\delta^*} .$$

We also have

$$T_3 = (p-1)T_7 + T_{13} + T_{10} \tag{93}$$

where

$$T_{13} = -2\left(1 - 2\frac{p-1}{k+2}\right) \xi^{\alpha\delta^* b_3 \cdots b_p} \xi_\alpha{}^\gamma{}_{b_3 \ldots b_p} R_{\gamma\delta^*} - \frac{p-1}{(k+1)(k+2)} \xi^{\alpha\delta^* b_3 \cdots b_p} \xi_{\alpha\delta^* b_3 \cdots b_p} R.$$

Our next aim is to make $T_3 \geqq 0$, and in preparation of this we introduce the scalar

$$G = \sup_\xi \frac{| \xi^{\alpha\delta_*} \xi^{\beta^*\gamma} G_{\alpha\beta^*\gamma\delta^*} |}{\xi^{\alpha\delta^*} \xi_{\alpha\delta^*}} ,$$

the analogous scalars H and K, and the scalar

$$L = \inf_\xi \frac{|\xi^\alpha \xi^{\beta*} R_{\alpha\beta*}|}{\xi^\alpha \xi_\alpha}.$$

THEOREM 6. *If on a compact V_{2k} with a Hermite-Kaehler metric the tensor $-R_{\alpha\beta*}$ is positive; and if*

$$G_{hijk} = 0 \quad \text{or} \quad H_{hijk} = 0, \tag{94}$$

or more generally if

$$(p-1)G < \left(1 - \frac{p-1}{k+1}\right)L \quad \text{or} \quad (p-1)H < \left(1 - \frac{p-1}{k+1}\right)L \tag{95}$$

for

$$1 \leqq p \leqq k, \tag{96}$$

then there exist no effective harmonic tensor of dimension p.

Also, if (95) *holds for all* (96), *then for the Betti numbers of V_{2k} we have*

$$B_{2l+1} = 0, \qquad B_{2l} = 1 \tag{97}$$

$$0 \leqq l \leqq k,$$

as in the case of complex projective space, which also satisfies all assumptions stated.

PROOF. We note that for $-R_{\alpha\beta*}$ positive, the term T_{10} is automatically $\geqq 0$. It then follows from our assumptions that T_3, as given by (92), is $\geqq 0$. Also, for a non-vanishing effective tensor we would have to have $T_3 > 0$ at some points, contrary to Lemma 5.

The conclusion (97) follows now directly from a highly important theorem of Hodge's, [4], 181, which we will quote as a lemma.

LEMMA 10 (Hodge). *On a compact V_{2k} wi:h Hermite-Kaehler metric we have*

$$B_0 \leqq B_2 \leqq B_4 \leqq \cdots \leqq B_{2[k/2]} \tag{98}$$

and

$$B_1 \leqq B_3 \leqq B_5 \leqq \cdots \leqq B_{2[k/2]+1}, \tag{99}$$

and if there are no effective tensors then we have the equalities

$$B_0 = B_2 = B_4 = \cdots = B_{2[k/2]}$$

$$B_1 = B_3 = B_5 = \cdots = B_{2[k/2]+1}.$$

More precisely, the number $B_{p+2} - B_p$, $(p + 2 \leqq k)$, is the number of linearly independent effective tensors of dimension $p + 2$.

With regards to this lemma we make several remarks. First, the total dimension of our space is $2k$, and the main conclusion of Theorem 6 has been stated only for p in (96). However, due to Poincaré duality this is sufficient. Second, for $p = 2$, the tensor h^{ij} is a harmonic tensor, and thus "creates" a

Betti unit for $p = 2$. It must *not* be an effective tensor lest it fall under Theorem 6, and thus "suffer destruction." But, why does Theorem 6 fail to operate? The answer is contained in Lemma 7. For ξ_{ij} non-effective, the terms (81) and (84) need not be zero, and for

$$\xi_{ij} \equiv h_{ij} \tag{100}$$

on P_{2k}, in spite of (95), we do *not* obtain $T_3 > 0$. But to some extent the situation seemingly continues to be a "delicate" one, since the direct calculation of T_3 gives $T_3 \equiv 0$ and not perchance $T_3 < 0$. Thus, seemingly, Lemma 5 still remains operative. But the conclusion from Lemma 5 is that the covariant derivative of (100) shall be 0. But this is relation (60) which was known to us from the outset and which indeed lies at the very root of the entire analysis at the termination of which it has just been re-obtained.

Our last remark describes a problem rather than a comment. Relation (98) implies in particular $B_{2l} \geqq 1$. But it was pointed out in [3] that the latter property by itself may be isolated from the complex-analytic setting altogether by simply presupposing the existence of a skew-tensor h_{ij} with property (60) and little more, in particular without any further tie-in to the tensor g_{ij} itself. Hodge's theorem in full and our application of it suggest the problem of finding a precise set of assumptions needed for securing the general relations (98) and (99) on a compact manifold. Finding such a set of assumptions would correspond to the classical process of identifying a space of one complex variable with a two-dimensional space of conformal flatness.

After these remarks we are turning to the tensor (71). The expression T_{13} is somewhat delicate due to the fact that

$$1 - 2\frac{p-1}{k+2} \geqq 0 \tag{101}$$

for

$$1 \leqq p \leqq \frac{k}{2} + 2 \tag{102}$$

only, and this is not enough for the Poincaré duality principle to operate. Hence we only obtain the following fragmentary result.

THEOREM 7. *If on our V_{2k} we have*

$$K_{hijk} = 0 \tag{103}$$

and $-R_{\alpha\beta^*}$ is positive, then for p in the partial interval (102) *we have* (97).

In order to obtain a result for all p we have to introduce a further restriction whose gist is that the space is not too far removed from an Einstein space $R_{\alpha\beta^*} = bg_{\alpha\beta^*}$. We introduce a tensor with the components

$$S_{\alpha\beta^*} = R_{\alpha\beta^*} - \frac{R}{2k} g_{\alpha\beta^*}$$

and we then can write

$$T_{13} = -\frac{1}{k}\left(1 - \frac{p-1}{(k+1)(k+2)}\right) R\xi^{\alpha\delta^* b_3 \cdots b_p} \xi_{\alpha\delta^* b_3 \cdots b_p} + 2\left(1 - 2\frac{p-1}{k+2}\right) \xi^{\alpha\delta^* b_3 \cdots b_p} \xi_\alpha{}^\gamma{}_{b_3 \cdots b_p} S_{\gamma\delta^*}.$$

If therefore we introduce the deviation

$$S = \sup_\xi \frac{|\xi^\alpha \xi^{\beta^*} S_{\alpha\beta^*}|}{\xi^\alpha \xi_\alpha}$$

we have

$$T_{13} \geqq 2\left\{\frac{|R|}{2k}\left(1 - \frac{p-1}{(k+1)(k+2)}\right) - \left|1 - 2\frac{p-1}{k+2}\right| \cdot |S|\right\} \xi^{\alpha\delta^* b_3 \cdots b_p} \xi_{\alpha\delta^* b_3 \cdots b_p}.$$

Now, for $\frac{k}{2} + 3 \leqq p \leqq k$ the bracket is

$$\geqq \frac{|R|}{2k+4} - \frac{4S}{k+2}$$

and thus we obtain the following theorem.

THEOREM 8. *If on our V_{2k} the Ricci tensor $-R_{\alpha\beta^*}$ is positive then relation* (97) *holds for all dimensions, if* (103) *holds and in addition we have*

$$S < \frac{|R|}{8};$$

or, more generally, for non-vanishing K_{hijk} if we have

$$K < \frac{4}{k+2}\left(\frac{|R|}{8} - S\right).$$

6. Non-compact spaces

We take a "non-compact" separable differentiable coordinate space with a Riemannian metric g_{ij} on it; or, more generally, such a space with a metric and non-symmetric connection, say. Being "non-compact" means that it requires a non-finite number of allowable coordinate systems by which to cover it. It should be noted that the underlying space as coordinate space might very well be a perfectly decent *compact* manifold to start with; but that the metric tensor on it might have some inadmissible peculiarities ("singularities" or "discontinuities" etc.) on isolated points or general lower-dimensional varieties etc.

The best way of dealing with such "exceptional" points is to omit them from the space structure altogether, and what is left over after this omission will as a rule be a non-compact space of the kind we have postulated. Now, the compactness of S_n was actually needed only once, namely for the proof of Lemma 1;

and in subsequent lemmas and theorems it had been carried only in order to make Lemma 1 operative, and also in order to validate the Hodge-de Rham "isomorphism" between harmonic integrals and Betti numbers. Thus, as regards harmonic integrals, we need only find a means of preserving the validity of Lemma 1. This was done in [1], p. 778, by a general but rather artificial definition which however was then verified in two interesting cases. We are going to re-state the more important one and also the extent to which our previous conclusions will remain in force.

For any tensor (19) we introduce the square norm

$$\varphi = \xi_{a_1 \dots a_p} \xi^{1a \cdots a_p}$$

as before, and we say that (19) has everywhere boundary values 0 if to every ϵ there exists a compact subset B_ϵ of S_n such that $\varphi < \epsilon$ for all points of the complementary subset $S_n - B_\epsilon$. If S_n is compact to start with, then we may choose B_ϵ to be all of S_n, and thus, in this sense, on a compact S_n everyone of our previously considered tensors has boundary values 0 at all (non-existent) boundary points. But in general we obtain

THEOREM 9. *All our previous statements on the non-existence of harmonic or pseudo-harmonic tensors, or of skew-tensors whose covariant derivative is likewise skew-symmetric, or on the non-existence of such tensors other than those satisfying relations* (32)*—whether in the Riemannian or the Hermitian case—remain in force for any one or all p, on a non-compact S_n provided the tensors envisaged are only those having boundary values* 0 *on S_n.*

PRINCETON UNIVERSITY

REFERENCES

1. BOCHNER, *Vector Fields and Ricci Curvature*, Bull. Amer. Math. Soc. **52** (1946), 776–797.
2. BOCHNER, *Curvature in Hermitian Metric*, Bull. Amer. Math. Soc. **53** (1947), 179–195.
3. BOCHNER, *Curvature and Betti numbers*, Ann. of Math. **49** (1948), 379–390.
4. W. V. D. HODGE, Theory and Applications of Harmonic Integrals, 1941.
5. J. A. SCHOUTEN AND D. J. STRUIK, Einfuehrung in die neueren Methoden der Differentialgeometrie, 1935, vol. I, 83 pp.
6. A. WEIL, *Sur la Théorie des formes différentielles attaché es á une variété analytique complexe*, Comment. Math. Helv. **20** (1947), 110–116.

ANNALS OF MATHEMATICS
Vol. 51, No. 2, March, 1950

EULER-POINCARÉ CHARACTERISTIC FOR LOCALLY HOMOGENEOUS AND COMPLEX SPACES*

BY S. BOCHNER

(Received April 6, 1949)

We will make some statements on the Euler-Poincaré characteristic of closed Riemannian spaces, both real and complex, and notably the latter, and they will all follow very simply and directly from the integral expression due to Allendoerfer-Weil, and Fenchel, and others, see [1], [2], [12], [10]. This expression evaluates the Euler characteristic in terms of the Riemann Christoffel curvature tensor pertaining to any positive-definite metric ground form on the space, but we will nevertheless be able to partially eliminate the reference to the curvature tensor or even to the ground form in some of our statements.

We think that the main result of this paper is Theorem 8 which states that if on a compact complex manifold of $2k$ real dimensions there are sufficiently many Abelian integrals of the first kind then

$$(-1)^k\chi \geqq 0. \tag{1}$$

This involves no explicit reference to a metric form on the manifold. However, in a refinement of the conclusion which separates the strict inequality

$$(-1)^k\chi > 0 \tag{2}$$

from the equality

$$\chi = 0 \tag{3}$$

it will be advisable to leave the reference to a certain metric in the statement proper, although by resorting to a cumbrous circumlocution it might be avoided as well. Actually, for $k = 1$, (algebraic curves), this circumlocution will be easy to state and it will be a novel expression for the classical Gauss-Bonnet formula in terms of Abelian integrals.

We might also point out that most of our statements will be concerned with the conclusion (1), and that we will compare this with and partially deduce from results of H. Hopf and H. Samelson [13], all of which are concerned with the conclusion

$$\chi \geqq 0. \tag{4}$$

The contrast between (1) and (4) will appear as a contrast between hyperbolic and elliptic geometry, as it were. We encountered it in several of our recent publications, and the present paper may be viewed as a further contribution to this topic; see [6] and [9].

Finally we wish to acknowledge that this paper owes its inception to many discussions had with Mr. Newton S. Hawley on the role of the Euler characteristic

* Written while engaged in a Project for the Office of Naval Research.

in general and for homogeneous spaces in particular and we wish to thank him for much enlightening information.

I. Locally Homogeneous Spaces

1. If S is a coordinate space of a fixed even dimension

$$n = 2k,$$

if it belongs to (at least) differentiability class C^3 and carries a positive definite metric form

$$ds^2 = g_{ij}\, dx_i\, dx_j, \tag{5}$$

and if it is closed (compact); then for the characteristic

$$\chi = \sum_{p=0}^{n} (-1)^p B_p$$

we have

$$\chi = \int_S K\, dv, \tag{6}$$

where the integrand K is a scalar function which apart from the numerical factor

$$[(8\pi)^k k!]^{-1} \tag{7}$$

can be written down as

$$| g |^{-1} \varepsilon^{\alpha_1\alpha_2\dots\alpha_n} R_{\alpha_1\alpha_2\beta_1\beta_2} R_{\alpha_3\alpha_4\beta_3\beta_4} \cdots R_{\alpha_{n-1}\alpha_n\beta_{n-1}\beta_n} \varepsilon^{\beta_1\beta_2\dots\beta_n}, \tag{8}$$

or alternately as

$$\varepsilon_{\alpha_1\dots\alpha_n} R^{\alpha_1\alpha_2}{}_{\beta_1\beta_2} R^{\alpha_3\alpha_4}{}_{\beta_3\beta_4} \cdots R^{\alpha_{n-1}\alpha_n}{}_{\beta_{n-1}\beta_n} \varepsilon^{\beta_1\dots\beta_n}, \tag{9}$$

and where dv is the intrinsic volume element $| g |^{1/2} dx_1 \cdots dx_n$.

2. Definition 1. We say that a Riemannian space S, whether closed or not, is homogenetic (with respect to the scalar function K) if the function (8) is a constant on it. A family of spaces $S, T, \dots$ will be called equigenetic (with respect to K) if they are homogenetic each and the constant value of the scalar function is the same for the several spaces.

For a closed homogenetic space we obviously have

$$\chi(S) = K \cdot v(S), \tag{10}$$

where $v(S)$ is the intrinsic volume $\int_S dv$, and for equigenetic spaces we have

$$\frac{\chi(S)}{v(S)} = \frac{\chi(T)}{v(T)} = \cdots = K. \tag{11}$$

Since $\chi(S)$ is an integer we obtain the following theorem.

THEOREM 1. *For a family of equigenetic closed spaces the characteristic has the same algebraic sign: either $\chi > 0$ for all of them, or $\chi = 0$, or $\chi < 0$.*

Also, if $\chi \neq 0$, then their intrinsic volumes are integer multiples of a positive lower bound.

3. DEFINITION 2. If two line-elements

$$g_{ij}(x)\, dx_i dx_j \tag{12}$$

$$h_{ij}(x)\, dx_i dx_j \tag{13}$$

are defined and belong to class C^2 in a neighborhood of the origin, then we will call them equimetric, if after an origin-preserving transformation of class C^3 in one of them, they become identically equal in a neighborhood of the origin. We call a Riemannian space homometric if the local line-elements of its ground form in all allowable coordinate systems are equimetric with one another, and hence with some "universal" line element representative of them. We finally call a family of spaces equimetric if they are each homometric, and their universal line-elements are equimetric.

Obviously, on a homometric space any scalar function which is expressible as a formula in the coefficients g_{ij} and its partial derivatives has a constant value, and on equimetric spaces this constant value is the same. This applies in particular to our expression (8) and hence we have

THEOREM 2. *Equimetric spaces are equigenetic and Theorem 1 applies.*

4. We now take a differentiable manifold S of class C^4 and on it a distance function $D(x, y)$ continuous for (x, y) in $S \times S$, with the customary properties: $D(x, y) = 0$ if and only if $x = y$, and $D(x, y) \leqq D(z, x) + D(z, y)$. Actually there is no need assuming that $D(x, y)$ is given for all points (x, y) of the entire $2n$-dimensional space $S \times S$ but only in some $2n$-dimensional neighborhood U of the diagonal set $x = y = x_0$. However for $D(x, y)$ in such a neighborhood U we require some differentiability property, namely that the squared function

$$\Omega(x, y) = (D(x, y))^2 \tag{14}$$

shall belong to class C^4 in the $2n$ local parameters obtained by products of local coordinate systems, and that for any x in S and "small" η we shall have

$$\Omega(x, x + \eta) \geqq A(x)\, (\eta_1^2 + \cdots + \eta_n^2). \tag{15}$$

In what follows, a distance function $D(x, y)$ shall have all properties enumerated thus far and such others as will be mentioned from time to time; and an "isometry" shall be a mapping which preserves the distance function or functions occurring.

DEFINITION 3. A space with a distance function will be called homometrized if corresponding to any two points P and Q there is an isometry between neighborhoods N_P, N_Q which maps P into Q. And two such spaces will be called equimetrized if, in addition, some neighborhood of one may be mapped isometrically onto a neighborhood of another.

The concepts of Definition 3 are closely related to those of Definition 2. In fact we have shown in [3] that a distance function gives rise to a non-negative symmetric tensor of class C^2, namely

$$g_{ij}(x) = \frac{\partial^2 \Omega(x, x + \eta)}{\partial \eta_i \, \partial \eta_j}\bigg|_{\eta=0}. \tag{16}$$

It is always non-negative definite, and under assumption (15) it is postive definite in a strict sense. Also, formula (16) is such that for homo- and equimetrized spaces, the resulting tensors (16) are homo- and equimetric, and thus we are led to

THEOREM 3. *Equimetrized spaces, if endowed with the ground form* (16), *are rendered equimetric and Theorem* 1 *holds.*

Thus in particular for two equimetrized closed spaces the Euler characteristic has the same algebraic sign.

A distance function gives rise to the tensor (16). Conversely, given a line-element, we can define lengths of arcs and hence, for a connected space, a distance function, which for near-by point is the geodesic distance, that is the length of the unique geodesic joining the points. Also, on applying the process (16) to this geodesic distance, the original tensor will be reobtained. However it must be pointed out that there are many other distance functions leading back to the same g_{ij}, for instance any $\lambda(D(x, y))$ where $\lambda(t)$ is a four times differentiable function in $0 \leqq t < \infty$ with $\lambda'(0) = 1$.

5. Consider a space S with a distance function on all of $S \times S$. If there is on S a transitive group of isometric homeomorphisms then S is homometrized by Definition 3. Now, if in particular the distance function $D(x, y)$ is locally geodesic then any homeomorphism preserving it is what in differential geometry is termed a "motion" for the given groundform, and thus we see that a Riemannian space with a transitive group of motions is homometric by Definition 2.

But the converse to the last statement holds only in some cases and fails in others, and the cases of failure are as important as those of holding. It apparently holds if the space is simply connected. In all cases known to us, a simply connected Riemannian space, closed or not, which is homometric is so by virtue of admitting a transitive group of motions, and it would be worth giving a proof that this must be so. However, if such a simply connected space $\tilde{S}$ with a transitive group of motions on it is not compact then it may, and in all typical cases will be the universal covering space of one or many compact spaces S each having the following relation to $\tilde{S}$ itself. Its fundamental group $\tilde{S}/S$ is a (discrete) sub-group of the group of all motions on $\tilde{S}$ and thus, by "projection" from $\tilde{S}$ into S, the latter space will be endowed with a Riemannian metric, of which, conversely, the metric on $\tilde{S}$ is a "periodic extension". Every local property of the groundform on

$\tilde{S}$ transfers to the groundform on S; thus, it is again homometric on S and in fact there are "local" groups of motion, that is local (in the original definition of Lie) groups of isometries for which the geodesic distance is invariant. However, one must not rashly assume that there will be a "global" transitive group of isometries on the space S in its entirety. To the contrary we will demonstrate in several ways that "as a rule" a compact space S just described will *not* admit any continuous group of motions on it, and in fact will only admit a finite number of isometries altogether, if any at all.—A compact space S of the kind just considered will be called *subordinate* to $\tilde{S}$.

The typical two dimensional example of subordinate spaces are the complex curves, that is closed Riemann surfaces of the theory of functions, of genus $p > 1$, all being subordinate to the unit circle

$$\tilde{S} : |z| < 1, \tag{17}$$

that is $x^2 + y^2 < 1$. The line-element on S is

$$ds^2 = \frac{dx^2 + dy^2}{(1 - x^2 - y^2)^2}, \tag{18}$$

or, in complex notation,

$$\tfrac{1}{2}\, ds^2 = \frac{dz\, d\bar{z}}{(1 - z\bar{z})^2} \tag{19}$$

(the factor 1/2 being due to a somewhat different normalization), and the transitive group of motions on $\tilde{S}$ are the Mobius transformations

$$z' = e^{i\alpha}\frac{z - z_0}{1 - \bar{z}_0 z}, \qquad |z_0| < 1. \tag{20}$$

It is known that the subordinate surfaces S consist of all complex curves of genus $p > 1$, and on the other hand it is known, compare [4] and [6], that none of the latter admits more than a finite number of isometries, or, alternately, of any complex homeomorphisms at all.

6. We will now illumine the subordination of spaces from a geometric viewpoint, and to that purpose we will axiomatize a phenomenon which to a considerable extent expresses the contrast between elliptic and hyperbolic spaces in general.

DEFINITION 4. We say that two Riemannian spaces S and V of the same dimension are antimetric if each is homometric and if their universal line-elements

$$g_{ij}\, dx_i\, dx_j \tag{21}$$

$$h_{ij}\, dx_i\, dx_j \tag{22}$$

can be so chosen that at the origin $x = 0$ we have

$$g_{ij} = h_{ij} \tag{23}$$

$$\frac{\partial g_{ij}}{\partial x_k} = \frac{\partial h_{ij}}{\partial x_k} = 0 \tag{24}$$

$$\frac{\partial g_{ij}}{\partial x_k \partial x_l} = -\frac{\partial h_{ij}}{\partial x_k \partial x_l}. \tag{25}$$

We note that (23) is an arbitrary normalization, and that (24) makes the two line-elements geodesic, and that thus (25) is the significant assumption. Owing to the formula

$$R_{hijk} = \frac{1}{2}\left(\frac{\partial^2 g_{hk}}{\partial x_i \partial x_j} + \frac{\partial^2 g_{ij}}{\partial x_h \partial x_k} - \frac{\partial^2 g_{hj}}{\partial x_i \partial x_k} - \frac{\partial^2 g_{ik}}{\partial x_h \partial x_j}\right) + g^{lm}([ij, m][hk, l] - [ik, m][hj, l]), \tag{26}$$

it follows from all assumptions that in suitable coordinate systems on S and V their curvature tensors have opposite algebraic signs

$$R_{hijk} = -R'_{hijk}, \tag{27}$$

and that therefore for the scalars (8) we have

$$K_S = (-1)^k K_V, \tag{28}$$

where as usual $n = 2k$.

Thus we immediately obtain the following conclusion

THEOREM 4. *If two compact spaces S and V are antimetric then for their Euler characteristics and intrinsic volumes we have the relation*

$$\frac{\chi(S)}{v(S)} = (-1)^k \frac{\chi(V)}{v(V)}. \tag{29}$$

In particular if $\chi(V) = 0$ then $\chi(S) = 0$ and if $\chi(S) > 0$ then $(-1)^k \chi(V) > 0$.

Now, for an application of this theorem we have the following situation in mind. There are given two antimetric spaces V and $\tilde{S}$, each admitting a transitive group of motions. The space V is compact, but the space $\tilde{S}$ is not compact and is simply connected and thus gives rise to subordinate spaces S. The space V ought to be restricted to being likewise simply connected, and in the majority of significant cases it will so be, but for the sake of a certain diversity we will not impose this restriction as a part of the following definition.

DEFINITION 5. A collection of spaces.

$$[V; \tilde{S}, S] \tag{30}$$

as just described will be called a geometric ensemble. Also we will call V elliptic, and the others hyperbolic, and within the ensemble the elliptic space on the one hand, and the hyperbolic spaces on the other hand will be counterparts of each other.

For instance, the spaces with the metric (18) or (19) are counterparts to the two-dimensional sphere with the line element

$$ds^2 = \frac{dx^2 + dy^2}{(1 + x^2 + y^2)^2}, \tag{31}$$

or, in complex notation, to the Gaussian sphere (complex projective curve) with the Fubini-Study line-element

$$\tfrac{1}{2}\, ds^2 = \frac{dz\, d\bar{z}}{(1 + z\bar{z})^2}. \tag{32}$$

Our purpose behind setting up Definition 5 is the following theorem.

THEOREM 5. *In a geometric ensemble* (30), *in addition to relation* (29) *we also have*

$$\chi(V) \geqq 0, \tag{33}$$

and hence

$$(-1)^k \chi(S) \geqq 0, \tag{34}$$

where $\chi(V) = 0$ *corresponds to* $\chi(S) = 0$, *and*

$$\chi(V) > 0 \tag{35}$$

corresponds to

$$(-1)^k \chi(S) > 0. \tag{36}$$

Also, for $\chi(V) > 0$ *we have*

$$v(S) \geqq \frac{1}{2^{n-1}}\, v(V). \tag{37}$$

PROOF. The additional assertion (33) on which the theorem hinges is due to Hopf and Samelson, see [13]. Finally, relation (37) follows from (29), if we consider that on the one hand relation (36) implies $|\chi(S)| \geqq 1$, and that on the other hand we have $\chi(V) \leqq 2^{n-1}$, since on a homogenous space V we have for the Betti numbers the restriction $B_p \leqq \binom{n}{p}$. The second half of Theorem 1 implies that in an ensemble (30) the numbers $v(S)$ have a positive lower bound, see also [15], and relation (37) gives such a bound specifically. In this connection it would be worth investigating whether the constant $c_n = 2^{-n+1}$ on the right side of (37) might not be replaced by some larger constant, possibly $c_n = 1$.

7. In defining antimetric elements (21), (22), we have postulated the validity of the relation (23)–(25) at the origin, but this selection of postulates has been a somewhat haphazard one. Actually for the needs of Theorem 4 somewhat less would have sufficed, and on the other hand, in all cases known, as exemplified by the pair of line-elements (18) and (31), a much more precise kinship is in evidence.

Usually, in suitable coordinate systems, the two line-elements are both analytic, and in the power series expansions

$$g_{ij} = \sum a_{ij;r_1\cdots r_n} x_1^{r_1} \cdots x_n^{r_n} \tag{38}$$

$$h_{ij} = \sum b_{ij;r_1\cdots r_n} x_1^{r_1} \cdots x_n^{r_n} \tag{39}$$

there occur no coefficients for which a total degree

$$r_1 + \cdots + r_n = N \tag{40}$$

is an odd integer, and those for which it is an even integer are in the relation

$$a_{ij,r_1\cdots r_n} = (-1)^{N/2} b_{ij;r_1\cdots r_n} . \tag{41}$$

That is, purely formally, one power series arises from the other on replacing the variables $x_1, \cdots, x_n$ by the variables $(-1)^{1/2}x_1, \cdots, (-1)^{1/2}x_n$, with both power series being real-valued nevertheless. For general n, this situation is exemplified by the pair of line elements

$$ds^2 = \frac{dx_1^2 + \cdots + dx_n^2}{(1 \pm (x_1^2 + \cdots + x_n^2))^2}, \tag{42}$$

where the plus sign belongs to the (elliptic) n-dimensional sphere V with the transitive group of orthogonal transformations on it, and the minus sign belongs to the hyperbolic interior

$$\tilde{S} : x_1^2 + \cdots + x_n^2 < 1, \tag{43}$$

with the transitive group of Moebius transformations on it.

If we write our previous two-dimensional line-element in the complex form (19), (32), then its multi-dimensional generalization is

$$\tfrac{1}{2}\, ds^2 = \frac{\sum_\alpha |dz_\alpha|^2 + (\sum_\alpha |z_\alpha|^2 \sum_\beta |dz_\beta|^2 - |\sum_\alpha \bar{z}_\alpha\, dz_\alpha|^2)}{(1 \pm \sum_\alpha |z_\alpha|^2)^2}, \tag{44}$$

where the plus sign belongs to complex projective space V and the minus sign to the neighborhood

$$\tilde{S} : |z_1|^2 + \cdots + |z_k|^2 < 1. \tag{45}$$

The transitive group of motions on V are the unitary transformations on complex projective space, and the motions on S are the complex fractional transformations of (45) into itself. The latter form a non-compact group, whereas the motion on V are a compact group. It should be noted that on V there are many more complex homeomorphisms present, namely all projective transformations, but they are a non-compact group, and the unitary transformations are a maximal compact subgroup of them.

The natural and significant generalizations of this complex ensemble are the so-called matrix spaces. Denote by $Z = (z_{\lambda\mu})$; $\lambda = 1, \cdots, l$; $\mu = 1, \cdots, m$, a rectangular matrix of complex variables and by $Z^* = (\bar{z}_{\mu\lambda})$ its adjoint matrix (that is, transposed conjugate), and consider the pair of line-elements

(46) $$\tfrac{1}{2}\, ds^2 = \text{trace}\, [(I_l \pm ZZ^*)^{-1}\, dZ(I_m \pm Z^*Z)^{-1}\, dZ^*],$$

where I_t is the square unit matrix of dimension t. If the $l \times m$ variables $z_{\lambda\mu}$ are independent complex variables, then the plus sign pertains to an (elliptic) complex Grassman variety which for $l = 1$, $m \geqq 1$, or $l \geqq 1$, $m = 1$ reduces to the complex projective space just discussed; a motion being a pair of unitary matrices of dimensions l and m respectively, see [14] and [15]. And the minus sign pertains to a hyperbolic counterpart, the covering space $\tilde{S}$ consisting of those points $(z_{\lambda\mu})$ whose matrix Z is such that the eigenvalues of the product matrix ZZ^* are all <1.

For the special cases $l = m$ two additional possibilities arise, one in which the matrix Z is a symmetric one, thus presenting only $\frac{1}{2}(l(l+1))$ independent variables, and another one in which the matrix Z is a skew-symmetric one thus giving only $\frac{1}{2}(l(l-1))$ complex variables.

In all these cases if we denote a pair of line-elements by

(47) $$g_{\alpha\beta^*}(z, \bar{z})\, dz_\alpha\, d\bar{z}_\beta$$

(48) $$h_{\alpha\beta^*}(z, \bar{z})\, dz_\alpha\, d\bar{z}_\beta\,,$$

then each coefficient $g_{\alpha\beta^*}$, $h_{\alpha\beta^*}$ is a power series in the $2k$ symbols $z_1, \cdots, z_1$ $\bar{z}_1, \cdots, \bar{z}_k$. It contains only those monomials

(49) $$z_1^{\lambda_1} \cdots z_k^{\lambda_k} \quad \bar{z}_1^{\mu_1} \cdots \bar{z}_k^{\mu_k}$$

for which

(50) $$\lambda_1 + \cdots + \lambda_k = \mu_1 + \cdots + \mu_k\,,$$

and furthermore

(51) $$h_{\alpha\beta^*}(z, \bar{z}) = g_{\alpha\beta^*}((-1)^{1/2}z, (-1)^{1/2}\bar{z}),$$

and these properties taken together imply, and are somewhat more precise than those stated for (38), (39).

These latter properties imply not only (24), but also the validity of

(52) $$\frac{\partial^3 g_{ij}}{\partial x_k\, \partial x_l\, \partial x_m} = \frac{\partial^3 h_{ij}}{\partial x_k\, \partial x_l\, \partial x_m} = 0$$

at the origin, and similarly for mixed partial derivatives of every odd order Now (52) and (24), if applied to (26) gives

(53) $$R_{hijk,m} = 0,$$

and

$$R'_{hijk,m} = 0,$$

at first at the origin, and due to homometry everywhere. Thus, all members of such a geometric ensemble are symmetric spaces in the definition of E. Cartan

and their realization in contrasting elliptic and hyperbolic versions has been known to Cartan in some aspects.

8. When considering the scalar (8), the contrast between "elliptic" and "hyperbolic" is expressed by the factor $(-1)^k$ which however establishes a distinction only for k odd and not for k even; furthermore, for $\chi = 0$, it certainly contributes nothing to recognizing a difference. However, a better discrimination is possible by examining the Gaussian sectional curvature

$$R_{ijkl}\lambda^i\mu^j\lambda^k\mu^l \tag{54}$$

or, what is usually more convenient, the Ricci mean curvature

$$-R_{ij}\xi^i\xi^j. \tag{55}$$

The latter translates the contrast between elliptic and hyperbolic into a straightforward distinction between "plus" and "minus" and it is less subject to incapacitation by the intrusion of zero-values.

Of course, a (flat) multi-torus is both elliptic and hyperbolic, the covering space $\tilde{S}$ coinciding with the flat Euclidean space, and in this case, (54) and (55) do vanish identically, and no distinction is offered since none exists. Now, take a genuinely elliptic space, say a sphere, and form its product with a torus. The product is not "genuinely" elliptic since its covering space is not compact, but it is elliptic within our liberal Definition 5, its simply-connected counterpart being the product of the counterpart of the sphere with a Euclidean space. Due to the product rule

$$\chi(V \times T) = \chi(V)\cdot\chi(T) \tag{56}$$

the characteristic of this elliptic space vanishes and so does its scalar (8) and thus no distinction is offered. However the quantities (54), (55) do not vanish identically. For the elliptic space they are everywhere non-negative and sometimes positive and for its counterpart, they are everywhere non-positive and sometimes negative.

In general it is heuristically plausible that for a simply-connected V we always have the strict inequality

$$-R_{ij}\xi^i\xi^j > 0 \tag{57}$$

throughout, although this might perhaps not be the case for (54) itself. Due to (27), relation (57) would imply the strict inequality

$$-R_{ij}\xi^i\xi^j < 0 \tag{58}$$

for $\tilde{S}$ and S, and whenever such is the case in a geometric ensemble an important consequence ensues. We have proved in [4], [6], that the validity of (58) on a compact S implies that there is at most a finite number of isometries on it, and this substantiates our previous remark in §5, that on a compact homometric space there may very well be transitively many local isometries, without there being transitively many or even infinitely many such isometries in the large.

9. We will now make sundry remarks. First of all, if T is a torus then (56) implies

$$\chi(V \times T) = 0. \tag{59}$$

Also, if we take the group invariant line elements on V and T and if on the product we put

$$ds^2 = ds'^2 + ds''^2 \tag{60}$$

then this line element is again invariant with respect to the product group of isometries, and hence (59) implies

$$K = 0. \tag{61}$$

However, the conclusion (61) and hence $\chi = 0$ can also be obtained from a merely local possibility of factoring off a torus, provided that metrically the local decomposition is a very precise one. We assume that locally there exist a system of coordinates $x_1, \cdots, x_n$ for which

$$ds^2 = dx_1^2 + ds'^2, \tag{62}$$

where ds'^2 depends on the variables $x_2, \cdots, x_n$ exclusively. Thus, g_{11} is a constant, $g_{ij} = 0$ for $2 \leqq j \leqq n$, and for $2 \leqq i, j \leqq n$, g_{ij} is independent of x_1. It is easy to see that then $R_{ijkl} = 0$ for all j, k, l. But in the sum (8), each addend contains a factor R_{1jkl}, and hence (61) holds. In [13], the authors also obtain $\chi = 0$ from some local phenomena involving toruses, and it would be an interesting task to relate our criterion to their more topological ones.

Our second remark concerns Cartan's condition (53). It obviously implies

$$K = \text{constant}, \tag{63}$$

since, on differentiating (8), (53) implies

$$K_{,i} = 0. \tag{64}$$

However it may be of some general interest to point out that on a compact S, in order to obtain (64) it is sufficient to assume that we have

$$R_{ijkl,a_1,a_2,\cdots,a_m} = 0 \tag{65}$$

for some sufficiently high m. We do not claim that (65) implies (53), although this might also be so, but only that (65) implies (64). In fact, if we form the $(m + k)^{\text{th}}$ covariant derivative of (8), and do it term-by-term by Leibnitz' rule, then (65) implies its vanishing, and then (64) will follow from the following lemma.

LEMMA 1. *On a compact S_n, of differentiability class C^∞ say, if for a scalar function φ we have*

$$\varphi_{,a_1,a_2,\cdots,a_m} = 0$$

for some m, then $\varphi = const.$

PROOF. On differentiating once more if necessary, we may assume that m is an even integer, $m = 2l$, and then Lemma 1 follows from

LEMMA 2. *On a compact S_n, if for a scalar φ we define the iterated Laplaceans* $\varphi^{(0)}(x) = \varphi(x)$,

$$\varphi^{(\lambda+1)}(x) = g^{ij}\varphi^{(\lambda)}{}_{,i,j}$$

and if for some $\lambda = l$ we have $\varphi^{(l)}(x) = 0$, then $\varphi = const.$
The proof follows by induction on l from the following lemma which we assume known, see [4] and [7].

LEMMA 3. *On a compact S_n, if for a scalar $\varphi(x)$ we have $g^{ij}\varphi_{,i,j}(x) = 0$,* (or $\geqq 0$, *say*), *then $\varphi = const.$ and if $g^{ij}\varphi_{,i,j} = const.$, then the latter constant is* 0.

10. Our last remark is of considerable geometric import, but essentially we will only state a problem without contributing towards its solution.

Take two Riemannian spaces. S_n, S_m, $n < m$. A differentiable mapping of S_n into S_m will be called equimetric, if it is locally one-one and if the groundform of S_n derives from the groundform of S_m by way of the mapping, that is

$$g_{ij} = h_{\alpha\beta}\frac{dy_\alpha}{dx_i}\frac{dy_{\beta'}}{dx_j}. \tag{66}$$

Obviously any distance function D_m on S_m which generates $h_{\alpha\beta}$ by formula (16) gives rise to a distance function D_n on S_n which in its turn generates g_{ij}. However, if D_m is the geodesic distance on S_m, the induced distance D_n is in general not again geodesic on S_n. For instance, in the "natural" imbedding of a sphere in Euclidean space, the induced distance is the distance by chords and not by spherical geodesics, and in fact, no compact S_n can be "geodesically" imbedded in a Euclidean E_m for any $m > n$ or even in real Hilbert space, $m = \infty$. But if we abide by equi-metric mapping, then in known cases many elliptic V_n can be so imbedded in some E_m, although apparently nothing is known about the possibility of such an imbedding from general assumptions. The known imbeddings of complex projective space (into *real* E_m of course, since into complex E_2 it is not possible) are highly non-obvious, and the general problem would be a very difficult one at best.

However a much more promising problem to investigate would be one referring to hyperbolic spaces. With regards to the two-dimensional $\tilde{S}$ with the metric (18), Hilbert has proved a theorem which if properly enunciated states that $\tilde{S}$ in its entirety cannot be equimetrically mapped into any E_m, $m \geqq 3$; and it strongly suggests itself that such an impossibility ought to be demonstrable for any hyperbolic space $\tilde{S}$, with $K \neq 0$ at least. In this connection, it should be pointed out that for a complex $\tilde{S}$ this conjecture has substance only if referring to real imbedding, since, as we have shown in [6], for complex imbedding it is even a local impossibility, at least for such complex line-elements as were introduced in §7.

II. COMPLEX SPACES

11. From now on we will be dealing with complex manifolds only, and our means of investigation will again be a metric groundform. However it will have the special features of an Hermite-Kaehler metric, that is, it will have the special form

$$\tfrac{1}{2}\, ds^2 = g_{\alpha\beta^*} dz_\alpha\, d\bar{z}_\beta \tag{67}$$

$(\alpha, \beta, \cdots = 1, \cdots, k)$ with the all-important condition

$$\frac{\partial g_{\alpha\beta^*}}{\partial z_\gamma} = \frac{\partial g_{\gamma\beta^*}}{\partial z_\alpha}. \tag{68}$$

For definitions and facts relating to such a metric we refer to [4], [5], and also [8].

Before turning to the subject matter proper, we are inserting a remark again leading to a problem. On a complex manifold, with or without a metric, the group of all complex homeomorphisms has properties analogous to those of a group of motions on a Riemannian space, and sometimes two parallel statements apply to the same space. Thus, if on a compact complex space with a Hermite-Kaehler metric we have

$$-R_{\alpha\beta^*}\zeta^\alpha \overline{\zeta^\beta} < 0 \tag{69}$$

then our first conclusion holds that there are at most a finite number of motions on it, whether complex analytic or only real differentiable ones. However, on the other hand, we have also proven that there is no one parametric group of complex homeomorphisms on it, and that under additional assumptions it is actually finite. This suggests the problem of investigating for a Hermite-Kaehler space under what additional assumptions, if any, the group of motions (at least its principal component) would be contained in the group of its complex homeomorphisms as such. We say "contained", because it certainly need not be equal to it, at least not on an elliptic space. For instance on complex projective space, with the plus-sign in (44), this line-element is certainly not invariant under *all* complex projective transformations, but as far as complex mappings are concerned, only under the subgroup of such like unitary transformations. And the problem suggested is to find out, for this and other spaces, whether it might not be invariant under some non-complex transformations as well.

12. Our next task is to transcribe the integrand (8) into conjugate complex variables z_α, $\bar{z}_\alpha$. We denote the integrand (8) without the normalizing factor (7), by A, introduce in addition to the differentials $dx_1, \cdots, dx_n$ another set of differentials $du_1, \cdots, du_n$, and set up the doubly external differential form

$$A\, dx_1 \cdots dx_n \qquad du_1 \cdots du_n\, . \tag{70}$$

It is easily seen that it is equal to the external power

$$(R_{i_1 i_2 j_1 j_2}\, dx_{i_1}\, dx_{i_2}\, du_{j_1}\, du_{j_2})^k. \tag{71}$$

If we now introduce the conjugate complex differentials

$$\begin{aligned} dz_\alpha &= dx_\alpha + (-1)^{1/2} dx_{k+\alpha}\,, & d\bar{z}_\alpha &= dx_\alpha - (-1)^{1/2} dx_{k+\alpha} \\ dw_\alpha &= du_\alpha + (-1)^{1/2} du_{k+\alpha}\,, & d\overline{w}_\alpha &= du_\alpha - (-1)^{1/2} du_{k+\alpha}\,, \end{aligned} \tag{72}$$

then (71) goes over into

$$4^k(R_{\alpha\beta^*\gamma\delta^*}\, dz_\alpha\, d\bar{z}_\beta\, dw_\gamma\, d\bar{w}_\delta)^k, \tag{73}$$

where the curvature components $R_{\alpha\beta^*\gamma\delta^*}$ pertain to the local parameters z_α, $\bar{z}_\alpha$. If now we carry out the exponentiation in (73) and recall that

$$\begin{aligned} dz_\alpha\, d\bar{z}_\alpha &= -2(-1)^{1/2} dx_\alpha\, dx_{k+\alpha} \\ dw_\alpha\, d\bar{w}_\alpha &= -2(-1)^{1/2} du_\alpha\, du_{k+\alpha} \end{aligned} \tag{74}$$

then we obtain

$$A = 4^{2k}(-1)^k B \tag{75}$$

where

$$\begin{gathered} B(z, \bar{z}) = \\ R_{\alpha_1\beta_1^*\gamma_1\delta_1^*} \cdots R_{\alpha_k\beta_k^*\gamma_k\delta_k^*} \varepsilon^{\alpha_1\cdots\alpha_k} \varepsilon^{\beta_1\cdots\beta_n} \varepsilon^{\gamma_1\cdots\gamma_k} \varepsilon^{\delta_1\cdots\delta_k}, \end{gathered} \tag{76}$$

and by inserting all numerical constants occurring we obtain for our previous scalar function K the final relation

$$K(x) = \frac{2^k(-1)^k B(z, \bar{z})}{\pi^k\, k!(\det |\, g_{\alpha\beta^*} |)^2}. \tag{77}$$

But the significant step is yet to come.

We take a neighborhood in the complex z-variables and we map it one-one into a neighborhood of complex variables $w_1, \cdots, w_m$, $m \geqq k$, by means of complex analytic functions

$$w_\sigma = f^\sigma(z_1, \cdots, z_k), \qquad \sigma = 1, \cdots, m. \tag{78}$$

If the w-neighborhood is endowed with a flat H.-K. metric

$$h_{\xi\eta^*}\, dw_\xi\, d\bar{w}_\eta, \qquad 1 \leqq \xi, \eta \leqq m, \tag{79}$$

then the mapping induces in the z-neighborhood a metric (67) whose curvature tensor is

$$R_{\alpha\beta^*\gamma\delta^*} = h_{\xi\eta^*} f^{\xi}_{,\alpha,\gamma} \overline{f^{\eta}_{,\beta,\delta}}, \tag{80}$$

see [5], p. 189. If we substitute this into the expression (76), and if we introduce the expression

$$D^{\sigma_1\cdots\sigma_k} = \sum\nolimits_{(\alpha),(\gamma)} \varepsilon^{\alpha_1\cdots\alpha_k} \varepsilon^{\gamma_1\cdots\gamma_k} f^{\sigma_1}_{,\alpha_1,\gamma_1} \cdots f^{\sigma_k}_{,\alpha_k,\gamma_k} \tag{81}$$

for all possible combination of indices

$$1 \leqq \sigma_1 \leqq m, \cdots, 1 \leqq \sigma_k \leqq m, \tag{82}$$

then we obtain the important relation

$$B = h_{\xi_1\eta_1^*} \cdots h_{\xi_k\eta_k^*} D^{\xi_1\cdots\xi_k} \overline{D_{\eta_1\cdots\eta_k}} \tag{83}$$

and since $h_{\xi\eta^*}$ is positive definite we hence obtain the following lemma

LEMMA 4. *In the given circumstances we have*

$$B > 0, \tag{84}$$

that is

$$(-1)^k K > 0, \tag{85}$$

unless we have

$$D^{\xi_1 \cdots \xi_k} = 0 \tag{86}$$

for all possible combination

$$1 \leqq \xi_j \leqq m, \qquad j = 1, \cdots, m, \tag{87}$$

in which exceptional case we have

$$B = (-1)^k K = 0. \tag{88}$$

It should be noted that in forming the expression (81), each of the m mapping functions $f^\sigma(z)$ is to be viewed as a scalar on the z-neighborhood, and that the symbol $f^\sigma_{,\alpha,\gamma}$ is its second covariant derivative relative to the (induced) metric (67).

13. The inference from Lemma 4 is as follows.

THEOREM 6. *If on a compact complex S_{2k}, with a H.-K. metric, in the neighborhood of every point the metric is obtainable from an equimetric immersion into a higher dimension flat complex Euclidean neighborhood, then we have*

$$(-1)^k \chi(S) \geqq 0, \tag{89}$$

and the equality

$$\chi(S) = 0 \tag{90}$$

holds if and only if in all neighborhoods we have identically

$$D^{\xi_1 \cdots \xi_k}(z, z) = 0 \tag{91}$$

for all $(\xi_1, \cdots, \xi_k)$.

In particular we have the following specific conclusion

THEOREM 7. *Let S_{2k} be a compact complex space with a H.-K. metric (67), and let $\tilde{S}$ be its universal covering space with the periodic extension of that metric on it.*

If then $\tilde{S}$ can be equimetrically mapped into a complex E_{2m}, then for S_{2k} we have relation (89), and (90) holds if and only if relation (86) hold identically for the imbedding functions.

This theorem is only a stepping stone toward the following one.

THEOREM 8. *If on a closed complex manifold S_{2k} there are given m complex analytic covariant vector fields*

$$\zeta_\alpha^{(\sigma)}(z), \qquad \sigma = 1, \cdots, m,$$

$m \geqq k$; if for each we have

$$\frac{\partial \zeta_\alpha^{(\sigma)}}{\partial z_\beta} = \frac{\partial \zeta_\beta^{(\sigma)}}{\partial z_\alpha} \tag{92}$$

and if for each point of S_{2k} *the matrix*

$$(93)\qquad | \zeta_\alpha^{(\sigma)} |_{\alpha=1,\cdots,k;\sigma=1,\cdots,m}$$

has maximal rank k; *then* $(-1)^k\chi \geqq 0$ *holds and* (90) *occurs if and only if we have*

$$(94)\qquad \varepsilon^{\alpha_1\cdots\alpha_k}\varepsilon^{\gamma_1\cdots\gamma_k}\zeta_{\alpha_1,\gamma_1}^{(\sigma_1)}\cdots\zeta_{\alpha_k,\gamma_k}^{(\sigma_k)} = 0$$

for all combinations

$$(95)\qquad 1 \leqq \sigma_j \leqq m$$

where $\zeta_{\alpha,\gamma}^{(\sigma)}$ *is the covariant derivative of* $\zeta_\alpha^{(\sigma)}$ *relative to the metric tensor*

$$(96)\qquad g_{\alpha\beta^*} = \sum_{\sigma=1}^m \zeta_\alpha^{(\sigma)}\overline{\zeta_\beta^{(\sigma)}}$$

PROOF. Starting with the given vector fields we introduce on S_{2k} the ground form (96). It is obviously non-negative and the rank of (93) being k makes it strictly positive. Also, (92) implies (68), thus making it into an H.-K. metric.

On the other hand, starting anew, we reason that property (92) allows us to set up the line integrals

$$(97)\qquad w_\sigma(z) = \int^z \zeta_\alpha^{(\sigma)} dz_\alpha ,$$

which are each an Abelian integral of the first kind on S_{2k} ; that is a complex analytic nowhere singular function on the covering space $\tilde{S}$ with additive periods corresponding to closed paths on S_{2k} itself. The rank of (93) being maximal means that there are sufficiently many of such integrals to permit a locally one–one complex imbedding of $\tilde{S}$ into E_{2m} . If now in E_{2m} we adopt the flat metric

$$(98)\qquad \sum_{\sigma=1}^m dw_\sigma d\overline{w}_\sigma$$

say, that is $h_{\xi\eta^*} = \delta_{\xi\eta}$, then (96) is the metric induced by this mapping, and Theorem 8 follows from Theorem 7.

Actually, the imbedding employed in the proof to Theorem 8 can be performed in two stages. First of all, the system of m Abelian integrals, if viewed on S_{2k} itself maps this space equimetrically into a certain flat torus T_{2m} , which latter torus in its turn derives its metric by "projection" from its universal covering space E_{2m} if endowed with the initial metric (98).

Finally we state for the record the following application of our preceding analysis to $k = 1$, thus giving the Gauss Bonnet formula for this case.

THEOREM 9. *On a complex* S_2 *of genus* $p > 1$ *we have*

$$(99)\qquad 2p - 2 = \frac{1}{\pi}\iint_{S_2} \frac{\partial^2 \log g}{\partial z \partial \bar{z}}\, dx\, dy = \frac{1}{4\pi}\iint_{S_2}\left(\frac{\partial^2 \log g}{\partial x^2} + \frac{\partial^2 \log g}{\partial y^2}\right) dx\, dy$$

where

$$g = \sum_{\rho=1}^p | \zeta^\rho |^2.$$

Herein $\zeta^1, \cdots, \zeta^p$ *is a basic system of complex Abelian differentials of the first kind. Also*

$$\frac{1}{2g}\frac{\partial^2 \log g}{\partial z \partial \bar{z}} = \frac{\sum_{\rho,\delta=1}^{p} \left| \zeta^\rho \frac{\partial \zeta^\delta}{\partial z} - \zeta^\delta \frac{\partial \zeta^\rho}{\partial z} \right|}{\sum_{\rho=1}^{p} | \zeta^\rho |^2}. \tag{100}$$

14. We have seen that the analytic condition (91), or (94), is necessary and sufficient for $\chi = 0$ and this raises the question of a direct geometric interpretation. We have none to offer except for a contributory remark.

If for the induced metric (67) the space S_{2k} is flat, that is

$$R_{\alpha\beta^*\gamma\delta^*} = 0, \tag{101}$$

then it is a torus and $\chi = 0$, and thus (91) holds. But conversely, (91) does not imply (101). To see that, take, for $k = 1$, a curve S_2 of genus $p > 1$, and map it into the previously introduced torus T_{2p}. Then take a curve of genus one, denoting it by U_2, and form the product

$$S_4 = S_2 \times U_2 . \tag{102}$$

It obviously can be mapped into the product

$$T_{2p+2} = T_{2p} \times U_2 \tag{103}$$

with its flat metric, and since $\chi(S_4) = 0$, due to $\chi(U_2) = 0$, condition (91) must be fulfilled. But we claim that the metric induced in (102) cannot be a flat one. In fact, by transition to universal covering spaces, (102) implies

$$\tilde{S}_4 = \tilde{S}_2 \times \tilde{U}_2 . \tag{104}$$

On the one hand, $\tilde{S}_2$ is the unit circle $| Z_1' | < 1$, and $\tilde{U}_2$ is the complex Z_2'-plane, and on the other hand, if S_4 were flat in an H.-K. metric then $\tilde{S}_4$ would be the complex (Z_1, Z_2)-plane. Now, (104) would indicate the possibility of a complex analytic transformation from the (Z_1, Z_2)-plane into $| Z_1' | < 1$, Z_2-*arbitrary*. In particular, there would exist a function $Z_1' = \varphi(Z_1, Z_2)$ defined in E_4 and yet bounded. Such a function must be a constant and the mapping would not be one–one as required.

15. In Theorem 6, relation (89) was obtained from

$$B \geqq 0, \tag{105}$$

and this in its turn followed from the nature of the expression (83). However (105) also holds if B is a limit of expressions (83), and thus we are led to the following theorem.

THEOREM 10. *For a compact complex* S_{2k} *with a H.-K. metric* (67) *we have* $(-1)^k\chi \geqq 0$, *provided that in the neighborhood of every point the tensor* $g_{\alpha\beta^*}$ *is the limit of a sequence of H.-K. tensors*

$$g^{(l)}_{\alpha\beta^*}, \qquad l = 1, 2, \cdots, \tag{106}$$

converging towards it uniformly with its second partial derivatives, such that each (106) *is obtainable from a complex imbedding into some* E_{2m}.

In certain cases it is also possible to retain a condition for $\chi = 0$, and we are stating the following

THEOREM 11. *We have* $(-1)^k\chi \geqq 0$ *if locally we can write*

$$(107) \qquad g_{\alpha\beta^*} = \sum_{\sigma=1}^{\infty} \frac{\partial f^{(\sigma)}}{\partial z_\alpha} \overline{\frac{\partial f^{(\delta)}}{\partial z_\beta}}$$

for a sequence of complex function $f^{(\sigma)}(z)$, $\sigma = 1, 2, \cdots$. *The sequence converging uniformly with its second derivatives.*

Also, we have $\chi = 0$, *if and only if* (91) *holds, where as before,* $D^{\xi_1\cdots\xi_k}$ *is to be formed by the expression* (81), *at present for all combinations* $1 \leqq \xi_j < \infty$, $j = 1, \cdots, k$.

For instance, for the curves S_2 of genus $p > 1$ with the metric (19) we have

$$(108) \qquad (1 - z\bar{z})^{-2} = \sum_{\sigma=0}^{\infty} (\sigma + 1) z^\sigma \bar{z}^\sigma = \sum_{\sigma=0}^{\infty} \left| \frac{\partial f^{(\sigma)}}{\partial z} \right|^2$$

where

$$(109) \qquad f^{(\sigma)}(z) = \frac{z^{\sigma+1}}{(\sigma + 1)^{1/2}}.$$

It is not hard to see that Theorem 11 applies, leading to $\chi < 0$, which of course agrees with $\chi = 2 - 2p$.

However it should be pointed out that the metric (19) which we have just induced from imbedding into Hilbert space, is radically different from the metric induced by imbedding into the space T_{2p} of Abelian integrals which latter imbedding produced the Gauss-Bonnet formula (99); since, as we have seen in [5] the metric (19) cannot be induced from an imbedding into finite-dimensional space. In other words, the "natural" metric (19), which is engendered by the geometric structure of the space is *not* the one leading to the explicit expression (99); if it were so leading, a large part of algebraic geometry in several variables would be much easier to manage.

16. Finally, we will consider imbedding into spaces of continuously many dimensions, as it were, and in this way establish a link with some view points of Part I of this paper.

As in [5], p. 194–195, we consider a domain D in the ordinary complex E_{2k}, and we assume that there acts on it a transitive group of complex homeomorphisms. The group is a Lie group and we will denote it by G. It is locally compact, and thus admits two Haar measures. Actually, the two measures are probably always identical, but we will not insist on this peculiarity. We select one of the two measures and denote the corresponding measure element by $d\theta$.

Since D lies in one coordinate system (and this fact is all-decisive in our present context), every element θ of G can be written in the form

$$(110) \qquad z_j' = f_j(z; \theta), \qquad j = 1, \cdots, k,$$

and we now set up the expression

$$(111) \qquad g_{\alpha\beta*}(z, \bar{z}) = \int_G \left(\sum_{j=1}^k \frac{\partial f_j(z, \theta)}{\partial z_\alpha} \frac{\overline{\partial f_j(z, \theta)}}{\partial z\beta} \right) d\theta,$$

assuming that the integral on G exists as an absolutely convergent L_1-function, together with the integral, for the first two partial derivatives, uniformly in the neighborhood of every point of D. Obviously it is a non-negative matrix of Hermite-Kaehler structure, and we are making the further assumption—which we are compelled to add explicitly—that it is a strictly positive one. We thus obtain an H.-K. metric

$$(112) \qquad \tfrac{1}{2}\, ds^2 = g_{\alpha\beta*}\, dz_\alpha\, d\bar{z}_\beta\,,$$

and it is not hard to deduce from the expression (111) that it is homometric with regard to the homeomorphisms of D.

Now, let Γ be a discontinuous subgroup of G and let the quotient $S = G/\Gamma$ be a compact complex space. The line-element (112) can be projected onto S, and for its characteristic we have thus the expression

$$(113) \qquad \chi(S) = \int_S \frac{2^k(-1)^k B(z, \bar{z})\, dv}{\pi^k k!(\det |\, g_{\alpha\beta*} \,|)^2},$$

with

$$(114) \qquad B(z, \bar{z}) = \int \cdots \int_G \sum_{j_1=1 \cdots j_k=1}^{k \quad k} |\, D^{j_1 \cdots j_k; \theta_1 \theta \cdots \theta_k} \,|\, d\theta_1 \cdots d\theta_k$$

and

$$(115) \qquad D^{j_1 \cdots j_k; \theta_1 \cdots \theta_k} = \varepsilon^{\alpha_1 \cdots \alpha_k} \varepsilon^{\gamma_1 \cdots \gamma_k} f_{j_1}(z, \theta_1)_{,\alpha_1,\gamma_1} \cdots f_{j_k}(z, \theta)_{,\alpha_k,\gamma_k}\,.$$

Hence we obtain

THEOREM 12. *For the compact complex space S subordinate to D we have $(-1)^k\chi \geqq 0$; and we have $\chi = 0$ if and only if*

$$(116) \qquad D^{j_1 \cdots j_k; \theta_1 \cdots \theta_k}(z, \bar{z}) = 0,$$

identically in all variables, indices and parameters occurring.

Actually, in most cases there will be a point z^0 of D, say the origin $z_1^0 = \cdots = z_k^0 = 0$, such that at this point

$$(117) \qquad \partial g_{\alpha\beta*}/\partial z_\gamma = 0.$$

This means that at this point condition (116) can also be written in the simplified form.

$$(118) \qquad \varepsilon^{\alpha_1 \cdots \alpha^k} \varepsilon^{\gamma_1 \cdots \gamma_k} \frac{\partial^2 f_{j_1}(z^0, \theta_1)}{\partial z_{\alpha_1} \partial \bar{z}_{\gamma_1}} \cdots \frac{\partial^2 f_{j_k}(z^0, \theta)}{\partial z_{\alpha_k} \partial \bar{z}_{\gamma_k}} = 0.$$

Now, due to the transitiveness of G on D, the validity of (116) at one point implies its validity at all other points, and thus we have the

COROLLARY. *If* (117) *holds at* z^0, *then* $\chi = 0$ *if and only if* (118) *holds for all combinations*

$$1 \leqq j_\lambda \leqq k, \quad \lambda = 1, \cdots, k; \qquad \theta_\lambda \varepsilon G, \quad \lambda = 1, \cdots, k.$$

Addendum. Almost automorphic vectors

(Received October 28, 1949)

The set-up of Theorem 8 can be generalized in a peculiar manner. Instead of assuming that the m vector fields $\zeta_\alpha^{(\sigma)}(z)$ with the property (92) are defined on the space S_{2k} itself, it suffices to assume more generally that they are defined on its covering space $\tilde{S}$ and that for any element T of the fundamental group Γ we have the relations

$$\zeta_\alpha^{(\sigma)}(Tz) = \sum_{\tau=1}^{m} C_\tau^\sigma(T)\zeta_\alpha^{(\tau)}(z), \qquad \sigma = 1, \cdots, m \tag{119}$$

where $\{C_\tau^\sigma\}$ is a unitary matrix. It would even suffice to assume only that our matrices are bounded in toto since any bounded representation of any group Γ is equivalent to a unitary one.

If now we introduce the tensor (96) on $\tilde{S}$, then it obviously is invariant under the elements of Γ; so that we are again dealing with an H.-K. metric on S_{2k} itself which is again locally derivable from a Euclidean imbedding. If therefore the matrix (93) has everywhere maximal rank k, then we again have conclusion (89), and (90) holds if and only if (91) holds. Also, the expressions entering (91) can be easily seen to be themselves invariant under the transformations (119), although this in itself is not necessary for the validity of the argument.

If we are only interested in obtaining conclusion (89) we can go one step further. If a system of vector fields satisfies the relation (119) then each individual member of the system is *almost automorphic* in the following sense. Any sequence of elements of the fundamental group Γ contains a subsequence $\{T_p\}$ such that the sequence $\zeta_\alpha^{(\sigma)}(T_p z)$ is uniformly convergent towards a limit on the entire $\tilde{S}$. In order to measure this uniformity it is necessary to have a metric on S which is invariant with respect to Γ but some such metric is always available to start with. Now, take a system of vector fields each of which is almost automorphic in accordance with the general definition just stated, and again define a metric tensor (96) on S. It is no longer invariant with respect to Γ but it is almost automorphic itself, that is, it is almost periodic with respect to the group Γ. Such metrics were already introduced in [4], and for the present context we note that by forming an average peculiar to almost periodic functions in general, a new metric will be obtained which will be invariant with respect to Γ and thus will exist on S_{2k} itself. The latter is locally a limit of metrics which are obtainable from a Euclidean imbedding each, so that the conclusion (89) can be again drawn, provided the limiting metric on S_{2k} is strictly positive definite everywhere. In order to secure this it is not sufficient to assume that the matrix (93) have maximal rank k everywhere on S, rather it is necessary to assume this with an appropriate "uniformity" clause attached to it, as for instance by demanding that the square sum of the determinants of rank k of the matrix (93) shall have a strictly positive lower bound uniformly on all of S.

PRINCETON UNIVERSITY

REFERENCES

1. C. B. ALLENDOERFER, *The Euler number of a Riemann manifold*, Amer. J. Math. 62 (1940), 243–248.
2. C. B. ALLENDOERFER and ANDRÉ WEIL, *The Gauss-Bonnet theorem for Riemannian polyhedra*, Trans. Amer. Math. Soc. 53 (1943), 101–129.
3. BOCHNER, *Differentiable and Riemann metric*, Duke Math. J. 4 (1938), 51–54.
4. BOCHNER, *Vector fields and Ricci curvature*, Bull. Amer. Math. Soc. 52 (1946), 776–797.
5. BOCHNER, *Curvature in Hermitian metric*, Bull. Amer. Math. Soc. 53 (1947), 179–195.
6. BOCHNER, *On compact complex manifolds*, J. Indian Math. Soc. n. s. 11 (1947), 1–21.
7. BOCHNER, *Curvature and Betti numbers*, Ann. of Math. 49 (1948), 379–390.
8. BOCHNER, *Curvature and Betti numbers, II*. Ann. of Math. 50 (1949), 77–93.
9. BOCHNER and DEANE MONTGOMERY, *Groups on analytic manifolds*, Ann. of Math. 48 (1947), 659–669.
10. S. S. CHERN, *Characteristic classes of Hermitian manifolds*, Ann. of Math. 47 (1946), 85–121.
11. C. EHRESMAN, *Sur la topologie de certains espaces homogènes*, Ann. of Math. 35 (1934), 396–443.
12. W. FENCHEL. *On total curvatures of Riemannian manifolds. I*, J. London Math. Soc. 15 (1940), 15–22.
13. H. HOPF and H. SAMELSON, *Ein Satz über die Wirkungsräume geschlossener Liescher Gruppen*, Comment. Math. Helv. 13 (1941), 240–251.
14. L. K. HUA, *On the theory of automorphic functions of a matrix variable I*. Amer. J. Math. 66 (1944), 470–488.
15. C. L. SIEGEL, *Some remarks on discontinuous groups*, Ann. of Math. 46 (1945), 708–718.

Annals of Mathematics
Vol. 52, No. 3, November, 1950

VECTOR FIELDS ON COMPLEX AND REAL MANIFOLDS

By S. Bochner

(Received January 6, 1950)

Previously, see [1]–[5], we have obtained statements linking Riemannian curvature to vector and tensor fields on a compact manifold, whether real or complex, and for the present some supplementary statements will be obtained in which curvature will not appear in the hypotheses.

1. On a compact complex V_{2k}, of real dimension $2k$, we consider a covariant vector field ξ_α, and a contravariant vector field η^α, both holomorphic, and we form the scalar contraction $\xi_\alpha \eta^\alpha$. It is again holomorphic, and if it is so on a compact space, then by the maximum modulus principle, it reduces to a constant

$$\xi_\alpha \eta^\alpha = c. \tag{1}$$

Thus for $k + 1$ vector fields

$$\xi_\alpha^{(j)}, \qquad j = 1, \cdots, k+1 \tag{2}$$

we obtain the system of relations

$$\xi_\alpha^{(j)} \eta^\alpha = c^j, \qquad j = 1, \cdots, k+1 \tag{3}$$

and from this we will deduce the following conclusion.

Theorem 1. *If on a compact complex V_{2k} there exist $k + 1$ analytic vector fields (2) such that for no system of constants*

$$(c^1, \cdots, c^{k+1}) \neq (0, \cdots, 0) \tag{4}$$

the determinant

$$| \xi_1^{(j)}, \cdots, \xi_k^{(j)}, \quad c^j |_{j=1,\cdots,k+1} \tag{5}$$

vanishes identically, then there exists on V_{2k} no contravariant analytic vector field η^α or tensor field $\eta^{\alpha_1 \cdots \alpha_p}$ other than zero. In this statement, "covariant" and "contravariant" may be interchanged.

Proof. If for a vector η^α we form the system (3) then, by our hypothesis, the constants c^j in it must all vanish. However our assumption also implies that the rank of the k by $k + 1$ matrix (2) must at some point have its maximal value k, and an analytic solution η^α of the system

$$\xi_\alpha^{(j)} \eta^\alpha = 0 \tag{6}$$

must then be zero, as claimed. For tensor fields $\eta^{\alpha_1 \cdots \alpha_p}$ we may now apply induction on p. Consider for instance a tensor field $\eta^{\alpha\beta}$. Each entity

$$\zeta^{(j)\beta} = \xi_\alpha^{(j)} \eta^{\alpha\beta}$$

is a contravariant vector field, and hence it vanishes identically by what we have just proved, but any solution $\eta^{\alpha\beta}$ of the system

$$\xi_\alpha^{(j)}\eta^{\alpha\beta} = 0, \qquad j = 1, \cdots, k+1$$

must vanish identically too.

2. The hypothesis of Theorem 1 may be replaced by an alternate one in which general constants no longer occur. By differentiation of (3) we obtain

$$\xi_\alpha^{(j)}\frac{\partial\eta^\alpha}{\partial z_\beta} + \frac{\partial\xi_\alpha^{(j)}}{\partial z_\beta}\eta^\alpha = 0 \qquad \beta = 1, \cdots, k; j = 1, \cdots, k+1,$$

and if we consider this as a system of equations in the $k(k+1)$ variables

$$\eta^\alpha, \quad \frac{\partial\eta^\alpha}{\partial z_\beta}$$

then it gives rise to a certain determinant

$$D_k\left(\xi_\alpha^{(j)}; \frac{\partial\xi_\alpha^{(j)}}{\partial z_\beta}\right) \tag{7}$$

which is a polynomial in the quantities inside the brackets. Now, an alternate hypothesis for Theorem 1 would be that this quantity (9) shall not vanish identically on V_{2k}, and we note that this would be an invariantive hypothesis in the sense that, for any point P_0, the non-vanishing of (9) is invariant with respect to a transformation of coordinates which leaves the point P_0 fixed. In fact, $D_k \neq 0$ at a point P_0 is a necessary and sufficient assumption for the conclusion that if for a system of power series

$$\eta^\alpha = a^\alpha + \textstyle\sum_{\beta=1}^k a_\beta^\alpha z_\beta + \text{(higher powers)}$$

with origin at P_0 we have

$$\zeta_\alpha^{(j)}\eta^\alpha = c^j + \text{(non-linear terms)}$$

then $a^\alpha = 0$, $a_\beta^\alpha = 0$.

3. For any $p \geqq 1$, for any two analytic tensor fields $\xi_{\alpha_1\cdots\alpha_p}$, $\eta^{\alpha_1\cdots\alpha_p}$ we again have

$$\xi_{\alpha_1\cdots\alpha_p}\eta^{\alpha_1\cdots\alpha_p} = c;$$

and if the two tensor fields are skew symmetric then for $p = k$ they are determined by the one component corresponding to $\alpha_1 = 1, \cdots, \alpha_k = k$, so that we then have

$$\xi_{1\cdots k}\eta^{1\cdots k} = c.$$

Any skew-tensor $\xi_{\alpha_1\cdots\alpha_k}$ determines a k-tuple Abelian integral of the first kind, and in keeping with the nomenclature of algebraic geometry we denote by p_g

(geometric genus) the number of linearly independent ones for constant complex coefficients. Now, for $p_g \geqq 2$ we have

$$\xi^{(j)}_{1\cdots k}\,\eta^{1\cdots k} = c^j, \qquad j = 1, 2,$$

and this implies

$$\eta^{1\cdots k} = 0. \tag{8}$$

We are thus led to the following theorem which for algebraic surfaces has been given by Picard [7].

THEOREM 1. *If the geometric genus of V_{2k} exceeds one, $p_g \geqq 2$, then the total group of complex homeomorphisms of V_{2k} cannot be transitive, not even locally so in the neighborhood of any point.*

PROOF. In fact the total group of such homeomorphisms is a complex Lie group (compare the discussion in [3]), and if it is locally transitive in the neighborhood of some point then Lie's own local theory implies that there exist k contravariant analytic vector fields $\eta^{\alpha}_{(j)}$; $j = 1, \cdots, k$, such that in the neighborhood of that point the determinant

$$|\eta^{\alpha}_{(j)}| \qquad \alpha, j = 1, \cdots, k$$

is $\neq 0$. However this determinant has the same transformation law as a quantity $\eta^{1\cdots k}$ which defines a skew symmetric tensor, and by (8) it must therefore vanish, which is contradictory.

Another obvious theorem (compare Picard l.c. and Painlevé [8]) is as follows.

THEOREM 3. *If $p_g \geqq 1$, and if there exists a one-parameter group of complex homeomorphisms (at least) then there must exist a skew symmetric analytic tensor $\zeta_{\alpha_1\cdots\alpha_{k-1}}$ which does not vanish identically.*

In fact, if $\xi_{\alpha_1\cdots\alpha_k}$ is skew and $\not\equiv 0$ and $\eta^{\alpha} \not\equiv 0$, then

$$\zeta_{\alpha_1\cdots\alpha_{k-1}} = \xi_{\alpha_1\cdots\alpha_k}\eta^{\alpha_k}$$

is $\not\equiv 0$.

4. We will now describe a connection with curvature by means of a metric which was first introduced in [2], and then utilized in [6] for the evaluation of the Euler-Poincaré characteristic.

If there are given r covarient analytic vector fields

$$\xi^{(j)}_{\alpha}, \qquad j = 1, \cdots, r, \tag{9}$$

with $r \geqq k + 1$ (we will later explain why not $r \geqq k$), and if we take any numerical positive definite Hermitian matrix A_{jl}, for instance the unit matrix $A_{jl} = \delta_{jl}$, then the expression

$$g_{\alpha\beta^*} = \sum\nolimits_{j,l} A_{jl}\,\xi^{(j)}_{\alpha}\,\overline{\xi^{(l)}_{\beta}} \tag{10}$$

defines a positive semi-definite Hermitian tensor field on V_{2k}. If we assume that the rank of the matrix (9) has its maximal value k at all points *without exception* then (10) is positive definite and may be used as a metric tensor. Next, if each vector field (9) satisfies the relation

$$\frac{\partial \xi_\alpha^{(j)}}{\partial z_\beta} = \frac{\partial \xi_\beta^{(j)}}{\partial z_\alpha}, \tag{11}$$

that is if it gives rise to a simple Abelian integral of the first kind

$$w_j = \int^z \xi_\alpha^{(j)}\, dz_\alpha, \tag{12}$$

then our metric tensor enjoys the crucial Kaehler property

$$\frac{\partial g_{\alpha\beta^*}}{\partial z_\gamma} = \frac{\partial g_{\gamma\beta^*}}{\partial z_\alpha} \tag{13}$$

and it derives from a locally Euclidean imbedding of our manifold V_{2k} into the Abelian variety of the coordinates (12) with the line element $\sum A_{jl}\, dw_j\, d\overline{w_l}$. We have shown in [2] that the Ricci tensor is then non-positive, having the value

$$-R_{\alpha\delta^*} = -\textstyle\sum_{jl} A_{jl}\, g^{\gamma\beta^*}\, \xi_{\alpha,\gamma}^{(j)}\, \overline{\xi_{\beta,\delta}^{(l)}}, \tag{14}$$

where $g^{\gamma\beta^*}$ is the inverse to (10) and $\xi_{\alpha,\gamma}^{(j)}$ is the covariant derivative of $\xi_\alpha^{(j)}$.

Now, we have proved in [1] that if on a compact complex V_{2k} with an Hermite-Kaehler metric the Ricci curvature is everywhere non-positive and at some point strictly negative then no contravariant analytic vector or tensor field may exist, which is the same conclusion as in Theorem 1. Thus, to a certain extent at least, our present Theorem 1 may be viewed as a particular case of a more extensive theorem in which a curvature hypothesis is the decisive assumption on which everything hinges.

5. On the face of it we might have admitted the possibility $r = k$ in the definition of (10) but this would have been illusory as the following theorem will imply.

THEOREM 4. *For $r \geqq k$, if there are given r vector fields (9) which are linearly independent for constant complex coefficients; if the matrix (9) has everywhere rank k and if (11) holds; then for $r \geqq k + 1$ the Ricci tensor (14) is not identically zero for any A_{jl}, but for $r = k$ it is identically zero for every A_{jl} and V_{2k} is topologically a torus.*

PROOF. (14) vanishes at a point if we there have

$$\xi_{\alpha,\gamma}^{(j)} = 0. \tag{15}$$

For $r \geqq k + 1$ we pick a point P_0 and form a linear combination

$$\xi_\alpha = \textstyle\sum_{j=1}^{r} a_j \xi_\alpha^{(j)} \tag{16}$$

with constants a_j not all zero, such that

$$\xi_\alpha(P_0) = 0.$$

Now (15) implies $\xi_{\alpha,\gamma} = 0$, that is

$$\frac{\partial \xi_\alpha}{\partial z_\beta} = \xi_\gamma \Gamma^\gamma_{\alpha\beta}, \tag{17}$$

and to this we may add the relations

$$\frac{\partial \xi_\alpha}{\partial \bar{z}_\beta} = 0. \tag{18}$$

But (17) and (18) taken together form a system of differential equations to which the classical uniqueness theorem applies and thus (16) would vanish identically, contrary to assumption.

For $r = k$, since, by assumption,

$$\det | \xi^{(j)}_\alpha | \not\equiv 0, \tag{19}$$

the following situation arises. The Abelian variety W as defined by (12) is topologically a torus and its complex analytic universal covering space is the whole open space. Also, V_{2k} maps into W locally topologically so that V_{2k} in its turn is an unbranched covering of W. Therefore V_{2k} itself is also a torus which is holomorphically covered by the total open space, and our assertion that its Ricci tensor (14) vanishes identically is an immediate consequence of the following lemma.

Every one-one analytic transformation

$$w_j = f_j(z_1, \cdots, z_k), \qquad j = 1, \cdots, k \tag{20}$$

of E_{2k} into itself which transforms a fundamental domain of multiperiodicity into another must be a linear transformation.

In fact, the functions (20) are such that for a period $(\omega_1, \cdots, \omega_k)$ we have

$$f_j(z_1 + \omega_1, \cdots, z_k + \omega_k) = f_j(z_1, \cdots, z_k) + \text{const.} \tag{21}$$

Therefore each partial derivative $\partial w_j / \partial z_l$ is multiperiodic in the ordinary sense and hence a constant.

6. We will now have a peculiar postscript to Theorem 1 which again will refer to the case $r = k$. As on previous occasions, see [1] and [6], we introduce the universal covering space $\tilde{V}$ of our compact manifold $V \equiv V_{2k}$ and the fundamental group $\Gamma = \tilde{V}/V$, and we consider on $\tilde{V}$ a set of vector fields

$$\xi^{(j)}_\alpha, \qquad j = 1, \cdots, k \tag{22}$$

which every element γ of Γ transforms into a linear combination

$$\xi'^{(j)}_\alpha = \sum_l a^j_{l(\gamma)} \xi^{(l)}_\alpha \tag{23}$$

where each matrix

$$\{a^j_{l(\gamma)}\} \tag{24}$$

is a unitary one.

Now, if η^α is defined on V then we can extend it periodically to all of $\tilde{V}$, see [3], and if we form the scalar functions

$$\varphi^{(j)} = \xi_\alpha^{(j)} \eta^\alpha \tag{25}$$

then each of these holomorphic functions on V is almost periodic relative to the group Γ in the following sense: any infinite sequence $\{\gamma_p\}$ contains an infinite subsequence $\{\delta_p\}$ such that the sequence of functions $\varphi^{(j)}(\delta_p P)$ is convergent towards a limiting function $\psi^{(j)}(P)$, as $p \to \infty$, uniformly in all of $\tilde{V}_1$ see [1]; and the "reverse" sequence $\psi^{(j)}(\delta_p^{-1} P)$ converges back towards the function $\varphi^{(j)}(P)$ from which we started. Now if $\varphi^{(j)}$ is holomorphic then each limiting function $\psi^{(j)}$ is holomorphic, and one at least may be chosen to attain the maximum of its absolute value. Thus it is a constant and so is the original function itself,

$$\varphi^{(j)} = c^j. \tag{26}$$

But η^α is assumed invariant with regard to Γ and (23) and (26) together imply

$$c^j = \textstyle\sum_l a_l^j(\gamma) c^l. \tag{27}$$

Thus, the vector $(c^1, \cdots, c^k)$ is an eigenvector of the entire group (24). If all its components are zero, and the rank of (22) is k somewhere, then (26) implies $\eta^\alpha = 0$. If, however, the vector is not zero, then after a change of basis in the linear space based on the k vectors (22) we may assume $c^1 \neq 0$, $c^2 = \cdots = c^k = 0$. This implies $a_1^1(\gamma) = 1$ and since the matrix is unitary it becomes fully reducible and the vector $\xi^{(1)}$ of the new basis is invariant with regard to Γ. Hence the following theorem.

THEOREM 5. *If there are given k vector fields* (22) *on an unbranched covering space $\tilde{V}$ of V then either a linear combination of them is one-valued on V itself, or there exists no contravariant vector field η^α on V.*

6. Finally we will make a statement on real manifolds. On such a manifold the maximal modulus principle applies to a solution of a Laplacean; and in order to have a Laplacean a positive definite symmetric tensor g_{ij} must be available to start with, that is to say the manifold must be given as a Riemannian space, and we will assume that it is so given. But on a real Riemannian space, whether differentiable or even real analytic, there is no distinction of "holomorphy" as between "covariant" and "contravariant" and the substitute for it is the distinction between a harmonic vector and a Killing vector. A harmonic vector has the two properties

$$\xi_{i,j} = \xi_{j,i} \tag{28}$$

$$g^{ij}\xi_{i,j} = 0, \tag{29}$$

and a Killing vector has the property

$$\eta_{i,j} = -\eta_{j,i} \tag{30}$$

which automatically implies

$$g^{ij}\eta_{i,j} = 0. \tag{31}$$

We now claim that in consequence of these properties the scalar

$$\varphi = g^{ab}\xi_a\eta_b \equiv \xi_a\eta^a$$

satisfies the Laplace equation

$$\Delta\varphi = 0 \tag{32}$$

where

$$\Delta\varphi = g^{rs}\varphi_{,r,s}\,. \tag{33}$$

In fact, on carrying out differentiations, (33) will be the sum of

$$T_1 = g^{rs}g^{ab}(\xi_{a,r,s}\eta_b + \xi_a\eta_{b,r,s})$$

and

$$T_2 = 2g^{rs}g^{ab}\xi_{a,r}\eta_{b,s}$$

and these two terms vanish separately. For T_2 this follows from (28) and (30) since these two relations transform it into its negative

$$-2g^{rs}g^{ab}\xi_{r,a}\eta_{s,b}$$

For T_1 we first of all obtain

$$T_1 = g^{rs}g^{ab}(\xi_{r,a,s}\eta_b - \xi_a\eta_{r,b,s}), \tag{34}$$

and if we put

$$\xi_{r,a,s} = \xi_{r,s,a} + \xi_\lambda R^\lambda_{r,a,s}$$

$$\eta_{r,b,s} = \eta_{r,s,b} + \eta_\mu R^\mu_{r,b,s}$$

then (34) is the sum of the term

$$g^{rs}g^{ab}(\xi_{r,s,a}\eta_b - \xi_a\eta_{r,s,b}) \equiv g^{ab}\eta_b(g^{rs}\xi_{r,s})_a - g^{ab}\xi_a(g^{rs}\eta_{r,s})_{1b}$$

which is zero on account of (29) and (31), and of the term

$$g^{ab}\xi_\lambda\eta_b g^{rs}R^\lambda_{r,a,s} - g^{ab}\xi_a\eta_\mu g^{rs}R^\mu_{r,b,s}$$

which vanishes identically.

But if (32) holds on a compact M_n then $\varphi = c$, and thus

$$\xi_i\eta^i = c,$$

and this leads to the following theorem.

THEOREM 6. *If on a compact Riemannian space, M_n there are given $n + 1$ harmonic vector fields $\xi_i^{(j)}$, $j = 1, \cdots, n + 1$, such that for no system of constant $(c^1, \cdots, c^{n+1}) \neq (0, \cdots, 0)$ the determinant*

$$|\, \xi_1^{(j)}, \cdots, \xi_n^{(j)}, c^j \,|\; j = 1, \cdots, n + 1$$

vanishes identically, or if the determinant

$$D_n\left(\xi_i^{(j)};\ \frac{\partial \xi_i^{(j)}}{\partial \xi_l}\right)$$

is different from 0 *on a dense set; then there exists on* M_n *no Killing vector other than zero. In this statement "harmonic vector" and "Killing vector" may be interchanged.*

PRINCETON UNIVERSITY

REFERENCES

1. BOCHNER, *Vector fields and Ricci curvature*. Bull. Amer. Math. Soc. 52 (1946), 776–797.
2. BOCHNER, *Curvature in Hermitian metric*. Bull. Amer. Math. Soc. 53 (1947), 179–195.
3. BOCHNER, *On compact complex manifolds*. J. Indian Math. Soc. XI (1947), 1–21.
4. BOCHNER, *Curvature and Betti numbers*. Ann. of Math., 49 (1948), 379–390.
5. BOCHNER, *Curvature and Betti numbers II*. Ann. of Math. 50 (1949), 77–93.
6. BOCHNER, *Euler-Poincaré characteristic for locally homogeneous and complex spaces*. Ann. of Math., 51 (1950), 241–261.
7. E. PICARD, *Mémoire sur la théorie des fonctions algébriques de deux variables*. Jour. de Math. (s. 4) 5 (1889), 135–319; in particular Chapter III.
8. P. PAINLEVÉ, Leçons sur la théorie des équations différentielles. Paris 1897, in particular pp. 255–288.

Reprinted from the Proceedings of the NATIONAL ACADEMY OF SCIENCES,
Vol. 37, No. 6, pp. 356–359. June, 1951

COMPLEX SPACES WITH TRANSITIVE COMMUTATIVE GROUPS OF TRANSFORMATIONS

BY S. BOCHNER

MATHEMATICS DEPARTMENT, PRINCETON UNIVERSITY

Communicated April 16, 1951

In algebraic geometry there is a theorem known to the effect that if on an algebraic variety there is given a transitive commutative Lie group of birational transformations then the variety can be uniformized by hyperelliptic functions of the same dimension as the variety,[1] and as a comment on this theorem we will give a proposition which is rather more comprehensive than itself.

THEOREM. *Let V_{2k}, $k \geq 1$, be a (non-algebraic, non-compact) complex coordinate space of 2k real dimensions. If, for $r \geq k$, there are on it r holomorphic contravariant vector fields*

$$\eta_p^{\alpha} \qquad \alpha = 1, \ldots, k; \ p = 1, \ldots, r, \tag{1}$$

such that the rank of the matrix (1) has everywhere its maximum value k ("transitivity") and that the bracket expressions

$$\varphi_{pq}^{\alpha} \equiv \eta_p^{\beta} \frac{\partial \eta_q^{\alpha}}{\partial z^{\beta}} - \eta_q^{\beta} \frac{\partial \eta_p^{\alpha}}{\partial z^{\beta}}$$

all vanish identically

$$\varphi_{pq}^{\alpha} = 0, \tag{2}$$

("commutativity"), then there are on the space k simple Abelian integrals of the first kind by which it is mapped holomorphically and locally one-one into the Euclidean E_{2k}. In particular,[2] if V_{2k} is compact it is a complex multitorus.

Proof: As on a previous occasion[3] we introduce the Hermitian tensorfield

$$g^{\alpha\beta^*} = \sum_{p=1}^{r} \eta_p{}^{\alpha}\overline{\eta_p{}^{\beta}}.$$

Due to the transitivity postulated, it is strictly positive definite and thus has an inverse $g_{\alpha\beta^*}$, and for the latter we claim the Kaehler property

$$\frac{\partial g_{\alpha\beta^*}}{\partial z_\gamma} - \frac{\partial g_{\gamma\beta^*}}{\partial z_\alpha} = 0. \qquad (3)$$

In fact,[3] equation (3) is equivalent with

$$\frac{\partial g^{\alpha\rho^*}}{\partial z_\beta} g^{\beta\sigma^*} - \frac{\partial g^{\alpha\sigma^*}}{\partial z_\beta} g^{\beta\rho^*} = 0$$

which in turn is equivalent with

$$\sum_{p,\, q} \overline{\eta_p{}^{\rho}}\, \overline{\eta_q{}^{\sigma}}\, \varphi_{pq}{}^{\alpha} = 0$$

and thus is certainly implied by (2).

We next claim that the given vector fields (1) are self-parallel in the constructed metric, that is

$$\frac{\partial \eta_p{}^{\alpha}}{\partial z^{\beta}} + \eta_p{}^{\gamma}\Gamma_{\gamma\beta}{}^{\alpha} = 0. \qquad (4)$$

In fact, due to [4]

$$\Gamma_{\beta\gamma}{}^{\alpha} = -\frac{\partial g^{\alpha\rho^*}}{\partial z_\beta} g_{\rho^*\gamma},$$

equation (4) is equivalent with

$$\frac{\partial \eta_p{}^{\alpha}}{\partial z_\beta} g^{\beta\rho^*} = \eta_p{}^{\gamma} \frac{\partial g^{\alpha\rho^*}}{\partial z_\gamma},$$

and hence with

$$\sum_q \overline{\eta_q{}^{\rho}} \varphi_{pq}{}^{\alpha} = 0,$$

and this is again implied by (2). It is actually equivalent with (2) but there is no need for verifying this. Now, this implies that in the Ricci relation

$$\eta^{\alpha}_{p,\, \beta,\, \gamma^*} - \eta^{\alpha}_{p,\, \gamma^*,\, \beta} = \eta_p{}^{\delta} R_{\delta\beta\gamma^*}{}^{\alpha}$$

the left side vanishes identically, and hence also the right side. But (1) has everywhere maximal rank, so that

$$R_{\delta\beta\gamma^*}{}^{\alpha} = 0,$$

and thus our space is a flat Kaehler space. Therefore, in the neighborhood of every point we can allowably normalize the metric tensor to

$$g_{\alpha\beta^*} = \delta_{\alpha\beta^*} \tag{5}$$

and therefore k linearly independent Abelian integrals of the stated kind can be introduced locally. However, since the normalization (5) can be introduced in every neighborhood, such Abelian integrals can be continued along every path and they remain linearly independent in the neighborhood of every point, which completes the proof of the theorem.

Actually, the normalization (5) shows that the contravariant vector fields

$$\zeta_\alpha^{\ r} = g_{\alpha\beta^*}\overline{\eta_r^{\ \beta}}$$

are likewise holomorphic and self-parallel, so that in particular we have

$$\frac{\partial \zeta_\alpha^{\ r}}{\partial z_\beta} - \frac{\partial \zeta_\beta^{\ r}}{\partial z_\alpha} = 0.$$

Therefore the r Abelian integrals

$$w^{\rho}(z) = \int_{z^0}^{z} \zeta_\alpha^{\ \rho}\, dz_\alpha$$

may be introduced, and if, for instance, the first k among them are linearly independent at the point z^0, then they will be so everywhere and offer the mapping claimed.

Remark.—We will now describe a general possibility for constructing a Kaehler metric (not necessarily a flat one) from given data by an old idea of M. Noether which actually only performs a holomorphic immersion into a complex projective space of sufficiently many dimensions and thus induces a Kaehler metric automatically.

Assume at first that we are given a number of scalar densities

$$\varphi_q(z), \qquad q = 1, \ldots, s \tag{6}$$

$s > k + 1$, each being a one-coordinate object which under a change of coordinates transforms itself by a multiplicative factor $\left(\frac{\partial(z)}{\partial(z')}\right)^g$ where g is any integer, positive or negative, the same for all. If then the s by $k + 1$ matrix

$$\left|\varphi_q, \frac{\partial \varphi_q}{\partial z_\alpha}\right| \qquad q = 1, \ldots, s; \quad \alpha = 1, \ldots, k, \tag{7}$$

has everywhere its maximal rank $k + 1$, then the equations

$$w_q = w_q(z)$$

will map our space V_{2k} holomorphically and locally regularly into the projective space P_{2s-2} of the points

$$w_1 : w_2 : \ldots : w_s.$$

Suppose now that we are given a holomorphic tensor density $\zeta^{\alpha_1 \ldots \alpha_m}_{\beta_1 \ldots \beta_n}$ of a given type $(m; n; g)$, with "symmetries" added. If the number of independent tensor components is N, then we denote this object by one cumulative index. Thus: ζ^A, and we envisage the presence of several such tensor-fields

$$\zeta_p{}^A, \quad A = 1, \ldots, N; \quad p = 1, \ldots, r \tag{8}$$

all of the same type, and with $p \geq N + 1$. Now, it is possible to verify that the N-dimensional square matrices of the matrix (8), whose number is $s = \binom{r}{N}$, are all densities of a certain weight; and if we denote them by (6) and impose the previous condition on the matrix (7) then a mapping into a projective space will ensue. The condition on the matrix (7) could probably be stated more simply in terms of the elements (6) itself, but we will not pursue this point any further.

[1] Due to Picard, Painlevé and others. For references see Castelnuovo, G., and Enriques, F., "Die Algebraischen Flächen, etc." *Encyclopädie der Mathematischen Wissenschaften*, Vol. III 21, Article III C 6b, pp. 744–746.

[2] Compare our "Vector Fields on Complex and Real Manifolds," *Ann. Math.*, **52**, 645–646 (1950).

[3] "Vectorfields and Ricci Curvature," *Bull. Am. Math. Soc.*, **52**, 796–797 (1946).

[4] "Curvature in Hermitian Metric," *Ibid.*, **53**, 180 (1947).

Reprinted from the Proceedings of the NATIONAL ACADEMY OF SCIENCES,
Vol. 37, No. 10, pp. 704–706. October, 1951

TENSORFIELDS AND RICCI CURVATURE IN HERMITIAN METRIC

BY S. BOCHNER

MATHEMATICS DEPARTMENT, PRINCETON UNIVERSITY

Communicated August 15, 1951

Previously[1] we have proved as follows.

THEOREM 1. *If on a compact Kaehler manifold the Ricci curvature is positive, $-R_{\alpha\beta^x} > 0$, then there exists no holomorphic tensorfield of covariant type $\zeta_{\alpha_1 \ldots \alpha_p}$; and for negative curvature there exists none of contravariant type $\zeta^{\beta_1 \ldots \beta_q}$.*

This theorem is unique among similar ones in that conclusions affecting tensors of arbitrarily many indices are obtained from restrictions on the Ricci tensor alone, and not on the curvature tensor in its entirety. In the present note we will extend this to tensors of mixed type

$$\zeta^{\beta_1 \ldots \beta_q}_{\alpha_1 \ldots \alpha_p} \tag{1}$$

as well, and also from Kaehler metric to hermitian metric in general. The previous proof to Theorem 1 will readily give the following proposition.

THEOREM 2. *For a holomorphic tensor (1) on a compact Kaehler manifold the expression*

$$-\sum_{r=1}^{p} \zeta^{\beta_1 \ldots \beta_q}_{\alpha_1 \ldots \lambda \ldots \alpha_p} R^{\lambda}_{\alpha_r} \zeta^{\alpha_1 \ldots \alpha_p}_{\beta_1 \ldots \beta_q} + \sum_{s=1}^{q} \zeta^{\beta_1 \ldots \mu \ldots \beta_q}_{\alpha_1 \ldots \alpha_p} R^{\beta_s}_{\mu} \zeta^{\alpha_1 \ldots \alpha_p}_{\beta_1 \ldots \beta_q}$$

cannot be ≥ 0 at all points and >0 at some points; and if it is 0 everywhere, then the tensor must have covariant derivative 0,

$$\zeta^{\beta_1 \ldots \beta_q}_{\alpha_1 \ldots \alpha_p, i} = 0. \tag{2}$$

Also, the assertion applies not only to tensorfields which are holomorphic, that is

$$\zeta^{\beta_1 \ldots \beta_q}_{\alpha_1 \ldots \alpha_p, \sigma^*} = 0, \tag{3}$$

but more generally to those satisfying the equations

$$g^{\sigma^*\rho} \zeta^{\beta_1 \ldots \beta_q}_{\alpha_1 \ldots \alpha_p, \sigma^*, \rho} = 0. \tag{4}$$

It should be noted that solutions of (4) are not so much "holomorphic," as rather "harmonic" in a certain sense, even when complex-valued, and that irrespective of curvature properties the number of linearly independent ones is always finite on a compact V_{2k}, by a general theorem.[3] An effective corollary to Theorem 3 is as follows.

THEOREM 3. *If at every point of our space we denote by M and m the algebraically largest and smallest eigen-value of the matrix $-R_{\alpha\beta^*}$ and if for $p \geq 0$, $q \geq 0$ we have $pm - qM \geq 0$ everywhere and*

$$pm - qM > 0 \tag{5}$$

somewhere, then there exists no solution of (4) for such p, q; and if we have $pm - qM = 0$ then any solution of (4) must also satisfy (2).

We note that (5) is fulfillable only if either $M \geq m \geq 0$, $M > 0$ or $0 \geq M \geq m$ $0 > m$, so that, in effect, Theorem 3 applies only to spaces whose Ricci curvature is either everywhere ≥ 0 or everywhere ≤ 0. In this respect it is no more general than Theorem 1, which it otherwise subsumes for $q = 0$ or $p = 0$, respectively. The most interesting conclusion from Theorem 3 arises for $M = m$.

THEOREM 4. *If a compact Kaehler manifold is an Einstein space*

$$-R_{\alpha\beta^*} = bg_{\alpha\beta^*},$$

then for $b > 0$ there is no holomorphic tensor (1) with $p > q$, and for $b < 0$ none with $p < q$; and those with $p = q$ must have covariant derivative 0.

Also, for $b = 0$, that is $R_{\alpha\beta^*} = 0$, every solution of (4) must have covariant derivative 0.

Thus, on complex projective space, complex Grassmann varieties, etc., there are no holomorphic tensors with more lower than upper indices, and on their hyperbolic counterparts, if compact, there are none with more upper than lower indices; and a space with vanishing Ricci curvature shows the behavior of a complex multi-torus.

In general hermitian metric (non-Kaehler) our assertions will remain in force, if we replace in the text the Ricci tensor $-R_{\alpha\beta^*}$ by the tensor

$$S_{\alpha\beta^*} = g^{\rho\sigma^*} R_{\alpha\beta^*\rho\sigma^*}, \tag{6}$$

which need no longer be equal to it, and if we retain condition (3) or (4) for our tensors (1), which conditions, however, are no longer implied by halomorphy component-wise.[1, 2] Thus, Theorem 3 remains literally in force if M, m are the maximum and minimum of (6), and, for instance, for $m > 0$ there is no solution of (4) with $p > 0$, $q = 0$.

[1] "Vector Fields and Ricci Curvature," *Bull. Am. Math. Soc.*, **52**, 776–797 (1946).

[2] "Curvature in Hermitian Metric," *Ibid.*, **53**, 179–195 (1947).

[3] "Tensorfields with Finite Bases," *Ann. Math.*, **53**, 400–411 (1951).

ANNALS OF MATHEMATICS
Vol. 53, No 3, May, 1951

TENSOR FIELDS WITH FINITE BASES

BY S. BOCHNER

(Received November 13, 1950)

We will establish the general principle (with applications) that any module of tensors or tensor densities of any mixed type on a compact manifold has a finite basis whenever the tensors are complex-holomorphic or, more generally, in the real case, are harmonic in a certain general manner. Our concept of "harmonic" is rather more general than the one current in the theory of cohomology and no such restrictions as skew-symmetry are required for us.

We will also operate, even more generally, on the universal covering space of the given one, and the finite bases will also occur for finite groupings of tensors on the covering space which reproduce themselves by arbitrary bounded representations of the fundamental group. Our consequences will be mainly analytical, but Theorem 8 also suggests the purely topological result of associating Betti numbers $\{B_r\}$ with every bounded irreducible representation of the fundamental group, the ordinary Betti numbers being those corresponding to the identity representation. This ought to apply to general complexes, and further statements bearing on this question will be made in a subsequent paper.

1. Holomorphic tensors

THEOREM 1. (i) *On a compact complex manifold* $V = V_{2k}$ *the module of all holomorphic tensor densities*

$$\xi^{\beta_1 \cdots \beta_q}_{\alpha_1 \cdots \alpha_p} \tag{1}$$

of any given type (p, q, W) *has a finite base; the integer* $W \gtrless 0$ *being the density weight.*

(ii) *Also, corresponding to any type, there are finite sets of points such that any tensor density vanishing on these points vanishes identically.*

Remark. Part (i) has been known previously only for vectors

$$\eta^\alpha \qquad (p = 0, q = 1, W = 0),$$

such being Lie generators of isomorphism, and the latter constituting a Lie group, see [9] and [5]; and for automorphic forms ($p = 0, q = 0$) on manifolds whose universal covering space is a complex cell, and in this case with an estimate for the dimension of the bases which our much more comprehensive approach cannot imitate, see [10], [13]. Part (ii) is a qualitative partial extension of the Riemann-Roch theorem.

PROOF. In the symbol (1) the lower and upper Greek indices are "unstarred" and this is meant to imply, see [5] and [4], that they shall transform themselves by use of holomorphic coefficients of the matrix

$$\frac{\partial(w_1, \cdots, w_k)}{\partial(z_1, \cdots, z_k)}, \tag{2}$$

and the density factor is meant to be the $-W^{\text{th}}$ power of the determinant of (2). But this does not imply and is not meant to presuppose that all components of (1) are holomorphic, unless such is expressly stated as it has been in the theorem.

We will show immediately afterwards that there always exists on V a (non-holomorphic) positive definite tensor

$$g^{\alpha\beta*} \tag{3}$$

of differentiability class C^∞ with the law of transformation

$$g'_{\alpha\beta*}(w, \bar{w}) = \frac{\partial z_\gamma}{\partial w_\alpha} \frac{\partial \bar{z}_\delta}{\partial \bar{w}_\beta} g_{\gamma\delta*}(z, \bar{z}) \tag{4}$$

and if we denote the value of its determinant by g then the quantity

$$g^W g^{\alpha_1\gamma_1^*} \cdots g^{\alpha_p\gamma_p^*} g_{\beta_1\delta_1^*} \cdots g_{\beta_q\delta_q^*} \xi^{\beta_1\cdots\beta_q}_{\alpha_1\cdots\alpha_p} \xi^{\delta_1\cdots\delta_q}_{\gamma_1\cdots\gamma_p} \tag{5}$$

is an ordinary (non-negative) scalar on V. For the present we will denote the object (1) simply by ξ and the quantity (5) by $|\xi|^2$. If Part (i) were false there would exist an infinite sequence of elements $\{\xi^{(\mu)}\}$ of the module such that any finite number of them were linearly independent. If we now take any sequence of points $\{P^\rho\}$ then by forming suitable linear combinations we could construct a new sequence

$$\{\xi^{(\rho)}\} \tag{6}$$

such that

$$\xi^{(r)}(P^\rho) = 0, \qquad \rho = 1, \cdots, r \tag{7}$$

$$|\xi^{(r)}| = 1 \tag{8}$$

for $r = 1, 2, 3, \cdots$. Due to the positive-definiteness of (3), the normalization (8) implies, in local coordinate systems, the uniform boundedness of the components (1) of the given sequence (6), locally. But since the components are holomorphic functions, they are then a fortiori also equi-uniformly continuous locally, and therefore a subsequence of (6) will converge uniformly towards a limiting tensor $\xi^{(\infty)}$ for which we would have

$$\xi^{(\infty)}(P^\rho) = 0, \qquad \rho = 1, 2, \cdots, \tag{9}$$

$$|\xi^{(\infty)}| = 1. \tag{10}$$

If now the sequence $\{P^\rho\}$ has been chosen everywhere dense in V, then (9) would imply $\xi^{(\infty)} \equiv 0$ in contradiction to (10), and hence the conclusion.

Next, if Part (ii) were false, then on choosing a sequence of points of the same description there would again be a sequence (6) satisfying (9) and (10), which again would be contradictory.

As for constructing a matrix (3) with the required properties we proceed as follows. With each point of V we can associate an allowable coordinate neighborhood U, containing the point, whose Euclidean pre-image is the sphere

$$S(r): z_1\bar{z}_1 + \cdots + z_k\bar{z}_k < r^2,$$

the origin corresponding to the point. We take the subneighborhood $U' \subset U$ whose pre-image is $S(r/3)$, and we now choose a *finite* number of neighborhoods $\{U_\sigma\}$ such that the subneighborhoods $\{U'_\sigma\}$ constitute a covering of V. With each U'_σ we associate an infinitely differentiable real function $\varphi_\sigma(t)$ in $0 \leqq t \leqq r$ which has the value 1 in $(0, r/3)$ and the value 0 in $(2r/3, r)$ and we now define a tensor $g_\sigma^{\alpha\beta^*}$ on V in the following way. On U^σ it has the value

$$\delta^{\alpha\beta}\varphi(z_1\bar{z}_1 + \cdots + z_k\bar{z}_k)$$

where $\delta^{\alpha\beta}$ is the Kronecker symbol, and on any coordinate neighborhood which is located entirely in $V - U'_\sigma$ it shall have value 0. This constitutes a covering, and since the values prescribed are compatible with (4) we can extend them to all allowable coordinate systems by using (4). And the sum $\sum_\sigma g_\sigma^{\alpha\beta^*}$ will then be a tensor (3) having all properties stated.

2. Harmonic tensors

THEOREM 2. *On a compact (real) five times differentiable manifold M_n we take all twice differentiable tensor densities*

$$\xi_{i_1\cdots i_p}^{j_1\cdots j_q}(x) \tag{11}$$

of any type (p, q, W). On denoting the components of (11) *in some fixed arrangement by one cumulative index, thus:*

$$\varphi_1(x), \cdots, \varphi_N(x), \tag{12}$$

where $N = n^{p+q}$, we take a module of our tensors which in some joint sufficiently small neighborhood of every point of M_n satisfy the same system of partial differential equations

$$h^{ij}\frac{\partial^2\varphi_A}{\partial x_i\partial x_j} + h_A^{Bi}(x)\frac{\partial\varphi_B}{\partial x_i} + h_A^B(x)\cdot\varphi_B = f_A(x), \qquad A = 1, \cdots, N, \tag{13}$$

with twice differentiable coefficients and positive definite h^{ij}.

Such a module has a finite basis. Also there are finite sets of points such that any element vanishing on them vanishes identically.

REMARK. Theorem 2 includes Theorem 1, since for a holomorphic tensor (1) we have the local equations

$$\left(\frac{\partial^2}{\partial x_\alpha^2} + \frac{\partial^2}{\partial y_\alpha^2}\right)\xi_A = 0 \qquad z_\alpha = x_\alpha + \sqrt{-1}\, y_\alpha \qquad \alpha = 1, \cdots, k$$

and by addition

$$\sum_{\alpha=1}^{k}\left(\frac{\partial^2\xi_A}{\partial x_\alpha^2} + \frac{\partial^2\xi_A}{\partial y_\alpha^2}\right) = 0$$

which is a system (13). On the other hand, Theorem 2 applies to skew-symmetric tensors $\xi_{i_1 \cdots i_p}$ which are harmonic in the particular definition of cohomology, and it furthermore applies to Killing vectors: $\xi_{i,j} + \xi_{j,i} = 0$, since for the latter we have

$$g^{jk}\xi_{i,j,k} + \xi_l R_i^l = 0,$$

where R_i^l is the Ricci tensor pertaining to g_{ij}. Thus we have a common source for the following phenomena of finiteness on compact manifolds: finiteness of Betti numbers, of the (Lie) group of isometrics in the real case, and of holomorphic isomorphisms in the complex case, and of bases for complex automorphic forms.

PROOF. If with the aid of some Riemannian ground matrix g^{ij} we proceed as in the proof to Theorem 1, we are left with verifying the following lemma.

LEMMA 1. *If a family of solutions of the system* (13) *in a joint neighborhood of the origin is uniformly bounded there, then in some smaller neighborhood their first partial derivatives are uniformly bounded, so as to make the functions themselves equi-uniformly continuous in the smaller neighborhood.*

A proof of this could probably be inferred from semi-topological considerations, see [11], but we will sketch a proof based on older procedures, see [12]. As on a previous occasion, see [1], p. 354, we introduce auxiliary variables

$$t_1, \cdots, t_N.$$

Putting

$$\Phi(x, t) = \sum_A t_A \varphi_A(x) \tag{14}$$

we obtain from (13) the relation

$$h^{ij}(x)\frac{\partial^2\Phi}{\partial x_i\,\partial x_j} + \sum_{A,B} t_A h_A^{Bi}\frac{\partial^2\Phi}{\partial x_i\,\partial t_B} + \sum_{A,B} t_A h_A^B \frac{\partial\Phi}{\partial t_B} = \sum_A t_A f_A(x) \tag{15}$$

and since $(\partial^2\Phi)/(\partial t_A\,\partial t_B) = 0$ we can add to the left side the term

$$\delta^{AB}(\partial^2\Phi)/(\partial t_A\,\partial t_B).$$

If we consider (14) as a function in the $n + N$ variables, then, for small t_A, equation (15) is again of elliptic type, and it is not hard to see that in order to prove Lemma 1 it suffices to prove that in the solution (14) of (15) the values of the functions maximize their partial derivatives locally. That is to say, if we revert to the notation of the lemma, it suffices to analyze it for $N = 1$, thus

$$h^{ij}(x)\frac{\partial^2\varphi}{\partial x_i\,\partial x_j} + h^i(x)\frac{\partial\varphi}{\partial x_i} + h(x)\varphi(x) = f(x). \tag{16}$$

For a sufficiently small neighborhood D_n of the origin with a suitably smooth boundary B_{n-1}, there exists a Green's function with the customary singularity such that, for $x \in D_n$, (16) implies

$$\varphi(x) = \int_{B_{n-1}} \varphi(\xi) \sum_{\alpha=1}^{n} G_\alpha(x; \xi) d\xi_1 \cdots d\hat{\xi}_\alpha \cdots d\xi_n \tag{17}$$
$$+ \int_{D_n} f(\xi) G(x; \xi)\, d\xi_1 \cdots d\xi_n$$

where each $G_\alpha(x; \xi)$ is a linear combination, with coefficients in the variables ξ, of the partial derivation $\partial G/\partial \xi_\beta$. Now on the right side of (17) we can form the first derivatives $\partial/\partial x_\beta$ by taking them under the integral sign and the conclusion follows.

We would like to remark that writing our tensor in the form (12) with one cumulative index only was not done solely for typographical purposes but with a possible generalization in mind. Recently, [8], we have generalized the notion of a tensor density to that of a "vectoroid" which may be always written in the form (12), and Theorems 1 and 2 will also be valid for such ones; the assumption in Theorem 2 remaining formally as stated.

3. Conglomerate and saltus functions

On an (oriented) compact complex V we take a *fixed* decomposition of it into simplices

$$S_1, \cdots, S_t, \tag{18}$$

each co-oriented with V itself. With each S_a, $a = 1, \cdots, t$, we associate a neighborhood U_a of its closure $\bar{S}_a$—which neighborhood may vary from case to case—and in U_a we take any arbitrary function φ_a, hereafter termed a "component", and we envisage an entire set of them:

$$\Phi = \{\varphi_a\}, \qquad a = 1, \cdots, t. \tag{19}$$

Such an object will be called a *conglomerate function*, see [2]. We next take all $(n - 1)$-dimensional faces occurring as boundaries of the simplices (18), each once, denoting them thus:

$$B_1, \cdots, B_s, \tag{20}$$

each in some fixed orientation. For each q, $1 \leqq q \leqq s$, B_q has a positive incidence with precisely one simplex (18) which we denote by $S_{a(q)}$, and a negative incidence with another one which we will denote by $S_{b(q)}$. We now form the (component) function

$$\delta_q(x) = \varphi_{a(q)} - \varphi_{b(q)} \tag{21}$$

which will be defined in a neighborhood N_q of $\bar{B}_q$, namely in

$$N_q = U_{a(q)} \cap U_{b(q)}, \tag{22}$$

and with these components we set up the (conglomerate) object

$$\Delta = \{\delta_q\}, \qquad q = 1, \cdots, s, \tag{23}$$

which will be called the saltus of (19), see again [2].

We may now start directly from an object (23), the components δ_q being defined in a neighborhood N_q of $\bar{B}_q$ each; and such an object will then be called a *saltus function* (per se) if and only if there is some conglomerate function (19) of which it is the saltus. It is very easy to give a necessary and sufficient condition for this to happen, the condition being that if we consider the object (23) as a conglomerate function in its own right on the chain (20) then its saltus in turn shall be identically zero. A function (19) has saltus 0—that is all $\delta_q = 0$—if and only if it corresponds to an "ordinary" one-valued function φ on all of V.

The conglomerate functions form a module (with complex coefficients) if we define linear composition by $\varphi_a = c'\varphi_a' + c''\varphi_a''$, and so do the saltus functions for themselves. Thus far the components φ_a, δ_q were analytically arbitrary and we are now going to impose analytic restrictions which will preserve the module character, and these restrictions will explicitly refer only to the saltus components whether they are viewed per se or as derived from (23). Our decisive restriction is as follows:

PROPERTY *I. Each component δ_q is holomorphic.*

We now take any *fixed* positive definite metric tensor $g_{\alpha\beta^*}$ which is differentiable sufficiently often and we set up the equation

$$g^{\alpha\beta^*} \frac{\partial^2 \Psi}{\partial z_\alpha \, \partial \bar{z}_\beta} = const. \tag{24}$$

Now, if the tensor has the Hermite-Kaehler property

$$\frac{\partial g_{\alpha\beta^*}}{\partial z_\gamma} = \frac{\partial g_{\alpha\beta^*}}{\partial z_\alpha}, \tag{25}$$

(that is to say if our manifold has such a tensor existing on it), then equation (24) is self-adjoint with respect to the volume element

$$dv = g \, dx_1 dy_1 \cdots dx_k dy_k$$

and any saltus function $\{\delta_q\}$ which is real analytic (and thus certainly if it satisfies property I) can be derived from a conglomerate function $\{\psi_a\}$ each component of which is a solution of (24). This was proven in [3], p. 796. If the metric is arbitrary Hermitian, equation (24) is not self-adjoint and the conclusion just stated would have to be re-examined. It probably holds in all cases, but in order to obviate the necessity of proving this for the present, we will make it into an additional postulate, thus:

PROPERTY *II. The saltus $\{\delta_q\}$ can be derived from a conglomerate function $\{\psi_a\}$ whose components are solutions of* (24).

A conglomerate or saltus function with properties I and II will be termed "*solvable*".

THEOREM 3. *Any module of solvable conglomerate functions $\{\Psi_a\}$ in which all components are solutions of* (24) *has a finite basis relative to its submodule in which all components are holomorphic.*

PROOF. Property I implies $\partial/\partial z_\gamma(\Psi_a - \Psi_b) = 0$. Therefore the object

$$\Psi_{\gamma *} = \frac{\partial \Psi}{\partial \bar{z}_\gamma} \tag{26}$$

is uniquely determined, and it also transforms itself according to

$$\Psi'_{\gamma *}(w, \bar{w}) = \frac{\partial \bar{z}_\delta}{\partial \bar{w}_\gamma} \cdot \Psi_{\delta *}(z, \bar{z}). \tag{27}$$

Now, on differentiating (24) we obtain for (26) the system of equations

$$g^{\alpha\beta *} \frac{\partial^2 \Psi_{\gamma *}}{\partial z_\alpha \partial \bar{z}_\beta} + \frac{\partial g^{\alpha\beta *}}{\partial \bar{z}_\gamma} \frac{\partial \Psi_{\beta *}}{\partial z_\alpha} = 0, \tag{28}$$

and by a slight adjustment of the proof to Theorem 1 we obtain the conclusion that the module of objects (26) has a finite basis. Now, $\Psi_{\gamma *} = 0$ means that $\Psi = \Psi_a$ is holomorphic and hence the theorem.

By application of property II it is not hard to deduce from Theorem 3 the following conclusion.

THEOREM 4. (i) *Any module of solvable saltus functions* $\{\Delta\}$ *contains a finite basis* Δ^ρ, $\rho = 1, \cdots, r$, *such that each* Δ *can be written in the form*

$$\Delta = \sum_{\rho=1}^{r} c_\rho \Delta^\rho + \Delta^h \tag{29}$$

where the remaining saltus Δ^h *is derivable from a conglomerate function* $\{\Psi_a\}$ *in which all* Ψ_a *are holomorphic.*

(ii) *Any module of solvable conglomerate functions* $\{\Phi\}$ *contains a finite basis* Φ^ρ, $\rho = 1, \cdots, r$, *such that each* Φ *can be written in the*

$$\Phi = \sum_{\rho=1}^{r} c_\rho \Phi^\rho + \Phi^h \tag{30}$$

where the residual conglomerate function Φ^h *has the same saltus as some solvable conglomerate function* Ψ*—not necessarily in the given module—whose components* Ψ_a *are all holomorphic.*

4. Singularities. Meromorphic conglomerate functions

We will now admit an arbitrary "exceptional" nowhere dense closed set E in V, and we note that a finite number of such sets is again such a set and that if a function δ_q in $N_q - E$ is holomorphic there (whether $N_q - E$ be connected or not) and has an analytic continuation into all of N_q then the continuation (which will be denoted by the same letter δ_q) is a unique one.

We now generalize the concept of a solvable conglomerate function $\{\varphi_a\}$ in the following manner: There is a set E depending on the function such that each φ_a is defined in $U_a - E$ for some neighborhood U_a of $\bar{S}_a$ and if we form the saltus components (21) in $N_q - E$ then each of them has an analytic continuation into all of N_q and the resulting object (23) is a solvable saltus function exactly as was defined before (that is before "singularities" were envisaged). We obviously obtain the following conclusion.

THEOREM 5. *Part* (ii) *of Theorem* 4 *also holds for solvable conglomerate functions with such singularities.*

The most interesting situation arises if we consider meromorphic functions on V. An object (19) is meromorphic if we may put

$$\varphi_a = \frac{f_a(z)}{g_a(z)} \tag{31}$$

with $f_a(z)$, $g_a(z)$ holomorphic in U; the resulting exceptional set E in V being then certainly nowhere dense in V. If a particular object Φ^h of this kind has the same saltus as an object Ψ with holomorphic components Ψ_a then $\Phi^0 \equiv \Phi^h - \Psi$ is a meromorphic conglomerate function with saltus 0, and is thus a meromorphic function on V of the ordinary kind; (but it should be noted that we require the representation (31) relative to a simplicial decomposition (18) which has been fixed beforehand and may not be refined without justification). We thus obtain the following statement.

THEOREM 6. *The module of solvable meromorphic conglomerate functions has a basis* $\{\Phi^\rho\}$ *such that*

$$\Phi = \sum_{\rho=1}^{r} c_\rho \Phi^\rho + \Phi^0 + \Psi, \tag{32}$$

where Φ^0 *is an ordinary meromorphic function and* Ψ *has holomorphic components.*

5. Meromorphic functions on covering spaces

We now take the universal covering space $\tilde{V}$ of V and denote the fundamental group by Γ. On $\tilde{V}$ we introduce a simplicial decomposition which is the "periodic extension" of the one given in V, see [5], and if an (ordinary) function φ on $\tilde{V}$ has the property that the differences

$$\varphi(\gamma P) - \varphi(P) = \varphi^\gamma(P), \qquad \gamma \in \Gamma \tag{33}$$

are holomorphic in V, then by projection onto V we obtain a conglomerate function having the decisive Property I there; and Property II which is then probably fulfilled automatically will be required explicitly. We may also admit "singularities" on a set $\tilde{E}$ on $\tilde{V}$ which is the periodic extension of such a set on V, and altogether the following theorem will ensue.

THEOREM 7. *Any module of functions* $\varphi(P)$ *on* $\tilde{V}$, *with or without "solvable" singularities, for which the differences* (33) *are all holomorphic in* $\tilde{V}$ *contain a finite base* $\varphi^\rho(P)$ *such that*

$$\varphi(P) = \sum_{\rho=1}^{r} c_\rho \varphi^\rho(P) + \varphi^h(P) \tag{34}$$

where $\varphi^h(P)$ *is a function on* $\tilde{V}$ *differing there from a holomorphic one by a difference for which the periods* (33) *are all* 0, *and thus exists on* V *itself.*

The module of meromorphic functions on $\tilde{V}$ with solvable singularities and holomorphic periods (33) *has a finite basis* φ^ρ *such that*

$$\varphi = \sum_{\rho=1}^{r} c_\rho \varphi^\rho + \varphi^0 + \psi \tag{35}$$

where φ^0 *is meromorphic on* V *and* ψ *is holomorphic on* $\tilde{V}$.

6. Tensor fields on covering spaces

In the theory of automorphic functions, one also envisages (scalar) densities which are not strictly periodic under the operations of the fundamental group Γ but reproduce themselves to within a numerical factor of absolute value 1, which factor is then automatically a character of the group Γ; and the statement on the existence of a finite basis also embraced this situation for any one given character. Now, we will extend this result in two directions, firstly by again admitting tensors of arbitrary type (p, q, W) and secondly by admitting not only a group character of Γ, which is a one-dimensional unitary matrix representation of it, but also any other finite-dimensional bounded representation of it, which then, after a minor change if need be, may be always assumed to be a unitary one. We have been commenting on this almost-automorphic (that is almost-periodic with respect to the group Γ) situation on several previous occasions, see [6], p. 259, and [7], and the present remarks are only one more contribution to this promising topic. Since our representation of Γ is to be multi-dimensional, several elements of the module of tensor fields of the given type have to be envisaged simultaneously, and it suggests itself to create a new object consisting of several tensors each, and then to introduce the module of such ones. In designating the tensor (1) we are now constrained to add to the indices (α, β) another one which will be purely enumerative and which will be denoted by λ, $\lambda = 1, \cdots, l$; and in order to simplify the writing of it all we now assume that we are dealing with the special type $(1, 0, 0)$ say, so that (1) is the vector ξ_α. The new object shall then be a set of such vectors

$$\zeta_\alpha = (\xi_{\alpha 1}, \cdots, \xi_{\alpha l}) \tag{36}$$

on $\tilde{V}$, and the module operation shall be

$$c\zeta_\alpha + c'\zeta'_\alpha = (\cdots, c\xi_{\alpha_\lambda} + c'\xi'_{\alpha_\lambda}, \cdots), \tag{37}$$

and for each $\gamma \in \Gamma$ we shall have

$$\xi_{\alpha_\lambda}(\gamma P) = \sum_{\mu=1}^{l} c_\lambda^\mu(\gamma)\xi_{\alpha\mu}(P), \tag{38}$$

where the matrix

$$\{c_\lambda^\mu(\gamma)\} \tag{39}$$

is unitary. We can now take over the proof to Theorem 1 by replacing the quantity (5) by the "norm"

$$\|\xi\|^2 = \sum_{\lambda=1}^{l} g^{\alpha\beta*}\xi_{\alpha\lambda}\overline{\xi_{\beta\lambda}} \tag{40}$$

and choosing the pointset $\{P^\rho\}$ as before, and similarly in the case of real manifolds, and altogether the following theorem will ensue.

THEOREM 8. *If on the universal covering space* $\tilde{V}$ *of* V *with fundamental group* Γ, *we denote the components of a tensor density* (1) *by a cumulative index thus:*

$$\xi_1, \cdots, \xi_N \tag{41}$$

and if for some finite integer l *we introduce sets of such tensors*

$$\{\xi_{A\lambda}\}, \qquad \lambda = 1, \cdots, l. \tag{42}$$

for which we have

$$\xi_{A\lambda}(\gamma P) = \sum_{\mu=1}^{l} c_\lambda^\mu(\gamma)\xi_{A\mu}(P), \tag{43}$$

$\gamma \in \Gamma$, *for some fixed unitary representation* (39) *of* Γ *then any module of such sets with holomorphic components has a finite basis.*

The conclusion also holds for a real M_n *and its covering space* $\tilde{M}_n$, *if the entire set of* N_l *components* $\xi_{A\lambda}$ *(in place of the previous system of* N *components* φ_A) *satisfies a system of equations of the form* (13).

7. Tensor fields with singularities

We will now make an extension of the statements concerning conglomerate functions from scalars to tensors, but the extension will be a very conditional one in that assumptions will be made on whose verifiability we are not able to comment at the present time.

We replace the scalar component φ by a tensor density

$$\varphi_{\varepsilon_1 \cdots \varepsilon_p}^{\eta_1 \cdots \eta_q} \tag{44}$$

of any type (p, q, W) with the same law of transformation as possessed by the object (1), but this time emphatically not having holomorphic components in the general case. We will again introduce a cumulative index which in the present context it will be convenient to exhibit as a superscript, and accordingly we will denote the object (44) thus:

$$\varphi = [\varphi^1, \cdots, \varphi^N]. \tag{45}$$

We will require the analogue to the partial derivative (26), but since we are starting out with a tensor we can introduce it only by way of a covariant derivation,

$$\varphi^A{}_{,\gamma^*} = \frac{\partial \varphi^A}{\partial \bar{z}_\gamma}; \tag{46}$$

therefore the underlying metric tensor $g_{\alpha\beta^*}$ must be such that for the affine connection based on it the covariant differentiation with respect to the complex conjugate variables shall not introduce any additional term of the form $\Gamma_{\beta\gamma^*}^A\varphi^B$; and the only way known to us of bringing this about is to stipulate that the metric have the Kaehler property (25) which we attempted to do without previously but apparently can no longer avoid presently. But a much more serious qualification is yet to come.

The analogue to (24), in order to be given in an invariant form, will have to be

$$g^{\beta*\alpha\cdot}\varphi^{A,}{}_{\beta*;\alpha} = 0. \tag{47}$$

If we write this out we obtain, locally, a set of relations

$$g^{\beta*\alpha}\frac{\partial^2\varphi^A}{\partial\bar{z}_\beta\,\partial z_\alpha} + h_B^{A\beta*}\frac{\partial\varphi^B}{\partial\bar{z}_\beta} = 0, \tag{48}$$

in which the derivatives $\partial\varphi^B/\partial z_\alpha$ themselves do not occur. This is very providential indeed, since on differentiating (48) with respect to $\bar{z}_\gamma$ we will obtain in analogy to relations (28), a set of relations of the form (13) for the quantities (46), and this will be just as convenient as in the case of scalars.

We now define a conglomerate tensor (φ_a^A) on our simplicial decomposition (18) to be a set of "component" tensors (45), just as in the case of scalars, and we now have again to stipulate our two properties. Property I again demands that the differences $\varphi_a^A - \varphi_b^A$ shall be holomorphic,

$$\frac{\partial}{\partial\bar{z}_\gamma}(\varphi_a^A - \varphi_b^A) = 0, \tag{49}$$

and this neither worse nor better than it was before. Property II again demands that every saltus $\delta_q^A = \varphi_{a\,(q)}^A - \varphi_{b\,(q)}^A$ occurring shall again be derivable from another conglomerate tensor (ψ_a^A) whose "components" satisfy the system of relation

$$g^{\beta*\alpha}\psi^A{}_{,\beta*,\alpha} = 0; \tag{50}$$

but whereas in the scalar case this was always automatically fulfilled for a Kaehler metric, and probably always even without one, we do not know whether such is the case for tensors as well, and we are introducing this property rather arbitrarily for what it is worth. But once it has been introduced the argument is the same as before, and we may state

THEOREM 9. *Theorems* 3, 4, 5, 6, 7 *also hold for solvable conglomerate tensor fields of any type.*

As a particular application we obtain the following theorem.

THEOREM 10. *On our* V_{2k} *we consider for* $1 \leqq p \leqq k$ *any external differential form*

$$\sum \xi_{\alpha_1\cdots\alpha_p}\,dz_{\alpha_1}\cdots dz_{\alpha_p} \tag{51}$$

in which the coefficients are meromorphic (that is holomorphic with singularities), and which in the neighborhood of each simplex S_a *is the external derivative of a similar differential form*

$$\sum \eta_{\alpha_1\cdots\alpha_{p-1}}\,dz_{\alpha_1}\cdots dz_{\alpha_{p-1}}, \tag{52}$$

and for which the conglomerate tensor formed with the components (52) *is a "solvable" one.*

The module of such forms (51) *has a finite basis relative to its smallest submodule containing all its exact forms and all its holomorphic forms.*

PRINCETON UNIVERSITY

REFERENCES

[1] BOCHNER, *Analytic mapping of compact Riemann spaces into Euclidean space*, Duke Math. 3 (1927), 339–354.

[2] BOCHNER, *Analytic and meromorphic continuation by means of Green's formula*, Ann. of Math. 44 (1943), 652–673.

[3] BOCHNER, *Vector fields and Ricci curvature*, Bull. Amer. Math. Soc. 52 (1946), 776–797.

[4] BOCHNER, *Curvature in Hermitian metric*, Bull. Amer. Math. Soc. 53 (1947), 179–195.

[5] BOCHNER, *On compact complex manifolds*, J. Indian Math. Soc. n.s. 11 (1947), 1–21.

[6] BOCHNER, *Euler-Poincaré characteristic for locally homogeneous and complex spaces*, Ann. of Math. 51 (1950), 241-261.

[7] BOCHNER, *Vector fields on complex and real manifolds*, Ann. of Math. 52 (1950), 642–649.

[8] BOCHNER, *A new viewpoint in differential geometry*, Canadian J. of Math., to appear.

[9] BOCHNER and DEANE MONTGOMERY, *Groups on analytic manifolds*, Ann. of Math. 48 (1947), 659–669.

[10] L. K. HUA, *On the theory of Fuchtian functions of several variables*, Ann. of Math. 47 (1946), 167–191.

[11] J. LERAY and J. SCHAUDER, *Topologie et equations fonctionelles*, Ann. Ecole norm (3) 51 (1934), 45–78.

[12] L. LICHTENSTEIN, *Nene Beiträge zur Theorie der linearen partiellen Differentialylei-chungen Zweiter Ordnung vom elliptischen Types*, Math. Zeit., 20 (1924), 194–212.

[13] C. L. SIEGEL, *Note on automorphic functions of several variables*, Ann. of Math. 43 (1942), 613–616.

A NEW VIEWPOINT IN DIFFERENTIAL GEOMETRY

S. BOCHNER

1. Introduction. The law of transformation for a general mixed tensor

$$\zeta^{i_1 \dots i_p}{}_{j_1 \dots j_q} \tag{1}$$

is somewhat complicated at first sight, but algebraically it can be stated succinctly in the following manner. If, without regard to the original position of various indices, we denote the components of the tensor in some fixed simple ordering by

$$u^1, \dots, u^N, \tag{2}$$

where

$$N = n^{p+q}, \tag{3}$$

then the law of transformation is the set of linear homogeneous relations

$$u'^A(y) = a^A_B u^B(x) \tag{4}$$

[summation over B from 1 to N] in which

$$\{a^A_B\} \equiv \{a^A_B(y; x)\} \tag{5}$$

is a square matrix of dimension N, and this matrix is the Kronecker product of p n-dimensional matrices

$$\left\{\frac{\partial y_i}{\partial x_j}\right\} \tag{6}$$

and q n-dimensional matrices

$$\left\{\frac{\partial x_j}{\partial y_i}\right\}. \tag{7}$$

Also, if (1) is not a "scalar" tensor but a tensor of weight W we must add to this as a further Kronecker factor the one-dimensional "matrix"

$$\left|\frac{\partial(x)}{\partial(y)}\right|^W, \tag{8}$$

or, if W is an integer, which it need not be, we may add the W one-dimensional factors

$$\frac{\partial(x)}{\partial(y)}, \tag{9}$$

if this be preferrred.

Received August 31, 1950.

If we set down relations (4) with unspecified matrices (5) and examine the properties needed for the erection of a calculus of tensors we find that we first of all require certain properties of "consistency" by which to validate relations (4) intrinsically, and to that purpose the self-explanatory properties

$$a_B^A(z;y)\,a_C^B(y;x) = a_C^A(z;x) \tag{10}$$

$$a_B^A(y;x) = \delta_B^A \tag{11}$$

are obviously sufficient; and the latter ones are verifiable from the mere chain property

$$\frac{\partial z_i}{\partial y_j}\,\frac{\partial y_j}{\partial x_k} = \delta_k^i \tag{12}$$

without using any other feature of (6), say. If, however, we proceed to introduce the interlocking concepts of "affine connection" and "covariant differentiation"—which, after all, are the heart of differential geometry— then the feature of tangentiality as manifest in the partial derivatives (6) is of the very essence, and the matrix (6) itself [or something resembling it] must of necessity be embodied in the calculus. We have found, however, in examining the nature of the ordinary affine connection L_{ij}^k that only one of its lower indices need be interacting with our special matrix (6), or (7) to be more precise, and we have correspondingly generalized it to an object

$$L_{Bi}^A \tag{13}$$

in which the other two indices A, B will be interacting with some other matrix (5), with only properties (10) and (11) presupposed known, which other matrix will be given in addition to, and quite independently from, the classical matrix (6) proper. This separation of algebraic from analytical pre-requisites, as it were, will not only clarify things but also imply a substantial generalization of the theory, and perhaps the most tangible advantage will be as follows.

If an ordinary tensor is symmetric, $\zeta_{ij} = \zeta_{ji}$, then we first of all may view it as an object (2) with $N = n^2$, but it should be noted that the "independent" components ζ_{ij} $(i \leqslant j)$, whose number is

$$N = \tfrac{1}{2}n(n+1),$$

are likewise an object of our kind, and it will have an affine connection entirely of its own to fit its own matrix structure. Similarly, a skew symmetric tensor $\zeta_{i_1 \ldots i_p}$ with $\binom{n}{p}$ independent components corresponding to $i_1 < i_2 < \ldots < i_p$ will be such an object. In particular, for $p = n$, we have *de facto* only one component $\zeta_{1 \ldots n}$, its matrix $a_1^1(y;x)$ being the Jacobian (9), so that from our approach it is entirely indistinguishable from an ordinary scalar density, and our affine connection (13), which in the present case has only components $A = B = 1$ is nothing else but the classical contraction L_{ji}^j, or something differing from it by a vector a_i as we shall see.

2. Vectoroids. We will introduce three positive integers n, N, ν, and their sizes will not be tied in explicitly in any manner whatsoever, although the contexts in which they will be given may imply dependencies per force. Indices $i, j, k, \ldots$ whether upper or lower will range from 1 to n, indices A, B, C from 1 to N, and indices $\alpha, \beta, \gamma, \ldots$ from 1 to ν.

We take an n-dimensional manifold M_n, compact or not (it need not even be separable), and on it a fixed covering by a family or arbitrary local coordinate systems—hereafter termed "allowable"—and we are defining a mathematical object, which we will call a *vectoroid*, in the following manner. In each allowable system it has N coordinate functions (2)—the number N will be called its dimension—and on the intersection of two allowable systems $U_1(x)$, $U_2(y)$ we have relations (4) for a matrix (5) for which (11) is satisfied on the intersection of any $U(x)$ with itself and (10) is satisfied on the intersection of any three systems $U_1(x)$, $U_2(y)$, $U_3(z)$.

An assemblage of matrices with properties (10), (11) will be called a (*matrix*) *structure*. In algebraic parlance we may say that a structure constitutes a matrix "representation" of the "groupoid" of coordinate transformations

$$y_i = f_i(x_i, \ldots, x_n) \qquad i = 1, \ldots, n \tag{14}$$

prevailing on intersections of pairs of allowable systems, and in topological parlance we may add that this generates a space of affinely interlocking N-dimensional linear fibers over M_n as base space and that a vectoroid is a continuous map of the base space into the full fiber bundle, the map of each base point being a point in the fiber over it.

As already stated in the introduction, any ordinary tensor of any weight may be interpreted as a vectoroid, and this includes, for $N = 1$, also scalars of any weight. For a scalar of weight 0, absolute scalar, that is, we have

$$a_1^1(y; x) = 1. \tag{15}$$

In the introduction we have also mentioned the possibility of reducing the dimension N whenever "symmetries" occur, and a general principle for so doing may be stated as follows.

Take a non-singular constant matrix λ_B^A, denoting its inverse by μ_B^A, and transform the structure (5) into

$$b_B^A(y; x) = \lambda_C^A\, a_D^C(y; x) \mu_B^D \tag{16}$$

and, correspondingly, the vectoroid u^A into

$$v^A = \lambda_C^A\, u^C. \tag{17}$$

If now it so happens that the new structure (16) fully decomposes into structures of dimension N_1, N_2, $N_1 + N_2 = N$, then in a suitably normalization of indices we may introduce the system of relations

$$\lambda_C^A\, u^C(x) = 0, \qquad A = N_1 + 1, \ldots, N \tag{18}$$

as "symmetry conditions" to be imposed on the original vectoroid (2); and if they happen to be fulfilled, then there exists a vectoroid of dimension N_1 only, having the components

$$v^1, \ldots, v^{N_1}, \tag{19}$$

and this new vectoroid may be said to be a fair substitute for the original one.

If a structure (5) has been given and is held fast, then the vectoroid (2) previously introduced is a "contravariant" one, and its "covariant" counterpart is an object t_A for which the law of transformation is

$$t'_A(y) = a_A^B(x; y) t_B(x), \tag{20}$$

or, what is the same

$$a_A^B(y; x)\, t'_B(y) = t_A(x). \tag{21}$$

If, however, we introduce the transposed inverse matrix

$$\bar{a}_B^A(y; x) = a_A^B(x; y) \tag{22}$$

then t_A can be easily identified with a vectoroid which is contravariant with regard to the new structure, and thus the contrast between contravariant and covariant is only a relative one.

Next, if we are given several of our matrix structures, equal or not, then any Kronecker product of them is again a structure, and a vectoroid pertaining to such a structure will sometimes be denoted by the symbol

$$t_{A_1 \ldots A_p}{}^{B_1 \ldots B_q}, \tag{23}$$

and then termed a *tensoroid*, whenever its structure is a product of q straight factors and p transposed inverse factors given. Also, if, for instance, in (23) the lower index A_1 and the upper index B_1 both pertain to the same structure, then we form the contraction

$$t_{A_2 \ldots A_p}{}^{B_2 \ldots B_q} = t_{CA_2 \ldots A_p}{}^{CB_2 \ldots B_q} \tag{24}$$

as usual. In particular, if t_A, u^A belong to the same structure then

$$t_A u^A \tag{25}$$

is an absolute scalar.

If in addition to the structure (5) we are given a second structure

$$b_\beta^\alpha \qquad \alpha, \beta = 1, \ldots, \nu, \tag{26}$$

if $\nu = N$, and if we are given a tensoroid $h_{A\alpha}$ for which

$$\det |h_{A\alpha}| \neq 0, \tag{27}$$

then there exists a tensoroid $h^{A\alpha}$ for which we have

$$h_{C\alpha} h^{C\beta} = \delta_\alpha^\beta, \quad h_{A\gamma} h^{B\gamma} = \delta_A^B; \tag{28}$$

and if furthermore the two structures are identical then the pair of tensoroids $h_{A\alpha}$, $h^{A\alpha}$, which we will now denote by

$$h_{AB}, \quad h^{AB}, \tag{29}$$

may be used for pulling indices pertaining to the structure (5) up and down as usual. We will always assume that the tensoroids (29) are symmetric, although for the shifting of indices alone this requirement would not be strictly needed.

Finally we note that there always exists a matrix structure defined by the Kronecker symbol, thus

$$b^{\alpha}_{\beta}(y; x) = \delta^{\alpha}_{\beta} \qquad \alpha, \beta, = 1, \ldots, \nu \tag{30}$$

and that a contravariant vectoroid pertaining to it has the law of transformation

$$u'^{\alpha}(y) = u^{\alpha}(x), \tag{31}$$

and thus consists of ν individual absolute scalars, and similarly for covariant vectoroids. If now we form the Kronecker product of an arbitrary structure (5) with the special structure (30), then a tensoroid $u^{A\alpha}$ pertaining to the product is a set of ν vectoroids u^A pertaining to (5) each, and any set of ν such vectoroids may be so viewed. In other words, from our general approach there is only a relative distinction between indices which are vectorial and those which are enumerative, and this merger of the two types of indices has some technical advantages.

3. Covariant differentiation. In the right side of (5) we have introduced both symbols y and x simultaneously and in a given order and in this way the consistency rules (10) and (11) have become self-explanatory. However, analytically, the individual components of the tensor (5) will be viewed as functions either of the variables $x = (x_1, \ldots, x_n)$ or of the variables $y = (y_1, \ldots, y_n)$ by themselves, depending on whether we view the intersection of the neighbourhoods $U_1(x)$, $U_2(y)$ as being part of the one or of the other; and in forming partial derivatives with respect to the variables x_i or y_j the variables exhibited will be those underlying.

If we are given a structure (5), and if the functions (5) and (14) belong to differentiability class C^1, then we define an affine connection

$$L^{B}_{Ai} \qquad A, B = 1, \ldots, N; i = 1, \ldots, n \tag{32}$$

by the law of transformation

$$a^{B}_{A}(y; x) \frac{\partial y_j}{\partial x_i} L'^{C}_{Bj}(y) = - \frac{\partial a^{C}_{A}(y; x)}{\partial x_i} + a^{C}_{B}(y; x)\, L^{B}_{Ai}(x), \tag{33}$$

and the following theorem can be easily verified.

THEOREM 1. *If t_A is a covariant vectoroid then the components*

(34) $$t_{A,i} = \frac{\partial t_A}{\partial x_i} - t_B L^B_{Ai}$$

constitute a tensoroid for the Kronecker product of (5) *and* (6).

Also, if (34) *is to constitute a tensoroid for arbitrary vectoroids* t_A *locally, then* (33) *must be satisfied.*

Furthermore, if u^A *is contravariant vectoroid then*

$$u^A{}_{,i} = \frac{\partial u^A}{\partial x_i} + u^B L^A_{Bi}$$

is a mixed tensoroid, that is to say, if L^B_{Ai} *is an affine connection for a structure* (5) *then* $-L^A_{Bi}$ *is one for the structure* (22).

It should be noted, however, that given a structure, the affine connection (32), if one exists, is determined up to an arbitrary additive tensor T^B_{Ai} only, so that associating $-L^A_{Bi}$ with the structure (22) is a deliberate act of normalization, pursuant to the fixed normalization of (32) for (5) itself.

If there is given a second structure (26) and if it has an affine connection $\Lambda^\beta_{\alpha i}$, then the Kronecker product of (5) and (26) admits the affine connection

(35) $$L^{(B\beta)}_{(A\alpha)i} = \delta^\beta_\alpha L^B_{Ai} + \delta^B_A \Lambda^\beta_{\alpha i},$$

and the resulting formula

(36) $$t_{A\alpha,i} = \frac{\partial t_{A\alpha}}{\partial x_i} - t_{B\alpha} L^B_{Ai} - t_{A\beta}\Lambda^\beta_{\alpha i}$$

is the generalization of a classical one; and similarly for mixed tensoroids.

For the special structure (30), relations (33) reduce to

(37) $$\frac{\partial y_j}{\partial x_i} \Lambda'^\gamma_{\alpha j} = \Lambda^\gamma_{\alpha i},$$

and that means that $\Lambda^\gamma_{\alpha i}$ is a vector with regard to the index i, with α and γ being enumerative indices. In particular (37) will be satisfied by

(38) $$\Lambda^\gamma_{\alpha i} = 0,$$

in which case (36) will reduce to the expression

(39) $$t_{A\alpha,i} = \frac{\partial t_{A\alpha}}{\partial x_i} - t_{B\alpha} L^B_{Ai},$$

but it must be stated that (38) is a deliberate "canonical" normalization for the solution of (37) and not the only one possible.

We are turning to the general expression (36). If we have

(40) $$t_{A\alpha,i} = 0$$

that is

$$\frac{\partial t_{A\alpha}}{\partial x_i} - t_{A\beta}\Lambda^{\beta}_{\alpha i} = t_{B\alpha} L^{B}_{Ai} \tag{41}$$

and if, for $\nu = N$, $t_{A\alpha}$ has a transposed inverse $t^{A\alpha}$, then (34) implies

$$L^{C}_{Ai} = t^{C\alpha} \frac{\partial t_{A\alpha}}{\partial x_i} - t^{C\alpha} t_{A\beta}\Lambda^{\beta}_{\alpha i}, \tag{42}$$

and thus L^{C}_{Ai} is uniquely determined by $\Lambda^{\beta}_{\alpha i}$. In particular for (38) we obtain

$$L^{C}_{Ai} = t^{C\alpha} \frac{\partial t_{A\alpha}}{\partial x_i}, \tag{43}$$

and this is a generalization from vectors to vectoroids of the known theorem that if there are given n independent parallel vector fields, the affine connection is uniquely determined hereby.

Finally, if we are also given a (non-symmetric) affine connection Γ^{k}_{ij} to fit the structure (6), then if everything given is twice differentiable, we can form the second covariant derivative

$$t_{A,i,j} = \frac{\partial t_{A,i}}{\partial x_j} - t_{B,i} L^{B}_{Aj} - t_{A,k} \Gamma^{k}_{ij}, \tag{44}$$

and we obtain

$$t_{A,i,j} - t_{A,j,i} = t_B L^{B}_{Aij}, \tag{45}$$

where

$$L^{B}_{Aij} = \frac{\partial L^{B}_{Aj}}{\partial x_i} - \frac{\partial L^{B}_{Ai}}{\partial x_j} + L^{B}_{Ci} L^{C}_{Aj} - L^{B}_{Cj} L^{C}_{Ai} + t_{A,k}(\Gamma^{k}_{ji} - \Gamma^{k}_{ij}) \tag{46}$$

is a tensoroid. We always have

$$L^{B}_{Aij} = - L^{B}_{Aji}$$

and for

$$\Gamma^{k}_{ij} = \Gamma^{k}_{ji}$$

we also have

$$L^{B}_{Aij,k} + L^{B}_{Ajk,i} + L^{B}_{Aki,j} = 0.$$

4. Remark on affine connections. Affine connections are a particular case of a general type of coefficients in invariant systems of linear partial differential operators, as we are going to describe briefly.

Consider for an ordinary scalar t the expression

$$c^{ij} \frac{\partial^2 t}{\partial x_i \partial x_j} + c^{k} \frac{\partial t}{\partial x_k} + ct \tag{47}$$

with $c^{ij} = c^{ji}$. If this is to be formally invariant under the transformation (14) then we must have

$$c^{ij}(x) = c'^{pq}(y) \frac{\partial x_i}{\partial y_p} \frac{\partial x_j}{\partial y_q}, \tag{48}$$

$$c^k(x) = c'^{pq}(y) \frac{\partial^2 x_k}{\partial y_p \partial y_q} + c'^r(y) \frac{\partial x_k}{\partial y_r}, \tag{49}$$

$$c(x) = c(y), \tag{50}$$

and it can be easily verified that the entire system of coefficients

$$\{c^{ij}, c^k, c\}$$

constitutes a vectoroid for an appropriate structure. However the sub-system $\{c^{ij}\}$ forms a vectoroid by itself, and if we insert some special values for this subset, and if we ignore the quantity c itself which is a scalar "by accident," then the remaining coefficients c^k transform themselves by relation (49) after the manner of an affine connection.

After this preliminary instance we will now set up a very comprehensive situation. If t_A is an unspecified vectoroid for a given structure (5) and if we demand that for some given integer l the system of partial differential operators

$$O_\alpha(t) = \sum_{0 \leqslant l_1 + \ldots + l_{\lambda_1} \leqslant l} c_\alpha^{Al_1 \ldots l_\lambda} \frac{\partial^{l_1 + \ldots l_\lambda} t_A}{\partial x_1^{l_1} \ldots \partial x_\lambda^{l_\lambda}}, \qquad \alpha = 1, \ldots, \nu, \tag{51}$$

shall form a vectoroid pertaining to some other structure (26), identically in all t_A, then the set of all coefficents occuring in all individual operators (51) form a vectoroid for a suitable structure. However the subset of those coefficients for which

$$l_1 + \ldots + l_\lambda = l \tag{52}$$

form a vectoroid by themselves, and if we insert some special values for this subset then the remaining "additional" coefficients transform in the manner of an affine connection.

In the particular case (34) we have $l = 1$ and the structure (26) is the Kronecker product of (5) and (6). Thus every index $\alpha = \alpha_0$ is a pair of indices $A = A_0$, $i = i_0$ and the highest coefficients corresponding to (52) have been specialized in the following "invariant" manner: if $A = A_0, i = i_0$ then $c_\alpha^{Al_1 \ldots l_\lambda} = 1$, otherwise $= 0$.

5. Compact manifolds. On a compact M_n we stipulate the following data, all of differentiability class C^2: a matrix structure (5), with an affine connection (32), a positive definite symmetric tensor (29) with

$$h^{AB}{}_{,i} = 0, \tag{53}$$

that is

$$\frac{\partial h^{AB}}{\partial x_i} + h^{CB} L^A_{Ci} + h^{AC} L^B_{Ci} = 0, \tag{54}$$

an ordinary (non-symmetric) affine connection

$$\Gamma^k_{ij}, \tag{55}$$

and an ordinary positive definite symmetric tensor g^{ij} which need *not* be related to (55).

We note that the Kronecker symbol δ^B_A is a mixed tensoroid for (5), and if for any mixed tensoroid t^B_A we adopt the definition

$$t^B_{A,i} = \frac{\partial t^B_A}{\partial x_i} - t^B_C L^C_{Ai} + t^C_A L^B_{Ci}, \tag{56}$$

as we will, then we obtain

$$\delta^B_{A,i} = 0, \tag{57}$$

and on recalling the relation

$$h_{AC}h^{BC} = \delta^B_A,$$

we now obtain from (53) and (57) the further inference

$$h_{AB,i} = 0. \tag{58}$$

Therefore, if we will employ the tensor (29) for shifting indices up and down, this operation of shifting will be commutative with covariant differentiation. Thus, if for a vectoroid t_A we form the scalar "square length"

$$\phi = h^{AB}t_At_B, \tag{59}$$

then we obtain

$$\frac{\partial \phi}{\partial x_i} \equiv \phi_{,i} = 2h^{AB}t_{A,i}t_B = t_{A,i}t^A = t_At^A{}_{,i}. \tag{60}$$

If we differentiate once more we obtain

$$\tfrac{1}{2}\phi_{,i,j} = h^{AB}t_{A,i}t_{B,j} + h^{AB}t_{A,i,j}t_B, \tag{61}$$

where of course

$$\phi_{,i,j} \equiv \frac{\partial^2 \phi}{\partial x_i \partial x_j} - \Gamma^k_{ij}\frac{\partial \phi}{\partial x_k}, \tag{62}$$

and $t_{A,i,j}$ is given by (41). And finally, if we introduce the invariant operator

$$\Delta\phi = g^{ij}\,\phi_{i,j} \equiv g^{ij}\frac{\partial^2 \phi}{\partial x_i \partial x_j} - g^{ij}\,\Gamma^k_{ij}\frac{\partial \phi}{\partial x_k}, \tag{63}$$

we obtain the equation

$$\tfrac{1}{2}\Delta\phi = g^{ij}h^{AB}t_{A,i}t_{B,j} + h^{AB}(g^{ij}t_{A,i,j})t_B = T_1 + T_2 \tag{64}$$

which we will now analyze.

As on previous occasions[1] we will use the lemma that, since M_n is compact,

[1]See our papers: *Vector fields and Ricci-curvature*, Bull. Amer. Math. Soc., vol. 52 (1946), 776-797, and *Curvature and Betti numbers*, Ann. of Math., vol. 49 (1948), 379-390 and II in vol. 50 (1949), 77-93.

we cannot have

(65) $$\Delta\phi \geqslant 0$$

on M_n unless we have

(66) $$\Delta\phi = 0.$$

Furthermore we have identically $T_1 \geqslant 0$, and $T_1 = 0$ if and only if $t_{A,i} = 0$. Thus, if for special reasons we have

(67) $$T_2 \geqslant 0,$$

then we must of necessity have

(68) $$T_2 = 0$$

and also

(69) $$t_{A,i} = 0.$$

In order to bring about (67) we introduce some fixed symmetric matrix S_{AB} and set up the system of partial differential equations

(70) $$g^{ij}t_{A,i,j} = S_A^B t_B (\equiv S_{AB}t^B),$$

and since this implies

(71) $$T_2 = S_{AB}t^A t^B,$$

we obtain the following conclusion.

THEOREM 2. *If on our compact M_n we are given a symmetric matrix S_{AB} which is positive definite then there is no vectoroid t_A other than* 0 *which satisfies the equations* (70).

If the matrix is only semi-definite then the only solutions of (70) *are those for which we have*

$$t_{A,i} = 0, \qquad S_A^B t_B = 0,$$

simultaneously.

If ζ_a is an ordinary vector field, and if it is harmonic (curl ζ = div ζ = 0) then it satisfies the equations

$$g^{ij}\zeta_{a,i,j} = - R_{ab}\zeta^b$$

where R_{ab} is the Ricci tensor based on g^{ij}, provided the affine connection is the ordinary one; and if the vector is a Killing vector ($\zeta_{i,j} + \zeta_{j,i} = 0$) then it satisfies the alternate equations

$$g^{ij}\zeta_{a,i,j} = R_{ab}\zeta^b.$$

Thus Theorem 2 includes the following statements made previously among others.[1] If $- R_{ab}$ is positive definite there exists no harmonic vector, and if it is negative definite there exists no Killing vector, and if it is semi-definite then in either case we must have

$$\zeta_{a,i} = 0, \qquad R_{ab}\zeta^b = 0,$$

simultaneously.

In the papers cited we also had results bearing on tensors of several indices. Some of them, in particular those referring to complex manifolds could also be subsumed under Theorem 2, but the analysis would have to be rather detailed and would not be profitable in the end. However, we would like to terminate with giving a sample of another type of Theorem[2] which is especially adapted to complex spaces and which actually gave us the impetus for devising the generalization from vector to vectoroid. The theorem is largely preanalytic in the sense that no affine connection will be involved or even assumed existing, and although the theorem originated in algebraic geometry[2] no algebraic structure will be involved either.

Let M_n be a compact space of n complex (that is $2n$ real) variables for which the functions (14) are holomorphic functions, that is local power series, from the complex variables x_i to the complex variables y_j, and let the matrix structure (5) be likewise holomorphic in either set of variables, and let t_A, u^A be holomorphic fields of vectoroids on all of M_n. The contraction (25) is an absolute scalar and also holomorphic, and being so on a compact manifold it must be a numerical constant

$$t_A u^A = c. \tag{72}$$

If now there are given $N + 1$ covariant vectoroids $t^\alpha_A (\alpha = 1, \ldots, N + 1)$ then a contravariant vectoroid u^A, if existing, has to satisfy a set of relations

$$t^\alpha_A u^A = c^\alpha, \qquad \alpha = 1, \ldots, N + 1 \tag{73}$$

and the following theorem ensues.[2]

THEOREM 3. *If on a compact complex M_n there exist $N + 1$ holomorphic vectoroids t_N, if the rank of the N by $N + 1$ matrix*

$$t^\alpha_N$$

is somewhere N and if for no set of constants c^α, not all zero, is the determinant

$$\det |t^\alpha_1, \ldots, t^\alpha_N, c^\alpha_N| \qquad \alpha = 1, \ldots, N + 1$$

identically 0, then there exists on M_n no holomorphic contravariant vectoroid u^A whatsoever.

If for each $\beta = 1, \ldots, N + 1$ we introduce the determinant

$$D^\beta = \det |t^1_A, \ldots, t^{\beta-1}_A, t^{\beta+1}_A, \ldots, t^{N+1}_A| \qquad A = 1, \ldots, N$$

then each D^β is a holomorphic density with the density factor

$$\det a^A_B(x; y) \qquad A, B = 1, \ldots, N$$

and the conditions of the theorem can also be stated in this way that these $N+1$ densities shall not be linearly dependent for constant complex coefficients.

Princeton University

[2] *Vector fields on complex and real manifolds*, Ann. of Math. vol. 52 (1950), 642-649.

Reprinted from Vol. II, PROCEEDINGS OF THE INTERNATIONAL CONGRESS OF MATHEMATICIANS, 1950
Printed in U.S.A.

LAPLACE OPERATOR ON MANIFOLDS

S. BOCHNER

During the last fifteen years or so the so-called Laplace-Beltrami operator has been steadily gaining in importance and prominence in many areas of analysis, and I shall try to summarize here some of its applications with which I am familiar.

We take a compact differentiable manifold M. A Laplacian on it has the form

$$\Delta f = \frac{1}{g^{1/2}} \frac{\partial}{\partial x_i}\left(g^{1/2} g^{ij} \frac{\partial f}{\partial x_j}\right) \tag{1}$$

$$\equiv g^{ij} \frac{\partial^2 f}{\partial x_i \partial x_j} - g^{ij}\Gamma^k_{ij} \frac{\partial f}{\partial x_k}$$

where g^{ij} is a positive definite tensor field (without any "singularities") and g is the determinant of its inverse g_{ij}. The quantity $g^{1/2}$ is a scalar density. If we replace it by any other scalar density, and, correspondingly, replace (1) by the more general expression

$$Lf = \frac{1}{b} \frac{\partial}{\partial x_i}\left(b g^{ij} \frac{\partial f}{\partial x_j}\right) \tag{2}$$

$$\equiv g^{ij} \frac{\partial^2 f}{\partial x_i \partial x_j} + \frac{1}{b} \frac{\partial}{\partial x_i}(b g^{ij}) \frac{\partial f}{\partial x_j},$$

then the latter operator is self-adjoint with regard to the volume element

$$dv = b\,dx_1 \cdots dx_n, \tag{3}$$

in the sense that for any two functions φ, ψ both differentiable twice, we have

$$\int_M (\varphi L\psi - \psi L\varphi)\, dv = 0.$$

Conversely, if we envisage the most general elliptic operator having in each coordinate system the form

$$g^{ij} \frac{\partial^2 f}{\partial x_i \partial x_j} + a^k \frac{\partial f}{\partial x_k} \tag{4}$$

and if we demand that it be self-adjoint, in the manner just stated, with respect to a prescribed volume element (3), then we must have

$$a^k \equiv \frac{1}{b} \frac{\partial}{\partial x_i}(b g^{ik})$$

and (4) has of necessity the form (2).

PART I. VECTOR FIELDS

We shall first very briefly touch upon an application to differential geometry in which the only property of the Laplacian used explicitly will be the following lemma.

LEMMA 1. *If for a scalar function $f(x)$ on our compact manifold we have*

$$\Delta f \geqq 0 \tag{5}$$

everywhere, then we must have equality everywhere, that is,

$$\Delta f \equiv 0. \tag{6}$$

Now, the expression (1) can be written as

$$\Delta f = g^{ij} f_{,i,j} \tag{7}$$

where $f_{,i,j}$ is the second covariant derivative as formed with the Christoffel symbols based on g_{ij}. If now we take an arbitrary vector field ξ_a on M, form its square length

$$f = g^{ab} \xi_a \xi_b \tag{8}$$

and apply the Laplacian to it, then we obtain

$$\tfrac{1}{2}\Delta f = g^{ij} g^{ab} \xi_{a,i} \xi_{b,j} + g^{ab}(g^{ij} \xi_{a,i,j}) \xi_b . \tag{9}$$

The first term on the right is always $\geqq 0$, and if by chance the second is also $\geqq 0$, then both must be 0 by the lemma. We now assume that ξ_a is the solution of an equation

$$g^{ij} \xi_{a,i,j} = A_a^c \xi_c \tag{10}$$

in which A_{ac} is a given symmetric tensor, so that the second term in (9) has then the value

$$A^{bc} \xi_b \xi_c , \tag{11}$$

and we are led to the following conclusion.

THEOREM 1. *If our symmetric tensor A_{ac} is strictly positive definite, then there is no solution of the differential equation* (10) *other than* $\xi_a = 0$.

If a vector field is harmonic (div $\xi = 0$ and curl $\xi = 0$), then it satisfies (10) with A_{ac} being $-R_{ac}$, where R_{ac} is the Ricci-tensor based on g_{ij}; if it is a Killing tensor ($\xi_{i,j} + \xi_{j,i} = 0$), then it satisfies it with $A_{ac} = R_{ac}$. We thus obtain the following theorem.

THEOREM 2. *If $-R_{ac}$ is positive definite (sphere, etc.), then there is no harmonic vector field, that is, the first Betti number is 0, and if $-R_{ac}$ is negative definite (constant negative curvature, etc.), then there is no Killing vector, and there exists no continuous group of isometries.*[1]

In the two-dimensional case there is even no continuous group of conformal homeomorphisms. In particular, for a closed Riemann surface of the classical type, of genus $p > 1$, there is only a finite number of conformal homeomorphisms, since by uniformizing it into the unit circle the surface acquires a non-Euclidean metric for which $-R_{ac} < 0$.[1]

More generally, in the case of several complex variables, no matter how many, if a compact manifold can be uniformized by a bounded domain in E_{2k}, then on such a compact manifold there exist only a finite number of complex isomorphisms. Algebraically put, if the complex domain is an algebraic variety and if we introduce the algebraic field of meromorphic functions on it, then the field has only a finite number of automorphisms.[2]

There are more complicated conclusions of this type involving higher Betti numbers and tensors of arbitrary order for both real and complex manifolds, compact ones or covering spaces of compact ones, but the principal point of reasoning is always Lemma 1 or a "localized" substitute thereof.[3]

PART II. ANALYTIC IMBEDDING

If we introduce the Hilbert space of L_2-functions relative to the volume element (3), then our formal operator (1) or, more generally, the formal operator (2) can be completed (that is, closed) to an operator which is self-adjoint, that is hypermaximal, in the technical sense,[4] and for the present we use only the following operational property of it.

It has a pure point-spectrum $\{-\rho_r\}$, and we thus obtain in our Hilbert space of L_2-functions a basis $\{\psi_r\}$ whose elements are twice differentiable solutions of the partial differential equations

$$-\Delta\psi_r = \rho_r\psi_r, \tag{12}$$

and we apply this in the following set-up.

Assume that M is a real analytic space. Nothing is known in general about the existence of analytic functions, scalars, or tensors, in the large. Assume, however, that we are also given the existence of an analytic positive definite tensor field g^{ij}, or what is the same, of an analytic Riemannian line-element

$$ds^2 = g_{ij}dx^i dx^j, \tag{13}$$

[1] *Vector fields and Ricci curvature*, Bull. Amer. Math. Soc. vol. 52 (1946) pp. 776–797.

[2] *On compact complex manifolds*, J. Indian Math. Soc. vol. 51 (1947) pp. 1–21.

[3] *Curvature and Betti numbers* (I) *and* II, Ann. of Math. vol. 49 (1948) pp. 379–390; vol. 50 (1949) pp. 77–93.

[4] *Analytic mapping of compact Riemann spaces into Euclidean space*, Duke Math. J. vol. 3 (1937) pp. 339–354.

without any singularities, that is. We then envisage the equations (12), and since all its coefficients are analytic, then by a theorem originally due to Hadamard, the functions ψ_r themselves are likewise analytic. Thus the one analytic tensor field g_{ij} known to exist generates a multitude of analytic scalar functions $\{\psi_r\}$, and it can be shown that the latter are dense in every differentiability class C^s on M in the following very precise sense.

LEMMA 2. *To any f in C^s there exists a sequence of finite linear combinations of the ψ_r which converge uniformly towards f, together with all partial derivatives of order $\leqq s$.*

By a general theorem of H. Whitney, our M_k can be topologically imbedded (C^1) in the Euclidean E_{2k+1}, and by applying Lemma 2 for $s = 1$ we see that by a small deformation Whitney's imbedding can be made into an analytic one.

Also, once this analytic imbedding has been achieved the previous statement for arbitrary s (including $s = \infty$) follows trivially, and not only for scalars, but for arbitrary tensors and tensor densities

$$\xi^{b_1 \cdots b_q}_{a_1 \cdots a_p}$$

as well. That is to say, the following conclusion holds.

THEOREM 3. *If on a compact analytic M_k there exists an analytic positive definite g_{ij}, then M_k can be mapped analytically topologically into the Euclidean E_{2k+1}, and in any type of scalars or tensors of differentiability class C^s, the analytic ones are C^s-dense.*[4]

PART III. STOCHASTIC PROCESSES[5]

We are returning to a differentiable M, and we are stating the operational properties of our operators much more fully than heretofore. The operator $-Lf$ has distinct eigenvalues λ_r such that

$$0 = \lambda_0 < \lambda_1 < \lambda_2 < \cdots \to \infty. \tag{14}$$

The smallest eigenvalue has multiplicity 1, and every other λ_r has a finite multiplicity μ_r at most. There is a complete orthonormal set of eigenfunctions $\varphi_{r\mu}$ pertaining to class C^2 (with φ_{01} being a constant) such that the equations

$$-L\varphi_r = \lambda_r \varphi_{r\mu}, \qquad \mu = 1, \cdots, \mu_r, \tag{15}$$

$r = 0, 1, \cdots$, are satisfied in the ordinary sense. If we put

$$g_r(x, y) = \sum_{\mu=1}^{\mu_r} \varphi_{r\mu}(x)\varphi_{r\mu}(y) \tag{16}$$

and introduce the kernel function

[5] In connection with parts III and IV see the author's paper *Quasi-analytic functions, Laplace operator, positive kernels*, Ann. of Math. vol. 51 (1950) pp. 68–91.

$$G(t; x, y) = \sum_{r=0}^{\infty} e^{-t\lambda_r} g_r(x, y) \tag{17}$$

for

$$0 < t < \infty, \tag{18}$$

then it has the important property

$$G(t; x, y) \geqq 0 \tag{19}$$

first established by Kolmogoroff in 1933,[6] and also the properties

$$\int_M G(t; x, y)\, dv = 1 \tag{20}$$

$$\int_M G(t_1 ; x, y) G(t_2 ; y, z)\, dv_y = G(t_1 + t_2 ; x, z), \tag{21}$$

which three properties make it into a so-called stationary stochastic process on M. Also, if for fixed y we denote $f(t; x, y)$ by $f(t; x)$, then it satisfies the "heat equation"

$$L_x f(t; x) = \frac{\partial f}{\partial t}, \tag{22}$$

the left side of (22) being uniquely determined by it, and if we take an arbitrary solution of (22) having boundary values $f(0; x) \equiv f(x)$, then its values for $t > 0$ are given by the formula

$$f(t; x) = \int_M f(0; y) G(t; x, y)\, dv_y .^7 \tag{23}$$

We next take in $0 \leqq \lambda < \infty$ a continuous function $\Phi(\lambda)$, $\Phi(0) = 0$, such that for every $t > 0$ we have a Stieltjes representation

$$e^{-t\Phi(\lambda)} = \sum e^{-\rho\lambda} d\gamma_t(\rho) \tag{24}$$

with $d\gamma_t(\rho) \geqq 0$ [that is, $\gamma_t(\rho)$ monotonely nondecreasing in $0 \leqq \rho \leqq +\infty$], and this amounts to assuming that we have $\Phi(\lambda) > 0$, $(-1)^{n-1}\Phi^{(n)}(\lambda) > 0$, $n = 1, 2, \cdots$, in $0 < \lambda < \infty$. If we now set up the expression

[6] A. Kolmogoroff, *Zur Theorie der stetigen zufälligen Prozesse*, Math. Ann. vol. 108 (1933) pp. 149–160.

[7] In the work of Kolmogoroff the self-adjointness of the operator is not required. To this compare K. Yosida, *Integration of Fokker-Planck's equation in a compact Riemannian space*, Arkiv för Matematik vol. 1 (1949) no. 9. However, under self-adjointness, the kernel with the properties (14)–(21) has the additional property of converging towards "identity" as the time variable tends to infinity. It would be interesting to decide to what extent self-adjointness is also necessary for this to take place. See also the author's note, *Diffusion equation and stochastic processes*, Proc. Nat. Acad. Sci. U. S. A. vol. 35 (1949) pp. 368–370.

$$G_\Phi(t; x, y) = \int_0^\infty e^{-t\Phi(\lambda_n)} g_n(x, y), \tag{25}$$

then owing to the representation

$$G_\Phi(t; x, y) = \int_0^\infty G(\rho; x, y) d\gamma_t(\rho) \tag{26}$$

it also satisfies the "stochastic" properties (19), (20), (21); and instead of (22) it satisfies the "heat equation"

$$-\Phi(-L)f = \frac{\partial f}{\partial t} \tag{27}$$

where $\Phi(-L)$ is the self-adjoint operator resulting from applying the numerical function $\Phi(\lambda)$ to the operator $-L$.

For $\Phi(\lambda) = \lambda$, G_Φ is the original kernel itself, and other very important cases arise on putting $\Phi(\lambda) = \lambda^\sigma$, $0 < \sigma < 1$. For instance, if M is the ordinary circle $-\pi \leqq x < \pi$, and (22) is

$$\frac{d^2 f}{dx^2} = \frac{\partial f}{\partial t}, \tag{28}$$

then we have $G(t; x, y) = G(t; x - y)$ where

$$G(t; x) = \left(\frac{\pi}{t}\right)^{1/2} \sum_{m=-\infty}^{\infty} e^{-(2\pi m - x)^2/t} \tag{29}$$

and

$$G_{\lambda^{1/2}}(t; x) = \frac{1 - e^{-2t}}{1 - 2e^{-t}\cos x + e^{-2t}}. \tag{30}$$

We now take on our M a fixed point P_0 and we consider on M all possible paths emanating from it, $x = P(t)$, $0 \leqq t \leqq 1$, $P(0) = P_0$, whether continuous or not. Now, the function $G(t; x, y)$, or any other one of our functions $G_\Phi(t; x, y)$, since it satisfies (19)–(21), can be used, by a fixed procedure, to introduce a probability (that is, a Lebesque measure of total measure 1) into the space of all such paths,[8] and we are going to state two properties of this probability, the first due to Paul Lévy,[9] and the second to Paul Lévy[9] and the author.[10]

(i) In the original case $\Phi(\lambda) = \lambda$, *and only in this case*, almost all paths are continuous in $0 \leqq t \leqq 1$.

(ii) However for $\Phi(\lambda) = \lambda$, and more generally for $\Phi(\lambda) = \lambda^\sigma$, $1/2 \leqq \sigma < 1$, almost all paths are *not* rectifiable, and thus in particular will have simple discontinuities only. More generally, if

[8] *Partial ordering in theory of stochastic processes*, Proc. Nat. Acad. Sci. U. S. A. vol. 36 (1950) pp. 439–443.

[9] P. Lévy, *Théorie de l'addition des variables aléatoires*, Paris, 1937, chapter VII, and in particular pp. 180, 199, 201.

[10] *Stochastic processes*, Ann. of Math. vol. 48 (1947) pp. 1014–1060, in particular pp. 1031–1037.

(31) $$\int_1^\infty \frac{\Phi(\lambda)}{\lambda^{3/2}}\, d\lambda < \infty,$$

then almost all paths are rectifiable, and if

(32) $$\int_1^\infty \frac{\Phi(\lambda)}{\lambda^{3/2}}\, d\lambda = \infty,$$

they are not.

Part IV. Quasi-analytic functions

We again assume that the space M and the coefficients of Lf are analytic or at least infinitely often differentiable. Any (continuous) function $f(x) \equiv f(0; x)$ has an expansion

(33) $$f(x) \sim \sum_{r=0}^{\infty} \sum_{\mu=1}^{\mu_r} c_{r\mu}\, \varphi_{r\mu},$$

and on putting

(34) $$f_r(x) = \sum_{\mu=1}^{\mu_r} c_{r\mu}\, \varphi_{r\mu}$$

we write for this

(35) $$f(x) \sim \sum_{r=0}^{\infty} f_r(x),$$

where each $f_r(x)$ is some eigenfunction

(36) $$-Lf_r(x) = \lambda_r f_r(x).$$

The solution $f(t; x)$ of (22) as given by (23) has the (convergent) expansion

(37) $$f(t; x) = \sum_{r=0}^{\infty} e^{-t\lambda_r} f_r(x),$$

and from (19) we conclude

(38) $$\max_{t,x} |\, f(t; x)\,| \leqq \max_x |\, f(0; x)\,|.$$

If L^n is the nth iterate of our operator, then (22) implies

(39) $$L_x^n f = \frac{\partial}{\partial t} L^{n-1} f = \frac{\partial^n f}{\partial t^n},$$

and this gives the relation

(40) $$\left| \frac{\partial^n f(t; x)}{\partial t^n} \right| \leq \max_x \left| L^n f(x) \right|$$

which we shall employ in connection with

(41) $$\left. \frac{\partial^n f(t; x)}{\partial t^n} \right|_{t=0} = L^n f(x)$$

in the following manner. For fixed $x = \xi$ we shall view $f(t;\xi)$ as a function of t in $0 \leqq t < \infty$, and to this function we shall apply Carleman's theorem on quasi-analytic functions. Also, $f(t;\xi)$ is a Dirichlet series in t, and if it vanishes in t, all its coefficients must be zero, and we therefore obtain the following lemma.

LEMMA 3. *If for an infinitely differentiable function* (35) *we have*

$$\sum_{n=0}^{\infty} \max_x \left| L^n f(x) \right|^{-1/n} = \infty \tag{42}$$

and if for a point $x = \xi$ we have

$$L^n f(\xi) = 0, \qquad n = 0, 1, 2, \cdots, \tag{43}$$

then at this point we have

$$f_r(\xi) = 0, \qquad r = 0, 1, 2, \cdots. \tag{44}$$

We call a pointset U on M a set of uniqueness if any eigenfunction of $-Lf$ vanishing on U must vanish identically, and in this way we obtain the following theorem.

THEOREM 4. *If in addition to* (42), (43) *holds for all points of a pointset of uniqueness, then $f(x)$ vanishes identically.*

For the operator $\Delta f = d^2 f/dx^2$ on the torus $-\pi \leqq x < \pi$, the eigenfunctions are

$$a \cos rx + b \sin rx = c \cos r(x - x_0),$$

and a pointset U is therefore a set of uniqueness if it contains infinitely many points or if it contains two points whose geodesic distance is an irrational part of 2π. Hence the following conclusion.

THEOREM 5. *If for an infinitely differentiable periodic function $f(x)$ we have*

$$\sum_{n=0}^{\infty} \max_x \left| \frac{d^{2n} f(x)}{dx^{2n}} \right|^{-1/n} = \infty \tag{45}$$

and if we have

$$\frac{d^{2n} f(\xi)}{d\xi^{2n}} = 0, \qquad n = 0, 1, \cdots \tag{46}$$

for infinitely many points, ξ, or for two points whose distance is irrational on the torus of length one, then

$$f(x) = 0. \tag{47}$$

Further remark: If we take a function $\Phi(\lambda)$ as in §4, then we may replace $-L$ by $\Phi(-L)$, and we obtain the following conclusion.

THEOREM 6. *If in addition to*

$$\sum_{n=0}^{\infty} \max_x \left| \Phi(-L)^n f \right|^{-1/n} = \infty \tag{48}$$

we have

$$\Phi(-L)^n f(\xi) = 0 \tag{49}$$

for all ξ of a uniqueness set, then

$$f(x) \equiv 0. \tag{50}$$

PART V. A THEOREM OF D. V. WIDDER AND P. LELONG

D. V. Widder[11] has proved the following. If a function $f(x)$ is infinitely differentiable in $0 \leqq x \leqq 1$ and if

$$(-1)^n \frac{d^{2n} f(x)}{dx^{2n}} \geqq 0 \quad \text{in} \quad 0 \leqq x \leqq 1, \tag{51}$$

then $f(x)$ is analytic in $0 \leqq x \leqq 1$.

More precisely, $f(x)$ can be analytically continued to an entire function $f(z)$, $z = x + iy$, and its order of magnitude is, roughly speaking, comparable to that of the function $\sin \pi z$, which is the standard function for satisfying (51), as it were.

Now P. Lelong[12] has generalized this to a Laplacian on a noncompact M. If we have, for a function $f(x)$, the condition

$$(-1)^n L^n f \geqq 0, \qquad n = 0, 1, \cdots, \tag{52}$$

then $f(x)$ is a real analytic function.

Also it can be continued into a complex analytic environment of M, $z_j = x_j + iy_j$, and Lelong has some statements on the size of this environment, and of the magnitude of $f(z_1, \cdots, z_k)$ in it.

PART VI. MEROMORPHIC FUNCTIONS ON ELLIPTIC SPACES

In the theorem of Widder and Lelong use has to be made of a principal solution of the Laplacian with the properties

$$\Delta_x H(x; \xi) = 0, \qquad H(x; \xi) = H(\xi; x)$$

among others, such a principal solution being the means for solving the boundary value problem of the ordinary kind.

On a compact manifold M_n there is only a generalized solution with

$$\Delta_x H(x; \xi) = -1,$$

[11] D. V. Widder, *Completely convex functions and Lidstone series*, Trans. Amer. Math. Soc. vol. 51 (1942) pp. 387–398.

[12] P. Lelong, *Sur les fonctions indéfiniment dérivables de plusieurs variables dont les Laplaciens successifs ont des signes alternés*, Duke Math. J. vol. 14 (1947) pp. 143–149.

and the corresponding type of boundary value problem, as first envisaged by Riemann, is the construction of harmonic functions having prescribed saltuses on designated pieces of hypersurfaces. The principal solution available is indeed suitable for treating this type of problem and one significant application is as follows.[1]

Suppose M_n is real analytic, say, and assume that it is covered by a finite number of functional elements $\{\varphi_a\}$, $a = 1, \cdots, N$, such that whenever two functional elements φ_a, φ_b overlap, their difference

$$\varphi_{ab} = \varphi_a - \varphi_b \tag{53}$$

shall be a solution of the Laplacian, that is

$$\Delta\varphi_{ab} = 0.$$

The individual elements φ_a by themselves need not be so, and furthermore the individual φ_a may have "singularities" on "exceptional" pieces of space, provided such singularities happen to cancel out from the differences φ_{ab} by ordinary analytic continuation. "Meromorphic" character expresses itself in this type of singularities, but there are singularities admissible for us which are much more general than meromorphic ones.

Now, the generalized principal solution can be used for constructing a new family of functional elements $f_{(a)}$ having the following properties: (i) each $f_{(a)}$ is defined where φ_a was defined, and has no singularities whatsoever; (ii) each $f_{(a)}$ is harmonic individually; and (iii) for all a, b we have

$$f_{(a)} - f_{(b)} = \varphi_a - \varphi_b .$$

By (iii) the differences

$$\varphi_a - f_{(a)} = \varphi_b - f_{(b)} = \varphi_c - f_{(c)} = \cdots$$

will then merge into *one* function F on M_n, and this function F will then have the same singularities as the given elements $\{\varphi_a\}$, the latter being "principal singular parts" of F, as it were.

We now make the special assumption that M_n is complex analytic, $n = 2k$; that the metric on it is Hermitian

$$ds^2 = g_{\alpha\bar\beta}\, dz_\alpha\, d\bar z_\beta ,$$

with the Kaehler property

$$\frac{\partial g_{\alpha\bar\beta}}{\partial z_\gamma} = \frac{\partial g_{\gamma\bar\beta}}{\partial z_\alpha},$$

in which case

$$\Delta\varphi \equiv g^{\alpha\bar\beta} \frac{\partial^2\varphi}{\partial z_\alpha\, \partial\bar z_\beta};$$

that the φ_a are complex-meromorphic, that is, $\partial\varphi/\partial\bar z_\alpha = 0$, and a fortiori $\Delta\varphi_a = 0$;

and that their differences φ_{ab} are complex holomorphic. This in itself does not yet sharpen any of the properties of the $f_{(a)}$ previously stated. If however we are adding the assumption: $-R_{\alpha\bar\beta} > 0$, (complex projective space, complex Grassmann varieties, etc.), then by the method of part I it can be proved that the vector field

$$\zeta_{\bar\alpha} = \frac{\partial f_{(a)}}{\partial \bar z_\alpha} \left(\equiv \frac{\partial f_{(b)}}{\partial \bar z_\alpha} \right)$$

is identically zero, so that each $f_{(a)}$ is not only harmonic but also complex-holomorphic and the following precise conclusion can be drawn:[1]

THEOREM 7. *If $-R_{\alpha\bar\beta}$ is positive definite (complex projective space, complex Grassmann varieties, etc.), and if the space is covered by a finite number of meromorphic functional elements $\{\varphi_a\}_{a=1,\cdots,N}$, and if the differences $\varphi_a - \varphi_b$ are holomorphic wherever φ_a, φ_b overlap; then there exists one joint meromorphic function Φ on M_{2k} having precisely the singularities of the given functional elements; meaning that $\Phi - \varphi_a$ is holomorphic for each a on the piece of space on which φ_a is defined.*

This generalizes the "trivial" theorem of the ordinary Gaussian sphere, $k = 1$, stating that there exists on it a rational function having at prescribed poles z_a prescribed principal parts

$$\varphi_a = \sum_{\nu=1}^{n_a} \frac{c_{a\nu}}{(z - z_a)^\nu},$$

$[\varphi = \sum_{\nu=1}^{n} c_\nu z^\nu \text{ for } z_a = \infty]$, the desired function Φ being then trivially the sum $\sum_a \varphi_a$. For $k > 1$ the construction of Φ as indicated above is very much more complicated than that, and its very existence is far from obvious.

PART VII. MEROMORPHIC FUNCTIONS ON MULTI-TORI

In Theorem 7, the restriction on $-R_{\alpha\bar\beta}$ to be positive definite is a very essential one even for $k = 1$, since already for the ordinary elliptic torus [having a flat metric with $R_{1\bar1} = 0$] the known theorem of Liouville for the existence of elliptic functions with given principal parts demands that the sum of the residues of the principal parts shall have value 0. Now, for $k = 1$, Liouville's condition is not only necessary but also sufficient and for arbitrary k it can be generalized in the following manner.[13]

We take a complex multi-torus M_{2k} whose complex universal covering space is the complex Euclidean E_{2k}, and on it a simplicial decomposition $\{S_a\}$, $a = 1, \cdots, N$, and for each S_a a complex-meromorphic element φ_a is a neighborhood of its closure such that whenever two simplices S_a, S_b intersect in a $(2k - 1)$-dimensional "oriented" face B_{ab}, the difference (53) is complex-holomorphic in a neighborhood of its closure. In order that there exist a meromorphic function

[13] *Analytic and meromorphic continuation by means of Green's formula*, Ann. of Math. vol. 44 (1943) pp. 652–673, especially Theorem 15 on p. 672.

Φ having the given φ_a's as principal parts as in Theorem 7, it is necessary and sufficient that for each $\alpha = 1, \cdots, k$ we have

$$\sum_{a,b} \int_{B_{a,b}} \varphi_{ab}\, d\zeta_\alpha\, \omega^{k-1} = 0,$$

where $\omega \equiv dz_1 d\bar{z}_1 + \cdots + dz_k d\bar{z}_k$ is the basic external form associable with the flat Euclidean metric given naturally.

Part VIII. Analytic completion

One of the most striking differences between functions in one and several complex variables is as follows. For $k = 1$, to any domain D there exists a function $f(z)$ which is analytic in D and in no larger domain; whereas for $k \geqq 2$ there are pairs of domains D, $\tilde{D}$, with $D \subset \tilde{D}$, such that any function of D is continuable into $\tilde{D}$. Also, perhaps the most significant statement bearing on the latter phenomenon is the following theorem which was first stated by F. Hartogs (virtually without a proof by himself). For $k \geqq 2$, if a bounded domain D has a boundary B consisting of one piece, and if a function $f(z_1, \cdots, z_k)$ is given in a neighborhood S of B, then it can be continued into $D + S$.

Now, analyticity means

$$\frac{\partial f}{\partial \bar{z}_\alpha} = 0, \qquad \alpha = 1, \cdots, k,$$

and thus our function in S satisfies in particular the Laplace equation

$$\frac{\partial^2 f}{\partial x_1^2} + \frac{\partial^2 f}{\partial y_1^2} + \cdots + \frac{\partial^2 f}{\partial x_k^2} + \frac{\partial^2 f}{\partial y_k^2} = 0$$

in $n = 2k$ variables, and also the equation

$$\frac{\partial^2 f}{\partial x_1^2} + \frac{\partial^2 f}{\partial y_1^2} = 0$$

in $m = 2$ variables, and in the paper cited before[13] we have given a sweeping generalization of Hartog's theorem to the following effect.

Theorem 8. *If a real analytic function $f(x_1, \cdots, x_n)$ is given in a neighborhood S of the connected boundary of a bounded domain D, then it can be continued into all of $D + B$ provided it is in S the joint solution of the Laplace equation*

$$\frac{\partial^2 f}{\partial x_1^2} + \cdots + \frac{\partial^2 f}{\partial x_n^2} = 0 \tag{54}$$

in all n-variables, and of some other (inhomogenous) linear partial differential equation with constant coefficients

$$\sum_{0 \leqq r_1 + \cdots + r_m \leqq R} a_{r_1 \cdots r_m} \frac{\partial^{r_1 \cdots r_m} f}{\partial x_1^{r_1} \cdots \partial x_m^{r_m}} = 0 \tag{55}$$

in the first m variables, with $m < n$.

In the paper cited we have assumed that the second operator (55) shall be of elliptic type in a certain sense, but this assumption can be abandoned altogether, as we shall show in a future publication, and we shall then also show that the first operator may be one of elliptic type with constant coefficients with which a suitable Green's kernel can be associated, as is, for instance, the case with the powers of the Laplacian Δf.

We also generalized Hartog's theorem in other directions and one of the conclusions obtained was as follows.

THEOREM 9. *If* $f(z_1, \cdots, z_k)$ *is analytic in a domain which is of the type of a cell, and if the set of the zeros of the function in the domain contains a bounding cycle of dimension* $2k - 2$, *then the function vanishes identically.*

PRINCETON UNIVERSITY,
PRINCETON, N. J., U. S. A.

Annals of Mathematics
Vol. 56, No. 3, November, 1952
Printed in U.S.A.

TENSOR-FIELDS IN NON-SYMMETRIC CONNECTIONS

By S. Bochner and K. Yano

(Received October 2, 1951)

In Section 1 of [3], the study was begun of the topic of "Curvature and Betti Numbers" for spaces on which a metric is interlocked with an affine connection which is not necessarily symmetric, but the formulas given there are mostly incorrect owing to the omission of certain terms which need not vanish for non-symmetric connections. In the present paper, we will rectify the formulas for such connections against the background of a study that will be rather more detailed than the one previously attempted.

1. We take an n-dimensional compact space V_n on which there is given a positive definite metric

$$ds^2 = g_{jk}\, dx^j\, dx^k \qquad (i, j, k, \cdots = 1, 2, \cdots, n) \tag{1.1}$$

and a metric connection E^i_{jk}, so that

$$g_{jk|l} \equiv \frac{\partial g_{jk}}{\partial x^l} - g_{sk} E^s_{jl} - g_{js} E^s_{kl} = 0, \tag{1.2}$$

where the solidus denotes covariant differentiation with respect to E^i_{jk}. The connection need not be symmetric, $E^i_{jk} \neq E^j_{kj}$, and the entity

$$S_{jk}{}^i = \tfrac{1}{2}(E^i_{jk} - E^i_{kj}) \tag{1.3}$$

will be called the torsion tensor. We define g^{is} by $g^{is}g_{sj} = \delta^i_j$, and we will use this to pull indices up and down, so that for instance

$$S^i{}_{jk} = g^{is} g_{kt} S_{sj}{}^t, \tag{1.4}$$

and we note that in virtue of (1.2) the pulling up and down of indices is commutative with covariant differentiation.

If we solve (1.2) and (1.3) with respect to E^i_{jk}, we obtain [6]

$$E^i_{jk} = \left\{ {i \atop jk} \right\} + S_{jk}{}^i - S^i{}_{jk} - S^i{}_{kj}, \tag{1.5}$$

where we have put

$$\left\{ {i \atop jk} \right\} = \frac{1}{2} g^{is} \left(\frac{\partial g_{sj}}{\partial x^k} + \frac{\partial g_{sk}}{\partial x^j} - \frac{\partial g_{jk}}{\partial x^s} \right), \tag{1.6}$$

and (1.5) shows that

$$\tfrac{1}{2}(E^i_{jk} + E^i_{kj}) = \left\{ {i \atop jk} \right\} - S^i{}_{jk} - S^i{}_{kj},$$

so that, the symmetric part of E^i_{jk} does not necessarily coincide with the Christoffel symbols $\left\{ {i \atop jk} \right\}$. In order that such be the case, we must have

$$S^i{}_{jk} + S^i{}_{kj} = 0$$

or

$$S_{ijk} + S_{ikj} = 0.$$

Thus, the covariant torsion tensor S_{ijk}, which is by definition antisymmetric in i and j, must be anti-symmetric in all indices. The converse being evident, we have

THEOREM 1. *The necessary and sufficient condition that the symmetric part of E^i_{jk} coincide with the Christoffel symbols $\left\{ {i \atop jk} \right\}$ is that the covariant components S_{ijk} of the torsion tensor be anti-symmetric in all indices.*

If, for example, for a tensor $T^i{}_{jk}$, we calculate $T^i{}_{jk|l|m} - T^i{}_{jk|m|l}$, then we find the Ricci formula:

$$T^i{}_{jk|l|m} - T^i{}_{ij|m|l} = T^s{}_{jk}E^i{}_{slm} - T^i{}_{sk}E^s{}_{jlm} - T^i{}_{js}E^s{}_{klm} - 2T^i{}_{jk|s}S_{lm}{}^s, \tag{1.7}$$

where

$$E^i{}_{jkl} = \frac{\partial E^i_{jk}}{\partial x^l} - \frac{\partial E^i_{jl}}{\partial x^k} + E^s_{jk}E^i_{sl} - E^s_{jl}E^i_{sk} \tag{1.8}$$

is the curvature tensor of the metric connection E^i_{jk}. The $E^i{}_{jkl}$ defined here differs by sign from that given in [3].

Applying the Ricci formula to g_{ij}, we find

$$-g_{sj}E^s{}_{ikl} - g_{is}E^s{}_{jkl} = 0,$$

and on putting

$$E_{ijkl} = g_{is}E^s{}_{jkl},$$

we obtain

$$E_{ijkl} = -E_{jikl} \qquad \text{and} \qquad E_{ijkl} = -E_{ijlk}. \tag{1.9}$$

It will be easily verified that the components $E^i{}_{jkl}$ of the curvature tensor satisfy, instead of the usual ones, the following Bianchi identities:

$$\begin{aligned} E^i{}_{jkl} + E^i{}_{klj} + E^i{}_{ljk} - 2(S_{jk}{}^i{}_{|l} + S_{kl}{}^i{}_{|j} + S_{lj}{}^i{}_{|k}) \\ + 4(S_{jk}{}^t S_{tl}{}^i + S_{kl}{}^t S_{tj}{}^i + S_{lj}{}^t S_{tk}{}^i) = 0, \end{aligned} \tag{1.10}$$

$$E^i{}_{jkl|m} + E^i{}_{jlm|k} + E^i{}_{jmk|l} - 2(E^i{}_{jtk}S_{lm}{}^t + E^i{}_{jtl}S_{mk}{}^t + E^i{}_{jtm}S_{kl}{}^t) = 0. \tag{1.11}$$

Furthermore, on putting

$$E^i_{jk} = \begin{Bmatrix} i \\ jk \end{Bmatrix} + \Omega_{jk}{}^i, \qquad \Omega_{jk}{}^i = S_{jk}{}^i - S^i{}_{jk} - S^i{}_{kj}, \tag{1.12}$$

we find

$$\Omega_{js}{}^s = 2S_{js}{}^s, \tag{1.13}$$

and

$$E^i{}_{jkl} = R^i{}_{jkl} + \Omega_{jk}{}^i{}_{;l} - \Omega_{jl}{}^i{}_{;k} + \Omega_{jk}{}^t\Omega_{tl}{}^i - \Omega_{jl}{}^t\Omega_{tk}{}^i, \tag{1.14}$$

where

$$R^i{}_{jkl} = \frac{\partial\begin{Bmatrix} i \\ jk \end{Bmatrix}}{\partial x^l} - \frac{\partial\begin{Bmatrix} i \\ jl \end{Bmatrix}}{\partial x^k} + \begin{Bmatrix} t \\ jk \end{Bmatrix}\begin{Bmatrix} i \\ tl \end{Bmatrix} - \begin{Bmatrix} t \\ jl \end{Bmatrix}\begin{Bmatrix} i \\ tk \end{Bmatrix},$$

and the semi-colon denotes covariant differentiation with respect to the Christoffel symbols introduced.

If we assume that S_{ijk} is anti-symmetric in all indices, then we have $\Omega_{jk}{}^i = S_{jk}{}^i$ and (1.14) becomes

$$E^i{}_{jkl} = R^i{}_{jkl} + S_{jk}{}^i{}_{;l} - S_{jl}{}^i{}_{;k} + S_{jk}{}^tS_{tl}{}^i - S_{jl}{}^tS_{tk}{}^i, \tag{1.15}$$

and this implies

$$E_{jk} = R_{jk} + S_{jk}{}^t{}_{;t} + S_{jl}{}^tS_{kt}{}^l, \tag{1.16}$$

where

$$E_{jk} = E^l{}_{jkl} \qquad \text{and} \qquad R_{jk} = R^l{}_{jkl}; \tag{1.17}$$

and due to

$$g^{jk}E_{ijkl} = g^{jk}E_{jilk} = E_{il},$$

we also have

$$g^{jk}E^i{}_{jkl} = E^i{}_l. \tag{1.18}$$

From (1.16) we deduce

$$\tfrac{1}{2}(E_{jk} + E_{kj}) = R_{jk} - S_j{}^{rs}S_{krs}, \tag{1.19}$$

$$\tfrac{1}{2}(E_{jk} - E_{kj}) = S_{jk}{}^t{}_{;t}, \tag{1.20}$$

and thus the tensor E_{jk} is not symmetric in general. But from (1.19) we have

$$E_{jk}\xi^j\xi^k = R_{jk}\xi^j\xi^k - (S_j{}^{rs}\xi^j)(S_{krs}\xi^k), \tag{1.21}$$

and hence

THEOREM 2. *In V_n with anti-symmetric torsion tensor, if $E_{jk} + E_{kj} = 0$, then $R_{jk}\xi^j\xi^k$ is non-negative.*

THEOREM 3. *In V_n with anti-symmetric torsion tensor, if $R_{jk}\xi^j\xi^k$ is non-positive, then $E_{jk}\xi^j\xi^k$ is also non-positive.*

2. A compact semi-simple group space is a good example of a metric space with torsion.

In such a space, there are given n linearly independent contravariant vectors $\overset{\cdot}{h}{}^i_a$ (a, b, c, d, e, f run from $\dot{1}$ to $\dot{n}$ and denote the vectors), for which [5]

$$\text{(2.1)} \qquad h^s_b \frac{\partial h^i_c}{\partial x^s} - h^s_c \frac{\partial h^i_b}{\partial x^s} = c_{bc}{}^a h^i_a,$$

where $c_{bc}{}^a$ are the constants of structure. Moreover, if we put

$$\text{(2.2)} \qquad g_{bc} = -c_{be}{}^f c_{cf}{}^e,$$

the quadratic form $g_{bc}z^b z^c$ is a positive definite one [4], and if we use g_{bc} and $g^{ab}(g^{ab}g_{bc} = \delta^a_c)$ for pulling indices up and down, we obtain, from the Jacobi identity

$$c_{ab}{}^e c_{ce}{}^f + c_{bc}{}^e c_{ae}{}^f + c_{ca}{}^e c_{be}{}^f = 0,$$

the relation

$$c_{bcd} \equiv c_{bc}{}^e g_{ed} = -c_{ab}{}^e c_{ce}{}^f c_{df}{}^a + c_{ab}{}^e c_{de}{}^f c_{cf}{}^a,$$

which shows that c_{bcd} is anti-symmetric in all indices b, c, and d.

If we now put

$$\text{(2.3)} \qquad g^{ij} = h^i_a h^j_b g^{ab},$$

then the determinant $|g^{ij}|$ is obviously different from zero, and if we define g_{jk} by $g^{ij}g_{jk} = \delta^i_k$, then

$$\text{(2.4)} \qquad g_{jk} = h^b_j h^c_k g_{bc},$$

where we have put

$$\text{(2.5)} \qquad h^b_j = g^{bc} g_{jk} h^k_c,$$

and consequently we have $h^i_a h^a_j = \delta^i_j$, $h^a_i h^i_b = \delta^a_b$.

The equation (2.4) shows that the quadratic differential form

$$\text{(2.6)} \qquad ds^2 = g_{jk}dx^j dx^k$$

is positive definite, and we introduce it as a metric on our space.

As usual, we now define an affine connection E^i_{jk} by demanding

$$h^i_{a|k} \equiv \frac{\partial h^i_a}{\partial x^k} + h^j_a E^i_{jk} = 0,$$

so that

$$\text{(2.7)} \qquad E^i_{jk} = h^i_a \frac{\partial h^a_j}{\partial x^k}.$$

From (2.1) and (2.7) we obtain

$$S_{jk}{}^{i} \equiv \tfrac{1}{2}(E^{i}_{jk} - E^{i}_{kj}) = \tfrac{1}{2}c_{bc}{}^{a}\, h^{b}_{j} h^{c}_{k} h^{i}_{a}, \tag{2.8}$$

so that

$$S_{jki} = \tfrac{1}{2}c_{bca}\, h^{b}_{j} h^{c}_{k} h^{a}_{i}, \tag{2.9}$$

and thus the covariant torsion tensor S_{jki} is anti-symmetric in all its indices

On the other hand, we have $g^{ij}{}_{|k} = (h^{i}_{a} h^{j}_{b} g^{ab})_{|k} = 0$, and consequently

$$g_{jk|l} = 0. \tag{2.10}$$

Thus, the affine connection E^{i}_{jk} is indeed metric, and everything stated in the preceding section will apply; firstly, (1.5) implies

$$E^{i}_{jk} = \begin{Bmatrix} i \\ jk \end{Bmatrix} + S_{jk}{}^{i}, \tag{2.11}$$

and the vectors h^{i}_{a} being absolutely parallel with respect to E^{i}_{jk}, the curvature tensor defined by (1.8) is identically zero:

$$E^{i}{}_{jkl} = 0 \qquad \text{and consequently} \qquad E_{jk} = 0. \tag{2.12}$$

On the other hand, equation (2.8) shows that

$$S_{jk}{}^{i}{}_{|l} = 0, \tag{2.13}$$

therefore, the first of the Bianchi identities (1.10) reduces to

$$S_{jk}{}^{t}S_{tl}{}^{i} + S_{kl}{}^{t}S_{tj}{}^{i} + S_{lj}{}^{t}S_{tk}{}^{i} = 0, \tag{2.14}$$

and, by virtue of (2.8), this is equivalent to the Jacobi identities.

But, if we denote by a semi-colon the covariant differentiation with respect to the Christoffel symbols, we obtain

$$S_{jk}{}^{i}{}_{|l} = S_{jk}{}^{i}{}_{;l}$$

by virtue of (2.14), and consequently

$$S_{jk}{}^{i}{}_{;l} = 0. \tag{2.15}$$

Thus, equation (1.15) gives

$$0 = R^{i}{}_{jkl} + S_{jk}{}^{t}S_{tl}{}^{i} - S_{jl}{}^{t}S_{tk}{}^{i}$$

or

$$R^{i}{}_{jkl} = S_{kl}{}^{t}S_{tj}{}^{i} \tag{2.16}$$

by virtue of (2.14), and hence also

$$R_{ijkl} = -S_{ijt}S_{kl}{}^{t}. \tag{2.17}$$

Thus, for the Ricci tensor, we have

$$R_{jk} = +S_{jst}S_k^{\ st}. \tag{2.18}$$

But, equations (2.2) and (2.9) show that

$$S_{jst}S_k^{\ st} = \tfrac{1}{4}g_{jk}, \tag{2.19}$$

and consequently we have

$$R_{jk} = \tfrac{1}{4}g_{jk} \tag{2.20}$$

which is a well-known formula.

3. As in [3], we will call a vector ξ_i pseudo-harmonic, if

$$\xi_{i|j} = \xi_{j|i} \quad \text{and} \quad g^{jk}\xi_{j|k} = \xi^k{}_{|k} = 0, \tag{3.1}$$

and for such a vector, the Ricci formula

$$\xi_{j|i|k} - \xi_{j|k|i} = -\xi_s E^s{}_{jik} - 2\xi_{j|s}S_{ik}^{\ \ s} \tag{3.2}$$

gives

$$\xi_{i|j|k} - \xi_{j|k|i} = +\xi_s E^s{}_{jki} - 2\xi_{j|s}S_{ik}^{\ \ s}$$

on account of $\xi_{j|i} = \xi_{i|j}$ and $-E^s{}_{jik} = +E^s{}_{jki}$. Also multiplying this equation by g_{jk} and summing with respect to the indices j and k, we find

$$g^{jk}\xi_{i|j|k} - \xi_s E^s{}_i + 2\xi_{r|s}S_i^{\ rs} = 0 \tag{3.3}$$

by virtue of (1.18) and $g^{jk}\xi_{j|k} = 0$.

Now, for a scalar $\phi(x)$, we have

$$g^{jk}\phi_{|j|k} = g^{jk}\frac{\partial^2\phi}{\partial x^j\partial x^k} - g^{jk}E^i_{jk}\frac{\partial\phi}{\partial x^i},$$

and hence from a theorem of E. Hopf [7] we obtain

LEMMA 1. *On a compact V_n, if, for a scalar field ϕ, we have*

$$\Delta\phi \equiv g^{jk}\phi_{|j|k} \geqq 0,$$

then we have

$$\Delta\phi = g^{jk}\phi_{|j|k} = 0.$$

Now, for ξ_i pseudo-harmonic, if $\phi = \xi^i\xi_i$, then

$$\Delta\phi = 2\xi^i g^{jk}\xi_{i|j|k} + 2g^{jk}\xi^i{}_{|j}\xi_{i|k},$$

and by (3.3), this is

$$\Delta\phi = 2\xi^i(\xi_s E^s{}_i - 2\xi_{r|s}S_i^{\ rs}) + 2g^{jk}\xi^i{}_{|j}\xi_{i|k},$$

and hence

$$\Delta\phi = (E_{jk} + E_{kj})\xi^j\xi^k - 2(S_{irs} + S_{isr})\xi^i\xi^{r|s} + (g_{rt}g_{su} + g_{ru}g_{st})\xi^{r|s}\xi^{t|u}. \tag{3.4}$$

Thus, from Lemma 1, we have

THEOREM 4. *On a compact V_n, if the matrix*

$$M = \begin{bmatrix} E_{jk} + E_{kj} & -(S_{irs} + S_{isr}) \\ -(S_{irs} + S_{isr}) & g_{rt}g_{su} + g_{ru}g_{st} \end{bmatrix}$$

defines a positive definite quadratic form in the variables ξ^i and $\xi^{rs} = \xi^{sr}$, then there exists no pseudo-harmonic vector other than zero vector.

If the matrix M defines a non-negative quadratic form, then every pseudo-harmonic vector ξ^i must satisfy

$$(E_{jk} + E_{kj})\xi^j\xi^k - 2(S_{irs} + S_{isr})\xi^i\xi^{r|s} + (g_{rt}g_{su} + g_{ru}g_{st})\xi^{r|s}\xi^{t|u} = 0.$$

If $S_{jk}{}^i$ vanishes, then our space is Riemannian and a pseudo-harmonic vector is an ordinary harmonic vector, and we have [1],

COROLLARY 1. *In a compact Riemannian space, if the form $R_{jk}\xi^j\xi^k$ is positive definite, there exists no harmonic vector other than zero vector. And if $R_{jk}\xi^j\xi^k$ is non-negative, then every harmonic vector must have vanishing covariant derivative, and then $R_{jk}\xi^j\xi^k = 0$.*

Now, if we have

$$E_{jk} + E_{kj} = 0 \quad \text{and} \quad S_{irs} + S_{isr} = 0,$$

then, for a pseudo-harmonic vector ξ^i, we have

$$\Delta\phi = 2\xi_{j|k}\xi^{j|k} \geqq 0,$$

from which

$$\xi_{j|k} \equiv \frac{\partial \xi_j}{\partial x^k} - \xi_i E^i_{jk} = 0$$

follows. Thus, there exist, in this case, at most n linearly independent (with constant coefficients) pseudo-harmonic vectors. Moreover, if such a pseudo-harmonic vector exists, it must satisfy

$$\xi_{j|k} = \xi_{j;k} - \xi_i S_{jk}{}^i = 0,$$

from which

$$\xi_{j;k} + \xi_{k;j} = 0.$$

The last equation shows that ξ^i is an ordinary Killing vector. Thus we have

COROLLARY 2. *In a compact V_n satisfying $E_{jk} + E_{kj} = 0$ and $S_{irs} + S_{isr} = 0$, a pseudo-harmonic vector must have vanishing covariant derivative with respect to the connection of the space, and consequently the number of linearly independent (with constant coefficients) pseudo-harmonic vectors is at most n. Moreover, if such a pseudo-harmonic vector exists, it is an ordinary Killing vector.*

Now, if $E_{jk} + E_{kj} = 0$ and $S_{irs} + S_{isr} = 0$, then, by Theorem 2, $R_{jk}\xi^j\xi^k$ is non-negative. Hence, by Corollary 1, an ordinary harmonic vector must have vanishing covariant derivative with respect to Christoffel symbols and satisfy

$$R_{jk}\xi^j\xi^k = S_{jrs}S_k^{\ rs}\xi^j\xi^k = 0.$$

Thus, if the rank of the matrix $\| S_{jrs}S_k^{\ rs} \|$ is n, we can conclude that there exists no ordinary harmonic vector. Thus we have

COROLLARY 3. *In a compact V_n satisfying $E_{jk} + E_{kj} = 0$ and $S_{irs} + S_{isr} = 0$, the ordinary harmonic vector must have vanishing covariant derivative with respect to the Christoffel symbols. Moreover, if the rank of the matrix $\| S_{jrs}S_k^{\ rs} \|$ is n, there exists no ordinary harmonic vector.*

A compact semi-simple group space falls under Corollary 2 and 3. On the other hand, in such a space, a pseudo-harmonic vector ξ_i can be written as

$$\xi_j = f_a(x)h_j^a,$$

and, by Corollary 2, it must have a vanishing covariant derivative with respect to the connection of the space; consequently $f_a(x)$ must be constants. Thus, we have

COROLLARY 4. *In a compact semi-simple group space, there exist n linearly independent pseudo-harmonic vectors, and any pseudo-harmonic vector is a linear combination with constant coefficients of these vectors.*

Moreover, from

$$0 = h^a_{j|k} = h^a_{j;k} - h^a_i S_{jk}^{\ \ i},$$

we have

$$h^a_{j;k} + h^a_{k;j} = 0,$$

and thus h_j^a are all ordinary Killing vectors and the space admits a simply transitive group of motions as well-known. Furthermore, due to (2.20), Corollary 3 implies the following known theorem.

COROLLARY 5. *There exists no ordinary harmonic vector in a compact semi-simple group space.*

Returning to a general compact orientable space V_n, we now assume $S_{jr}^{\ \ r} = 0$ (this is satisfied if S_{jki} is anti-symmetric in all indices) and in this case relation (3.3) is not only necessary but also sufficient for a vector to be pseudo-harmonic.

In fact, for any vector v^i, we have

$$v^i_{\ |i} = v^i_{\ ;i} + v^j\Omega_{ji}^{\ \ i} = v^i_{\ ;i} + 2v^jS_{ji}^{\ \ i},$$

from which

$$v^i_{\ |i} = \frac{1}{\sqrt{g}}\frac{\partial\sqrt{g}v^i}{\partial x^i} \tag{3.5}$$

by virtue of the assumption $S_{ji}{}^{i} = 0$, where g is the determinant formed with g_{jk}. Thus, for any vector v^i, we have

$$\int v^i{}_{|i}\, dv = 0, \tag{3.6}$$

the integral being taken over the whole space, where dv is the volume element:

$$dv = \sqrt{g}\, dx^1\, dx^2 \cdots dx^n.$$

Thus we have first

$$\int g^{jk}(\xi^i \xi_i)_{|j|k}\, dv = 0$$

for any vector field ξ^i, or

$$\int (\xi^{i|j}\xi_{i|j} + \xi^i g^{jk}\xi_{i|j|k})\, dv = 0. \tag{3.7}$$

Next we have

$$\int (\xi^i{}_{|j}\xi^j)_{|i}\, dv = \int (\xi^i{}_{|j|i}\xi^j + \xi^i{}_{|j}\xi^j{}_{|i})\, dv = 0$$

or

$$\int (\xi^i{}_{|i|j}\xi^j + E_{jk}\xi^j\xi^k - 2\xi^i{}_{|s}S_{ji}{}^{s}\xi^j + \xi^i{}_{|j}\xi^j{}_{|i})\, dv = 0 \tag{3.8}$$

by virtue of the Bianchi identity:

$$\xi^i{}_{|j|k} - \xi^i{}_{|k|j} = \xi_s E^i{}_{sjk} - 2\xi^i{}_{|s}S_{jk}{}^{s}. \tag{3.9}$$

On the other hand, we have

$$\int (\xi^i{}_{|i}\xi^j)_{|j}\, dv = 0$$

or

$$\int \xi^i{}_{|i|j}\xi^j + \xi^i{}_{|i}\xi^j{}_{|j})\, dv = 0. \tag{3.10}$$

Forming (3.8)–(3.10), we find

$$\int (E_{jk}\xi^j\xi^k - 2S_{irs}\xi^i\xi^{r|s} + \xi^{i|j}\xi_{j|i} - \xi^i{}_{|i})\, dv = 0. \tag{3.11}$$

Next forming (3.7)–(3.11), we have

$$\int [\xi^i(g^{jk}\xi_{i|j|k} - \xi_s E^s{}_i + 2S_{irs}\xi^{r|s}) + \tfrac{1}{2}(\xi^{i|j} - \xi^{j|i})(\xi_{i|j} - \xi_{j|i}) + \xi^i{}_{|i}\xi^j{}_{|j}]\, dv = 0. \tag{3.12}$$

Thus, if the vector ξ_i satisfies (3.3) then we must have

$$\xi_{i|j} = \xi_{j|i} \quad \text{and} \quad \xi^i{}_{|i} = 0,$$

and hence we have [9]

THEOREM 5. *In a compact orientable V_n with $S_{jr}{}^r = 0$, a necessary and sufficient condition that a vector ξ_i be a pseudo-harmonic vector is that the ξ_i satisfy* (3.3).

4. We shall call a vector ξ^i pseudo-Killing vector if it satisfies the condition [3]

$$(4.1) \qquad \xi_{i|j} + \xi_{j|i} = 0 \quad \text{and automatically} \quad \xi^i{}_{|i} = 0.$$

For such a vector, the Ricci formula (3.2) gives

$$-\xi_{i|j|k} - \xi_{j|k|i} = +\xi_s E^s{}_{jki} - 2\xi_{j|s} S_{ik}{}^s$$

by virtue of $\xi_{j|i} = -\xi_{i|j}$, and hence

$$(4.2) \qquad g^{jk}\xi_{i|j|k} + \xi_s E^s{}_i - 2\xi_{r|s} S_i{}^{rs} = 0$$

by virtue of $g^{jk}\xi_{j|k} = 0$.

Therefore, for $\phi = \xi^i\xi_i$, we have

$$(4.3) \quad \Delta\phi = -[(E_{jk} + E_{kj})\xi^j\xi^k - 2(S_{irs} - S_{isr})\xi^i\xi^{r|s} - (g_{rt}g_{su} - g_{ru}g_{st})\xi^{r|s}\xi^{t|u}],$$

and hence, from Lemma 1, we obtain

THEOREM 6. *On a compact V_n, if the matrix*

$$M' = \begin{bmatrix} E_{jk} + E_{kj} & -(S_{irs} - S_{isr}) \\ -(S_{irs} - S_{isr}) & -(g_{rt}g_{su} - g_{ru}g_{st}) \end{bmatrix}$$

defines a negative definite quadratic form in the variables ξ^i and $\xi^{rs} = -\xi^{sr}$, then there exists no pseudo-Killing vector other than zero vector.

If the matrix M' defines a non-positive quadratic form, then every pseudo-Killing vector ξ^i must satisfy

$$(E_{jk} + E_{kj})\xi^j\xi^k - 2(S_{irs} - S_{isr})\xi^i\xi^{r|s} - (g_{rt}g_{su} - g_{ru}g_{st})\xi^{r|s}\xi^{t|u} = 0.$$

If $S_{jk}{}^i$ vanishes, then our space becomes Riemannian and pseudo-Killing vector becomes ordinary Killing vector, and we have [1]

COROLLARY 1. *In a compact Riemannian space, if the form $R_{jk}\xi^j\xi^k$ is negative definite, then there exists no Killing vector other than zero vector. And if $R_{jk}\xi^j\xi^k$ is non-positive, then every Killing vector must have vanishing covariant derivative and we then have $R_{jk}\xi^j\xi^k = 0$.*

Now, for $S_{ir}{}^r = 0$, (3.7) and (3.11) imply

$$(4.4) \quad \int [\xi^i(g^{jk}\xi_{i|j|k} + \xi_s E^s{}_i - 2S_{irs}\xi^{r|s}) + \tfrac{1}{2}(\xi^{i|j} + \xi^{j|i})(\xi_{i|j} + \xi_{j|i}) - \xi^i{}_{|i}\xi^j{}_{|j}]\, dv = 0,$$

and thus, if the vector ξ_i satisfies (4.2), and $\xi^i{}_{|i} = 0$, then we must have

$$\xi_{i|j} + \xi_{j|i} = 0,$$

and hence we have [9]

THEOREM 7. *In a compact orientable V_n with $S_{jr}{}^r = 0$, a necessary and sufficient condition that a vector ξ_i be a pseudo-Killing vector is that ξ_i satisfy* (4.2) *and $\xi^i{}_{|i} = 0$.*

Also, if, more specifically, S_{jki} is anti-symmetric in all indices, then we have

$$\xi_{i|j} + \xi_{j|i} = \xi_{i;j} + \xi_{j;i},$$

and hence we have

THEOREM 8. *In a space with anti-symmetric torsion tensor, the pseudo-Killing vector and the ordinary Killing vector coincide.*

5. We shall call an anti-symmetric tensor $\xi_{i_1 i_2 \cdots i_p}$ pseudo-harmonic [3], if it satisfies the conditions:

$$\xi_{[i_1 i_2 \cdots i_p | r]} = 0 \tag{5.1}$$

or

$$\xi_{i_1 i_2 \cdots i_p | r} = \xi_{r i_2 \cdots i_p | i_1} + \xi_{i_1 r i_3 \cdots i_p | i_2} + \cdots + \xi_{i_1 i_2 \cdots i_{p-1} r | i_p} \tag{5.2}$$

and

$$g^{rs} \xi_{i_1 i_2 \cdots i_{p-1} r | s} = 0, \tag{5.3}$$

where the brackets [ ] denote the anti-symmetric part.

The generalization of (3.3) then is

$$\begin{aligned} g^{rs}\xi_{i_1 i_2 \cdots i_p | r | s} &- \textstyle\sum_{s=1}^{p} \xi_{i_1 \cdots (s) \cdots i_p} E^l{}_{i_s} \\ &+ \textstyle\sum_{s<t}^{i \cdots p} \xi_{i_1 \cdots (s) \cdots (t) \cdots i_p} (E^l{}_{i_s i_t}{}^m - E^l{}_{i_t i_s}{}^m) \\ &+ 2 \textstyle\sum_{s=1}^{p} \xi_{i_1 \cdots (s) \cdots i_p | m} S_{i_s}{}^{lm} = 0. \end{aligned} \tag{5.4}$$

If we put $\phi = \xi^{i_1 i_2 \cdots i_p} \xi_{i_1 i_2 \cdots i_p}$, we then obtain

$$\Delta\phi = K^{(p)}_{ijkl}\, \xi^{iji_3 \cdots i_p} \xi^{kl}{}_{i_3 \cdots i_p} - 2S^{(p)}_{ijrst} \xi^{iji_3 \cdots i_p} \xi^{rs}{}_{i_3 \cdots i_p}{}^{|t} + G_{rstuvw}\, \xi^{rsi_3 \cdots i_p | t} \xi^{uv}{}_{i_3 \cdots i_p}{}^{|w},$$

where

$$K^{(p)}_{ijkl} = \frac{p}{2}(E_{ik} g_{lj} - E_{jk} g_{li} - E_{il} g_{kj} + E_{jl} g_{ki}) - \frac{p(p-1)}{2}(E_{iklj} - E_{jkli} - E_{ilkj} + E_{jlki}),$$

$$S^{(p)}_{ijrst} = \frac{p}{2}(S_{irt} g_{js} - S_{jrt} g_{is} - S_{ist} g_{jr} + S_{jst} g_{ir}),$$

$$G_{rstuvw} = (g_{ru} g_{sv} - g_{rv} g_{su}) g_{tw}.$$

Thus, from Lemma 1, we obtain

THEOREM 9. *On a compact V_n, if the matrix*

$$N = \begin{bmatrix} K^{(p)}_{ijkl} & -S^{(p)}_{ijrst} \\ -S^{(p)}_{ijrst} & G_{rstuvw} \end{bmatrix}$$

defines a positive definite quadratic form in the variables $\xi^{ij} = -\xi^{ji}$ and $\xi^{rst} = -\xi^{srt}$, then there exists no pseudo-harmonic tensor of pth order other than zero tensor.

If the matrix N defines a non-negative quadratic form, then any pseudo-harmonic tensor $\xi_{i_1 i_2 \cdots i_p}$ must satisfy

$$K^{(p)}_{ijkl}\,\xi^{iji_3\cdots i_p}\xi^{kl}{}_{i_3\cdots i_p} - 2S^{(p)}_{ijrst}\,\xi^{iji_3\cdots i_p}\xi^{rs}{}_{i_3\cdots i_p}{}^{|t} + G_{rstuvw}\,\xi^{rsi_3\cdots i_p|t}\xi^{uv}{}_{i_3\cdots i_p}{}^{|w} = 0.$$

For $S_{jk}{}^i = 0$, we have [10]

COROLLARY 1. *In a compact Riemannian space, if the form*

$$\frac{1}{2}K^{(p)}_{ijkl}\,\xi^{ij}\xi^{kl} = pR_{ij}\,\xi^{ia}\xi^{j}{}_a + \frac{p(p-1)}{2}R_{ijkl}\,\xi^{ij}\xi^{kl}$$

in the variables $\xi^{ij} = -\xi^{ji}$ is positive definite, then there exists no harmonic tensor of the pth order other than zero tensor. And if $K^{(p)}_{ijkl}\xi^{ij}\xi^{kl}$ is non-negative, then every harmonic tensor must have vanishing covariant derivative.

In this connection, we shall note another consequence of Corollary 1.

We assume that

$$0 < A \leqq \frac{-R_{ijkl}\,\xi^{ij}\xi^{kl}}{\xi^{ij}\xi_{ij}} \leqq B$$

for any anti-symmetric tensor ξ^{ij}, A and B being constants.

If we put

$$\xi^{ij} = p^i q^j - p^j q^i,$$

p^i and q^i being two unit vectors which are mutually orthogonal, then we get, from (5.5),

$$\tfrac{1}{2}A \leqq -R_{ijkl}\,p^i q^j p^k q^l \leqq \tfrac{1}{2}B,$$

$-R_{ijkl}\,p^i q^j p^k q^l$ being the sectional curvature.

Now, for $n-1$ unit vectors $q^i_{(1)}, q^i_{(2)}, \cdots, q^i_{(n-1)}$ which are orthogonal to p^i and to each other, we have

$$-R_{ijkl}\,p^i p^j p^k p^l = 0, \quad \tfrac{1}{2}A \leqq -R_{ijkl}\,p^i q^j_{(a)} p^k q^l_{(a)} \leqq \tfrac{1}{2}B, \qquad (a = 1, 2, \cdots, n-1)$$

and from this and

$$p^j p^l + \textstyle\sum_{a=1}^{n-1} q^j_{(a)} q^l_{(a)} = g^{jl},$$

we obtain

$$\tfrac{1}{2}(n-1)A \leqq R_{ik}\,p^i p^k \leqq \tfrac{1}{2}(n-1)B. \tag{5.6}$$

From (5.6), we have

$$R_{ij}\xi^i\xi^j \geqq \tfrac{1}{2}(n-1)A\xi^i\xi_i$$

for any vector ξ^i and consequently

$$R_{ij}\xi^{ia}\xi^j{}_a \geqq \tfrac{1}{2}(n-1)A\xi^{ij}\xi_{ij} \tag{5.7}$$

for any anti-symmetric tensor ξ^{ij}, and, from (5.5), we have

$$R_{ijkl}\xi^{ij}\xi^{kl} \geqq -B\xi^{ij}\xi_{ij}. \tag{5.8}$$

Thus, from (5.7) and (5.8), we find

$$pR_{ij}\xi^{ia}\xi^j{}_a + \frac{p(p-1)}{2}R_{ijkl}\xi^{ij}\xi^{kl} \geqq \frac{p}{2}[(n-1)A-(p-1)B]\xi^{ij}\xi_{ij},$$

and consequently, if

$$\frac{A}{B} > \frac{p-1}{n-1}, \tag{5.9}$$

then, the form $K^{(p)}_{ijkl}\xi^{ij}\xi^{kl}$ will be positive definite.

But, as we have

$$\frac{1}{2} > \frac{p-1}{n-1} \quad \text{for} \quad p = 1, 2, \cdots, \left[\frac{n}{2}\right],$$

we can say that if

$$\frac{A}{B} = \frac{1}{2} \quad \text{and consequently} \quad A = \frac{1}{2}B,$$

then

$$pR_{ij}\xi^{ia}\xi^j{}_a + \frac{p(p-1)}{2}R_{ijkl}\xi^{ij}\xi^{kl}$$

is positive definite for $p = 1, 2, \cdots, \left[\frac{n}{2}\right]$. Thus we have

COROLLARY 2. *In a compact Riemannian space, if the curvature tensor satisfies*

$$0 < \frac{1}{2}B \leqq \frac{-R_{ijkl}\xi^{ij}\xi^{kl}}{\xi^{ij}\xi_{ij}} \leqq B \tag{5.10}$$

for any anti-symmetric tensor ξ^{ij}, B being a constant, then all the pth Betti numbers vanish $(p = 1, 2, \cdots, n-1)$.

This result may be compared with that of H. E. Rauch [8].

As a generalization of Theorem 5, one can prove

THEOREM 10. *In a compact orientable metric space whose torsion tensor $S_{jk}{}^i$ satisfies $S_{jr}{}^r = 0$, a necessary and sufficient condition that an anti-symmetric tensor $\xi_{i_1 i_2 \cdots i_p}$ be a pseudo-harmonic tensor is that it satisfy* (5.4).

6. We shall call pseudo-Killing tensor an anti-symmetric tensor $\xi_{i_1 i_2 \cdots i_p}$ which satisfies the condition:

$$\xi_{i_1 i_2 \cdots i_p | r} = \xi_{[i_1 i_2 \cdots i_p | r]} \tag{6.1}$$

or

$$\xi_{i_1 i_2 \cdots i_p | r} = -\frac{1}{p}\left[\xi_{r i_2 \cdots i_p | i_1} + \xi_{i_1 r i_3 \cdots i_p | i_2} + \cdots + \xi_{i_1 i_2 \cdots i_{p-1} r | i_p}\right] \tag{6.2}$$

and automatically

$$g^{rs}\xi_{i_1 i_2 \cdots i_{p-1} r | s} = 0. \tag{6.3}$$

By a method analogous to that used in the preceding Section, we obtain, from (6.2),

$$\begin{aligned} & g^{rs}\xi_{i_1 i_2 \cdots i_p | r | s} + \frac{1}{p}\sum_{s=1}^{p} \xi_{i_1 i_2 \cdots s \cdots i_p} E^{l}{}_{i_s} \\ & \quad - \frac{1}{p}\sum_{s<t}^{1\cdots p} \xi_{i_1 \cdots s \cdots t \cdots i_p}(E^{l}_{i_s i_t}{}^{m} - E^{l}_{i_t i_s}{}^{m}) \\ & \quad - \frac{2}{p}\sum_{s=1}^{p} \xi_{i_1 \cdots s \cdots i_p | m} S_{i_s}{}^{lm} = 0. \end{aligned} \tag{6.4}$$

Now, if we put $\phi = \xi^{i_1 i_2 \cdots i_p}\xi_{i_1 i_2 \cdots i_p}$ and calculate $\Delta\phi$, then, as in the preceding Section, we obtain

$$\begin{aligned} \Delta\phi = & -\frac{1}{p} K^{(p)}_{ijkl}\,\xi^{iji_3 \ldots i_p}\xi^{kl}{}_{i_3 \cdots i_p} + \frac{2}{p} S^{(p)}_{ijrst}\,\xi^{iji_3 \ldots i_p}\xi^{rs}{}_{i_3 \cdots i_p}{}^{|t} \\ & + G_{rstuvw}\,\xi^{rsi_3 \ldots i_p | t}\xi^{uv}{}_{\cdot \cdots i}{}^{|w}. \end{aligned}$$

Thus, by virtue of Lemma 1, we have

THEOREM 11. *On a compact V_n, if the matrix*

$$N' = \begin{bmatrix} K^{(p)}_{ijkl} & -S_{ijrst} \\ -S_{ijrst} & -pG_{rstuvw} \end{bmatrix}$$

defines a negative definite quadratic form in the variables $\xi^{ij} = -\xi^{ji}$ and $\xi^{rst} = -\xi^{srt}$, then there exists no pseudo-Killing tensor of the pth order other than zero tensor.

If the matrix N' defines a non-positive quadratic form then any pseudo-Killing tensor $\xi_{i_1 i_2 \cdots i_p}$ must satisfy

$$K^{(p)}_{ijkl}\,\xi^{iji_3 \ldots i_p}\xi^{kl}{}_{i_3 \cdots i_p} - 2S^{(p)}_{ijrst}\,\xi^{iji_3 \ldots i_p}\xi^{rs}{}_{i_3 \cdots i_p}{}^{|t} - pG_{rstuvw}\,\xi^{rsi_3 \ldots i_p | t}\xi^{uv}{}_{i_3 \cdots i_p}{}^{|w} = 0.$$

For $S_{jk}{}^{i} = 0$, we have [10]

COROLLARY 1. *In a compact Riemannian space, if the form*

$$K^{(p)}_{ijkl}\,\xi^{ij}\xi^{kl} = pR_{ij}\,\xi^{ia}\xi^{j}{}_{a} + \frac{p(p-1)}{2} R_{ijkl}\,\xi^{ij}\xi^{kl}$$

in the variable $\xi^{ij} = -\xi^{ji}$ *is negative definite, then there exists no Killing tensor of the* p^{th} *order other than zero tensor. And if* $K^{(p)}_{ijkl}\xi^{ij}\xi^{kl}$ *is non-positive, then every Killing tensor must have vanishing covariant derivative.*

Also, Theorem 7 has the following generalization:

THEOREM 12. *In a compact orientable* V_n *with* $S_{jr}{}^{r} = 0$, *a necessary and sufficient condition that an anti-symmetric tensor* $\xi_{i_1 i_2 \cdots i_p}$ *be a pseudo-Killing tensor is that it satisfy* (6.4) *and* $\xi^{i}{}_{i_2\cdots i_p|i} = 0$.

7. Following E. Hopf [7], we obtain the following generalization of Lemma 1:

LEMMA 2. *On a compact* V_n, *if, for a scalar* ϕ *and a vector* A_s, *we have*

$$\Delta\phi + 2A^s\phi_{|s} \equiv g^{jk}\phi_{|j|k} + 2A^s\phi_{|s} \geqq 0,$$

then we have

$$\Delta\phi + 2A^s\phi_{|s} = 0.$$

Now, for ξ_i pseudo-harmonic, if $\phi = \xi^i\xi_i$, then

$$\Delta\phi + 2A^s\phi_{|s} = (E_{jk} + E_{kj})\xi^j\xi^k - 2(S_{irs} + S_{isr} - g_{ir}A_s - g_{is}A_r)\xi^i\xi^{r|s} + (g_{rt}g_{su} + g_{ru}g_{st})\xi^{r|s}\xi^{t|u}.$$

Thus, from Lemma 2, we have

THEOREM 13. *On a compact* V_n, *if the matrix*

$$P = \begin{bmatrix} E_{jk} + E_{kj}, & -(S_{irs} + S_{isr} - g_{ir}A_s - g_{is}A_r) \\ -(S_{irs} + S_{isr} - g_{ir}A_s - g_{is}A_r), & g_{rt}g_{su} + g_{ru}g_{st} \end{bmatrix}$$

where A_s *is a certain vector, defines a positive definite quadratic form in the variables* ξ^i *and* $\xi^{rs} = \xi^{sr}$, *then there exists no pseudo-harmonic vector other than zero vector.*

If the matrix P defines a non-negative quadratic form, then every pseudo-harmonic vector must satisfy

$$(E_{jk} + E_{kj})\xi^j\xi^k - 2(S_{irs} + S_{isr} - g_{ir}A_s - g_{is}A_r)\xi^i\xi^{r|s} + (g_{rt}g_{su} + g_{ru}g_{st})\xi^{r|s}\xi^{t|u} = 0.$$

Next, for ξ_i pseudo-Killing, if $\phi = \xi^i\xi_i$, then

$$\Delta\phi + 2A^s\phi_{|s} = -[(E_{jk} + E_{kj})\xi^j\xi^k - 2(S_{irs} - S_{isr} + g_{ir}A_s - g_{is}A_r)\xi^i\xi^{r|s} - (g_{rt}g_{su} - g_{ru}g_{st})\xi^{r|s}\xi^{t|u}].$$

Thus, from Lemma 2, we have

THEOREM 14. *On a compact* V_n, *if the matrix*

$$P' = \begin{bmatrix} E_{jk} + E_{kj}, & -(S_{irs} - S_{irs} + g_{ir}A_s - g_{is}A_r) \\ -(S_{irs} - S_{isr} + g_{ir}A_j - g_{is}A_r), & -(g_{rt}g_{su} - g_{ru}g_{st} \end{bmatrix}$$

where A_s *is a certain vector, defines a negative definite quadratic form in the variables* ξ^i *and* $\xi^{rs} = -\xi^{sr}$, *then there exists no pseudo-Killing vector other than zero vector.*

If the matrix P' defines a non-positive quadratic form, then every pseudo-Killing vector E^j must satisfy

$$(E_{jk}+E_{kj})\xi^j\xi^k - 2(S_{irs} - S_{isr} + g_{ir}A_s - g_{is}A_r)\xi^i\xi^{r|s} - (g_{rt}g_{su} - g_{ru}g_{st})\xi^{r|s}\xi^{t|u} = 0.$$

In the particular case where the torsion tensor satisfies an equation of the form

$$S_{irs} - S_{isr} + g_{ir}A_s - g_{is}A_r = 0, \tag{7.1}$$

we have

COROLLARY 1. *On a compact V_n with the torsion tensor S_{ijk} satisfying an equation of the form* (7.1), *if $E_{jk}\xi^j\xi^k$ is negative definite, there exists no pseudo-Killing vector other than zero vector.*

If $E_{jk}\xi^j\xi^k$ is non-positive, then every pseudo-Killing vector ξ^i must have vanishing covariant derivative with respect to the affine connection of the space.

It is to be noted that if $S_{jk}{}^i$ has the form $S_{jk}{}^i = \delta^i_j\phi_k - \delta^i_k\phi_j$, then (7.1) is satisfied.

Such an unspecified vector field A_s can also be introduced in a suitable fashion in the matrices occurring in Theorems 9 and 11, without altering the wording of the theorems otherwise.

PRINCETON UNIVERSITY AND
INSTITUTE FOR ADVANCED STUDY

REFERENCES

1. BOCHNER, S. *Vector fields and Ricci curvature.* Bull. Amer. Math. Soc., 52, (1946) 776–797.
2. BOCHNER, S. *Curvature and Betti numbers.* Ann. of Math., 49, (1948) 379–390.
3. BOCHNER, S. *Curvature and Betti numbers, II.* Ann. of Math., 50, (1949) 77–93.
4. CARTAN, E. and J A. SCHOUTEN. *On the geometry of the group manifold of simple and semi-simple groups.* Proc. van Wettens., Amsterdam, 26, (1926) 803–815.
5. EISENHART, L. P. Continuous groups of transformations. Princeton University Press, 1933.
6. HAYDEN, H. A. *Subspaces of a space with torsion.* Proc. London Math. Soc., 34, (1932) 27–50.
7. HOPF, E.: *Elementare Bemerkungen über die Lösungen partieller Differentialgleichungen zweiter Ordnung vom elliptischen Typus.* Preuss. Akad. Wiss. Sitzungsber., 19, (1927) 149–152.
8. RAUCH, H. E. *A contribution to differential geometry in the large.* Ann. of Math., 54, (1951) 38–55.
9. YANO, K. *On harmonic and Killing vector fields.* Ann. of Math., 55, (1952) 38–45.
10. YANO, K. *Some remarks on tensor fields and curvature.* Ann. of Math., 55, (1952) 328–347.

Reprinted from the Proceedings of the NATIONAL ACADEMY OF SCIENCES
Vol. 41, No. 7, pp. 485–490. July, 1955

STATIONARY SPACE-TIME IN GENERAL RELATIVITY

BY S. BOCHNER

PRINCETON UNIVERSITY

Communicated May 20, 1955

In the notation of the recent book of A. Lichnerowicz[1] we take a V_4 with an everywhere regular hyperbolic local line element which, somewhat differently from the book, we denote by

$$\sum_{\alpha,\,\beta=0}^{3} \bar{g}_{\alpha\beta}\, dx^{\alpha}\, dx^{\beta}. \tag{1}$$

We assume that V_4 is stationary, as there defined, and this implies that V_4 is the topological product of V_3 with R: $(-\infty < t < \infty)$ and that on V_3 the line element (1) induces certain data which on V_3 are linked by the relation

$$\Delta\xi = -\xi R + \frac{1}{4}\xi^3 H^2, \tag{2}$$

with which we will be concerned exclusively. Underlying it is a positive-definite Riemannian metric,

$$\sum_{i,\,j=1}^{3} g_{ij}\, dx^i\, dx^j, \tag{3}$$

which is induced by expression (1). The quantity ξ is a scalar function which is positive throughout,

$$\xi > 0, \tag{4}$$

and $\Delta\xi$ is the Laplacean $g^{ij}\xi_{,i,\,j}$; R is another scalar, which is denoted in Lichnerowicz with indices because it is induced by the Ricci tensor $\bar{R}_{\alpha\beta}$ pertaining to expression (1) and which is such that, if

$$\bar{R}_{\alpha\beta} = 0$$

throughout, then also

$$R = 0 \tag{5}$$

thoughout; and, finally,

$$H^2 = H_{ij}H^{ij} \geq 0, \tag{6}$$

where H_{ij} is a certain skew tensor.

Definition 1: We will say that V_4 is *empty* if on V_3 we have

$$H = 0, \qquad \xi = \text{const.}, \tag{7}$$

and we will say that it is *semi-empty* if only

$$H = 0. \tag{8}$$

Remark 1: The space is empty if and only if expression (1) is locally Euclidean, but this will not concern us. Also, relation (2), always on a V_3, also occurs in some field theories with suitable interpretation of the symbols, whether the theory be based on a V_4 or on V_k, $k \geq 5$, and relation (7) or (8) always implies a certain manner of *emptiness.*

In what follows, we are assuming that V_3 is C^2, g_{ij} is C^1, ξ is C^2, R is C^0, and H_{ij} is C^0.

Important criteria for the occurrence of relation (7), or perhaps only (8), are due to, among others,[1] Levi-Civita, R. Serini, Ch. Racine, Lichnerowicz, Einstein and Pauli, and Yvonne Fourès and Lichnerowicz (for a recent result in a related but different direction see a paper by A. H. Taub[2]), and it is the purpose of the present paper to comment on one of these criteria by way of a generalization. The generalization as such has been given before[3] in our work on "Curvature and Betti numbers" but will be restated for the present context.

THEOREM 1. *Let V_3 be compact. (i) If we assume*

$$-\int_{V_3} \xi R\, dv \geq 0, \tag{9}$$

then V_4 is semi-empty, and equality holds:

$$-\int_{V_3} \xi R\, dv = 0. \tag{10}$$

(ii) If, more precisely,

$$-R \geq 0, \tag{11}$$

then V_4 is empty, and, in fact, equation (5) holds.

Proof: If we integrate equation (2), we obtain

$$\int_{V_3} \Delta\xi\, dv = -\int_{V_3} \xi R\, dv + \frac{1}{4}\int_{V_3} \xi^3 H^2\, dv. \tag{12}$$

On the other hand, since V_3 is compact, we have[4]

$$\int_{V_3} \Delta\xi \, dv = 0, \tag{13}$$

and relations (6), (9), (12), and (13) imply (8) and (10) as claimed. If, more precisely, relation (11) holds, then (2) implies $\Delta\xi \geq 0$, and the desired conclusion $\xi =$ const. follows now either from the formula[4]

$$\int_{V_3} (\xi + C) \cdot \Delta\xi \, dv = -\int_{V_3} g^{ij} \, \xi_{,i,j} \, dv \tag{14}$$

or from the following local theorem due to E. Hopf.

LEMMA. *In a co-ordinate neighborhood V, if a function $\xi(P)$ of class C^2 satisfies the inequality $\Delta\xi \geq 0$ and if there is a fixed point P^0 in V such that $\xi(P) \leq \xi(P^0)$ everywhere in V, then we must have $\xi(P) = \xi(P^0)$ everywhere in V.*

We are now turning to comment on Theorem 1, in case V_3 is not (necessarily) compact.

Assumption 1: There is given on V_3 a group $\Gamma:(\gamma)$ of one-to-one continuous transformations (with fixed points, perhaps) having the following property. There exists a compact set S^0 in V_3 such that, corresponding to any point P^0 in V_3, there is an element γ^0 in Γ such that the transformed point $\gamma^0 P^0$ lies in S^0.

Definition 2: Let $f(P)$ be a continuous function on V_3, either a scalar or, more generally, any tensor density. We call it "almost periodic" (or almost automorphic), relative to the given group Γ, if every infinite sequence of elements $\{\gamma_n\}$ in Γ contains an infinite subsequence $\{\gamma_{n_p}\}$, $p = 1, 2, \ldots$, having the following property. Corresponding to any point, there is a co-ordinate neighborhood in which the sequence of functions $f(\gamma_{n_p}P)$ converges uniformly in each component, and, if we denote the limit function by $f^*(P)$, then, reversely, the sequence of functions $f^*(\gamma_{n_p}^{-1}P)$ converges, in the same fashion, *toward the function $f(P)$ from which we have started.*

Remark 2: The italicized part of definition 2 makes it unnecessary to require that (the space V_3 shall have uniform structure and that) the sequences shall converge uniformly in all of V_3.

It is easily seen that sums, products, and contractions are again almost periodic and that for an almost periodic scalar $\xi(P)$ the number

$$m = \sup_P \xi(P) \tag{15}$$

is finite (and also the inf.).

THEOREM 2. *Under assumption* 1, *if the functions*

$$g_{ij}, g^{ij}, \xi, \xi_{,i}, \xi_{,i,j}, R, H_{ij}, \tag{16}$$

are all almost periodic with respect to Γ, and if we have $R \leq 0$, then V_4 is empty.

Proof: Relation (11) implies $\Delta\xi \geq 0$. Now, if the supremum (15) is attained, that is, if there is a point P^0 for which $\xi(P^0) = m$, then the lemma implies $\xi =$ const. everywhere. But relations (2), (6), and (11) then also imply $H = 0$ as claimed.

If, however, it is not attained, then there exists a sequence of points $\{P_n\}$ for which $\xi(P_n) \to m$. By requirement 1 there exist elements $\gamma_n \in \Gamma$ such that $\gamma_n P_n \to S^0$. But S^0 is compact, and if we apply definition 2 judiciously, we may assume that the sequence $\{\gamma_n\}$ has the following properties simultaneously. We have $\gamma_n P_n \to P^0$, where P^0 is a point of S^0, and the 7 sequences (see expressions [16]),

$$g_{ij}(\gamma_n P_n),\ g^{ij}(\gamma_n P_n),\ \xi(\gamma_n P_n), \ldots,$$

all converge uniformly in the neighborhood of every point of V_3. If we denote the limit functions by asterisks, $g_{ij}{}^*$, g^{*ij}, ξ^*, R^*, etc., then we again have

$$\Delta^*\xi^* = -\xi^* R^* + \frac{1}{4}\xi^{*3}H^{*2},$$

and the function ξ^* does attain its supremum. Therefore, ξ^* is constant; but, by the reverse part of definition 1, the function ξ is likewise a constant, and this completes the proof of the theorem.

If V_3 is the Euclidean E_3 and Γ the group of all translations, then assumption 1 is fulfilled, with S being any one single point of E_3. Also, in this case a function $\xi(x)$, $x = (x^1, x^2, x^3)$, is almost periodic in the sense of definition 2 if it is almost periodic in the sense of Bohr,[5] that is, if and only if it is a limit, uniformly in E_3, of finite sums

$$\sum_\lambda a_{\lambda_1\lambda_2\lambda_3} e^{i(\lambda_1 x^1 + \lambda_2 x^2 + \lambda_3 x^3)}$$

for arbitrary real numbers $\lambda_1, \lambda_2, \lambda_3$. Theorem 2 now implies the following:

THEOREM 3. *Assume that V_3 is the Euclidean E_3 and that we are given one global line element* (3) *on it. If the individual functions*

$$g^{ij}(x),\ g_{ij}(x),\ \frac{\partial g_{ij}}{\partial x^k},\ \frac{\partial^2 g_{ij}}{\partial x^k \partial x^l},\ \xi,\ \frac{\partial \xi}{\partial x^k},\ \frac{\partial^2 \xi}{\partial x^k \partial x^l},\ R,\ H_{ij},$$

are all almost periodic in the sense of Bohr, and if $R \leq 0$, then V_4 is empty space.

An almost periodic function $f(x)$ on a group has a mean $M\{f(x)\}$ which on E_3 can be evaluated as

$$M\{f(x)\} = \lim_{T\to\infty} \frac{1}{(2T)^3} \int_{-T}^{T}\int_{-T}^{T}\int_{-T}^{T} f(x^1, x^2, x^3)\, dx^1\, dx^2\, dx^3,$$

say. Now, for a domain D in E_3 with sufficiently smooth boundary B, Stokes's formula gives

$$\int_D \Delta\xi \cdot dv = \int_B \frac{\partial \xi}{\partial n}\, d\omega; \tag{17}$$

but if D is the cube $-T \leq x^i \leq T$, $i = 1, 2, 3$, then, for $T \to \infty$,

$$\int_B |d\omega| = 0(T^2), \tag{18}$$

and

$$\frac{d\xi}{dn} = 0(1), \tag{19}$$

and therefore

$$M\{\Delta\xi\} = 0. \tag{20}$$

However, relation (2) implies

$$M\{\Delta\xi\} = -M\{\xi R\} + \frac{1}{4}M\{\xi^3H^2\}, \tag{2.}$$

and

$$f \geq 0 \quad \text{and} \quad Mf = 0 \quad \text{imply} \quad f = 0. \tag{22}$$

Hence the following conclusion.

THEOREM 4. *If in Theorem* 3 *we replace relation* (11) *by the weaker assumption*

$$-M\{\xi R\} \geq 0, \tag{23}$$

then V_3 is semi-empty.

Remark 3: If in Theorems 3 or 4 the line element (3) is not global but only local, then we must assume that functions (16) are almost periodic in the sense of Bohr, and the conclusions remain in force. For Theorem 3 this is a direct consequence of Theorem 2, and for Theorem 4 this follows from the fact that, if we put

$$\Delta\xi \cdot dv = \Delta\xi\sqrt{g}\,dx^1\,dx^2\,dx^3,$$

then the integrand is almost periodic in the sense of Bohr, and similarly for the other integrands in equation (21).

A meaningful extension of Theorem 4 to a V_3 with a general group Γ is not easy to obtain; but we will make a statement, a somewhat technical one, for the case in which the topological quotient V_3/Γ is itself a (compact) manifold, which we will denote by U_3. In this case the group Γ is countable, and we enumerate its elements as a sequence $\gamma^0, \gamma^1, \gamma^2, \ldots$, in which γ^0 is the identity, and we decompose V_3 into a sequence of disjoint "fundamental domains" $U^0, U^1, U^2, \ldots$, which are such that γ^n transforms U^0 into U^n. They are not really "domains," that is, open sets, but their closures are (compact) closures of domains. From relation (2) we obtain

$$\int_{U^m}\Delta\xi\,dv = -\int_{U^m}\xi R\,dv + \frac{1}{4}\int_{U^m}\xi^3H^2\,dv, \tag{24}$$

and we note that for any function $f(P)$ on V_3 we have

$$\int_{U^m} f(P)\,dv(P) = \int_{U^0} f(\gamma^mP)\,dv(\gamma^mP). \tag{25}$$

Now, the integrand on the right side is the product of $dx^1\,dx^2\,dx^3$ with $f(\gamma^mP)$ $(g(\gamma^mP))^{1/2}$, and, if the last function is almost periodic in our sense, then it is almost periodic on Γ in the sense of von Neumann, uniformly for P in U^0. Such a function has, again, an almost periodic mean $M\{\quad\}$, again with property (22), and equation (24) implies

$$M_\gamma\left\{\int_{U^0}\Delta\xi\,dv\right\} = -M_\gamma\left\{\int_{U^0}\xi R\,dv\right\} + \frac{1}{4}M_\gamma\left\{\int_{U^0}\xi^3H^2\,dv\right\}. \tag{26}$$

Hence the following conclusion.

THEOREM 5. *If it is known that, in analogy to equation* (13), *we have*

$$M_\gamma\{\textstyle\int_{U^0}\Delta\xi\,dv\} = 0, \tag{27}$$

then the assumption

$$M_\gamma\{\textstyle\int_{U^0}\xi R\,dv\} \leq 0$$

implies that V_4 is semi-empty.

S. BOCHNER

This theorem is useless unless equation (27) can be established, and, if one wishes to prove this by imitating the procedure used for Theorem 4, then the following features have to be verified, subject to assumptions. It is possible to arrange the ordering $\{\gamma^m\}$ in such a manner that for a sequence of integers $T_n \to \infty$ we have

$$M\{f(P)\} = \lim_{n\to\infty} \frac{1}{T_n} \sum_{m=1}^{T_n} f(\gamma^m P) \tag{28}$$

uniformly for P in U^0. Also, the union of sets

$$\sum_{m=0}^{T_n} U^m$$

constitutes a piece of space whose actual boundary consists of several sufficiently smooth surfaces whose two-dimensional induced surface area is $o(T_n)$ as $n \to \infty$.

Remark 4: Finally, we note that, with the aid of the mean (27), we may even obtain the statement on emptiness itself without requiring almost-periodicity at all. Assume that for a certain mean (28) we may go over from equation (25) to equation (26); that relation (22) holds; and that we have not only equation (27) but also

$$M_\gamma\{\textstyle\int_{U^0} \xi\cdot\Delta\xi\, dv\} = -M_\gamma\{\textstyle\int_{U^0} g^{ij}\, \xi_i\xi_j\cdot dv\}$$

in analogy to equation (14); then $-R \geq 0$ implies that V_4 is empty.

[1] *Theories relativistes de la gravitation et de l'électromagnétisme* (Paris: Masson & Cie, 1955).

[2] "Empty Space-Times Admitting a Three Parameter Group of Motions," *Ann. Math.*, **53,** 472–490, 1951.

[3] "Vector Fields and Ricci Curvature," *Bull. Am. Math. Soc.*, **52,** 776–797, 1946, esp. pp. 782–83; K. Yano and S. Bochner, *Curvature and Betti Numbers* (Princeton, N.J.: Princeton University Press, 1953), pp. 182–185.

[4] "Remark on the Theorem of Green," *Duke J.*, **3,** 334–338, 1937.

[5] "Beiträge zur Theorie der fastperiodischen Funktionen. II. Funktionen mehnerer Variablen," *Math. Ann.*, **96,** 383–409, 1926.

Salomon Bochner

CURVATURE AND BETTI NUMBERS IN REAL AND COMPLEX VECTOR BUNDLES *

Introduction.

The title of the present paper is intended to indicate its topic, but in fact no Betti numbers or similar invariants will be introduced or even suggested.

In differential geometry of a manifold M_n any concept or statement involving affine connection, Riemannian metric, curvature tensor, etc. can also be interpreted as an attribute not of M_n itself but of a certain vector bundle erected over it « tangentially ». Now, starting from this observation it suggests itself to envisage vector bundles other than the classical ones, and an outline of the theory of differential geometry for vector bundles in general, based on tensor calculus, was given in our previous paper [3]. Since this paper was written, the topic of vector bundles has acquired considerable actuality, and in the present paper a certain continuation to the previous one, which may be of interest to recent trends, will be presented.

In [3] we dealt mostly with local questions but two global theorems were stated as well. Of these the first—it is there numbered Theorem 2—, makes a statement referring to « Curvature and Betti numbers », and in a sense it is a very comprehensive statement too, as we will see. But it requires some comment and explication in order to recognize how comprehensive in scope it really is, and it is the purpose of the present paper to contribute such explications. We will not be concerned with re-obtaining

* Done with the support of the United States Air Force through the Office of Scientific Research of the Air Force Research and Development Command.

specific previous results of ours in enlarged form, but instead we will draw up fairly comprehensive propositions, under which recent differential-geometric statements made in Kodaira [7], [8] will be subsumed systematically.

Syllogistically we could take the paper [3] as known, and start out with the continuation to its Theorem 2 directly. However, it will be more convenient to repeat essential prerequisites in a suitable fashion, but we must emphasize that their repetition is not designed to supersede the original paper [3] as such.

1. - Tensor algebra in vector bundles.

1.1. As in [3] we are describing an N-dimensional vector bundle over an n-dimensional base space M_n by a system of transition matrices

$$\{a^A_B(y;x)\}\,, \qquad A, B = 1, \dots, N \tag{1.1.1}$$

between any two intersecting elements in a fixed covering of M_n by a system of neighborhoods. We will call such a system (1.1.1) a *matrix structure*, and it is subject to the consistency properties

$$a^A_B(z;y)\; a^B_C(y;x) = a^A_C(z;x) \tag{1.1.2}$$

[summation over B] and

$$a^A_B(x;x) = \delta^A_B \tag{1.1.3}$$

[$=1$, if $A = B$, and $=0$ if $A \neq B$].

A generic point of the vector bundle constructed by (1.1.1) is either a *contravariant vectoroid*

$$u^A = \{u^1, \dots, u^N\} \tag{1.1.4}$$

with the transition relations

$$u'^A(y) = a^A_B(y;x)\, u^B(x)\ ; \tag{1.1.5}$$

or, alternately, a *covariant vectoroid*

$$t_A = \{t_1, \ldots, t_N\} \tag{1.1.6}$$

with the law of transition

$$t'_A(y) = a_A^B(x; y)\, t_B(x) \ . \tag{1.1.7}$$

For (1.1.7) we can also write

$$a_A^B(x; y)\, t'_B(y) = t_A(x) \ ,$$

and we further note as follows. With any structure (1.1.1) we can associate a « dual » one, namely

$$\bar{a}_A^B(y; x) \equiv a_B^A(x; y) \ , \tag{1.1.8}$$

with « duality » being an involution, and what is covariant for one of the two structures (1.1.1) and (1.1.8) is contravariant for the other.

1.2. An ordinary co-, or contravariant vector ξ_i, η^j is such a vectoroid pertaining to the classical matrix structure

$$\left\{ \frac{\partial y_j}{\partial x_i} \right\} \ , \tag{1.2.1}$$

and a mixed tensor, in the ordinary sense,

$$\xi_{i_1 \ldots i_p \,\cdots\cdots}^{\cdots\cdots\, j_1 \ldots j_q} \ , \qquad p + q = s \ , \tag{1.2.2}$$

is a covariant vectoroid pertaining to the matrix structure which is the Kronecker product of p factors (1.2.1) and q dual factors.

Quite generally, if we are given s matrix structures of the form (1.1.1),

$$\{a_{B_r}^{A_r}(y; x)\} \ , \qquad r = 1, \ldots, s \ , \tag{1.2.3}$$

of various dimensions N_r, with no connection whatsoever between them stipulated, then their Kronecker product is again a structure, and a covariant vectoroid pertaining to it, if denoted by the symbol

$$(1.2.4) \qquad t_{A_1 \dots A_s} ,$$

obeys the law of transformation

$$(1.2.5) \qquad t'_{A_1 \dots A_s}(y) = a_{A_1}^{B_1}(x; y) \dots a_{A_s}^{B_s}(x; y)\, t_{B_1 \dots B_s}(x) .$$

Inasmuch as (1.2.4) pertains to a Kronecker product of various matrix structures we will also call it a covariant *tensoroid*, and will also consider contravariant and mixed tensoroids

$$t^{B_1 \dots B_s}_{\dots\dots\dots} , \qquad t^{\dots\dots\dots\; B_1 \dots B_q}_{A_1 \dots A_p\; \dots\dots\dots}$$

in an obvious sense. But we are emphasizing that any tensoroid is also a vectoroid for a matrix structure suitably chosen.

1.3. If with any structure (1.1.1) we form the determinant

$$(1.3.1) \qquad a_1^1(y; x) = \det \left| a_A^B(y; x) \right|_{A, B = 1, \dots N} ,$$

then this is a new structure of dimension 1. The vectoroids pertaining to it might be called « densities », but the distinction between contravariant and covariant must be observed for densities too.

Structures of the form (1.3.1) are already known in the classical theory. An object (1.2.2) is there called a tensor of weight W, for $W > 0$, if it is a covariant vectoroid pertaining to the Kronecker product of p factors (1.2.1), q dual factors, and W one-dimensional factors

$$\det \left| \frac{\partial y_j}{\partial x_i} \right| .$$

We remark that if we are given N vectoroids

$$(1.3.2) \qquad \{ t_A^{(a)} \} , \qquad a = 1, \dots, N ,$$

for the general structure (1.1.1), then the quantity

$$t_1 = \det \left| t_A^{(a)} \right|_{A,\, a=1, \dots, N}$$

is a covariant density pertaining to (1.3.1).

Finally we remark that we can always introduce the one-dimensional trivial structure

$$a_1^1(y; x) = 1 \, . \tag{1.3.3}$$

Its vectoroids are the ordinary scalar functions on M_n, and there is no difference between covariant and contravariant then.

1.4. If we are given the structure (1.1.1) and a second structure

$$\{ b_D^C(y; x) \} \, , \tag{1.4.1}$$

of dimension M, then the $(N+M)$-dimensional matrix

$$\begin{Bmatrix} a_B^A(y; x) \, , & 0 \\ 0 \, , & b_D^C(y; x) \end{Bmatrix} \tag{1.4.2}$$

is a structure again. We will call it the *logical sum* of the two structures, and we will require it for the complex cases only, and not before.

2. - Tensor analysis in vector bundles.

2.1. From now on we assume that with each matrix structure introduced there is associated a fixed *affine connection* which we will denote by the symbol

$$L_{Bi}^A \tag{2.1.1}$$

usually. It satisfies the law of transformation

$$a_A^B(y; x) \frac{\partial y_j}{\partial x_i} L'^C_{Bj}(y) = - \frac{\partial a_A^C(y; x)}{\partial x_i} + a_B^C(y; x) \, L_{Ai}^B(x) \, , \tag{2.1.2}$$

and for a given vector bundle none need exist, globally. But if one does exist, as we will always assume, the following uniqueness statement can be made.

If M^A_{Bi} is a second affine connection, then the difference

$$T^A_{Bi} = M^A_{Bi} - L^A_{Bi} \tag{2.1.3}$$

satisfies the relations

$$a^B_A(y;x)\,\frac{\partial y_j}{\partial x_i}\;T'^C_{Bj}(y) = a^C_B(y;x)\;L'^B_{Ai}(x)\;, \tag{2.1.4}$$

and is thus a tensoroid, mixed as the position of the indices indicates, pertaining to a product of two factors (1.1.1) and one factor (1.2.1). —Conversely, for any affine connection L^A_{Bi} and an arbitrary such tensoroid T^A_{Bi} the sum

$$L^A_{Bi} + T^A_{Bi} \tag{2.1.5}$$

is an affine connection again.

2.2. If (2.1.1) satisfies (2.1.2) then

$$-\,L^B_{Ai} \tag{2.2.1}$$

satisfies the same relation for the dual structure (1.1.8). Thus the latter has also an affine connection. By what we have just seen, the latter is not uniquely determined as such, but we are stipulating that it shall be the expression (2.2.1) itself. A similar « normalization » shall apply in other similar cases as well.

2.3. The structure (1.3.1) has also an affine connection, namely

$$L^1_{1i} = L^A_{Ai}\,(= \sum_{A=1}^{N} L^A_{Ai})\;. \tag{2.3.1}$$

We will not make use of this fact, and we are mentioning it only because it occurs in the classical theory already.

2.4. If we denote the affine connection of (1.2.3) by

$$L^{A_r}_{B_r i}\,, \tag{2.4.1}$$

then the expression

$$\sum_{r=1}^{s} \delta^{A_1}_{B_1} \cdots \delta^{A_{r-1}}_{B_{r-1}} L^{A_r}_{B_r i}\, \delta^{A_{r+1}}_{B_{r+1}} \cdots \delta^{A_s}_{B_s} \tag{2.4.2}$$

is an affine connection for the Kronecker product of the factors (1.2.3). We will denote it by the symbol

$$L^{A_1 \dots A_s}_{B_1 \dots B_s i}\,, \tag{2.4.3}$$

and call it the « Kronecker sum » of the connections (2.4.1).

2.5. If t_A is a vectoroid for (1.1.1) then the expression

$$t_{A,\,i} = \frac{\partial t_A}{\partial x_i} - t_C L^C_{Ai} \tag{2.5.1}$$

is a tensoroid for the Kronecker product of (1.1.1) and (1.2.1). As usual, we will call it the *covariant derivative* of t_A. By applying previous rules we have

$$t^B_{.,\,i} = \frac{\partial t^B}{\partial x_i} + t^C L^B_{Ci}\,, \tag{2.5.2}$$

and quite generally

$$t^{B_1 \dots B_q}_{A_1 \dots A_p,\,i} = \frac{\partial}{\partial x_i} t^{B_1 \dots B_q}_{A_1 \dots A_p} \tag{2.5.3}$$

$$- \sum_{\varrho=1}^{p} t^{B_1 \dots B_q}_{A_1 \dots C_\varrho \dots A_p} L^{C_\varrho}_{A_\varrho i} + \sum_{\sigma=1}^{q} t^{B_1 \dots C_\sigma \dots B_q}_{A_1 \dots A_p} L^{B_\sigma}_{C_\sigma i}\,.$$

2.6. By our general assumption there is also an affine connection pertaining to the structure (1.2.1) and we denote it by

$$L^p_{qr}\,. \tag{2.6.1}$$

Since a covariant derivative of a tensoroid is again a tensoroid we can, by the use of (2.6.1), form new covariant derivatives successively.

It is easy to verify that we have

$$t_{A,i,j} - t_{A,j,i} = t_C L^C_{ij} + t_{A,k} (L^k_{ij} - L^k_{ji}) \tag{2.6.2}$$

where, by definition,

$$L^B_{Aij} = \frac{\partial L^B_{Aj}}{\partial x_i} - \frac{\partial L^B_{Ai}}{\partial x_j} + L^B_{Ci} L^C_{Aj} - L^B_{Cj} L^C_{Ai} \ . \tag{2.6.3}$$

The last expression is a tensoroid, and we will call it the *curvature tensoroid*, in analogy to the classical case. It obviously has the property

$$L^B_{Aij} = - L^B_{Aji} \ . \tag{2.6.4}$$

We are now adding the assumption that the affine connection (2.6.1) shall be symmetric in the lower indices

$$L^p_{qr} = L^p_{rq} \ , \tag{2.6.5}$$

and we would like to point out that we know of no really satisfactory generalization of this symmetry requirement to affine connection as defined for structures in general.

The assumption (2.6.5) has the following consequences. Firstly, we have

$$L^B_{Aij,\,k} + L^B_{Aki,\,j} + L^B_{Ajk,\,i} = 0 \ , \tag{2.6.6}$$

and we note this down for its own sake, although it will be of no importance to us. What will be of importance is the fact that (2.6.2) now reduces to

$$t_{A,i,j} - t_{A,j,i} = t_C L^C_{Aij} \ , \tag{2.6.7}$$

and that more generally we have

$$\begin{aligned} & t^{B_1 \dots B_q}_{A_1 \dots A_p,\, i,\, j} - t^{B_1 \dots B_q}_{A_1 \dots A_p,\, i,\, j} \\ & = \sum_{\varrho=1}^{p} t^{B_1 \dots B_q}_{A_1, \dots C_\varrho \dots A_p} L^{C_\varrho}_{A_\varrho ij} - \sum_{\sigma=1}^{q} t^{B_1 \dots C_\sigma \dots B_q}_{A_1 \dots A_p} L^{B_\sigma}_{C_\sigma ij} \ . \end{aligned} \tag{2.6.8}$$

In the last formula the symbol

$$L^{B_r}_{A_r ij} \tag{2.6.9}$$

is the curvature tensoroid for the affine connection of (2.4.1) of the Kronecker factor (1.2.3). If now we denote the curvature tensoroid for the Kronecker product by

$$L^{B_1 \dots B_s}_{A_1 \dots A_s ij} \; , \tag{2.6.10}$$

then by substituting the Kronecker sum (2.4.2) in (2.6.3) it turns out that (2.6.10) is again a Kronecker sum, namely

$$\sum_{r=1}^{s} \delta^{B_1}_{A_1} \dots \delta^{B_{r-1}}_{A_{r-1}} L^{B_r}_{A_r ij} \delta^{B_{r+1}}_{A_{r+1}} \dots \delta^{B_s}_{A_s} \; . \tag{2.6.11}$$

Also, for the dual to a matrix, the curvature tensoroid changes the algebraic sign, and all this together gives (2.6.8).

2.7. We will also assume that for each matrix structure (1.1.1) there is given a tensoroid

$$g_{AB} \tag{2.7.1}$$

pertaining to the Kronecker product of (1.1.1) with itself, and that it has the following properties. It is symmetric,

$$g_{AB} = g_{BA} \; , \tag{2.7.2}$$

non-singular,

$$\det |g_{AB}|_{A, B = 1, \dots, N} \neq 0 \; , \tag{2.7.3}$$

and it has covariant derivative 0,

$$g_{AB, i} = 0 \; . \tag{2.7.4}$$

Since, always,

$$g_{AB, i, j} - g_{AB, j, i} = g_{CB} L^{C}_{Aij} + g_{AC} L^{C}_{Aij} \; , \tag{2.7.5}$$

if we put

$$L_{ABij} = g_{AC} L^C_{Bij} \, , \tag{2.7.6}$$

then (2.7.2) and (2.7.5) imply

$$L_{ABij} = - L_{BAij} \, , \tag{2.7.7}$$

as in the classical theory again. This relation is a companion to (2.6.4) but a somewhat subtler one, and we emphasize that we do *not* have anything like

$$L_{ABij} = L_{ijAB} \, ,$$

because a symbol L_{ijAB} is *not* definable, as far as we know.

Furthermore, assumption (2.7.3) implies the existence of a tensoroid

$$g^{AB} \tag{2.7.8}$$

which is again symmetric, and uniquely determined, such that

$$g_{AC} \, g^{CB} = \delta^B_A \, . \tag{2.7.9}$$

It has again the property

$$g^{AB}_{\cdot\cdot,i} = 0 \, , \tag{2.7.10}$$

and the pair of tensoroids (2.7.1) and (2.7.8) can be used for pulling indices up and down as in the classical theory.

Thus for instance

$$\xi_A \eta^A = \xi^A \eta_A = g^{AB} \xi_A \eta_B = g_{AB} \xi^A \eta^B \, .$$

2.8. By the general assumption made in 2.7 there also exists in the structure (1.2.1) a positive definite symmetric tensor g_{pq} for which

$$g_{pq,i} = 0 \tag{2.8.1}$$

relative to the affine connection (2.6.1). But since we have assumed that the latter is symmetric, see (2.6.5), it follows that (2.6.1) is the familiar RIEMANN-CHRISTOFFEL symbol, and in order to emphasize this we will sometimes denote it by

(2.8.2) $$\Gamma^p_{qr} \ .$$

2.9. If, for each r,

(2.9.1) $$g_{A_r B_r}$$

is a « fundamental » tensoroid pertaining to (1.2.3) then the object

(2.9.2) $$g_{(A)(B)} = g_{A_1 B_1} \cdots g_{A_s B_s}$$

is such a tensoroid for the Kronecker product of these structures.

In fact covariant derivation obeys LEIBNIZ' rule for the derivative of a product, and thus, in particular, the needed relation

$$g_{(A)(B),\, i} = 0$$

is a consequence of

$$g_{A_r B_r,\, i} = 0 \ , \qquad r = 1, \ldots, s \ .$$

In a product of structures the fundamental tensoroid will be always chosen to be the expression (2.9.2).

3. - A GENERAL THEOREM.

3.1. If $\varphi(P)$ is a scalar on M_n then its LAPLACEAN

(3.1.1) $$\Delta\varphi = g^{ij}\, \varphi_{,i,j}$$

is in particular a divergence

(3.1.2) $$g^{ij} \lambda_{i,j}$$

where $\lambda_i = \varphi_{,i}$. But for a compact M_n we have, for any vector field λ_i,

$$\int_{M_n} g^{ij} \lambda_{i,j} \, dv = 0 \quad , \tag{3.1.3}$$

and hence the following conclusion.

LEMMA 3.1.1. (i) *On a compact M_n, for any scalar $\varphi(P)$,* if

$$\Delta\varphi \geq 0 \tag{3.1.4}$$

throughout, then also

$$\Delta\varphi = 0$$

throughout.

(ii) *More generally, for any vector field λ_i, if*

$$g^{ij}\lambda_{i,j} \geq 0$$

throughout, then also

$$g^{ij}\lambda_{i,j} = 0$$

throughout.

For a non-compact M_n there is the following local theorem, due to *E.* Hopf, which however, as it stands, is only applicable to $\Delta\varphi$ (or a somewhat more general elliptic operator of second order) but not to the divergence of a vector in general.

LEMMA 3.1.2. *In a coordinate neighborhood V, if a function $\varphi(P)$ satisfies the inequality $\Delta\varphi \geq 0$ and if there is a fixed point P^0 in V such that $\varphi(P) \leq \varphi(P^0)$ everywhere in V, then we must have $\varphi(P) = \varphi(P^0)$ everywhere in P^0* .

3.2. We now take a matrix structure (1.1.1) for which all the assumptions enumerated in Section 2 are satisfied, and for a vectoroid ξ_A in it we form the scalar

$$\varphi = g^{AB}\xi_A\xi_B \equiv \xi_A\xi^A \; . \tag{3.2.1}$$

Obviously

$$\frac{1}{2}\,\Delta\varphi = T_1 + T_2 \ , \tag{3.2.2}$$

where

$$T_1 = g^{rs} g^{AC} \xi_{A,\,r} \xi_{C,\,s} \tag{3.2.3}$$

and

$$T_2 = g^{AC}(g^{rs}\xi_{A,\,r,\,s})\,\xi_C \equiv (g^{rs}\,\xi_{A,\,r,\,s})\,\xi^A \ . \tag{3.2.4}$$

Now, we always have

$$T_1 \geq 0 \ , \tag{3.2.5}$$

and thus, in order to obtain (3.1.4) we only have to secure

$$T_2 \geq 0 \ , \tag{3.2.6}$$

To this end we assume that ξ_A satisfies a certain equation (or rather that its components satisfy a system of equations) of the form

$$g^{rs}\xi_{A,\,r,\,s} = g^{EC}\,S_{AE}\xi_C \tag{3.2.7}$$

that is

$$g^{rs}\xi_{A,\,r,\,s} = S_{A\,\cdot}^{\cdot\,C}\,\xi_C \ , \tag{3.2.8}$$

where S_{AE} is a given tensoroid on the product of (1.1.1) with itself. We do *not* assume that this tensoroid is symmetric, so that

$$S_{AE} \neq S_{EA}$$

may well occur.

Now, obviously

$$T_2 = S^{AC}_{\cdot\ \cdot}\,\xi_A\xi_C \tag{3.2.9}$$

and hence, by part (i) of Lemma 3.1.1, the following conclusion.

THEOREM 3.2.1. *On a compact M_n, if a vectoroid ξ_A satisfies an equation* (3.2.8) *in which the given tensoroid S_{AC} is positive definite, in the sense that*

$$(3.2.10) \qquad S^{AC}_{\cdot\cdot}\, t_A t_C > 0 \quad \textit{for} \quad g^{AC}_{\cdot\cdot}\, t_A t_C > 0$$

then ξ_A must be identically 0.

If the matrix S_{AC} is only positive semi-definite, then the only solutions of (3.2.8) *are those for which*

$$(3.2.11) \qquad \xi_{A,\,i} = 0 \;, \qquad S_{A\cdot}^{\cdot\, C}\, \xi_C = 0$$

simultaneously.

By applying Lemma 3.1.2 the following generalization arises.

THEOREM 3.2.2. *On a non-compact M_n, the entire conclusion holds if the vectoroid has « boundary values* 0 *» in the following sense. To each $\epsilon > 0$ there corresponds a compact subset M_n^{ε} of M_n such that*

$$(3.2.12) \qquad g^{AB} \xi_A \xi_B < \varepsilon^2$$

for all points outside M_n^{ε}.

Theorem 3.2.2 subsumes Theorem 3.2.1, because on a compact M_n we can choose $M_n^{\varepsilon} = M_n$, and (3.2.12) holds always for the set $M_n - M_n^{\varepsilon}$ because this is the empty set now.

3.3. There is another, more important generalization to a non-compact M_n, which we have been stressing from the very beginning of our work (see, for instance, [1], p. 782/3, and [4], p. 182/4) and which we are now going to describe in our latest version as introduced in [5], p. 487.

(i) M_n is not (necessarily) compact, and there is given on M_n a group $\Gamma : (\gamma)$ of one-to-one continuous transformations (with fixed points, perhaps) having the following property. There exists a compact set $M^0{}_n$ in M_n such that corresponding to any point P^0 in M_n there is an element γ^0 in Γ such that the transformed point $\gamma^0 P^0$ lies in $M^0{}_n$.

(ii) There is given on M_n a fixed covering by neighborhoods $\{U\}$ such that each element γ of Γ transforms the family $\{U\}$ into itself, and all matrix structures (1.1.1) which we shall introduce shall refer to this one fixed system of neighborhoods always.

(iii) Let $f(P)$ be a continuous function on M_n, scalar, or having components (tensoroids, affine connection, etc.). We call it « almost periodic » (or, « almost automorphic ») relative to the given group Γ, if every infinite sequence of elements $\{\gamma_n\}$ in Γ contains an infinite sub-sequence $\{\gamma_{n_p}\}$, $p = 1, 2, \ldots$, having the following property. Corresponding to any point, there is a neighborhood U of it in which the sequence of functions $f(\gamma_{n_p} P)$ converges uniformly, in each of its components, and if we denote the limit function by $f^*(P)$, then reversely, the sequence of functions $f^*(\gamma_{n_p}^{-1} P)$ converges, in the same fashion, *towards the function $f(P)$ from which we have started.*

REMARK 3.3.1. As already remarked in [5], the italicized part of requirement (iii) makes it unnecessary to impose any kind of «uniform structure» on the space M_n globally and to require that our sequences shall converge uniformly with regard to such a structure globally.

We can now conclude as follows, always using Lemma 3.1.2, see [5], p. 487.

THEOREM 3.3.1. *Theorem 3.2.1 also holds on a non-compact M_n with a group $\Gamma:(\gamma)$ as described, if the quantities*

$$g^{rs},\ g^{AB},\ \xi_A,\ \xi_{A,i},\ \xi_{A,i,j} \tag{3.3.1}$$

are almost automorphic as described.

REMARK 3.3.2. It should be observed that we do not require that the structure coefficients (1.1.1), (1.2.1), or the affine connections (2.1.1), (2.6.1) shall be almost automorphic by themselves. It follows however from the law of transformations (1.1.5) that if there are « sufficiently many » almost automorphic vectoroids ξ^A then the coefficients (1.1.1) will be almost automorphic themselves, and it follows then from the equation (2.5.2) that if there are sufficiently many almost automorphic derivatives $\xi^A{}_{.,j}$ then the coefficients L^A_{Bi} will be likewise so.

Theorems 3.2.2 and 3.3.1 are two generalizations of Theorem 3.2.1 but in different directions. But they can technically be brought together in one statement by the following device.

In requirement (i) for almost automorphy we do *not* require that the pointset $M^0{}_n$ shall be a compact subset of M_n but only a *relatively closed* subset of it.

We call a function $f(P)$ *almost modular* if it satisfies technically the same requirement (iii) as before, and a theorem which includes all previous ones, is now as follows.

THEOREM 3.3.2. *The conclusions of Theorem* 3.2.1 *are also valid if the quantities* (3.3.1) *are almost modular and* ξ_A *has boundary values* 0 *in the relatively closed subset* $M^0{}_n$ *of* M_n *in the following sense. To each* $\epsilon > 0$ *there is a compact subset* $M_n{}^\varepsilon$ *of* $M^0{}_n$ *such that* (3.2.12) *is valid in* $M^0{}_n - M_n{}^\varepsilon$.

4. - SPECIFIC VERSIONS OF THE THEOREM.

The heart of the preceding theorems is the assumption (3.2.8) with the requirement (3.2.10), and we are now going to give, in successive stages, sufficient conditions for verifying these.

4.1. We assume that our matrix (1.1.1) is the Kronecker product of two matrices

$$\{a^C_A(y;\, x)\}\ ,\qquad \{b^D_B(y;\, x)\} \tag{4.1.1}$$

and we denote their fundamental tensors by

$$g_{AC} \quad \text{and} \quad h_{BD} \tag{4.1.2}$$

respectively, so that in particular

$$g_{AC,\,i} = 0\ ,\qquad h_{BD,\,i} = 0\ .$$

Correspondingly we now denote our previous vectoroid ξ_A by the symbol ξ_{AB}. In keeping with 2.9 we now have

$$\varphi = g^{AC} h^{BD} \xi_{AB} \xi_{CD} , \tag{4.1.3}$$

and for the equations (3.2.8) and (3.2.9) we now write

$$g^{rs} \xi_{AB, r, s} = S_{A \cdot B \cdot}^{\cdot C \cdot D} \xi_{CD} \tag{4.1.4}$$

and

$$T_2 = S^{ABCD}_{\cdot \cdot \cdot \cdot} \xi_{AB} \xi_{CD}$$

respectively.

We now assume, by explicit assumption, that we are given a certain additive decomposition

$$S_{A \cdot B \cdot}^{\cdot C \cdot D} = P_{A \cdot B \cdot}^{\cdot C \cdot D} + \delta_{A \cdot}^{\cdot C} Q_{B \cdot}^{\cdot D} \tag{4.1.5}$$

where $\delta_{A \cdot}^{\cdot C}$, as usual, $= 1$ for $A = C$, and $= 0$ for $A \neq C$. We then have

$$T_2 = P_{A \cdot B \cdot}^{\cdot C \cdot D} \xi^{AB}_{\cdot \cdot} \xi_{CD} + Q_{B \cdot}^{\cdot D} \xi^{AB}_{\cdot \cdot} \xi_{AD} , \tag{4.1.6}$$

and the purpose of this decomposition will reveal itself in the following further assumption.

We assume that the first structure (4.1.1) is replaced by the Kronecker product of s equal factors, so that our tensoroid can now be denoted by

$$\xi_{A_1 \ldots A_s B} , \tag{4.1.7}$$

and (4.1.3) is to be replaced by

$$g^{A_1 C_1} \ldots g^{A_s C_s} h^{BD} \xi_{A_1 \ldots A_s B} \xi_{C_1 \ldots C_s D} . \tag{4.1.8}$$

We are assuming that the decomposition (4.1.6) remains in force for every s so that explicitly

$$T_2 = Q_{B \cdot}^{\cdot D} \xi^{A_1 \ldots A_s B}_{\cdot \, \cdot\cdot \, \cdot \, \cdot} \xi_{A_1 \ldots A_s D} + T_2^P , \tag{4.1.9}$$

where

$$(4.1.10) \qquad T_2^P = P_{A_1 \dots A_s \cdot \, \cdot\cdot \, \cdot B \cdot}^{\cdot \, \cdot\cdot \, \cdot C_1 \dots C_s \cdot D} \xi^{A_1 \dots A_s B}_{\cdot \; \cdot \cdot \; \cdot \; \cdot} \xi_{C_1 \dots C_s D} \cdot$$

Herein the matrix $Q_{B\cdot}^{\cdot D}$, which depends only on the second structure (4.1.1), shall be fixed, that is, independent of s; whereas the matrix in the second term, that is in term (4.1.10) shall be a Kronecker sum

$$\sum_{r=1}^{s} \delta_{A_1}^{C_1} \dots \delta_{A_{r-1}}^{C_{r-1}} P_{A_r \cdot B \cdot}^{\cdot C_r \cdot D} \delta_{A_{r+1}}^{C_{r+1}} \dots \delta_{A_s}^{C_s}$$

in which the decisive factors

$$P_{A_r \cdot B \cdot}^{\cdot C_r \cdot D}$$

shall be the same for all r. Thus

$$(4.1.11) \quad T_2^P = \sum_{r=1}^{s} P_{A_r \cdot B \cdot}^{\cdot C_r \cdot D} \xi^{A_1 \dots A_s B}_{\cdot \; \cdot\cdot \; \cdot \; \cdot} \xi_{A_1 \dots A_{r-1} C_r A_{r+1} \dots A_s D} \cdot$$

Next we assume that our tensoroid (4.1.7) is, with regard to its indices $A_1, \dots, A_s$ (either) symmetric (in all of them, or perhaps, only anti-symmetric in all of them). We then have

$$(4.1.12) \qquad T_2 = sP_{A \cdot B \cdot}^{\cdot C \cdot D} \xi^{A A_2 \dots A_s B}_{\cdot \; \cdot \; \cdot\cdot \; \cdot \; \cdot} \xi_{C A_1 \dots A_s D}$$

$$+ Q_{B \cdot}^{\cdot D} \xi^{A_1 A_2 \dots A_s B}_{\cdot \; \cdot \; \cdot\cdot \; \cdot \; \cdot} \xi_{A_1 A_2 \dots A_s D} \cdot$$

And finally we assume that $P_{A \cdot B \cdot}^{\cdot C \cdot D}$ is positive definite in the sense that we have

$$(4.1.13) \qquad P_{A \cdot B \cdot}^{\cdot C \cdot D} t^{AB}_{\cdot \; \cdot} t_{CD} > 0$$

whenever

$$t^{AB} t_{AB} > 0 \; .$$

Then, for sufficiently large s we have $T_2 > 0$ and hence the following conclusion.

THEOREM 4.1.1. *If on a compact M_n with structures* (4.1.1), *a tensoroid* (4.1.7) *pertaining to* s *Kronecker factors* (1.1.1) *and one factor* (1.4.1) *is symmetric* (*or antisymmetric*) *relative to the* s *factors and satisfies a differential equation*

$$g^{rs}\xi_{A_1 \ldots A_s B, r, s} = S_{A_1 \ldots A_s \cdot \ldots \cdot B \cdot}^{\cdot \ \ldots \ \cdot C_1 \ldots C_s \cdot D} \xi_{C_1 \ldots C_s D},$$

of the given description, then, in consequence of the positivity assumption (4.1.13) *there is an integer* s_0 *such that for* $s \geqslant s_0$ *the tensoroid* (4.1.7) *is identically* 0.

The iteration of the structure (1.1.1) in this theorem goes back to KODAIRA [8].

4.2. We now assume in particular that the second structure in (4.1.1) is the classical structure (1.2.1) so that for $s = 1$ our tensoroid (4.1.7) is a tensoroid

$$\xi_{Ai} \ . \tag{4.2.1}$$

If it were not for the index A, this would be an « ordinary » vector, and we now prefer to view the structure (1.1.1) as something superimposed from the outside on it. That is to say, we again view (4.2.1) as a vector in the classical sense, but one whose nature has been modified by erecting a vector bundle on the base space over which (4.2.1) is defined « in reality. » In short we will view it as a *bundle-valued vector*, and subject it to restrictions which are *formally* the same as if the bundle were not present at all.

Specifically we call this bundle-valued vector *harmonic* if it satisfies the conditions

$$\xi_{Ai, i} = \xi_{Aj, i} \tag{4.2.2}$$

$$g^{ij}\xi_{Ai, j} = 0 \ , \tag{4.2.3}$$

as usual, but, and this is a very important « but », the covariant differentiations shall be done in exactly the manner as laid down in Section 2, remembering, to that purpose, that (4.2.1) is a tensoroid after all.

Therefore we have, by (4.2.2),

$$g^{rs}\,\xi_{Ai,\,r,\,s} = g^{rs}\,\xi_{Ar,\,i,\,s}\ ,$$

and herein we will put, by (2.6.8),

$$\xi_{Ar,\,i,\,s} = \xi_{Ar,\,s,\,i} + \xi_{Cr}\,L^{C}_{Ars} + \xi_{Ap}\,R^{p}_{r\,is}\ ,$$

were $L^{C}_{Ar\,s}$ is of course the curvature tensoroid of the bundle, and $R^{p}_{r\,is}$ the curvature tensor of the base space. Since, however (4.2.3) implies

$$g^{rs}\,\xi_{Ar,\,s,\,i} = 0\ ,$$

we obtain

$$g^{rs}\,\xi_{Ai,\,r,\,s} = P_{A\,.\,i\,.}^{\,.\,C\,.\,j}\,\xi_{Cj} + Q_{i\,.}^{\,.\,j}\,\xi_{Aj} \tag{4.2.4}$$

where

$$P_{A\,.\,i\,.}^{\,.\,C\,.\,j} = g^{js}\,L^{C}_{Ais} \equiv L^{C\,.\,.\,j}_{\,.\,Ai} \tag{4.2.5}$$

(the last is a symbol introduced ad hoc), and

$$Q_{q\,.}^{\,.\,p} = g^{rs}\,R^{p}_{rqs} = R^{p}_{q}\ .$$

the latter being the Ricci tensor in M_n, suitably normed.

Now, our old sum formula (2.6.11) implies now immediately the corresponding sum formula in (4.1.11), and hence the following conclusion from Theorem 4.1.1.

Theorem 4.2.1. *If on a compact M_n the curvature in a bundle* (1.1.1) *is positive in the sense that*

$$L^{C}_{Aij}\,t^{Ai}t_{Cs}\,g^{sj} > 0$$

for

$$t^{Ai}\,t_{Ai} > 0$$

then for sufficiently large s *there is no symmetric multi-bundle-*

valued vector $\xi_{A_1 \dots A_s i}$, *which is harmonic in the sense that*

(4.2.6) $$\xi_{A_1 \dots A_s i, j} = \xi_{A_1 \dots A_s j, i}$$

(4.2.7) $$g^{ij} \xi_{A_1 \dots A_s i, j} = 0 \ ,$$

other than $\equiv 0$.

Next, for any $p(\leqslant n)$ we say that a bundle-valued tensor

(4.2.8) $$\xi_{A i_1 \dots i_p}$$

is harmonic, if, as usual, it is anti-symmetric in the base space indices $i_1, \dots, i_p$ and

(4.2.9) $$\xi_{A i_1 \dots i_p, j} = \sum_{q=1}^{p} \xi_{A i_1 \dots j \dots i_p, i_q}$$

and

(4.2.10) $$g^{ij} \xi_{A i_1 \dots i_{p-1} i, j} = 0 \ .$$

In this case we have

$$T_2 = T_2^P + T_2^Q$$

where

$$\frac{1}{p} T_2^Q = R^t_{a_1} \xi_{A t a_2 \dots a_p} \xi^{A a_1 \dots a_p}$$

$$+ (p-1) R^{u\,.\,.\,t}_{a_2 a_1} \xi_{A t u b_3 \dots b_p} \xi^{A a_1 a_2 b_3 \dots b_p}$$

and

$$\frac{1}{p} T_2^P = P_{A\,.\,i\,.}^{\,.\,C\,.\,j} \xi_{C j a_2 \dots a_p} \xi^{A i a_2 \dots a_p} \ ,$$

where the tensor in the last expression has the same value (4.2.5) as before. Hence the following statement.

THEOREM 4.2.2. *The conclusion in Theorem* 4.2.1 *also holds for a multi-bundle-valued harmonic tensor*

(4.2.11) $$\xi_{A_1 \ldots A_s i_1 \ldots i_p}$$

of any dimension p.

Finally one last generalization.

THEOREM 4.2.3. *The conclusion even holds for a multi-bundle-valued harmonic tensor*

(4.2.12) $$\xi_{A_1 \ldots A_s B i_1 \ldots i_p}$$

of any dimension p, *in which the index* B *refers to some arbitrary fixed structure* (1.4.1) *added multiplicatively to the* s *factors* (1.1.1), *and the* p *factors* (1.2.1).

In fact, always referring to the sum formula (2.6.11) for curvature it is easily seen that the additional structure (1.4.1) contributes to the value of T_2^p a term which is independent of the principal structure (1.1.1) and of the integer s, and is thus likewise outweighed for s sufficiently large.

4.3. We note that the specific theorems of this section can be extended to a non-compact M_n, if the data (3.3.1) are almost-modular, where of course now the ξ_A in 3.3.1 is to be the tensoroid (4.2.11) and (4.2.12) respectively.

We also note that the conclusions would hold again, if, alternately, instead of bundle-valued harmonic tensors we would envisage such Killing tensors, again formally defined as before, see [4].

Finally we note that for an orientable compact M_n it is also true for bundle-valued tensors that conversely the solution of the second order differential equation (4.2.4) and its analogue for arbitrary p is harmonic again, that is, satisfies the equations (4.2.9) and (4.2.10) separately again. See [4], pp. 74-76.

5. - COMPLEX VECTOR BUNDLES.

We will make a transition from real to complex numbers both in the base space and in the vector fiber over it, and although the two transitions will be closely parallel, there will be a certain difference between them nevertheless.

5.1. With regard to the base space M_n we assume that it is a manifold in the ordinary sense on which a holomorphic neighborhood structure is defined. Its true dimension is then an even number, $n = 2k$, and, locally, the true real parameters can now, for analytical purposes, be replaced by complex parameters and their conjugates which together we will denote as in [1] by

$$\{x_\alpha ; x_{\beta *}\}, \qquad 1 \leq \alpha, \beta \leq k \tag{5.1.1}$$

$$x_{\beta *} = \overline{x_\beta} .$$

We will also employ the notation

$$x_{k+\beta} = x_{\beta *} ,$$

and then denote the parameters (5.1.1.) by the symbol

$$\{x_i\} , \qquad i = 1, \dots, n \,(= 2k) , \tag{5.1.2}$$

with Roman indices, formally as before. The classical matrix (1.2.1) has now, in the notation (5.1.1), the form

$$\begin{pmatrix} \dfrac{\partial y_\alpha}{\partial x_\beta} , & 0 \\ 0 , & \dfrac{\partial y_{\gamma *}}{\partial x_{\delta *}} \end{pmatrix} , \tag{5.1.3}$$

a transformation of parameters being given by holomorphic functions $y_\alpha = y_\alpha(x_1, \dots, x_k)$, and their conjugate complex ones.

5.2. On the other hand, whether the base space M_n is real or holomorphic (n being even or odd) we can always introduce a matrix structure

$$\{a^A_B(y ; x)\} , \tag{5.2.1}$$

of any dimension N in which the coefficients are complex-valued now and the vectoroids u^A, t_A have components which are complex numbers too. No holomorphy is involved at all. Of course, if the base space does have holomorphic structure, then we can, if we choose to, formulate the requirement that the coefficients (5.2.1) or some other quantities occurring shall be holomorphic (or, say, anti-holomorphic) functions then. But we wish to emphasize that any such requirement will be explicitly made and is not to be taken for granted otherwise.

It is readily seen that the entire tensor algebra and analysis carries over to a complex structure, whether the base space is real or holomorphic, and for the parameters (5.1.1) in the latter case.

5.3. With any complex structure (5.2.1) we can associate the conjugate complex structure

$$\{ a^{A*}_{B*}(y;\, x) \} , \tag{5.3.1}$$

where

$$a^{A*}_{B*}(y;\, x) = \overline{a^{A}_{B}(y;\, x)} , \tag{5.3.2}$$

and also the logical sum of the two structures, thus

$$\begin{Bmatrix} a^{A}_{B}(y;\, x) , & 0 \\ 0 , & a^{C*}_{D*}(y;\, x) \end{Bmatrix} . \tag{5.3.3}$$

Conjugation of structures has many obvious properties and we only mention the following ones. It is an involution, that is, the conjugate of a conjugate is the original structure, and, furthermore, the Kronecker product of conjugates is the conjugate of the Kronecker product.

If t_A is a vectoroid for (5.2.1), and if we put

$$t_{A*} = \overline{t_A} \tag{5.3.4}$$

then t_{A*} is a vectoroid for (5.3.1) and the object

$$t_{\mathfrak{A}} = \{ t_A ,\, t_{C*} \} \tag{5.3.5}$$

is a vectoroid for (5.3.3). There are in the bundle (5.3.3) vectoroids for which (5.3.4) is not fulfilled, but if a vectoroid (5.3.5) does have this property, then we say that it is « self-adjoint. » In general we say that an object with one or several indices is self-adjoint, if « starring » all indices simultaneously, will turn its value into the conjugate complex one. Thus, for instance, the matrix (5.3.3) itself, if we denote it for a moment by

$$\{ a^{\mathfrak{A}}_{\mathfrak{B}} (y; x) \} \tag{5.3.6}$$

is obviously self-adjoint. A scalar is self-adjoint if and only if it is real-valued. In what follows we will start out with objects which are self-adjoint, and all operations with them will be such that any resulting object will be self-adjoint again.

If the base space is real, and if L^A_{Bi} is a (complex-valued) affine connection pertaining to (5.2.1) then the enlarged object $L^{\mathfrak{A}}_{\mathfrak{B}i}$ with the additional components

$$L^{A^*}_{B^*i} = \overline{L^A_{Bi}} \tag{5.3.7}$$

$$L^A_{B^*i} = 0 = L^{A^*}_{Bi} \tag{5.3.8}$$

is an affine connection for (5.3.3), and is self-adjoint. But if the base space is holomorphic we must also « star » the space index, if self-adjointness is to be attained, and thus (5.3.7) is to be replaced by

$$L^{A^*}_{B^*i} = \overline{L^A_{Bi^*}} \ . \tag{5.3.9}$$

In what follows only self-adjoint affine connections which the property (5.3.8) shall be associated with the structure (5.3.6).

5.4. Next, the fundamental tensor $g_{\mathfrak{A}\mathfrak{B}}$ pertaining to (5.3.6) shall be of the following kind:

$$g_{\mathfrak{A}\mathfrak{B}} = g_{\mathfrak{B}\mathfrak{A}} \tag{5.4.1}$$

$$g_{AB} = 0 = g_{A^*B^*} \tag{5.4.2}$$

$$g_{A^*B} = \overline{g_{AB^*}} \tag{5.4.3}$$

$$g_{AB^*} t^A t^{B^*} > 0 \quad \text{if} \quad t^A t^{A^*} > 0 \ . \tag{5.4.4}$$

Of course, we are again demanding

$$g_{\mathfrak{A}\mathfrak{B},i} = 0 \ . \tag{5.4.5}$$

The inverse to such a tensoroid has the same properties.

For a complex base space relation (5.4.5) implies again

$$g_{CB^*}\, L^{C}_{A\alpha\beta^*} = -\, g_{AD^*}\, L^{D^*}_{B^*\alpha\beta^*} \ , \tag{5.4.6}$$

in addition to

$$L^{A}_{Bij} = -\, L^{A}_{Bji} \tag{5.4.7}$$

and to self-adjointness. It can be easily seen that this has the following consequence.

LEMMA 5.4.1. *The quantity*

$$g^{\varrho\sigma^*}\, g^{AB^*}\, g^{\alpha\beta^*}\, L^{C}_{A\alpha\sigma^*}\, t_{C\varrho}\, t_{B^*\beta^*} \tag{5.4.8}$$

is real-valued.

5.5. From now on the *base space* shall be *holomorphic*. Relation (5.4.5) gives in particular

$$\frac{\partial g_{AB^*}}{\partial x_\alpha} = g_{CB^*}\, L^{C}_{A\alpha} + g_{AD^*}\, L^{D^*}_{B^*\alpha} \ , \tag{5.5.1}$$

and this suggests introducing also the assumption

$$L^{A}_{B\alpha^*} = 0 = L^{A^*}_{B^*\alpha} \ , \tag{5.5.2}$$

because in this case relations (5.5.1) can be inverted thus:

$$L^{A}_{B} = g^{AC^*}\, \frac{\partial g_{C^*B}}{\partial x_\alpha} \ . \tag{5.5.3}$$

Now, a glance at relations (2.1.2) will show that (5.5.2) is possible only if

$$\frac{\partial a_B^A(y;\,x)}{\partial x_{a^*}} = 0 \,, \tag{5.5.4}$$

that is if the structure matrix is a holomorphic one. However, in general, relations (5.5.2) and (5.5.3) will not be needed, and assumption (5.5.4) will not be made. The only exception is the base space itself. Since it is holomorphic, the equations (5.5.4) are satisfied for (5.1.3) automatically, but this does not yet imply (5.5.2) « by itself ». However, for the base space we have required that the connection shall be a symmetric one, and in this case, the earlier requirement (5.3.8) does bring about the further requirement (5.5.2) by symmetry. Thus, for the base space our requirements add up to the assumption that it shall be a Kaehler space, no less.

Non-symmetric connections in the base space have been envisaged before (see [2], [4], and [6]), but in the theorem we wish to bring they would be rather unsuitable, apparently.

5.6. If the index A refers to a complex structure (5.2.1) and $\alpha_1, \ldots, \alpha_p$ to the « unstarred » parameters (5.1.1), then the « partial » tensoroid

$$t_{A\alpha_1 \ldots \alpha_p}$$

shall be called harmonic if it is antisymmetric in $\alpha_1, \ldots, \alpha_p$ and

$$t_{A\alpha_1 \ldots \alpha_p, \varrho} = \sum_{q=1}^{p} t_{A\alpha_1 \ldots \alpha_{q-1}\varrho\alpha_{q+1} \ldots \alpha_p, \alpha_q} \tag{5.5.5}$$

$$g^{\varrho\sigma^*} t_{A\alpha_1 \ldots \alpha_{p-1}\varrho, \sigma^*} = 0 \,. \tag{5.5.6}$$

THEOREM 5.5.1. *If on a compact Kaehler base space there is given a principal complex structure* (5.2.1) *and some other secondary complex structure* (1.4.1) *both with linear connections and fundamental tensoroids as stipulated in this section, but with no other holonomy requirements imposed; and if the affine connection of the principal structure is positive, in the sense that the (real-valued) form* (5.4.8) *is positive definite, then for sufficiently large s, any tensoroid*

(5.5.7) $$\xi_{A_1 \dots A_s B a_1 \dots a_p}$$

which is harmonic relative to the space indices $a_1, \dots, a_p$ *in the manner stated must be* $\equiv 0$.

In (5.5.7) *each index* A_r *refers to the principal complex structure and* B *to the secondary structure, and* (5.5.7) *is assumed either symmetric or antisymmetric in the indices* $A_1, \dots, A_s$.

Proof. Following KODAIRA [7] we are setting up not a LAPLACEAN but a divergence. For instance, for $s = 1$ and $p = 1$ it is

(5.5.8) $$g^{\varrho\sigma^*} g^{AC^*} h^{BD^*} g^{\alpha\beta^*} (\xi_{AB\alpha,\,\varrho}\, \xi_{C^* D^* \beta^*})_{,\,\sigma^*} \,,$$

and part (ii) of Lemma 3.1.1 applies to it. The expression (5.5.8) has the value $T_1 + T_2$ where

$$T_1 = g^{\varrho\sigma^*} \xi_{AB\alpha,\,\varrho}\, \xi^{AB\alpha}_{\dots,\,\sigma^*} \geq 0$$

and

$$T_2 = g^{\varrho\sigma^*} \xi_{AB\alpha,\,\varrho,\,\sigma^*}\, \xi^{AB} \,.$$

But from this point, the analysis of the expression T_2 is the same as in Section 4 for the real data.

THEOREM 5.5.2. *If we restrict the concept of a harmonic bundle-valued tensor, by adding to the requirements* (5.5.5), (5.5.6), *the further requirement*

(5.5.9) $$g^{\varrho\sigma^*} \xi_{A a_1 \dots a_p,\,\sigma^*,\,\varrho} = 0$$

then the conclusion of theorem 5.5.1 *also holds on a non-compact* M_n, *for almost modular data as stated in Section* 4.3.

We have formulated the last theorem because in a rather particular case of Theorem 5.5.2, which we had anticipated in [1], p. 792/3, the additional requirement (5.5.9) was satisfied automatically, as there utilized, and in the light of recent trends in the theory of automorphic and modular functions, statements on « Curvature and Betti numbers » for almost modular situations might become acute eventually.

REFERENCES

[1] S. Bochner, *Vector Fields and Ricci Curvature.* « Bulletin American Mathem. Society » 52 (1946), 776-797.

[2] — — *Curvature and Betti Numbers II.* « Annals of Mathematics » 50 (1949), 77-93.

[3] — — *A New Viewpoint in Differential Geometry.* « Canadian Journal of Mathematics » 3 (1951), 460-470.

[4] K. Yano and S. Bochner, *Curvature and Betti Numbers.* Princeton, 1953.

[5] S. Bochner, *Stationary Space-Time in General Relativity.* « Proceedings National Academy of Sciences », Wash. 41 (1955), 485-490.

[6] S. I. Goldberg, *Tensor Fields and Curvature in Hermitian Manifolds with Torsion.* « Annals of Mathematics » 63 (1956), 64-76.

[7] K. Kodaira, *On a differential-geometric method in the theory of Analytic Stacks.* « Proceedings of the National Academy of Sciences » 39 (1953), 1268-1273.

[8] — — *On Kaehler Varieties of Restricted Type.* « Annals of Mathematics » 60 (1954), 28-48.

BOCHNER, S.
1961
Annali di Matematica pura ed applicata
(IV), Vol. LIII, pp. 57-62

Vector fields on Riemannian spaces with boundary.

by S. BOCHNER (a Princeton, U.S.A.)

To Enrico Bompiani on his scientific Jubilee.

Summary. - *Non-existence of harmonic and* KILLING *vectors on a Riemannian space with boundary. Our statements are more detailed than some previously given by* K. YANO.

1. Introduction.

The present paper attaches itself to recent papers by K. YANO, [1]. In a subsequent paper, [2], YANO extended his statements to tensor fields, but we will be content to deal with vector fields only.

On the whole, YANO envisaged only vector fields of which it is assumed, to start with, that on the boundary each of them is either purely tangential or purely perpendicular. In our variant on his reasoning this assumption will not be made to start with, although our sufficient conditions will approach this assumption eventually.

We note that our analysis will on the face of it be somewhat different from YANO's. YANO employs on the boundary the «invariant» quantities of VAN DER WAERDEN and BORTOLOTTI. To us however it seemed more suggestive to operate with the classical geodesic coordinates, which have also been studied by BOMPIANI for other purposes, see for instance [3].

2. Geodesic coordinates.

In what follows the indices i, j, k, l will run from 1 to n, the last value n included, but the indices p, q, r, s will run only from 1 to $n-1$, the value n not being included.

In a neighborhood V^0 of the origin in real E^n: $(x^1, \dots, x^n)$ we denote by D^0 the half-neighborhood $x^n < 0$ and by B^0 its boundary $x^n = 0$. We take in V^0 a line element in its geodesic form

$$ds^2 = g_{pq}(x^1, \dots, x^{n-1}, x^n)dx^p dx^q + (dx^n)^2 \tag{1}$$

and we recall, see [4], p. 57, that on B^0 the second fundamental form and the mean curvature are given by

$$H_{pq} = -\frac{1}{2}\partial_n g_{pq}, \qquad H = g^{pq}H_{pq} = H^p_p. \tag{2}$$

Also for the CHRISTOFFEL symbols we have

$$\Gamma^n_{nn} = 0, \qquad \Gamma^p_{nn} = 0, \qquad \Gamma^n_{pn} = 0, \qquad \Gamma^n_{pq} = H_{pq}, \qquad \Gamma^r_{pn} = -H^r_p$$

$$2\Gamma^p_{qr} = g^{ps}(\partial_r g_{qs} + \partial_q g_{rs} - \partial_s g_{qr}). \tag{3}$$

Now, let ξ_i be a (covariant) vector field of class C^1 on D^0, with boundary values of class C^1 on B^0. On B^0, each such vector has a normal part which is a scalar on B^0, and a tangential part which is a vector on B^0. In our special coordinates, the normal component is simply the last component ξ_n itself, and it also has the value ξ^n, but in the invariant idiom it is the contraction $\xi_i N^i$ where N^i is the outer normal unit vector on B^0. For the tangential part the components are numerically $\xi_1, \ldots, \xi_{n-1}$, but in order to emphasize their meaning we will also denote them by $t_1, \ldots, t_{n-1}$. Also we will make use of the symbol t, and by div $(t \cdot \xi_n)$ we will understand «invariantly» the divergence in B^0 of the vector field which arises by multipliyng the components of the tangential part t_p with the scalar $\xi_i N^i$.

Now, on D^0, or on B^0, we have

$$\nabla_n \xi_n = \partial_n \xi_n, \tag{4}$$

$$\nabla_p \xi_n = \partial_p \xi_n + H^q_p \xi_q \equiv \partial_p \xi_n + H^q_p t_q, \tag{5}$$

$$\nabla_n \xi_p = \partial_n \xi_p + H^q_p \xi_q \equiv \partial_n \xi_p' + H^q_p t_q, \tag{6}$$

$$\begin{aligned} \nabla_q \xi_p &= \partial_q \xi_p - \xi_r \Gamma^r_{qp} - \xi_n H_{pq} \\ &\equiv \nabla'_q t_p - \xi_n H_{pq}. \end{aligned} \tag{7}$$

In (7), the operator ∇'_q indicates covariant differentiation in the $(n-1)$-dimensional space B^0.

3. Harmonic vectors.

All our conclusions will be predicated on the behavior of a certain integrand on $[B^0]$ and by definition the integrand is the quantity

$$J = (\xi^i \nabla_j \xi_i) N^j. \tag{8}$$

Due to our normalization (1) this first of all reduces to

$$J = \xi^n \nabla_n \xi_n + t^p \nabla_n \xi_p. \tag{9}$$

We now consider the separate (though parallel) cases in which ξ_i is a harmonic vector field or a KILLING vector field.

If ξ_i is harmonic then by definition

$$\nabla_i \xi_j = \nabla_j \xi_i \tag{10}$$

and

$$\nabla^i \xi_i = 0. \tag{11}$$

From the latter we obtain, see (7) and (2),

$$\nabla_n \xi_n = - g^{pq} \nabla_p \xi_q = - g^{pq} (\nabla'_p t_q - \xi_n H_{pq}) = - \operatorname{div} t + \xi_n H,$$

so that, by (9),

$$J = - \xi^n \cdot \operatorname{div} t + (\xi^n)^2 H + t^p \nabla_n \xi_p .$$

Now, (10) implies in particular $\nabla_n \xi_p = \nabla_p \xi_n$, and thus by (5),

$$J = - \xi^n \cdot \operatorname{div} t + (\xi^n)^2 H + t^p \partial_p \xi_n + H_{pq} t^p t^q . \tag{12}$$

Furthermore on B^0 we also have the identity

$$t^p \partial_p \xi_n = \operatorname{div} (t \cdot \xi_n) - \xi_n \cdot \operatorname{div} t, \tag{13}$$

and if we substitute this in (12) we finally obtain

$$J = J_0 + \operatorname{div} (t \cdot \xi_n) \tag{14}$$

where

$$J_0 = - 2\xi_n \cdot \operatorname{div} t + (\xi_n)^2 H + H^{pq} t_p t_q . \tag{15}$$

In order to utilize all this we take in an oriented Riemannian manifold V of class C^r, with r sufficiently large, a domain D with compact closure whose boundary B consists of one or several closed hypersurfaces. The latter shall be sufficiently smooth so that in the neighborhood of every point a line element of the form (1) can be introduced.

By STOKES' theorem we have generally

$$\int_D \nabla^j (\xi^i \nabla_j \xi_i) dv = \int_B J d\omega.$$

Now, if ξ_i is harmonic in D then, (see [5], Chapter II),

$$\int_D \nabla^j (\xi^i \nabla_j \xi_i) dv = \int_D (\nabla^j \xi^i \cdot \nabla_j \xi_i + R_{ij} \xi^i \xi^j) dv. \tag{16}$$

On the other had we have

$$\int_B J d\omega = \int_B J_0 d\omega,$$

where J_0, which already occurs in the second paper [1], is defined by (15), the reason being that

$$\int_B \operatorname{div}(t \cdot \xi_i N^i) d\omega = 0.$$

In fact, B is oriented and has boundary 0, and by STOKES' theorem the integral for any divergence is 0. Hence the following conclusion.

THEOREM 1. - *If $R_{ij} \geq 0$ and ξ_i is harmonic in D, and if on the boundary B of D one of the following alternate sets of conditions is satisfied:*

(i) $$t_p = 0, \qquad H \leq 0$$

(ii) $$\xi_n = 0, \qquad H^{pq} t_p t_q \leq 0$$

(iii) $$\operatorname{div} t = 0, \qquad H^{pq} t_p t_q \leq 0, \quad H \leq 0,$$

then in D

$$R^{ij}\xi_i\xi_j = 0, \qquad \nabla_i \xi_j = 0, \tag{17}$$

and, in cases (ii) *and* (iii), *on B,*

$$H^{pq} t_p\, t_q = 0. \tag{18}$$

Also, if $R_{ij} > 0$, then (17) *is possible only if $\xi_i = 0$, that is no such vector field other than 0 exists.*

4. **Killing vectors.**

For a KILLING vector ξ^i we have

$$\nabla_j \xi_i = - \nabla_i \xi_j$$

and, automatically,

$$\nabla^i \xi_i = 0.$$

In particular $\nabla_n \xi_n = 0$, and (9) is to be replaced by

$$J = t^p \nabla_n \xi_p .$$

Next, due to $\nabla_n \xi_p = - \nabla_p \xi_n$, because of (5),

$$J = - t^p \partial_p \xi_n - H_{pq} t^p t^q,$$

then, because of (13),

$$J = J_0 - \operatorname{div} (t \cdot \xi_n)$$

where now

$$J_0 = \xi_n \cdot \operatorname{div} t - H_{pq} t^p t^q.$$

Finally, (16) is to be replaced by

$$\int_D \nabla^j (\xi^i \nabla_j \xi_i) dv = \int_D (\nabla^j \xi^i \cdot \nabla_j \xi_i - R_{ij} \xi^i \xi^j) d\omega$$

and the conclusion is as follows.

THEOREM 2. - *If* $R_{ij} \leq 0$, *and* ξ^i *is a Killing vector, and if on* B *one of the following alternate sets of conditions is satisfied:*

(i) $t^p = 0$

(ii) $\xi^n = \text{constant}, \qquad H_{pq} t^p t^q \geq 0$

(iii) $\operatorname{div} t = 0, \qquad H_{pq} t^p t^q \geq 0,$

then, on D, (17) *holds, and, in cases* (ii) *and* (iii), *on* B, (18) *holds.*

Thus, if $R_{ij} < 0$, *then there is no such Killing vector vector other than* 0.

Remark. Both in theorem 1 and theorem 2, if B consists of several closed hypersurfaces, then it suffices if on each hypersurface one of the three alternate conditions holds, not necessarily the same.

5. A counter example.

Case (i) of theorem 2 does not implicate H_{pq}. However, in theorem 1 say, the condition $t_p = 0$ by itself, or also $\xi_n = 0$ by itself would not suffice. In fact, take in the ordinary (x, y) plane the annulus

$$1 < x^2 + y^2 < 2$$

and in it the vector field

$$\xi_1 = \frac{x}{x^2 + y^2}, \qquad \xi_2 = \frac{y}{x^2 + y^2}.$$

It is harmonic, and on the boundary its tangential component is 0. But, $\partial_1 \xi_2 \neq 0$, contrary to (17). Similarly,

$$\xi_1 = \frac{y}{x^2 + y^2}, \qquad \xi_2 = \frac{-x}{x^2 + y^2}$$

is harmonic, with normal component 0 on the boundary, and again $\partial_1 \xi_2 \neq 0$.

BIBLIOGRAPHY

[1] KENTARO YANO, *Harmonic and Killing vector fields in compact orientable Riemannian spaces with boundary*, Annals of Mathematics 69 (1959), 588-597. Also, *Sur les vecteurs harmoniques et vecteurs de Killing dans un espace de Riemann à frontière*, Comptes Rendues Paris, 247 (1958), 1085-1087.

[2] KENTARO YANO, *Harmonic and Killing tensor fields in Riemannian spaces with boundary*, Journal of the Mathematical Society of Japan, 10 (1958), 430-437.

[3] E. BOMPIANI, *Spazi Riemanniani luoghi di varietà totalmente geodetiche*, Rendiconti di Palermo, 48 (1924), 121-134.

[4] L. P. EISENHART, *Riemannian Geometry*, Princeton University Press, 1926.

[5] K. YANO and S. BOCHNER, *Curvature and Betti numbers*, 1953.

ANNALS OF MATHEMATICS
Vol. 75, No. 1, January, 1962
Printed in Japan

INFINITE LINEAR PSEUDOGROUPS OF TRANSFORMATIONS

BY S. BOCHNER AND R. C. GUNNING

(Received March 28, 1961)

0. Introduction

The complex analytic functions of one complex variable and the complex analytic mappings in several complex variables can be viewed as mappings between real euclidean spaces, defined by a system of linear partial differential equations with constant coefficients, the Cauchy-Riemann equations [3]. This system of differential equations has the property that the composition of any two solutions is again a solution — a property of fundamental importance in function theory. We may ask whether there are any other such systems of linear differential equations; the present paper is devoted to a discussion of this problem. Our conclusion, briefly stated, is that there are essentially no other such differential equations, so that this property is characteristic of complex analytic functions in a sense. More precisely, we list in Theorems 1 and 2 all the families of mappings, defined by linear partial differential equations with constant coefficients, closed under composition, and irreducible in a natural sense; we do not, of course, distinguish families which differ merely by a real linear change of coordinates. The list includes, besides the complex analytic mappings and the trivial families of mappings, only three classes of linear transformations: the set of all real linear transformations, all complex linear transformations, and all quaternion linear transformations.

The general problem of classifying the families of local mappings defined by systems of differential equations and closed under composition was treated extensively by E. Cartan; these are the infinite continuous groups (or pseudogroups in a more modern terminology) of his papers [4] and [5]. The problem which we are discussing is really the special case of Cartan's general problem in which the differential equations defining the pseudogroup are linear and have constant coefficients; but the present results seem not to occur among those stated by Cartan himself. Cartan's classification, first of all, was restricted to families of complex analytic transformations, while we consider merely C^∞ real transformations; and Cartan's general definition of equivalence (as in Chapter III of [4]) is much weaker than ours. The difference lies in the fact that we are treating a comparatively simple case of the remarkably difficult general problem. This simplification manifests itself in two principal ways. First, it suffices, in our approach, to use the classification of associative algebras, rather than the more difficult classification of general Lie groups by means of their Lie algebras; and secondly, the problems of analyzing the prolongation of differential systems, and the integrability conditions for systems of differential equations, can be avoided altogether. Our results may be viewed as an introductory chapter to the general theory, simple enough to indicate the deep interest of the problem in its entirety.

1. Terminology and definitions

Let Φ be the set of all local C^∞ mappings of an n-dimensional euclidean space E_n into itself. An element $F \in \Phi$ is thus a mapping from some open subset $U \subset E_n$ into another open subset $V \subset E_n$; and in terms of a fixed coordinate system in E_n the mapping can be described by components in the form

$$F\colon (x_1, \cdots, x_n) \longrightarrow (f_1(x), \cdots, f_n(x)) \,, \tag{1}$$

where $f_1(x), \cdots, f_n(x)$ are C^∞ real-valued functions in U.

To each point x in the domain of any mapping $F \in \Phi$ there can be associated the values of the partial derivatives of the components of the mapping F at that point, the values

$$\xi^i_{j_1\cdots j_s}(x) = \frac{\partial^s f_i(x)}{\partial x_{j_1} \cdots \partial x_{j_s}} \;; \tag{2}$$

these values are of course symmetric in the lower indices $j_1, \cdots, j_s$. The result is a set of mappings from the domain of F into the spaces of symmetric tensors $\xi^i_{j_1\cdots j_s}$. To say that the components of the mapping F are solutions of some system of linear partial differential equations of order t means that the image tensors (2) are subject to some conditions of the form

$$\textstyle\sum_{s=1}^{t} \sum_{i,j_1,\cdots,j_s=1}^{n} c^{\nu i}_{j_1\cdots j_s} \xi^i_{j_1\cdots j_s}(x) = 0 \,, \qquad \nu = 1, 2, 3, \cdots, \tag{3}$$

for some constants $c^{\nu i}_{j_1\cdots j_s}$. The equations (3) are of course not uniquely determined, since they can be replaced by suitable linear combinations of themselves; the invariant form of condition (3) is just this: the values $\xi^i_{j_1\cdots j_s}(x)$ for all points x in the domain of the function must lie in some fixed linear subspace of the vector space of all such tensors $\xi^i_{j_1\cdots j_s}$. To express this condition more formally, introduce the complex linear spaces $\mathcal{S}(s)$ consisting of all (constant) tensors $\xi^i_{j_1\cdots j_s}$, $1 \leqq j_1, \cdots, j_s \leqq n$, which are fully symmetric in the lower indices $j_1, \cdots, j_s$; and introduce further the direct sums of these linear spaces, $\mathcal{T}(t) = \sum_{s=1}^{t} \mathcal{S}(s)$. To any linear subspace $\mathcal{A} \subset \mathcal{T}(t)$ there corresponds the subset $\Phi(\mathcal{A}) \subset \Phi$ consisting of all local mappings $F \in \Phi$ such that for all points x in the domain of the mapping F the associated tensors (2) lie in the subspace $\mathcal{A} \subset \mathcal{T}(t)$; the set $\Phi(\mathcal{A})$ will be called the *set of mappings defined by the linear differential condition* $\mathcal{A}$. The index t can be considered as the *degree* of the differential condition $\mathcal{A}$. If actually $\mathcal{A} \subset \mathcal{S}(s)$, then the differential condition $\mathcal{A}$ will be called *homogeneous of degree* s. One further notational convention should be introduced here. Since the linear space $\mathcal{T}(t)$ is the direct sum of the subspaces $\mathcal{S}(s)$, $1 \leqq s \leqq t$, there are the natural projection mappings $\pi(s)$: $\mathcal{T}(t) \rightarrow \mathcal{S}(s)$, which amount to a consideration of only the homogeneous terms of degree s of the tensors in $\mathcal{T}(t)$. Then to each linear subspace $\mathcal{A} \subset \mathcal{T}(t)$, there correspond the various projections $\mathcal{A}(s) = \pi(s)(\mathcal{A}) \subset \mathcal{S}(s)$; obviously $\mathcal{A} \subset \mathcal{A}(1) + \cdots + \mathcal{A}(t)$, but the containment is generally proper. The subspaces $\mathcal{A}(s)$ will be called the *homogeneous parts* of the linear differential condition $\mathcal{A}$.

A non-singular linear change of coordinates in the underlying euclidean space E_n can often be used to simplify the form of a given linear differential condition $\mathcal{A}$. In particular, suppose that the change of coordinates is given by

$$(4) \qquad x_i^* = \sum_{j=1}^{n} g^{ij} x_j \ , \qquad x_i = \sum_{j=1}^{n} g_{ij} x_j^* \ ,$$

where (g^{ij}), (g_{ij}) are inverse matrices; and suppose that an element $F \in \Phi$ can be written in these two coordinate systems in the equivalent forms

$$(x_1, \cdots, x_n) \longrightarrow (f_1(x), \cdots, f_n(x)) \ ; \quad (x_1^*, \cdots, x_n^*) \longrightarrow (f_1^*(x), \cdots, f_n^*(x)) \ .$$

Then a straightforward use of the chain rule for differentiation shows that the tensor (2) and its equivalent in the coordinates $(x_1^*, \cdots, x_n^*)$ are related by the formula

$$(5) \qquad \xi^{*i}_{j_1 \cdots j_s} = \sum_{k, l_1, \cdots, l_s = 1}^{n} g^{ik} \xi^k_{l_1 \cdots l_s} g_{l_1 j_1} \cdots g_{l_s j_s} \ .$$

Notice in particular that for $s = 1$, in which case the tensor ξ^i_j can be viewed as a matrix ξ, condition (5) can be written in matrix form

$$(6) \qquad \xi^* = G \xi G^{-1} \ ,$$

where $G = (g^{ij})$ is the matrix of the transformation (4); thus a linear transformation in E_n has the effect of an inner automorphism on $\mathcal{S}(1)$ when the latter is viewed as a matrix algebra.

A subset $\Psi \subset \Phi$ of local mappings will be called *reducible to dimension* r, $0 < r < n$, if after a suitable non-singular linear change of coordinates in E_n all the transformations $F \in \Psi$ have coordinate functions (1) such that the components $f_1(x), \cdots, f_r(x)$ depend only on the variables $x_1, \cdots, x_r$; the subset Ψ will be called *irreducible* if it is not reducible to any dimension r, $0 < r < n$. To characterize this condition of reducibility in another manner, introduce the subspaces $\mathcal{R}_r^* \subset \mathcal{S}(1)$ defined as follows:

(7) $\mathcal{R}_r^* = \{\xi_j^i \in \mathcal{S}(1) \mid \xi_j^i = 0 \quad \text{if } 1 \leqq i \leqq r \text{ and } r+1 \leqq j \leqq n\}$;

and let $\mathcal{R}_r$ denote any one linear subspace of $\mathcal{S}(1)$ which can be transformed to $\mathcal{R}_r^*$ by some change of coordinates (6). Recall from (2) that for a transformation $F \in \Phi(\mathcal{R}_r^*)$, $\partial f_i(x)/\partial x_j = \xi_j^i(x) \in \mathcal{R}_r^*$, hence $\partial f_i(x)/\partial x_j = 0$ whenever $1 \leqq i \leqq r$ and $r+1 \leqq j \leqq n$. It is therefore obvious that a subset $\Psi \subset \Phi$ is reducible to dimension r if and only if $\Psi \subset \Phi(\mathcal{R}_r)$. In particular, if $\mathcal{A} \subset \mathcal{T}(t)$ is a linear subspace such that $\mathcal{A}(1) \subset \mathcal{R}_r$ then $\Phi(\mathcal{A})$ is reducible to dimension r.

For any two mappings $F, G \in \Phi$, with $F\colon U \to V$ and $G\colon V \to W$, the composite mapping $H = G \circ F\colon U \to W$ is also an element of Φ; in this sense the full set Φ is closed under composition. A subset $\Psi \subset \Phi$ will be called a *linear pseudogroup* of local C^∞ mappings in E_n if:

(1) Ψ is closed under composition of mappings, in the above sense; and

(2) $\Psi = \Phi(\mathcal{A})$ for some linear differential condition $\mathcal{A}$.

The degree of $\mathcal{A}$ will be called the degree of the associated linear pseudogroup. The problem which we are to consider here can then be described as that of classifying the irreducible linear pseudogroups in E_n; the reducible linear pseudogroups will be mentioned but briefly.

2. Linear pseudogroups of first degree

As noted before, the tensors ξ_j^i in $\mathcal{S}(1)$ can be viewed as $n \times n$ matrices; this identifies $\mathcal{S}(1)$ with the full algebra of $n \times n$ real matrices. A linear differential condition $\mathcal{A} \subset \mathcal{S}(1)$ is then a linear subspace of the full matrix algebra.

LEMMA 1. *A subset $\Phi(\mathcal{A}) \subset \Phi$ defined by a linear differential condition $\mathcal{A}$ of first degree is a linear pseudogroup if and only if $\mathcal{A}$ is a subalgebra of the full matrix algebra $\mathcal{S}(1)$; it is, moreover, an irreducible pseudogroup precisely when $\mathcal{A}$ is an irreducible subalgebra of the full matrix algebra $\mathcal{S}(1)$.*†

PROOF. If $F\colon U \to V$, $G\colon V \to W$, $H = G \circ F\colon U \to W$ are mappings in Φ then on setting

$$\xi_j^i(x) = \frac{\partial f_i(x)}{\partial x_j}, \qquad \eta_j^i(x) = \frac{\partial g_i(x)}{\partial x_j}, \qquad \zeta_j^i(x) = \frac{\partial h_i(x)}{\partial x_j},$$

it follows from the chain rule for differentiation that

(8) $$\zeta_j^i(x) = \sum_{k=1}^n \eta_k^i(F(x))\xi_j^k(x) .$$

Now if $\mathcal{A} \subset \mathcal{S}(1)$ is a subalgebra and $F, G \subset \Phi(\mathcal{A})$, then $\xi_k^i(x)$, $\eta_j^k(y) \in \mathcal{A}$ for any points $x \in U$, $y \in V$, and from (8) it follows that $\zeta_j^i(x) \in \mathcal{A}$ for all

† For the concepts and results needed from the theory of linear algebra, see Albert's book ([1] in the bibliography at the end of this paper); specific references to this book will be given as the occasion arises.

$x \in U$ so that $H \in \Phi(\mathcal{A})$; thus $\Phi(\mathcal{A})$ is a linear pseudogroup. Conversely suppose that $\Phi(\mathcal{A})$ is a linear pseudogroup, and consider two matrices $\xi^i_j, \eta^i_j \in \mathcal{A}$; then the linear transformations

$$f_i(x) = \sum_{j=1}^n \xi^i_j x_j \,, \qquad g_i(x) = \sum_{j=1}^n \eta^i_j x_j$$

are clearly contained in $\Phi(\mathcal{A})$, so that their composition

$$h_i(x) = \sum_{j,k=1}^n \xi^i_k \eta^k_j x_j$$

is contained in $\Phi(\mathcal{A})$. This in turn shows that the matrix product $\sum_{k=1}^n \xi^i_k \eta^k_j$ lies in $\mathcal{A}$, so that $\mathcal{A}$ is indeed a subalgebra of $\mathcal{S}(1)$.

In defining reducibility of a subset $\Psi \subset \Phi$ of local transformations it was noted that Ψ is reducible to dimension r precisely when $\Psi \subset \Phi(\mathcal{R}_r)$; hence for a linear differential condition $\mathcal{A}$ of first degree, it is clear that $\Phi(\mathcal{A})$ is reducible to dimension r precisely when $\mathcal{A} \subset \mathcal{R}_r$. Rephrasing this condition, and recalling the definition of $\mathcal{R}_r$, the subset $\Phi(\mathcal{A}) \subset \Phi$ is reducible if and only if the subalgebra $\mathcal{A} \subset \mathcal{S}(1)$ can be mapped into some subalgebra $\mathcal{R}^*_r \subset \mathcal{S}(1)$ as defined in (7), for $0 < r < n$, by an inner automorphism (6) of the matrix algebra $\mathcal{S}(1)$. The latter condition is just the usual definition of reducibility of a subalgebra of a full matrix algebra (cf. [1, p. 113]); hence $\Phi(\mathcal{A})$ is irreducible if and only if $\mathcal{A}$ is irreducible.

REMARK. An obvious generalization of the above, as noted earlier, is the consideration of a set of local C^∞ transformations which are solutions of some system of partial differential equations, no longer necessarily a linear system. If the system of equations involves only the first derivatives, then the analogue of Lemma 1 is the assertion that an irreducible such pseudogroup corresponds to an irreducible subgroup of the real general linear group; the subgroups involved may be algebraic subgroups, if the differential equations are polynomials in the various derivatives $\xi^i_j = \partial y_i / \partial x_j$, or analytic subgroups, if these equations are (real) analytic, and so on. Further investigations in this direction would necessitate, in place of the present Lemma 2, an analogous classification of such irreducible matrix groups, as given, for example, in [6], or in [10].

The determination of the linear pseudogroups thus reduces to the purely algebraic problem of the classification of subalgebras of the full matrix algebra; the results are well known, and will be summarized in the following lemma. As notational preliminary introduce the 2×2 real matrix $J = \begin{pmatrix} 0 & 1 \\ -1 & 0 \end{pmatrix}$ and its k-fold direct sum

$$(9) \qquad J_k = J \oplus \cdots \oplus J = \begin{pmatrix} J & 0 & \cdots & 0 \\ 0 & J & \cdots & 0 \\ \cdots & \cdots & \cdots & \cdots \\ 0 & 0 & \cdots & J \end{pmatrix},$$

which is a $2k \times 2k$ real matrix; and introduce the 4×4 real matrices

$$(10) \qquad \mathcal{I} = \begin{pmatrix} 0 & 1 & 0 & 0 \\ -1 & 0 & 0 & 0 \\ 0 & 0 & 0 & -1 \\ 0 & 0 & 1 & 0 \end{pmatrix}, \quad \mathcal{J} = \begin{pmatrix} 0 & 0 & 1 & 0 \\ 0 & 0 & 0 & 1 \\ -1 & 0 & 0 & 0 \\ 0 & -1 & 0 & 0 \end{pmatrix},$$

$$\mathcal{K} = \begin{pmatrix} 0 & 0 & 0 & 1 \\ 0 & 0 & -1 & 0 \\ 0 & 1 & 0 & 0 \\ -1 & 0 & 0 & 0 \end{pmatrix}$$

and their k-fold direct sums $\mathcal{I}_k$, $\mathcal{J}_k$, $\mathcal{K}_k$ defined analogously to (9), these being $4k \times 4k$ real matrices.

LEMMA 2. *If $\mathcal{A}$ is an irreducible subalgebra of the full $n \times n$ real matrix algebra $\mathcal{S}(1)$ then, after a suitable inner automorphism* (6) *of the matrix algebra, $\mathcal{A}$ can be brought into one of the following forms*:

(i) $\mathcal{A} = 0$;

(ii) $\mathcal{A} = \mathcal{S}(1)$;

(iii) $\mathcal{A}$ = *the algebra of all matrices $A \in \mathcal{S}(1)$ which commute with J_k, (where $n = 2k$)*;

(iv) $\mathcal{A}$ = *the algebra of all matrices $A \in \mathcal{S}(1)$ which commute simultaneously with $\mathcal{I}_k$, $\mathcal{J}_k$, $\mathcal{K}_k$, (where $n = 4k$)*.

PROOF. Let $\mathcal{B} = \{B \in \mathcal{S}(1) \mid BA = AB \text{ for all } A \in \mathcal{A}\}$. Then [1, p. 117, Theorem 2] $\mathcal{A}$ is a simple algebra, $\mathcal{B}$ is a division algebra, and $\mathcal{A}$ is isomorphic to the direct product $\mathcal{M}_k \times \mathcal{B}^{-1}$ of a full algebra $\mathcal{M}_k$ of $k \times k$ matrices and the reciprocal algebra $\mathcal{B}^{-1}$ to $\mathcal{B}$. (The reciprocal algebra $\mathcal{B}^{-1}$ is isomorphic to $\mathcal{B}$, except that the order of multiplication is reversed in the isomorphism; cf., [1, p. 3].) Moreover [1, p. 118, Theorem 3] $\mathcal{B}$ is fully decomposable into a number of equivalent irreducible components, say $\mathcal{B} = \mathcal{D} \oplus \cdots \oplus \mathcal{D}$, where now $\mathcal{D}$ is an irreducible representation of the division algebra $\mathcal{B}$; and the algebra $\mathcal{A}^* = \{\mathcal{A} \in \mathcal{S}(1) \mid AB = BA \text{ for all } B \in \mathcal{B}\}$ is a simple algebra isomorphic to the direct product $\mathcal{M}_l \times \mathcal{B}^{-1}$ for some l. Note that $\mathcal{A} \subset \mathcal{A}^*$, hence $\mathcal{A}^*$ is irreducible since $\mathcal{A}$ is irreducible by hypothesis. Now the dimension of the full matrix algebra in which a simple algebra is irreducibly represented is uniquely determined by the structure of the algebra (as in [1, p. 117, Theorem 2]); so since $\mathcal{A}$ and $\mathcal{A}^*$ are irreducibly represented in $\mathcal{S}(1)$ and are both similar to $\mathcal{B}^{-1}$ it follows that $\mathcal{A}$ and $\mathcal{A}^*$ are isomorphic. Since $\mathcal{A} \subset \mathcal{A}^*$, it is further clear that $\mathcal{A} = \mathcal{A}^*$. Thus an irreducible subalgebra $\mathcal{A} \subset \mathcal{S}(1)$ can be described as the commutator algebra of some subalgebra of $\mathcal{S}(1)$ of the form $\mathcal{D} \oplus \cdots \oplus \mathcal{D}$, where $\mathcal{D}$ is an irreducible representation of a real division algebra. Now the only real division algebras, except the real numbers themselves, are the complex numbers and the quaternion numbers [1, p. 146, Theorem 28]. After an inner automorphism of the ambient matrix algebra (note Lemma 4, p. 112, in [1]), the complex numbers as a real matrix algebra are generated by the 2×2 identity matrix and J; and the quaternions are generated by the 4×4 identity matrix and the matrices $\mathcal{I}$, $\mathcal{J}$, $\mathcal{K}$. The lemma then follows as stated.

THEOREM 1. *The only first degree irreducible linear pseudogroups of C^∞ local mappings in E_n are the following*:

(i) *the set of constant mappings*;

(ii) *the set Φ of all C^∞ local mappings*;

(iii) *the set of all complex analytic mappings of k complex variables, in a complex structure on euclidean $n = 2k$ space*;

(iv) *the set of all linear mappings of euclidean $n = 4k$ space defined by $4k \times 4k$ real matrices A which commute simultaneously with $\mathcal{I}_k$, $\mathcal{J}_k$, $\mathcal{K}_k$.*

PROOF. By Lemma 1, any irreducible linear pseudogroup of first degree is of the form $\Phi(\mathcal{A})$ for some irreducible matrix subalgebra $\mathcal{A}$; it is then merely necessary to examine the classification of irreducible matrix algebras as recalled in Lemma 2. The trivial matrix algebras (i) and (ii) of Lemma 2 yield the trivial linear pseudogroups (i) and (ii) listed above. Now consider the algebra (iii) of Lemma 2. After an appropriate choice of coordinates in euclidean $n = 2k$ space the local mappings (1) of the pseudogroup $\Phi(\mathcal{A})$ will consist of those mappings such that the matrix $\xi^i_j(x) = \partial f_i(x)/\partial x_j$ commutes with the matrix J_k for all points x in the domain of the mapping. Since J_k is a direct sum of 2×2 matrix blocks, the matrix $\xi^i_j(x)$ can be decomposed correspondingly into 2×2 blocks, and the condition is that each block commute with J. The latter condition, in turn, is easily seen to amount to the usual Cauchy-Riemann equations, hence $\Phi(\mathcal{A})$ is precisely the set of complex analytic local mappings defined by these equations. Finally consider the algebra (iv) of Lemma 2. As in the preceding case, after a suitable choice of coordinates in the euclidean $n = 4k$ space, the pseudogroup $\Phi(\mathcal{A})$ will consist of those local mappings such that the matrix $\xi^i_j(x) = \partial f_i(x)/\partial x_j$ can be decomposed into 4×4 blocks each of which commutes simultaneously with the matrices $\mathcal{I}$, $\mathcal{J}$, $\mathcal{K}$. For notational convenience consider the block $1 \leqq i, j \leqq 4$. Recalling the definitions (10) a straightforward calculation shows that the functions $f_i(x)$ are subject to the differential conditions

$$
\begin{aligned}
&\frac{\partial f_1}{\partial x_1} = \frac{\partial f_2}{\partial x_2} = \frac{\partial f_3}{\partial x_3} = \frac{\partial f_4}{\partial x_4} \\
&\frac{\partial f_1}{\partial x_2} = -\frac{\partial f_2}{\partial x_1} = \frac{\partial f_3}{\partial x_4} = -\frac{\partial f_4}{\partial x_3} \\
&\frac{\partial f_1}{\partial x_3} = -\frac{\partial f_2}{\partial x_4} = -\frac{\partial f_3}{\partial x_1} = \frac{\partial f_4}{\partial x_2} \\
&\frac{\partial f_1}{\partial x_4} = \frac{\partial f_2}{\partial x_3} = -\frac{\partial f_3}{\partial x_2} = -\frac{\partial f_4}{\partial x_1} .
\end{aligned}
\tag{11}
$$

Introducing the complex variables $z_1 = x_1 + ix_2$, $z_2 = x_3 - ix_4$, $w_1 = f_1 + if_2$, $w_2 = f_3 - if_4$, it follows immediately from (11) that w_1, w_2 are holomorphic functions of the complex variables z_1, z_2 in the relevant domains. But it also follows from (11) that

$$
\frac{\partial w_1}{\partial z_1} = \frac{\overline{\partial w_2}}{\partial z_2} \quad \text{and} \quad \frac{\partial w_1}{\partial z_2} = -\frac{\overline{\partial w_2}}{\partial z_1} . \tag{12}
$$

But since the functions w_1, w_2 are holomorphic, (12) can only hold when the partial derivatives are constants, hence when the original functions w_1, w_2 are linear. The functions $f_1, \cdots, f_4$ are therefore linear, and the matrix ξ^i_j of the linear transformation commutes with $\mathcal{I}_k, \mathcal{J}_k, \mathcal{K}_k$. This completes the proof.

REMARKS. The linear transformations of the form (iv) can be written more simply in terms of the complex coordinates $z_1, z_2, \cdots, z_{2k}, w_1, w_2, \cdots, w_{2k}$ introduced in the proof above. Equation (12) shows that the first 4×4 block in the matrix representing the linear transformation must be of the form $\begin{pmatrix} a & b \\ -\bar{b} & \bar{a} \end{pmatrix}$, and that the remaining blocks have a similar form. It is convenient to introduce the column vectors

$$
Z_1 = \begin{pmatrix} z_1 \\ z_3 \\ \cdots \\ z_{2k-1} \end{pmatrix}, \quad Z_2 = \begin{pmatrix} z_2 \\ z_4 \\ \cdots \\ z_{2k} \end{pmatrix}, \quad W_1 = \begin{pmatrix} w_1 \\ w_3 \\ \cdots \\ w_{2k-1} \end{pmatrix}, \quad W_2 = \begin{pmatrix} w_2 \\ w_4 \\ \cdots \\ w_{2k} \end{pmatrix};
$$

it is then clear that the transformation can be written in matrix form

$$
\begin{aligned}
W_1 &= AZ_1 + BZ_2 \\
W_2 &= -\bar{B}Z_1 + \bar{A}Z_2 .
\end{aligned}
\tag{13}
$$

This group was investigated by E. Cartan in his paper [7]. It is interesting to note that on changing the minus sign into a plus sign in (13), the corresponding group becomes reducible, when considered as a real linear group.

3. Linear pseudogroups of higher degree

Each linear subspace $\mathcal{A} \subset \mathcal{T}(t) = \mathcal{S}(1) \oplus \cdots \oplus \mathcal{S}(t)$ determines, as discussed in § 1, a subset $\Phi(\mathcal{A}) \subset \Phi$ of local C^∞ mappings. Now to each element $F \in \Phi(\mathcal{A})$ and to each point x in the domain of F we can further associate the tensor $(\xi^i_j(x), \cdots, \xi^i_{j_1 \cdots j_t}(x)) \in \mathcal{T}(t)$ with components given by (2); the set of all tensors so arising, for all $F \in \Phi(\mathcal{A})$ and all suitable points x, generate a linear subspace $\mathcal{A}^* \subset \mathcal{A}$, which we shall call the *realization* of the original linear subspace $\mathcal{A}$. It is clear from this definition that $\mathcal{A}^* \subset \mathcal{A}$ and that $\Phi(\mathcal{A}^*) = \Phi(\mathcal{A})$; indeed $\mathcal{A}^*$ could equally well have been defined as the smallest linear subspace of $\mathcal{A}$ such that $\Phi(\mathcal{A}) = \Phi(\mathcal{A}^*)$.

In the case $t = 1$ a consideration of linear transformations, as in the proof of Lemma 1, demonstrated that $\mathcal{A}^* = \mathcal{A}$; this is in general not true for $t > 1$. It is clear that the problem of calculating $\mathcal{A}^*$ from $\mathcal{A}$ is really a problem in local differential equation theory: to determine the full space of functions F which are locally required to satisfy the differential equations corresponding to the condition that $F \in \Phi(\mathcal{A})$. A detailed discussion of such topics would follow the lines developed by E. Cartan, in Chapter I of his paper [4] or in his book [8]. These topics were taken up again recently by Kuranishi in [9]. For our present purposes, however, such considerations are not at all necessary; the linearity condition, together with the group composition condition, are quite restrictive, and a more detailed study of the structure of possible realizations $\mathcal{A}^*$ need not be taken up. Therefore we shall simply consider the linear spaces $\mathcal{A}^*$ themselves, rather than the initial spaces $\mathcal{A}$.

To each tensor $(\xi^i_j, \cdots, \xi^i_{j_1 \cdots j_t}) \in \mathcal{A}^* \subset \mathcal{T}(t)$ associate its first component $\xi^i_j \in \mathcal{S}(1)$; the set of all such tensors forms a particular linear subspace of $\mathcal{S}(1)$ which has been called $\mathcal{A}^*(1)$. For any mapping $F \in \Phi(\mathcal{A}^*)$ it is clear that $F \in \Phi(\mathcal{A}^*(1))$ as well; thus $\Phi(\mathcal{A}^*) \subset \Phi(\mathcal{A}^*(1))$.

LEMMA 3. *If $\Phi(\mathcal{A}) = \Phi(\mathcal{A}^*)$ is an irreducible linear pseudogroup defined by a linear differential condition $\mathcal{A}$, then $\mathcal{A}^*(1)$ is an irreducible subalgebra of the full matrix algebra $\mathcal{S}(1)$.*

PROOF. Select any two tensors $\xi^i_j, \eta^i_j \in \mathcal{A}^*(1)$. In consequence of the definition of $\mathcal{A}^*(1)$ there will be finitely many mappings $F^\nu \in \Phi(\mathcal{A}^*)$ such that

$$\xi^i_j = \sum_\nu \partial f^\nu_i(x_\nu)/\partial x_j$$

for suitable points x_ν. Upon replacing these mappings by simple translates of themselves if necessary, we can assume that the points x_ν are one and the same point x; and then the mapping $F = \sum_\nu F_\nu \in \Phi(\mathcal{A}^*)$, with the usual vector addition of mappings into euclidean space, is such that $\xi^i_j = \partial f_i(x)/\partial x_j$. Letting $y = F(x)$ there exists, by the same argument, a mapping $G \in \Phi(\mathcal{A}^*)$ such that $\eta^i_j = \partial g_i(y)/\partial y_j$. Then as in (8) it follows from the chain rule for differentiation that $\mathcal{A}^*(1)$ is a subalgebra of $\mathcal{S}(1)$. It was noted earlier that a subset $\Psi \subset \Phi$ is reducible to dimension r precisely when $\Psi \subset \Phi(\mathcal{R}_r)$; it is therefore clear that $\mathcal{A}^*(1) \subset \mathcal{R}_r$ implies that $\Phi(\mathcal{A}) \subset \Phi(\mathcal{A}^*(1))$ is reducible to dimension r. Consequently if $\Phi(\mathcal{A})$ is irreducible, $\mathcal{A}^*(1)$ can lie in no subspace of the form $\mathcal{R}_r$; and, as in Lemma 1, this implies that $\mathcal{A}^*(1)$ is irreducible as a subalgebra of $\mathcal{S}(1)$.

It now follows, from this lemma and the remarks in the preceding paragraphs, that any irreducible linear pseudogroup of degree $t > 1$ must be contained, either properly or improperly, in one of the irreducible linear pseudogroups of degree 1 listed in Theorem 1. In cases (i) and (iv) of Theorem 1, the transformations are linear, hence all the tensors $\xi^i_{j_1 \cdots j_s}(x)$ of degree $s > 1$ are automatically zero; in other words, $\mathcal{A}^*(s) = \mathcal{A}^* \cap \mathcal{S}(s) = 0$ whenever $s > 1$. Thus these cases need be considered no further. In case (ii) of Theorem 1, $\mathcal{A}^*(1) = \mathcal{S}(1)$. In case (iii) of Theorem 1 all of the local mappings are complex analytic in a suitable complex structure, thus they can be written as mappings $(z_1, \cdots, z_k) \to (w_1, \cdots, w_k)$ in the complex structure, and the derivatives (2) can be replaced by the complex analytic derivatives $\partial^s w_i / \partial z_{j_1} \cdots \partial z_{j_s}$ to determine the (complex) vector space $\mathcal{A}^*$. In these terms, $\mathcal{A}^*(1) = \mathcal{S}(1)$, where $\mathcal{S}(1)$ is now taken to be the full matrix algebra over the complex numbers. Consequently, the latter two cases (ii) and (iii) of Theorem 1, have the same form, except that the underlying field is the real number field in case (ii) and the complex number field in case (iii); with this observation, the two cases can be discussed simultaneously.

LEMMA 4. *If $\Phi(\mathcal{A}) = \Phi(\mathcal{A}^*)$ is an irreducible linear pseudogroup defined by a linear differential condition $\mathcal{A}$, then either:*

(i) $\mathcal{A}^*(2) = 0$, *or*

(ii) $\mathcal{A}^* \cap \mathcal{T}(2) = \mathcal{T}(2)$, *the full space of tensors $(\xi^i_j, \xi^i_{j_1 j_2})$ over the appropriate ground field.*

PROOF. If $F: U \to V$, $G: V \to W$, $H = G \circ F: U \to W$ are local mappings in Φ and $\xi^i_{j_1 \cdots j_s}$, $\eta^i_{j_1 \cdots j_s}$, $\zeta^i_{j_1 \cdots j_s}$ are the tensors associated to F, G, H respectively as in (2), then from the chain rule for differentiation it follows that in addition to (8),

$$\zeta^i_{j_1 j_2}(x) = \sum_{k=1}^n \eta^i_k(F(x))\xi^k_{j_1 j_2}(x) + \sum_{k_1, k_2 = 1}^n \eta^i_{k_1 k_2}(F(x))\xi^{k_1}_{j_1}(x)\xi^{k_2}_{j_2}(x) \;. \tag{14}$$

Then upon applying the argument used in the proof of Lemma 3 it follows that whenever $(\xi^i_j, \xi^i_{j_1 j_2})$ and $(\eta^i_j, \eta^i_{j_1 j_2})$ are tensors belonging to the space $\mathcal{A}^* \cap \mathcal{T}(2)$, the composed tensor $(\zeta^i_j, \zeta^i_{j_1 j_2})$ defined by

$$\begin{aligned} \zeta^i_j &= \sum_{k=1}^n \eta^i_k \xi^k_j \\ \zeta^i_{j_1 j_2} &= \sum_{k=1}^n \eta^i_k \xi^k_{j_1 j_2} + \sum_{k_1, k_2 = 1}^n \eta^i_{k_1 k_2} \xi^{k_1}_{j_1} \xi^{k_2}_{j_2} \;, \end{aligned} \tag{15}$$

also belongs to the space $\mathcal{A}^* \cap \mathcal{T}(2)$. This is, of course, a vector space as well, so that $(t\xi^i_j, t\xi^i_{j_1j_2}) \in \mathcal{A}^* \cap \mathcal{T}(2)$ for all numbers t in the field. Substituting this tensor into (15) in place of $(\xi^i_j, \xi^i_{j_1j_2})$, it follows clearly that both

$$(16) \qquad \left(\sum_{k=1}^n \eta^i_k \xi^k_j, \sum_{k=1}^n \eta^i_k \xi^k_{j_1j_2}\right) \quad \text{and} \quad \left(0, \sum_{k_1,k_2=1}^n \eta^i_{k_1k_2}\xi^{k_1}_{j_1}\xi^{k_2}_{j_2}\right)$$

must belong to $\mathcal{A}^* \cap \mathcal{T}(2)$ separately. If $\mathcal{A}^*(2) = 0$ there is of course nothing to prove; and it has been noted that in all of the remaining cases, $\mathcal{A}^*(1)=\mathcal{S}(1)$, the full space of matrices over the relevant field. In particular, by taking for ξ^i_j the identity matrix, it follows from (16) that for any tensor $(\eta^i_j, \eta^i_{j_1j_2}) \in \mathcal{A}^* \cap \mathcal{T}(2)$, both $(0, \eta^i_{j_1j_2})$ and $(\eta^i_j, 0)$ are contained in $\mathcal{A}^* \cap \mathcal{T}(2)$ separately. One further application of (16) shows that, whenever $(\eta^i_j, \eta^i_{j_1j_2}) \in \mathcal{A}^* \cap \mathcal{T}(2)$, then

$$(17) \qquad \left(0, \sum_{k,k_1,k_2=1}^n \alpha^i_k \eta^k_{k_1k_2} b^{k_1}_{j_1} b^{k_2}_{j_2}\right) \in \mathcal{A}^* \cap \mathcal{T}(2)$$

for arbitrary matrices α^i_k, b^k_j. It is then a simple algebraic observation that $\mathcal{A}^* \cap \mathcal{T}(2)$ must contain all symmetric tensors $(0, \eta^k_{k_1k_2})$, from which the lemma follows as stated. (For completeness, the last assertion can be demonstrated as follows. If $\mathcal{A}^*(2) \neq 0$, there is some entry $\eta^r_{pq} = \eta^r_{qp} \neq 0$; in particular, assume that either $p = q$ or that $p \neq q$ and all terms $\eta^r_{pp} = 0$. Then setting $\alpha^i_k = \delta^i_p \delta^r_k$, $b^k_j = (\delta^k_p + \delta^k_q)(\delta^\sigma_j + \delta^\tau_j)$ in (17), where δ^i_j is the Kronecker symbol, yields a tensor $(0, \zeta^i_{j_1j_2}) \in \mathcal{A}^* \cap \mathcal{T}(2)$ such that $\zeta^\rho_{\sigma\tau} = \zeta^\rho_{\tau\sigma} \neq 0$ but all other entries $\zeta^i_{j_1j_2} = 0$. These tensors form a basis for $\mathcal{S}(2)$, and thus follows the assertion.)

LEMMA 5. *If $\Phi(\mathcal{A}) = \Phi(\mathcal{A}^*)$ is an irreducible linear pseudogroup defined by a linear differential condition $\mathcal{A}$ such that $\mathcal{A}^*(2)\neq 0$, then $\mathcal{A}^* \cap \mathcal{T}(t)=\mathcal{T}(t)$, over the appropriate field, for all t.*

PROOF. The result of Lemma 4 provided the desired condition for $t = 2$. We shall only consider the case $t = 3$ in detail here; the general case follows just the same line of argument, but notational complications arise which are not worth the trouble of handling. With the notation as in the proof of Lemma 4, it follows that, whenever $(\xi^i_{j_1\cdots j_s})$, $(\eta^i_{j_1\cdots j_s}) \in \mathcal{A}^* \cap \mathcal{T}(3)$, then the composed tensor $(\zeta^i_{j_1\cdots j_s})$ defined by (15) and

$$(18) \qquad \begin{aligned} \zeta^i_{j_1j_2j_3} &= \sum_{k=1}^n \eta^i_k \xi^k_{j_1j_2j_3} + \sum_{k_1,k_2,k_3=1}^n \eta^i_{k_1k_2k_3}\xi^{k_1}_{j_1}\xi^{k_2}_{j_2}\xi^{k_3}_{j_3} \\ &\quad + \sum_{k_1,k_2=1}^n \eta^i_{k_1k_2}[\xi^{k_1}_{j_1j_2}\xi^{k_2}_{j_3} + \xi^{k_1}_{j_1j_3}\xi^{k_2}_{j_2} + \xi^{k_1}_{j_2j_3}\xi^{k_2}_{j_1}] \end{aligned}$$

also belongs to $\mathcal{A}^* \cap \mathcal{T}(3)$. Upon substituting $(t\xi^i_{j_1j_s})$ in (18) in place of $(\xi^i_{j_1j_s})$, it follows as in Lemma 4 that $\mathcal{A}^* \cap \mathcal{T}(3)$ contains separately the tensors $(0, 0, \xi^i_{j_1j_2j_3})(0, \xi^i_{j_1j_2}, 0)$, and $(\xi^i_{j_1}, 0, 0)$, and hence any linear combinations of these. As noted, the tensors $(0, \xi^i_{j_1j_2}, 0)$ and $(\xi^i_j, 0, 0)$ have arbitrary values. But then (18) shows that there must be some tensor with $\zeta^i_{j_1j_2j_3} \neq 0$. A repetition of the argument of Lemma 4 shows then that the terms $(0, 0, \zeta^i_{j_1 j_2 j_3})$ can be arbitrary symmetric tensors, thus completing the proof.

THEOREM 2. *The only irreducible linear pseudogroups of C^∞ local mappings in E_n are, in addition to those listed in Theorem 1, the following, (all of second degree):*

(v) *the set of all real linear mappings;*

(vi) *the set of all complex linear mappings in a suitable complex structure on euclidean $n = 2k$ space.*

PROOF. The irreducible linear pseudogroups of first degree having been listed in Theorem 1, it suffices to consider a linear differential condition $\mathcal{A}$ of degree $t \geqq 2$. As in Lemma 3, the term $\mathcal{A}^*(1)$ is the full matrix algebra over either the real or complex numbers. Then by Lemma 4 either $\mathcal{A}^*(2) = 0$ or $\mathcal{A}^* \cap \mathcal{T}(t) = \mathcal{T}(t)$; the condition $\mathcal{A}^*(2) = 0$ means that the mappings in $\Phi(\mathcal{A}) = \Phi(\mathcal{A}^*)$ are linear, and thus leads to cases (v), (vi) as stated in the theorem. If $\mathcal{A}^*(2) \neq 0$, then Lemma 5 shows that $\mathcal{A}^* \cap \mathcal{T}(t) = \mathcal{T}(t)$, and there are consequently no further conditions imposed. This leads once again to cases (ii), (iii) of Theorem 1.

PRINCETON UNIVERSITY

BIBLIOGRAPHY

1. A. A. ALBERT, Structures of Algebras, Amer. Math. Soc. Colloquium Publications, vol. 24, 1939.
2. V. BARGMANN, *Irreducible unitary representations of the Lorentz group*, Ann. of Math., 48 (1947), 568-640.
3. S. BOCHNER and W. T. MARTIN, Several Complex Variables, Princeton Univ. Press, Princeton, 1948.
4. E. CARTAN, *Sur la structure des groupes infinis de transformations*, Ann. École Norm. Sup., 21 (1904), 153-206; 22 (1905), 219-308.
5. ———, *Les sous-groupes des groupes continus de transformations*, Ann. École Norm. Sup., 25 (1908), 57-194; *Les groupes de transformations continus, infinis, simples*, Ann. École Norm. Sup., 26 (1909), 93-161.
6. ———, *Les groupes projectifs qui ne laissent invariante aucune multiplicité plane*, Bull. Soc. Math. France. 41 (1913), 53-96; *Les groups projectifs continus réels qui ne laissent invariante aucune multiplicité plane*, J. Math. pures et appl., 10 (1914), 149-186.
7. ———, *Sur les groupes linéaires quaternioniens*, Vierteljahr. Naturf. Gesellschaft Zürich, 85 (1940), 191-203.
8. ———, Les systèmes différentiels extérieurs et leurs applications géométriques, Hermann, Paris, 1945.
9. M. KURANISHI, *On E. Cartan's prolongation theorem of exterior differential systems*, Amer. J. Math., 79 (1957), 1-47.
10. H. WEYL, *Theorie der Darstellung kontinuierlicher halb-einfacher Gruppen durch lineare Transformationen*, Math. Z., 23 (1925), 271-309; 24 (1926), 328-395.

Section H

Differential Equations

Some of Bochner's papers can conveniently be grouped together under the heading of differential equations. Some of these papers were concerned with analogues of the analytic continuation of functions of several complex variables for functions that are the solutions of various systems of partial differential equations.

Über Sturm-Liouvillesche Polynomsysteme.

Von

S. Bochner in München.

Wir betrachten irgendeine Differentialgleichung der Form

$$p_0(x)y'' + p_1(x)y' + p_2(x)y + \lambda y = 0. \tag{1}$$

Die Koeffizienten $p_0(x)$, $p_1(x)$, $p_2(x)$ sind irgendwelche reell- oder komplexwertige Funktionen der Variablen x, von denen wir in erster Linie nur anzunehmen brauchen, daß sie in einem gemeinsamen Intervall J der x-Achse definiert sind; und λ bedeutet einen Parameter, der aller komplexen Werte fähig ist.

Wir setzen nunmehr voraus, daß es eine Folge von Parameterwerten

$$\lambda_0, \lambda_1, \lambda_2, \ldots, \lambda_n, \ldots \tag{2}$$

und eine Folge von Polynomen

$$P_0(x), P_1(x), P_2(x), \ldots, P_n(x), \ldots \tag{3}$$

gibt, von denen das n-te, $P_n(x)$, von *genau* n-tem Grade ist, und die so beschaffen sind, daß das Polynom $P_n(x)$ im Intervall J der Gleichung (1) für den Wert $\lambda = \lambda_n$ genügt:

$$p_0(x)P_n''(x) + p_1(x)P_n'(x) + p_2(x)P_n(x) + \lambda_n P_n(x) = 0. \tag{4}$$

Ein solches System von Polynomen wollen wir ein Sturm-Liouvillesches Polynomsystem nennen. Von solcher Art sind die wichtigsten Polynomsysteme der mathematischen Physik: die Legendreschen (allgemeiner: die hypergeometrischen), die Laguerreschen und die Hermiteschen Polynome (vgl. etwa: Courant-Hilbert, Methoden der mathematischen Physik I, Kap II, § 8 und § 10).

Es ist eine sehr leichte Aufgabe, die Gesamtheit *aller* Sturm-Liouvilleschen Polynomsysteme aufzustellen. Da aber diese Aufstellung in der einschlägigen Literatur nicht vorzukommen scheint, sei es gestattet, sie im folgenden näher auszuführen.

Gegeben seien eine Gleichung (1) und dazugehörige Eigenwerte (2) und Eigenfunktionen (3). Setzt man in (4) $n = 0$, so ergibt sich

$$p_2(x) P_0(x) + \lambda_0 P_0(x) = 0;$$

da nach Voraussetzung $P_0(x)$ eine *von Null verschiedene* Konstante sein soll, erhalten wir daraus

$$p_2(x) = -\lambda_0,$$

also daß die Funktion $p_2(x)$ eine Konstante sein muß. Wir können ohne wesentliche Einschränkung annehmen, daß $p_2(x) = 0$, sonst ersetze man die Gleichung (1) durch

$$p_0(x) y'' + p_1(x) y' + \lambda y = 0 \tag{5}$$

und betrachte (5) für die Eigenwerte $\lambda_n' = \lambda_n + p_2$. Aus

$$p_1(x) P_1'(x) + \lambda_1 P_1(x) = 0$$

folgt weiterhin

$$p_1(x) = -\frac{1}{P_1'(x)} \lambda_1 P_1(x), \tag{6}$$

also daß $p_1(x)$ eine lineare Funktion in x ist,

$$p_1(x) = \delta + \varepsilon x,$$

und aus

$$p_0(x) P_2''(x) = -p_1(x) P_2'(x) - \lambda_2 P_2(x)$$

folgt

$$p_0(x) = \alpha + \beta x + \gamma x^2.$$

Wir brauchen also nur noch die Gleichung

$$(\alpha + \beta x + \gamma x^2) y'' + (\delta + \varepsilon x) y' + \lambda y = 0 \tag{7}$$

daraufhin zu untersuchen, wann sie ein Sturm-Liouvillesches Polynomsystem erzeugt.

Transformiert man die Variable x in eine lineare Verbindung $x = a + b\xi$ mit beliebigen komplexen Koeffizienten a und b, $b \neq 0$, so entsteht aus (7) eine Gleichung derselben Bauart (in der Variablen ξ) und falls die Gleichung (7) das Polynomsystem (3) erzeugt, erzeugt die neue Gleichung das Sturm-Liouvillesche Polynomsystem

$$Q_n(\xi) = P_n(a + b\xi).$$

Zur weiteren Diskussion der Gleichungen (7) genügt es daher, sich auf einige „Normalformen" zu beschränken, aus denen alle anderen durch passende lineare Transformationen der Variablen x hervorgehen. Es sind dies:

$$\begin{aligned}
&1. && (\delta + \varepsilon x) y' + \lambda y = 0,\\
&2. && y'' + (\delta + \varepsilon x) y' + \lambda y = 0,\\
&3. && x y'' + (\delta + \varepsilon x) y' + \lambda y = 0,\\
&4. && x^2 y'' + (\delta + \varepsilon x) y' + \lambda y = 0,\\
&5. && x(1-x) y'' + (\delta + \varepsilon x) y' + \lambda y = 0.
\end{aligned}$$

Die Normalform 1. entspricht dem Fall $\alpha = \beta = \gamma = 0$, die Normalform 2. dem Fall $\alpha \neq 0, \beta = \gamma = 0$, und die Normalform 3. dem Fall $\beta \neq 0, \gamma = 0$. Ist $\gamma \neq 0$, so muß man zwei Unterfälle unterscheiden, je nachdem die zwei Wurzeln der Gleichung $\alpha + \beta x + \gamma x^2 = 0$ gleich oder verschieden sind. Im ersten Unterfall kann man durch eine passende lineare Transformation der Variablen der gemeinsamen Wurzel den Wert 0 erteilen und demnach unsere Gleichung auf die Normalform 4. reduzieren; im zweiten Unterfalle kann man durch eine lineare Transformation von x erreichen, daß die zwei Wurzeln die Werte 0 und 1 erhalten und somit unsere Differentialgleichung in die Normalform 5. übergeht.

Um von der Gleichung (7) bei vorgegebenen Koeffizienten $\alpha, \beta, \gamma, \delta, \varepsilon$ zu entscheiden, ob sie ein Sturm-Liouvillesches Polynomsystem erzeugt, kann man in folgender Weise verfahren. Man setze die Funktion $y(x)$ als Potenzreihe an:

$$y(x) = a_0 + a_1 x + a_2 x^2 + a_3 x^3 + \dots \tag{8}$$

$$y'(x) = 1 a_1 + 2 a_2 x + 3 a_3 x^2 + \dots \tag{9}$$

$$y''(x) = 2 \cdot 1 \cdot a_2 + 3 \cdot 2 \cdot a_3 x + \dots \tag{10}$$

und führe die Reihen (8) bis (10) in die Gleichung (7) ein. Das gibt für die Koeffizienten $a_0, a_1, a_2, \dots$ die folgenden Bedingungsrelationen, welche den Charakter einer dreigliedrigen Rekursionsformel haben

$$\begin{aligned}(\nu+2)(\nu+1)\alpha a_{\nu+2} + [(\nu+1)\nu\beta + (\nu+1)\delta] a_{\nu+1}\\ + [\nu(\nu-1)\gamma + \nu\varepsilon + \lambda] a_\nu = 0 \qquad (\nu = 0, 1, 2, 3, \dots).\end{aligned} \tag{11}$$

In allen Einzelfällen, auf die wir die Diskussion der allgemeinen Gleichung (7) zurückführen werden, verschwindet in (11) entweder der Koeffizient von $a_{\nu+2}$ oder der von $a_{\nu+1}$, so daß die Rekursionsformel jeweils nur zweigliedrig ist. Dadurch wird es jeweils möglich zu entscheiden, ob man zu jedem n, $n = 0, 1, 2, 3, \dots$, durch passende Wahl von λ, $\lambda = \lambda_n$, das Rekursionsschema durch eine Folge von Zahlen $a_0^{(n)}, a_1^{(n)}, a_2^{(n)}, \dots$ befriedigen kann, für welche $a_{n+1}^{(n)} = a_{n+2}^{(n)} = a_{n+3}^{(n)} = \dots = 0$ und $a_n^{(n)} \neq 0$. Ist so etwas möglich, dann bildet die Polynomfolge

$$P_n(x) = \sum_{\mu=0}^{n} a_\mu^{(n)} x^\mu$$

ein Sturm-Liouvillesches System. Unter Umständen (allerdings, wie wir sehen werden, nur unter sehr speziellen) kann *eine* Gleichung *mehrere* Sturm-Liouvillesche Polynomsysteme erzeugen. Hierbei sehen wir zwei Polynomsysteme, $P_n(x)$ und $Q_n(x)$, als verschieden an, falls für mindestens einen Wert n_0 von n, $n_0 > 0$, die Polynome $P_{n_0}(x)$ und $Q_{n_0}(x)$ sich nicht nur um einen konstanten Faktor unterscheiden.

Wie man der Relation (6) entnimmt, ist für $\varepsilon = 0$ auch $\delta = 0$, und (7) gibt dann

$$(\alpha + \beta x + \gamma x^2) P_n''(x) = -\lambda_n P_n(x).$$

Da jetzt $P_0(x)$ und $P_1(x)$ zum Eigenwert $\lambda = 0$ gehören und die Gleichung (7) nur zwei linear unabhängige Lösungen haben kann, sind die Eigenwerte der anderen Polynome, z. B. der von $P_3(x)$, von Null verschieden, und daraus folgt $\gamma \neq 0$. In den Normalformen 1. bis 3. werden wir demzufolge den Fall $\varepsilon = 0$ ausschließen.

Die Normalform 1. kann man durch den Übergang von x zu $\delta + \varepsilon x$ auf die Gestalt

$$x y' + \lambda y = 0$$

reduzieren. In dieser endgültigen Form erzeugt sie, als einziges, für $\lambda_n = -n$ das System aller Potenzen

$$P_n(x) = x^n.$$

Die Normalform 2. kann man vermöge der Transformation $\xi = \delta \sqrt{-\frac{1}{2\varepsilon}} + \sqrt{-\frac{\varepsilon}{2}}\, x$ auf die Gestalt

$$\frac{d^2 y}{d\xi^2} - 2\xi \frac{dy}{d\xi} + \lambda y = 0,$$

d. h. auf die Gestalt

$$y'' - 2xy' + \lambda y = 0$$

zurückführen. Diese Gleichung erzeugt bekanntlich die Hermiteschen Polynome

$$P_n(x) = \sum_{\varrho=0}^{\left[\frac{n}{2}\right]} (-1)^\varrho \binom{2\varrho}{\varrho} \binom{n}{2\varrho} (2x)^{n-2\varrho}.$$

Die Normalform 3. verwandelt sich durch die Transformation $x = -\frac{1}{\varepsilon}\xi$ in die allgemeine Laguerresche Gleichung

$$\xi \frac{d^2 y}{d\xi^2} + (\delta - \xi) \frac{dy}{d\xi} + \lambda y = 0,$$

die wir folgendermaßen schreiben:

$$xy'' + (k+1-x)y' + \lambda y = 0,$$

wobei k irgendeine reelle oder komplexe Zahl sein mag. Die Rekursionsformel (11) lautet jetzt

$$(\nu+1)(\nu+k+1)a_{\nu+1} + (\lambda-\nu)a_\nu = 0.$$

Unsere Gleichung erzeugt für $k \neq -1, -2, -3, \ldots$ als (einziges) Sturm-Liouvillesches System die Polynome

$$P_n^{(k)}(x) = \frac{1}{n!}\sum_{\varrho=0}^{n}(-1)^\varrho \varrho! \binom{n}{\varrho}\binom{n+k}{\varrho} x^{n-\varrho}.$$

Für $k = 0$ sind dies gewöhnliche und für allgemeines k verallgemeinerte Laguerresche Polynome.

Von der Normalform 4. nehmen wir den Sonderfall $\delta = 0$ vorweg. Unsere Gleichung schreiben wir jetzt

$$x^2 y'' + (k+1)xy' + \lambda y = 0,$$

und wir erhalten die Rekursionsformel

$$[\nu(\nu+k)+\lambda]a_\nu = 0.$$

Wollen wir erreichen, daß $a_n \neq 0$, so müssen wir setzen

$$\lambda_n = -n(n+k).$$

Für diesen Wert des Parameters gibt es nun tatsächlich eine polynomiale Lösung unserer Gleichung, nämlich

$$P_n(x) = x^n.$$

Diese Lösung ist nur dann nicht die einzige, falls für ein ganzzahliges m aus dem Intervall $0 \leqq m \leqq n-1$

$$m(m+k) + \lambda_n = 0, \tag{12}$$

in welchem Falle alle Polynome

$$x^n + b_n x^m$$

für beliebige Werte von b_n unsere Gleichung lösen. Die Relation (12) ist gleichbedeutend mit

$$m^2 + mk - n^2 - nk = 0,$$

$$(m-n)(m+n+k) = 0,$$

$$m = -n-k.$$

Sie kommt also dann und nur dann vor, falls k einen der Werte -1 $-2, -3, \ldots$ hat.

Interessanter als der eben behandelte Sonderfall ist der allgemeine Fall $\delta \neq 0$. Durch eine lineare Transformation der Variablen können wir erreichen, daß δ den Wert -1 annimmt:

$$x^2 y'' + ((k+1)x - 1)y' + \lambda y = 0. \tag{13}$$

Die Rekursionsformel lautet jetzt

$$(\nu+1)a_{\nu+1} = [\nu(\nu+k) + \lambda]\, a_\nu.$$

Für $k = -1, -2, -3, \ldots$ gibt es Paare von nicht negativen ganzen Zahlen m und n, $m < n$, für welche

$$m(m+k) = n(n+k),$$

und für jedes solche n gibt es keine polynomiale Lösung von (13) von *genau* n-tem Grade. Für jeden sonstigen Wert von k erzeugt unsere Gleichung *ein* Polynomsystem

$$P_n^{(k)}(x) = \sum_{\varrho=0}^{n} \varrho!\binom{n}{\varrho}\binom{-n-k}{\varrho} x^{\varrho}.$$

Für $k = 1$ sind dies Polynome, welche in der Theorie der Besselschen Funktionen eine wichtige Rolle spielen. Die Besselsche Funktion $J_{n+\frac{1}{2}}(x)$, für $n = 0, 1, 2, 3, \ldots$, beträgt[1])

$$\frac{e^{i\left[x-\frac{\pi}{2}(n+1)\right]}}{\sqrt{2\pi x}} P_n^{(1)}\left(\frac{1}{2ix}\right) + \frac{e^{-i\left[x-\frac{\pi}{2}(n+1)\right]}}{\sqrt{2\pi x}} P_n^{(1)}\left(\frac{-1}{2ix}\right).$$

Die Normalform 5. schreiben wir folgendermaßen:

$$x(1-x)y'' + (k - (s+1)x)y' + \lambda y = 0.$$

Ihre Rekursionsformel lautet:

$$(\nu+1)(\nu+k)a_{\nu+1} = [\nu(\nu+s) - \lambda]\, a_\nu.$$

In erster Linie setzen wir voraus: $k \neq 0, -1, -2, -3, \ldots$. Für $\lambda_n = n(n+s)$ gibt es dann immer ein Polynom n-ten Grades

$$P_n(x) = G(n+s, -n, k; x),$$

welches unserer Gleichung genügt, wobei wir unter $G(a, b, c; x)$ die hypergeometrische Reihe

$$1 + \frac{ab}{1\cdot c}x + \frac{a(a+1)b(b+1)}{1\cdot 2\cdot c(c+1)}x^2 + \ldots$$

verstehen. Diese Polynome sind dann und nur dann für alle n von *genau* n-tem Grade, d. h. sie bilden dann und nur dann ein Sturm-Liouvillesches

[1]) Vgl. z. B. G. N. Watson, Theory of Bessel Functions (1922), p. 53, Formel (1).

System, falls $s \neq -1, -2, -3, \ldots$. Es bedeute nunmehr k einen der bisher ausgeschlossenen Werte $0, -1, -2, -3, \ldots$. Wir ersetzen die Variable x durch die lineare Verbindung $1-x$, wodurch unsere Gleichung übergeht in

$$x(1-x)y'' + ((s+1-k) - (s+1)x)y' + \lambda y = 0.$$

Die neue Gleichung fällt unter den eben besprochenen Typus, sofern nicht $s+1-k$ einen der Werte $0, -1, -2, -3, \ldots$ hat.

Es bleiben also nur noch die Fälle übrig:

$$k = 0, -1, -2, -3, \ldots, \qquad s = -1, -2, -3, \ldots \quad \text{mit} \quad s < k.$$

Diese sehr speziellen Fälle wollen wir aber nicht näher diskutieren, da hierbei keine neuen Besonderheiten zutage treten würden.

(Eingegangen am 1. Juni 1928.)

ANNALS OF MATHEMATICS
Vol. 47, No. 2, April, 1946

LINEAR PARTIAL DIFFERENTIAL EQUATIONS, WITH CONSTANT COEFFICIENTS

S. BOCHNER

(Received September 14, 1945)

We will derive by a simple method some elementary properties of solutions of systems of linear partial differential equations with constant coefficients. In particular, we will obtain a general theorem on removable singularities. No use will be made of Green's functions or other source functions. Accordingly, our results will be stated for equations in general, although most of them will be of consequence only for equations of elliptic or similar type.

CHAPTER I. DIFFERENTIAL EQUATIONS

For fixed n, we consider an operator of the form

$$\Lambda f \equiv \sum_{r_1+\cdots+r_n \leqq N_0} a_{r_1\cdots r_n} \frac{\partial^{r_1+\cdots+r_n} f}{\partial x_1^{r_1} \cdots \partial x_n^{r_n}}, \tag{1}$$

that is a *finite* sum of the form

$$af + b_1 \frac{\partial f}{\partial x_1} + \cdots + b_n \frac{\partial f}{\partial x_n} + c\frac{\partial^2 f}{x_1^2} + \cdots .$$

The coefficients are all constants. The variables $x_1, \cdots, x_n$ are real, the coefficients may be complex. The integer N_0, if it is the smallest possible one, will be called the order of Λf. Whenever an entire system of operators will be introduced, as will be the case in Chapter II, the order N_0 will be the smallest integer admissible for all operators occurring.

If we consider the differential equation

$$\Lambda f = 0, \tag{2}$$

then for $N \geqq N_0$ we will say that $f(x)$ is a *strict* solution of class C^N in an open set D, if $f(x)$ is defined and belongs to differentiability class C^N in D, and if it satisfies the given equation at every point of D.

We will say that $f(x)$ is a *weak* solution of class C^N in D, if it is defined almost everywhere in D and Lebesque integrable in every compact subset of D, and if corresponding to any point x^0 in D there exist a neighborhood $U = U(x^0)$, such that in $U, f(x)$ is a weak limit of strict solutions of class C^N in U. In other words, there exist a sequence of functions $\{f^{(k)}(x)\}$, $k = 1, 2, \cdots$, each of which is defined, and a strict solution of class C^N in U, such that for every bounded measurable function $\psi(x)$ in U, we have

$$\lim_{k\to\infty} \int_U f^{(k)}(x)\psi(x)\, dv_x = \int_U f(x)\psi(x)\, dv_x ,$$

$[dv_x = dx_1 \cdots dx_n]$.

In what follows, the symbol D' will invariably denote an open subset of D whose closure in space is a bounded subset of D. Furthermore, any function $\varphi(x)$ in D which vanishes outside some D' will be called a *testing function*.

We will start with a very simple theorem.

THEOREM 1. *A function $f(x)$ of class C^N in D is a strict solution of* (2), *if and only if we have*

$$\int_D f \cdot \mathbf{\Lambda}\varphi \cdot dv_x = 0 \tag{3}$$

for every testing function of class C^N.

PROOF. If f and φ both belong to C^N, and if φ is a testing function, then by Stokes' theorem we have

$$\int_D f \cdot \Lambda\varphi \cdot dv_x = \int_D \varphi \cdot \Lambda f \cdot dv_x .$$

Thus, (3) is equivalent with

$$\int_D \varphi \cdot \Lambda f \cdot dv_x = 0. \tag{4}$$

However, for given f we will have relation (4) holding for all testing functions of any class C^M, if and only if $\Lambda f = 0$, as asserted.

LEMMA 1. *If $\{U\}$ is a covering of D by a system of neighborhoods, if D' is a subset of D as described before, and if $\varphi(x)$ is a testing function of class C^N which vanishes outside D', then there exist a finite number of neighborhoods $U_1, \cdots, U_r$ out of the given covering, and corresponding functions $\varphi_1, \cdots, \varphi_r$ of class C^N, such that φ_ρ vanishes outside U_ρ, $\rho = 1, \cdots, r$, and that*

$$\varphi(x) = \varphi_1(x) + \cdots + \varphi_r(x) \tag{5}$$

in all of D.

This lemma is a familiar tool in the theory of differential equations and of differentiable manifolds.[1]

LEMMA 2. *If $\{U\}$ is a covering of D and if we assume that relation* (3) *holds for every testing function of class C^N which vanishes outside some U, then it holds for all testing functions of class C^N.*

PROOF. Follows from (5).

THEOREM 2. *If $f(x)$ is a weak solution of class C^N, then*

$$\int_D f \cdot \Lambda\varphi \cdot dv_x = 0 \tag{6}$$

holds for all testing functions of class C^N.

PROOF. If $f^{(k)}(x)$ approximates weakly to f on U, then by Theorem 1 we have

$$\int_U f^{(k)}(x) \cdot \Lambda\varphi \cdot dv_x = 0 \tag{7}$$

[1] See S. BOCHNER, *Remark on the theorem of Green*, Duke Journal, 3 (1937), 334–338.

for every testing function in U. Letting $k \to \infty$, we obtain (7) for $f(x)$ itself, and by Lemma 2, for D instead of U.

By combining Theorem 2 with one half of Theorem 1, we obtain

THEOREM 3. *If $f(x)$ is a weak solution of class C^N, and if $f(x)$ happens to belong to differentiability class C^M, $M \geqq N_0$, then $f(x)$ is a strict solution of class C^M.*

Our next aim is to invert Theorem 2. This will be done by using so-called h-averages.[2] If the boundary of D' has a distance $\geqq \rho$ from the boundary of D, and if $f(x)$ in D is integrable in every compact subset of D, then for x in D' we can form the h-average

$$f_h(x) = \frac{1}{(2h)^n} \int_{-h}^{h} \cdots \int_{-h}^{h} f(x_1 + t_1, \cdots, x_n + t_n)\, dv_t,$$

if $h < \rho$. If $h < \rho/M$, $M \geqq 1$, we can iterate the process of forming the h-average M times. The Mth iterate will be denoted by $f_{h,M}(x)$ and also called the h-average of order M.

THEOREM 4. *If $f(x)$ is integrable and (6) holds for $\varphi \in C^N$ and D' is an open subset of D as before, then for every $M \geqq 1$ and h sufficiently small we have relation*

$$\int_{D'} f_{h,M}(x) \cdot \Lambda\varphi \cdot dv_x = 0 \tag{8}$$

for every testing function φ of class C^N in D'.

PROOF. If $\varphi(x)$ is a testing-function in D', then for each sufficiently small $t = (t_1, \cdots, t_n)$, the function $\varphi(x - t)$ is a testing-function in D. Therefore by assumption (6) we have

$$\int_D f(x) \cdot \Lambda\varphi(x - t) dv_x = 0,$$

and by a translation of coordinates we hence obtain

$$\int_{D'} f(x + t) \cdot \Lambda\varphi(x) \cdot dv_x = 0$$

for t sufficiently small. If we integrate this with respect to t, we obtain (8) for $M = 1$ and small h, and by iteration for $M \geqq 1$.

THEOREM 5. *If $f(x)$ is a strict solution of class C^N in D, then in D', $f_{h,M}(x)$ is a strict solution for $M \geqq 1$ and small h.*

PROOFS. Theorems 4 and 1.

THEOREM 6. *If $f(x)$ is a weak solution of (1) in D, then in D', $f_{h,M}(x)$ is a strict solution for $M \geqq N_0 + 1$, and small h.*

PROOF. If $f(x)$ is integrable, then $f_h(x)$ is continuous, and $f_{h,M}(x)$ belongs to class C^{M-1}. Now, if $f(x)$ is a weak solution, then by Theorems 2 and 4 we have relation (8). For $M \geqq N_0 + 1$, $f_{h,M}$ is therefore a strict solution by Theorem 1.

[2] For the role of h-averages in the calculus of varations, see J. W. CALKIN, and C. B. MORREY, *Functions of several variables and absolute continuity*, Duke Journal, 4 (1940), 170-186 and 187-215.

THEOREM 7. *If $f(x)$ is integrable and if* (6) *holds for* $\varphi \in C^N$, *then for every* M, *no matter how large,* $f(x)$ *is a weak solution of* (1) *of class* C^M.

PROOF. By Theorem 4, we have relation (8), and by Theorem 1 it hence follows that $f_{h,M}$ is a strict solution for $M \geqq N_0 + 1$. However, for fixed $M, f_{h,M}(x)$ converges weakly to $f(x)$ in D' as $h \to 0$, and this proves the theorem.

THEOREM 8. *If $f(x)$ is a weak solution of class C^N, then it is also a weak solution of class C^M, for all $M \geqq N$.*

PROOF. Theorems 2 and 7.

We also note

THEOREM 9. *If $f(x)$ is a weak solution in D, then $f_{h,M}$ is a weak solution in D', for $M \geqq 1$.*

Finally we point out a special theorem.

THEOREM 10. *Any weak solution of the Laplace equation*

$$\frac{\partial^2 f}{\partial x_1^2} + \cdots + \frac{\partial^2 f}{\partial x_n^2} = 0$$

is also (after correction on a null-set) a strict solution, that is, a harmonic function.

PROOF. If a sequence $f^{(k)}(x)$ is weakly convergent, then in particular the norms $\int_{D'} | f^{(k)}(x) | \, dx$ are bounded in k. Now, if $f(x)$ is harmonic in D, then at each point x^0 in D, $f(x^0)$ is equal to the $(n - 1)$-dimensional average over the boundary of the sphere with center at x^0. From this it follows that $f(x^0)$ is also equal to the n-dimensional average over the interior of the sphere. From this it follows that every weakly convergent sequence of harmonic functions in D is boundedly convergent in every compact subset. From the Poisson integral it then follows that the sequence is also uniformly convergent and that so is also the sequence of their partial derivatives of every order. Thus the limit function is likewise harmonic.

CHAPTER II. SYSTEMS OF EQUATIONS

We will next consider systems of equations. If

$$\Lambda^1 f, \cdots, \Lambda^r f \tag{9}$$

are a fixed system of operators of the type considered before, and if

$$L^1 g, \cdots, L^r g$$

is a variable system of such operators, then

$$Lf \equiv L^1(\Lambda^1 f) + L^2(\Lambda^2 f) + \cdots + L^r(\Lambda^r f) \tag{10}$$

is again an operator of this kind. We will say that (10) has been *induced* by the system (9), and we will call it an *induced* operator.

If $f(x)$ is a strict common solution of the system of equations

$$\Lambda^\rho f = 0, \qquad \rho = 1, \cdots, r, \tag{11}$$

and if it has derivatives of sufficiently high order, then it is also a solution of the induced equation

$$Lf = 0. \tag{12}$$

Thus by Theorem 7 we obtain

THEOREM 11. *If $f(x)$ is a common weak solution of the system* (11), *then it is also a weak solution of every induced equation.*

We will now add a second generalization. We will assume that the symbol $f(x)$ represents not one function but a finite set of function

$$f(x) = (f_1(x), \cdots, f_s(x)), \tag{13}$$

the integer s having no arithmetical connection with the dimension n of the space. An operator Λf shall be an expression of the form

$$\Lambda_1 f_1 + \cdots + \Lambda_s f_s,$$

where each $\Lambda_\sigma f_\sigma$ is an operator of our original type. It should be noted that although Λf operates on a vector function of s components, its value is a one-component function. If the notions of integrability, differentiability, weak convergence, h-average, etc. are applied to each component separately; and if the integral

$$\int f \cdot \Lambda\varphi \cdot dv_x$$

which occurs in several theorems is replaced by

$$\int (f_1 \cdot \Lambda_1\varphi + \cdots + f_s \cdot \Lambda_s\varphi) dv_x$$

where $\varphi(x)$ is a one-component function; then previous theorems will also apply to vector-functions, and to systems of equations as well.

APPLICATIONS. As an application of Theorem 11, we consider for $n = 3$ the system of equations

$$\operatorname{div} f = 0, \qquad \operatorname{rot} f = 0, \tag{14}$$

that is the system

$$\frac{\partial f_i}{\partial x_j} - \frac{\partial f_j}{\partial x_i} = 0, \qquad \frac{\partial f_1}{\partial x_1} + \frac{\partial f_2}{\partial x_2} + \frac{\partial f_3}{\partial x_3} = 0. \tag{15}$$

By the known relation $\Delta f = \operatorname{grad} \operatorname{div} f - \operatorname{rot} \operatorname{rot} f$, the system (14) induces the equations

$$\Delta f_1 = 0, \qquad \Delta f_2 = 0, \qquad \Delta f_3 = 0,$$

and thus by Theorems 10 and 11 we obtain the theorem of H. Weyl.[3]

[3] The method of orthogonal projection in potential theory, Duke Journal, 7 (1940), p. 412.

THEOREM 12. *For $n = 3$, every weak solution of* (14) *has derivatives of all orders.*

Similar to this is

THEOREM 13. *For $n = 2$, a weak solution of the Cauchy-Riemann equations*

$$u_x - v_y = 0, \qquad u_y + v_x = 0 \tag{16}$$

is a strict solution, and thus $u + iv$ is analytic in $x + iy$.

The reason being that equations (16) induce the equations $\Delta u = 0$, $\Delta v = 0$.

Our approach also throws light on the classical theorem of Morera which we will derive in a general set-up. As before, we take any space dimension n, and we introduce for any $s \geqq 1$, expressions

$$\Lambda^i f = \Lambda_1^i f_1 + \cdots + \Lambda_s^i f_s; \qquad i = 1, \cdots, n. \tag{17}$$

We denote by N_0 the precise order of the system and we assume that $f(x)$ belongs to class C^{N_0} in an open set D.

If we will assume that a function f also belongs to C^{N_0+1}, then by Stokes theorem the integral

$$\int_S \sum_{i=1}^n (-1)^{i-1} \Lambda^i(f(x))\, dx_1 \cdots dx_{i-1}\, dx_{i+1} \cdots dx_n \tag{18}$$

will have the same value as

$$\int_B \left(\sum_{i=1}^n \frac{\partial}{\partial x_i} \Lambda^i f \right) dv_x , \tag{19}$$

where B is a domain in D, and S is its boundary, the latter being sufficiently smooth. Now, if the surface integral vanishes for all $(n - 1)$-dimensional surfaces S, then f is a strict solution of

$$\sum_{i=1}^n \frac{\partial}{\partial x_i} \Lambda^i f = 0. \tag{20}$$

We now claim that the conclusion also holds if f belongs only to C^{N_0}.

THEOREM 14. *If the system* (17) *is of order N_0 and f belongs to C^{N_0} and if* (18) *vanishes over all spheres, then f is a weak solution of* (20).

PROOF. If (18) vanishes for all spheres S, then for fixed S it vanishes for translated spheres $S - (t)$, for sufficiently small t. Or, if we replace x by $x + t$, we see that (18) vanishes for $f(x + t)$ in place of $f(x)$. If we then integrate with respect to t under the integral, we arrive at the vanishing of (18) for $f_h(x)$ instead of $f(x)$. But if $f(x)$ belongs to C^{N_0}, then $f_h(x)$ belongs to C^{N_0+1}, and thus $f_h(x)$ is a strict solution of (20). Therefore, $f(x)$ itself is a weak solution of (20), as asserted.

For instance, if u and v are continuous functions in (x, y), and $\alpha, \beta, \gamma, \delta$ are constants, and

$$\int_S (\alpha u + \beta v)\, dx + (\gamma u + \delta v)\, dy = 0$$

for all closed curves in a domain, then (u, v) are a weak solution of

$$\frac{\partial}{\partial y}(\alpha u + bv) - \frac{\partial}{\partial x}(\gamma u + \delta v) = 0.$$

In particular, if

$$\int_S f\,dz \equiv \int_S (u + iv)(dx + i\,dy) = 0,$$

then we have

$$\frac{\partial}{\partial x}(iu - v) - \frac{\partial}{\partial y}(u + iv) = 0$$

and this equation decomposes into the Cauchy-Riemann equations (16). However, a weak solution of the latter equations is also a strict solution, and in this way we obtain a peculiar new proof for the theorem of Morera.

Another curious little theorem for arbitrary n is as follows.

THEOREM 15. *If $f(x)$ is continuous in D, and if*

$$\int_S f dx_2\, dx_3 \cdots dx_n = 0$$

for every spherical hypersurface, then $(\partial f/\partial x_1)$ exists and is equal to 0.

PROOF. By Theorem 14, $f(x)$ is a weak solution of $(\partial f/\partial x_1) = 0$, and $f_h(x)$ is a strict solution. Thus $f_h(x)$ is constant in x_1, and since in the present case $f_h(x)$ converges uniformly to $f(x)$, the latter function is also constant in x_1.

Finally we note the following theorem.

THEOREM 16. *If it is known that f belongs to C^1 in D, and that*

$$\int_S \sum_{i=1}^{n} (-1)^{i-1} \frac{\partial f}{\partial x_i}\, dx_1 \cdots dx_{i-1}\, dx_{i+1} \cdots dx_n = 0$$

for all spheres S, then f is harmonic.

CHAPTER III. REMOVABLE SINGULARITIES

Let A be an arbitrary bounded measurable set in n-space and for $0 < \epsilon < \infty$ let A_ϵ denote the ϵ-neighborhood of A, that is, the union of all open spheres of radius ϵ with centers at points of A. As shown elsewhere,[1] there is in entire space a function $Q_\epsilon(x)$ of class C^∞ having the following properties: (i) $|Q_\epsilon(x)| \leqq 1$; (ii) $Q_\epsilon(x) = 0$ in $A_{2\epsilon}$; (iii) $Q_\epsilon(x) = 1$ outside $A_{3\epsilon}$; and (iv) for every multi-index $(k_1, \cdots, k_n)$ there exist a constant which is independent of x and ϵ such that

$$\left| \frac{\partial^{k_1+\cdots+k_n} Q_\epsilon(x)}{\partial x_1^{k_1} \cdots \partial x_n^{k_n}} \right| \leqq \frac{C}{\epsilon^{k_1+\cdots+k_n}}.$$

Now, let D be a bounded open set, let D_0 be an open subset, and let A be the difference $D - D_0$. We will look upon A as an "exceptional set" in D. Let

(21) $$\Lambda^\rho f = 0, \qquad \rho = 1, \cdots, r,$$

be a system of equations for the vector function

$$f = (f_1, \cdots, f_s), \tag{22}$$

and let N_0 be the order of the system (21). Let f be a given weak solution of (21) in D_0 and let it be integrable in D_0. We complete the vector function f on the exceptional set by assigning there the values 0, and we are posing the problem of deciding under what conditions the completed function will be a weak solution of (21) in all of D.

If φ is a testing function in D, then

$$\varphi(x) \cdot Q_\epsilon(x)$$

will be a testing function of the same class in D_0. Since f was assumed to be a weak solution of (21) in D_0, we have

$$\int_D (f_1 \cdot \Lambda_1(\varphi Q_\epsilon) + \cdots + f_s \cdot \Lambda_s(\varphi Q_\epsilon))\, dv_x = 0. \tag{23}$$

If we take into consideration that the partial derivatives of φ of order $\leqq N_0$ are bounded in D, and if we make use of all properties of $Q_\epsilon(x)$, it is not hard to see that for x in D we have

$$| \Lambda_\sigma^\rho(\varphi Q_\epsilon) - \Lambda_\sigma^\rho(\varphi) | \leqq \frac{C_0}{\epsilon^{N_0}}, \tag{24}$$

where C_0 is independent of x and ϵ. If we use property (iii) and compare (23) and (24) we next obtain the decisive inequality

$$\left| \int_D (f_1 \cdot \Lambda_1^\rho \varphi + \cdots + f_s \cdot \Lambda_s^\rho \varphi)\, dv_x \right| \leqq \frac{C_0}{\epsilon^{N_0}} \int_{A_{3\epsilon}} (|f_1| + \cdots + |f_s|)\, dv_x .$$

By using Theorem 7 we now arrive at the following theorem.

THEOREM 17. *If a vector function f is a weak solution of a system (21) of order N_0 in a bounded open set $D_0 = D - A$, where D is a larger bounded open set and A an exceptional set in it; if B_ϵ is the ϵ-neighborhood of A in D_0 and $v(\epsilon)$ its Lebesque measure; and if we denote by $T(\epsilon)$ the strong average of f over B_ϵ, that is*

$$T(\epsilon) = \frac{1}{v(\epsilon)} \int_{B_\epsilon} (|f_1| + \cdots + |f_s(x)|)\, dv_x ; \tag{25}$$

then adding the values $f \equiv 0$ on A will produce a weak solution of (21) in all of D provided we have

$$T(\epsilon) = o\left(\frac{\epsilon^{N_0}}{v(\epsilon)}\right) \tag{26}$$

as $\epsilon \to 0$.

If the set A has (n-dimensional) Lebesque measure 0, it makes no difference by which values we complete f on the points of A, since a weak solution is determined only up to a null-set. If furthermore the system (21) is such that every weak solution is automatically a strict solution, then the given vector f in D_0

will automatically determine its additional values on A, and in this case our theorem is truly a theorem on removable singularities.

If A is a sufficiently smooth pointset of dimension m, $0 \leqq m < n$, then

$$v(\epsilon) = O(\epsilon^{n-m}) \tag{27}$$

and (26) is implied by

$$T(\epsilon) = o(\epsilon^{N_0+m-n}). \tag{28}$$

For $N_0 + m = n$ this is $T(\epsilon) = o(1)$ and thus the requirement is that f shall vanish in the strong average when x approaches the exceptional set A. If $N_0 + m > n$, a stringent mode of vanishing is required, whereas for $N_0 + m < n$ a certain latitude of unboundedness is not precluded *a priori*.

If we take an integer M such that $0 < M < N$, if we introduce the derivatives

$$f_{\sigma,\mu_1\cdots\mu_n} = \frac{\partial^{\mu_1+\cdots+\mu_n} f_\sigma}{\partial x_1^{\mu_1} \cdots \partial x_{x_n}^{\mu_n}} \tag{29}$$

for

$$0 \leqq \mu_1 + \cdots + \mu_n \leqq M,$$

and if f belongs to class C^M, then we can set up the averages

$$T_M(\epsilon) = \frac{1}{v(\epsilon)} \int_{B_\epsilon} \left(\sum_{\sigma=1}^{s} \sum_{\mu} |f_{\sigma,\mu}| \right) dv_x . \tag{30}$$

Very often relation (26) can be replaced by

$$T_M(\epsilon) = c\left(\frac{\epsilon^{N_0-M}}{v(\epsilon)} \right) \tag{31}$$

and similarly (28) by

$$T_M(\epsilon) = o\ (\epsilon^{N_0-M+m-n}). \tag{32}$$

This replacement will certainly be admissible if there exists a system of equations

$$L(f_\sigma ; f_{\sigma,\mu}) = 0 \tag{33}$$

of order $N_0 - M$ such that a combination of the systems (33) and (29) will induce the original system (21). In particular for $M = N_0 - 1$ we thus obtain

$$T_{N_0-1}(\epsilon) = o(\epsilon^{m-n+1}),$$

and for a hypersurface A of dimension $n - 1$ this will be fulfilled whenever f and its derivatives of order $\leqq N_0 - 1$ are approaching values 0 as x approaches A.

Chapter IV. A Uniqueness Theorem

We will draw a peculiar conclusion from the last theorem.

Theorem 18. *If D_0 is a bounded domain and if B is an $(n - 1)$-dimensional piece of hypersurface on the boundary of D_0 and B_ϵ is its ϵ-neighborhood in D_0 ; if* *the system* (21) *of order* N_0 *is such that every weak solution is automatically analytic in the real variables* $x_1, \cdots, x_n$; *and if*

$$T(\epsilon) = o\left(\frac{\epsilon^{N_0}}{v(\epsilon)}\right),$$

or more specifically if

$$T(\epsilon) = o(\epsilon^{N_0-1}),$$

then $f(x) \equiv 0$.

If (21) *is inducible from a system of order* $N_0 - M$ *as described in Chapter III, and if*

$$T^M(\epsilon) = o(\epsilon^{N_0-M-1}),$$

then again $f(x) \equiv 0$.

We can take a domain D_1 which borders on B from the outside. If we put $A = B + D_1$, $D = D_0 + A = D + B + D_1$, we may apply Theorem 17 and the conclusion is that $f(x)$, if completed by values 0 in A, is a weak solution in all of D. By our specific assumption, $f(x)$ is analytic in D, and since it vanishes in an open subset D_1, it vanishes identically, as asserted.

Appendix

Operators with non-constant coefficients

Some of our results, notably the theorem on removable singularities, will remain valid if the coefficients a_r which occur in our operators are general functions of $(x_1, \cdots, x_n)$. We introduce an expression of the form

$$\sum_{r_1+\cdots+r_n \leqq N_0} a_{r_1} \cdots {}_{r_n}(x) \frac{\partial^{r_1+\cdots+r_n} f}{\partial_{x_1}^{r_1} \cdots \partial_{x_n}^{r_n}}$$

in which the coefficients $a_r(x)$ have bounded continuous derivatives of order $\leqq N_0$ in the given domain D; or of any larger order N and in some larger domain, if the context will so require. The given operator will be denoted by $\bar{\Lambda}f$; whereas by Λf we will denote its formal adjoint, that is the operator

$$\sum_{r_1+\cdots+r_n \leqq N_0} (-1)^{r_1+\cdots+r_n} \frac{\partial^{r_1+\cdots+r_n} (a_r f)}{\partial_{x_1}^{r_1} \cdots \partial_{x_n}^{r_n}}.$$

If f belongs to C^{N_0} and φ is a testing function of class C^{N_0}, we have

$$\int_D (f \cdot \Lambda\varphi - \varphi \cdot \bar{\Lambda}f)\, dv_x = 0,$$

and we will say that $f(x)$ is a weak solution of the equation

$$\bar{\Lambda}f = 0$$

in D if it is defined almost everywhere and Lebesque integrable in every compact subset of D and if for every testing function φ of class C^{N_0} we have

$$\int_D f \cdot \Lambda\varphi\, dv_x = 0. \tag{34}$$

A similar definition holds for a vector function being a solution of an equation

$$\bar{\Lambda}_1(f_1) + \cdots + \bar{\Lambda}_s(f_s) = 0.$$

In testing relation (34) we may restrict ourselves to functions φ from C^N, for some $N \geqq N_0$, since for every testing function φ of class C^{N_0} there exist a sequence of testing functions $\{\varphi_n\}$, each of class C^N, such that

$$\Lambda\varphi_n \longrightarrow \Lambda\varphi.$$

Furthermore, the fundamental facts about induced equations remain in force, since the formal adjoint of $\bar{L}(\bar{\Lambda}f)$ is $\Lambda(Lf)$ and (34) implies

$$\int_D f \cdot \Lambda(L\varphi)\, dv_x = 0.$$

It must be pointed out that in general no smoothing process in the nature of an h-average will be available for our weak solutions, and we cannot claim that a weak solution is a weak limit of strict solutions. However, the proofs in Chapters III and IV do not depend on the latter facts and we thus may state

THEOREM 19. *Theorems* 17 *and* 18 *are also valid for operators with non-constant coefficients.*

PRINCETON UNIVERSITY

A HOMOGENEOUS DIFFERENTIAL SYSTEM OF INFINITE ORDER WITH NONVANISHING SOLUTION

S. BOCHNER AND D. V. WIDDER

1. Introduction. The only solution of the homogeneous differential system

$$\frac{d^{2k-1}}{dt^{2k-1}} [t^k f(t)] = 0, \tag{1}$$

$$f(0+) = f(+\infty) = 0 \tag{2}$$

is $f(t) \equiv 0$. This may be seen by observing that none of the fundamental solutions of equation (1), t^n $(n = -k, -k+1, \cdots, k-2)$, satisfies the boundary conditions (2). We now replace equation (1) by another of infinite order, retaining the boundary conditions (2), and investigate the possibility of a nonvanishing solution. We let k become infinite in equation (1) after introduction of a factor. Set

$$L_{k,t}[f(x)] = \frac{(-t)^{k-1}}{k!(k-2)!} \frac{d^{2k-1}}{dt^{2k-1}} [t^k f(t)] \tag{3}$$

and consider the system

$$\lim_{k\to\infty} L_{k,t}[f(x)] = 0, \qquad 0 < t < \infty, \tag{4}$$

with the boundary conditions (2). Since the differential operator (3) serves to invert the Stieltjes transform,[1]

$$f(x) = \int_0^\infty \frac{\phi(t)}{x+t} dt, \tag{5}$$

$$\lim_{k\to\infty} L_{k,t}[f(x)] = \phi(t),$$

it is clear that the only solution of the system (4), (2) of the form (5) must vanish identically. Accordingly if the limit (4) exists boundedly,[2] for example, there is no nonvanishing solution of the system. However, we shall show that if the limit (4) merely exists at each point of the interval then there are many nonvanishing solutions of the system. One very simple solution is $f(t) = t(1+t)^{-2}$. We shall find

Presented to the Society, September 4, 1947; received by the editors June 25, 1947.

[1] See *The Laplace transform* by D. V. Widder, Princeton University Press, 1946, p. 345.

[2] D. V. Widder, loc. cit. p. 373.

it convenient to make an exponential change of variable which will change equation (4) into

$$\lim_{n\to\infty} D\prod_{k=1}^{n}\left(1-\frac{D^2}{k^2}\right)y(x) = 0,$$

where D indicates differentiation with respect to x. The system in question becomes

$$(6)\qquad (\sin \pi D)y(x) = 0, \qquad -\infty < x < \infty,$$

$$(7)\qquad y(-\infty) = y(+\infty) = 0,$$

and a nonvanishing solution is

$$y(x) = e^{-x}(1+e^{-x})^{-2}.$$

2. Existence of a nonvanishing solution.

THEOREM 1. *The function*

$$h(x) = \frac{e^{-x}}{(1+e^{-x})^2}$$

satisfies the system (6), (7).

To prove this let us introduce the following notation:

$$E_n(D) = D\prod_{k=1}^{n}\left(1-\frac{D^2}{k^2}\right),$$

$$g(x) = \frac{1}{1+e^{-x}},$$

$$c_n = \frac{(2n+1)!}{n!n!}.$$

We shall show by induction that

$$(8)\qquad E_n(D)g(x) = c_n[h(x)]^{n+1}.$$

This is true for $n=0$ since $g'(x)=h(x)$. Assume equation (8) true when n is replaced by $n-1$ and differentiate both sides twice with respect to x,

$$\frac{d^2}{dx^2}E_{n-1}(D)g(x) = -c_{n-1}\left\{\frac{2n(2n+1)e^{-(n+1)x}}{(1+e^{-x})^{2n+2}} + \frac{n^2e^{-nx}}{(1+e^{-x})^{2n}}\right\}.$$

If this equation is divided by n^2 and subtracted from equation (8) for

$n-1$ we obtain

$$
\begin{aligned}
E_n(D)g(x) &= \frac{c_{n-1}}{n^2}\,\frac{2n(2n+1)e^{-(n+1)x}}{(1+e^{-x})^{2n+2}} \\
&= c_n h(x)^{n+1}. \qquad (9)
\end{aligned}
$$

By one more differentiation we have

$$
E_n(D)h(x) = c_n(n+1)h(x)^n h'(x). \qquad (10)
$$

Since $h(x)$ is an even function it is clear that $E_n(D)h(x)$ vanishes at $x=0$. For each $x\neq 0$ the right-hand side of equation (10) is the general term of a convergent series and hence tends to zero with n. The test ratio of the series is

$$
\lim_{n\to\infty}\frac{c_n}{c_{n-1}}\left(\frac{n+1}{n}\right)h(x) = 4h(x),
$$

and this is clearly less than unity. Thus

$$
(\sin \pi D)h(x) = 0, \qquad -\infty < x < \infty.
$$

It is clear by inspection that $h(\pm\infty)=0$. This proves the theorem.

3. Derivatives of the solution.

THEOREM 2. *Every even derivative of the function $h(x)$ of Theorem 1 satisfies the system* (6), (7).

By equation (9) we have $E_n(D)g^{(p)}(x)=c_nD^ph(x)^{n+1}$, $p=0, 1, 2, \cdots$. Without computing the pth derivative of $h(x)^{n+1}$ completely we can obtain an upper bound for it as follows. It is clear that

$$
D^p h(x)^{n+1} = D^p \frac{1}{(e^{x/2}+e^{-x/2})^{2n+2}} = \frac{a_p(x)}{(e^{x/2}+e^{-x/2})^{2n+p+2}}, \qquad (11)
$$

where

$$
a_p(x) = \sum_{j=0}^{p} c_{p,j}e^{(j-p/2)x}. \qquad (12)
$$

Here the coefficients $c_{p,j}$ need not be determined. If

$$
|c_{p,j}| \leqq A_pC_{p,j}, \qquad j=0, 1, \cdots, p,
$$

then

$$
|a_p(x)| \leqq A_p(e^{x/2}+e^{-x/2})^p, \qquad -\infty < x < \infty. \qquad (13)
$$

By differentiation of equation (12) we have

$$|a_p'(x)| \leqq A_p \frac{p}{2}(e^{x/2}+e^{-x/2})^p. \tag{14}$$

By induction on p we obtain by differentiating equation (11) that

$$a_{p+1}(x) = a_p'(x)(e^{x/2}+e^{-x/2}) - 2^{-1}(2n+p+2)a_p(x)(e^{x/2}-e^{-x/2}).$$

Then inequalities (13) and (14) give

$$|a_{p+1}(x)| \leqq A_p(n+p+1)(e^{x/2}+e^{-x/2})^{p+1}$$

or

$$A_{p+1} = (n+p+1)A_p.$$

For $p=0$, $a_p(x)=1$, and $A_0=1$. Hence

$$A_p = \frac{(n+p)!}{n!}$$

and

$$\begin{aligned} |D^p[h(x)]^{n+1}| &\leqq \frac{(n+p)!}{n!}[h(x)]^{n+1} \\ |E_n(D)h^{(p)}(x)| &\leqq c_n\frac{(n+p+1)!}{n!}[h(x)]^{n+1}. \end{aligned} \tag{15}$$

As in §2 the right-hand side of inequality (15) tends to zero with $1/n$ for each $x\neq 0$. Moreover if p is even, the left-hand side is an odd function and consequently vanishes at the origin. Hence $h^{(p)}(x)$ satisfies equation (6) for even p.

Since

$$\begin{aligned} h^{(p)}(x) &= E_0(D)g^{(p)}(x) = c_0D^p[h(x)], \\ |h^{(p)}(x)| &\leqq c_0p!h(x), \end{aligned} \tag{16}$$

it follows that $h^{(p)}(x)$ vanishes at $\pm\infty$, and the theorem is proved.

4. Further solutions. We can now obtain a very large class of solutions of the system (6), (7).

THEOREM 3. *If the function*

$$f(w) = \sum_{k=0}^{\infty} a_k w^k$$

is an even function of order one and minimal type and if $h(x)$ *is the function of Theorem* 1, *then the function*

$$F(x) = \sum_{k=0}^{\infty} a_k h^{(k)}(x)$$

satisfies the system (6), (7).

Let $\epsilon_0, \epsilon_1, \cdots$ be a sequence of complex numbers tending to zero. Then the most general function of order one and minimal type takes the form[3]

$$f(w) = \sum_{k=0}^{\infty} \frac{(\epsilon_k w)^k}{k!} w^k.$$

Since $f(w)$ is even, $\epsilon_k = 0$ for even k. Clearly $F(x)$ is also even, and $E_n(D)F(x)$ is odd for each n, and vanishes at the origin. That is, $F(x)$ satisfies equation (6) at $x=0$.

If $x \neq 0$ we may determine a positive number ϵ less than unity and such that

$$4e^{-x} < (1 + e^{-x})^2(1 - \epsilon). \tag{17}$$

Now determine an integer N so large that $|\epsilon_k| \leqq \epsilon$ when $k \geqq N$. We have at once that

$$E_n(D)F(x) = \sum_{k=0}^{\infty} \frac{\epsilon_k^k}{k!} E_n(D)h^{(k)}(x). \tag{18}$$

Inequality (15) is sufficient to establish the uniform convergence of the series (18) and thus to validate the term-by-term differentiation. By Theorem 2 it is clear that

$$\lim_{n\to\infty} \sum_{k=0}^{N} \frac{\epsilon_k^k}{k!} E_n(D)h^{(k)}(x) = 0.$$

Moreover, by inequality (15)

$$\left| \sum_{k=N}^{\infty} \frac{\epsilon_k^k}{k!} E_n(D)h^{(k)}(x) \right| \leqq c_n[h(x)]^{n+1} \sum_{k=N}^{\infty} \frac{(n+k+1)!}{n!} \frac{\epsilon^k}{k!}$$

$$\leqq \frac{(n+1)\, c_n}{(1-\epsilon)^{n+2}} [h(x)]^{n+1}.$$

But this dominating function is again the general term of a series

[3] See, for example, L Bieberbach, *Lehrbuch der Funktionentheorie*, vol. 2, 1931, p. 235.

with test ratio

$$\lim_{n\to\infty} \frac{c_{n+1}}{c_n} \frac{h(x)}{1-\epsilon} = \frac{4h(x)}{1-\epsilon}.$$

This ratio is less than 1 by (17), so that $F(x)$ satisfies (6) for all x.

Finally, from inequality (16)

$$|F(x)| \leqq c_0 h(x) \sum_{k=0}^{\infty} |\epsilon_k|^k,$$

and $F(x)$ also satisfies equations (7). This completes the proof.

5. **A counter-example.** We can now show that the result of Theorem 3 is best possible in a certain sense.

THEOREM 4. *The word "minimal" in Theorem 3 cannot be replaced by the word "normal".*

To prove this it will be sufficient to exhibit an even entire function

$$f(w) = \sum_{k=0}^{\infty} a_k w^k$$

of order 1 and type t (arbitrarily small) such that the function

$$F(x) = \sum_{k=0}^{\infty} a_k h^{(k)}(x)$$

will fail to satisfy equation (6) for some x. We choose $a_k = (it)^k/k!$ for even k and $a_k = 0$ for odd k. Then $f(w) = \cos(tw)$ is a function with the properties desired. Moreover the function $F(x)$ becomes

$$F(x) = \frac{h(x+it) + h(x-it)}{2} = \text{Re } h(x+it).$$

Simple computations give

$$\begin{aligned} &-\frac{E_n(D)F(x)}{(n+1)c_n} \\ &= \text{Re}\left\{\frac{(e^{x/2}-e^{-x/2})\cos(t/2) + i(e^{x/2}+e^{-x/2})\sin(t/2)}{[(e^{x/2}+e^{-x/2})\cos(t/2) + i(e^{x/2}-e^{-x/2})\sin(t/2)]^{2n+3}}\right\}. \end{aligned} \tag{19}$$

If we define an argument α and a modulus λ as follows,

$$\tan\alpha = \frac{e^{x/2}-e^{-x/2}}{e^{x/2}+e^{-x/2}} \tan\frac{t}{2}, \qquad \lambda^2 = e^x + e^{-x} + 2\cos t$$

equation (19) becomes

$$(20)\qquad -\frac{\lambda^{2n+3}E_n(D)F(x)}{(n+1)c_n} = \operatorname{Re}\left\{\frac{(e^{x/2}-e^{-x/2})\cos(t/2)+i(e^{x/2}+e^{-x/2})\sin(t/2)}{e^{i\alpha(2n+3)}}\right\}.$$

If we choose x so near zero that $\lambda<2$, it is clear that $\lambda^{2n+3}/[(n+1)c_n]$ approaches zero with $1/n$. If in addition we choose x so near zero that $\alpha=(2\pi)/N$, where N is an odd integer (which is seen to be possible from the definition of α), then the right-hand side of equation (20) reduces to

$$(e^{x/2}-e^{-x/2})\cos(t/2)\neq 0$$

for infinitely many values of n. Hence for such x

$$\lim_{n\to\infty} E_n(D)F(x)\neq 0,$$

and our result is established.

PRINCETON UNIVERSITY AND
HARVARD UNIVERSITY

Reprinted from the Proceedings of the NATIONAL ACADEMY OF SCIENCES,
Vol. 38, No. 3, pp. 227–30. March, 1952

PARTIAL DIFFERENTIAL EQUATIONS AND ANALYTIC CONTINUATIONS

BY S. BOCHNER

MATHEMATICS DEPARTMENT, PRINCETON UNIVERSITY

Communicated January 28, 1952

In Euclidean E^n we will say that an operator with constant coefficients

$$\Delta f = \sum_{p_1+\ldots+p_n \leq 2h} a_{p_1\ldots p_n} \frac{\partial^{p_1+\ldots+p_n} f}{\partial x_1^{p_1}\ldots\partial x_n^{p_n}} \tag{1}$$

of any even order is *elliptic semihomogeneous* (in a generalized sense) if it possesses a Green's function $G(x)$ which is defined and *analytic* in E^n except at the origin, and if there is an r_0 such that for any $r_1 \geq 0, \ldots, r_n \geq 0, r_1 + \ldots + r_n \geq r_0$ we have at infinity

$$\frac{\partial^{r_1+\ldots+r_n} G}{\partial x_1^{r_1}\ldots\partial x_n^{r_n}} = 0(|x|^{-n-1}), \qquad |x| \to \infty. \tag{2}$$

We have recently demonstrated[1] that such a Green's function exists if the polynomial

$$\sum_{p_1+\ldots+p_n \leq 2h} a_{p_1\ldots p_n}(it_1)^{p_1}\ldots(it_n)^{p_n}$$

and both its highest and lowest homogeneous parts occurring, are all three different from zero for any real $t_1, \ldots, t_n$ not all zero.

THEOREM 1. *In E_n, if D is a bounded domain having a connected boundary B, and if U is a neighborhood of B, and if an analytic function $f(x_j)$ in U satisfies an elliptic semihomogeneous equation*

$$\Delta f = 0 \tag{3}$$

in all its variables, and also an arbitrary (non-homogeneous, non-elliptic) equation

$$\Lambda f = \sum_{q_1+\ldots+q_m \leq M} b_{q_1\ldots q_m} \frac{\partial^{q_1+\ldots+q_m} f}{\partial x_1^{q_1}\ldots\partial x_m^{q_m}} = 0 \tag{4}$$

in fewer than all variables,

$$m < n,$$

the latter having again constant coefficients; then the function has an analytic continuation into all of $D + U$.

Proof: This theorem is an outgrowth of the theorem of Hartogs for complex variables, and it was already previously[2] given in the present version, and in more refined ones, but only for the operator (1) being the ordinary Laplacean and the operator (4) being elliptic in its variables, more or less.

The Green's function of (1) gives rise to a Green's integral

$$g(x_j) = \int_B \sum_{\alpha=1}^{n} (X_\alpha \cdot \sigma^\alpha), \tag{5}$$

$$\sigma^\alpha = (-1)^{\alpha-1} d\xi_1 \ldots d\xi_{\alpha-1} d\xi_{\alpha+1} \ldots d\xi_n,$$

$$X_\alpha = \sum_{\mu_1+\ldots+\mu_n \leq 2h-1} \frac{\partial^{\mu_1+\ldots+\mu_n} f(\xi)}{\partial \xi_1^{\mu_1} \ldots d\xi_n^{\mu_n}} G^\alpha_{\mu_1 \ldots \mu_n}(x - \xi), \tag{6}$$

where (i) each $G^\alpha_{\mu_1 \ldots \mu_n}(\xi)$ is a linear combination with constant coefficients of derivatives of $G(\xi)$, (ii) for fixed x and any solution $f(\xi)$ of (3) we have

$$\frac{\partial X_1}{\partial \xi_1} + \ldots + \frac{\partial X_n}{\partial \xi_n} = 0 \tag{7}$$

and (iii) if B_0 bounds a domain in which $f(\xi)$ is a solution of (3) then for x in the domain we have $g(x_j) = f(x_j)$, that is

$$f(x_j) = \int_{B_0} \sum_\alpha X_\alpha \cdot \sigma^\alpha. \tag{8}$$

Now, for the proof of the theorem we may assume that B is a piecewise differentiable simplicial closed surface, and if we take a layer around it which is bounded by two "parallel" surfaces, and apply (8) to the layer then we obtain in U a Laurent decomposition

$$f(x_j) = f_1(x_j) - f_2(x_j), \tag{9}$$

where f_1 is analytic in $D + U$ and f_2 in $(E^n - D) + U$, and where for x in D, or in $E^n - D$, respectively, both are represented by the "same" integral (5). Our next step is to show that both are again solutions of the "accessory" equation (4),

$$\Lambda f_1 = 0, \qquad \Lambda f_2 = 0, \tag{10}$$

and as previously[2] we do this in the following manner. We view the integral (5) as an operator, obviously distributive, from solutions of the elliptic equation (3) in the neighborhood of B to functions $g(x)$ in $E^n - B$, denoting it then by $g(x) = L(f(\xi); x)$, and due to the fact that in (6) the variable x occurs only in the combinations $x - \xi$ this operator is commutative with partial differentiations as well, so that for any Λ with constant coefficients we have

$$\Lambda_x g(x) = L(\Lambda_\xi f(\xi); x). \tag{11}$$

If in particular $\Lambda f(\xi) = 0$ then also $\Lambda g(x) = 0$ and hence (10). It will be worth while to state the reason behind (11) in a comprehensive lemma, and we again introduce the differential form $\sigma^{\alpha\beta}$ which for $\alpha < \beta$ is defined as

$$(-1)^{\alpha+\beta-1} d\xi_1 \ldots d\xi_{\alpha-1} d\xi_{\alpha+1} \ldots d\xi_{\beta-1} d\xi_{\beta+1} \ldots d\xi_n,$$

for $\alpha = \beta$ is 0, and for $\alpha > \beta$ is $-\sigma^{\beta\alpha}$.

LEMMA 1. *If B is a (non-closed) surface with boundary C, and if in a neighborhood of $B + C$ we are given a solution $f(\xi)$ of an elliptic equation (3), and if we introduce the functional (11); then for any differential operator (4) with constant coefficients we have*

$$\Lambda_x g(x) - L(\Lambda_\xi f(\xi);\, x) = \int_C \sum_{\alpha,\,\beta=1}^{n} Y_{\alpha\beta} \cdot \sigma^{\alpha\beta},$$

where each $Y_{\alpha\beta}$ is a bilinear form in partial derivatives of $G(x - \xi)$, and in partial derivatives of $f(\xi)$ of order $\leq (2h - 1) + (M - 1)$.

For the conclusion of Theorem 1 it suffices to show that the function f_2 in (9) is analytic everywhere. We have already obtained $\Lambda f_2 = 0$, and from inspection of the integral (5) it follows that at infinity f_2 has the same behavior (2) as $G(x)$ itself, with another r_0 of course. Thus our conclusion follows from the following lemma which can be proved by Fourier analysis.

LEMMA 2. *If an analytic function $f(x_j)$ in an exterior of a sphere $|x| > R$ satisfies an equation (4) in fewer than all variables, and at infinity has the behavior (2), then $f(x_j)$ is a polynomial of degree $\leq M - 1$ in all its variables and is thus analytic everywhere.*

For the proof of the Laurent decomposition (9) instead of applying (8) to a layer around all of B we might have used as previously[2] the following "local" argument.

LEMMA 3. *If we consider the integral (5) for any union of simplices B and any $f(\xi)$ on it, and if in a neighborhood of point of a simplex, $f(\xi)$ is a solution of (3), then in the neighborhood of this point the function $g(x)$ can be continued analytically across B from both sides, and the difference of $g(x)$ on both sides is $f(x)$ itself,*

$$g_+(x) - g_-(x) = f(x).$$

With this lemma the following sharpening of Theorem 1 ensues.

THEOREM 2. *If the piecewise differentiable surface B is a union of the closures of differentiable simplices $B_1, \ldots, B_r$; if in a neighborhood U_ρ of each (non-closed) B_ρ there is a given analytic solution f_ρ of (3) and (4); if each f_ρ together with its partial derivatives of order $\leq (2h - 1) + (M - 1)$ has continuous boundary values on the boundary of B_ρ; and if on each $(n - 2)$-dimensional edge common to two simplices all these boundary values are the same for the two functions; then in the domain D bounded by B there exists an analytic function $f(x)$ which in each U_ρ coincides with the function $f_\rho(x)$ given.*

Another conclusion is as follows.

THEOREM 3. *If a surface B with a boundary C does not decompose E^n, and if in a neighborhood of B there is given a solution of (3) and (4) for which* *the partial derivatives (or only their boundary values) of order* $\leq (2h - 1) + (M - 1)$ *are 0 on C; then f(x) itself is* $= 0$.

A full account with results in other directions will be given elsewhere.

[1] "Zeta Functions and Green's Functions for Linear Partial Differential Operators of Elliptic Type with Constant Coefficients," *Ann. Math.* (to appear).

[2] "Analytic and Meromorphic Continuation by Means of Green's Formula," *Ibid.*, **44**, 652–673 (1943).

Annals of Mathematics
Vol. 57, No. 1, January, 1953
Printed in U.S.A.

ZETA FUNCTIONS AND GREEN'S FUNCTIONS FOR LINEAR PARTIAL DIFFERENTIAL OPERATORS OF ELLIPTIC TYPE WITH CONSTANT COEFFICIENTS

S. Bochner

In Memory of Harald Bohr

(Received November 28, 1951)

We will investigate multi-dimensional series of the form

$$\sum\nolimits'_{(n)} \frac{Q_g(n_j)}{T(n_j)^s} c^{2\pi i(n,x)} \tag{1}$$

$$(n, x) = n_1x_1 + \cdots + n_kx_k , \tag{2}$$

in which $T(n_j)$ is an arbitrary polynomial of elliptic (will be defined) type of any even degree, and Q_g is any further polynomial and we will demonstrate that these series really have sums, for any real variables x_j and all complex s; they they are C^∞ and even analytic in the x_j, and entire functions in s; except that for $(x_j) \neq (0)$—and of course for congruent points—there are some simple poles in the s-plane of the kind familiar from the theory of Zeta functions, Epstein Zeta functions, and such like expansions. There are no "functional equations" present however, at least not in an analytically fruitful manner, unless one chooses to interpret Theorem 11 as embodying a functional equation incipiently; and some remarks tending to substantiate such an interpretation for Theorem 11 have been presented in a separate note, see [1].

Actually, in our theorem proper, $T(n_j)$ and also $Q(n_j)$ will not be polynomials at all, but rather more general functions of non-oscillatory behavior reminiscent of such, and it will be syllogistically advisable to have these generalizations undertaken in the body of the statements, but we will not attempt to assess the gains from such generalizations in terms of concretely novel consequences thus resulting.

If with a linear operator with constant coefficients

$$\Lambda f = (-1)^h \sum_{0 \leqq r_1 + \cdots + r_k \leqq 2h} a_{r_1 \cdots r_k} \frac{\partial^{r_1 + \cdots + r_k} f}{\partial x_1^{r_1} \cdots \partial x_k^{r_k}} \tag{3}$$

we associate the polynomial

$$T(t_j) = (-1)^h \sum\nolimits_{(r)} a_{r_1 \cdots r_k} (2\pi i t_1)^{r_1} \cdots (2\pi i t_k)^{r_k}, \tag{4}$$

then the series (1), for $Q_g = 1$, is in the nature of a multi-periodic Green's functions pertaining to the sth power $\Lambda^s f$ of the operator (3) itself, the exponentiation to be made in the sense of operator theory abstractly. These general powers, especially for s complex, are far from being differential operators themselves, but the outcome of our analysis will be that their Green's functions behave in very much the same way as for $s = 1$, nevertheless.

I. Zeta Functions

We will operate in the Euclidean E_k, $k \geqq 2$, and we will always use the abbreviations

$$(5)\quad t = (t_1^2 + \cdots + t_k^2)^{\frac{1}{2}}, \qquad x = (x_1^2 + \cdots + x_k^2)^{\frac{1}{2}}, \qquad n = (n_1^2 + \cdots + n_k^2)^{\frac{1}{2}},$$

etc. A function will be denoted by $\varphi(t_1, \cdots, t_k)$ or more briefly by $\varphi(t_j)$, and if, denoted by $\varphi(t)$, say, then this will often imply that it is a function of the (positive) radius t only.

DEFINITION 1. We will say that a function $\varphi(t_j)$ is *strictly of type* δ,

$$\delta \lessgtr 0,$$

if $\varphi(t_j)$ is defined everywhere in E_k and is infinitely differentiable there; if it is 0 in a certain neighborhood of the origin $0 < t < t^0$; and, what is decisive, if for the function itself and for everyone of its partial derivatives

$$(6)\qquad D^p\varphi = \frac{\partial^{p_1+\cdots+p_k}}{\partial t_1^{p} \cdots \partial t_k^{p_k}}, \qquad p_1 + \cdots + p_k = p$$

we have

$$(7)\qquad D^p\varphi = 0(t^{\delta-p}), \qquad t \to \infty.$$

We will say that $\varphi(t_j)$ is *practically* of type δ, if it is defined and infinitely differentiable in a "big" outer shell $t^0 < t < \infty$, with $t^0 < 1$, and there satisfies (7) as before.

If $\varphi(t_j)$ is [practically] of type δ then

$$(8)\qquad \varphi(t_j)\cdot t_1^{\delta_1} \cdots t_k^{\delta_k}$$

is [practically] of type $\delta + \mathrm{Re}(\delta_1 + \cdots + \delta_k)$. Furthermore, if $\varphi(t_j)$ is practically of type δ, and if we take a function $\chi(t)$ which is infinitely differentiable in $0 < t < \infty$, has value 0 in $0 < t < t^1$ for $t^0 < t^1 < 1$, and has value 1 in $1 \leqq t < \infty$; then the product

$$(9)\qquad \tilde{\varphi}(t_j) = \chi(t)\varphi(t_j)$$

is a function strictly of type δ, for which

$$(10)\qquad \tilde{\varphi}(t_j) = \varphi(t_j), \qquad 1 \leqq t < \infty.$$

In particular, we have

$$(11)\qquad \tilde{\varphi}(n_1, \cdots, n_k) = \varphi(n_1, \cdots, n_k)$$

for integers

$$(12)\qquad (n_1, \cdots, n_k) \neq (0, \cdots, 0),$$

and this "smoothing" of a function which is practically of type δ to one which is strictly so, will be a leading step in our analysis.

If $\varphi(t_j)$ is strictly of type δ than for $\varepsilon > 0$ we may set up the Fourier transform

$$\begin{aligned} F^{\varepsilon}(x_j) &= \int_{E_k} e^{i(x,t)-\varepsilon t^2}\varphi(t_j)\,dv_j, \\ (x,t) &= x_1t_1 + \cdots + x_kt_k, \end{aligned} \tag{13}$$

and we may carry out the partial differentiations leading to

$$x^{2g}F^{\varepsilon}(x_j) = \int_{E_k} e^{i(x,t)}\Delta_t^g\{e^{-\varepsilon t^2}\varphi(t_j)\}\,dv_t \tag{14}$$

where $g \geqq 0$ is any integer, and

$$\Delta_t \equiv -\sum_{j=1}^{k}\frac{\partial^2}{\partial t_j^2}. \tag{15}$$

If we let $\varepsilon \to 0$ in the right side of (14) we formally obtain

$$\int_{E_k} e^{i(x,t)}\Delta_t^g\varphi(t_j)\,dv_t, \tag{16}$$

and by (7) this is absolutely convergent whenever

$$2g > \delta + k, \tag{17}$$

and we claim that in this case the formal limit is an actual one. The best way of verifying this is to proceed more generally in the following set-up. For the moment we denote the summability factor in (13) not by $e^{-\varepsilon t^2}$ but by $e^{-\varepsilon^2 t^2}$, and we then replace it by $K(\varepsilon t)$, where $K(t)$ has the following properties. It is continuous in $0 \leqq t < \infty$ with $K(0) = 1$; it is infinitely differentiable in $0 < t < \infty$; for $t \to \infty$ we have

$$\frac{d^pK(t)}{dt^p} = 0(t^{-q}) \tag{18}$$

in all $p \geqq 0$, $q \geqq 0$; and in the neighborhood of $t = 0$ we have, for $p > 0$,

$$\frac{d^pK(t)}{dt^p} = o(t^{-p}). \tag{19}$$

Since $\varphi(t_j)$ vanishes in the neighborhood of the origin we can perform what corresponds to the transition from (13) to (14); and after carrying out the partial differentiation on the right side of (14), since we have

$$\int_{E_k} e^{i(x,t)}K(\varepsilon t)\Delta^g\varphi(t_j)\,dv_t \to \int_{\delta_k} e^{i(x,t)}\Delta^g\varphi(t_j)\,dv_t$$

for (17), what is left to do is to verify that for

$$p + q = 2g, \qquad p > 0$$

we have

$$\varepsilon^p \int_{E_k} | K^{(p)}(\varepsilon t) | \cdot | D^q \varphi(t_j) | \, dv_t \to 0. \tag{20}$$

Now the left side is majorized by

$$\varepsilon^p \int_{t_0}^{\infty} | K^{(p)}(\varepsilon t) | \, t^{\delta-q+k-1} \, dt = \varepsilon^{2g-\delta-k} \int_{\varepsilon t^0}^{\infty} | K^{(p)}(t) | \, t^{\delta-q+k-1} \, dt$$
$$= \int_{\varepsilon t^0}^{1} + \int_{1}^{\infty}$$

and the estimate (20) will follow if in the first part we substitute (19) and in the second (18).

Thus $F^{\varepsilon}(x_j)$ has a limit $F(x_j)$ for $x \neq 0$, continuously in the neighborhood of every point. Also, due to (14), for large g, it is small at infinity, and we can set up the Poisson summation formula

$$\sum\nolimits_{(n)} e^{-\varepsilon n^2} \varphi(n_j) e^{2\pi i(n,x)} = \sum\nolimits_{(n)} F^{\varepsilon}(x_j + 2\pi i \, n_j), \tag{21}$$

and the right side has a limit as $\varepsilon \to 0$ and hence also the left side. Finally, if we differentiate (13) with respect to an x_j, then the new integrand is strictly of type $\delta + 1$ and all the statements just made again apply, and altogether the following theorem may be recorded.

THEOREM 1. (i) *If $\varphi(t_j)$ is strictly of type δ for any δ then the limit*

$$F(x) = \lim_{\varepsilon \to 0} \int_{E_k} e^{i(x,t) - \varepsilon t^2} \varphi(t_j) \, dv_t \tag{22}$$

exists for $x \neq 0$ uniformly in the neighborhood of any point. The function $F(x)$ is C^{∞}, for $x \neq 0$, and its partial derivatives of any mixed order can be formed under the integral sign.

(ii) *If $\varphi(t_j)$ is practically of type δ, then the limit*

$$G(x) = \lim_{\varepsilon \to 0} \sum\nolimits_{(n)}' e^{2\pi i(x,n) - \varepsilon n^2} \varphi(n_j) \tag{23}$$

exists for $(x_j) \neq (0)$ on the torus

$$V: -\tfrac{1}{2} \leqq x_j < \tfrac{1}{2}, \qquad j = 1, \cdots, k \tag{24}$$

and is C^{∞} and its derivatives can be formed under the summation sign.

(iii) *Also, both in (22) and (23), the function e^{-t^2} may be replaced by a kernel $K(t)$ as previously described, and in particular by e^{-t}.*

We consider a polynomial

$$T(t_j) = T_{2h}(t_j) + (\text{lower powers}) \tag{25}$$

of an even degree $2h$ with complex coefficients, and we call this polynomial, or the associated operator (3), elliptic if

$$T_{2h}(t_j) > 0, \qquad 0 < t < \infty \tag{26}$$

and also (in the non-homogeneous case)

$$T(t_j) \neq 0, \tag{27}$$

the latter again for $0 < t < \infty$, or only in a shell $t^0 < t < \infty$, with $t^0 < 1$.

For $k \geqq 3$ such an outer shell is simply connected and it then is possible to continue univalently $T(t_j)^s$, for any complex s, from its obvious values at infinity into the entire shell, by monodromy. For $k = 2$ however this must be stipulated explicitly (all this in the inhomogeneous case only).

If now we introduce some further polynomial $Q_g(t_j)$, homogeneous say, then part (ii) of Theorem 1 readily implies the following conclusion.

THEOREM 2. *If $T(t_j)$ is elliptic, then for any complex $s = \sigma + i\sigma'$ the limit*

$$\zeta(s; x_j) = \lim_{\varepsilon\to 0} \sum{}' e^{2\pi i(x,n)-\varepsilon n^2} \frac{Q_g(n_j)}{T(n_j)^s} \tag{28}$$

exists for $x \neq 0$ on (24), uniformly in the neighborhood of every point. It is C^∞, and derivatives can be formed under the summation sign.

We are next turning to the dependence on s, and we will describe it in a very general set-up.

DEFINITION 2. We will call a function $\varphi(t_j\,; s)$ a Zeta kernel if: (i) it is defined for t_j in a "big" shell $t^0 < t < \infty$, $t^0 < 1$, and for s in a domain D of the complex s-plane; (ii) it is C^∞ in the t_j and holomorphic in s; (iii) all derivatives

$$D_t^p \varphi(t_j\,; s) = \frac{\partial^{p_1+\cdots+p_k}\varphi}{\partial t_1^{p_1} \cdots \partial t_k^{p_k}} \tag{29}$$

are continuous in $(t_j\,; s)$ and holomorphic in s; and, what is decisive, (iv) corresponding to each point s^0 of D there exists a neighborhood N of s^0 and a real number $\delta = \delta(N)$ such that we have

$$D_t^p\, \varphi(t_j\,; s) = 0(t^{\delta-p}), \qquad t \to \infty, \tag{30}$$

uniformly for s in N, for every $p \geqq 0$.

THEOREM 3. (i) *If $\varphi(t_j\,; s)$ is a Zeta kernel, then the junction*

$$\zeta(s; x_j) = \lim_{\varepsilon\to 0} \sum_{(n)} e^{2\pi i(n,x)-\varepsilon n^2}\, \varphi(n_j\,; s), \tag{31}$$

which by Theorem 2 exists for $x \neq 0$ and s in D, is holomorphic in D for every $x \neq 0$.

(ii) *Also, for $x = 0$, the limit*

$$\lim_{\varepsilon\to 0}\left[\sum_{(n)}{}' e^{-\varepsilon n^2}\varphi(n_j\,; s) - \int_{t\geqq 1} e^{-\varepsilon t^2}\varphi(t_j\,; s)\, dv_t\right] \tag{32}$$

exists, for s in D, and is again holomorphic there.

PROOF. Our kernel $\varphi(t_j\,; s)$ is practically of type δ locally majorizèdly in s. We now take a fixed factor $\chi(t)$ as before, and introduce the function

$$\tilde{\varphi}(t_j\,; s) = \chi(t)\, \varphi(t_j\,; s), \tag{33}$$

and if we set down the relations

$$F^{\varepsilon}(x_j;s) = \int_{E_k} e^{i(x,t)-\varepsilon t^2}\tilde{\varphi}(t_j;s)\,dv_t \tag{34}$$

$$x^{2g}F^{\varepsilon}(x_j;s) = \int_{E_k} e^{i(x,t)}\Delta_t^g\{e^{-\varepsilon t^2}\tilde{\varphi}(t_j;s)\}\,dv_t \tag{35}$$

$$\sum_{(n)} e^{-\varepsilon n^2}\tilde{\varphi}(n_j;s)e^{2\pi i(n,x)} = \sum_{(n)} F^{\varepsilon}(x_j+2\pi i n_j;s), \tag{36}$$

then all statements previously made with regard to relations (13), (14), (21) will remain valid, majorizèdly in the holomorphic variable s. This immediately proves part (i) of Theorem 3, and it also establishes part (ii) but with

$$\lim_{\varepsilon\to 0}\left[\sum_{(n)}{}' e^{-\varepsilon n^2}\tilde{\varphi}^{(n_j;s)} - \int_{E_k} e^{-\varepsilon t^2}\tilde{\varphi}(t_j;s)\,dv_t\right] \tag{37}$$

instead of (32). However, due to (10), (11), for fixed ε, the difference between the expressions (37) and (32) is

$$\int_{t\leqq 1} e^{-\varepsilon t^2}\tilde{\varphi}(t_j;s)\,dv_t \equiv \int_{t^0\leqq t\leqq 1} e^{-\varepsilon t^2}\chi(t)\varphi(t_j;s)\,dv_t,$$

and for $\varepsilon\to 0$ this also has a limit, and the limiting function is holomorphic in s; which completes the proof of part (ii) in the wording given.

THEOREM 4. (i) *The function*

$$\zeta(s;x_j) = \lim_{\varepsilon\to 0}\sum{}' e^{2\pi i(x,n)-\varepsilon n^2}\frac{Q_g(n_j)}{T(n_j)^s}$$

of Theorem 2 is, for $x\neq 0$, an entire function in s.

(ii) *For $x=0$, the Dirichlet series*

$$\sum_{(n)}{}' \frac{Q_g(n_j)}{T(n_j)^s} = \zeta(s) \tag{38}$$

and the associated integral

$$\int_{t\geqq 1}\frac{Q^g(t_j)}{T(t_j)^s}\,dv_t = H(s) \tag{39}$$

are both absolutely convergent in the right half plane

$$\sigma > \frac{k+g}{2h} \tag{40}$$

and the holomorphic functions thus defined are such that their difference has an analytic continuation into the entire plane, that is

$$\zeta(s) - H(s) = \text{entire function}. \tag{41}$$

(iii) *Taken separately, the functions* $\zeta(s)$, $H(s)$ *have meromorphic continuations into the entire s-plane. If* $T(n_j)$ *is homogeneous, then* $H(s)$ *and* $\zeta(s)$ *have (at most) one simple pole at the obvious point*

$$s = \frac{k+g}{2h} \tag{42}$$

the residue being

$$\frac{1}{2h}\int_{S_{k-1}} \frac{Q_g(\alpha_1, \cdots, \alpha_k)\, d\omega_\alpha}{T(\alpha_1, \cdots, \alpha_k)^{(k+g)/2h}} \tag{43}$$

where S_{k-1} *is the unit sphere*

$$S_{k-1} : \alpha_1^2 + \cdots + \alpha_k^2 = 1, \tag{44}$$

and $d\omega_\alpha$ *is the natural* $(k-1)$*-dimensional volume element on it.*

If $T(n_j)$ *is non-homogeneous, there is first of all again a simple pole at the point* (42) *with the residue*

$$\frac{1}{2h}\int_{S_{k-1}} \frac{Q_g(\alpha_j)\, d\omega_\alpha}{T_{2h}(\alpha_j)^{(k+g)/2h}} \tag{45}$$

where $T_{2h}(t_j)$ *is the highest part of* $T(t_j)$; *and there may also appear other poles, all simple ones, at the points*

$$s = \frac{k+g-m}{2h}, \qquad m = 1, 2, 3, 4, \cdots \tag{46}$$

which are placed at equal distances to the left of the first one.

PROOF. Part (i) follows from part (i) of Theorem 3 on putting $\varphi(t_j\,;\,s) = Q_g \cdot T^{-s}$, and part (ii) follows again from part (ii) of Theorem 3 on realizing that in the half-plane (40) the difference (41) is identical with

$$\lim_{\varepsilon\to 0}\left[\sum{}' e^{-\varepsilon n^2}\frac{Q_g(n_j)}{T(n_j)^s} - \int_{t\geqq 1} e^{-\varepsilon t^2}\frac{Q_g(t_j)}{T(t_j)^s}\, dv_t\right]$$

which latter expression has the continuation claimed. As regards part (iii), in the homogeneous case we have, in (40),

$$H(s) = \int_{t=1}^{\infty} \frac{t^{g+k-1}}{t^{2hs}}\, dt \cdot \int_{S_{k-1}} \frac{Q_g(\alpha_j)}{T(\alpha_j)^s}\, d\omega_\alpha\,;$$

but in this product the first factor is $(2hs - g - k)^{-1}$, and the second factor is an entire function, and hence the conclusion.

If T is not homogeneous, we put

$$T = T_{2h} + H_{2h-1} = T_{2h}\left(1 + \frac{P_{2h-1}}{T_{2h}}\right)$$

and if we substitute this in the integral (39) we obtain, at first formally,

$$H(s) = \int_{t \geqq 1} \frac{Q_g(t_j)}{T_{2h}^s} \sum_{n=0}^{\infty} \binom{-s}{n} \frac{P_{2h-1}^n}{T_{2h}^n} dv_t$$

$$= \sum_{n=0}^{\infty} \binom{-s}{n} \int_{t \geqq 1} \frac{Q_g P_{2h-1}^n}{T_{2h}^{s+n}} dv_t .$$

Now, the nth term of the sum is of the kind just discussed with s replaced by $s + n$, and Q_g replaced by $Q_g P_{2h-1}^n$, and since the last expression is an inhomogeneous polynomial of degree $\leqq (2h - 1)n$, we see that for $n > 0$, the term has at most simple poles at the points

$$s + n = \frac{g + \nu + k}{2h}, \qquad 0 \leqq \nu \leqq (2h - 1)n,$$

and is holomorphic otherwise. Finally, corresponding to any bounded domain of the s-plane we can find an N such that on putting

$$\left(1 + \frac{P_{2h-1}}{T_{2h}}\right)^{-j} = \sum_{n=0}^{N} \binom{-s}{n} \left(\frac{P_{2h-1}}{T_{2h}}\right)^n + R_N(t_j\,;\, s)$$

the remainder

$$\int_{t \geqq \frac{1}{2}} \frac{Q_g(t_j)}{T_{2h}^s} R_N(t_j\,;\, s)\, dv_t$$

will be absolutely uniformly convergent for s in D, and thus will be holomorphic, compare [11], p. 270. And from all this our assertion follows.

II. Green's Functions

If on our torus (24) we are given two continuous functions

(47) $$H(x_j) \sim \sum \gamma_{n_1 \cdots n_k} e^{2\pi i(n,x)}; \qquad \varphi(x_j) \sim \sum \alpha_{n_1 \cdots n_k} e^{2\pi i(n,x)},$$

then

(48) $$\int_V H(x_j)\varphi(y_j - x_j)\, dv_x = \sum_{(n)} \gamma_n \alpha_n e^{2\pi i(n,y)}.$$

Now, if for an elliptic operator (3) we set up the functions

(49) $$G^{\varepsilon}(x_j) = \sum_{(n)}{}' e^{-\varepsilon n^2} \frac{e^{2\pi i(n,x)}}{T(n_j)}$$

then we obviously have

(50) $$\Lambda G^{\varepsilon}(x_j) = \sum_{(n)}' e^{-\varepsilon n^2} e^{2\pi i(n,x)},$$

and (48) implies, for any continuous $\varphi(x_j)$,

$$\begin{aligned} \int_V \Lambda G^\varepsilon(x_j)\cdot\varphi(y_j - x_j)\,dv_x &= \textstyle\sum_n' e^{-\varepsilon n^2}\gamma_{n_1\cdots n_k} e^{2\pi i(n,y)} \\ &= \textstyle\sum_n - \gamma_{0\cdots 0} \\ &= \varphi^\varepsilon(y_j) - \int_V \varphi(x_j)\,dv_x, \end{aligned} \tag{51}$$

where

$$\lim\nolimits_{\varepsilon\to 0}\varphi^\varepsilon(y_j) = \varphi(y_j) \tag{52}$$

uniformly on V. Hence the following theorem.

THEOREM 5. *For any continuous $\varphi(x_j)$ we have*

$$\varphi(y_j) - \int_V \varphi\,dv = \lim\nolimits_{\varepsilon\to 0}\int_V \Lambda G^\varepsilon(x_j)\varphi(y_j - x_j)\,dx_x, \tag{53}$$

the limit existing.

But now assume that $\varphi(x_j)$ vanishes in a neighborhood of the origin and put $y_j = 0$. On the one hand we then have $\varphi(y_j) = \varphi(0) = 0$, and on the other hand, by Theorem 1 or 3, we may then take the limit $\varepsilon\to 0$ under the integral sign, and altogether we obtain

$$\int_V (\Lambda G(x_j) + 1)\varphi(-x_j)\,dv_x = 0, \tag{54}$$

where

$$G(x_j) = \lim\nolimits_{\varepsilon\to 0}\sum_{(n)}' e^{-\varepsilon n^2}\frac{e^{2\pi i(n,x)}}{T(n_j)}. \tag{55}$$

From (54) we easily conclude as follows.

THEOREM 6. *For the function* (55) *we have*

$$\Lambda G(x_j) = -1 \tag{56}$$

for $(x_j) \neq (0)$ on the torus.

We are going to obtain the analogue of Theorems 5, 6 for non-periodic functions in all of E_k. In this case one must introduce an integral instead of a series and the point $t = 0$ is a singularity which interferes with the analysis. Because of that we are unable to reach non-homogeneous operators in unrestricted generality but only the following sub-class of them which however includes the homogeneous ones in their totality, at any rate.

We will call a polynomial (or operator) semi-homogeneous if the lowest powers occurring are of an even degree and nowhere zero, that is if we have

$$T(t_j) = T_{2h} + T_{2h-1} + \cdots + T_{2h_0} \tag{57}$$

where

$$T_{2h_0}(\alpha_j) \neq 0, \text{ for } \alpha_1^2 + \cdots \alpha_k^2 = 1.$$

Note that $h_0 = 0$ is not excluded so that a polynomial with non-vanishing constant term is admissible.

We introduce a function $L(x, t)$ which for $h_0 = 0$ is itself zero and otherwise has the value

$$L(x, t) = \sum_{l=0}^{2h_0-1} \frac{(2\pi i(x, t))^l}{l!}$$

(see Bochner [2], and [3], Chapter VI, and Herglotz [8]), and we then introduce the part functions

$$(58) \qquad \begin{aligned} F_0(x_j) &= \int_{t\leqq 1} \frac{e^{2\pi i(x,t)} - L(x, t)}{T(t_j)} \, dv_t \\ F_1^\varepsilon(x_j) &= \int_{t\geqq 1} e^{-\varepsilon t^2} \frac{e^{2\pi i(x,t)}}{T(t_j)} \, dv_t, \end{aligned}$$

and we note as follows. $F_0(x_j)$ is C^∞ (and actually analytic) and it can be differentiated freely under the integral sign so that in particular

$$\Lambda F_0(x_j) = \int_{t\leqq 1} e^{2\pi i(x,t)} \, dv_t.$$

The second part differs from a function

$$\int_{E_k} \frac{\chi(t)}{T(t_j)} e^{-\varepsilon t^2 + 2\pi i(x,t)} \, dv_t = \int_{t^0 \leqq t < \infty}$$

in which $\chi(t)$ is a smoothing factor as previously introduced, by the addend

$$\int_{t^0 \leqq t \leqq t^1} \frac{\chi(t)}{T(t_j)} e^{-\varepsilon t^2 + 2\pi i(x,t)} \, dv_t,$$

which is likewise analytic and differentiable; and thus by Theorem 1 we conclude that F_1^ε converges with all its derivatives towards a function $F_1(x_j)$, boundedly in the neighborhood of every point not the origin.

If now we introduce the function

$$F^\varepsilon(x_j) = F_0(x_j) + F_1^\varepsilon(x_j)$$

and its limit

$$(59) \qquad F(x_j) = F_0(x_j) + \lim_{\varepsilon\to 0} F_1^\varepsilon(x_j),$$

then for any continuous function

$$\varphi(x_j) \sim \int_{E_k} e^{2\pi i(x,t)} \gamma(t_j) \, dv_t$$

which vanishes outside a finite interval we have, by convolution,

$$\int_{E_k} \Lambda F^\varepsilon(x_j)\varphi(y_j - x_j) \, dv_t = \int_{t<1} \gamma(t_i) e^{2\pi i(y,t)} \, dv_t + \int_{1<t<\infty} \gamma(t_j) e^{-\varepsilon t^2} e^{2\pi i(y,t)} \, dv_t$$

and hence we conclude as follows.

THEOREM 7. *If an operator is elliptic semi-homogeneous, then for a continuous function $\varphi(x_j)$ which vanishes outside a bounded set we have*

$$\varphi(y_j) = \lim_{\varepsilon\to 0} \int_{E_k} \Lambda F^{\varepsilon}(x_j)\varphi(y_j - x_j)\, dv_{x_j},$$

the limit existing. Also

$$F(x_j) = 0 \text{ for } x \neq 0.$$

III. Spherical coordinates and analyticity

It is known (for the most general result see [13]) that any solution of an elliptic equation, if differentiable sufficiently often is automatically analytic. Therefore our function $\zeta(s; x_j)$ is for $s = 1$ not only C^∞ as stated in Theorem 2 but also analytic; and we are going to prove the non-obvious theorem that this likewise holds for arbitrary values of s, real or complex, at least in the homogeneous case. Our analysis would cover the semi-homogeneous operators as well, but we will not insist on verifying this. In order to appreciate the meaning of the results note that for $Q_0 = 1$, the function $\zeta(s; x_j)$ is the Green's function for the operational s^{th} power $\Lambda^s f$ of the operator given, and that for general s such a power no longer has the structure of some differential expression, or of some similar expression given "explicitly". Our leading theorem will be Theorem 8, and its proof will be given at the end of this Section, after several lemmas and other theorems will have been dealt with first.

THEOREM 8. *If $T(t_j)$ is elliptic homogeneous, then the function*

$$\lim_{\varepsilon\to 0} \sum_{(n)}{}' \frac{e^{2\pi i(n,x)}}{T(n_j)^s} e^{-\varepsilon n^2} \tag{60}$$

is, for every complex s, analytic in (x_j) for $x \neq 0$.

More generally, if $\varphi(\alpha_j)$ is defined and analytic in the unit sphere (44) *and if for some complex $\gamma(= -s)$ we put*

$$\varphi(t_j) = t^\gamma \varphi(\alpha_j) \tag{61}$$

where $(t; \alpha_j)$ are spherical coordinates in $E_k - (0)$,

$$t_j = t\alpha_j, \qquad t = (t_1^2 + \cdots + t_k^2)^{\frac{1}{2}}, \qquad \alpha_1^2 + \cdots + \alpha_k^2 = 1, \tag{62}$$

then the function

$$G_\varphi(x_j) = \lim_{\varepsilon\to 0} \sum_n{}' e^{-\varepsilon n^2} \varphi(n_j) e^{2\pi i(n,x)} \tag{63}$$

is analytic for $(x_j) \neq (0)$.

The method of proof we are going to use was foreshadowed in Hobson [9], and it is also closely allied to "spherical summability" previously studied by ourselves, see [4]. It will be convenient to introduce the symbol

$$l = \frac{k-2}{2} \tag{64}$$

for permanent use; and in addition to the spherical coordinates (62) we are also introducing coordinates

$$(65)\qquad x_j = x\xi_j, \qquad x = (x_1^2 + \cdots + x_k^2)^{\frac{1}{2}}, \qquad \xi_1^2 + \cdots + \xi_k^2 = 1$$

so that the general Fourier transform

$$(66)\qquad F(x_j) = \frac{1}{(2\pi)^{l+1}} \int_{E_k} e^{i(x,t)} \varphi(t_j)\, dv_t$$

will be formally written as

$$(67)\qquad F(x;\xi_j) = \frac{1}{(2\pi)^{l+1}} \int_{t=0}^{\infty} dt \int_{S_{k-1}} e^{ixt(\xi,\alpha)} \varphi(t\alpha_j) t^{k-1}\, d\omega_\alpha .$$

Next, for an "arbitrary" function $\varphi(t_j)$ we introduce the expansion in spherical harmonics,

$$(68)\qquad \varphi(t\alpha_j) = \textstyle\sum_{n=0}^{\infty} \varphi_n(t;\alpha_j)$$

which for $k = 2$ is the expansion of

$$\varphi(t \cos\theta,\, t \sin\theta)$$

into the Fourier series

$$\tfrac{1}{2}A_0(t) + \textstyle\sum_{n=1} (A_n(t) \cos n^\theta + B_n(t) \sin n^\theta).$$

For $k \geqq 3$ we have, compare [5], p. 325,

$$(69)\qquad \varphi_n(t;\xi_j) = \frac{\Gamma(l)(n+l)}{2\pi^{l+1}} \int_{S_{k-1}} P_n^l(\cos\gamma)\varphi(t\alpha_j)\, d\omega_\alpha,$$

$$(70)\qquad \cos\gamma = (\xi, \alpha) = \textstyle\sum_{j=1}^{k} \xi_j\alpha_j,$$

where $P_n^l(\cos\gamma)$ is definable by the generating relation

$$(1 - 2w\cos + w^2)^{-l} = \textstyle\sum_0^\infty P_n^l(\cos\gamma) w^n,$$

say. If now we substitute (69) in (68) and then (68) in (67), and if we utilize the addition theorem

$$(71)\qquad e^{iz\cos\gamma} = 2^l\,\Gamma(l) \sum_{n=0}^{\infty} i^n (n+l) \frac{J_{n+l}(z)}{z^l} P_n^l(\cos\gamma)$$

for $k \geqq 3$, or the "degenerate" relation

$$(72)\qquad e^{iz\cos\gamma} = J_0(z) + \textstyle\sum_1^\infty (i^n e^{-in\gamma} + (-i)^n e^{in\gamma}) J_n(z)$$

for $k = 2$, compare [15], p. 368 and p. 22, then we end up, for $(x_j) \neq (0)$, with the expansion

$$(73)\qquad [F(x\xi_j) \equiv]\ F(x;\xi_j) = \textstyle\sum_{n=0}^{\infty} F_n(x;\xi_j)$$

where

$$(74)\qquad F_n(x;\xi_j) = \frac{i^n}{x^k} \int_0^\infty \varphi_n\left(\frac{t}{x};\xi_j\right) J_{l+n}(t) t^{l+1}\, dt,$$

compare [1], and it is this expansion from which further conclusions will be drawn.

For any $n \geqq 0$, an nth spherical can be defined either in Euclidean E_k as a homogeneous polynomial of degree n which is a solution of the Laplacean

$$(75) \qquad -\Delta\varphi \equiv \frac{\partial^2 \varphi}{\partial t_1^2} + \cdots + \frac{\partial^2 \varphi}{\partial t_k^2} = 0,$$

or it can be defined on the S_{k-1} as a (twice differentiable) solution of the equation

$$(76) \qquad L_\alpha \varphi_n(\alpha_j) = n(n + k - 2)\, \varphi_n(\alpha_j)$$

where $L_\alpha \varphi$ is the "curvilinear" Laplacean on the S_{k-1} as induced by (75). From either description it follows that for any $n \geqq 0$, spherical harmonics form a vector space. Therefore, if $\varphi_n(t;\, \xi_j)$ is an nth harmonic for each t, then so is an integral

$$\int_0^\infty \varphi_n(t;\, \xi_j) K(t)\, dt,$$

if convergent. Therefore, for every $x > 0$, (74) is an nth harmonic again and (73) constitutes an expansion of $F(x_j)$ into such harmonics, the dual one to (68) as it were,—but this fact as such will be of no particular consequence to the present context, although in other contexts it is of considerable interest indeed, see [1]. Also, the expansion in spherical harmonics is valid literally whenever the series is absolutely-uniformly convergent, say, and this will be so under our assumptions. The other relations will be likewise valid automatically, and we will make no further comment on that account.

Turning again to algorithms we note that for a homogeneous function (61) the separation of variables continues itself into the expansion (68), thus

$$(77) \qquad t^\gamma \varphi(\alpha_j) = \textstyle\sum_0^\infty t^\gamma \varphi_n(\alpha_j)$$

and if in this case we add the convergence factor $e^{-\varepsilon t}$, thus,

$$(78) \qquad F^\varepsilon(x_j) = \frac{1}{(2\pi)^{l+1}} \int_{E_k} \varphi(t_j) e^{-\varepsilon t + i(t,x)}\, dv_t\,,$$

then we correspondingly have

$$(79) \qquad F^\varepsilon(x_j) = \textstyle\sum_{n=0}^\infty F^\varepsilon_n(x;\, \xi_j)$$

where

$$(80) \qquad F^\varepsilon_n(x;\, \xi_j) = \frac{\varphi_n(\xi_j) i^n}{x^{\gamma+k}} \int_0^\infty e^{-\varepsilon(t/x)} J_{l+n}(t) t^{l+1}\, dt.$$

The last integral however converges heuristically towards

$$2^{\gamma+l+1} \frac{\Gamma(\frac{1}{2}(\gamma + n + k))}{\Gamma(\frac{1}{2}(n - \gamma))},$$

and this leads to the following conclusion, at first heuristic.

THEOREM 9. *Heuristically, if* $\varphi(\alpha_j)$ *is an arbitrary function on the unit sphere* (44), *and if we extend it in* $E_k - (0)$ *to a homogeneous function* $\varphi(t_j)$ *of an arbitrary weight* γ, *real or complex; then the Fourier integral*

$$F(x;\, \xi_j) = \frac{1}{(2\pi)^{l+1}} \int_{E_k} \varphi(t_j) e^{i(t,x)}\, dv_t \tag{81}$$

is again a homogeneous function, but of weight $-\gamma - k$, *thus*

$$F(x;\, \xi_j) = x^{-\gamma-k} \psi^{\gamma}(\xi_j), \tag{82}$$

and if on (44) *we introduce the expansion in spherical harmonics*

$$\varphi(\xi_j) = \textstyle\sum_0^{\infty} \varphi_n(\xi_j) \tag{83}$$

then the corresponding expansion of the new function is

$$\psi^{\gamma}(\xi_j) = 2^{\gamma+l+1} \sum_{n=0}^{\infty} \frac{\Gamma(\frac{1}{2}(\gamma + n + k))}{\Gamma(\frac{1}{2}(n - \gamma))}\, i^n \varphi_n(\xi_j). \tag{84}$$

If for a moment we put $-\gamma = s$, $\gamma + k = s'$, then we obtain $s' = k - s$, and this relation is the one which dominates "functional equations" so-called, compare again [1].

As for making the heuristic statements rigorous, we first of all note that for $\operatorname{Re}(\nu + \mu) > 0$ we have

$$\begin{aligned}&\int_0^{\infty} e^{-\varepsilon t} J_{\nu}(t) t^{\mu-1}\, dt \\ &\quad = \frac{\Gamma(\mu + \nu)}{2^{\nu}(1 + \epsilon^2)^{\frac{1}{2}(\nu+\mu)}\Gamma(\nu + 1)} F\left(\frac{\mu + \nu}{2}, \frac{1 - \mu + \nu}{2}, \nu + 1; \frac{1}{1 + \epsilon^2}\right),\end{aligned} \tag{85}$$

see [15], p. 385, and hence the following lemma.

LEMMA 1. *For given* n_0, $n_0 = 0, 1, 2, 3, \cdots$, *if*

$$\operatorname{Re}\gamma > -k - n_0 \tag{86}$$

and $x > 0$, *then the limit relation*

$$\lim_{\varepsilon \to 0} \int_0 e^{-\varepsilon(t/x)} J_{l+n}(t) t^{\gamma+l+1}\, dt = 2^{\gamma+l+1} \frac{\Gamma(\frac{1}{2}(\gamma + n + k))}{\Gamma(\frac{1}{2}(n - \gamma))} \tag{87}$$

holds, uniformly in a neighborhood of x, *and for* $n = n_0, n_0 + 1, n_0 + 2, \cdots$, *each. Also*

$$\int_0^{\infty} e^{-\varepsilon(t/x)} J_{l+n}(t) t^{\gamma+l+1}\, dt = 0(n^{\operatorname{Re}\gamma+k}) \tag{88}$$

as $n \to \infty$, *uniformly for* $0 < \varepsilon < \infty$, $0 < x < \infty$.

PROOF. We put in (85), $\nu = l + n$, $\mu = \gamma + l + 2$ and if we first assume that γ is real then in the hypergeometric function all terms are positive for $n > |\gamma| + 1$, and for the remaining values of n only a finite number of coefficients are non-negative altogether. Hence we can let $\varepsilon \to 0$ in (85) and the right side of (85) converges towards the right side in (87) if we apply, as we may, the known formula

$$F(a, b, c; 1) = \frac{\Gamma(c)\Gamma(c - a - b)}{\Gamma(c - a)\Gamma(c - b)}$$

in conjunction with the duplication formula for gamma functions. This establishes (87) in the neighborhood of any point x, and (88) follows easily from the positiveness, previously stated, of the coefficients in the expansion. If now γ acquires an imaginary part, then this modifies each term in each hypergeometric function by a factor having a fixed numerical bound and this then validates (87) by Abel's theorem and (88) by inspection.

LEMMA 2. *If $\varphi(\xi_j) \in C^{2g}$ on the unit sphere, for some integer $g = 0, 1, 2, \cdots$, then for its harmonics we have*

$$\varphi_n(\xi_j) = 0(n^{-2g+2l}), \qquad n \to \infty, \tag{89}$$

uniformly in ξ_j; and the estimates depend only on the bounds for the partial derivatives of order $\leqq 2g$, and no other properties of $\varphi(\xi_j)$.

Also, for any partial differential operator D^q of total order q on the sphere we have in a similar fashion

$$D^q\varphi_n(\xi_j) = 0(n^{-2q+2l+q}), \qquad n \to \infty. \tag{90}$$

PROOF. The polynomial $P_n^l(\cos\gamma)$ is a solution of (76) in both the variables ξ_j and α_j separately,

$$LP_n^l(\cos\gamma) = n(n + k - 2)P_n^l(\cos\gamma). \tag{91}$$

On the other hand for any two functions $\chi_1(\alpha_j)$, $\chi_2(\alpha_j)$ of class C^{2g} on S_{k-1} we have, for any integer g,

$$\int_{S_{k-1}} \chi_1(\alpha)\cdot L^g\chi_2(\alpha)\, d\omega_\alpha = \int_{S_{k-1}} L^g\chi_1(\alpha)\cdot\chi_2(\alpha)\, d\omega_\alpha,$$

and if we put here $\chi_1 = P_n^l$, $\chi_2 = \varphi$ then (69) implies, for $\varphi\varepsilon C^{2g}$, and $k \geqq 3$

$$\varphi_n(\xi_j) = \frac{\Gamma(l)(n + l)}{2\pi^{l+1}[n(n + k - 2)]^g} \int_{S_{k-1}} P_n^l(\cos\gamma)\cdot L^g\,\varphi(\alpha_j)\, d\omega_\alpha. \tag{92}$$

However, for real γ, we have

$$P_n^l(\cos\gamma) = 0(n^{2l-1}) \tag{93}$$

see [14], p. 166, and therefore (102) implies (89), as claimed. In the "exceptional" case $k = 2$, that is $l = 0$, we do not have the estimate (93), but the weaker estimate $0(1)$; however the factor $n + l$ which appears in (71) is no longer present in (72) and the final estimate is formally the same.

For the second half of the lemma we note that $P_n^l(\cos\gamma)$ is a trigonometric polynomial of order n, and in a (real) neighborhood of any point on the sphere, a qth derivative

$$D_\xi^q P_n^l(\cos\gamma)$$

is then maximized by the ordinary first q derivatives of the trigonometric polynomial with regard to the angle γ as an independent variable. However, by the theorem of S. Bernstein, such a derivative is maximized by the size of the polynomial itself, with the factor n^q added, and hence the conclusion.

LEMMA 3. *If $\varphi(\xi_j)$ is analytic on the (real) unit sphere*

$$\xi_1^2 + \cdots + \xi_k^2 = 1, \tag{94}$$

then there exists a complex neighborhood

$$|\zeta_1 - \xi_1|^2 + \cdots + |\zeta_k - \xi_k|^2 < \rho^2 \tag{95}$$

of it, and an $\varepsilon > 0$, such that for the harmonics of $\varphi(\xi_j)$ we have the estimate

$$\varphi_n(\zeta_1, \cdots, \zeta_k) = 0(e^{-\varepsilon n}), \qquad n \to \infty \tag{96}$$

uniformly in the entire complex neighborhood. Also for a majorized class of analytic function $\varphi(\xi_j)$ the estimate (96) *holds uniformly in them.*

The converse also holds.

PROOF. The converse is trivial, since if we have (96) in (95) then the sum in (83) is uniformly convergent in the complex neighborhood (95), and $\varphi(\xi_j)$ is analytic.

Now, if $\varphi(\xi_j)$ is given analytic, then from (92) we deduce

$$\varphi_n(\xi_j) = 0(n^{-2g+1}) \cdot \int_{S_{k-1}} |P_n^l(\cos\gamma)| \cdot L^g \varphi(\alpha_j)\, d\omega_\alpha,$$

uniformly in all $g \geqq 1$ and uniformly in all complex angles

$$\cos\gamma = \zeta_1\alpha_1 + \cdots + \zeta_k\alpha_k$$

with $\zeta_1, \cdots, \zeta_k$ being complex numbers of any kind. Now, due to the analyticity of $\varphi(\xi_j)$ on (94), there exists an $A > 0$ such that

$$|L^g\varphi(\alpha_j)| \leqq (2g)!A^{2g}$$

for α_j on (94). Furthermore, since $P_n^l(\cos\gamma)$ is a trigonometric polynomial of dimension n, it follows from (93) that for each $\eta > 0$, no matter how small, there exists a neighborhood (95) such that in this neighborhood we have

$$|P_n^l(\cos\gamma)| \leqq A_1\, e^{\eta n} n^k,$$

with A_1 independent of η. Altogether we conclude that for $n \geqq 1$, $g \geqq 1$ we have

$$|\varphi_n(\xi_j)| \leqq \frac{(2g)!B^{2g}\, e^{\eta n}}{n^{2g}}\, n^k \tag{97}$$

for some $B > 0$ which is fixed, any η which is sufficiently small, and a neighborhood (95) depending on that η only. Now, the right side in (97) is

$$\leqq A_2 \left(\frac{n}{2gB}\right)^{-2g} e^{\eta n} \cdot n^k$$

where A_2 and B are both fixed, and for

$$\frac{n}{e} \leqq 2gB < \frac{n}{e} + 1$$

this is majorized by

$$A_3 n^k e^{-(n/Be)} e^{\eta n} = A_3 n^k e^{-((1/Be)-\eta)n}$$

with A_3, B fixed. If now we put, $\eta = \varepsilon$, $(Be)^{-1} = 3\varepsilon$, then we obtain the estimate (96) in a neighborhood (95) as claimed.

Now, for the quotient of gamma functions occurring in (84) we have

$$\frac{\Gamma(\frac{1}{2}(\gamma + n + k))}{\Gamma(\frac{1}{2}(n - \gamma))} = 0(n^{\gamma+\frac{1}{2}k}), \qquad n \to \infty \tag{98}$$

except for the values

$$\gamma = -k, \quad -k - 1, \quad -k - 2, \quad \cdots, \tag{99}$$

which make the numerator infinite, and from a combination of Lemmas 2, 3 we obtain the following conclusion.

THEOREM 10. *Let γ be a complex number different from* (99). *If a function $\varphi(\alpha_j)$ on* (94) *is C^{2g}, with*

$$2g > \operatorname{Re} \gamma + \tfrac{3}{2}k - 1 \tag{100}$$

then the series

$$\psi^\gamma(\xi) = 2^{\gamma+l+1} \sum_{n=0}^{\infty} \frac{\Gamma(\frac{1}{2}(\gamma + n + k))}{\Gamma(\frac{1}{2}(n - \gamma))} i^n \varphi_n(\xi) \tag{101}$$

is uniformly convergent, and the sum is continuous. Also, if for some $q^0 > 0$ we have

$$2g > \operatorname{Re} \gamma + \tfrac{3}{2}k - 1 + q^0, \tag{102}$$

then $\psi^\gamma(\xi_j)$ belongs to C^{q^0}, and if $\varphi(\xi_j) \in C^\infty$, then $\psi^\gamma(\xi_j) \in C^\infty$.

If $\varphi(\xi_j)$ is analytic, then $\psi^\gamma(\xi_j)$ is analytic.

Also if the expansion (83) *begins with $n = n_0$ then only the values $\gamma = -k - n$, $n \geqq n_0$, need be excluded.*

We are now going to utilize Lemma 1, and if we consider that a mixed partial derivative of a homogeneous function is again homogeneous, then the following theorem ensues.

THEOREM 11. *For $\operatorname{Re} \gamma > -k$, if $\varphi(\alpha_j) \in C^{2g}$, where* (100) *holds, then the integral*

$$F^\varepsilon(x_j) = \frac{1}{(2\pi)^{l+1}} \int_{E_k} \varphi(t_j) e^{-\varepsilon t + i(t,x)} \, dv_t \tag{103}$$

has a limit, as $\varepsilon \to 0$, *boundedly in the neighborhood of every point* $(x_j) \neq 0$, *and the limit is*

$$F(x_j) = x^{-\gamma-k} \psi^{\gamma}(\xi_j). \tag{104}$$

If (102) *holds, then* $F(x_j) \in C^{q^0}$; *if* $\varphi(t_j) \in C^{\infty}$, *then* $F(x_j) \in C^{\infty}$; *and if* $\varphi(t_j)$ *is analytic, then so is* $F(x_j)$.

If the harmonic expansion of $\varphi(\alpha_j)$ *begins with* $n = n_0$, *then all this is also valid for*

$$\operatorname{Re} \gamma > -k - n_0 ,$$

provided the integral (103) *be interpreted as*

$$\lim_{\eta \to 0} \int_{t > \eta},$$

this limit existing.

We now wish to turn to series instead of integrals. But the formula (21) can be made to apply only if the right side converges, and in view of (104) this will only occur if $\operatorname{Re} \gamma > 0$, at first. And if we again use our lemmas for precise estimates, the following conclusion obtains.

THEOREM 12. (Continuation of Theorem 11). (i) *For* $\operatorname{Re} \gamma > 0$, *the limit*

$$G(x) = \lim_{\varepsilon \to 0} \sum\nolimits_{(n)}' e^{-\varepsilon n} \varphi(n_j) e^{2\pi i (n,x)} \tag{105}$$

likewise exists, uniformly in the neighborhood of any point $(x_j) \neq 0$ *on the torus; and it is continuous,* C^{q^0}, C^{∞}, *or analytic, respectively.*

(ii) *Such derivatives of the function* (105) *as have been stated existing may be formed under the summation sign.*

(iii) *In the neighborhood of the origin we have*

$$G(x) = (2\pi)^{l+1} x^{-\gamma-k} \psi^{\gamma}(\xi_j) + \text{(remainder)} \tag{106}$$

where the remainder is continuous, C^{q^0}, C^{∞}, *or analytic there.*

We are now finally ready to prove Theorem 8 as stated at the beginning of this section. We take an analytic function $\varphi(\alpha_j)$ on S_{k-1} and we introduce in E_k the homogeneous function

$$\varphi^{\gamma}(t_j) = t^{\gamma} \varphi(\alpha_j) \tag{107}$$

for variable γ and we correspondingly use the notation

$$H_{\gamma}(x_j) = \lim_{\varepsilon \to 0} \sum\nolimits_{(n)}' e^{-\varepsilon n} \psi^{\gamma}(n_j) e^{2\pi i (n,x)}. \tag{108}$$

Now, for $\operatorname{Re} \gamma > 0$ we have just established that this is analytic in the x_j, and we still have to show that this is likewise so for arbitrary γ. Now, it follows from Theorem 1, that for arbitrary γ our function is at any rate C^{∞} (—see in particular part (iii) of Theorem 1—), and that it can be differentiated formally, and if in particular we apply the ordinary Laplacean then we obtain

$$\Delta_x H_{\gamma-2}(x_j) = 4\pi^2 H_{\gamma}(x).$$

The analyticity in the x_j follows now by "backward" induction from γ onto $\gamma - 2$ from the lemma that if $f(x_j) \in C^2$, and $g(x_j)$ is analytic, and if

$$\Delta_x f(x_j) = g(x_j)$$

holds then $f(x_j)$ is likewise analytic. This is the "easiest" case of the general theorem that a solution of an elliptic differential equation with analytic coefficients is itself analytic; and in the present case it follows from the familiar formula

$$f(x) = \frac{1}{\omega_k} \int_{B_{k-1}} \sum_{j=1}^{k} \left(\frac{\partial K(x - \xi)}{\partial \xi_j} f(\xi) - K(x - \xi) \frac{\partial f}{\partial \xi_j} \right) (-1)^{j-1}$$

$$\cdot d\xi_1 \cdots d\xi_{j-1}\, d\xi_{j+1} \cdots d\xi_k - \int_{B_k} K(x - \xi) g(\xi)\, dv_\xi ,$$

in which $K(x)$ is the Newtonian kernel, by noting that the surface integral is always analytic no matter what $f(x_j)$ is, and that the volume integral is analytic whenever $g(x_j)$ is analytic by a general lemma, a proof of which will be found in [6], p. 42. This then completes the proof of theorem 8.

IV. Asymptotic Behavior at the Origin

We will use spherical harmonics for proving the following supplement to Theorem 1.

THEOREM 13. *If $\varphi(t_j)$ is strictly of type δ as in Theorem* 1, *then for both the integral and the series we have the estimate*

$$F(x),\ G(x) = \begin{cases} 0(x^{-\delta-k}), & \text{if } \delta + k > 0 \\ 0(\log 1/x), & \text{if } \delta + k = 0, \end{cases}$$

in the neighborhood of the origin.

More precisely, the approximating functions $F^\varepsilon(x)$, $G^\varepsilon(x)$ satisfy this estimate uniformly in $\varepsilon > 0$.

We will reduce this theorem to a lemma from which conclusions could be drawn for functions $\varphi(t_j)$ not necessarily C^∞ but only C^{2q}, for q sufficiently large depending on δ; and not necessarily vanishing in the neighborhood of the origin as our concept of a function being *strictly* of type δ presupposes, but only vanishing of a sufficiently high order again depending on δ. But this kind of refinement, together with making the estimates more precise, will be deferred to another occasion.

By inserting the factor $e^{-\varepsilon t^2}$ in the integral (66) we are led to estimating the expression

$$(109) \qquad \frac{1}{x^k} \int_0^\infty e^{-\varepsilon(t^2/x^2)}\, \varphi_n\!\left(\frac{t}{x}; \xi_j\right) J_{l+n}(t) t^{l+1}\, dt$$

as an approximation to (74). And if in (109) we suppress the reference to the variables ξ_j, and make a change of coordinates $u = t^{\frac{1}{2}}$, then on the basis of the expansion (73) and the formula (36) the reader will have no difficulty in reducing the theorem to the following lemma.

LEMMA 4. *Assumptions. Let* $\psi(u)$ *in* $0 < u < \infty$ *belong to class* C^r *for some* $r \geqq 0$, *and for some* $u^0 > 0$ *let*

$$\left|\frac{d^p\,\psi(u)}{du^p}\right| \leqq \begin{cases} Au^{\frac{1}{2}\delta-p} & u_0 \leqq u < \infty \\ Au^{\frac{1}{2}\tau-p}, & 0 < u \leqq u_0 \end{cases} \tag{110}$$

$0 \leqq p \leqq r$, *for some real* δ, *and for some real* τ *for which*

$$\tau + k > 0. \tag{111}$$

Put

$$\psi^\varepsilon(u) = e^{-\varepsilon u}\,\psi(u), \qquad 0 < \varepsilon < \infty \tag{112}$$

and introduce for $0 < x < \infty$ *the expression*

$$M_n^\varepsilon(x) = \frac{1}{x^k}\int_0^\infty \psi^\varepsilon\left(\frac{u}{x^2}\right) u^{\frac{1}{2}l} J_{l+n}(u^{1/2})\,du \qquad n = 0, 1, 2, 3, \cdots. \tag{113}$$

CONCLUSIONS. (i) *If*

$$r > \delta + l + 2, \tag{114}$$

and $x^0 > 0$ *is given, then there exists a constant*

$$B = B(k, \delta, \tau, r, x^0, u^0) \tag{115}$$

such that

$$(n+1)^{-r}\,|\,M_n^\varepsilon(x)\,| \leqq \begin{cases} \dfrac{BA}{x^{\delta+k}} & \text{if } \delta + k > 0 \\ BA\log\dfrac{1}{x} & \text{if } \delta + k = 0 \\ BA & \text{if } \delta + k < 0, \end{cases} \tag{116}$$

for x *in* $0 < x < x^0$.

(ii) *If also a number* $N > 0$ *is given and if we also have*

$$\tau > r - (l+2), \qquad \tau > N - k, \qquad r \geqq N - l, \tag{117}$$

then there exists a further constant

$$B = B(k, \delta, \tau, r, x^0, u^0, N) \tag{118}$$

such that

$$(n+1)^{-r}\,|\,M_n^\varepsilon(x)\,| \leqq \frac{BA}{x^N} \tag{119}$$

for x *in* $x^0 < x < \infty$.

(iii) *Also, in either case the limit*

(120) $$M_n(x) = \lim_{\varepsilon\to 0} M_n^{\varepsilon}(x)$$

exists, and of course satisfies the estimates too.

PROOF OF THE LEMMA. It is not hard to see that if there exist constants (115), (118) for some finite values x^0, u^0 then they also exist for any other ones and that it therefore suffices to put

$$x^0 = 1, \qquad u^0 = 1,$$

as we will do. Furthermore the factor $e^{-\varepsilon u}$ in (112) can be replaced by a much more general factor $K(\epsilon u)$, that is

$$\psi^{\varepsilon}(u) = \psi(u)\, K(\varepsilon u),$$

with $K(u)$ having the following properties. It is defined and continuous in the closed interval $0 \leqq u < \infty$; and in the open interval $0 < u < \infty$ it has r continuous derivatives, the r being the same as in the theorem, such that

(121) $$\left| u^p \frac{d^p K(u)}{du^p} \right| \leqq A, \qquad 0 < u < \infty,$$

for $0 \leqq p \leqq r$; for $0 \leqq p \leqq r - 1$ we have

(122) $$\frac{d^p K(u)}{du^p} = 0(u^{-\frac{1}{2}\delta-\frac{1}{2}l-\frac{3}{4}}) \qquad \text{as } u \to \infty,$$

and finally, $K(0) = K(+0) = 1$. In particular, we may put $K(u) = e^{-\varepsilon^{\lambda}}$, for any $\lambda > 0$, but we can also take for it a Riesz kernel

$$K(u) = \begin{cases} (1 - u^{\alpha})^{\beta}, & 0 < u < 1 \\ 0, & 1 < u < \infty \end{cases}$$

for suitable $\alpha > 0$, $\beta > 0$ say. And our analysis will also show automatically that the value of the resulting limit $M_n(x)$ is independent of the particular kernel employed, because, after a certain stage, it will be permissible to make the direct substitution $\varepsilon = 0$ by which the kernel occurring will be eliminated in the end.

For the proof proper, we introduce in (113) the function

$$S_{\nu}(u) = u^{\frac{1}{2}\nu} J_{\nu}(u^{\frac{1}{2}}),$$

for which the relation

$$S_{\nu}(u) = 2 \frac{d}{du} S_{\nu+1}(u)$$

holds, see [15], p. 46, and if we carry out r partial differentiation, which are justified by (121) and (122), we obtain, except for a factor $(-2)^r$ the value

(123) $$\frac{1}{x^k} \int_0^{\infty} \frac{d_r}{du^r} \left\{ \psi^{\varepsilon}\left(\frac{u}{x^2}\right) u^{-n/2} \right\} u^{\frac{1}{2}(l+n+r)} J_{l+n+r}(u^{1/2})\, du.$$

If we herein carry out the differentiation by the product rule, we obtain $r + 1$ terms each having the form

$$\frac{1}{r}\int_0^\infty \frac{d^p}{du^p}\left\{\psi^\varepsilon\left(\frac{u}{x^2}\right)\right\} u^{-\frac{1}{2}n-q} u^{\frac{1}{2}(l+n+r)} J_{l+n+r}(u^{\frac{1}{2}})\,du, \tag{124}$$

where

$$p \geqq 0, \qquad q \geqq 0, \qquad p + q = r, \tag{125}$$

except for a factor which is maximized by n^q, which in its turn is $\leqq n^r$. This has been accounted for by the factor $(n + 1)^{-r}$ on the left side of (116) and (119), and from now on we have only to show that (124) is maximized by the right side in (116) and (119) respectively.

Since we have

$$J_{l+n}(t) = \begin{cases} 0(t^{l+n}), & 0 \leqq t < 1 \\ 0(1), & 1 \leqq t < \infty, \end{cases}$$

uniformly in n, (124) is majorized by the sum of

$$\frac{1}{x^{l+r}}\int_{x^{-2}} \left|\frac{d^p\psi^\varepsilon(u)}{du^p}\right| u^{\frac{1}{2}(l-r)+r}\,du \tag{126}$$

and of

$$x^n\int_0^{x^{-2}} \left|\frac{d^p\psi^\varepsilon(u)}{du^p}\right| u^{l+p+\frac{1}{2}n}\,du$$

which latter term is in turn majorized by

$$\int_0^{x^{-2}} \left|\frac{d^p\psi^\varepsilon(u)}{du^p}\right| u^{l+p}\,du. \tag{127}$$

We now claim that uniformly in $0 < \varepsilon < \infty$ the totality of terms (126), (127) for all combinations (125) is in turn maximized by the totality of terms

$$\int_0^{x^{-2}} \left|\frac{d^p\psi(u)}{du^p}\right| u^{l+p}\,du, \tag{128}$$

$$\frac{1}{x^{l+r}}\int_{x^{-2}}^\infty \left|\frac{d^p\psi(u)}{du^p}\right| u^{\frac{1}{2}(l-r)+p}\,du \tag{129}$$

in which the convergence factor no longer occurs. In fact if we put

$$p = \lambda + \mu, \qquad \lambda \geqq 0, \qquad \mu \geqq 0 \tag{130}$$

then

$$\frac{d^p\psi^\varepsilon(u)}{du^p} = \sum_{\lambda=0}^{p}\binom{p}{\lambda}\frac{d^\lambda\chi(u)}{du^\lambda}\frac{d^\mu}{du^\mu}\{e^{-\varepsilon\mu}\}$$

and due to

$$\frac{d^{\mu}(e^{-\varepsilon u})}{du^{\mu}} \leqq C(r)u^{-\mu},$$

see (121), this is indeed so.

This leaves us with the task of actually estimating (128) and (129) only. We will first treat (128) for $x < 1$ $(= x^0)$, and to that end we put

$$\int_0^{x^{-2}} = \int_0^1 + \int_1^{x^{-2}}$$

and we introduce herein the two parts of assumption (120), for $u^0 = 1$. This leads to

$$\int_0^1 u^{\frac{1}{2}\tau+l}\,du + \int_1^{x^{-2}} u^{\frac{1}{2}\delta+l}\,du,$$

and the first part is $0(1)$ by (111), and the second is $0(x^{-\delta-k})$ if $\delta + k > 0$; $= 0(\log^{(1/x)})$ if $\delta + k = 0$; and $0(1)$ if $\delta + k < 0$; all as claimed in (116). As regards (129) for $x \leqq 1$, we note that it is majorized by

$$\frac{1}{x^{l+r}} \int_{x^{-2}}^{\infty} u^{\frac{1}{2}(\delta+l-r)}\,du,$$

and for $r > \delta + l + 2$, as we have stipulated in (114), this integral is first of all finite, and the value of the expression is $0(x^{-\delta-k})$, always, even if $\delta + k \leqq 0$. This completes the proof of part (i).

For part (ii) we note that for $x \geqq 1$, (128) is majorized by

$$\int_0^{x^{-2}} u^{\frac{1}{2}\tau+l}\,du = 0(x^{-\tau-k})$$

and this is indeed $0(x^{-N})$, if $\tau \geqq N - k$. Term (129) however must be decomposed into

$$\int_{x^{-2}}^1 + \int_1^{\infty}.$$

For the first part we have

$$\frac{1}{x^{l+r}} \int_{x^{-2}}^1 u^{\frac{1}{2}(\tau+l-r)}\,du = 0(x^{-l-r}) = 0(x^{-N}).$$

provided $\tau + l - r > -2$ and also $r \geqq N - l$, which was so stipulated; and for the second we have

$$\frac{1}{x^{l+r}} \int_1^{\infty} u^{\frac{1}{2}(\delta+l-r)}\,du = 0(x^{-l-r})$$

provided $\delta + l - r < -2$, which was stipulated.

Finally for part (iii) of the lemma we write (123) as

$$\int_0^\infty H_n(x;\, \varepsilon;\, u)\, du,$$

and our proof has demonstrated that there is a function $H_n(x;\, u) \geqq 0$ such that

$$|\, H_n(x;\, \varepsilon;\, u)\,| \leqq H_n(x;\, u) \qquad \lim_{\varepsilon\to 0} H_n(x;\, \varepsilon;\, u) = H_n(x;\, u)$$

and that the function

$$\tilde{M}_n(x) = \int_0^\infty H_n(x;\, u)\, du$$

likewise satisfies the estimates (116) and (119) and this completes the proof of the lemma.

We will terminate with a proposition which will be of consequence in another context.

THEOREM 14. *If $T(t_j)$ is semi-homogeneous elliptic then the Green's functions for the entire E_k as defined by (59) has the following property: corresponding to any $N > 0$ there exists an integer $q_0 = q_0(N)$, such that for $q \geqq q_0$ we have*

$$\frac{\partial^q F(x_i)}{\partial x_1^{q_1} \cdots \partial x_k^{q_k}} = 0(x^{-N}), \qquad q_1 + \cdots + q_k = q,$$

as $x \to \infty$. For instance, $q \geqq N + 2h_0$ is satisfactory.

PROOF. First of all, for $q \geqq 2h_0$ we have

$$D^q(F(x_j)) = \int_{t\leqq 1} \frac{Q(t_j)}{T(t_j)}\, e^{2\pi i(t,x)}\, dv_t + \lim_{\varepsilon\to 0} \int_{t\geqq 1} e^{-\varepsilon t}\, \frac{Q(t_j)}{T(t_j)}\, e^{2\pi i(t,x)}\, dv_t\,,$$

where

$$Q(t_j) = \Pi_j(2\pi i\, t_j)^{q_j},$$

with the corrective term $L(x, t)$ no longer present, and it is easy to see that the total expression can also be written as

$$\lim_{\varepsilon\to 0} \int_{E_k} e^{-\varepsilon t}\, \frac{Q(t_j)}{T(t_j)}\, e^{2\pi i(t,x)}\, dv_t\,.$$

But now we can introduce our smoothing factor $\chi(t)$ and write this as the sum of

$$\lim_{\varepsilon\to 0} \int_{E_k} e^{-\varepsilon t} \chi(t)\, \frac{Q(t_j)}{T(t_a)}\, e^{2\pi i(t,x)}\, dv_t$$

and

$$\int_{E_k} (1 - \chi(t))\, \frac{Q(t_j)}{T(t_j)}\, e^{2\pi i(t,x)}\, dv_t\,,$$

and apply part (ii) of Lemma 4 separately. For the first addend we have $\delta = q_0 - 2h$, $\tau = -\infty$, and it therefore is $0(x^{-N})$ for any N no matter how large.

For the second, however, we have $\delta = -\infty$, $\tau = 2h_0$, and it is easy to verify that the value of q given will secure the estimate asserted.

PRINCETON UNIVERSITY

REFERENCES

[1] BOCHNER, *Theta relations with spherical harmonics*. Proc. Nat. Acad. Sci. U. S. A., 37 (1951), 804–808.

[2] BOCHNER, *Darstellung reell-variabler und analytischer Funktionen durch verallgemeinerte Fourier- und Laplace-Integrale*. Math. Ann., 97(1927), 637–662.

[3] BOCHNER, Fouriersche Integrale, 1932.

[4] BOCHNER, *Summation of multiple Fourier series by spherical means*. Trans. Amer. Math. Soc., 40(1936), 175–207.

[5] BOCHNER and A. E. TAYLOR, *Some theorems on quasi-analyticity for functions in several variables*. Amer. J. Math., 61(1939), 303–329.

[6] BOCHNER and W. T. MARTIN, Several complex variables, 1948.

[7] L. GÅRDING, *Le problème de Dirichlet pour les equations aux dérivées partielles elliptiques homogènes a coefficients constants*. C. R. Acad. Sci. Paris, 230(1950), 1030–1032.

[8] G. HERGLOTZ, *Ueber die Integration linearer, partieller Differentialgleichungen mit konstanten Koeffizienten*. Berichte Sächsische Akad. der Wiss. Leipzig, 1926, pp. 287–318.

[9] E. W. HOBSON, *On the evaluation of a certain surface integral, and its applications to the expansion, in series, of the potential of ellipsoids*. Proc. London Math. Soc., 24(1893), 80–96.

[10] F. JOHN, *The fundamental solution of linear elliptic differential equations with analytic coefficients*. Communications on Pure and Applied Mathematics, 3(1950), 273–304

[11] E. LANDAU, Handbuch der Primzahlen, 1909, Vol. I.

[12] S. MINAKSHISUNDARAM and Å. PLEIJEL, *Some properties of the Eigenfunctions of the Laplace operator on Riemann surfaces*. Canadian J. Math., 3(1949), 242–256.

[13] I. G. PETROVSKY, *Sur l'analycitée des solutions des systemès d'equations differentielles*, Rec. Math., (Mat. Sbornik) N. S., 5(1939), 3–70.

[14] G. SZEGÖ, Orthogonal Polynomials, 1939.

[15] G. N. WATSON, Bessel Functions, 1922.

Weak solutions of linear partial differential equations.

By S. BOCHNER.

We are returning to the topic of our paper : Linear partial differential equations with constant coefficients (*Ann. Math.*, t. 47, 1946, p. 202-212) and of the first part of Chapter VIII in the book Bochner-Martin, *Several complex variables*, but in the present paper the coefficients will be variable throughout. We will call a function in a domain " locally integrable " if it is defined almost everywhere, is measurable, and is integrable in every compact subset. More general " distributions " would not be suitable to the present context.

In Euclidean E_n we consider all differential symbols

$$D_{r_1\ldots r_n} = \frac{\partial^{r_1+\ldots+r_n}}{\partial x_1^{r_1}\ldots\partial x_n^{r_n}} \qquad (r_1 \geqq 0, \ \ldots, \ r_n \geqq 0). \tag{1}$$

We say that a symbol is *lower* than another,

$$D_{s_1\ldots s_n} < D_{r_1\ldots r_n} \tag{2}$$

if

$$s_1 \leqq r_1, \quad s_2 \leqq r_2 \quad \ldots, \quad s_n \leqq r_n. \tag{3}$$

We introduce a *finite non-vacuous* set

$$\mathcal{D} = \{ D_{r_1\ldots r_n} \}, \tag{4}$$

and we say that it is *complete* if with any element it contains *all* lower ones.

For given $p \geqq 1$ we say that the set

$$\mathcal{D}_p = \{ D_{s_1\ldots s_n} \} \tag{5}$$

is a divisor of order p of $\mathfrak{D}$, if it contains the identity $D_{0\ldots 0}$, is complete, and if corresponding to any element $D_{r_1\ldots r_n}$ in (4) with $r_1+\ldots+r_n \geqq p$ there exists in (5) a lower element $D_{s_1\ldots s_n}$ with $s_1+\ldots+s_n=r_1+\ldots+r_n-p$. If in (4) we always have

$$r_1+\ldots+r_n \leqq N \tag{6}$$

then the set of all symbols $\{D_{s_1\ldots s_n}\}$ with $s_1+\ldots+s_n \leqq N-p$ plus $D_{0\ldots 0}$ is such a $\mathfrak{D}_p$. However frequently it is too large, and, in our theorems, the smaller the divisor the better the result.

If $f(x_1, \ldots, x_n)$ is suitably differentiable then with (4) we associate the absolute value

$$|\mathfrak{D}f(x)| = \sum_{\mathfrak{D}} |D_{r_1\ldots r_n}f(x)|,$$

and we note that the absolute value

$$|\mathfrak{D}_p f(x)| = \sum_{\mathfrak{D}_p} |D_{s_1\ldots s_n}f(x)| \tag{7}$$

will be smaller if $\mathfrak{D}_p$ is smaller.

We will consider operators

$$Tf = \sum_{\mathfrak{D}} a_{r_1\ldots r_n}(x)\, D_{r_1\ldots r_n} f \tag{8}$$

with coefficients belonging to differentiability class C^∞ in a given domain S and their formal transposed

$$'Tf = \sum_{\mathfrak{D}} (-1)^{r_1+\ldots+r_n} D_{r_1\ldots r_n}(a_{r_1\ldots r_n} f). \tag{9}$$

The smallest number N satisfyring (6) is their order. A testing function in S is a function φ in C^∞ with compact support. If T is any operator then $\psi = T\varphi$ is again a testing function. Also, for any two operators T^1, T^2 we have

$$(T^1 T^2)\varphi = T^1(T^2\varphi), \tag{10}$$

and this in fact holds for any function which is suitably differentiable.

If φ is a testing function and $f \in C^N$ then we have

$$\int_S \varphi \, . \, Tf \, dv = \int_S f \, . \, {}'T\varphi \, dv. \tag{11}$$

On the basis of this we say that $f(x)$ is a *strong* solution of

$$Tf = 0 \tag{12}$$

if it is C^N and satisfies (12) pointwise. It is a *weak* solution, if it is locally integrable and

$$\int_S f \, . \, {}'T\varphi \, dv = 0 \tag{13}$$

for all testing functions φ. More generally if we are given several operators $T^1, \ldots, T^\mu$, and locally integrable functions $(f^1, \ldots, f^\mu)$ then we say that they are a weak solution of

$$T^1 f^1 + \ldots + T^\mu f^\mu = 0 \tag{14}$$

if we have

$$\sum_{\lambda=0}^{\mu} \int_S f^\lambda \, . \, {}'T^\lambda \varphi \, dv = 0 \tag{15}$$

for all φ. Obviously a strong solution is a weak solution. If in a weak solution each f belongs to the corresponding C^N then it is a strong solution.

An inhomogenous equation

$$T^1 f^1 + \ldots + T^\mu f^\mu = f^{\mu+1} \tag{16}$$

will be interpreted as the homogenous equation

$$T^1 f^1 + \ldots + T^\mu f^\mu - D_{0\ldots0} f^{\mu+1} = 0. \tag{17}$$

Thus if $f^{\mu+1}$ is locally integrable then $(f^1, \ldots, f^\mu)$ is a weak solution of (16) if by addition of $f^{\mu+1}$ it becomes a weak solution of (17), that is if

$$\sum_{\lambda} \int f^\lambda \, . \, {}'T^\lambda \varphi \, dv = \int f^{\mu+1} \, . \, \varphi \, dv \tag{18}$$

for all φ.

THEOREM 1. — *For any* T^0, *a weak solution of* (14) *is also a weak solution of*

$$(T^0T^1)f^1+(T^0T^2)f^2+\ldots+(T^0T^\mu)f^\mu=0. \tag{19}$$

In fact we have

$$\sum_\lambda \int f^\lambda.'(T^0T^\lambda)\varphi\,dv=\sum_\lambda \int f^\lambda.'T^\lambda('T^0\varphi)\,dv.$$

However $'T^0\varphi$ is again a testing function, and therefore the sum is 0 by definition (15) of a weak solution for (14).

THEOREM 2. — *If for some operators* $T_1, \ldots, T_{r-1}$ *we are given a weak solution* $(f_1, \ldots, f_r)$ *of the system of equations*

$$f_s=T_{s-1}f_{s-1} \qquad (2\leqq s\leqq r), \tag{20}$$

then for any further operator T_r *we have*

$$\int_S f_1('T_1'T_2\ldots'T_r\varphi)\,dv=\int_S f_r.'T_r\varphi\,dv \tag{21}$$

for any testing function φ.

In fact, since $'T_r\varphi$ is again a testing function then the relation $f_r=T_{r-1}f_{r-1}$ implies that the right side in (21) has the value

$$\int_S f_{r-1}.'T_{r-1}'T_r\varphi\,dv.$$

If we herein " substitute " $f_{r-1}=T_{r-2}f_{r-2}$ then we obtain for this

$$\int_S f_{r-2}.'T_{r-2}'T_{r-1}'T_r\varphi\,dv,$$

and so forth.

Theorem 2 easily leads to the following theorem.

THEOREM 3. — *Let there be given a table of locally integrable functions*

$$\{f_r^\lambda\} \qquad (1\leqq\lambda\leqq\mu,\ 1\leqq r\leqq r_\lambda), \tag{22}$$

which are known to be joint weak solutions of a system of equation

$$f_s^\lambda=T_{s-1}^\lambda f_{s-1}^\lambda \qquad (1\leqq\lambda\leqq\mu,\ 2\leqq s\leqq r_\lambda), \tag{23}$$

and let $T_{r_\lambda}^\lambda$, $1\leqq\lambda\leqq\mu$, *be further operators.*

(i) *The functions* f_1^λ, $\lambda = 1, \ldots, \mu$, *are a weak solution of the equation*

$$\sum_\lambda T^\lambda_{r_\lambda} T^\lambda_{r_\lambda - 1} \ldots T^\lambda_1 f^\lambda_1 = 0 \tag{24}$$

if and only if the functions $f^\lambda_{r_\lambda}$, $\lambda = 1, \ldots, \mu$, *are a weak solution of*

$$\sum_\lambda T^\lambda_{r_\lambda} f^\lambda_{r_\lambda} = 0. \tag{25}$$

(ii) *If in the table* (22) *we have*

$$f^1_1 = f^2_1 = \ldots = f^\mu_1 = f \tag{26}$$

and if we put

$$T = \sum_\lambda T^\lambda_{r_\lambda} T^\lambda_{r_\lambda - 1} \ldots T^\lambda_1, \tag{27}$$

then f *is a weak solution of*

$$Tf = 0 \tag{28}$$

if and only if $f^\lambda_{r_\lambda}$, $\lambda = 1, \ldots, \mu$, *is a weak solution of* (25).

Remark 1. — If $f(x)$ is suitably differentiable then we can *define* the elements of the table (22) by (26) and (23), and thus, if $f(x)$ is a strong solution of (28), and the order of the addends in (27) is not larger than in T itself, then automatically, we can replace (28) by the system of equations (23) and (25). However, for a weak solution of (28) there is no reason to assume that, starting with the functions (26), the operations (23) can be carried out " weakly ", and the existence of the table (22) must be stipulated.

THEOREM 4. — *Let* A *be a set in* E_n. *There exists a family of functions* $\varphi_\varepsilon(x_1, \ldots, x_n)$ in E_n, $0 < \varepsilon \leq 1$, *with the following properties.* $\varphi_\varepsilon \in C^\infty$ in E_n. *If, for any* $\eta > 0$, *we denote by* A_η *the* η-*neighborhood of* A, *then*

$$\varphi_\varepsilon = 0, \quad \text{for} \quad x \varepsilon A_{\frac{1}{10}\varepsilon}, \qquad \text{and} \qquad = 1 \quad \text{for} \quad x \varepsilon E_n - A_\varepsilon, \tag{29}$$

and

$$|D_{r_1 \ldots r_n} \varphi_\varepsilon(x_1, \ldots, x_n)| \leq \frac{C_{r_1 + \ldots + r_n}}{\varepsilon^{r_1 + \ldots + r_n}}, \tag{30}$$

$r_1 \geq 0, \ldots, r_n \geq 0$, *where* C_r *is independent of* ε.

Proof. — (The construction given in "Bochner-Martin" is faulty. The following alternate one was given to us by John W. Woll, graduate student at Princeton University.) Take a function $\psi(t)$ which is defined and C^∞ in $0 \leqq t < \infty$, $\psi(t) = 1$ for $0 \leqq t \leqq \frac{1}{2}$, $\psi(t) = 0$ for $1 \leqq t < \infty$, and $\psi(t) \geqq 0$ for $0 \leqq t < \infty$.

Put

$$K(\xi_1, \ldots, \xi_n) = \frac{\psi(\xi_1^2 + \ldots + \xi_n^2)}{\int_{E_n} \psi(t_1^2 + \ldots + t_n^2)\, dv_t}.$$

Denote by $\varphi^\eta(x_1, \ldots, x_n)$ the function in E_n which is $= 1$ for x in $E_n - A_{4\eta}$ and $= 0$ for x in $A_{4\eta}$, and introduce the smoothing

$$\psi_\eta(x_1, \ldots, x_n) = \frac{1}{\eta^n} \int_{E_n} \varphi^\eta(x_1 + \xi_1, \ldots, x_n + \xi_n)\, K\left(\frac{\xi_1}{\eta}, \ldots, \frac{\xi_n}{\eta}\right) dv_\xi.$$

Finally put

$$\varphi_\varepsilon(x_1, \ldots, x_n) = \psi_{\frac{1}{10}\varepsilon}(x_1, \ldots, x_n).$$

We take in E_n an open set S and a relatively closed subset A, and we assume that the open set

$$S^0 = S - A$$

is not empty. If we are given any locally integrable function $f(x)$ in S^0 then we will denote by $\tilde{f}(x)$ a function in S which is its "completion" by value 0, that is

$$\tilde{f}(x) = \begin{cases} f(x) & (x \text{ in } S^0), \\ 0 & (x \text{ in } A). \end{cases}$$

Also, for any indices λ, r, $\tilde{f}_r^\lambda$ will be the same type of completion of the function f_r^λ.

THEOREM 5. — *Let $(f^1, \ldots, f^\mu)$ be a weak solution of*

$$T^1 f^1 + \ldots + T^\mu f^\mu = 0 \tag{31}$$

in S^0, and let N^λ be the order of T^λ, $\lambda = 1, \ldots, \mu$.

If, for $\varepsilon \to 0$, *we have*

(32) $$\int_{S^0 \cap A_\varepsilon} |f^\lambda(x)|\, dv_x = o(\varepsilon^{N^\lambda}) \qquad (\lambda = 1, \ldots, \mu),$$

then $\{\tilde{f}^\lambda\}$ *is a weak solution of*

(33) $$T^1 \tilde{f}^1 + \ldots + T^\mu \tilde{f}^\mu = 0$$

in S.

Proof. — Let φ be a testing function in S. We have to verify that

(34) $$\sum_\lambda \int_S \tilde{f}^\lambda \,.\, {}'T^\lambda \varphi \, dv$$

is zero. We write for this expression the sum

$$\sum_\lambda \int_S \tilde{f}^\lambda \,.\, {}'T^\lambda(\varphi(1-\varphi_\varepsilon))\, dv_x + \sum_\lambda \int_S \tilde{f}^\lambda \,.\, {}'T^\lambda(\varphi\varphi_\varepsilon)\, dv_x,$$

and since $\varphi\varphi_\varepsilon$ is a testing function in S^0 the second expression has value 0. On the other hand, since $\varphi_\varepsilon = 0$ outside A_ε and $\tilde{f} = 0$ on A the first expression is

(35) $$\sum_\lambda \int_{S^0 \cap A_\varepsilon} f^\lambda \,.\, {}'T^\lambda(\varphi(1-\varphi_\varepsilon))\, dv_x.$$

Now, (30) implies

$${}'T^\lambda(\varphi(1-\varphi_\varepsilon)) = O(\varepsilon^{-N^\lambda}) \qquad (\varepsilon \to 0),$$

and assumption (32) therefore implies that the integral (35) tends to 0 as $\varepsilon \to 0$, which proves the theorem. An easy generalization of it is the following one.

THEOREM 6. — *If the functions* (22) *satisfy the equations* (23) *and* (24) *weakly in* S^0, *if* N_r^λ *is the order of* T_r^λ *and if*

(36) $$\int_{S^0 \cap A_\varepsilon} |f_r^\lambda|\, dv_x = o(\varepsilon^{N_r^\lambda}),$$

for all (λ, r) *then the functions* $\tilde{f}_r^\lambda$ *satisfy the equations weakly in* S.

We now choose a representation (27) in which $N_r^\lambda \leq p$ for some $p \geq 1$. By doing this judiciously, and recalling remark 1, the following conclusion can be drawn.

THEOREM 7. — *If $f(x)$ in S^0 is a strong solution of*

$$\sum_{\mathfrak{O}} a_{r_1 \ldots r_n}(x) D_{r_1 \ldots r_n} f(x) = 0, \tag{37}$$

and if for some divisor $\mathfrak{O}_p$ of $\mathfrak{O}$ we have

$$\int_{S_0 \cap A_\varepsilon} |\mathfrak{O}_p(f(x))| \, dv_x = o(\varepsilon^p),$$

then $f(x)$ is a Weak solution of the same equation in all of S.

We will say that A is a semi-smooth variety of dimension $n-q$, and then denote it by B_{n-q}, if it has topological dimension $n-q$ and if its η-neighborhood A_η, for small $\eta > 0$, has n-dimensional measure $o(\eta^q)$. The following conclusion from Theorem 7 is obvious.

THEOREM 8. — *If $f(x)$ in S^0 is a strong solution of (37) and if we have*

$$|\mathfrak{O}_p f(x)| = o(\varepsilon^{p-q}) \qquad (\varepsilon \to 0) \tag{38}$$

where ε is the distance of the point x from a semi-smooth variety B_{n-q}, uniformly in all x, then $f(x)$ can be continued into a weak solution of the equation in $S^0 + B_{n-q}$.

If we decompose a neighborhood U in E_n by an $(n-1)$-dimensional semi-smooth variety B_{n-1} into a part U^+ and U^-; if $f(x)$ is a strong solution of (37) in U^+, and if we have

$$|\mathfrak{O}_p f(x)| = o(\varepsilon^{p-1}), \tag{39}$$

uniformly in x, then by putting $f(x) = 0$ in U^- (and also B_{n-1}) a weak solution in all of U arises. For $p = 1$ the condition is

$$|\mathfrak{O}_1 f(x)| \to 0, \tag{40}$$

as x approaches the boundary B_{n-1}.

Remark 2. — Call an equation (37) " properly elliptic " if every weak solution (after rectification of values on a set of measure 0) is a strong solution, and if a solution which is zero in a neighborhood of a point vanishes identically. For solutions of such equations the theorem implies that if the functions and its derivatives of order $\leqq N-1$ have boundary values 0 for a semi-smooth piece of the boundary, then the function vanishes identically. If one has information about partial derivatives of order $\leqq N-p$ only, then the stronger limiting behavior (39) must be assumed known.

Remark 3. — For fixed q, the bound ε^{p-q} is smaller the larger p is, and thus one might think that Theorem 8 is " better " the smaller p is. Such however is not the case always. In fact, for $q=1$, if B_{n-1} is very smooth then the largest value $p=N$, that is the condition

$$f(x)=o(\varepsilon^{N-1}),$$

is the best, because, if B_{n-1} is very smooth and if a function and its derivatives of order $\leqq N-p$ are as small as $o(\varepsilon^{p-1})$, then, by application of the ordinary mean value theorem it turns out that the function and derivatives of order $\leqq N-(p+1)$ are $o(\varepsilon^{p+1-1})$ automatically. Thus, in this case, Theorem 5 is still the best, even though Theorem 6, which is based on it, appears to be a rather more sophisticated version of the proposition underlying it.

The situation however is no longer so, if we consider a lower-dimensional B_{n-q}, $q\geqq 2$, no matter how smooth. It does not decompose U and the resulting theorem is not about continuation across a boundary but about removing a lower dimensional variety of (alleged) irregularities. The bound $o(\varepsilon^{p-q})$ is now a large quantity if $p<q$, the larger the smaller p is. Theorem 8 states that if the function and certain partial derivatives are not too large in the vicinity of B_{n-2} then, in the properly elliptic case, they are in fact quite well-behaved and there are no singularities at all. And in the non-elliptic case it remains a weak solution of its equation across the exceptional set, anyway. For such cases, the theorem is indeed better the smaller p.

Remark 4. — Returning to remark 2 we not that for $q=1$, that is $A = B_{n-1}$, if the boundary B_{n-1} is very un-smooth then smaller values of p might be of advantage again. Assume that it is any pointset of topological dimension $n - 1$ say, which separates U into U^+ and U^{-1}, and that it is possible to construct a sequence of open set A_ε containing it (they need *not* be ε-neighborhoods) whose n-dimensional volume is $o(\varepsilon^r)$, for some positive value $r > 0$, and a family of functions φ_ε with properties (29) and (30) as before. We would then obtain the conclusion that

$$|\mathfrak{D}_p f(x)| = o(\varepsilon^{p-r})$$

implies, for the properly elliptic case, the idendical vanishing of $f(x)$, and this theorem is then indeed the better the smaller p is, provided that information about the boundary behaviour of partial derivatives of suitably high order is available.

SYSTÊMES DES ÉQUATIONS AUX DÉRIVÉES PARTIELLES QUI SE RATTACHENT AUX FORMULES DE GREEN (*)

Communication de

S. BOCHNER

(Université de Princeton)

Dans un espace Euclidien E_n à points réels $\xi = (\xi_1, \ldots, \xi_n)$, $x = (x_1, \ldots, x_n)$, etc, nous introduisons une forme différentielle extérieure de dimension n-1,

$$\Omega(f(\xi);\, \xi - x) = \Sigma_{(\alpha)}\, \Omega_{\alpha_1 \ldots \alpha_{n-1}}\, d\xi_{\alpha_1} \wedge d\xi_{\alpha_2} \wedge \ldots \wedge d\xi_{\alpha_{n-1}}, \tag{1}$$

chaque composant $\Omega_{\alpha_1 \ldots \alpha_{n-1}}$ étant une expression finie

$$\sum_{p_1 \cdots p_n} G_{p_1 \cdots p_n}(\xi - x) \frac{\partial^{p_1 + \cdots + p_n} f(\xi)}{\partial \xi_1^{p_1} \ldots\ldots \partial \xi_n^{p_n}} \tag{2}$$

à une fonction indéterminée $f(\xi) \equiv f(\xi_1, \ldots, \xi_n)$. Les coefficients sont des fonctions déterminées de deux points ξ, x qui sont définies pour $\xi \neq x$ et ne dépendent que de la différence $\xi - x$. Toutes les constantes et toutes les fonctions qui se rencontrent peuvent prendre des valeurs complexes, et il sera sous-entendu que toutes les fonctions introduites sont C^∞ dans les ensembles où elles sont définies.

La dérivée extérieure de (1) est le produit de l'élément de volume

$$(dv_\xi \equiv)\, d\xi = d\xi_1 \wedge d\xi_2 \wedge \ldots \wedge d\xi_n \tag{3}$$

par un composant $\Omega_{12\ldots n}$ qui est encore une expression de la forme (2). En introduisant la famille des opérateurs

$$Tf(\xi) = \sum_{(q)} c_{q_1 \cdots q_n} \frac{\partial^{q_1 + \cdots + q_n} f(\xi)}{\partial \xi_1^{q_1} \ldots\ldots \partial \xi_n^{q_n}} \tag{4}$$

(sommes finies) à coefficients constants, nous représentons $\Omega_{1\ldots n}$ par une somme finie

$$\sum_{j=1}^{k} G_j(\xi - x)\, T_j f(\xi) \tag{5}$$

quelconque à des coefficients qui sont intégrables même à l'origine où ils ne sont pas définis :

$$\int_{0 < |\xi| < 1} |G_j(\xi)|\, d\xi < \infty \qquad j = 1, \ldots, k; \tag{6}$$

l'existence d'une représentation (5) de telle sorte étant une hypothèse que l'on impose expressément.

* This research was supported by the United States Air Force, through the Office of Scientific Research of the Air Research and Development Command.

Ces données préparées, nous introduisons le système des équations

$$(7) \qquad T_j f(x) = \rho_j(x) \qquad j = 1, \dots, k ;$$

les conditions d'intégrabilité

$$(8) \qquad T_j \rho_\ell(x) = T_\ell \rho_j(x) \qquad 1 < j, \ell < k ;$$

qui y correspondent, et l'intégrale de Green

$$(9) \qquad g(x) = \int_{B_{n-1}} \Omega(f(\xi); \xi - x) - \int_{B_n} \sum_j G_j(\xi - x) \cdot \varphi_j(\xi)\, d\xi$$

qui s'y rattache. L'ensemble B_n est un domaine borné, et B_{n-1} est sa frontière suffisamment régulière. Nous introduisons immédiatement un voisinage U de B_{n-1}, l'adhérence $\bar{B}_n = B_n + B_{n-1}$, et le voisinage

$$\tilde{B}_n = B_{n-1} \cup U \cdot \bar{B}_n \cup U$$

de $\bar{B}_n$.

<u>Théorème 1</u> - <u>Si $f(x)$ est donnée dans $\tilde{B}_n$ et les fonctions</u>

$$(10) \qquad \rho_1(x), \dots, \rho_k(x)$$

<u>y sont définies par (7), l'intégrale (9) est zéro en dehors de $\bar{B}_n$: $g(x) = 0$ pour $x \notin \bar{B}_n$</u>

C'est une conséquence directe du théorème de Stokes.

<u>Théorème 2</u> - <u>Etant donné un système de fonctions (10) dans $\tilde{B}_n$ tel que dans U il existe une fonction $f(x)$ qui y satisfait les relations (7), alors, pour chaque $x \notin B_{n-1}$, pour des vecteurs h suffisamment petits on a</u>

$$(11) \qquad g(x+h) = \int_{B_{n-1}} \Omega(f(\xi + h); \xi - x) - \int_{B_n} \Omega_j\, G_j(\xi - x) \cdot \rho_j(\xi + h)\, d\xi,$$

<u>ce qui montre que $g(x)$ est C^∞ pour $x \notin B_{n-1}$ et y entraine la relation</u>

$$(12) \qquad T^x g(x) = \int_{B_{n-1}} \Omega(T^\xi f(\xi); \xi - x) - \int_{B_n} \sum_j G_j(\xi - x) \cdot T\rho_j(\xi)\, d\xi,$$

<u>pour un opérateur T de la forme (4) quelconque</u>.

<u>Démonstration</u> - D'abord, en remplaçant ξ par $\xi - h$, on obtient la formule (11) non pour la figure (B_n, B_{n-1}) mais pour la figure (B_n^h, B_{n-1}^h) où B_n^h résulte de B_n en le déplaçant par $-h$ et B_{n-1}^h est la frontière de B_n^h. Pour h suffisamment petit, $B_{n-1}^h - B_{n-1}$ est la frontière d'un domaine dans U et d'après le Théorème 1 on peut remplacer les deux figures l'une par l'autre, q.e.d.

Si l'on applique le Théorème 2 à des fonctions $f(x)$, $\rho_j(x)$, qui sont liées par (7) dans l'ouvert $\tilde{B}_n$ entier et l'on fait attention au Théorème 1 on peut arriver à la conclusion suivante :

Théorème 3 - Sur chaque ouvert A de E_n, l'intégrale de Green (9) donne naissance à une fonctionnelle

$$g(x) = Lf(x) \tag{13}$$

qui y est une application distributive de la classe des fonctions C^∞ en soi-même (non bi-univoque en général).

Pour un sous-ouvert B_n dont l'adhérence est contenue dans A on a

$$\int_{B_{n-1}} \Omega(f(\xi);\xi-x) - \int_{B_n} \sum G_j(\xi-x).T_j f(\xi)\,d\xi = \begin{cases} Lf(x), & x \in B_n \\ 0, & x \notin \bar{B}_n \end{cases} \tag{14}$$

En outre, pour tous les opérateurs (4) on a l'égalité

$$L\,T f(x) = T\,L f(x)$$

Le concept de la fonctionnelle (13) va nous mettre en état d'énoncer le théorème suivant.

Théorème 4 - Etant donnés un système des opérateurs

$$T_1, \ldots T_k \tag{16}$$

qui se rattache à une intégrale de Green et un système des fonctions $\rho_1, \ldots, \rho_k$ qui est défini dans $\tilde{B}_n$, et y satisfait aux conditions d'intégrabilité (8), si l'on peut résoudre le système des équations

$$T_j f(x) = \rho_j, \qquad j = 1, \ldots, k, \tag{17}$$

dans un voisinage de la frontière B_{n-1} de B_n on peut aussi résoudre le système

$$T_j g(x) = L\rho_j, \qquad j = 1, \ldots, K \tag{18}$$

dans le domaine B_n.

Démonstration - En mettant $T = T_j$ dans (12) on obtient, en faisant usage de (8),

$$T_j g = \int_{B_{n-1}} \Omega(\rho_j(\xi);\xi-x) - \int_{B_n} \sum_\ell G_\ell(\xi-x).T_\ell \rho_j(\xi)\,d\xi ,$$

et d'après (14), pour $f(\xi) = \rho_j(\xi)$, la somme des deux intégrales a la valeur $L\rho_j$ dans B_n.

A l'ordinaire, on n'envisage que des intégrales de Green dont la fonctionnelle associée est l'identité,

$$Lg = g , \tag{19}$$

et c'est le cas principal sans doute. Nous allons ajouter des énoncés qui s'y réfèrent, et pour indiquer que l'on est dans le cas (19) on parlera non de l'intégrale de Green mais de la formule de Green plus précisément.

En premier lieu, le Théorème 4 se modifie en version suivante.

Théorème 5 - Etant donnés un système des opérateurs (16) qui se rattachent à une formule de Green

$$(20) \qquad \int_{B_{n-1}} \Omega(f(\xi); \xi - x) - \int_{B_n} \sum_j G_j(\xi - x) . T_j f(\xi) d\xi = \begin{cases} f(x), & x \in B_n \\ 0, & x \notin \bar{B}_n, \end{cases}$$

et un système de fonctions ρ_j qui est défini dans $\tilde{B}_n$ et satisfait aux conditions d'intégrabilité (8), s'il existe une solution f(x) du système des équations (17) dans un voisinage de la frontière B_{n-1} il existe aussi une (autre) solution g(x) dans le domaine B_n.

Les deux solutions différentes f(x), g(x) du même système (17) empiètent l'une sur l'autre dans l'ouvert $B_n \cap U$ qui s'étend le long de la frontière entière B_{n-1} en dedans de B_n, et la différence $h(x) = f(x) - g(x)$ y satisfait au système homogène

$$(21) \qquad T_j h(x) = 0.$$

Si la frontière B_{n-1} se compose de morceaux séparés, on ne doit pas attendre que les parties de h(x) provenant des morceaux intérieurs de la frontière aient des traits de comportement déterminables dans le domaine entier B_n. Mais pour le morceau extérieur de la frontière cela peut arriver. Il existe des systèmes des équations (21) ayant la propriété suivante (découverte pour la première fois par Hartogs, voir [1], [2], [3]). Si la frontière B_{n-1} d'un domaine borné B_n est d'un seul tenant, une solution du système (21) dans un voisinage de B_{n-1} peut se prolonger en tant que solution dans tout le domaine B_n. D'où la conclusion suivante.

Théorème 6 - Si le système $\{T_j\}$ du Théorème 5 possède aussi cette propriété, au cas d'une frontière B_{n-1} d'un seul tenant la solution f(x) donnée dans U peut être prolongée en tant que solution dans tout l'ouvert $\tilde{B}_n$. Plus généralement, au cas d'une frontière à plusieurs morceaux, il y a dans B_n une solution qui coincide avec f(x) dans un voisinage du morceau extérieur.

Dans le théorème prochain on ajoutera une hypothèse explicite sur des opérateurs appartenant à une formule de Green (sans la propriété de Hartogs précitée).

Hypothèse H - On peut représenter l'espace donné E_n par un produit des sous-espaces Euclidiens

$$(22) \qquad E = E_{n_1} \times E_{n_2} \times \ldots \times E_{n_k}, \qquad n_1 + \ldots + n_k = n,$$

de telle sorte que pour une fonction $\varphi(x_1, \ldots, x_n)$ sur un ouvert de E_n qui ne dépend que des coordonnées de E_{n_j} on a

$$(23) \qquad T_\ell \varphi = 0, \qquad \ell \neq j,$$

$j = 1, \ldots, k.$

Par exemple, le système des équations

$$\frac{\partial f}{\partial x_m} = \rho_m , \qquad m = 1, \ldots , n ,$$

qui appartient à une formule de Green

$$\gamma_n f(x) = \int_{B_{n-1}} f \frac{\partial G}{\partial \xi_m} \sigma^m - \int_{B_n} \sum_{m=1}^{n} \frac{\partial f}{\partial \xi_m} \frac{\partial G}{\partial \xi_m} d\xi$$

classique satisfait à l'hypothèse H. Il en est de même pour le système

$$\frac{\partial f}{\partial \bar{z}_\alpha} = \rho_\alpha (x_1, y_1, \ldots x_k, y_k) , \qquad \alpha = 1, \ldots , k$$

$$\frac{\partial}{\partial \bar{z}} = \frac{1}{2} \left(\frac{\partial}{\partial x} - i \frac{\partial}{\partial y} \right)$$

qui appartient à une formule de Green

$$(2\pi i)^k f(z, \bar{z}) = \int_{B_{2k-1}} f \frac{\partial V}{\partial \zeta_\alpha} d\zeta_\alpha \wedge \omega^{k-1} - \frac{1}{k} \int_{B_{2k}} \sum_\alpha \frac{\partial f}{\partial \bar{\zeta}_\alpha} \frac{\partial V}{\partial \zeta_\alpha} \omega^k ,$$

$$\omega = d\bar{\zeta}_1 \wedge d\zeta_1 + \ldots + d\bar{\zeta}_k \wedge d\zeta_k , \qquad \zeta = \xi + i\eta$$

$$V = \frac{1}{1-k} \left(\sum_{\alpha=1}^{k} (\zeta_\alpha - z_\alpha)(\bar{\zeta}_\alpha - \bar{z}_\alpha) \right)^{1-k} , \qquad k \geq 2.$$

Nous remarquons que l'article présent est une élaboration systématique des énoncés faits par rapport à ces deux cas spéciaux dans [1] p. 664/5.

Théorème 7 - Sous l'hypothèse H, si un système de fonctions (10) satisfait aux conditions d'intégrabilité (8) dans un ouvert de E_n, alors dans tout sous-ouvert borné D qui est un produit

$$(24) \qquad D = D_1 \times D_2 \times \ldots \times D_k , \qquad D_j \subset E_{n_j} , \quad j = 1, \ldots , K$$

il existe une solution f du système (17).

Démonstration - Pour une fonction $\varphi(x)$ à support compact dans un ouvert D, la formule de Green se réduit à

$$(25) \qquad \varphi(x) = \int_D \sum_j G_j (\xi - x) . T_j \varphi(\xi) . d\xi .$$

En dénotant les points de E_{n_j} par x_j, ξ_j, nous prenons pour chaque j des voisinages bornés fixes quelconques

$$A_\ell^j \subset E_{n_\ell} , \qquad \ell \neq j ,$$

et nous introduisons la fonction

$$H_j(\xi_j) = \int_{\substack{\Pi A_\ell^j \\ \ell \neq j}} G_j(\xi) \prod_{\ell \neq j} d\xi_\ell . \tag{26}$$

Si D_j est un ouvert dans E_{n_j} et $\varphi_j(x_j)$ est une fonction à support compact dans D_j, en appliquant l'égalité (25) à l'ouvert

$$D = A_1^j \times \ldots \times A_{j-1}^j \times D_j \times A_{j+1}^j \times \ldots \times A_k^j$$

on obtient d'après (23)

$$\varphi_j(x_j) = \int_D G_j(\xi - x)\, T_j \varphi_j(\xi_j)\, d\xi .$$

Donc

$$\varphi_j(x_j) = \int_{D_j} H_j(\xi_j - x_j)\, T_j \varphi_j(\xi_j)\, d\xi_j ,$$

ce qui équivaut à

$$\varphi_j(x_j) = \int_{D_j} H_j(x_j - \xi_j)\, T_j^* \varphi_j(\xi_j)\, d\xi_j , \tag{27}$$

l'opérateur T^* étant, pour un opérateur (4) quelconque, l'opérateur transposé

$$\sum_{q_1 \ldots q_n} (-1)^{q_1 + \cdots + q_n} c_{q_1 \ldots q_n} \frac{\partial^{q_1 + \cdots + q_n}}{\partial \xi_1^{q_1} \ldots \partial \xi_n^{q_n}} .$$

La relation (27) étant valable pour toutes les fonctions d'épreuve φ, on en peut déduire un résultat intermédiaire.

Lemme - Pour un ouvert borne D_j fixe, la transformée

$$f(x_j) = \int_{D_j} H_j(\xi_j - x_j)\, \rho(\xi_j)\, d\xi_j \tag{28}$$

satisfait à l'équation

$$T_j f = \rho(x) \tag{29}$$

dans D_j.

Les conditions d'intégrabilité (8) permettent d'étendre le système donné $\{\rho_j\}$ en un système

$$\{\rho_{p_1 \ldots p_j}\} \qquad 1 \leq j \leq k , \qquad 1 \leq p_\ell \leq k , \tag{30}$$

qui est symétrique dans tous les indices et dans lequel, pour $j \geqslant 2$, on a

$$\rho_{p_1 \ldots p_j} = T_{p_1} \rho_{p_2 \ldots p_j} \tag{31}$$

En dénotant l'opérateur (28) par

$$f = \mathcal{T}_j \rho ,$$

on a

$$T_p . \mathcal{T}_q = \mathcal{T}_q . T_p , \qquad 1 \leqslant p \leqslant q \tag{32}$$

$$T_p . \mathcal{T}_p = \underline{\text{identité}} , \tag{33}$$

et l'on vérifie aisément que la fonction

$$f(x_1, \ldots, x_n) = \sum_{1 \leqslant j \leqslant k} (-1)^j \sum_{p_1 < p_2 < \ldots < p_j} \mathcal{T}_{p_1} . \mathcal{T}_{p_2} \ldots \mathcal{T}_{p_j} (\rho_{p_1 \ldots p_j}) \tag{34}$$

satisfait au système (7).

Remarque - A partir de la relation (27) on peut généraliser immédiatement les raisonnements dans la direction suivante.

Dans les opérateurs $T \equiv T_j$ les coefficients ne sont plus des constantes mais des fonctions dans E_n, T^* étant

$$\sum (-1)^{q_1 + \cdots + q_n} \frac{\partial^{q_1 + \cdots + q_n} (c_{q_1 \cdots q_n} f)}{\partial \xi_1^{q_1} \ldots \partial \xi_n^{q_n}} .$$

On remplace chaque fonction $H(\xi - x) \equiv H_j(\xi - x)$ par une fonction $H(\xi ; x)$ telle que, pour D, D' bornés, on a

$$\int_D |H(\xi ; x)| dx < M(D, D') < \infty \qquad \xi \in D'.$$

Toutes les fonctions données ρ_j et toutes les fonctions cherchées f sont intégrables (seulement), et toute relation

$$Tf = \rho$$

est prise au sens faible :

$$\int_D f . T^* \varphi \, dx = \int_D f . \varphi \, dx$$

pour toutes les fonctions d'épreuves φ.

Etant donné un système des fonctions $\{\rho_j\}$ on sait qu'il satisfait aux conditions d'intégrabilité (8) si on peut l'étendre en un système (30) pour lequel les relations (31) sont satisfaites faiblement.

L'énoncé même est le suivant. Etant donné un système $\{T_j\}$, si l'on y peut associer des fonctions $\{H_j(\xi, x)\}$ pour lesquelles on a

$$\varphi(x) = \int_{D_j} H_j(x ; \xi_j) \,.\, T_j^* \varphi(\xi_j)\, d\xi_j$$

pour des fonctions d'épreuve, alors la transformée

$$\mathcal{T}_j = \int_{D_j} H_j(\xi_j ; x)\, \rho\, (\xi_j)\, d\xi_j$$

satisfait à $T_j \mathcal{T}_j \rho = \rho$, et étant données les relations d'intégrabilité, la fonction (34) satisfait au système (7) au sens faible.

BIBLIOGRAPHIE

[1] S. BOCHNER, Analytic and meromorphic continuation by means of Green's formula, Annals of Mathematics, 44 (1943), 652-673.

[2] - Partial differential equations and analytic continuations, Proceedings of the National Academy of Sciences, Washington, 38 (1952), 227-230.

[3] - Green's formula and analytic continuation, in Contributions to the theory of partial differential equations, Annals of Mathematics Study, 33, Princeton, 1954.

INTERVENTIONS

Monsieur SCHWARTZ: la condition "G_j est indéfiniment dérivable" (qui exprime une condition d'ellipticité du système des T_j) est-elle nécessaire ?

Monsieur BOCHNER répond qu'elle l'est probablement tandis que la sommabilité des G_j au voisinage de l'origine ne l'est sans doute pas.

Monsieur DELSARTE fait observer que le Professeur BOCHNER a employé à plusieurs reprises la détermination d'intégrale de Cauchy pour "intégrale de Green". Or les intégrales employées par Monsieur BOCHNER sont des intégrales de bord, étendues à des variétés de dimension n-1, dans C^n. Il est bien connu, que les intégrales de Cauchy et leurs diverses généralisations utilisées dans la théorie des fonctions de plusieurs variables complexes sont en général des intégrales étendues à des variétés à n dimensions réelles. D'où provient la différence ?

Monsieur BOCHNER répond qu'il peut y avoir réduction de la dimension des variétés sur lesquelles se calcule les intégrales, lorsque ces variétés sont suffisamment régulières.

Monsieur FANTAPPIE fait observer que, parmi les opérateurs L permutables avec les dérivations il y a aussi des opérateurs non locaux, exemple les translations. Il signale aussi que parmi les équations aux dérivées partielles qui ne sont pas à coefficients constants, une classe particulièrement remarquable pour la physique, est celle des équations invariantes par rapport à un groupe, ce groupe permet de les intégrer dans bien des cas ; notamment si le groupe est compact, on peut résoudre le problème des valeurs propres et des fonctions propres.

STURM-LIOUVILLE AND HEAT EQUATIONS WHOSE EIGEN-FUNCTIONS ARE ULTRASPHERICAL POLYNOMIALS OR ASSOCIATED BESSEL FUNCTIONS

BY

S. Bochner

Princeton University

1. Introduction.

1.1. This paper intends to be "analysis" and not "probability" but the problem is stochastic in origin, and we will use stochastic parlance for describing it now, and afterwards occasionally.

All heat equations whose boundary conditions are not too singular have this in common that their Green's functions are everywhere non-negative, and this fact is the very basis for interpreting propagation of heat as a random diffusion of particles, alternately.

In the simplest classical instances of the heat equation, the Green's kernel is not only just positive but possesses a feature of positivity which is much more penetrating than that. The effect of this is that not only the "ordinary" (Gauss-) Brownian process but also all the other infinitely subdivisible homogeneous stochastic processes are associable with it. Now, we have shown ([1], Chapter IV) that a certain subclass of the other homogeneous processes (—all symmetric stable processes among them—) exist automatically whenever a Gaussian process is definable and we have termed them "subordinate" to the Gaussian therefore. Now, the (symmetric) Poisson process is not so "subordinate" to the Gaussian, and is in fact in a certain sense even "supraordinate" to it; and, as far as is known to us, the Poisson process does not have analogues for heat equations in general.

We have found however, that a Poisson process exists for the two equations

$$(1.1.1)\qquad (1-x^2)\frac{\partial^2 f}{\partial x^2}-(2\gamma+1)x\frac{\partial f}{\partial x}=\frac{\partial f}{\partial t}, \qquad -1\leqq x\leqq 1,$$

$$(1.1.2)\qquad \frac{\partial^2 f}{\partial x^2}+\frac{2\gamma}{x}\frac{\partial f}{\partial x}=\frac{\partial f}{\partial t} \qquad 0\leqq x<\infty,$$

for *arbitrary real* $\gamma\geqq 0$, for a rather natural boundary condition, and this is the topic we will be dealing with. Equation (1.1.1) involves ultra-spherical polynomials, and although previously introduced in [2] will be treated anew. Equation (1.1.2) involves Bessel functions, and although it can be viewed and even analyzed as a limiting case of (1.1.1) will be treated on its own. Also, we will frequently refer to (1.1.1) as the "*discrete*" case, and to (1.1.2) as the *continuous* case.

We emphasize that our equations are *one*-dimensional with respect to the space variable x but have certain "singularities" at the endpoints of the x-interval over which they are envisaged. For the *particular* values of the parameter

$$\gamma=\frac{k-1}{2}, \qquad k=1,2,3,\cdots, \tag{1.1.3}$$

(and only for those) our equations are obtainable from certain k-dimensional ones which although higher-dimensional are less "singular" nevertheless. The latter equations also admit Poisson processes, and for general reasons in fact; and to that compare the comment to Theorem 3.6.2 of the present paper and also the paper [3] previously instigated by us.

Equation (1.1.2) brings to mind the equation

$$\frac{\partial^2 f}{\partial x^2}+\frac{2\gamma}{x}\frac{\partial f}{\partial x}=\Delta_y f \tag{1.1.4}$$

which is being searchingly analyzed by the Maryland school (A. Weinstein, J. B. Diaz, M. H. Martin, G. S. S. Ludford, H. F. Weinberger, E. K. Blum). However, in (1.1.4) it is the variable x which corresponds to the time variable in (1.1.2), and there seems to be no overlap in the topics thus far.

2. The Sturm-Liouville Equation.

2.1. In what follows we will consider on a certain space $R:(x)$ a certain vector family of (usually real-valued) functions $\{f(x)\}$ and with each $f(x)$ there will be associated a certain two-point function $f(x;y)$ on the product space $(x,y)\in(R,R)$.

Definition 2.1.1 We will say that the association $f(x)\rightarrow f(x;y)$ has property P if whenever for a function we have

$$f(x)\geqq 0 \quad \text{for all } x\in R \tag{2.1.1}$$

then for the associated function we also have

$$f(x;y)\geqq 0 \quad \text{for all } (x,y)\in(R,R). \tag{2.1.2}$$

2.2. For fixed $\gamma \geqq 0$ we are introducing the eigenvalues

$$\lambda_n = n(n+2\gamma), \qquad n=0,1,2,\cdots. \tag{2.2.1}$$

In one technical detail the parameter $\gamma=0$ is different from the parameters $\gamma>0$, and will be usually singled out therefore.

The function $P_n^{(\gamma)}(x)$, $n=0,1,2,\cdots$ shall be the known polynomial solution of

$$(1-x^2)P''(x)-(2\gamma+1)xP'(x) = -\lambda_n P(x), \quad n=0,1,2,\cdots, \tag{2.2.2}$$

and for our purpose we are standardizing it by the normalization

$$P_n^{(\gamma)}(1)=1 \tag{2.2.3}$$

which is not the customary one (compare [4], p. 80).

Frequently, though not always, we will omit the parameter γ as superscript or subscript, as we have already done in (2.2.2), but it must be emphasized that omitting the parameter does *not* mean that its value has been specialized (to $\gamma=\frac{1}{2}$, say).

We will now list some properties of our polynomials, but the decisive one will be stated later. The normalization (2.2.3) implies

$$|P_n(x)| \leqq 1, \qquad -1 \leqq x \leqq 1 \tag{2.2.4}$$

and also

$$P_n(-1) = (-1)^n. \tag{2.2.5}$$

If we put for $\gamma>0$,

$$\rho_n = \frac{\gamma+n}{\gamma}\,\frac{\Gamma(n+2\gamma)}{n!\,\Gamma(2\gamma)},$$

then we have the "generating" relation

$$\frac{1-w^2}{(1-2wx+w^2)^{2\gamma+1}} = \sum_{n=0}^{\infty} \rho_n w^n P_n(x), \qquad 0 \leqq w < 1, \tag{2.2.6}$$

and we note that $\rho_n = 0(n^{2\gamma})$.

For the weight function

$$\nu(x) = \frac{1}{c_{\gamma+\frac{1}{2}}}\int_{-1}^{x}(1-\xi^2)^{\gamma-\frac{1}{2}}\,d\xi, \qquad \nu(+1)=1 \tag{2.2.7}$$

$$c_\delta = \int_{-1}^{1}(1-\xi^2)^{\delta-1}\,d\xi$$

we have the orthogonality relation

$$\int_{-1}^{1} P_m(x)P_n(x)\,d\nu(x) = \frac{\delta_{mn}}{\rho_m}. \tag{2.2.8}$$

For functions with finite L_2-norm

(2.2.9) $$\| f \| = \left(\int_{-1}^{1} f(x)^2 \, d\nu(x) \right)^{\frac{1}{2}}$$

we have the expansion

(2.2.10) $$f(x) \sim \sum_{n=0}^{\infty} \rho_n a_n P_n(x)$$

(2.2.11) $$a_n = \int_{-1}^{1} f(x) P_n(x) \, d\nu(x)$$

and the Parseval relation

(2.2.12) $$\| f \|^2 = \sum_{0}^{\infty} \rho_n a_n^2.$$

More generally, if we take another such expansion

(2.2.13) $$g(x) \sim \sum \rho_n b_n P_n(x),$$

then

(2.2.14) $$\int_{-1}^{1} f(x) g(x) \, d\nu(x) = \sum \rho_n a_n b_n.$$

The series (2.2.10) converges in square mean, and if it happens to converge a.e. then the sum function is $g(x)$ a.e. In particular, if

(2.2.15) $$\sum_{0}^{\infty} \rho_n \, | a_n | < \infty$$

then, due to (2.2.4), we have

(2.2.16) $$f(x) = \sum_{0}^{\infty} \rho_n a_n P_n(x)$$

a.e.

Next, (2.2.2) implies

(2.2.17) $$(2\gamma + 1) P_n'(1) = \lambda_n P_n(1).$$

and therefore $P_n'(1) > 0$. On the other hand, the function

(2.2.18) $$\frac{dP_n^{(\gamma)}(x)}{dx}$$

satisfies again (2.2.2) but for $(\gamma + 1, n - 1)$ instead of (γ, n). By the mean value theorem we have

$$\frac{1 - P_n^{(\gamma)}(x)}{1 - x} = \frac{dP_n^{(\gamma)}(\xi)}{d\xi}, \qquad x \leqq \xi \leqq 1,$$

and if we now apply (2.2.3) and (2.2.4) to the function (2.2.18) and recall (2.2.17) we obtain

$$(2.2.19) \qquad 0 \leqq \frac{1-P_n^{(\gamma)}(x)}{1-x} \leqq \lim_{x\to 1} \frac{1-P_n^{(\gamma)}(x)}{1-x} = \frac{\lambda_n}{2\gamma+1}.$$

2.3. If in (2.2.2) we replace $P(\cos x/n)$ by $T(x)$, divide the equation by n^2 and let $n \to \infty$ we formally obtain the equation

$$(2.3.1) \qquad T'' + \frac{2\gamma}{x} T' = -T, \qquad 0 \leqq x < \infty.$$

Furthermore it is known ([4], p. 186) that, in our normalization, we have

$$\lim_{n\to\infty} P_n^{(\gamma)}(\cos x/n) = T^{(\gamma)}(x) \equiv T(x),$$

where

$$(2.3.2) \qquad T(x) \equiv 2^{\gamma-\frac{1}{2}}\Gamma(\gamma+\tfrac{1}{2}) x^{\frac{1}{2}-\gamma} J_{\gamma-\frac{1}{2}}(x),$$

$$(2.3.3) \qquad T(0) = 1, \qquad |T(x)| \leqq 1,$$

and $J_\nu(x)$ is the Bessel function. We might therefore expect that (2.3.2) is a solution of (2.3.1) and the classical differential equation for $J_\nu(x)$ does indeed so imply.

The full analogue to (2.2.2) is the equation

$$(2.3.4) \qquad T'' + \frac{2\gamma}{x} T' = -\alpha^2 T$$

with the continuous eigenvalues $0 \leqq \alpha < \infty$. The corresponding eigenfunction, if denoted by $T_\alpha^{(\gamma)}(x)$, is simply the function $T^{(\gamma)}(\alpha x)$ so that the specific symbol $T_\alpha^{(\gamma)}(x)$ can be dispensed with. The analogue to (2.2.19) is

$$(2.3.5) \qquad 0 \leqq \frac{1-T^{(\gamma)}(x)}{x^2} \leqq \lim_{x\downarrow 0} \frac{1-T^{(\gamma)}(x)}{x^2} = \frac{1}{2\gamma+1}.$$

The "discrete" coefficient ρ_n is to be replaced by the derivative of the function

$$(2.3.6) \qquad \mu(\alpha) = \frac{\alpha^{2\gamma+1}}{2^{\gamma+\frac{1}{2}}\Gamma(\gamma+\frac{3}{2})}, \qquad 0 \leqq \alpha < \infty,$$

and the latter occurs in the formulas

$$(2.3.7) \qquad f(x) \sim \int_0^\infty \phi(\alpha) T(\alpha x)\, d\mu(\alpha)$$

$$(2.3.8) \qquad \phi_f(\alpha) \equiv \phi(\alpha) \sim \int_0^\infty f(x) T(x\alpha)\, d\mu(x)$$

$$\int_0^\infty f(x)^2 d\mu(x) = \int_0^\infty \phi(\alpha)^2 d\mu(\alpha), \tag{2.3.9}$$

which are analogues to (2.2.10), (2.2.11) and (2.2.12). It is readily seen that they are modifications of the Hankel inversion formulas

$$f_0(x) \sim \int_0^\infty \phi_0(\alpha) J_{\gamma-\frac{1}{2}}(\alpha x)(\alpha x)^{\frac{1}{2}} d\alpha$$

$$\phi_0(\alpha) \sim \int_0^\infty f_0(x) J_{\gamma-\frac{1}{2}}(x\alpha)(x\alpha)^{\frac{1}{2}} dx$$

and the Parseval relation

$$\int_0^\infty f_0(x)^2 dx = \int_0^\infty \phi_0(\alpha)^2 d\alpha,$$

and we can therefore state as follows ([5], p. 178). If we introduce the L_2-norm

$$\| f \| = \left(\int_0^\infty f(x)^2 d\mu(x)\right)^{\frac{1}{2}}, \tag{2.3.10}$$

then for any $f \varepsilon L_2$ the integral (2.3.8) exists as

$$\underset{A\to\infty}{\text{limit in mean}} \int_0^A f(x) T(x\alpha) d\mu(x),$$

say. The limit function $\phi(\alpha)$ is again in L_2 and (2.3.7) and (2.3.9) hold. If, by chance, we have

$$\int_0^\infty |\phi(\alpha)| d\mu(\alpha) < \infty, \tag{2.3.11}$$

then, due to (2.3.3), the integral (2.3.7) can be evaluated in the ordinary sense. Also, if for another pair of functions $g(x)$, $\psi(\alpha)$ in L_2 we have

$$g(x) \sim \int_0^\infty \psi(\alpha) T(x\alpha) a\mu(\alpha) \tag{2.3.12}$$

then for any $f \varepsilon L_2$, we have

$$\int_0^\infty f(x) g(x) d\mu(x) = \int_0^\infty \phi(\alpha)\psi(\alpha) d\mu(\alpha), \tag{2.3.13}$$

and this generalizes (2.3.9) again.

We are listing the formula

$$\int_0^\infty e^{-t\alpha} T(\alpha x) d\mu(\alpha) = \frac{t 2^{\gamma+\frac{1}{2}}\Gamma(\gamma+1)}{(t^2+x^2)^{\gamma+1}\pi^{\frac{1}{2}}}, \qquad t > 0,$$

([6], p. 386, formula (6)) as an analogue to (2.2.6), if one be wanted, and the (self-inversive) pair of formulas

$$(2.3.14) \qquad (2t)^{-(\gamma+\frac{1}{2})} \int_0^\infty T(\alpha x) e^{-x^2/4t} \, d\mu(x) = e^{-t\alpha^2}$$

$$\int_0^\infty T(x\alpha) e^{-t\alpha^2} \, d\mu(\alpha) = (2t)^{-(\gamma+\frac{1}{2})} e^{-x^2/4t}$$

which are indigenous to the continuous case only.

2.4. THEOREM 2.4.1. *For $\gamma > 0$ we have*

$$(2.4.1) \quad 1/c_\gamma \int_0^\pi P_n(xy + (1-x^2)^{\frac{1}{2}}(1-y^2)^{\frac{1}{2}} \cos\theta) \sin^{2\gamma-1}\theta d\theta = P_n(x)P_n(y)$$

and

$$(2.4.2) \quad 1/c_\gamma \int_0^\pi T((x^2 + y^2 - 2xy\cos\theta)^{\frac{1}{2}}) \sin^{2\gamma-1}\theta d\theta = T(x)T(y).$$

For $\gamma = 0$ we have $T(x) = \cos x$, $P_n(\cos x) = \cos nx$ and the substitute for these relations is

$$(2.4.3) \qquad \cos x \cos y = \tfrac{1}{2}[\cos(x+y) + \cos(x-y)].$$

These are the properties crucial for us. Formula (2.4.1) is due to Gegenbauer, and is stated in [6], p. 369, second formula from the bottom of the page. Formula (2.4.2) is due to Sonine, and is stated in [6], as formula (17) on page 367, (in which put $m = 0$). The emphasis in these formulas lies for us *only* in the fact that the averaging weight in them is *non-negative*, whether it be the continuous density $(\sin\theta)^{2\gamma-1}$ in (2.4.1), (2.4.2) or the discontinuous weight in (2.4.3). Theorem 2.4.1 obviously implies the following one.

THEOREM 2.4.2. *If, for* (2.2.15), *we associate with the function*

$$(2.4.4) \qquad f(x) = \sum_0^\infty \rho_n a_n P_n(x)$$

the function

$$(2.4.5) \qquad f(x;y) = \sum_0^\infty \rho_n a_n P_n(x) P_n(y),$$

or, for (2.3.11), *with the function*

$$(2.4.6) \qquad f(x) = \int_0^\infty \phi(\alpha) T(\alpha x) d\mu(\alpha)$$

the function

$$(2.4.7) \qquad f(x;y) = \int_0^\infty \phi(\alpha) T(\alpha x) T(\alpha y) d\mu(\alpha),$$

then we have property P in the sense of definition 2.1.1.

The transition from $f(x)$ to $f(x;y)$ as fixed in the theorem is a *displacement operator* in the sense of J. Delsarte [7], [8] and B. M. Levitan [9], but neither Delsarte nor Levitan have been concerned with "positivity" as we are.

2.5. On a space $R:(x)$ we take a sequence of functions $\{\chi_n(x)\}$, and a vector space of functions presentable by expansions

$$f(x) \sim \sum a_n \chi_n(x). \tag{2.5.1}$$

For a short while we are admitting complex values for functions and coefficients, defining the two-point functions as

$$f(x;y) \sim \sum a_n \chi_n(x)\, \overline{\chi_n(y)}, \tag{2.5.2}$$

but we are defining property P literally as in definition 2.1.1.

For instance, if $\{\chi_n(x)\}$ is a sequence of exponentials $\{e^{imx}\}$, say $m = 0, \pm 1, \pm 2, \cdots$, then

$$f(x;y) = f(x-y), \tag{2.5.3}$$

and property P holds. In general we can state as follows.

THEOREM 2.5.1. *If on a group space $R:(x)$ with multiplication xy we take a sequence of one-dimensional bounded representations*

$$\chi_0(x), \chi_1(x), \chi_2(x), \cdots, \tag{2.5.4}$$

and define $f(x)$ and $f(x;y)$ by (2.5.1) *and* (2.5.2) *then property P holds.*

More generally, any $\chi_n(x)$ may be the "adjusted" trace

$$\chi(x) = (1/r) \sum_{\alpha=1}^{r} \phi_{\alpha\alpha}(x) \tag{2.5.5}$$

of an irreducible finite-dimensional bounded representation

$$\{\phi_{\alpha\beta}\}, \alpha, \beta = 1, \cdots, r, \tag{2.5.6}$$

(for equal or non-equal r).

In fact, we have formally

$$f(x;y) = M_s f(xsy^{-1}s^{-1}) \tag{2.5.7}$$

where M_s is the almost periodic average, and our theorem applies to such functions (2.5.1) for which this formal relation (2.5.7) is a rigorous one.

If all representations used are one-dimensional then

$$M_s f(xsy^{-1}s^{-1}) = M_s f(xy^{-1}ss^{-1}) = M_s f(xy^{-1}) = f(xy^{-1}),$$

and for a commutative group this gives indeed (2.5.3).

THEOREM 2.5.2. *Furthermore, if we take the real parts*

$$P_n(x)=\tfrac{1}{2}\{\chi_n(x)+\overline{\chi_n(x)}\}=\tfrac{1}{2}\{\chi_n(x)+\chi_n(x^{-1})\}\tag{2.5.8}$$

of the function (2.5.4) *then property* P *also holds for the associated expansion*

$$f(x)\sim\sum a_nP_n(x),\qquad f(x;y)\sim\sum a_nP_n(x)P_n(y).\tag{2.5.9}$$

In fact, relation (2.4.3) now generalizes to

$$f(x;y)=\tfrac{1}{2}M_s\{f(xsy^{-1}s^{-1})+f(xsys^{-1})\}.\tag{2.5.10}$$

We note that ultraspherical polynomials $P_n^{(\gamma)}(x)$ do not subsume under Theorem 2.5.2, or an obvious adaptation thereof, not even in the case (1.1.3)—except for $k=1$—even though in this case they are zonal functions on the sphere $\xi_1^2+\cdots+\xi_{k+1}^2=1$, as well known.

2.6. Returning to an arbitrary space $R:(x)$, we consider a sequence of real functions $P_0(x),P_1(x),P_2(x),\cdots$ which is such that for the vector families of finite linear sums

$$f(x)=\sum_{n=0}^{N}a_nP_n(x),\qquad f(x;y)=\sum_{n=0}^{N}a_nP_n(x)P_n(y)$$

property P holds. If also

$$P_0(x)=1,\tag{2.6.1}$$

then it follows easily that

$$a\leqq f(x)\leqq b\quad\text{implies}\quad a\leqq f(x;y)\leqq b\tag{2.6.2}$$

for any finite real numbers $\alpha<b$. If now also each $P_n(x)$ is bounded and if we put

$$\rho_n=\inf_x P_n(x),\qquad \sigma_n=\sup_x P_n(x)$$

then (2.6.2) implies

$$\rho_n\leqq P_n(x)P_n(y)\leqq\sigma_n.$$

Therefore $\rho_n^2\leqq\sigma_n$, $\sigma_n^2\leqq\sigma_n$, and this implies $-1\leqq-\rho_n\leqq\sigma_n\leqq1$, that is,

$$|P_n(x)|\leqq 1.$$

Thus we see that the estimate (2.2.4) for ultraspherical polynomials "follows" from property P, once the latter has been established. Another "motivation" for the estimate (2.2.4) has been given by I. Schur and G. Szegö (see [4], p. 159).

2.7. Next, we are going to make a statement on our two-point functions (2.4.5) and (2.4.7) in which property P is not involved.

In what follows the space R will be either the interval $-1 \leqq x \leqq 1$ or the half line $0 \leqq x < \infty$ and the volume element "dv" on R will be the measure element "$d\nu$" or "$d\mu$" respectively.

THEOREM 2.7.1. *In both the discontinuous and the continuous case the expression*

$$\int_R f(x;y)g(y)dv \tag{2.7.1}$$

is a "convolution" in the following sense.

If we are given a function (2.2.16) *in* L_2 *subject to condition* (2.2.15) *and another function* (2.2.13) *in* L_2 *then* (2.7.1) *has the value*

$$\sum_{n=0}^{\infty} \rho_n a_n b_n P_n(x) \tag{2.7.2}$$

If we are given a function (2.3.7) *in* L_2 *subject to condition* (2.3.11) *and another function* (2.3.12) *in* L_2 *then* (2.7.1) *has the value*

$$\int_0^{\infty} \phi(\alpha)\psi(\alpha)T(\alpha x)d\mu(\alpha). \tag{2.7.3}$$

In fact, in the discrete case, for fixed x we may view $f(x;y)$ as a function in L_2. Its Fourier coefficients are $a_nP_n(x)$ and the value (2.7.2) is then the right side in (2.2.14). In the continuous case, we deduce, in a similar manner, the value (2.7.3) from (2.3.13).

3. The Heat and Diffusion Equation.

3.1. If we apply the operator

$$\Delta_x f = (1-x^2)\frac{d^2f}{dx^2} - (2\gamma+1)x\frac{dx}{df} \qquad -1 \leqq x \leqq 1 \tag{3.1.1}$$

to a function (2.2.10) then we obtain formally

$$\Delta f \sim -\sum_{0}^{\infty} \lambda_n \rho_n a_n P_n(x), \tag{3.1.2}$$

and subject to convergence assumptions this holds rigorously. Now if a system of coefficients $\{a_n\}$ is such that we have not only

$$\sum \rho_n a_n^2 < \infty \tag{3.1.3}$$

but also

$$\sum \lambda_n^2 \rho_n a_n^2 < \infty \tag{3.1.4}$$

then (3.1.2) is again an expansion of a function in L_2 and we may now define an (unbounded) self-adjoint operator Δf on the real Hilbert space as follows. It applies to precisely those functions $f(x)$ in L_2 for which we have not only (3.1.3) but also (3.1.4) and Δf is *defined* to be the function in L_2 whose expansion is (3.1.2).

In a similar fashion if we apply the operator

$$\Delta_x f = \frac{d^2 f}{dx^2} + \frac{2\gamma}{x}\frac{df}{dx}, \qquad 0 \leqq x < \infty \tag{3.1.5}$$

to an expansion (2.3.7) then we formally obtain

$$\Delta f \sim -\int_0^\infty \alpha^2 \phi(\alpha) T(\alpha x)\, d\mu(\alpha), \tag{3.1.6}$$

and here again we define the operator Δf to apply to those functions $f(x) \in L_2$ for which we have

$$\int_0^\infty \phi(\alpha)^2 d\mu(\alpha) < \infty \tag{3.1.7}$$

and

$$\int_0^\infty \alpha^4 \phi(\alpha)^2 d\mu(\alpha) < \infty \tag{3.1.8}$$

simultaneously. We wish to point out that there is a minor difference between the two cases which does not affect results materially and it is this. Relation (3.1.3) implies (3.1.4), but this is not so for (3.1.7) and (3.1.8) since we may have $\int_0^\infty \alpha^4 \phi(\alpha)^2\, d\mu(\alpha) < \infty$ and $\int_0^\infty \phi(\alpha)^2\, d\mu(\alpha) = \infty$.

3.2. We are now denoting either of the two equations (1.1.1) or (1.1.2) by

$$\Delta_x f = \frac{\partial f}{\partial t} \qquad 0 < t < \infty \tag{3.2.1}$$

and we are considering solutions of the following kind only. The solution $f(x,t)$ shall be an object $f_t(x)$ which for each $t > 0$ is an element of L_2 (with regard to x) and for each fixed t the operator $\Delta_x f$ shall apply to it in the manner introduced. Also the Banach valued function $f_t(.)$ shall have a derivative in norm

$$\lim_{h \to 0} \left\| \frac{f_{t+h} - f_t}{h} - \frac{\partial f_t}{\partial t} \right\| = 0 \tag{3.2.2}$$

and (3.2.1) shall be a relation between the Banach valued functions resulting. (Compare to this approach [10], pp. 132-138.)

In the discrete case we therefore have

$$f_t(x) \sim -\sum_0^\infty \rho_n a_n(t) P_n(x) \tag{3.2.3}$$

$$\sum_0^\infty \rho_n \lambda_n^2 a_n(t)^2 < \infty \tag{3.2.4}$$

$$\Delta_x f_t(x) \sim \sum \rho_n \lambda_n a_n(t) P_n(x), \tag{3.2.5}$$

and (3.2.2) implies that for each n, $a_n(t)$ has a derivative in the ordinary sense and that

$$\frac{\partial f_t}{\partial t} \sim -\sum_0^\infty \rho_n a_n'(t) P_n(x), \tag{3.2.6}$$

where

$$\sum \rho_n (a_n'(t))^2 < \infty. \tag{3.2.7}$$

Finally, by uniqueness of Fourier expansions, (3.2.1) is equivalent with

$$-\lambda_n a_n(t) = a_n'(t), \qquad n = 0, 1, 2, \cdots \tag{3.2.8}$$

and this is equivalent with

$$a_n(t) = a_n e^{-\lambda_n t}, \qquad n = 0, 1, 2, \cdots \tag{3.2.9}$$

therefore. Also

$$\sum_0^\infty \rho_n a_n^2 e^{-2\lambda_n t} < \infty, \qquad 0 < t < \infty, \tag{3.2.10}$$

and it is easily verified that $f_t(x)$ is a solution of (3.2.1) in our sense if and only if there is a system of coefficients $\{a_n\}$, subject to restriction (3.2.10), for which

$$f_t(x) = \sum_0^\infty \rho_n a_n e^{-\lambda_n t} P_n(x). \tag{3.2.11}$$

Also $f_t(x)$ has a limit in norm as $t \to 0$, and the limit element has Fourier coefficients $\{a_n^0\}$, if and only if the expansion (3.2.11) is formally convergent as $t \to 0$ towards a Fourier series, that is if we have (3.1.3), and $a_n^0 = a_n$ then.

In the continuous case the analogues to (3.2.3), (3.2.4) and (3.2.5) are

$$f_t(x) \sim \int_0^\infty \phi_t(\alpha) T(\alpha x) \, d\mu(\alpha) \tag{3.2.12}$$

$$\int_0^\infty \alpha^4 \phi_t(\alpha)^2 \, d\mu(\alpha) < \infty \tag{3.2.13}$$

(3.2.14) $$\Delta_x f_t(x) \sim -\int_0^\infty \alpha^2 \phi_t(\alpha) T(\alpha x) d\mu(\alpha),$$

and $\phi_t(\alpha)$ is itself a Banach valued function now, whose values are elements of L_2 themselves. Thus, for each t, $\phi_t(\alpha)$ is determined up to an α-set of measure 0 only. Due to the Parseval relation, requirement (3.2.2) is equivalent with the requirement that $\phi_t(\alpha)$ shall have a derivative in norm with respect to t, so that

(3.2.15) $$\frac{\partial f_t}{\partial t} \sim \int_0^\infty \frac{\partial \phi_t(\alpha)}{\partial t} T(\alpha x) d\mu(\alpha),$$

and equation (3.2.1) itself now becomes the equation

(3.2.16) $$-\alpha^2 \phi_t(\alpha) = \frac{\partial \phi_t(\alpha)}{\partial t}$$

in which both sides are L_2-elements for each t. The exceptional α-sets of measure 0 may be such that their union for all t will not be a two-dimensional set of measure 0 in the (α, t)-plane at first. But, by a general reasoning, [11], they can be so adjusted that $\phi_t(\alpha)$ will become a measurable function in the (α, t) plane, and that for almost all α, (3.2.16) will become a "true" equality for almost all t, in the sense that the relation

(3.2.17) $$\phi_t(\alpha) = \phi(\alpha) e^{-\alpha^2 t}$$

will ensue. We have the following statement therefore. $f_t(x)$ is a solution of (3.2.1) in our sense if and only if we have

$$f_t(x) = \int_0^\infty \phi(\alpha)^2 e^{-\alpha^2 t} T(\alpha x) d\mu(\alpha)$$

where $\phi(\alpha)$ is measurable and

$$\int_0^\infty \phi(\alpha)^2 e^{-2\alpha^2 t} d\mu(\alpha) < \infty, \qquad 0 < t < \infty.$$

Also, $f_t(x)$ has a limit in L_2-norm, as $t \to 0$, and the limit if $f(x)$, if and only if $\phi(\alpha)$ is a Fourier transform, that is if (3.1.7) holds, and it is the transform of the limit element then.

3.3. THEOREM 3.3.1. *If we introduce the functions*

(3.3.1) $$K(x, y; t) = \sum_0^\infty \rho_n e^{-t\lambda_n} P_n(x) P_n(y)$$

(3.3.2) $$K(x, y; t) = \int_0^\infty e^{-t\alpha^2} T(\alpha x) T(\alpha y) d\mu(\alpha)$$

$0 < t < \infty$, *then, in either case, for any* $f \in L_2$ *the integral*

$$f_t(x) = \int_R K(x, y; t) f(y) dv(y) \tag{3.3.3}$$

is the solution of (3.2.1) *with boundary element* f, *as described.*

This follows directly from Theorem 2.7.1. Thus $K(x, y; t)$ is a Green's kernel of (3.2.1) and the theorem establishes a precise meaning within which it so is. We note that in the continuous case the kernel has the value

$$K(x, y; t) = (2t)^{-(\gamma+\frac{1}{2})} \exp\left(-\frac{x^2 + y^2}{4t}\right) T\left(\frac{ixy}{2t}\right) \tag{3.3.4}$$

(see [6], formula (1) on page 395), but in the discrete case no such simple formula is known to us.

THEOREM 3.3.2. *The kernel has the following properties.*

(3.3.5) (i) $K(x, y; t) \geqq 0$

(3.3.6) (ii) $\int_R K(x, y; t) dv(y) = 1,$

(3.3.7) (iii) $\int_R K(x, y; r) K(y, z; s) dv(y) = K(x, z; r + s)$

(iv) *If* $f \in L_2$ *and if we introduce the "transform"* (3.3.3), *then* $\| f_t \| \leqq \| f \|$, *and* $f_t \to f$ *in* L_2*-norm.*

Proof. Property (iv) is obvious. Property (iii) follows easily from Theorem 2.7.1. Property (ii) in the discrete case results from term-by-term integration of (3.3.1); and in the continuous case, if we use the value (3.3.4) it follows from a ready made formula again, see [6], formula (4) on page 394. Property (i) in the continuous case is made manifest by the expression (3.3.4), and in the discrete case we are going to deduce it from property P systematically.

3.4. We consider the normed ring of bounded sequences

$$C: (c_0, c_1, c_2, \cdots) \tag{3.4.1}$$

with norm

$$\| C \| = \sup_n | c_n | \tag{3.4.2}$$

and operations

$$\alpha C' + \beta C'' = \{\alpha c_n' + \beta c_n''\} \tag{3.4.3}$$

$$C'C'' = \{c_n' c_n''\}, \tag{3.4.4}$$

and we recall that a sequence $C^k = \{c_n{}^k\}$, $k = 1, 2, \cdots$ is weakly convergent to C if $\| C^k \| \leqq M$ and $c_n = \lim_{k \to \infty} c_n{}^k$.

For fixed γ we consider the vector space of all absolutely convergent expansions

$$f(x) = \sum_0^\infty \rho_n a_n P_n(x) \tag{3.4.5}$$

and for each C we associate with it the expansion

$$Cf(x) = \sum_0^\infty \rho_n c_n a_n P_n(x). \tag{3.4.6}$$

Now, we call the sequence (3.4.1) *positive-definite* if this transformation is positive, that is if

$$f(x) \geqq 0 \text{ implies } Cf(x) \geqq 0. \tag{3.4.7}$$

The set of all such elements C will be denoted by P. Obviously P is a semi-ring ($\alpha \geqq 0, \beta \geqq 0$) and is weakly closed. If we consider the expansion

$$e^{tc_n} = \sum_0^\infty \frac{t^k}{k!} c_n{}^k$$

the following conclusion ensues.

THEOREM 3.4.1. *If $\{c_n\} \epsilon P$ then also $\{e^{tc_n}\} \epsilon P$ for any $t \geqq 0$.*

On the other hand Theorem 2.4.2 implies as follows

THEOREM 3.4.2. *For every y in $-1 \leqq y \leqq 1$, the sequence $\{c_n = P_n(y)\}$ belongs to P.*

This will lead to the following statement.

THEOREM 3.4.3. *In the discrete case, for any $\gamma > 0$, if $\sigma \geqq 0$ and if $G(B)$ is a countably additive non-negative set function on the Borel sets $\{B\}$ of the half-closed interval $-1 \leqq y < 1$, and if we put*

$$\mu_n = \sigma \lambda_n + \int_{-1}^{1-0} \frac{1 - P_n(y)}{1 - y} dG(y), \qquad n = 0, 1, 2, \cdots \tag{3.4.8}$$

then the sequence

$$C: \ c_n = e^{-t\mu_n} \tag{3.4.9}$$

belongs to P for any $t > 0$.

Special cases are

$$c_n = e^{-t\lambda_n} \tag{3.4.10}$$

and

$$c_n = e^{-t(1 - P_n(y))}. \tag{3.4.11}$$

Proof. By Theorems 3.4.1 and 3.4.2 our assertion holds for $\mu_n = P_n(y)$, and, if we multiply c_n by e^{-t}, also for (3.4.11). If we now replace t by $t(2\gamma+1)(1-y)^{-1}$ then we obtain

$$\mu_n = \frac{1-P_n(y)}{1-y}, \tag{3.4.12}$$

and if we let $y\uparrow 1$ and recall (2.2.19) then we obtain (3.4.10). On the other hand, in (3.4.12) we may replace t by $t\cdot dG(y)$, and, since P is "multiplicative" we obtain any finite sum

$$\sum_{p=1}^{q}\frac{1-P_n(y_p)}{1-y_p}\,dG(y_p).$$

But the integral in (3.4.8) is the limit of such approximating sums, whence the conclusion.

By formula (2.2.6) the sum $\sum \rho_n w^n P_n(x)$ is $\geqq 0$ for $0 < w < 1$. By Theorems 3.4.1 and 3.4.3 we have therefore

$$\sum_{n=0}^{\infty} \rho_n w^n e^{-t\lambda_n} P_n(x) P_n(y) \geqq 0$$

for $0 \leqq w < 1$. If we let $w \to 1$, the series converges to the series (3.3.1), and this completes the proof of Theorem 3.3.2, finally.

3.5. Because of the closure properties of P, Theorem 3.4.1 implies the following one.

THEOREM 3.5.1. *If $H(y)$ is bounded monotonely increasing in $-1 \leqq y \leqq 1$ then the sequence*

$$c_n = \int_{-1}^{1} P_n(y)\,dH(y) \tag{3.5.1}$$

is in P.

Also, such a representation is unique, as known from the theory of the moment problem.

THEOREM 3.5.2. (1) *Conversely any element in P has such a representation. Also, if*

$$\sum \rho_n |c_n| < \infty \tag{3.5.2}$$

then

$$H(y) = \int_{-1}^{y} h(y)\,d\nu(y), \qquad h(y) \geqq 0.$$

Proof. If (3.5.2) holds then the function

(3.5.3) $$h(y)=\sum_0^\infty \rho_n c_n P_n(y), \qquad c_n=\int_{-1}^1 P_n(y)h(y)d\nu(y),$$

has the following property. If we are given a function (2.2.16) subject to (2.2.15) then we have $\int_{-1}^1 f(y)h(y)d\nu(y)=\sum_0^\infty \rho_n a_n c_n=\sum_0^\infty \rho_n a_n c_n P_n(0)$, and thus, by definition of P, $f(x)\geqq 0$ implies

$$\int_{-1}^1 f(y)h(y)d\nu(y)\geqq 0.$$

However the functions (2.2.16) subject to (2.2.15) include all ordinary polynomials, and thus $h(y)\geqq 0$ as claimed. For general $\{c_n\}\epsilon P$ we first introduce the dampened sequence

$$c_n(t)=c_n e^{-t\lambda_n}$$

$t>0$, which by Theorem 3.4.3 is in P again. However, it is subject to (3.5.2), and thus, by what we have already proven, we have

$$c_n e^{-t\lambda_n}=\int_{-1}^1 P_n(y)d_yH(y;t).$$

Now, let $t\to 0$, and apply Helly's theorem.

In [2] we established a more general version of Theorem 3.5.1, in that the boundedness of the sequence $\{c_n\}$ was not at all made part of the definition of a positive-definite sequence, and the requirement (3.4.7) was stipulated for polynomials only. Technically this is a rather non-trivial generalization, but only technically so, since (3.5.3) implies

(3.6.1) $$|c_n|\leqq c_0=\int_R dH(y),$$

once (3.5.3) has been arrived at, and in the present more stochastic discussion it is somewhat a side-issue, we think.

3.6. Turning to the continuous case, we now define as follows. A function $\psi(\alpha)$ in $0\leqq\alpha<\infty$ is positive definite, if $\psi(\alpha)$ is continuous and bounded, $|\psi(\alpha)|\leqq M$, and if

(3.6.2) $$\int_0^\infty \phi(\alpha)T(\alpha x)d\mu(\alpha)\geqq 0, \qquad \int_0^\infty |\phi(\alpha)|\,d\mu(\alpha)<\infty$$

implies

(3.6.3) $$\int_0^\infty \psi(\alpha)\phi(\alpha)T(\alpha x)d\mu(\alpha)\geqq 0.$$

Theorem 2.4.2 again implies that for each y in $0\leqq y<\infty$, $\psi(\alpha)\equiv \mathrm{T}(\alpha y)$

is in P, and (2.3.5) again implies that $e^{-t\alpha^2}$ is in P for each t. But this follows also directly from formula (2.3.14) as an application of the (easy) first half of the following theorem.

THEOREM 3.6.1. *A function is in P if it can be represented as an integral*

$$\psi(\alpha) = \int_0^\infty T(\alpha y) dH(y) \tag{3.6.4}$$

with a bounded non-decreasing function $H(y)$ in $0 \leqq y < \infty$.

Conversely, every element $\psi(\alpha)$ in P can be so represented.

The converse is a little harder to prove than in Theorem 3.5.2, but we will omit the details nevertheless. This "omission" does not create a "gap" in the following theorem and in later theorems because for these theorems it is admissible to *define* the class P by the integral representation (3.6.4) so as to make the "converse" immaterial. On the other hand we note that for validity of the converse as stated it is not even necessary to postulate the boundedness of $\psi(\alpha)$ to start with. Rather, it suffices to assume that $\psi(\alpha)$ is a continuous function for which (3.6.2) implies (3.6.3) whenever $\phi(\alpha)$ is 0 outside some finite interval.

The analogue to Theorem 3.4.3 is the following.

THEOREM 3.6.2. *If we put*

$$\chi(\alpha) = \sigma\alpha^2 + \int_{0+}^\infty \frac{1 - T(\alpha y)}{y^2} dG(y) \tag{3.6.5}$$

where $\sigma \geqq 0$ and $G(B) \geqq 0$, $\int_{0+}^\infty dG(y) < \infty$, then

$$e^{-t\chi(\alpha)} \tag{3.6.6}$$

belongs to P for each $t > 0$. Particular cases are

$$e^{-t\alpha^2} \tag{3.6.7}$$

and

$$e^{-t(1-T(\alpha y))} \tag{3.6.8}$$

in each $y \geqq 0$.

We also note that for $\gamma = 0$, that is $T(x) = \cos x$, a function $\psi(\alpha)$ is positive definite in the present sense if and only if the function $\psi_0(\alpha) = \psi(|\alpha|)$ in $-\infty < \alpha < \infty$ is positive-definite in the "classical" sense, that is

$$(3.6.9)\qquad \sum_{m,n=0}^{N} \psi_0(\alpha^m - \alpha^n)\rho_m\bar{\rho}_n \geqq 0.$$

More generally, if γ has a value (1.1.3), then $\psi(\alpha)$ is positive definite in the present sense if and only if the "radial" function

$$\psi_0(\alpha) \equiv \psi_0(\alpha_1, \cdots, \alpha_k) = \psi((\alpha_1^2 + \cdots + \alpha_k^2)^{\frac{1}{2}})$$

in E_k satisfies (3.6.9) for any points $\alpha^0, \alpha^1, \cdots, \alpha^N$ in E_n and any $\rho_0, \cdots, \rho_N$.

In the discrete case the situation is not so. It is true that for $\gamma = 0$ the sequence $\{c_n\}$ is positive definite in the present sense if and only if the matrix $c_{m,n} = c_{|m-n|}$, $-\infty < m, n < \infty$ is positive definite in the ordinary sense; but for other values (1.1.3) no such connection with multi-dimensional positive-definiteness is in evidence.

4. Homogeneous Stochastic Processes.

4.1. In the discrete case we now consider sequences $\{c_n\}$ in P for which

$$(4.1.1)\qquad c_0 = 1.$$

Under the restriction (3.5.2) if we introduce the functions

$$(4.1.2)\qquad h(x) = \sum \rho_n c_n P_n(x), \qquad h(x,y) = \sum \rho_n c_n P_n(x) P_n(y)$$

then we have

$$(4.1.3)\qquad h(x) \geqq 0, \qquad \int_R h(x)\,dv = 1$$

$$(4.1.4)\qquad h(x,y) \geqq 0, \qquad \int_R h(x,y)\,dv(y) = 1.$$

Also, if for two sequences $\{c_n'\}$, $\{c_n''\}$ we introduce the product $\{c_n = c_n' c_n''\}$ then for the corresponding two-point functions we have

$$(4.1.5)\qquad h(x,z) = \int_R h'(x,y)h''(y,z)\,dv_y.$$

Furthermore, if the restriction (3.5.2) is not available then we can bring it about by introducing the approximating sequences

$$(4.1.6)\qquad c_{\epsilon n} = e^{-\epsilon\lambda_n} c_n, \qquad 0 < \epsilon < 1,$$

which are themselves in P. If we denote the corresponding two point functions by $h_\epsilon(x,y)$ then it is appropriate to introduce the set-function, with respect to one of the variables only,

$$(4.1.7)\qquad H_\epsilon(x,B) = \int_B h_\epsilon(x;y)\,dv_y,$$

and let $\epsilon \to 0$ afterwards. If we omit the subscript ϵ, then this function has the properties

$$H(x,B) \geqq 0, \qquad H(x,R)=1 \tag{4.1.8}$$

$$\int_R H'(x, dB_y)H''(y,B_z) = H(x,B_z). \tag{4.1.9}$$

If we now let $\epsilon \downarrow 0$, then it is not hard to verify that (4.1.7) converges, for each x, towards a function $H(x,B)$ for which (4.1.8) holds, and which is such that the convolution rule (4.1.9) likewise holds, the integral existing suitably.

The interpretation of all this is as follows. The function $h(x)$ itself is a probability density on R, on the face of it, but it is not stochastically meaningful by itself. The really meaningful object is $h(x,y)$ which is the density for a probability *re*-distribution on R, and the object $H(x,B)$ is its non-density generalization properly. In the Euclidean case one has $h(x,y)=h(y-x)$, and even more generally $H(x,B)=H(B-x)$, where $H(B)=H(0,B)$, and the illusion arises as if one-variable object were the primary one. But this is an "accident" of Euclidean space, which can be duplicated on commutative group spaces in general, but nowhere else.

In the continuous case, if $\phi(\alpha) \in P$, $\phi(0)=1$, then subject to restriction (2.3.11), we put

$$h(x) = \int_0^\infty \phi(\alpha)T(\alpha x)d\mu(\alpha), \qquad h(x,y) = \int_0^\infty \phi(\alpha)T(\alpha x)T(\alpha y)d\mu(\alpha).$$

Otherwise we put first $\phi_\epsilon(\alpha)=e^{-\epsilon\alpha^2}\phi(\alpha)$, and all the relations (4.1.3)-(4.1.5) and (4.1.7)-(4.1.9) are as before.

4.2. We now wish to find out, in the discrete case say, when a two-point function $h(x,y)$ with the properties (4.1.4) will have a representation (4.1.2), more or less. Since the system $\{P_n(x)\}$, $n=0,1,\cdots$, is complete on R, the two-fold system $\{P_m(x)P_n(y)\}$, $m,n=0,1,1,\cdots$, is complete on $R \times R$. A general function $h(x,y)$ has therefore an expansion

$$\sum_{m=0}^\infty \sum_{n=0}^\infty \rho_m\rho_n c_{m,n} P_m(x)P_n(x),$$

but for "special" functions $h(x,y)$ only the diagonal terms $m=n$ may occur. The conditions (4.1.4) have very little bearing on this restriction, and the latter must be imposed additionally therefore. To this end we are introducing the operation

$$g(x) = \int_R h(x,y)f(y)dv(y) \tag{4.2.1}$$

from $f(.)$ to $g(.)$ and we note that conditions (4.1.4) express the following requirement. If $f(y) \geqq 0$ then $g(x) \geqq 0$, and in this case also

$$\int_R g(x)dv(x) = \int_R f(y)dv(y)\,; \tag{4.2.2}$$

but in general

$$\int_R |g(x)|\,dv(x) \leqq \int_R |f(y)|\,dv(y). \tag{4.2.3}$$

However, we can also ignore this feature of positivity, and view (4.2.1) as an operation from L_2 into L_2. Now, in the "special" case (4.1.2) we have

$$g(x) \sim \sum \rho_n c_n a_n P_n(x),$$

and this can be expressed as follows. It is a bounded self-adjoint operator on the real Hilbert-space L_2, and, what is decisive, this operator is commutative with the Laplace operator $\Delta_x f$ as introduced in 3.1. Also

$$\| g \| \leqq \| f \|.$$

In keeping with a certain nomenclature already established in the theory of probability, we will describe this feature of commutativity by saying that our "transition matrix" $h(x,y)$ is a "homogeneous" one, and "homogeneity" is meant to connote a certain group-like "invariance" with respect to those data by which the commutativity with the "basic" operator $\Delta_x f$ is brought about. See [1], pp. 97-98.

4.3. *Definition* 4.3.1. A *homogeneous stochastic process* is defined as follows. In the discrete case it is a family of elements $\{c_n(t)\}$ in P, $0 \leqq t < \infty$, with the properties: $c_n(r+s) = c_n(r)c_n(s)$, $c_n(0) = 1$, $c_n(t)$ is continuous in t. In the continuous case it is a family of functions $\phi(\alpha, t) \in P$, $0 \leqq t < \infty$ such that $\phi(\alpha, r)\phi(\alpha, s) = \phi(\alpha, r+s)$, $\phi(\alpha, 0) = 1$, $\phi(\alpha, t)$ is continuous in (α, t).

The corresponding two-point functions $h(x,y;t)$, or more generally $H(x, B_y; t)$, $0 < t < \infty$, have the properties (in addition to "measurability" requirements)

$$h(x,y;t) \geqq 0, \qquad \int_R h(x,y;t)\,dv_y = 1 \tag{4.3.1}$$

$$\int_R h(x,y;t)h(y,z;s)\,dv_y = h(x,z;r+s) \tag{4.3.2}$$

$$\lim_{t \downarrow 0} \int_R h(x,y;t)f(y)\,dv_y = f(x) \qquad (L_2\text{-norm}) \tag{4.3.3}$$

or

$$H(x, B_y; t) \geqq 0 \qquad H(x, R_y; t) = 1 \tag{4.3.4}$$

$$\int_R H(x, dB_y; r) H(y, B_z; s) = H(x, B_z; r+s) \tag{4.3.5}$$

$$\lim_{t \downarrow 0} \int_R f(y) d_y H(x, B_y; t) = f(x) \qquad (L_2\text{-norm}). \tag{4.3.6}$$

THEOREM 4.3.1. (i) *If* $\{c_n(t)\}$ *is a homogeneous stochastic process then we can put*

$$c_n(t) = \exp\{-t\sigma\lambda_n - t \int_{-1}^{1-0} \frac{1-P_n(y)}{1-y} dG(y)\} \tag{4.3.7}$$

where $\sigma \geqq 0$, $G(B) \geqq 0$, $\int_{-1}^{1-0} dG < \infty$. *In the continuous case we can write*

$$\phi(\alpha, t) = \exp\{-t\sigma\alpha^2 - t \int_{0+}^{\infty} \frac{1-T(\alpha y)}{y^2} dG(y)\} \tag{4.3.8}$$

where $G(B) \geqq 0$, $\int_{0+}^{\infty} dG < \infty$.

(ii) *Conversely, any such expression* (4.3.7) *or* (4.3.8) *constitutes a homogeneous stochastic process.*

(iii) *The integral in* (4.3.7) *is* $o(n^2)$ *as* $n \to \infty$, *and in* (4.3.8) *it is* $o(\alpha^2)$ *as* $\alpha \to \infty$, *and in either case,* σ *and* $G(B)$ *are uniquely determined.*

Part (ii) is nothing but a restatement of Theorems 3.4.3 and 3.6.2. Part (i) and (iii), interesting as they are, have very little to do with property P, and we will not dwell on them. But we are referring to [1], Chapter 3, where the (multidimensional) Euclidean cases have been analyzed in systematic fashion, and to [2], where the case of ultraspherical polynomials has been sketched.

For $\gamma = 0$, (4.3.8) is the "classical" stochastic process on the straight line $-\infty < x < \infty$,

$$\phi(\alpha, t) = \exp\{-t\sigma\alpha^2 - t \int_{0+}^{\infty} \frac{1-\cos \alpha y}{y^2} dG(y)\}, \tag{4.3.9}$$

when specialized to symmetric distributions. Also, (4.3.7) is then

$$c_n(t) = \exp\{-t\sigma n^2 - t \int_{0+}^{\pi} \frac{1-\cos ny}{y^2} dG(y)\} \tag{4.3.10}$$

and this is the "adaption" of (4.3.9) to random variables on the (compact) torus $-\pi \leqq x < \pi$ as we have elaborated in [1], p. 74.

4.4. The particular processes (3.4.11) and (3.6.8) which were so directly descendent from property P correspond to the (symmetric) Poisson process; and, in our analysis, the processes (3.4.10) and (3.6.7) which correspond to the Gaussian process were obtained by formal closure properties of the class P from those. In order to assess the relative primacy of these two basic processes the question is in order whether membership in P would not imply the Poisson process from the Gaussian process conversely, and the answer to this question is in the negative. We will give a reasoning for the continuous case only, but it would undoubtedly be forthcoming in the (less obvious) discrete case as well. For variable t, the family of positive definite functions $\{e^{-t\alpha^2}\}$ is already closed with respect to multiplication and (weak) limits and the only closure operation remaining is (positive) addition combined with such. It is easily reasoned that the only positive-definite functions thus ensuing are of the form

$$\phi(\alpha) = \int_0^\infty e^{-t\alpha^2} d\rho(t), \tag{4.4.1}$$

$d\rho(t) \geqq 0$, $\int_0^\infty d\rho(t) = 1$, and the question before us is to decide whether a function $e^{-u(1-T(\alpha y))}$, $u > 0$, $y > 0$, is representable thus. Now, if it were so, then $1 - T(\alpha y)$ would have to be monotonely non-decreasing in $0 < \alpha < \infty$. But the formula

$$\frac{d}{dy}(y^{-\nu}J_\nu(y)) = -y^{-\nu}J_{\nu+1}(y)$$

contradicts this, since $J_{\nu+1}(y)$ is negative somewhere.

Since the Poisson process (3.4.11) is not presentable by the integral (4.4.1) it is of course natural to inquire which homogeneous processes so are, and a very general answer to this question was given in [1], pp. 91-99, which we will here reproduce.

We take in $0 \leqq y < \infty$ a function $\nu(y)$ having the following properties. It is continuous in $0 \leqq y < \infty$ with $\nu(0) = 0$, and differentiable in $0 < y < \infty$, and the derivative $\frac{d\nu}{dy}$ is completely monotone in $0 < y < \infty$. The general formula for such a function is

$$\nu(y) = cy + \int_{0+}^\infty (1 - e^{-ty}) d\rho(t), \tag{4.4.2}$$

where $c \geqq 0$, and $\rho(t)$ is monotonely non-decreasing in $0 < t < \infty$ with

$$\int_{0+}^1 t d\rho(t) + \int_1^\infty d\rho(t) < \infty.$$

If this is so, then for each $u > 0$ the function $e^{-u\nu(y)}$ is completely monotone in $0 \leqq y < \infty$ and has the value 1 for $y = 0$, and we therefore have a representation

$$(4.4.3) \qquad e^{-u\nu(y)} = \int_0^\infty e^{-tr} d_r\gamma(r;u)$$

where

$$d_r\gamma(r;u) \geqq 0, \qquad \int_0^\infty d_r\gamma(r;u) = 1.$$

Such a function $\gamma(r;u)$ we called a *subordinator,* and we called it a *proper* subordinator if $\gamma(0+;u) = \gamma(0;u)$ for $0 < u < \infty$. Also the necessary and sufficient condition for a subordinate to be proper is: $\nu(y) \to \infty$ as $y \to \infty$.

THEOREM 4.4.1. *If we are given a density $h(x,y;t)$ with the properties (4.3.1)-(4.3.3) then for any proper subordinator $\gamma(r;u)$ the function*

$$(4.4.4) \qquad h_\nu(x,y;t) = \int_0^\infty h(x,y;r)\, d_r\gamma(r;t)$$

is of the same kind. For a non-density $H(x, B_y;t)$ we can correspondingly form

$$H(x,B_y;t) = \int_0^\infty H(x,B_y;t)\, d_r\gamma(r;t)$$

and the subordinator need not be proper now.

Also, in the discrete case, if we have a (suitably) convergent expansion

$$(4.4.5) \qquad h(x,y;t) = \sum_0^\infty e^{-\mu_n t}\rho_n P_n(x) P_n(y)$$

$\mu_n \geqq 0$, then

$$(4.4.6) \qquad h_\nu(x,y;t) = \sum_0^\infty \; e^{-\nu(\mu_n)t}\rho_n P_n(x) P_n(y);$$

and, in the continuous case,

$$(4.4.7) \qquad h(x,y;t) = \int_0^\infty e^{-\chi(\alpha)t} T(\alpha x) T(\alpha y)\, d\mu(\alpha)$$

will go over into

$$(4.4.8) \qquad h_\nu(x,y;t) = \int_0^\infty e^{-\nu(\chi(\alpha))t} T(\alpha x) T(\alpha y)\, d\mu(\alpha).$$

If we apply this in particular to our kernels (3.3.1) to (3.3.2) we obtain functions

$$(4.4.9) \qquad \sum_0^\infty \rho_n e^{-\nu(n(n+2\gamma))t} P_n(x) P_n(y)$$

$$\int_0^\infty e^{-\nu(\alpha^2)t}T(\alpha x)T(\alpha y)\,d\mu(\alpha), \tag{4.4.10}$$

for any function $\nu(y)$ of the form (4.4.2) provided also $\nu(y)\to\infty$ as $y\to\infty$, which will insure absolute convergence, say.

The "simplest" functions $\nu(y)$ are y^ρ, $0<\rho<1$, and they correspond to the symmetric stable processes so-called, and we are stating for emphasis that we have

$$\sum_0^\infty \rho_n e^{-(n(n+2\gamma))^\rho t}P_n(x)P_n(y)\geqq 0 \tag{4.4.11}$$

$$\int_0^\infty e^{-\alpha^{2\rho}t}T(\alpha x)T(\alpha y)\,d\mu(\alpha)\geqq 0, \tag{4.4.12}$$

in any $0<\rho\leqq 1$.

We also recall that the kernels (4.4.9) and (4.4.10) are the Green's kernels of the "diffusion" equation

$$\nu(-\Delta)_x f=\frac{\partial f}{\partial t} \tag{4.4.13}$$

where $\nu(-\Delta)$ is the operational function $\nu(y)$ of the non-negative self-adjoint operator $-\Delta$, as defined in Hilbert space theory. Especially, the kernels (4.4.11) and (4.4.12) pertain to the equation

$$(-\Delta)_x{}^\rho f=\frac{\partial f}{\partial t}$$

which we have introduced first in [12] in this context.

Actually, of course, if we are given the more general homogeneous stochastic process, and if we denote (4.3.8) by $\phi(\alpha,t)=e^{-t\nu(\alpha)}$, and (4.3.7) by $c_n(t)=e^{-t\nu(\lambda_n)}$, where $\nu(y)$ is a non-negative function to suit the context, whether $\nu(y)$ be of the form (4.4.2) or not; then of course the corresponding function $h(x,y;t)$—or $H(x,B_y;t)$—will be the Green's kernel of the equation (4.4.13) again. The difference however, is this. With the functions (4.4.2) we can set up the equation (4.4.13) "always"; that is, whenever a Laplacean can be defined suitably, and the construction (4.4.4) is then a "universal" one. But the larger class of functions $\nu(y)$ envisaged depends on the individual set-up, and must be determined from case to case. In our present two cases the larger class even includes the Poisson processes, which we think is a rare enough occurrence. But what the actual stochastic consequences of this occurrence are would have to be examined yet.

REFERENCES

[1] S. Bochner, Harmonic analysis and the theory of probability, California University Press, 1955.

[2] ———, "Positive zonal functions on spheres," Proc. Nat. Acad. Sc., Washington, vol. 40 (1954), pp. 1141-1147.

[3] G. A. Hunt, Positive transformation on group spaces, 1954, 34 pp., Cornell University, Contract AF 18 (600)-685.

[4] G. Szegö, Orthogonal polynomials, 1949.

[5] S. Bochner, Fouriersche Integrale, Chelsea, 1948.

[6] G. N. Watson, Bessel Function, 1922.

[7] J. Delsarte, "Sur une extension de la formule de Taylor," Journal de Math. Pur. Appl. (9), vol. 17 (1936), pp. 213-231.

[8] ———, "Une extension nouvelle de la théorie de fonctions presque periodique de Bohr," Acta Mathem., vol. 69 (1938), pp. 259-317.

[9] B. M. Levitan, "The application of Generalized Displacement Operators to Linear Differential Equations of the Second Order," Translation No. 59 (1950), American Math. Soc.

[10] S. Bochner and K. Chandrasekharan, Fourier Transforms, Princeton, 1948.

[11] S. Bochner and J. von Neumann, "On compact solutions of Operational-Differential equations," Annals of Math., vol. 36 (1935), pp. 255-291.

[12] S. Bochner, "Diffusion equation and Stochastic processes," Proc. Natl. Acad. Sc., Washington, vol. 35 (1949), pp. 368-370.

Acknowledgments

The American Mathematical Society gratefully acknowledges the kindness of these institutions in granting the following permissions:

Accademia Nazionale Dei Lincei:

> *Su un teorema di Frobenius Pern Le Funzioni di Jacobi*, Rendiconti di questa Accademia, vol. IX, series VIII, fasc. 6, 1950, pp. 327–331.

ACTA Scientiarum Mathematicarum (Szeged):

> *Lattice points and Fourier expansions*, Acta Sci. Math. 12 (1950), 1–15.

Akademie der Wissenschaften zu Göttingen:

> *Abstrakte Funktionen und die Besselsche Ungleichung Nachrichten von der Gesellschaft der Wissenschaften zu Göttingen*, Math. Phys. (1933), 178–180.

American Council of Learned Societies:

> Bochner, Salomon, *Dictionary of Scientific Biography*, vol. 17, supp. II, pp. 88–90.

American Scientist:

> *Role of Mathematics in the Rise of Mechanics*, American Scientist 50 (1962), 294–311.

Annali di Matematica:

> *Theorem of Morera in several variables*, Annali di Matematica, serie IV, Tomo XXXIV, 1953.
>
> *Vector fields on Riemannian spaces with boundary*, Annali di Matematica Pura ed Applicata, serie IV, Tomo LIII, 1961.

Bayerische Akademie der Wissenschaften:

Über die Struktur von Fourierreihen fastperiodischer Funktionen, Sitzungsberichte der Bayerischen Akademie der Wisenschaften, 1928, pp. 181–190.

Berliner Mathematische Gesellschaft E.V.:

Über Fourierreihen von fastperiodischen Funktionen, Sitzungsberichte BMG 26. Jahrgang.

B. G. Teubner, GMBH:

Orthogonale Systeme analytischer und harmonischer Funktionen, Jahresbericht der Deutschen Mathematiker-Vereinigung Bd. 30, 1921, pp. 100–101.

Beitrag zur absoluten Konergenz fastperiodischer Fourierreihen, Deutsche Mathematiker, Vereiningung, Jahresbericht, Band 39, 1930, pp. 52–54.

Kurven Endlicher Länge, Deutsche Mathematiker, Vereiningung, Jahresbericht, Band 40, 1931, pp. 245–249.

Calcutta Mathematical Society:

General analytical setting for the general limit theory of probability, Golden Jubilee Commemoration volume (1958–1959), pp. 111–128.

Canadian Mathematical Society:

A new viewpoint in differential geometry, Canad. J. Math. 3 (1951), 460–470.

Centre National de la Recherche Scientifique-Publications:

Systémes des équations aux dérivées partielles qui se rattachent aux formules de Green, La Theorie des Equations aux Dérivées Partielles, CNRS (1956), 19–27.

Duke University Press:

Fourier series of L_2-Functions, Duke Math. J. 16 (1949), 579–583.

F. Centrale des Reviues Dunod-Gauthier Villars:

Function rings and analytic measures, J. Math. Pures Appl. (1963), 209–221.

Weak solutions of linear partial differential equations, J. Math. Pures Appl. 35 (1956), 194–202.

Sur les fonctions presque periodiques de Bohr, Anal. Math., 1–3.

Friends of Fondren Library, Rice University:

Frontispiece appearing in Part 4, from the cover of The Flyleaf, 33 (1983).

Fundamenta Mathematicae:

Integration von Funktionen, Deren Werte die Elemente Eines Vektorraumes Sind, Fund. Math. (1933), 262–276.

Eine Bemerkung zum Satz von Fubini, Fund. Math. (1933), 277–280.

Absolute-Additive Abstrakte Mengenfunktionen, Fund. Math. (1933), 211–213.

Hungarian Academy of Sciences:

Über Faktorenfolgen Für Fouriersche Reihe, Acta Litterarum Ac Scientiaru, Tom IV, Fasc. III, vol. 28, 1929, pp. 125–129.

Indian Mathematical Society:

On compact complex manifolds, J. Indian Math. Soc. 11 (1947), 1–21.

Remarks on Gaussian sums and Tauberian theorems, J. Indian Math. Soc. 15 (1951), 97–104.

Algebraic and linear dependence of automorphic functions in several variable, J. Indian Math. Soc. 16 (1952), 1–6.

Connection between functional equations and modular relations, J. Indian Math. Soc. 16 (1952), 99–102.

Hartogs' theorem in complex spaces with singularities, J. Indian Math. Soc. 16 (1952), 138–146.

Theorems on analytic continuation which occur in the study of Riemann's functional equation, J. Indian Math. Soc. 21 (1957), 127–147.

Institut Mittage-Leffler:

Hauptlösungen von Differenzenglichungen, Acta Math. 51 (1927), 1–21.

Abstrakte fastperiodisch Funktionen, Acta Math. 61 (1933), 150–184.

Fastperiodische Lösungen der Wellwngleichung, Acta Math. 62 (1933), 227–237.

Institute of Mathematical Statistics:

A limit theorem for the periodogram, Ann. Math. Stat. 29 (1958).

International Philosophical Quarterly:

Aristotle's physics and todays physics, Internat. Philos. Quart. 4 (1964), 217–244.

History of Science Society, ISIS:

The significance of some basic mathematical conceptions for physics, ISIS 54 (1963), 179–205.

Johns Hopkins University Press:

Bounded analytic functions in several variables and multiple Laplace integrals, 59 (1937), 732–738.

Some theorems of quasi-analyticity for functions of several variables, Amer. J. Math. 61 (1939), 303–328.

Gauss summability of trigonometric integrals, Amer. J. Math. 71 (1949), 50–59.

London Mathematical Society:

Properties of Fourier series of almost periodic functions, Proc. London Math. Soc. (2) 26 (1926), 432–452.

Remarks on the theorems of Picard-Landau and Picard-Schottky, J. London Math. Soc. 1 (1925), 100–103.

Properties of Fourier series of almost periodic functions, J. London Math. Soc. 1 (1926), 132–133.

Notes on two theorems of Norbert Weiner, J. London Math. Soc. 1 (1926), 240–244.

Remarks on the integration of almost periodic functions, J. London Math. Soc. 8 (1933), 250–254.

Homogeneous systems of differential equations with almost periodic coefficients, J. London Math. Soc. 8 (1933), 279–288.

Remark on the Euclidean algorithm, J. London Math. Soc. 9 (1933).

An extension of a Tauberian theorem on series with positive terms, J. London Math. Soc. 9 (1933), 141–148.

On a theorem of M. Riesz for Fourier series, J. London Math. Soc. 14 (1939), 62–73.

Properties of Fourier series of almost periodic functions, Proc. London Math. Soc (2) 26 (1926), 432–452.

Lunds Universitets Matematiska Institution

Bessel functions and modular relations of higher type and hyperbolic differential equations, Communications du Seminaire mathematique de L'universite de Lund Tome Supplementaire, 1952, pp. 12–20.

MacMillan Publishing Company:

Space, Dictionary of the History of Ideas (Philip P. Wiener, ed.), vol. IV, pp. 295–307.

Continuity and discontinuity in nature and knowledge, Dictionary of the History of Ideas (Philip P. Wiener, ed.), vol. I, pp. 492–504.

Symmetry and asymmetry, Dictionary of the History of Ideas (Philip P. Wiener, ed.), vol. IV, 345–353.

Mathematics in cultural history, Dictionary of the History of Ideas (Philip P. Wiener, ed.), vol. III, pp. 177–185.

Infinity, Dictionary of the History of Ideas (Philip P. Wiener, ed.), vol. II.

McGraw-Hill Publishing Company:

Mathematics, McGraw-Hill Encyclopedia of Science and Technology, 1987, pp. 175–180.

Mathematical Association of America:

Fourier series came first, Amer. Math. Monthly 86 (1979), 197–199.

Mathematical reflections, Amer. Math. Monthly 81 (1974), 827–852.

National Academy Press:

John Von Neumann 1903–1957: *A biographical memoir*, Biographical Memoirs, vol. 32, National Academy Press, Washington, DC, 1958.

General almost automorphy, Proc. Nat. Acad. Sci. U.S.A. 72 (1975), 3815–3818.

Oxford University Press:

Summation over lattice points in K-space, Quart. J. Math. 19 (1948), 238–248.

Summations over lattice points in K-space (*A supplemental note*), Quart. J. Math. Oxford Ser. (2) 1 (1950), 80.

Princeton University Press:

Green's formula and analytic continuation, Ann. of Math. Stud. 33 (1954), 1–14.

Contributions to the theory of Riemann surfaces (L. Alfors et al., eds.), Princeton University Press, Princeton, NJ, 1953, pp. 189–201.

(*frontispiece* #3) *from Problems in Analysis: A symposium in honor of Salomon Bochner*, Princeton University Press, Princeton, NJ, 1970.

Plato and the one-ness of knowledge, The Role of Mathematics in the Rise of Science, Princeton University Press, Princeton, NJ, 1966.

Structure of complex spaces, Contributions to the theory of Riemann surfaces, Princeton University Press, Princeton, NJ 1953, 189–201.

Inversion formulae and unitary transformations, Ann. of Math. 35 (1934), 111–115.

Distribution functions and positive-definite functions, Ann. of Math. 35 (1934), 252–257.

On compact solutions of operational-differential equations. I, Ann. of Math. 36 (1935), 255–291.

Summation of derived Fourier Series—An application to Fourier expansions on compact Lie groups, Ann. of Math. 37 (1936), 345–356.

A converse of Poisson's theorem in the theory of probability, Ann. of Math. 37 (1936), 816–822.

Singularities of composite functions in several variables, Ann. of Math. 38 (1937), 293–302.

A theorem on analytic continuation of functions in several variables, Ann. of Math. 39 (1938), 14–19.

Closure theorems for translations, Ann. of Math. 39 (1938), 287–300.

Linear functionals on certain spaces of abstractly-valued functions, Ann. of Math. 39 (1938), 913–944.

Additive set functions on groups, Ann. of Math. 40 (1939), 769–799.

Finitely additive integral, Ann. of Math. 41 (1940), 495–504.

Hadamard's theorem for Dirichlet series, Ann. of Math. 41 (1940), 711–714.

Additive set functions and vector lattices, Ann. of Math. 42 (1940), 316–324.

Hilbert distances and positive definite functions, Ann. of Math. 42 (1941), 647–656.

On a theorem of Tannaka and Krein, Ann. of Math. 43 (1942), 56–58.

Absolutely convergent Fourier expansions for non-commutative normed rings, Ann. of Math. 43 (1942), 409–418.

Analytic and meromorphic continuation by means of Green's formula, Ann. of Math. 44 (1943), 652–673.

Group invariance of Cauchy's formula in several variables, Ann. of Math. 45 (1944), 686–707.

Boundary values of analytic functions in several variables and of almost periodic functions, Ann. of Math. 45 (1944), 708–722.

Compact groups of differentiable transformations, Ann. of Math. 46 (1945), 372–381.

Groups of differentiable and real or complex analytic transformations, Ann. of Math. 46 (1945), 685–694.

Formal Lie groups, Ann. of Math. 47 (1946), 192–201.

Linear partial differential equations with constant coefficients, Ann. of Math. 47 (1946), 202–212.

Locally compact groups of differential transformations, Ann. of Math. 47 (1946), 639–653.

Distributive order-preserving operations in partially ordered vector sets, Ann. of Math. 48 (1947), 168–179.

Groups on analytic manifolds, Ann. of Math. 48 (1947), 659–669.

Stochastic processes, Ann. of Math. 48 (1942), 1014–1061.

Curvature and Betti numbers, Ann. of Math. 48 (1948), 379–390.

On the localization property for multiple Fourier series, Ann. of Math. 49 (1948), 966–978.

Curvature and Betti numbers. II, Ann. of Math. 50 (1949), 77–93.

Quasi-analytic functions, Laplace operator, positive kernels, Ann. of Math. 51 (1950), 68–91.

Euler-Poincaré characteristic for locally homogeneous and complex spaces, Ann. of Math. 515 (1950), 241–261.

Vector fields on complex and real manifolds, Ann. of Math. 52 (1950), 642–649.

Some properties of modular relations, Ann. of Math. 53 (1951), 332–363.

Tensor fields with finite bases, Ann. of Math. 53 (1951), 400–411.

Tensor fields with finite bases, Ann. of Math. 56 (1952), 504–519.

Zeta functions and Green's functions for linear partial differential operators of elliptic type with constant coefficients, Ann. of Math. 57 (1953), 32–56.

Length of random paths on general homogeneous spaces, Ann. of Math. 57 (1953), 309–319.

Complex spaces with singularities, Ann. of Math. 57 (1953), 490–516.

Partial ordering in the theory of Martingales, Ann. of Math. 62 (1955), 162–169.

On Riemann's functional equation, Ann. of Math. 63 (1956), 336–360.

On Riemann's functional equation with multiple gamma factors, Ann. of Math. 67 (1958), 29–41.

A special integral transformation in Euclidean space, Ann. of Math. 68 (1958), 150–158.

Strong law of large numbers for sequences of almost periodic functions, Ann. of Math. 68 (1959), 718–732.

Indefinite linear pseudogroups of transformations, Ann. of Math. 75 (1962), 93–104.

Contributions to Fourier analysis, Ann. of Math. 25 (1950), 3–45.

Analytic functions with almost periodic coefficients, Ann. of Math. (2) 35 (1934), 152–161.

Rice University Press:

The rise of functions, Complex Analysis II: Proceedings of the Conference on Complex Analysis, Rice University, 26–29 March 1969, Rice Univ. Stud., vol. 56, Rice Univ. Press, Houston, TX, 1970, pp. 3–23.

Singularities and discontinuities, Proceedings of the Converence on Complex Analysis, Rice University, March 1972, Rice Univ. Stud., vol. 59, Rice Univ. Press, Houston, TX, 1973.

The emergence of analysis in the Renaissance and after, Rice Univ. Stud., vol. 64, Rice Univ. Press, Houston, TX, 1978, pp. 11–56.

Einstein between centuries, Rice Univ. Stud., vol. 65, Rice Univ. Press, Houston, TX, 1979, pp. 1–54.

Seminario Matematico:

Curvature and Betti numbers in real and complex vector bundles, Rendiconti del Seminario Matematico 15 (1955–56), 225–253.

S. Hirzel Verlag Stuttgart:

Über einen arithmetischen Satz der Röntgenkristallographie, Probleme Der Modernen Physik, 1928, pp. 88–94.

Springer-Verlag:

Über orthogonale Systeme analytischer Funktionen, Math. Z. 14 (1922), 180–207.

Konvergenzsatze für Fourierreihen grenzperiodischer Funktionen, Math. Z. 27 (1928), 187–211.

Ein Satz über lineare Operationen, Math. Z. 29 (1929), 737–744.

Über Stürm-Liouvillesche Polynomsysteme, Math. Z. 29 (1929), 730–736.

Allgemeine lineare Differenzengleichungen mit asymptotish konstanten Koeffizienten, Math. Z. 33 (1931), 426–450.

Ein Konvergenzsatz für mehrvariablige Fouriersche Integrale, Math. Z. 34 (1931), 440–447.

Limitierung mehrfacher Folgen nach dem Verfahren der arithmetischen Mittel, Math. Z. 35 (1932), 122–126.

Ein Satz von Landau und Ikehara, Math. Z. 37 (1933), 1–9.

Simultankonvergenz von mehrfachen Fourierschen Reihen und Integralen, Math. Z. 50 (1933), 1–6.

Green-Goursat theorem, Math. Z. 63 (1955), 230–242.

Beitrage zue Theorie der fastperiodischen Funktionen, I. *Teil. Funktionen einer Variablen*, Math. Ann. 96 (1926), 119–147.

Beitrage zue Theorie der fastperiodishen Funktionen, II. *Teil. Funktionen Mehrerer Variablen*, Math. Ann. 96 (1926), 383–409.

Fortsetzung Riemannscher Flachen, Math. Ann. 98 (1928), 406–421.

Darstellungen reellvariabler und analytischer Funktionen durch verallgemeinerte Fourier und Laplace-Integrale, Math. Ann. 97 (1927), 632–662.

Der Dimensionsbegriff für Punktmengen in kartesischen Räumen, Math. Ann. 102 (1930), 768–784.

Über gewisse Differential und allgemeinere Gleichungen, deren lösungen fastperiodisch sind I. *Teil. Der Existenzsatz*, Math. Ann. 102 (1929), 489–504.

Über gewisse Differential und allgemeinere Gleichungen, deren lösungen fastperiodisch sind. II. *Teil. Der Beschränktheitrassatz*, Math. Ann. 103 (1930), 588–597.

Über gewisse Differential und allgemeinere Gleichungen, deren lösungen fastperiodisch sind. III. *Teil. Systeme von Gleichungen*, Math. Ann. 104 (1931), 579–587.

Die Poissonsche Summationsformel in mehreren Veränderlichen, Math. Ann. 106 (1932), 56–63.

Monotone Funktionen, Stieljessche Integrale und harmonische Analyse, Math. Ann. 108 (1933), 378–410.

Hartogs' theorem in Euclidean space and a related Cousin theorem on the torus, Contributions to Function Theory, Tata Institute of Fundamental Research, 1960.

Potential-theoretic approach to analytic Bohr almost periodic functions in half-planes, Math. Ann. 150 (1963), 150–155.

A formal approach to stochastic stability, Zeitschrift für Wahrscheinlichkeitstheorie und Verwandte Geibiete 31 (1975), pp. 187–198.

Positivity of the heat kernel for ultraspherical polynomials and similar functions, Arch. Rational Mech. Anal. 70 (1979), 211–217.

Tôhoku Mathematical Journal:

Some general convergence theorems, Tôhoku Math. J. 42 (1936), 191–194.

The University of California Press:

Stationarity, boundedness, almost periodicity of random-valued functions, Proceedings of the Third Berkeley Symposium on Mathematical Statistics and Probability, The University of California Press, Berkeley, CA, pp. 7–27.

University of Maryland Book Center:

Sturm-Liouville and heat equations whose eigen-functions are ultraspherical polynomials or associated Bessel functions, Proceedings of the Conference on Differential Equations (J. B. Diaz and L. E. Payne, eds.).

The University of Michigan Press:

Functions in one complex variable as viewed from the theory of functions in several variables, Lectures on Functions of a Complex Variable (Wilfred Kaplan et al., eds.), The University of Michigan Press, Ann Arbor, MI, pp. 315–333.

Walter de Gruyter & Company:

Über eine Klasse singulärer Integralgleichungen, Sonderausgabe Aus Den Sitzungsberichten Der Preussischen Akademie Der Wissenschaften, Phys-Math Klasse, 1930, XXII.

Umkehrsätze für allgemeine Limitierungsverfahren, Sonderausgabe Aus Den Sitzungsberichten Der Preussischen Akademie Der Wissenschaften, Phys-Math Klasse, 1933, III.

Spektralzerlegung linearer Scharen unitärer Operatoren, Sopnderausgabe Aus Den Sitzungsberichten Der Preussischen Akademie Der Wissenschasften, Phys-Math Klasse, 1933, IX.